Nonlinear Finite Element Methods

T0181608

Peter Wriggers

Nonlinear Finite Element Methods

 Springer

Prof. Dr. Peter Wriggers
Leibniz Universität Hannover
Fakultät für Maschinenbau
Institut für Kontinuumsmechanik
Appelstr. 11
30167 Hannover
Germany
wriggers@ikm.uni-hannover.de

ISBN: 978-3-642-09002-8   e-ISBN: 978-3-540-71001-1

Cover design: WMXDesign GmbH, Heidelberg

Printed on acid-free paper

9 8 7 6 5 4 3 2 1

springer.com

Preface

During the last decade, the method of finite elements (FEM) has proven to be a universal tool for the analysis of complex structures in engineering. With increasing computing power, different nonlinear problems can be treated nowadays.

This book describes, besides the physical and mathematical background of the finite element method, special discretization techniques and algorithms which have to be applied for nonlinear problems of solid mechanics. This includes the necessary basics of continuum mechanics, constitutive equations for engineering materials and variational principles as well as the matrix formulation of FEM with respect to different configurations and the development of algorithms for the solution of the resulting nonlinear algebraic equation systems. Furthermore, time dependent problems are discussed for nonlinear dynamical systems and constitutive equations. Since nonlinear problems exhibit singularities also formulations are included which facilitate a computation of limit and bifurcation points.

In addition, element formulations will be derived for nonlinear truss-, beam- and shell structures and applied in an exemplary fashion. In the same way, different discretization techniques for three-dimensional solids will be developed.

Adaptive methods become more significant for a save and efficient application of the finite element method. Hence the basic procedures of adaptive techniques will be described and formulated such that they can be applied to physically and geometrically nonlinear problems. Additionally, the formulation and treatment of contact problems is included. Associated discretization techniques and algorithms are presented for large sliding in contact interfaces.

The book is intended for graduate students of mechanical and civil engineering who want to familiarize themselves with numerical methods applied to problems in solid mechanics. This applies also to PhD-students and engineers working in industry who need further background information on the application of finite elements to nonlinear problems. Due to that several examples are included in the text for a deeper understanding of the formulations and algorithms.

Numerical results stem from scientific collaboration with my former PhD students E. Boerner, A. Boersma, R. Eberlein, C. S. Han, S. Löhnert, S.

Meynen, T. Raible, S. Reese, A. Rieger, J. Sansour, O. Scherf, H. Spiess and H. Tschöpe which often enough resulted in joint work in which new papers or reports were written. I would like to express my appreciation to all of them since they helped with their constructive comments and criticisms to improve the text.

A new chapter has been added in which my former PhD student Joze Korelc, now full professor at the University of Ljubjana, discusses the methods and merits of automatic code development and creation for nonlinear finite element analysis.

Finally I like to mention the German Science Council (DFG) which supported my work on nonlinear finite element methods through different projects over the years. The results of this work can be found at many different places throughout the book. Last but not least, I like to thank the Springer Verlag for the pleasant collaboration during the last years.

Hannover, June 2008   *Peter Wriggers*

Contents

1. Introduction

Driven by the development of powerful and inexpensive computers, the field of computer aided engineering emerged. It provides predictive tools as well as insights in complex engineering processes. Hence engineers working in many different application areas demand numerical simulation tools for their investigations which some years ago where only accessible by experiments. Modeling of engineering problems leads in many cases to ordinary and partial differential equations which often are of nonlinear nature. A powerful tool to solve these differential equations is the finite element method which was developed over the last 50 years. Applications in engineering include frames, shells and continua in structural analysis within the disciplines of civil, mechanical or aerospace engineering, for an historical overview see e.g. Felippa (2000). Furthermore, the method is used to solve heat conduction problems as well as electrical and magnetic field problems and last but not least simulations of fluids can be carried out using finite elements. The first book covering this wider field of finite element applications was written by Zienkiewicz and Cheung (1967).

Based on this demand, a number of all-purpose finite element codes were developed which can be applied for the solution of many different problems. Additionally, a huge number of special purpose codes have been developed which are tailored for a specific engineering application. Many of these program codes solve nonlinear problems. Often the theoretical background and the associated solution algorithms of the codes are not completely transparent for the user who then is not informed properly and thus has difficulties in judging and assessing the results obtained in the nonlinear analysis. Nonlinear simulations can lead to solutions which are non-unique, e.g. localization can occur in structural applications as well as limit points or bifurcation. Convergence of the numerical analysis is not always obtained in nonlinear applications. Often no mathematical error analysis is available for nonlinear problems. Due to these circumstances, the user of a nonlinear finite element code needs, besides practical experience, a good theoretical background regarding the finite element method and the underlying theory. This book was written to provide the necessary background for problems stemming mainly from solid mechanics.

Nonlinearities which occur in practical applications in civil engineering
are of different nature. For example, in the area of steel constructions, elasto-
plastic analyzes are necessary to compute the limit loads of truss, frame or
shell structures. In case of cable constructions, geometrically nonlinear effects have to be included to describe the large displacements. In concrete
constructions or soil mechanics, complicated nonlinear material laws have to
be considered for a realistic description of the engineering problem. Also an-
choring constructions in which concrete and steel are highly nonlinear due
to possible frictional sliding at interfaces. During the manufacturing of concrete, heat is generated due to chemical reactions which lead together with the
change of the constitutive parameters to heat induced stresses. This process
is a thermomechanical coupled one and can only be realistically described by
a nonlinear model.

Many applications and constructions in mechanical engineering can only
be successfully simulated by nonlinear methods. Among these simulations
are the computation of bearing capacities of rubber bearings or forming processes. All of them include finite deformation analysis and nonlinear constitutive equations. Finally, the numerical simulation of crash problems can be
mentioned as a complex nonlinear problem which is applied widely in the
automotive industry.

All mentioned applications require large numerical finite element models
with several thousand up to ten million degrees of freedom. Thus, besides the
correct formulations of the problem in the continuum mechanics setting, it is
also necessary to provide efficient and robust methods for the solution.

Since the solution methods have to be adjusted with respect to the type
of nonlinearity, the main nonlinearities will be discussed which are related to
solid mechanics.

- **Geometrical nonlinearity** occurs in problems where large displacements
  and rotations have to be considered like in structural elements as cables,
  frames membranes or shells but the strains are still small. Geometrical
  nonlinearity is sufficient in many cases to predict singular points in stability
  analysis.
- **Finite deformations** can occur in problems like metal forming or tyre mechanics. Here not only the displacements are large but also the strains. Contrary to geometrically nonlinear applications, in which only small strains
  occur, problem with finite deformations include arbitrarily large strains.
- **Physical nonlinearity**: Many materials depict nonlinear behaviour.
  Among these are visco-elastic polymers or steel, concrete and soil which
  show elasto-plastic responses. The material behaviour is characterized by
  a nonlinear response function between stresses and strains or by a set of
  evolution equations.
- **Stability problems** can be subdivided into structural mechanics into
  two classes: geometrical and material instability. Geometrical instability

includes bifurcations like buckling of frames or shells but can also be connected to limit points which indicate snap-through behaviour of a structure. Material instabilities come along with necking or shear bands in metals but also geo-materials. The origin of these instabilities lies either in an instability of the equilibrium equations or in the loss of positive definiteness of the acoustic tensor which is related to the incremental constitutive tensor of the material. Both instabilities react in a very sensitive way to imperfections.

— **Nonlinear boundary conditions**: Problems which are characterized by nonlinearities stemming from the boundary are associated with contact between two bodies or deformation dependent loading.

— **Coupled problems** occur when different interacting fields which describe e.g. solids, heat conduction in solids or fluids are needed to formulate a complex physical problem. Examples are thermomechanical coupling, fluid-structure-interaction or problems in which chemical reactions, heat generation and conduction and mechanical stresses have to be coupled to model an engineering process, like the design of a new material. In all cases, nonlinearities in each of the different field equations have to be considered.

During the last years, many breakthroughs were achieved worldwide for each of these different areas which result in better approximative behaviour of the finite elements and in better and more efficient algorithms. Goal of these developments is the design of methods for nonlinear problems which are robust, accurate and efficient. With the achievement of these goals, finite element methods can more safely be applied to nonlinear problems in engineering. Today's problems with small to moderate strains can be solved for elastic and inelastic materials using standard finite element codes. Problems with finite deformations, material instabilities and contact cannot be solved routinely with existing software. Here research is still necessary to enhance the existing programs and tools.

This book includes the basic relations needed for the mathematical modelling of an engineering problem in solid mechanics and the algorithmic treatment used for its numerical simulation. The latter will be investigated with respect to robustness, accuracy and efficiency. Since there are numerous different nonlinear problems and associated solution methods, only an introduction can be provided for this extensive field.

The book is subdivided into ten further chapters which include the following subject areas:

— **Chapter 2** contains an introduction to the major nonlinearities which occur in solid mechanics. Based on simple examples, which can be solved analytically, different phenomena are discussed which are treated in general form in-depth in the following chapters.

— The foundation of nonlinear continuum mechanics is summarized in **Chap. 3** to establish a basis for an unified treatment of finite element formulations. Different strain measures for finite deformations are introduced, after that the associated stress tensors are presented together with the general balance laws and related weak forms. The field of material theory is so large that this area cannot be treated in-depth in this monograph. Hence the constitutive material equations is restricted to hyperelastic and elasto-plastic materials for finite deformations and does present visco-elastic and visco-plastic constitutive laws for small strains only.

— **Chapter 4** discusses the finite element approximations of the nonlinear equations derived in Chap. 3. First the shape functions are described which can be applied for one-, two- and three-dimensional problems. The approximation is based on the isoparametric concept which will be discussed in the light of its excellent suitability for nonlinear problems. The chapter contains also the discretization and matrix formulation of the nonlinear weak forms within the classical displacement formulation leading to nonlinear algebraic equation systems. Finally, linearizations of the nonlinear matrix equations are presented which are needed within the solution algorithms.

— **Chapter 5** discusses algorithms for the solution of nonlinear equation systems which stem from the discretization of the partial differential equations of continuum mechanics. Besides the classical NEWTON method, arc-length procedures are also included. Furthermore, methods for the direct and iterative solution of linear equations systems, which have to be solved within the iterative solution procedures, are subject of this chapter. Finally algorithms for parallel computers with distributed memory are discussed.

— Explicit and implicit algorithms for time dependent problems are described in **Chap. 6**. These are classical and modern methods for time integration of the equations of motion used in nonlinear applications and integration methods for time dependent constitutive equations. The latter are derived for visco-elastic, visco-plastic and elasto-plastic constitutive equations. The presented algorithms can be applied to problems with small and large strains.

— The foundation for the treatment of stability problems is subject of **Chap. 7**. Here the necessary algorithms for investigations regarding the stability behaviour of structures are discussed. Besides the classical buckling analysis, the computation of singular points using extended systems is also presented. The chapter closes with algorithms for path switching needed in post-buckling analysis.

— A new and relevant aspect of modern finite element methods is considered in **Chap. 8**. The topic is automatic error control using error estimators and indicators. The associated adaptive algorithms are derived for elastic and elasto-plastic applications. Examples depict the behaviour of different methods.

— Nonlinear structural members which can be formulated by a one- or two-parametric description are presented in **Chap. 9**. Objects using a one-parametric description are trusses, beams and axi-symmetrical shells including axi-symmetrical membranes. A two-parametric description is applied to model nonlinear shells. The weak forms and associated finite element discretizations are derived for all formulations in a geometrically exact manner.

— Special two- and three-dimensional solid elements are contained in **Chap. 10**. These are constructed especially for incompressible materials and for thin solids with bending behaviour. All elements are formulated either in the initial or current configuration. Special mixed element, enhanced elements and stabilized elements are developed and compared regarding their advantages and disadvantages.

— **Chapter 11** discussed contact problems undergoing finite deformations since many technical problems include contact constraints, especially when large deformations have to be considered. Contact kinematics as well as different possibilities to solve the associated variational inequalities are presented. This chapter includes also constitutive equations for friction within the contact interface and two-dimensional finite element discretization.

Within Chaps. 4–11, different matrix formulations based on the finite element method are derived. Numerical simulations of special examples are presented to compare different formulations and to illustrate the general behaviour of finite element approximations and associated solution algorithms.

The book is written as a textbook for nonlinear applications of the finite element method in solid mechanics. Due to that the chapters have to be read in a chronological order. To obtain a general overview Chaps. 2–6 have to be studied. Chapters 7–11 contain more specialized applications of the method. These chapters can be studied on its own, in case that the reader possesses already the corresponding knowledge. This is also true for Chaps. 4–6 in case that the reader has already the necessary background in continuum mechanics.

2. Nonlinear Phenomena

Numerous different nonlinearities can occur in solid mechanics which are either of geometrical or of physical nature. The treatment of associated problems demands a large bandwidth of methods and algorithms which will be discussed in the following chapters. Based on introductory examples, different phenomena of nonlinear behaviour will be described to introduce the reader the nature of the problems. Deliberately simplified mechanical models are used, which are just complicated enough to represent the desired nonlinear feature. All solutions can still be solved analytically which helps to understand the problem. However, engineering problems cannot be formulated with such simplified models. Due to that numerical methods have to be applied for real world applications.

2.1 Geometrical Nonlinearity

In structural analysis, it is usually sufficient to consider only small deformations and strains since many parts of the structure can only undergo small strains to maintain their usability. With this restriction, a linear constitutive equation can be introduced when elastic deformations are present. However, even under this assumption, there are many problems which depict large displacements or rotations, such as cables, beams or shells. Such problems require a nonlinear theory which includes the geometry in an exact way. Some examples which represent different geometrically nonlinear behaviour are discussed in the following.

2.1.1 Large Displacements of a Rigid Beam

The first example for geometrically nonlinear behaviour is a rigid beam of length l, see Fig. 2.1a which is supported by an elastic rotational spring with stiffness c at its left end.

Equilibrium at the deformed system yields directly, see Fig. 2.1b,

$$F\,l\cos\varphi = c\,\varphi\,.\tag{2.1}$$

Equation (2.1) relates the force F in a nonlinear way to the beam rotation φ. The nonlinearity stems from the change of geometry in the equilibrium

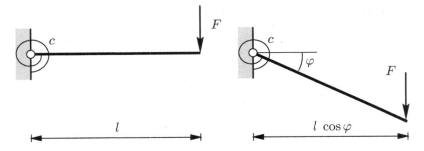

Fig. 2.1a System and loading   **Fig. 2.1b** Undeformed system

equation. Hence this type of behaviour is known as geometrical nonlinearity. For small rotations φ the approximation $\cos\varphi \to 1$ is obtained. With this the linear solution $F = c\varphi/l$ can be derived from (2.1). Figure 2.2 shows the increase of the force as a function of the rotation φ for both cases. One observes clearly that the linear solution deviates from the exact geometrically nonlinear one for large rotations.

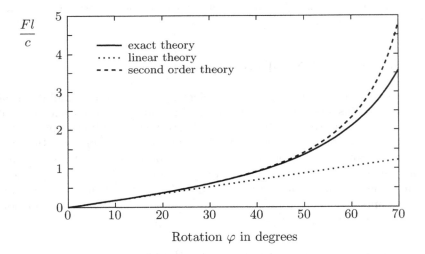

Fig. 2.2 Force versus rotation

Often the so-called second order theories are applied to include nonlinear effects in the mechanical model. The idea is to descibe the nonlinear terms using a TAYLOR series which is terminated after the second term. For the presented example, the relation of the second order theory

$$F = \frac{c\hat{\varphi}}{l\left(1 - \frac{\hat{\varphi}^2}{2}\right)} \tag{2.2}$$

is obtained with $\cos \varphi \approx 1 - \frac{\hat{\varphi}^2}{2}$. This equation approximates the exact solution up to a rotation of $\hat{\varphi} \approx \pi/3$ very well, see Fig. 2.2. The solution (2.2) deviates, however, for larger rotations from the exact solution.

In case that the flexibility of the beam has to be considered too, we would have to include also nonlinear strain–displacement relations for the beam. The associated equations are derived in Sect. 6.2. Due to the fact that they are quite complicated, an analytical solution cannot be derived for the nonlinear elastic beam.

2.1.2 Large Displacements of an Elastic System

In this example, the influence of flexibility of a structure is investigated in the context of geometrically nonlinear behaviour. Let us consider two horizontal elastic springs which have a linear force–displacement relation. The structure is loaded by a point force F, see Fig. 2.3a.

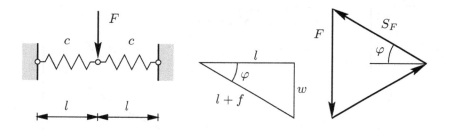

Fig. 2.3a System and loading   **Fig. 2.3b** Geometry and equilibrium

To obtain the load–deflection curve of the force F with respect to the vertical displacement w, the kinematical relation between the vertical displacement w and the elongation f of the spring has to be formulated as well as the equilibrium and the constitutive law for the spring. The kinematical relation is given by, see Fig. 2.3b,

$$w^2 + l^2 = (l + f)^2 \quad \longrightarrow \quad f = l \left[\sqrt{1 + \left(\frac{w}{l}\right)^2} - 1 \right] \tag{2.3}$$

and

$$\sin \varphi = \frac{w}{l + f}. \tag{2.4}$$

Equilibrium follows from Fig. 2.3b by considering symmetry as

$$S_F \sin \varphi = \frac{F}{2}. \tag{2.5}$$

Furthermore, the constitutive equation for the spring is assumed to be linear

$$S_F = c\,f \tag{2.6}$$

where c is the spring stiffness and S_F denotes the force in the spring. Inserting (2.4) in (2.5) yields with (2.6)

$$c\,f\,\frac{w}{l+f} = \frac{F}{2}\,. \tag{2.7}$$

This equation can be reformulated using (2.3) in terms of the vertical displacement w

$$\frac{w}{l}\left[1 - \frac{1}{\sqrt{1+\left(\frac{w}{l}\right)^2}}\right] = \frac{F}{2\,c\,l}\,. \tag{2.8}$$

The associated load–deflection curve is depicted in Fig. 2.4.

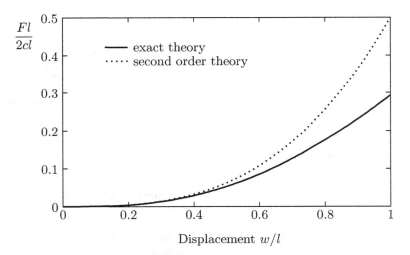

Fig. 2.4 Load–deflection curve

The load–deflection curve has a horizontal tangent at zero displacement. For that reason, this example cannot be formulated in terms of a linearized theory. However, it is possible to derive a second order theory for small values of w ($w/l \ll 1$) by a TAYLOR series expansion of the square root in (2.8): $1/\sqrt{1+\left(\frac{w}{l}\right)^2} \approx 1 - \frac{1}{2}\left(\frac{w}{l}\right)^2$. This approach yields the cubical polynomial

$$\frac{\hat{w}}{l}\left[\frac{1}{2}\left(\frac{\hat{w}}{l}\right)^2\right] = \frac{F}{2\,c\,l}\,, \tag{2.9}$$

which approximates the exact solution for $w/l < 0.4$ well, see Fig. 2.4.

Remark 2.1: This example depicts, besides large displacements w, also large elongations (strains) in the spring. For most technically relevant materials, a linear relation between forces and elongation does not exist in such a case. For example, for steel bars, plastic deformations for larger strains has to be considered, see Sect. 2.2. For rubber bands, the spring stiffness depicts a nonlinear characteristic, see Chap. 3. In the example, a linear relationship is assumed in (2.6), which can be found in real springs. Thus only the effect of geometrically nonlinear behaviour was discussed, omitting here nonlinear effects stemming from the material.

2.1.3 Bifurcation Problem

The solution of a nonlinear problem is not always unique. This feature will be discussed by means of the stability problem described in Fig. 2.5a. This example is equivalent to the first one; only the load acts now in horizontal direction. Formulating equilibrium at the deformed system, see Fig. 2.5b, yields

$$F l \sin \varphi = c \varphi \quad \longrightarrow \quad \frac{\varphi}{\sin \varphi} = \frac{F l}{c} . \qquad (2.10)$$

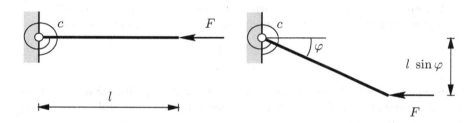

Fig. 2.5a System and loading   **Fig. 2.5b** Deformed system

This equation has multiple solutions. The trivial solution is $\varphi = 0$, which is valid for all values of F. For $\frac{Fl}{c} > 1$ ($|\varphi| \geq |\sin \varphi|$), there exist two more solutions which are depicted in Fig. 2.6. In total, three solutions of (2.10) exist for $\frac{Fl}{c} > 1$. Hence the solution is no longer unique. The point ($\frac{Fl}{c} = 1$) at which the three different solutions start is known as bifurcation point. An essential question is now which solution path will be followed by the system when the load is increased beyond the bifurcation point. The answer for that is provided by the theory of stability. For the case at hand, it can be shown that the trivial solution is instable. The physical meaning of this is that for a small disturbance of the trivial solution $\varphi = 0$ equilibrium is lost and the system will change to a new stable equilibrium state. Usually large deformations and even dynamical effects occur in such situation. In technical construction, such behaviour leads most of the times to a total collapse. Hence the identification of such instable solution is of great practical importance. The other solutions of (2.10) – shown in Fig. 2.6 by a – are stable

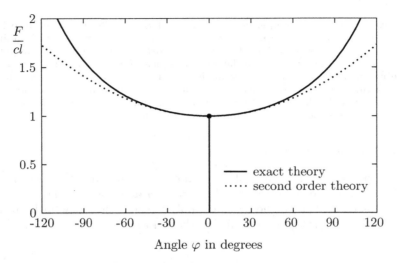

Fig. 2.6 Load-deflection curve

and insensitive against small disturbances. An approximation of (2.10) using second order theory yields with $\sin \varphi \approx \hat{\varphi} - \hat{\varphi}^3 / 6$ and $1 / (1 - x) \approx 1 + \hat{x}$

$$1 + \frac{\hat{\varphi}^2}{6} = \frac{Fl}{c} . \tag{2.11}$$

This equation reproduces the same behaviour as (2.10), since three solutions are also present for $\frac{Fl}{c} > 1$. (A more general formulation will be discussed later in Chaps. 5 and 7.)

Remark 2.2: Often, in practical applications, it is only of interest to compute the bifurcation point of a structure under a given loading. In its vicinity, see Fig. 2.6, it can be assumed that φ is small and hence approximate $\sin \varphi \approx \hat{\varphi}$. From (2.10), the linear homogeneous equation

$$(Fl - c) \hat{\varphi} = 0 \tag{2.12}$$

is obtained which is either fulfilled trivially for $\hat{\varphi} = 0$ or non-trivially for $F = F_c = \frac{c}{l}$. One calls F_c the critical load (F_c is eigenvalue of the eigenvalue problems (2.12)). The critical load F_c is equivalent to the load related to the bifurcation point in the exact equation (2.10).

2.1.4 Snap-Through Problem

In this example, the system depicted in Fig. 2.7a is considered which consists of two trusses – modelled as springs of length L_0 – under the action of a vertical point force F. The springs are supported at their left and right end, respectively.

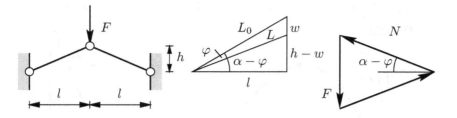

Fig. 2.7a System and loading   **Fig. 2.7b** Geometry and equilibrium

With the kinematical relations $(h - w)^2 + l^2 = L^2$ and $h^2 + l^2 = L_0^2$, see Fig. 2.7b, the length change of the spring is as follows

$$ f = L - L_0 = l \left[\sqrt{1 + \left(\frac{h - w}{l} \right)^2} - \sqrt{1 + \left(\frac{h}{l} \right)^2} \right] . \qquad (2.13)$$

Equilibrium with regard to the deformed system can be written as

$$ N \sin(\alpha - \varphi) = -\frac{F}{2} , \qquad (2.14)$$

see Fig. 2.7b with the normal force N. This relation leads with $\sin(\alpha - \varphi) = (h - w)/L$ to

$$ N \frac{h - w}{L} = -\frac{F}{2} . \qquad (2.15)$$

The spring characteristic is linear, hence $N = c f$, see also Remark 2.1. Inserting (2.13) into (2.15) yields the nonlinear relation between force F and displacement w

$$ c\,(h - w) \frac{L - L_0}{L} = -\frac{F}{2} \implies \frac{w - h}{l} \left[1 - \frac{L_0}{l \sqrt{1 + \left(\frac{w-h}{l} \right)^2}} \right] - \frac{F}{2\,c\,l} = 0 . $$
$$ (2.16) $$

The associated load–deflection curve is plotted for $L_0 / l = 1.25$ in Fig. 2.8. The load increases until point D and decreases afterwards for increasing displacement w. The latter situation cannot be reached in case of a static loading. To model this process in a physically correct way a dynamic process should be introduced once point D is passed. The system cannot not be in static equilibrium until point E is reached for a load larger or equal than the load at D. The associated process, in which the system changes from one equilibrium state to another instantaneous, is called snap-through. Due to that, point D is called snap-through point. Since the solution is not unique at such point (equilibrium can be found for load F in D or E which is associated with two different displacements w) a snap-through problem is a

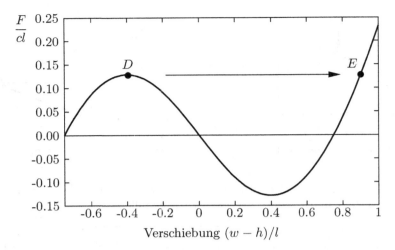

Fig. 2.8 Load–deflection curve

stability problem. Furthermore, the load is limited at point D which leads to the notion of a limit point.

Depending on the geometry and the loading, snap-through behaviour can be observed in many technical structures like trusses, beams or shells. Usually snap-through is connected with a total failure of the structure. However, there are also applications which rely on snap-through behaviour. One of such problems is related to the opening of, e.g. a glass of jam. Here a loud noise of the lid proves that the glass was not opened before and that the contents are untouched. The noise is related to a snap-through of a thin shell (the lid) following from a change of the internal pressure. The deformations related to this process are very small, thus the lid is not damaged.

From the previous examples, it can be concluded that geometrical nonlinearity leads to numerous different phenomena, which can occur individually or in combined form. In principle, always nonlinear behaviour of a new structural design has to be considered until it can be proven that a linearized treatment of the problem is adequate.

Remark 2.3: The application of approximate theories (e.g. second order theories) is not necessary for the present examples since the geometrically exact equations can be easily analyzed. Approximate theories always make sense when the exact formulation of a problem is too complex and the approximate theory enables an analytical treatment. If, however, a computational treatment is chosen – as in this monograph – then the exact relations should always be used since the resulting nonlinear equations can be solved on a computer without difficulties. Due to this, we will abstain from the derivation of approximate theories in the following text.

2.2 Physical Nonlinearity

Within the treatment of geometrically nonlinear problems, which have been considered so far, only linear elastic stress–strain relations, such as the linear characteristic for the elongation of a spring, were used. This is a good approximation for many materials but holds only under certain restrictive assumptions like small strains. Simple examples show that this is not always true. An elongated rubber band depicts, e.g. with increasing elongation a greater stiffness; furthermore a metal wire suffers permanent deformation under bending which is due to elasto-plastic material behaviour. Permanent deformations occur in the last case once a limiting stress is exceeded. This behaviour is also called plastic flow and the limiting stress is known as yield stress.

As an example for elasto-plastic deformations, the system depicted in Fig. 2.9a is considered. It consists of two bars with same cross section but different materials. Generally, elasto-plastic material behaviour is assumed as shown in Fig. 2.9b where the stress is limited by the yield stress σ_y. The material data for YOUNG modulus are chosen as $E_1 = 2\,E_2 = 2\,E$; furthermore the yield stresses of the two materials are different $\sigma_{y1} = 3\,\sigma_{y2} = 3\,\sigma_y$ where the subscript is related to the bars.

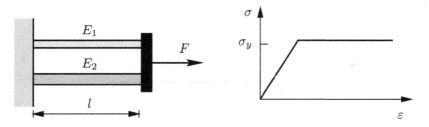

Fig. 2.9a System and loading   **Fig. 2.9b** Material behaviour

Under the assumption that the loading is so small that the yield stress will not be reached in one of the bars, a purely elastic response occurs. In that case, equilibrium can be formulated

$$N_1 + N_2 = F \quad \longrightarrow \quad \sigma_1 + \sigma_2 = \frac{F}{A}, \tag{2.17}$$

where the normal force N_i in bar i is related to the stress by $N_i = A\,\sigma_i$. The kinematical relations yield

$$u_1 = u_2 = u\,, \qquad \epsilon = \frac{u}{l} \tag{2.18}$$

for the displacements in normal direction u_i and the associated strain ϵ. Finally, the linear elastic material law of HOOKE relates stresses to strains

$$\sigma_i = E_i \, \epsilon = E_i \, \frac{u}{l} \, . \tag{2.19}$$

A combination of all equations results in a linear relation between force F and displacement u:

$$E_1 A \, \frac{u}{l} + E_2 A \, \frac{u}{l} = F \quad \longrightarrow \quad u = \frac{F \, l}{(E_1 + E_2)A} \, . \tag{2.20}$$

The stresses in both bars follow as

$$\sigma_i = E_i \, \frac{F}{(E_1 + E_2)A} \quad \longrightarrow \quad \sigma_i = E_i \, \frac{F}{3EA} \, . \tag{2.21}$$

Note that bar 2 starts to yield for $F = 3 \, A \, \sigma_y$ while bar 1 is still elastic. The associated displacement is $u = \sigma_y \, l \, / \, E$. In case of a further increase of the load, the constant normal force $N_2 = A \, \sigma_y$ in bar 2 has now to be considered, see Fig. 2.9b. This results, analogous to (2.20), in a displacement of

$$u = \left(\frac{F}{A} - \sigma_y \right) \frac{l}{2E} \, . \tag{2.22}$$

Finally, bar 1 yields with $N_{1\,min} = 3 \, A \, \sigma_y$ at an applied force of $F = 4 \, A \, \sigma_y$. After that, it is not possible to increase the load further and the load–displacement diagram in Fig. 2.10 is obtained.

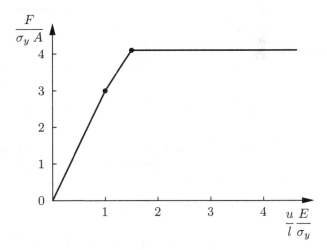

Fig. 2.10 Load–displacement diagram

Again a nonlinear relation between load and displacement is observed as in the previous examples. However, here this nonlinear behaviour stems from the elasto-plastic material law defined in Fig. 2.9b.

In many technical applications, nonlinear behaviour is observed which results from a combination of geometrical and physical nonlinearities. Examples are deep drawing of metal sheets, bulk forming processes or car crashes. The relevant treatment is considered in the following chapters.

2.3 Nonlinearity Due to Boundary Conditions

Another but different source for nonlinearities is related to special boundary constraints. One major cause of such nonlinear behaviour are boundary constraints which change with the deformation state of a system (e.g. during the increase of a prescribed load). These occur when one body comes into contact with another one during a deformation process. Here a penetration of one body into the other is ruled out and the contact zone between the two bodies changes depending on the load level.

The essential characteristics of contact problems are discussed using the following simple model. We consider to bars with stiffness EA, see Fig. 2.11a. The system is subjected to a point force F. Both bars are separated by a gap δ. We look for the displacement and stress state in the bars on the condition that one bar cannot penetrate the other one. This condition leads with Fig. 2.11a to an inequality for the displacements

$$u_1 - u_2 \leq \delta. \tag{2.23}$$

Here the "less" sign is correct in the case that bar 1 does not touch bar 2. The "equal" sign is true for contact of the bars.

In case of $u_1 - u_2 < \delta$ displacement u_2 is zero. Thus the displacement of bar 1 is

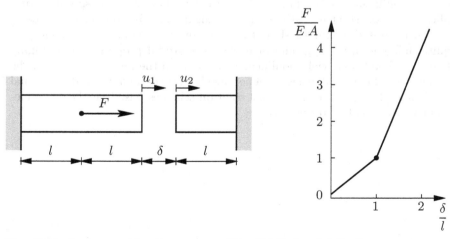

Fig. 2.11a System and loading   Fig. 2.11b Load deflection curve

$$u_1 = \frac{F\,l}{EA}\,. \tag{2.24}$$

An increase of the force F such that $F > EA\frac{\delta}{l}$ is fulfilled leads to contact of bars 1 and 2, then equation $u_1 - u_2 = \delta$ is valid. With the displacements of both bars at $x = 2\,l$

$$u_1 = \frac{F\,l}{EA} + \frac{N_1\,2\,l}{EA} \quad \text{and} \quad u_2 = -\frac{N_2\,l}{EA} \tag{2.25}$$

and the condition that the normal force at the contact point has to be equal for both bars $(N_1(2l) = N_2(2l) = N)$ follows from $u_1 - u_2 = \delta$:

$$\frac{F\,l}{EA} + 3\frac{N\,l}{EA} = \delta \longrightarrow N = \frac{1}{3}\left(EA\frac{\delta}{l} - F\right). \tag{2.26}$$

Using (2.25) and (2.26) yields, in case of contact, a relation between force F and displacement u_1

$$F = EA\left(3\frac{u_1}{l} - 2\frac{\delta}{l}\right). \tag{2.27}$$

This equation includes, for $u_1 = \delta$, the limiting case of the beginning contact in which $N = 0$ is obtained from (2.26). Hence the displacement u_1 of bar 1 can also be computed from (2.24). Figure 2.11b depicts the resulting nonlinear load–deflection curve where the nonlinear behaviour stems only from the contact mechanism.

Basically, geometrical constraints and equilibrium equations have to be considered in contact formulations which are not differentiable since the system can assume two different states of being in contact or being not in contact. This is reflected by the kink in the load deflection curve.

Since contact nonlinearities are linked in many technical applications to further nonlinearities, like finite or inelastic deformations, it is especially complicated to construct robust end efficient algorithms for contact problems.

All discussed examples show that the source of nonlinear behaviour is quite different. Besides geometrical effects, material properties or changing boundary conditions yield nonlinear behaviour. In the following chapter, the underlying theoretical basis is generalized for two- and three-dimensional solids and the necessary numerical solution schemes based on the finite element method will be developed.

3. Basic Equations of Continuum Mechanics

This chapter contains a summary of the continuum mechanics background which is needed for the finite element formulation of solid mechanics and structural problems. The kinematical relations, the balance laws with their weak forms and the constitutive equations are described in detail in this chapter.

Kinematical relations will be formulated for the current and the referential description of motion. Based on that strain measures will be introduced. Variational formulations will be derived which are basis for nonlinear finite element methods. Isotropic hyperelastic material behaviour will be discussed as an example for nonlinear constitutive laws which can be applied to describe large strains not only in three-dimensional solids but also in structural elements like trusses or shells. Furthermore, inelastic material behaviour will be treated for small and finite strain applications within the framework of classical material equations.

Since this book is devoted to nonlinear finite element formulations, the underlying theory of continuum mechanics cannot always be treated in the necessary depth. Hence several extensive derivations are not presented, but at such point the relevant literature will be cited. The reader who needs a more in-depth treatment of continuum mechanics is referred to standard books, e.g. Truesdell and Toupin (1960), Truesdell and Noll (1965), Eringen (1967), Malvern (1969), Becker and Bürger (1975), Altenbach and Altenbach (1994), Chadwick (1999) or Holzapfel (2000) for the basics on continuum mechanics, Ogden (1984) for the theory of elasticity and Marsden and Hughes (1983) or Ciarlet (1988) for the mathematical background of the theory of elasticity.

3.1 Kinematics

The kinematical relations concern the description of the deformation and motion of a body, the derivation of strain measures and the time derivatives of kinematical quantities. All kinematical relations are needed within the constitutive equations and the weak formulation of balance laws.

3.1.1 Motion and Deformation Gradient

In this section, the motion and deformation of homogeneous bodies are considered. Here the continuum approach is applied in which a body B is described in a formal way by a set of continuously distributed points $P \in B$, also called particles or material points, which occupy a region within the EUCLIDEAN point space $\mathbb{E}^3$. The configuration of a body B is a one-to-one mapping $\varphi\colon B \longrightarrow \mathbb{E}^3$, which places the particles of B in $\mathbb{E}^3$. With this definition, the location of a particle X from B is given for the configuration φ as $\mathbf{x} = \varphi(X)$. Thus, the placement of a body B is described by $\varphi(B) = \{\varphi(X) \,|\, X \in B\}$ and called configuration $\varphi(B)$ of body B.

The motion of body B is then given as a one-parametric series of configurations $\varphi_t\colon B \to \mathbb{E}^3$. The location of a particle X at time $t \in \mathbb{R}^+$ yields

$$\mathbf{x} = \varphi_t(X) = \varphi(X,t). \tag{3.1}$$

This equation describes a curve in $\mathbb{E}^3$ for a particle X. $\mathbf{X} = \varphi_0(X)$ defines the reference configuration of body B, with $\mathbf{X}$ being the location of the particle X for this configuration. Thus from (3.1)

$$\mathbf{x} = \varphi(\varphi_0^{-1}(\mathbf{X}), t) \tag{3.2}$$

can be deduced.

Remark 3.1: It is not necessary that the body assumes the reference configuration at any time. Since the reference configuration can be chosen arbitrarily, it is often assumed for practical purposes that the configuration of body B at the beginning of the deformation (initial configuration) is equivalent to the reference configuration. However, there are applications like isoparametric interpolation functions within finite element formulations for which reference configurations will be defined which are purely fictitious.
Usually it is not necessary to distinguish between $\mathbf{X}$ and X. Then the notation simplifies and instead of (3.2), the relation

$$\mathbf{x} = \varphi(\mathbf{X}, t) \tag{3.3}$$

is obtained, where $\mathbf{X}$ represents particle X in the reference configuration B. Based on this, the placements $\mathbf{x}$ and $\mathbf{X}$ can be formulated as position vectors in $\mathbb{E}^3$ with respect the origin $\mathbf{O}$, see Fig. 3.1. Point X is defined in the reference configuration by the position vector $\mathbf{X} = X_A \, \mathbf{E}_A$. $\mathbf{E}_A$ defines an orthogonal base system in the reference configuration with origin $\mathbf{O}$. Hence (3.3) can be written in terms of components of the vector as

$$x_i = \varphi_i(X_A, t). \tag{3.4}$$

If the motion is characterized with respect to the material coordinates $\{X_1, X_2, X_3\}$, this is called material or referential description. In the material description – often also referred to as LAGRANGIAN description of motion – one follows the movement of a particle of body B in time.

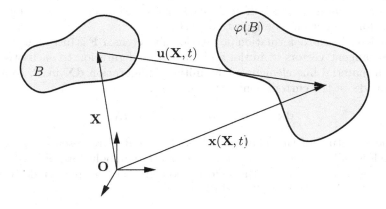

Fig. 3.1 Motion of body B

In the following, capital letters are used as indices for components of vectors and tensors with respect to the basis $\mathbf{E}_A$ of the reference configuration where X_A are the LAGRANGIAN coordinates of the particle X.

Another possibility is the use of the spatial coordinates $\{\, x_1\,,x_2\,,x_3\,\}$ when the motion of body B has to be described. In this formulation, attention is paid to a point in space and the change of the motion with time t at this point. This description is called current, spatial or EULERIAN description of motion. Small letters are used for indices of vectors and tensors which are related to the basis $\mathbf{e}_i$ of the current or spatial configuration. The quantities x_i are the spatial coordinates of X.

For simplicity, an orthogonal cartesian basis will be assumed in the following. This is in accordance with the formulation of numerical methods based on FEM in which often isoparametric interpolations are used which rely on an orthogonal base system. A more general description using curvilinear coordinates is just technical but leads eventually to a significantly more complex formulation.

The equations of continuum mechanics can be formulated with respect to the deformed or undeformed configurations of a body. From the theoretical point of view, there is no difference or preference whether the equations are related to the initial or current configuration. Thus the configuration can be chosen freely. However, physical implications as in the theory of plasticity have to be taken into account, see e.g. (Lubliner (1990), p. 453) when selecting a certain description. Additionally, this selection can have consequences regarding the selected numerical method. Here differences with respect to efficiency can be observed which will be discussed in later chapters.

Since it is not clear from the outset which formulation is preferable, the following strain measures will be derived for the reference configuration B as well as for the current configuration $\varphi(B)$.

To describe the deformation process locally, a tensor $\mathbf{F}$ is introduced which relates tangent vectors of initial and current configuration to each other. It maps a material line element of the initial configuration $\mathbf{dX}$ in B, to a line element $\mathbf{dx}$ of the current configuration $\varphi(B)$.

$$\mathbf{dx} = \mathbf{F}\,\mathbf{dX} \quad \text{or} \quad dx_i = F_{iA}\,dX_A\,. \tag{3.5}$$

By the structure of this equation, it is clear that $\mathbf{F}$ represents a gradient. Hence $\mathbf{F}$ is called deformation gradient. From the symbolic form $\mathbf{F} = \partial\,\mathbf{x}\,/\,\partial\,\mathbf{X}$ follow the components of the deformation gradient as partial derivatives $\partial x_i\,/\,\partial X_A = x_{i,A}$. With (3.3) and (3.4)

$$\mathbf{F} = \operatorname{Grad}\varphi(\mathbf{X}, t) = F_{iA}\,\mathbf{e}_i \otimes \mathbf{E}_A = \frac{\partial x_i}{\partial X_A}\,\mathbf{e}_i \otimes \mathbf{E}_A \tag{3.6}$$

is obtained. The matrix formulation of $\mathbf{F}$ yields

$$[F_{iA}] = \begin{bmatrix} x_{1,1} & x_{1,2} & x_{1,3} \\ x_{2,1} & x_{2,2} & x_{2,3} \\ x_{3,1} & x_{3,2} & x_{3,3} \end{bmatrix}\,. \tag{3.7}$$

Since the gradient in (3.6) is a linear operator, the local transformation in (3.5) is also linear. To maintain the connection of B during the deformation process, the mapping (3.5) has to be one-to-one which excludes a singularity of $\mathbf{F}$. The latter condition can be recast in the form

$$J = \det\mathbf{F} \neq 0\,, \tag{3.8}$$

where J defines a determinant named after JACOBI. Furthermore, to exclude a self penetration of the body, the following constraint has to be fulfilled by the deformation gradient: $J > 0$. Since $\mathbf{F}$ cannot be singular, the inverse $\mathbf{F}^{-1}$ exists, which can be applied to invert relation (3.5)

$$\mathbf{dX} = \mathbf{F}^{-1}\,\mathbf{dx}\,. \tag{3.9}$$

The inverse of the deformation gradient has the following form

$$\mathbf{F}^{-1} = (F^{-1})_{iA}\,\mathbf{E}_A \otimes \mathbf{e}_i \quad \text{with} \quad (F^{-1})_{iA} = \frac{\partial X_A}{\partial x_i}\,. \tag{3.10}$$

Here $\mathbf{X}$ is given by $\mathbf{X} = \varphi^{-1}(\mathbf{x})$.

Knowing the deformation gradient allows to express further transformations of differential quantities between B and $\varphi(B)$. The transformation of surface area elements between B and $\varphi(B)$ is given by the formula of NANSON (see e.g. Ogden (1984), p. 88)

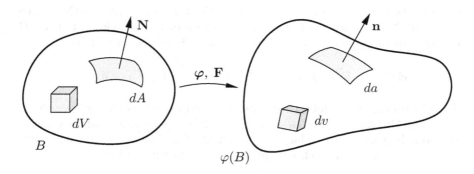

Fig. 3.2 Transformation of differential elements

$$\mathbf{da} = \mathbf{n}\,da = J\,\mathbf{F}^{-T}\,\mathbf{N}\,dA = J\,\mathbf{F}^{-T}\,\mathbf{dA}\,. \tag{3.11}$$

In this equation, $\mathbf{n}$ is the normal vector of the surface of the deformed body $\varphi(B)$ and $\mathbf{N}$ is the normal vector in B, see Fig. 3.2. J is the JACOBI determinant, defined in (3.8) and da and dA are the area elements of the associated configurations, respectively.

The transformation between volume elements of initial and current configuration is provided by the relation

$$dv = J\,dV\,. \tag{3.12}$$

By introducing a displacement vector $\mathbf{u}(\mathbf{X}, t)$ as difference between the position vectors of current and initial configuration

$$\mathbf{u}(\mathbf{X}, t) = \varphi(\mathbf{X}, t) - \mathbf{X}, \tag{3.13}$$

the deformation gradient (3.6) can be written as follows

$$\mathbf{F} = \text{Grad}\,[\,\mathbf{X} + \mathbf{u}(\mathbf{X}, t)\,] = \mathbf{1} + \text{Grad}\,\mathbf{u} = \mathbf{1} + \mathbf{H}\,. \tag{3.14}$$

The tensor $\mathbf{H} = \text{Grad}\,\mathbf{u}$ is called displacement gradient.

3.1.2 Strain Measures

In this section, different strain measures are discussed which are used in forthcoming formulations. The first strain tensor referred to the initial configuration B is defined by

$$\mathbf{E} := \frac{1}{2}\,(\,\mathbf{F}^T\mathbf{F} - \mathbf{1}\,) = \frac{1}{2}\,(\,\mathbf{C} - \mathbf{1}\,) \tag{3.15}$$

and called GREEN-LAGRANGE strain tensor. The tensor $\mathbf{C} := \mathbf{F}^T\,\mathbf{F}$ in (3.15) is the right CAUCHY-GREEN tensor which expresses the square of the line element $\mathbf{dx}$ by the material line element $\mathbf{dX}$: $\mathbf{dx} \cdot \mathbf{dx} = \mathbf{dX} \cdot \mathbf{C}\,\mathbf{dX}$. Hence the strain $\mathbf{E}$ describes the change of the square of the line elements from B to $\varphi(B)$. In component form, the strain tensor $\mathbf{E}$ can be written as

$$\mathbf{E} = E_{AB}\,\mathbf{E}_A \otimes \mathbf{E}_B \qquad \text{with} \quad E_{AB} = \frac{1}{2}\left(F_{iA}\,F_{iB} - \delta_{AB} \right).$$

The KRONECKER symbol δ_{AB} denotes the components of the unit tensor $\mathbf{1}$. Often the GREEN-LAGRANGE strain tensor $\mathbf{E}$ is expressed in analytical investigations by using the displacement gradient. This is actually not necessary when a numerical approach is applied. We obtain with (3.14)

$$\mathbf{E} = \frac{1}{2}\left(\mathbf{H} + \mathbf{H}^T + \mathbf{H}^T\mathbf{H} \right). \tag{3.16}$$

The higher order term $\mathbf{H}^T\mathbf{H}$ depicts the nonlinear character of the GREEN-LAGRANGE strain tensor. Within the geometrically linear theory, this term is neglected by assuming that the displacement gradient is of small order ($\|\mathbf{H}\| \ll 1$). In that case, the strain tensor $\mathbf{E}$ reduces to the linear strain measure ε

$$\varepsilon = \frac{1}{2}\left(\mathbf{H} + \mathbf{H}^T \right) = \frac{1}{2}\left(u_{A,B} + u_{B,A} \right) \mathbf{E}_A \otimes \mathbf{E}_B. \tag{3.17}$$

Exercise 3.1: Determine the deformation gradient (3.6) and the GREEN-LAGRANGE strain tensor (3.16) for a plane deformation. The deformation map is given by $\mathbf{x} = \mathbf{X} + \mathbf{u}(X_1, X_2)$, see Fig. 3.3.

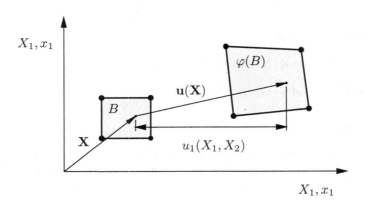

Fig. 3.3 Plane deformation

Solution: $\mathbf{F}$ is computed from (3.7) based on the components of $\mathbf{x}$

$$x_1 = X_1 + u_1(X_1, X_2),$$
$$x_2 = X_2 + u_2(X_1, X_2),$$
$$x_3 = X_3.$$

This yields the matrix form of $\mathbf{F}$

$$[F_{iA}] = \begin{bmatrix} 1 + u_{1,1} & u_{1,2} & 0 \\ u_{2,1} & 1 + u_{2,2} & 0 \\ 0 & 0 & 1 \end{bmatrix}.$$

The GREEN-LAGRANGE strain tensor $\mathbf{E}$ follows with $\mathbf{F}$ from (3.15). However, it is also possible to compute $\mathbf{E}$ with (3.16) but more time-consuming. By matrix multiplication

$$\left[\mathbf{F}^T\mathbf{F}\right] = \begin{bmatrix} (1 + u_{1,1})^2 + u_{2,1}^2 & (1 + u_{1,1})u_{1,2} + (1 + u_{2,2})u_{2,1} & 0 \\ (1 + u_{1,1})u_{1,2} + (1 + u_{2,2})u_{2,1} & (1 + u_{2,2})^2 + u_{1,2}^2 & 0 \\ 0 & 0 & 1 \end{bmatrix}$$

is obtained. Hence the components of the GREEN-LAGRANGE strain tensor are given by

$$E_{11} = u_{1,1} + \frac{1}{2}\left(u_{1,1}^2 + u_{2,1}^2\right),$$

$$E_{22} = u_{2,2} + \frac{1}{2}\left(u_{2,2}^2 + u_{1,2}^2\right),$$

$$E_{12} = \frac{1}{2}\left(u_{1,2} + u_{2,1}\right) + \frac{1}{2}\left(u_{1,1}\,u_{1,2} + u_{2,2}\,u_{2,1}\right).$$

The components E_{33}, E_{13} and E_{23} are zero for a plane deformation. The first term on the right hand side is related to the linear part of the strain measure.

The GREEN-LAGRANGE strain measure is often used in nonlinear structural engineering applications. Mostly, this strain measure is applied for problems with large displacements but small strains (e.g. within beam or shell theory), since it can describe arbitrary rigid body motions correctly. These are defined by

$$\mathbf{x}_R = \mathbf{Q}\,\mathbf{X} + \mathbf{c},$$

where $\mathbf{c}$ is a constant vector describing a translation and $\mathbf{Q}$ is constant proper orthogonal tensor which rotates $\mathbf{X}$. By inserting this motion in (3.6), $\mathbf{F}_R = \mathbf{Q}$ is derived. Due to the property of the orthogonal tensor $\mathbf{Q}$ that $\mathbf{Q}^T\mathbf{Q} = \mathbf{1}$, it follows immediately that $\mathbf{E} = \mathbf{0}$ for a rigid body motion.

A generalization of (3.15) can be found e.g. in Ogden (1984) and is defined by

$$\mathbf{E}^\alpha = \frac{1}{\alpha}\left(\mathbf{U}^\alpha - \mathbf{1}\right), \quad \alpha \in \mathbb{R}. \tag{3.18}$$

This strain tensor is referred to the initial configuration B. It is constructed such that its linearization yields the classical linear strain measure (3.17). For $\alpha = 0$, the strain tensor

$$\mathbf{E}^{(0)} = \ln\mathbf{U} \tag{3.19}$$

follows which is known under the name of HENCKY.

The equivalent to the generalized strain measure (3.18) which is formulated with respect to the current configuration $\varphi(B)$ is given by

$$\mathbf{e}^\alpha = \frac{1}{\alpha}(\mathbf{V}^\alpha - \mathbf{1}), \quad \alpha \in \mathbb{R}. \tag{3.20}$$

The poplar decomposition of the deformation gradient was used within the definition of the strain measures (3.18) and (3.20). This splits the deformation gradient in a multiplicative way in a proper orthogonal rotation tensor $\mathbf{R}$ (with $\mathbf{R}^{-1} = \mathbf{R}^T$) and the symmetrical stretch tensors $\mathbf{U}$ and $\mathbf{V}$, see e.g. (Ogden 1984, p. 92):

$$\mathbf{F} = \mathbf{R}\,\mathbf{U} = \mathbf{V}\,\mathbf{R}, \tag{3.21}$$

$$F_{iB}\,\mathbf{e}_i \otimes \mathbf{E}_B = (R_{iA}\mathbf{e}_i \otimes \mathbf{E}_A)(U_{CB}\,\mathbf{E}_C \otimes \mathbf{E}_B),$$

$$F_{iB}\,\mathbf{e}_i \otimes \mathbf{E}_B = (V_{ik}\mathbf{e}_i \otimes \mathbf{e}_k)(R_{mB}\,\mathbf{e}_m \otimes \mathbf{E}_B).$$

Due to the orthogonality of $\mathbf{R}$, the right CAUCHY-GREEN tensor can be written as $\mathbf{C} = \mathbf{F}^T\,\mathbf{F} = \mathbf{U}^T\,\mathbf{R}^T\,\mathbf{R}\,\mathbf{U} = \mathbf{U}^T\,\mathbf{U} = \mathbf{U}^2$. The last result follows from the symmetry of $\mathbf{U}$. With this result, the GREEN-LAGRANGEstrain tensor (3.15) can be written as $\mathbf{E} = \frac{1}{2}(\mathbf{U}^2 - \mathbf{1})$ which is included in (3.18) for the special case of $\alpha = 2$.

When using the strain measures (3.18) and (3.20), attention has to be paid to the fact that it is only possible to compute $\mathbf{U}$ or $\mathbf{V}$ via a spectral decomposition (e.g. in case of $\alpha = 0.5$ the square root $\mathbf{U}^{1/2}$ of the right stretch tensor $\mathbf{U}$ has to be calculated which can only be achieved by spectral decomposition). Spectral decomposition of $\mathbf{U}$ and $\mathbf{V}$ is provided by

$$\mathbf{U} = \sum_{i=1}^{3} \lambda_i\,\mathbf{N}_i \otimes \mathbf{N}_i, \qquad \mathbf{V} = \sum_{i=1}^{3} \lambda_i\,\mathbf{n}_i \otimes \mathbf{n}_i, \tag{3.22}$$

where λ_i are the principal values of the stretch tensors, also called principal stretches. They are equal for $\mathbf{U}$ and $\mathbf{V}$. The eigenvectors $\mathbf{N}_i$ of $\mathbf{U}$ are related to the reference configuration. The eigenvectors $\mathbf{n}_i$ of $\mathbf{V}$ are referred to the spatial configuration. The eigenvectors $\mathbf{n}_i$ can be obtain from $\mathbf{N}_i$ via the rotation $\mathbf{n}_i = \mathbf{R}\,\mathbf{N}_i$. This result follows directly from (3.21): $\mathbf{V} = \mathbf{R}\,\mathbf{U}\,\mathbf{R}^T = \sum_{i=1}^{3} \lambda_i\,(\mathbf{R}\,\mathbf{N}_I) \otimes (\mathbf{R}\,\mathbf{N}_I)$ using $(3.22)_2$.

Since $\mathbf{C} = \mathbf{U}^2$, it can be easily shown that the spectral decomposition of the right CAUCHY-GREEN tensor is given by

$$\mathbf{C} = \sum_{i=1}^{3} \lambda_i^2\,\mathbf{N}_i \otimes \mathbf{N}_i. \tag{3.23}$$

For the practical computation of the spectral decomposition we refer to Appendix A.1.5.

Exercise 3.2: Compute the principal stretches of a plate which is loaded in plane. The deformation gradient $\mathbf{F}$ is given by

$$\mathbf{F} = \begin{bmatrix} 3 & -1 & 0 \\ 2 & 2 & 0 \\ 0 & 0 & 1 \end{bmatrix}.$$

Furthermore, the right stretch tensor $\mathbf{U}$ and the rotation tensor $\mathbf{R}$ have to be determined.

Solution: The right CAUCHY-GREEN deformation tensor can be written as

$$\mathbf{C} = \mathbf{F}^T \mathbf{F} = \mathbf{U}^2$$

by using the polar decomposition (3.21). The eigenvalues of $\mathbf{C}$ are the square of the eigenvalues of $\mathbf{U}$. Hence it is sufficient to compute the eigenvalues of $\mathbf{C}$, which will be denoted here by λ_i^2. The eigenvalues follow as the zero values of the determinant: $\det\left[\mathbf{C} - \lambda^2 \mathbf{1}\right]$. With the given deformation gradient $\mathbf{F}$,

$$\mathbf{C} = \begin{bmatrix} 13 & 1 & 0 \\ 1 & 5 & 0 \\ 0 & 0 & 1 \end{bmatrix}$$

is obtained and furthermore

$$\det\left[\mathbf{C} - \lambda^2 \mathbf{1}\right] = \begin{vmatrix} (13 - \lambda^2) & 1 & 0 \\ 1 & (5 - \lambda^2) & 0 \\ 0 & 0 & (1 - \lambda^2) \end{vmatrix}$$

is deduced. The associated characteristic polynomial has the form

$$(1 - \lambda^2)\,(\lambda^4 - 18\,\lambda^2 + 64) = 0,$$

and yields the solutions

$$\lambda_{1,2}^2 = 9 \pm \sqrt{17}, \quad \lambda_3^2 = 1.$$

Observe that the third spatial direction (trivial solution $\lambda_3^2 = 1$) is completely decoupled from the other solutions.

To compute $\mathbf{U}$, the eigenvalues λ_i^2 have to be inserted in the homogeneous equation system $\left[\mathbf{C} - \lambda_i^2\,\mathbf{1}\right]\mathbf{N}_i = \mathbf{0}$. This provides, after normalising ($\|\mathbf{N}_i\| = 1$), the eigenvectors of $\mathbf{U}$

$$\mathbf{N}_1 = \left\{ \begin{matrix} 0.993 \\ 0.122 \\ 0.0 \end{matrix} \right\}, \quad \mathbf{N}_2 = \left\{ \begin{matrix} -0.122 \\ 0.993 \\ 0.0 \end{matrix} \right\}, \quad \mathbf{N}_3 = \left\{ \begin{matrix} 0.0 \\ 0.0 \\ 1.0 \end{matrix} \right\}.$$

Now $\mathbf{U}$ follows as

$$\mathbf{U} = \lambda_1\,\mathbf{N}_1 \otimes \mathbf{N}_1 + \lambda_2\,\mathbf{N}_2 \otimes \mathbf{N}_2 + \lambda_3\,\mathbf{N}_3 \otimes \mathbf{N}_3.$$

The dyadic products $\mathbf{N}_i \otimes \mathbf{N}_i$ are given in matrix formulation by $\mathbf{N}_i\,\mathbf{N}_i^T$. Thus the last equation can be written as

$$\mathbf{U} = \lambda_1\,\mathbf{N}_1\,\mathbf{N}_1^T + \lambda_2\,\mathbf{N}_2\,\mathbf{N}_2^T + \lambda_3\,\mathbf{N}_3\,\mathbf{N}_3^T = \begin{bmatrix} 3.60 & 0.17 & 0 \\ 0.17 & 2.23 & 0 \\ 0 & 0 & 1 \end{bmatrix}.$$

With this result $\mathbf{U}$ is known. The associated rotation tensor $\mathbf{R}$ is computed from the relation $\mathbf{R} = \mathbf{F}\,\mathbf{U}^{-1}$. The corresponding matrix multiplication yields after inversion of $\mathbf{U}$

$$\mathbf{R} = \begin{bmatrix} 0.86 & -0.51 & 0 \\ 0.51 & 0.86 & 0 \\ 0 & 0 & 1 \end{bmatrix}.$$

The computation of $\mathbf{R}$ and $\mathbf{U}$ can be simplified when the knowledge of a plane deformation state is used explicitly. The in-plane rotation is described by the ansatz

$$\mathbf{R} = \begin{bmatrix} \cos\theta & \sin\theta & 0 \\ -\sin\theta & \cos\theta & 0 \\ 0 & 0 & 1 \end{bmatrix}$$

as a rotation of two orthogonal eigenvectors. The rotation is described completely by the still unknown angle θ. Since the relation $\mathbf{R}^T = \mathbf{R}^{-1}$ holds, $\mathbf{U}$ can be computed with (3.21) from $\mathbf{U} = \mathbf{R}^T\,\mathbf{F}$. The symmetry condition $\mathbf{U} = \mathbf{U}^T$

$$U_{12} = R_{11}\,F_{12} + R_{21}\,F_{22} \equiv R_{12}\,F_{11} + R_{22}\,F_{21} = U_{21}$$

yields an equation for the unknown angle θ

$$\tan\theta = \frac{F_{12} - F_{21}}{F_{11} + F_{22}},$$

leading in this example to the angle $\theta = -31.0$. Hence the rotation tensor $\mathbf{R}$ is known and the stretch tensor follows from $\mathbf{U} = \mathbf{R}^T\,\mathbf{F}$.

As another special case of the generalized strain measures (3.18) and (3.20), the so-called ALMANSI strain tensor

$$\mathbf{e} := \mathbf{e}^{(-2)} = \frac{1}{2}\,(\,1 - \mathbf{V}^{-2}\,) = \frac{1}{2}\,(\,1 - \mathbf{b}^{-1}\,) = \frac{1}{2}\,(\,1 - \mathbf{F}^{-T}\,\mathbf{F}^{-1}\,) \qquad (3.24)$$

is obtained with $\alpha = -2$ from (3.20). In this equation, the left CAUCHY-GREEN tensor

$$\mathbf{b} := \mathbf{F}\,\mathbf{F}^T = \mathbf{V}\,\mathbf{R}\,\mathbf{R}^T\,\mathbf{V}^T = \mathbf{V}^2 \qquad (3.25)$$

was introduced. This tensor is related to the spatial configuration and will be of significance in later chapters in which constitutive equations are formulated and implemented in a numerical scheme.
The generalized strain measures (3.18) and (3.20) can be written based on the spectral decomposition (3.22) of the stretch tensors $\mathbf{U}$ and $\mathbf{V}$ as

$$\mathbf{E}^\alpha = \frac{1}{\alpha}\sum_{i=1}^{3}(\lambda_i^\alpha - 1)\,\mathbf{N}_i \otimes \mathbf{N}_i \quad \text{and} \quad \mathbf{e}^\alpha = \frac{1}{\alpha}\sum_{i=1}^{3}(\lambda_i^\alpha - 1)\,\mathbf{n}_i \otimes \mathbf{n}_i. \qquad (3.26)$$

With this result, the GREEN-LAGRANGE and the ALMANSI strain tensor are obtained for the special case $\alpha = 2$ and $\alpha = -2$ in terms of the principal stretches

$$\mathbf{E} = \sum_{i=1}^{3} \frac{1}{2}(\lambda_i^2 - 1)\, \mathbf{N}_i \otimes \mathbf{N}_i\,, \quad \text{and} \quad \mathbf{e} = \sum_{i=1}^{3} \frac{1}{2}(1 - \lambda_i^{-2})\, \mathbf{n}_i \otimes \mathbf{n}_i\,. \quad (3.27)$$

Remark 3.2: The principal directions are known before hand in several applications – such as e.g. truss structures or axi-symmetric membranes with isotropic constitutive behaviour – which is shown, e.g. in Sect. 9.3. In such cases, a formulation of the strain measures in principal stretches is advantageous. Furthermore, it is much simpler to fit constitutive equations for elastic materials undergoing finite strains to experimental data when they are formulated in principal strains, see Exercise 3.6. Hence, also in these cases, a formulation in principal stretches should be applied.

For motions which are constraint by special conditions, it is often possible to incorporate these constraint conditions directly into the kinematical relations. In case of incompressibility which plays a prominent role in rubber materials and metal plasticity, the constraint condition $\det \mathbf{F} = J = 1$ has to be fulfilled. The following multiplicative decomposition of the deformation gradient

$$\mathbf{F} = J^{\frac{1}{3}}\, \widehat{\mathbf{F}}, \qquad \widehat{\mathbf{F}} = J^{-\frac{1}{3}}\, \mathbf{F} \qquad (3.28)$$

was suggested in Flory (1961). It preserves a priori the volume of $\widehat{\mathbf{F}}$ (isochoric motion), since $\det \widehat{\mathbf{F}} \equiv 1$.

By inserting (3.28) in (3.15), a relation between the isochoric part of the right CAUCHY–GREEN deformation tensor $\widehat{\mathbf{C}}$ and $\mathbf{C}$ is obtained

$$\widehat{\mathbf{C}} = \widehat{\mathbf{F}}^T \widehat{\mathbf{F}} = J^{-\frac{2}{3}}\, \mathbf{F}^T\, \mathbf{F} = J^{-\frac{2}{3}}\, \mathbf{C}\,. \qquad (3.29)$$

The multiplicative split of $\mathbf{F}$ in a volume changing part (J) and a volume preserving part ($\widehat{\mathbf{F}}$) in the nonlinear theory corresponds to an additive decomposition of the strain tensor in the geometrically linear theory in a deviator $\mathbf{e}_D$ and a volumetric part

$$\boldsymbol{\epsilon} = \mathbf{e}_D + \frac{1}{3}\,\mathrm{tr}\,\boldsymbol{\epsilon}\,\mathbf{1}\,. \qquad (3.30)$$

Exercise 3.3: The deformation gradient given in Exercise 3.2 has to be decomposed in its isochoric and dilatoric part.
Solution: Using (3.28) yields with $J = \det \mathbf{F} = 8$ to the split

$$\mathbf{F} = J^{\frac{1}{3}}\, \widehat{\mathbf{F}} = 2 \begin{bmatrix} 1.5 & -0.5 & 0 \\ 1 & 1 & 0 \\ 0 & 0 & 0.5 \end{bmatrix}\,.$$

3.1.3 Transformation of Vectors and Tensors

Knowledge regarding the transformation between differential quantities in the current and reference configuration is essential for many theoretical derivations and their applications in finite element methods. Tangent fields and one forms which are related to the current configuration can be expressed in terms of quantities in the reference configuration. With the notation introduced in Marsden and Hughes (1983) this is called *pull back*. Conversely a *push forward* relates tangent fields and one forms referred to the reference configuration to the current configuration $\varphi(B)$. For a detailed mathematical background, see e.g. Marsden and Hughes (1983).

Tangent fields or one forms are connected to the base vectors, see Appendix A.2.6. For a covariant gradient of a scalar field $G(\mathbf{X}) = g(\mathbf{x}) = g[\varphi(\mathbf{X})]$ relation

$$\operatorname{Grad} G \;=\; \mathbf{F}^T \operatorname{grad} g \Longleftrightarrow \frac{\partial G}{\partial X_A} = \frac{\partial g}{\partial x_i}\frac{\partial x_i}{\partial X_A}, \tag{3.31}$$

$$\operatorname{grad} g \;=\; \mathbf{F}^{-T} \operatorname{Grad} G \tag{3.32}$$

can be derived. In an analogous way, the transformation for the covariant gradient of the vector field $\mathbf{W}(\mathbf{X}) = \mathbf{w}\,(\mathbf{x}) = \mathbf{w}\,[\varphi(\mathbf{X})]$ is obtained

$$\operatorname{Grad} \mathbf{W} = \operatorname{grad} \mathbf{w}\,\mathbf{F} \Longleftrightarrow \operatorname{grad} \mathbf{w} = \operatorname{Grad} \mathbf{W}\,\mathbf{F}^{-1}. \tag{3.33}$$

As an application, the deformation gradient is computed from a displacement field $\mathbf{u}\,[\varphi(\mathbf{X})]$ which is referred to the current configuration. With (3.14) and (3.33) it follows

$$\begin{aligned}
\mathbf{F} &= 1 + \operatorname{Grad} \mathbf{u} & \mid \mathbf{F}^{-1} \\
1 &= \mathbf{F}^{-1} + \operatorname{Grad} \mathbf{u}\,\mathbf{F}^{-1}, \\
\Longrightarrow \mathbf{F}^{-1} &= 1 - \operatorname{grad} \mathbf{u}.
\end{aligned} \tag{3.34}$$

Hence the inverse of the deformation gradient can be obtained directly with displacements which are referred to the current configuration. This result will be applied later in formulations of the finite element method.

A typical application of a *pull back* operation to tensors is given by the transformation of the ALMANSI strain tensor to the GREEN-LAGRANGE strains using (3.15) and (3.24)

$$\mathbf{E} = \mathbf{F}^T \frac{1}{2}\left(1 - \mathbf{F}^{-T}\mathbf{F}^{-1}\right)\mathbf{F} = \mathbf{F}^T \mathbf{e}\,\mathbf{F}, \tag{3.35}$$

which of course does not change the physical meaning of the strain measure. It only chances the configuration.

Remark 3.3: Initial and current configuration are often parameterized in numerical methods by the introduction of convective coordinates. These can be thought as lines which are carved on the body B, see Appendix A.1.2, especially Fig. A.1. In this parametrisation, it is assumed that the cartesian coordinates $\{X_A\}$ and $\{x_i\}$ can be represented as functions of the convective coordinates $\{\Theta^j\}$. Using convective coordinates, the tangent vector can be computed in each point $\mathbf{X}$ in B as

$$\mathbf{G}_j = \frac{\partial \mathbf{X}}{\partial \Theta^j} = \mathbf{X}_{,j}. \tag{3.36}$$

This is also true for a point which is described with respect to the current configuration $\boldsymbol{\varphi}(\mathbf{X}, t)$ in $\varphi(B)$

$$\mathbf{g}_j = \frac{\partial \boldsymbol{\varphi}(\mathbf{X}, t)}{\partial \Theta^j} = \boldsymbol{\varphi}_{,j}. \tag{3.37}$$

Using the chain rule

$$\mathbf{g}_j = \frac{\partial \boldsymbol{\varphi}(\mathbf{X}, t)}{\partial \mathbf{X}} \frac{\partial \mathbf{X}}{\partial \Theta^j} = \mathbf{F}\,\mathbf{G}_j \tag{3.38}$$

is derived from both previous relations. This means that tangent vectors transform like line elements $\mathbf{dx}$ and $\mathbf{dX}$, see (3.5). With (3.38), it is possible to describe the deformation gradient by the tangent vectors as follows

$$\mathbf{F} = \mathbf{g}_i \otimes \mathbf{G}^i. \tag{3.39}$$

The tangent vectors are covariant vectors which are connected to their contravariant counter parts (one forms) by $\mathbf{g}_i \cdot \mathbf{g}^k = \delta_i{}^k$. Using (3.38) relation

$$\mathbf{F}\,\mathbf{G}_i \cdot \mathbf{A}\,\mathbf{G}^k = \delta_i{}^k \longrightarrow \mathbf{A} = \mathbf{F}^{-T} \tag{3.40}$$

is deduced, where $\mathbf{A} = \mathbf{F}^{-T}$ denotes the transformation tensor for the contravariant base vectors. Hence the transformation $\mathbf{g}^k = \mathbf{F}^{-T}\,\mathbf{G}^k$ is valid.

The covariant or contravariant base vectors can serve as basis for a vector or tensor. Once this basis is known, it is relatively easy to perform *pull back* or *push forward* operations, see Appendix A.2.6. For the GREEN-LAGRANGE strain tensor, this leads with $\mathbf{F}^T\mathbf{F} = (\mathbf{G}^i \otimes \mathbf{g}_i)(\mathbf{g}_k \otimes \mathbf{G}^k)$ to

$$\begin{aligned} \mathbf{E} &= \frac{1}{2}\left(g_{ik} - G_{ik}\right)\mathbf{G}^i \otimes \mathbf{G}^k \tag{3.41} \\ &= \frac{1}{2}\left(g_{ik} - G_{ik}\right)\mathbf{F}^T\,\mathbf{g}^i \otimes \mathbf{F}^T\,\mathbf{g}^k = \mathbf{F}^T\left[\frac{1}{2}\left(g_{ik} - G_{ik}\right)\mathbf{g}^i \otimes \mathbf{g}^k\right]\mathbf{F}. \end{aligned}$$

This result is equivalent to the *pull back* operation in (3.35).

3.1.4 Time Derivatives

The dependence of the deformation $\boldsymbol{\varphi}(\mathbf{X}, t)$ on time t has to be considered in nonlinear problems in case that the constitutive behaviour is history dependent (e.g. in plasticity or visco-elasticity) or in case that the complete process is of dynamical nature. In such applications, time derivatives are needed which will be derived here for kinematical quantities.

The velocity of a material point with respect to the reference configuration is defined by the material time derivative

$$\mathbf{v}(\mathbf{X}, t) = \frac{D\boldsymbol{\varphi}}{Dt} = \frac{\partial \boldsymbol{\varphi}(\mathbf{X}, t)}{\partial t} = \dot{\boldsymbol{\varphi}}(\mathbf{X}, t). \tag{3.42}$$

In the current configuration, the velocity $\hat{\mathbf{v}}$ of a particle which assumes point $\mathbf{x}$ at time t in $\varphi(B)$ is given by

$$\hat{\mathbf{v}}(\mathbf{x}, t) = \hat{\mathbf{v}}(\varphi(\mathbf{X}, t), t) = \mathbf{v}(\mathbf{X}, t). \tag{3.43}$$

The acceleration is given in an analogous way by the second derivative with respect to time

$$\mathbf{a} = \ddot{\boldsymbol{\varphi}}\left(\mathbf{X}, t\right) = \dot{\mathbf{v}}\left(\mathbf{X}, t\right). \tag{3.44}$$

Based on this definition, the acceleration can be determined with reference to the current configuration. With (3.43) and the chain rule, it yields

$$\hat{\mathbf{a}} = \dot{\hat{\mathbf{v}}} = \frac{\partial}{\partial t}\left[\hat{\mathbf{v}}\left(\boldsymbol{\varphi}(\mathbf{X}, t), t\right)\right] = \frac{\partial \hat{\mathbf{v}}}{\partial t} + \operatorname{grad}\hat{\mathbf{v}}\,\hat{\mathbf{v}}. \tag{3.45}$$

The first term is called local part and the second term is called convective part of the acceleration. The local time derivative is computed holding the current position $\mathbf{x}$ fixed. The time derivative (3.45) is of significance in fluid mechanics.

The time derivative of the deformation gradient $\mathbf{F}$ yields with (3.6), (3.42) and (3.33)

$$\dot{\mathbf{F}} = \operatorname{Grad}\dot{\boldsymbol{\varphi}}\left(\mathbf{X}, t\right) = \operatorname{Grad}\mathbf{v} = \operatorname{grad}\hat{\mathbf{v}}\,\mathbf{F}. \tag{3.46}$$

In this equation, the spatial velocity gradient $\operatorname{grad}\hat{\mathbf{v}}$ occurs which is often denoted by l. It can be written with (3.46) as

$$\mathbf{l} = \dot{\mathbf{F}}\,\mathbf{F}^{-1}. \tag{3.47}$$

Equation (3.46) can now be used to compute the time derivative of the GREEN-LAGRANGE strain tensor (3.15)

$$\dot{\mathbf{E}} = \frac{1}{2}\left(\dot{\mathbf{F}}^T\,\mathbf{F} + \mathbf{F}^T\,\dot{\mathbf{F}}\right). \tag{3.48}$$

Using (3.47) in (3.46) yields the time derivative of $\mathbf{E}$

$$\dot{\mathbf{E}} = \mathbf{F}^T\frac{1}{2}\left(\mathbf{l} + \mathbf{l}^T\right)\mathbf{F} = \mathbf{F}^T\,\mathbf{d}\,\mathbf{F}. \tag{3.49}$$

This equation has an equivalent structure as (3.35) and hence denotes a *pull back* of the symmetric spatial velocity gradient $\mathbf{d} = \frac{1}{2}\left(\mathbf{l} + \mathbf{l}^T\right)$ to the reference configuration.

Finally, the convective time derivative of a spatial tensor is considered which is also called LIE derivative. The LIE derivative is defined for a spatial tensor $\mathbf{g}(\mathbf{x}, t)$ with covariant basis by

$$\mathcal{L}_v\,\mathbf{g} := \mathbf{F}\left\{\frac{\partial}{\partial t}\left[\mathbf{F}^{-1}\,\mathbf{g}\,\mathbf{F}^{-T}\right]\right\}\mathbf{F}^T. \tag{3.50}$$

This means that tensor $\mathbf{g}$ must be transformed first to the reference configuration by a *pull back* operation. Here the material time derivative can be computed and afterwards the resulting quantity is related to the current configuration by a *push forward* operation.

The analogous rule for the LIE derivative of a spatial tensor $\hat{\mathbf{g}}$ with contravariant basis is given by

$$\mathcal{L}_v \hat{\mathbf{g}} := \mathbf{F}^{-T} \left\{ \frac{\partial}{\partial t} \left[\mathbf{F}^T \hat{\mathbf{g}} \mathbf{F} \right] \right\} \mathbf{F}^{-1} . \tag{3.51}$$

Using this relation, the LIE derivative of the ALMANSI strain tensor

$$\mathcal{L}_v \mathbf{e} = \mathbf{F}^{-T} \left\{ \frac{\partial}{\partial t} \left[\mathbf{F}^T \mathbf{e} \mathbf{F} \right] \right\} \mathbf{F}^{-1} = \mathbf{F}^{-T} \dot{\mathbf{E}} \mathbf{F}^{-1} \tag{3.52}$$

is obtained which leads to

$$\dot{\mathbf{E}} = \mathbf{F}^T \mathcal{L}_v \mathbf{e} \, \mathbf{F} . \tag{3.53}$$

A comparison with (3.49) shows that the LIE derivative of the ALMANSI strain tensor is equivalent to the symmetric spatial velocity gradient $\mathbf{d}$.

Exercise 3.4: In finite deformation plasticity, often a multiplicative split of the deformation gradient in an elastic and an inelastic part is postulated: $\mathbf{F} = \mathbf{F}_e \mathbf{F}_p$, see Fig. 3.8. Compute the LIE derivative of the spatial strain measure $\mathbf{b}_e = \mathbf{F}_e \mathbf{F}_e^T$.

Solution: The *pull back* of $\mathbf{b}_e$ to the reference configuration leads with (3.50) to

$$\mathbf{F}^{-1} \mathbf{b}_e \mathbf{F}^{-T} = \mathbf{F}^{-1} (\mathbf{F} \mathbf{F}_p^{-1} \mathbf{F}_p^{-T} \mathbf{F}^T) \mathbf{F}^{-T} = \mathbf{F}_p^{-1} \mathbf{F}_p^{-T} . \tag{3.54}$$

The subsequent time derivative yields

$$\frac{\partial}{\partial t} \mathbf{F}_p^{-1} \mathbf{F}_p^{-T} = \dot{\mathbf{F}}_p^{-1} \mathbf{F}_p^{-T} + \mathbf{F}_p^{-1} \dot{\mathbf{F}}_p^{-T} . \tag{3.55}$$

This equation can be rewritten by using the identity $\mathbf{F}_p \mathbf{F}_p^{-1} = \mathbf{1}$, which results in $\dot{\mathbf{F}}_p^{-1} = -\mathbf{F}_p^{-1} \dot{\mathbf{F}}_p \mathbf{F}_p^{-1}$,

$$\frac{\partial}{\partial t} \mathbf{F}_p^{-1} \mathbf{F}_p^{-T} = -\mathbf{F}_p^{-1} \dot{\mathbf{F}}_p \mathbf{F}_p^{-1} \mathbf{F}_p^{-T} - \mathbf{F}_p^{-1} \mathbf{F}_p^{-T} \dot{\mathbf{F}}_p^{T} \mathbf{F}_p^{-T} . \tag{3.56}$$

The final transformation to the spatial configuration by a *push forward* operation yields

$$\mathcal{L}_v \mathbf{b}_e = \mathbf{F} \left(\frac{\partial}{\partial t} \mathbf{F}_p^{-1} \mathbf{F}_p^{-T} \right) \mathbf{F}^T = -\mathbf{F}_e \left(\dot{\mathbf{F}}_p \mathbf{F}_p^{-1} + \mathbf{F}_p^{-T} \dot{\mathbf{F}}_p^{-T} \right) \mathbf{F}_e^T . \tag{3.57}$$

With the definition of a plastic velocity gradient $\tilde{\mathbf{L}}_p = \dot{\mathbf{F}}_p \mathbf{F}_p^{-1}$ in the intermediate plastic configuration analogous to (3.47), see Sect. 3.3.2, the LIE derivative of $\mathbf{b}_e$ is obtained as

$$\mathcal{L}_v \mathbf{b}_e = -2 \mathbf{F}_e \frac{1}{2} \left(\tilde{\mathbf{L}}_p + \tilde{\mathbf{L}}_p^T \right) \mathbf{F}_e^T . \tag{3.58}$$

This result can be interpreted as a *push forward* of the symmetric part of the plastic velocity gradient from the intermediate plastic configuration to the current configuration.

3.2 Balance Equations

This section contains the differential equations which describe the local balance equations such as balance of mass, balance of linear and angular momentum as well as the first law of thermodynamics. These equations represent the fundamental relations of continuum mechanics. A detailed derivation of these equations can be found in e.g. Truesdell and Toupin (1960), Truesdell and Noll (1965), (Malvern (1969), Chap. 5), Marsden and Hughes (1983) and Holzapfel (2000).

3.2.1 Balance of Mass

In this section, only processes are considered in which the mass of a system is conserved. This means that the change of mass has to be zero ($\dot{m} = 0$). Hence an infinitesimal mass element in initial and current configuration has to be equal which leads with $dm(\mathbf{X}) = \rho_0 \, dV$ and $dm(\mathbf{x}) = \rho \, dv$ to

$$\rho \, dv = \rho_0 \, dV \,. \tag{3.59}$$

Here ρ_0 and ρ are the densities in initial and current configuration, respectively. With (3.12), the volume elements dV and dv can be transformed leading to the LAGRANGIAN description of the mass balance

$$\rho_0 = J \, \rho \,. \tag{3.60}$$

For completeness, the rate form of mass continuity in spatial form is presented as

$$\dot{\rho}(\mathbf{x}\,,t) + \rho(\mathbf{x}\,,t) \, \mathrm{div}\,\mathbf{v}(\mathbf{x}\,,t) = 0 \tag{3.61}$$

which follows from the evaluation of

$$\dot{m} = \frac{D}{Dt} \int_B \rho(\mathbf{x}\,,t)\,dv = 0 \,.$$

3.2.2 Balance of Linear and Angular Momentum

The linear momentum or the translational momentum is given in the current and initial configuration with (3.59) by

$$\mathbf{L} = \int_{\varphi(B)} \rho\,\mathbf{v}\,dv = \int_B \rho_0\,\mathbf{v}\,dV \tag{3.62}$$

for the continuous case. The balance of linear momentum reads: *The change of linear momentum* $\mathbf{L}$ *in time (material time derivative) is equal to the sum of all external forces (volume and surface forces) acting on body B.*

Mathematically, this statement can be expressed by

$$\dot{\mathbf{L}} = \int_{\varphi(B)} \rho\,\bar{\mathbf{b}}\,dv + \int_{\varphi(\partial B)} \mathbf{t}\,da. \tag{3.63}$$

$\rho\,\bar{\mathbf{b}}$ defines the volume force (e.g. gravitational force). $\mathbf{t}$ is the stress vector acting on the surface of the body. With CAUCHY's theorem which relates the stress vector $\mathbf{t}$ to the surface normal $\mathbf{n}$ via the linear mapping

$$\mathbf{t} = \boldsymbol{\sigma}\,\mathbf{n}, \quad t_i = \sigma_{ik}\,n_k, \quad \left\{\begin{array}{c} t_1 \\ t_2 \\ t_3 \end{array}\right\} = \left[\begin{array}{ccc} \sigma_{11} & \sigma_{12} & \sigma_{13} \\ \sigma_{21} & \sigma_{22} & \sigma_{23} \\ \sigma_{31} & \sigma_{32} & \sigma_{33} \end{array}\right] \left\{\begin{array}{c} n_1 \\ n_2 \\ n_3 \end{array}\right\} \tag{3.64}$$

(here presented in direct tensor notation, sum- and matrix formulation) the stress vector can be expressed in terms of a stress tensor $\boldsymbol{\sigma}$. Using now the divergence theorem, see Appendix A.2.8, the local balance equation of linear momentum is derived from (3.63). With reference to the current configuration $\varphi(B)$ relation

$$\operatorname{div}\boldsymbol{\sigma} + \rho\,\bar{\mathbf{b}} = \rho\dot{\mathbf{v}}, \quad \sigma_{ik,i} + \rho\,\bar{b}_k = \rho\,\dot{v}_k \tag{3.65}$$

is obtained. The stress tensor $\boldsymbol{\sigma}$ is called CAUCHY stress tensor. $\rho\dot{\mathbf{v}}$ describes the inertial forces which can be neglected in case of purely static investigations.

The angular momentum with reference to a point O given by $\mathbf{x}_0$ is defined with respect to the current and initial configuration with (3.59) as

$$\mathbf{J} = \int_{\varphi(B)} (\varphi - \mathbf{x}_0) \times \rho\mathbf{v}\,dv = \int_B (\varphi - \mathbf{x}_0) \times \rho_0\,\mathbf{v}\,dV. \tag{3.66}$$

The balance of angular momentum can be phrased as follows: *The change in time (material time derivative) of angular momentum* $\mathbf{J}$ *with respect to a point* O *is equal to the sum of all moments stemming from external volume and surface forces with respect to point 0.*

$$\dot{\mathbf{J}} = \int_{\varphi(B)} (\varphi - \mathbf{x}_0) \times \rho\,\bar{\mathbf{b}}\,dv + \int_{\varphi(\partial B)} (\varphi - \mathbf{x}_0) \times \mathbf{t}\,da. \tag{3.67}$$

This equation yields after some manipulations the local balance of angular momentum which simply demands the symmetry of the CAUCHY stress tensor

$$\boldsymbol{\sigma} = \boldsymbol{\sigma}^T, \quad \sigma_{ik} = \sigma_{ki}. \tag{3.68}$$

Observe that the balance of linear and angular momentum leads, in the special case of nonexisting external forces, to conservation of linear and angular momentum

$$\dot{\mathbf{L}} = \mathbf{0} \Leftrightarrow \mathbf{L} = \text{const.}, \tag{3.69}$$

$$\dot{\mathbf{J}} = \mathbf{0} \Leftrightarrow \mathbf{J} = \text{const.} \tag{3.70}$$

3.2.3 First Law of Thermodynamics

Another balance law which postulates the conservation of energy in a thermo-dynamical process is known as the first law of thermodynamics. It reads: *The change in time (material time derivative) of the total energy E is equal to the sum of the mechanical power P of all external loads plus the heat supply Q*

$$\dot{E} = P + Q\,. \tag{3.71}$$

The mechanical power due to volume and surface loads is given by

$$P = \int_{\varphi(B)} \rho\,\bar{\mathbf{b}} \cdot \mathbf{v}\,dv + \int_{\varphi(\partial B)} \mathbf{t} \cdot \mathbf{v}\,da\,. \tag{3.72}$$

The heat supply

$$Q = - \int_{\varphi(\partial B)} \mathbf{q} \cdot \mathbf{n}\,da + \int_{\varphi(B)} \rho\,r\,dv \tag{3.73}$$

consists of a conduction through the surface of the body which is described by the heat flux vector $\mathbf{q}$ and the surface normal $\mathbf{n}$ and a distributed inner heat source r (specific heat supply).

The total energy is composed of the kinetic energy

$$K = \int_{\varphi(B)} \frac{1}{2}\rho\mathbf{v} \cdot \mathbf{v}\,dv \tag{3.74}$$

and the internal energy

$$U = \int_{\varphi(B)} \rho u\,dv\,. \tag{3.75}$$

u is the specific internal energy. Inserting all these relations into equation $\dot{E} = P + Q$ yields after several manipulations the local form of the first law of thermodynamics

$$\rho\,\dot{u} = \boldsymbol{\sigma} \cdot \mathbf{d} + \rho\,r - \mathrm{div}\,\mathbf{q}\,. \tag{3.76}$$

The term $\boldsymbol{\sigma} \cdot \mathbf{d}$ is called specific stress power.

In the framework of constitutive theory, the free HELMHOLTZ energy ψ is often introduced by the relation

$$\psi = u - \eta\,\theta\,. \tag{3.77}$$

Here η denotes the entropy of the system and θ is the absolute temperature. With this definition, the first law of thermodynamics can be recast as

$$\rho\,\dot{\psi} = \boldsymbol{\sigma} \cdot \mathbf{d} + \rho\,r - \mathrm{div}\,\mathbf{q} - \dot{\eta}\,\theta - \eta\,\dot{\theta}\,. \tag{3.78}$$

The special case that neither heat is supplied to an elastic body nor external forces act on the body leads to conservation of total energy

$$\dot{E} = \dot{K} + \dot{U} = 0 \Leftrightarrow E = \mathrm{const.} \tag{3.79}$$

3.2.4 Introduction of Different Stress Tensors

Equations (3.65) and (3.68) are referred to the current configuration. Often it is desirable to relate all quantities to the initial configuration B. For this purpose, further stress tensors have to be introduced. Since a given stress vector does not change when referred to the current or initial configuration, the following transformation can be performed using NANSON'S formula (3.11) for surface elements

$$\int_{\partial\varphi(B)} \boldsymbol{\sigma}\,\mathbf{n}\,da = \int_{\partial B} \boldsymbol{\sigma}\,J\,\mathbf{F}^{-T}\,\mathbf{N}\,dA = \int_{\partial B} \mathbf{P}\,\mathbf{N}\,dA\,, \qquad (3.80)$$

which defines the first PIOLA–KIRCHHOFF stress tensor $\mathbf{P}$. Observe that the first PIOLA-KIRCHHOFF stress can be written in terms of the CAUCHY stress

$$\mathbf{P} = J\,\boldsymbol{\sigma}\,\mathbf{F}^{-T} \qquad P_{Ai} = J\,\sigma_{ik}(F_{Ak})^{-1}\,. \qquad (3.81)$$

The spatial stress tensor $\boldsymbol{\sigma}$ in (3.81) is multiplied only from one side by $\mathbf{F}$, hence the tensor $\mathbf{P}$ is a two field tensor with one basis referred to the current and the other to the initial configuration.

Naturally, it is simpler to work in the initial configuration with symmetrical stress tensors the second PIOLA-KIRCHHOFF stress was introduced. This tensor results from a complete transformation of the CAUCHY stress to the initial configuration of B

$$\begin{aligned}
\mathbf{S} &= \mathbf{F}^{-1}\,\mathbf{P} = J\,\mathbf{F}^{-1}\,\boldsymbol{\sigma}\,\mathbf{F}^{-T}\,, & (3.82)\\
S_{AB} &= (F_{iA})^{-1}\,P_{Bi} = J\,(F_{iA})^{-1}\,\sigma_{ik}(F_{Bk})^{-1}\,. & (3.83)
\end{aligned}$$

$\mathbf{S}$ does not represent a stress which can be interpreted physically. Hence it is a pure mathematical quantity which however plays a prominent role in constitutive theory, since $\mathbf{S}$ is work conjugated to the GREEN-LAGRANGE strain tensor (3.15).

Besides the CAUCHY stress tensor $\boldsymbol{\sigma}$ often the so-called KIRCHHOFF stress tensor $\boldsymbol{\tau}$ is introduced which results from a *push forward* of the second PIOLA-KIRCHHOFF stress tensor $\mathbf{S}$ to the current configuration

$$\boldsymbol{\tau} = \mathbf{F}\,\mathbf{S}\,\mathbf{F}^{T}\,, \qquad \boldsymbol{\tau} = J\,\boldsymbol{\sigma}\,. \qquad (3.84)$$

Spectral decomposition can also be applied to the stress tensors. It yields for the CAUCHY, the KIRCHHOFF and the first and second PIOLA-KIRCHHOFF stress tensors

$$\begin{aligned}
\boldsymbol{\sigma} &= \sum_{i=1}^{3} \sigma_i\,\mathbf{m}_i \otimes \mathbf{m}_i\,, & \boldsymbol{\tau} &= \sum_{i=1}^{3} \tau_i\,\mathbf{m}_i \otimes \mathbf{m}_i\,, \\
\mathbf{P} &= \sum_{i=1}^{3} P_i\,\mathbf{m}_i \otimes \mathbf{M}_i\,, & \mathbf{S} &= \sum_{i=1}^{3} S_i\,\mathbf{M}_i \otimes \mathbf{M}_i\,.
\end{aligned} \qquad (3.85)$$

3.2.5 Balance Equations with Respect to Initial Configuration

With the first PIOLA-KIRCHHOFF stress, the local balance of linear momentum (3.65) can be recast with respect to the initial configuration as

$$\text{DIV}\,\mathbf{P} + \rho_0\,\bar{\mathbf{b}} = \rho_0\,\dot{\mathbf{v}}, \tag{3.86}$$

where DIV denotes the divergence operation with respect to the initial configuration. Furthermore, the use of (3.81) in the balance of angular momentum (3.68) yields

$$\mathbf{P}\,\mathbf{F}^T = \mathbf{F}\,\mathbf{P}^T. \tag{3.87}$$

From this it is clear that the first PIOLA–KIRCHHOFF stress tensor is non-symmetric. Using (3.82), the balance of angular momentum yields the symmetry of the second PIOLA-KIRCHHOFF stress tensor: $\mathbf{S} = \mathbf{S}^T$.

Transformation of the first law of thermodynamics (3.76) to the initial configuration can be obtained with the transformation of the stress power using (3.49)

$$J\boldsymbol{\sigma}\cdot\mathbf{d} = \mathbf{F}\,\mathbf{S}\,\mathbf{F}^T \cdot \mathbf{F}^{-T}\dot{\mathbf{E}}\,\mathbf{F}^{-1} = \mathbf{S}\cdot\dot{\mathbf{E}} \tag{3.88}$$

and (3.59) as

$$\rho_0\,\dot{u} = \mathbf{S}\cdot\dot{\mathbf{E}} - \text{DIV}\,\mathbf{Q} + \rho_0\,R. \tag{3.89}$$

Here the heat source R and the heat flux vector $\mathbf{Q}$ are referred to the initial configuration. The stress power (3.88) can be written with (3.84) or (3.15) as

$$\mathbf{S}\cdot\dot{\mathbf{E}} = \frac{1}{2}\mathbf{S}\cdot\dot{\mathbf{C}} = \boldsymbol{\tau}\cdot\mathbf{d}. \tag{3.90}$$

Here the first two terms are related to the initial configuration whereas the last term is referred to the current configuration.

3.2.6 Time Derivatives of Stress Tensors

The time derivative of stress tensors is of significance for the statement of incremental forms of constitutive equations. For stresses which are referred to the initial configuration (e.g. the second PIOLA-KIRCHHOFF stress tensor $\mathbf{S}$), the derivative with respect to time is given by the material time derivative

$$\dot{\mathbf{S}} = \frac{\partial \mathbf{S}(\mathbf{X}, t)}{\partial t}. \tag{3.91}$$

Time derivatives for stress tensors like the CAUCHY stress tensor $\boldsymbol{\sigma}$ which are related to the current configuration are computed according to (3.45)

$$\dot{\boldsymbol{\sigma}} = \frac{\partial \boldsymbol{\sigma}}{\partial t} + \text{grad}\,\boldsymbol{\sigma}\,\mathbf{v}, \tag{3.92}$$

$$\dot{\sigma}_{ik} = \frac{\partial \sigma_{ik}}{\partial t} + \frac{\partial \sigma_{ik}}{\partial x_l}\,v_l. \tag{3.93}$$

It can easily be shown, see e.g. Truesdell and Toupin (1960), that the material time derivative of the CAUCHY stress tensor is not objective, but objectivity is an inevitable prerequisite for the formulation of constitutive equations. Hence numerous time derivatives were formulated – so-called objective time derivatives – which can be applied to compute stress rates. The LIE derivative of a stress tensor provides an objective stress rate, see, e.g. Truesdell and Toupin (1960) or Marsden and Hughes (1983). It is given for the KIRCHHOFF stress tensor using (3.50) as

$$\mathcal{L}_v \boldsymbol{\tau} = \mathbf{F} \left\{ \frac{\partial}{\partial t} \left[\mathbf{F}^{-1} \boldsymbol{\tau} \mathbf{F}^{-T} \right] \right\} \mathbf{F}^T . \tag{3.94}$$

With $\dot{\mathbf{F}}^{-1} = -\mathbf{F}^{-1} \dot{\mathbf{F}} \mathbf{F}^{-1}$ and some algebraic manipulations,

$$\mathcal{L}_v \boldsymbol{\tau} = \dot{\boldsymbol{\tau}} - \mathbf{l}\boldsymbol{\tau} - \boldsymbol{\tau} \mathbf{l}^T = \overset{\triangle}{\boldsymbol{\tau}} \tag{3.95}$$

can be derived using (3.47). The term $\overset{\triangle}{\boldsymbol{\tau}}$ is also called OLDROYD stress rate, see, e.g. Marsden and Hughes (1983). It is equivalent to the LIE derivative of the KIRCHHOFF stress tensor. Observe that the LIE derivative of $\boldsymbol{\tau}$ is obtained as *push forward* of the material time derivative of the second PIOLA-KIRCHHOFF stress if (3.84) is employed in (3.94)

$$\mathcal{L}_v \boldsymbol{\tau} = \mathbf{F} \dot{\mathbf{S}} \mathbf{F}^T . \tag{3.96}$$

Another objective stress rate called the JAUMANN stress rate is applied in many formulations of elasto-plastic material behaviour at finite strains. This rate is defined by

$$\overset{\triangledown}{\boldsymbol{\tau}} = \dot{\boldsymbol{\tau}} - \mathbf{w} \boldsymbol{\tau} + \boldsymbol{\tau} \mathbf{w} . \tag{3.97}$$

where $\mathbf{w} = \frac{1}{2} (\mathbf{l} - \mathbf{l}^T) = -\mathbf{w}^T$ is the skew symmetric part of the spatial velocity gradient. Since $\mathbf{l} = \mathbf{d} + \mathbf{w}$ is valid, the LIE derivative of $\boldsymbol{\tau}$ can be written with (3.95) as

$$\mathcal{L}_v \boldsymbol{\tau} = \overset{\triangledown}{\boldsymbol{\tau}} - \mathbf{d} \boldsymbol{\tau} - \boldsymbol{\tau} \mathbf{d} . \tag{3.98}$$

This relates the JAUMANN stress rate to the LIE derivative (3.94).

By the exchange of the deformation gradient $\mathbf{F}$ by the rotation tensor for the polar decomposition $\mathbf{R}$ in the LIE derivatives above further objective stress rates can be defined. An example is given by

$$L_{\mathbf{v}}^R(\boldsymbol{\tau}) = \dot{\boldsymbol{\tau}} - \boldsymbol{\Omega} \boldsymbol{\tau} + \boldsymbol{\tau} \boldsymbol{\Omega} \quad \text{with} \quad \boldsymbol{\Omega} = \dot{\mathbf{R}} \mathbf{R}^T . \tag{3.99}$$

This stress rate is called GREEN-NAGHDI stress rate. In case that $\mathbf{d} \equiv \mathbf{0}$, it can be shown that the JAUMANN stress rate is identical to the GREEN-NAGHDI stress rate since $\mathbf{w} = \boldsymbol{\Omega}$.

3.3 Constitutive Equations

The kinematical relations and balance laws derived so far are not sufficient to solve a boundary or initial value problem in continuum mechanics. For a complete set of equations, a constitutive equation has to be formulated which characterizes the material response of a solid body.

The constitutive theory describes, in relation to the nature of the task, either the microscopic or the macroscopic behaviour of a material. For most materials like steel or concrete which are used in technical applications, a macroscopic description is sufficient. In that case the functional dependence of stresses or heat flux with respect to the motion or temperature has to be considered. Since real materials can exhibit very complex behaviour, approximations have to be applied within the derivation process of constitutive equations. These, however, have to be extensive enough to cover all effects observed in experimental investigations. Furthermore, basic principles from mechanics have to be obeyed to obtain theoretically sound constitutive equations. These principles, which are listed in the following, can contribute on their part to a simplification of the constitutive equations.

Using the principle of determinism, a decision will be made with regard to independent and dependent variables which occur in the constitutive equations. Classically motion and temperature are chosen as unknowns. The principle of equipresence demands the same set of variables for all constitutive equations. By the principle of local action, the material functions are restricted to a pointwise dependence on the deformation gradient, the temperature and its gradient. Finally, the invariance of constitutive equations with respect to rigid body motions is postulated. This specifies the form of the material law, e.g. the deformation gradient $\mathbf{F}$ is exchanged as variable by the right stretch tensor $\mathbf{U}$ or the right CAUCHY-GREEN tensor $\mathbf{C}$.

Another essential restriction for constitutive equations is provided by the second law of thermodynamics. The second law of thermodynamics postulates that heat cannot flow itself from a system with low temperature to a system with a higher temperature. Another physical observation is that a substance with equally distributed temperature which is free of heat sources can only receive mechanical energy but not release it. These observations lead two inequalities, see Truesdell and Noll (1965, S.295), which contain mathematical statements regarding the local entropy production and the entropy production as a result of heat conduction. An essential postulate states that for closed systems the entropy always increases ($d\eta > 0$) within an irreversible process. With this the direction of process has to be considered. Since only the weaker form of the second law of thermodynamics is needed one inequality is sufficient, see e.g. Malvern (1969, S. 255). By introducing the absolute temperature $\theta : (\theta > 0)$, the entropy production is given by

$$\Gamma \equiv \frac{d}{dt} \int\limits_{\phi(B)} \rho\,\eta\,dv - \int\limits_{\phi(B)} \frac{\rho\,r}{\theta}\,dv + \int\limits_{\phi(\partial B)} \frac{1}{\theta}\,\mathbf{q}\cdot\mathbf{n}\,.\,da. \qquad (3.100)$$

The postulate that entropy production Γ is always larger than zero, $\Gamma \geq 0$, leads by addition of the energy balance (3.76) to the second law of thermodynamics

$$\rho \dot{\eta} \geq \frac{\rho r}{\theta} - \mathrm{div}\left(\frac{\mathbf{q}}{\theta}\right). \tag{3.101}$$

With the introduction of the free HELMHOLTZ energy (3.77), $\psi = e - \eta\,\theta$, the so-called reduced form of the second law of thermodynamics can be defined by using (3.76)

$$\rho\left(\dot{\theta}\eta + \dot{\psi}\right) - \boldsymbol{\sigma}\cdot\mathbf{d} + \frac{1}{\theta}\mathbf{q}\cdot\mathrm{grad}\,\theta \leq 0. \tag{3.102}$$

The free HELMHOLTZ energy ψ denotes the part of inner energy which performs work at constant temperature. The free HELMHOLTZ energy is relevant for the construction of constitutive equations, since its derivation with respect to a strain measure yields the stresses.

With inequalities (3.101) and (3.102), the irreversibility of processes can be described in which mechanical energy is transformed to heat energy (e.g. in case of friction or inelastic deformations).

The material form of (3.101) is derived in the same way as the first law of thermodynamics, leading to

$$\rho_R \dot{\eta} \geq \rho_R \frac{R}{\theta} - \mathrm{Div}\left(\frac{\mathbf{Q}}{\theta}\right). \tag{3.103}$$

Some special cases of thermodynamical processes can now be stated: (1) supply of heat energy is excluded as well in the interior as over the surface of the body ($R = 0\,,\mathbf{q} = \mathbf{0}$); such process is called *adiabatic*. (2) A process in which the temperature in the body is kept constant ($\theta = const.$) is known as *isothermal* process.

3.3.1 Elastic Material

Purely elastic material behaviour is discussed in this section under the assumption of so-called GREEN elasticity, also named hyper elasticity, see, e.g. Ogden (1984, Chap. 4). This description is valid for many materials – like, e.g. foam or rubbers – which undergo finite deformations. In case of small strains, these constitutive equations reduce to the classical law of HOOKE known from the linear theory of elasticity.

The constitutive equation for the second PIOLA-KIRCHHOFF stress tensor can be derived from the potential ψ in case of a hyper elastic material. ψ describes the strain energy stored in the body (for this reason it is called strain energy function). The derivative of ψ with respect to the right CAUCHY-GREEN tensor yields

$$\mathbf{S} = 2\,\rho_0\,\frac{\partial\psi(\mathbf{C})}{\partial\mathbf{C}}\,, \qquad S_{AB} = 2\,\rho_0\,\frac{\partial\psi(C_{CD})}{\partial C_{AB}}. \tag{3.104}$$

The exclusive dependence of ψ on $\mathbf{C}$ is substantiated by the general principles as discussed in the last section. For a more detailed treatment, see e.g. Truesdell and Noll (1965), Malvern (1969) and Ogden (1984).

The material behaviour is independent on directions for technically important classes of isotropic materials which includes, e.g. steel, aluminium, rubber or concrete. With this assumption, it is possible to specify the general function ψ in (3.104). By introducing isotropy groups, a function can be derived which only depends upon the invariants of the strain tensors, see e.g. Ogden (1984). For the right CAUCHY-GREEN deformation tensor $\mathbf{C} = \mathbf{F}^T \mathbf{F}$ and the left CAUCHY-GREEN deformation tensors $\mathbf{b} = \mathbf{F} \mathbf{F}^T$, which have due to their definition the same invariants, it follows

$$\psi(\mathbf{C}) = \psi(I_C, II_C, III_C) = \psi(I_b, II_b, III_b) = \psi(\mathbf{b}). \qquad (3.105)$$

The invariants I_C, II_C and III_C are defined in Appendix A.1.5.

The strain energy function ψ can be written with $\mathbf{C} = \mathbf{U}^2$ or $\mathbf{b} = \mathbf{V}^2$ as a function of the right ($\mathbf{U}$) or left ($\mathbf{V}$) stretch tensor

$$\bar{\psi}(\mathbf{U}) = \bar{\psi}(I_U, II_U, III_U). \qquad (3.106)$$

With the relations, see also Appendix A.1.5,

$$\begin{array}{rcl} I_C & = & \lambda_1^2 + \lambda_2^2 + \lambda_3^2, \\ II_C & = & \lambda_1^2 \lambda_2^2 + \lambda_2^2 \lambda_3^2 + \lambda_3^2 \lambda_1^2, \\ III_C & = & \lambda_1^2 \lambda_2^2 \lambda_3^2, \end{array} \qquad (3.107)$$

the invariants can be expressed in terms of the principal stretches, see e.g. Truesdell and Noll (1965). Thus the strain energy function assumes the form

$$\psi(\mathbf{C}) \equiv \psi(\mathbf{b}) = \psi(\lambda_1^2, \lambda_2^2, \lambda_3^2) \qquad (3.108)$$

when using the principal stretches λ_i^2 of $\mathbf{C}$ or $\mathbf{b}$.

This description of the material by an isotropic tensor function (3.105), see e.g. Ogden (1984), leads by using the chain rule to the following relation between the second PIOLA-KIRCHHOFF stresses and the right CAUCHY-GREEN tensor

$$\mathbf{S} = 2\rho_0 \left[\left(\frac{\partial \psi}{\partial I_C} + I_C \frac{\partial \psi}{\partial II_C} \right) \mathbf{1} - \frac{\partial \psi}{\partial II_C} \mathbf{C} + III_C \frac{\partial \psi}{\partial III_C} \mathbf{C}^{-1} \right]. \qquad (3.109)$$

In this equation, the relations

$$\frac{\partial I_C}{\partial \mathbf{C}} = \mathbf{1}, \quad \frac{\partial II_C}{\partial \mathbf{C}} = I_C \mathbf{1} - \mathbf{C}, \quad \frac{\partial III_C}{\partial \mathbf{C}} = III_C \mathbf{C}^{-1} \qquad (3.110)$$

were applied. They describe the derivatives of the invariants of a tensor with respect to the tensor itself.

Exercise 3.5: Relate the constitutive equation (3.109) to the current configuration and express the CAUCHY stress tensor in terms of the left CAUCHY-GREEN tensors.

Solution: With (3.83) $\boldsymbol{\sigma} = J^{-1} \mathbf{F} \mathbf{S} \mathbf{F}^T$ is obtained and hence by considering (3.104)

$$\boldsymbol{\sigma} = 2\rho \mathbf{F} \frac{\partial \psi(\mathbf{C})}{\partial \mathbf{C}} \mathbf{F}^T$$

follows. This yields with (3.109)

$$\boldsymbol{\sigma} = 2\rho \left[\left(\frac{\partial \psi}{\partial I_C} + I_C \frac{\partial \psi}{\partial II_C} \right) \mathbf{F} \mathbf{F}^T - \frac{\partial \psi}{\partial II_C} \mathbf{F} \mathbf{C} \mathbf{F}^T + III_C \frac{\partial \psi}{\partial III_C} \mathbf{F} \mathbf{C}^{-1} \mathbf{F}^T \right].$$

Since the invariants of $\mathbf{C}$ and $\mathbf{b}$ are equal, the required result

$$\boldsymbol{\sigma} = 2\rho \left[\left(\frac{\partial \psi}{\partial I_b} + I_b \frac{\partial \psi}{\partial II_b} \right) \mathbf{b} - \frac{\partial \psi}{\partial II_b} \mathbf{b}^2 + III_b \frac{\partial \psi}{\partial III_b} \mathbf{1} \right]$$

is obtained with $\mathbf{F} \mathbf{C}^{-1} \mathbf{F}^T = \mathbf{1}$. By comparison with (3.109), it can be shown that the equation is equivalent to

$$\boldsymbol{\sigma} = 2\rho \mathbf{b} \frac{\partial \psi(\mathbf{b})}{\partial \mathbf{b}}. \tag{3.111}$$

This relation can be used to compute the CAUCHY stress tensor directly for a given strain energy function ψ of an isotropic material.

Constitutive equations of the form (3.109) are still very complex because ψ can be an arbitrary function of the invariants. This can result in a large number of constitutive parameters which are not easily determined by experiments. Hence it is desirable to formulate material functions in nonlinear elasticity with a minimum number of constitutive parameters. First formulations can be found in Mooney (1940) for incompressible rubber materials and Rivlin (1948) who postulated[1]

$$W(I_C, II_C) = c_1 (I_C - 3) + c_2 (II_C - 3). \tag{3.112}$$

For a complete formulation of an incompressible problem, the constraint ($J - 1 = 0$) has to be introduced which can be achieved by using the method of LAGRANGE multipliers.

For an in-depth discussion of further strain energy functions for rubber materials, reference is made to the overview article by Ogden (1972). A very good fit to experimental data is provided by a generalized strain energy function which was developed by Ogden (1972). An extension to compressible materials like foams is possible by introducing further the term $g(J)$

$$W(\lambda_k) = \sum_{i=1}^{r} \mu_i K_i (\lambda_k) + g(J) \text{ with } K_i(\lambda_k) = \frac{1}{\alpha_i} (\lambda_1^{\alpha_i} + \lambda_2^{\alpha_i} + \lambda_3^{\alpha_i} - 3). \tag{3.113}$$

---

[1] In this equation and all following equations, the specific strain energy $\rho_0 \psi$ is written as W to simplify notation.

The construction of this constitutive equation using principal stretches is motivated by the generalized strain measures (3.18) and (3.20). The constitutive parameters μ_i and α_i have to be determined from experiments. However when fitting the experimental data, the condition has to be obeyed that the constitutive equation has to reduce to HOOKE law for small strains. Furthermore, mathematical investigations regarding existence of solutions restrict the choice of the constitutive parameters further, see e.g. Marsden and Hughes (1983), Ogden (1984) and Ciarlet (1988). The associated conditions are

$$\sum_{i=1}^{r} \mu_i \alpha_i = 2\mu \quad \text{and} \quad \mu_i \alpha_i > 0. \tag{3.114}$$

The first restriction produces the effect that the strain energy function (3.113) yields at $\lambda_k = 1$ the constitutive tensor of the classical linear theory (parameter μ corresponds to the shear modulus). The second restriction is related to the existence of solutions in finite elasticity, see Ogden (1972). Parameters determined from experiments have to obey both restrictions. To further fulfil polyconvexity which ensures existence of solutions in finite elasticity, the inequalities $\mu_i > 0$ and $\alpha_i > 1$ or $\mu_i < 0$ and $\alpha_i < 1$ have to be met by the constitutive parameters, see Marsden and Hughes (1983). This demand is stronger than the second restriction in (3.114). Usually the constitutive parameters μ_i and α_i of real materials which obey restriction $(3.114)_2$ also fulfil the stronger restriction of polyconvexity.

The strain energy function of the MOONEY-RIVLIN material can be obtained from (3.113) by using $r = 2$, $c_1 = \frac{1}{2}\mu_1$, $c_2 = -\frac{1}{2}\mu_2$, and $\alpha_1 = 2$, $\alpha_2 = -2$. This leads with $g(J) = 0$ to

$$W(\lambda_i) = \frac{\mu_1}{2}(\lambda_1^2 + \lambda_2^2 + \lambda_3^2 - 3) - \frac{\mu_2}{2}(\lambda_1^{-2} + \lambda_2^{-2} + \lambda_3^{-2} - 3) \text{ with } \mu_1 - \mu_2 = \mu. \tag{3.115}$$

Neo-Hooke Material. The choice of $W(\lambda_k) = \mu_1 K_1(\lambda_k) + g(J)$ with $\alpha_1 = 2$ and $\mu_1 = \mu$ yields the special case of a compressible NEO-HOOKE material which can be transformed with $(3.108)_1$ to the form

$$W(I_C, J) = g(J) + \frac{1}{2}\mu(I_C - 3). \tag{3.116}$$

The function $g(J)$ in (3.113) and (3.116) has to be convex for compressible materials. Furthermore, the growth conditions

$$\lim_{J \to +\infty} W \to \infty \quad \text{and} \quad \lim_{J \to 0} W \to \infty \tag{3.117}$$

have to be fulfilled by the strain energy W. The latter conditions can be physically interpreted in the way that stresses have to approach $-\infty$ for a volume going to zero and $+\infty$ for a volume approaching ∞. These growth conditions also play a role in the mathematical treatment of finite elasticity,

e.g. for questions regarding existence and uniqueness of solutions, see e.g.
Marsden and Hughes (1983), Ciarlet (1988).

In Ciarlet (1988), a special ansatz was chosen for the compressible part
in (3.116) to fulfil the growth conditions

$$g(J) = c\,(J^2 - 1) - d \ln J - \mu \ln J \quad \text{with} \quad c > 0, d > 0. \tag{3.118}$$

Further choices for the function $g(J)$ and a related discussion can be found
in Doll and Schweizerhof (2000).

By inserting relation (3.118) for $g(J)$ into (3.116), a constitutive relation
for the second PIOLA-KIRCHHOFF stress tensor is obtained. With the special
choice of $c = \Lambda / 4$ and $d = \Lambda / 2$, it follows

$$\mathbf{S} = \frac{\Lambda}{2}\,(\,J^2 - 1\,)\,\mathbf{C}^{-1} + \mu\,(\,\mathbf{1} - \mathbf{C}^{-1}\,). \tag{3.119}$$

The constitutive parameters Λ and μ are known as the LAMÉ constants.
It should be remarked that the constitutive relation for the second PIOLA-
KIRCHHOFF stress tensor does not have anything in common with a linear
relation between $\mathbf{S}$ and $\mathbf{E}$ which is often used in engineering applications of
structural members, see also Remark 3.4.

Equation (3.119) can be referred to the current configuration by express-
ing the second PIOLA-KIRCHHOFF stress tensor by the CAUCHY stress tensor
via $\boldsymbol{\sigma} = J^{-1}\,\mathbf{F}\,\mathbf{S}\,\mathbf{F}^T$. After some algebraic manipulations, relation

$$\boldsymbol{\sigma} = \frac{\Lambda}{2\,J}\,(\,J^2 - 1\,)\,\mathbf{1} + \frac{\mu}{J}\,(\,\mathbf{b} - \mathbf{1}\,) \tag{3.120}$$

is obtained where all terms are living in the spatial configuration.

Remark 3.4: Nonlinear elastic material behaviour is often described in engineering literature by a linear relation between the second PIOLA-KIRCHHOFF stress
tensor and the GREEN-LAGRANGE strain tensor (ST. VENANT material)

$$\mathbf{S} = \Lambda\,\text{tr}\,\mathbf{E}\,\mathbf{1} + 2\,\mu\,\mathbf{E}. \tag{3.121}$$

This constitutive equation corresponds to HOOKE'S laws of the infinitesimal theory
of elasticity with the LAMÉ constants Λ and μ (these constants can be converted
to the modulus of elasticity $E = \frac{(3\Lambda + 2\mu)\,\mu}{\Lambda + \mu}$ and POISSON'S ratio $\nu = \frac{\Lambda}{2\,(\Lambda + \mu)}$).

Generally, it can be shown that the constitutive equation (3.121) is restricted
to deformations with large displacements and finite rotations but small strains. ST.
VENANT'S law depicts major deficiencies in the compressible range: in the limit case
of the compression of a body to volume "0" the stress $\boldsymbol{\sigma}$ approaches zero instead of
$\lim_{J \to 0} \boldsymbol{\sigma} \to -\infty$. With such behaviour, the material equation provided in (3.121) is
not applicable for general simulations of solids within the finite deformation range.
However, it can be successfully used for large deflection analysis of thin structural
members like beams or shells.

Split in Isochoric and Volumetric Parts. In finite elasticity of foam or rubber materials, the deformation is split into a volumetric part represented by J and an isochoric part described by $\widehat{\mathbf{C}}$, see (3.29), since both parts can depict different material behaviour, see e.g. Lubliner (1985). This split is also useful when quasi-incompressible materials are described by special numerical formulations – like mixed methods – since the split permits a different treatment of the incompressible part.

One possibility for the formulation of the constitutive equation is given by an additive split of the strain energy function in its volumetric and isochoric parts: $W(\widehat{\mathbf{C}}, J) = \hat{W}(\widehat{\mathbf{C}}) + U(J)$. This leads for the strain energy function introduced in (3.116) to

$$W(\widehat{\mathbf{C}}, J) = U(J) + \frac{1}{2}\mu\left(I_{\widehat{C}} - 3\right). \tag{3.122}$$

Here the term $U(J)$ is different from $g(J)$: $U(J) = \frac{K}{4}\left(J^2 - 1\right) - \frac{K}{2}\ln J$, since the third term in the sum in (3.118) disappears and the LAMÉ constant Λ has to be exchanged by the modulus of compression K.

The second PIOLA-KIRCHHOFF stresses are computed via

$$\mathbf{S} = 2\frac{\partial W}{\partial \mathbf{C}} = 2\frac{\partial \hat{W}}{\partial \widehat{\mathbf{C}}}\frac{\partial \widehat{\mathbf{C}}}{\partial \mathbf{C}} + 2\frac{\partial U}{\partial J}\frac{\partial J}{\partial \mathbf{C}}. \tag{3.123}$$

For an explicit evaluation of this equation, the derivatives $\partial J/\partial\mathbf{C}$ and $\partial\widehat{\mathbf{C}}/\partial\mathbf{C}$ are needed. These are given by

$$\frac{\partial J}{\partial \mathbf{C}} = \frac{\partial\sqrt{\det\mathbf{C}}}{\partial\mathbf{C}} = \frac{1}{2}J\mathbf{C}^{-1}, \tag{3.124}$$

see also $(3.110)_3$, and with (3.29) by

$$\begin{aligned}
\frac{\partial\widehat{\mathbf{C}}}{\partial\mathbf{C}} &= \frac{\partial(J^{-\frac{2}{3}}\mathbf{C})}{\partial\mathbf{C}} = \frac{\partial J^{-\frac{2}{3}}}{\partial\mathbf{C}}\otimes\mathbf{C} + J^{-\frac{2}{3}}\frac{\partial\mathbf{C}}{\partial\mathbf{C}} \\
&= J^{-\frac{2}{3}}\left(\mathbb{E} - \frac{1}{3}\mathbf{C}^{-1}\otimes\mathbf{C}\right) =: \mathbb{P}
\end{aligned} \tag{3.125}$$

$$\frac{\partial\hat{C}_{EF}}{\partial C_{AB}} = J^{-\frac{2}{3}}\left(\mathbb{E}_{ABEF} - \frac{1}{3}C_{AB}^{-1}C_{EF}\right) =: \mathbb{P}_{ABEF}.$$

The fourth order unit tensor $\mathbb{E}_{ABEF} = \frac{1}{2}\left(\delta_{AE}\delta_{BF} + \delta_{AF}\delta_{BE}\right)$ follows from $\partial\mathbf{C}/\partial\mathbf{C}$. The explicit form of (3.123) can now be stated as

$$\begin{aligned}
\mathbf{S} &= \mathbb{P}\left[2\frac{\partial\hat{W}}{\partial\widehat{\mathbf{C}}}\right] + \frac{\partial U}{\partial J}J\mathbf{C}^{-1} = \mathbf{S}_{ISO} + \mathbf{S}_{VOL}, \\
S_{AB} &= \mathbb{P}_{ABEF}\,2\frac{\partial\hat{W}}{\partial\hat{C}_{EF}} + \frac{\partial U}{\partial J}J C_{AB}^{-1}.
\end{aligned} \tag{3.126}$$

For the special choice of the strain energy function (3.122), the volumetric and isochoric stresses

$$\mathbf{S}_{ISO} = \mu\,\mathbb{P}\,[\mathbf{1}]\,, \qquad \mathbf{S}_{VOL} = \frac{K}{2}\,(\,J^2 - 1\,)\,\mathbf{C}^{-1} \qquad (3.127)$$

can be specified. For the isochoric part of the second PIOLA-KIRCHHOFF stress tensor, the explicit form: $\mathbf{S}_{ISO} = \mu\,J^{-\frac{2}{3}}\,(\mathbf{1} - \frac{1}{3}\,\mathrm{tr}\mathbf{C}\,\mathbf{C}^{-1}\,)$ can be deduced.

The transformation to the current configuration yields for the KIRCHHOFF stress tensor introduced in (3.84)

$$\begin{aligned}
\boldsymbol{\tau} &= \mathbf{F}\,\mathbf{S}\,\mathbf{F}^T = \mathbf{F}\left\{ 2\,\mathbb{P}\left[\frac{\partial \widehat{W}}{\partial \widehat{\mathbf{C}}}\right] + \frac{\partial U}{\partial J}\,J\,\mathbf{C}^{-1} \right\}\mathbf{F}^T \\[2mm]
&= \mathbf{F}\left\{ 2\,\frac{\partial \widehat{W}}{\partial \widehat{\mathbf{C}}} - \frac{1}{3}\left(\frac{\partial \widehat{W}}{\partial \widehat{\mathbf{C}}}\cdot\mathbf{C}\right)\mathbf{C}^{-1} \right\}\mathbf{F}^T + \frac{\partial U}{\partial J}\,J\,\mathbf{1}\,. \qquad (3.128)
\end{aligned}$$

By defining the operator $\mathrm{dev}(\bullet) = (\bullet) - \frac{1}{3}\,\mathrm{tr}(\bullet)\,\mathbf{1}$, equation (3.128) can be written as

$$\boldsymbol{\tau} = J\,p\,\mathbf{1} + \mathrm{dev}\,\widehat{\boldsymbol{\tau}} = \boldsymbol{\tau}_{vol}\,\mathbf{1} + \boldsymbol{\tau}_{iso}\,. \qquad (3.129)$$

This relation depicts clearly the split of the stress tensor into a volumetric and an isochoric part. The following definitions were used in (3.129)

$$p = \frac{\partial U}{\partial J} \qquad \text{and} \qquad \widehat{\boldsymbol{\tau}} = \widehat{\mathbf{F}}\,2\,\frac{\partial \widehat{W}}{\partial \widehat{\mathbf{C}}}\,\widehat{\mathbf{F}}^T\,. \qquad (3.130)$$

It is easily seen from Exercise 3.5 that the second term of the last equation can also be formulated as $\widehat{\boldsymbol{\tau}} = \widehat{\mathbf{b}}\,2\,\partial \widehat{W}\,/\,\partial \widehat{\mathbf{b}}$. Here the definition $\widehat{\mathbf{b}} = J^{-\frac{2}{3}}\,\mathbf{b}$ has to be applied analogously to (3.29), see also Miehe (1994).

Formulation with Respect to Principal Stretches. In case that the elastic strain energy is given in terms of the principal stretches $\lambda_1\,,\lambda_2\,,\lambda_3$, see (3.113), the second PIOLA-KIRCHHOFF stresses is computed with (3.104) from

$$\mathbf{S} = 2\,\frac{\partial W(\lambda_k)}{\partial \mathbf{C}} = 2\sum_{i=1}^{3}\frac{\partial W}{\partial \lambda_i}\,\frac{\partial \lambda_i}{\partial \mathbf{C}}\,. \qquad (3.131)$$

Here the derivative $\partial W\,/\,\lambda_i$ is ascertainable directly from w when w is given as a function of the principal stretches, see e.g. Ogden (1984, p. 482). The partial derivative $\partial \lambda_i\,/\,\partial \mathbf{C}$ which occurs when the chain rule is applied in (3.131) can be determined from the eigenvalue problems $(\,\mathbf{C} - \lambda_i^2\,\mathbf{1}\,)\,\mathbf{N}_i = \mathbf{0}$. Using the results provided in Simo and Taylor (1991),

$$\frac{\partial \lambda_i}{\partial \mathbf{C}} = \frac{1}{2\,\lambda_i}\,\mathbf{N}_i \otimes \mathbf{N}_i \qquad (3.132)$$

can be written. Thus the second PIOLA-KIRCHHOFF stresses follow as

$$\mathbf{S} = \sum_{i=1}^{3} \frac{1}{\lambda_{(i)}} \frac{\partial W}{\partial \lambda_{(i)}} \mathbf{N}_{(i)} \otimes \mathbf{N}_{(i)} = \sum_{i=1}^{3} S_{(i)} \mathbf{N}_{(i)} \otimes \mathbf{N}_{(i)} . \qquad (3.133)$$

In this equation no sum is carried out over i. This is denoted by the bracket around the index.

A comparison with the spectral decomposition (3.23) of the right CAUCHY-GREEN strain measure shows that $\mathbf{S}$ and $\mathbf{C}$ have the same eigenvectors. This fact is consistent with the restriction to isotropic material behaviour. Since the eigenvectors $\mathbf{n}_i$ of the current configuration can be obtained via a pure rotation of the eigenvectors $\mathbf{N}_i$: $\mathbf{n}_i = \mathbf{R}\,\mathbf{N}_i$, it is simple to transform (3.133) to the spatial configuration. The KIRCHHOFF stress tensor follows with $\boldsymbol{\tau} = \mathbf{F}\,\mathbf{S}\,\mathbf{F}^{T}$ as

$$\boldsymbol{\tau} = \sum_{i=1}^{3} \lambda_{(i)} \frac{\partial W}{\partial \lambda_{(i)}} \mathbf{n}_{(i)} \otimes \mathbf{n}_{(i)} = \sum_{i=1}^{3} \tau_{(i)} \mathbf{n}_{(i)} \otimes \mathbf{n}_{(i)} . \qquad (3.134)$$

By comparison with (3.133), the principal values of the KIRCHHOFF stresses can be related to the principal values of the second PIOLA-KIRCHHOFF stresses via $\tau_i = \lambda_i^2\, S_i$.

For numerical computations, it is necessary to transform the constitutive equations (3.133) and (3.134) to a cartesian coordinate system. This transformation can be performed via the relations $\mathbf{N}_I = \mathbf{D}\,\mathbf{E}_I$ and $\mathbf{n}_i = \mathbf{D}^{\varphi}\,\mathbf{e}_i$, respectively. The transformation matrices are defined by $\mathbf{D} = \mathbf{N}_J \otimes \mathbf{E}_J$ and $\mathbf{D}^{\varphi} = \mathbf{n}_j \otimes \mathbf{e}_j$. Using the component form of the transformation tensors $\mathbf{D}$ (the $D_{IK} = \mathbf{E}_I \cdot \mathbf{N}_K$ denote the directional cosines of the eigenvectors $\mathbf{N}_I$ with respect to the cartesian basis $\mathbf{E}_I$), (3.133) can be written as

$$\mathbf{S} = \sum_{i=1}^{3} S_{(i)}\, D_{(i)\,J}\, D_{(i)\,K}\, \mathbf{E}_J \otimes \mathbf{E}_K . \qquad (3.135)$$

Hence the components of the second PIOLA-KIRCHHOFF stress tensor are given by

$$S_{JK} = \sum_{i=1}^{3} S_{(i)}\, D_{(i)\,J}\, D_{(i)\,K} . \qquad (3.136)$$

Analogous relations are valid for the KIRCHHOFF stresses

$$\tau_{jk} = \sum_{i=1}^{3} \tau_{(i)}\, D^{\varphi}_{(i)\,j}\, D^{\varphi}_{(i)\,k}, \qquad (3.137)$$

where the transformation matrix $\mathbf{D}^{\varphi}$ has to be used with the components $D^{\varphi}_{ik} = \mathbf{e}_i \cdot \mathbf{n}_k$.

Remark 3.5:

a. A closed form for the eigenvector basis $\mathbf{N}_{(i)} \otimes \mathbf{N}_{(i)}$ can be found in Morman (1987)

$$\mathbf{N}_{(i)} \otimes \mathbf{N}_{(i)} = \frac{\lambda_i^2}{(\lambda_i^2 - \lambda_j^2)(\lambda_i^2 - \lambda_k^2)} \left[\mathbf{C} - (I_C - \lambda_i^2)\mathbf{1} + III_C\,\lambda_i^{-2}\,\mathbf{C}^{-1} \right] . \quad (3.138)$$

Here the indices i, j, k have to be exchanged in a cyclic way by indices 1, 2, 3. This description was applied in, e.g. Simo and Taylor (1991) for a finite element implementation of the constitutive equation (3.131). The associated incremental form is relatively complicated.

b. For numerical computation, a matrix formulation of the stresses is often chosen in which the stresses are put in vector form (VOIGT notation), see Chap. 4. For the transformation of (3.136), the relation

$$\mathbf{S} = \left\{ \begin{matrix} S_{11} \\ S_{22} \\ S_{33} \\ S_{12} \\ S_{23} \\ S_{31} \end{matrix} \right\} = \left[\begin{matrix} D_{11}^2 & D_{21}^2 & D_{31}^2 \\ D_{12}^2 & D_{22}^2 & D_{32}^2 \\ D_{13}^2 & D_{23}^2 & D_{33}^2 \\ D_{11}\,D_{12} & D_{21}\,D_{22} & D_{31}\,D_{32} \\ D_{12}\,D_{13} & D_{22}\,D_{23} & D_{32}\,D_{33} \\ D_{13}\,D_{11} & D_{23}\,D_{21} & D_{33}\,D_{31} \end{matrix} \right] \left\{ \begin{matrix} S_1 \\ S_2 \\ S_3 \end{matrix} \right\} = \mathbf{D}\,\bar{\mathbf{S}} \quad (3.139)$$

is obtained considering that the stress vector, which represents the eigenvalues, has only three components.

Exercise 3.6: The first PIOLA-KIRCHHOFF stresses have to be specified for a NEO-HOOKE-, a MOONEY-RIVLIN- and an OGDEN-material under the assumption of incompressibility and formulated for a uniaxially loaded bar and a bi-axially loaded plate. Starting point is the strain energy function (3.113) with three terms. The following set of parameters should be used which was determined by Ogden (1972) based on the experimental work reported in Treloar (1944): $\mu_1 = 6.3$, $\mu_2 = 0.013$, $\mu_3 = -0.1$ and $\alpha_1 = 1.3$, $\alpha_2 = 5.0$, $\alpha_3 = -2.0$.

Solution: For simple uniaxial loading of a rod, the principal values can be stated directly using (3.134) since for incompressibility the CAUCHY stress tensor is equivalent to the KIRCHHOFF stress tensor. However, in the case of incompressibility, the unknown pressure p has to be computed from the constraint condition $J = 1$. The CAUCHY stresses are given by

$$\sigma_i = \lambda_i \frac{\partial W}{\partial \lambda_i} + p. \quad (3.140)$$

The stresses orthogonal to the axis of the rod are zero for uniaxial loading which leads to the conditions

$$\sigma_2 = \lambda_2 \sum_{i=1}^{3} \mu_i \lambda_2^{\alpha_i - 1} + p = 0,$$

$$\sigma_3 = \lambda_3 \sum_{i=1}^{3} \mu_i \lambda_3^{\alpha_i - 1} + p = 0.$$

From the incompressibility constraint $J = \lambda_1 \lambda_2 \lambda_3 = 1$ and the assumption that the stretches orthogonal to the axis of the rod are equal $\lambda_2 = \lambda_3$, it follows that $\lambda_2 = \lambda_3 = \lambda_1^{-\frac{1}{2}}$. Inserting this result in the previous relation yields

$$p = -\sum_{i=1}^{3} \mu_i \lambda_1^{-\frac{1}{2}\alpha_i}$$

and leads finally with (3.140) to

$$\sigma_1 = \sum_{i=1}^{3} \left[\mu_i \lambda_1^{\alpha_i} - \lambda_1^{-\frac{1}{2}\alpha_i} \right] . \tag{3.141}$$

The component P_1 of the first PIOLA-KIRCHHOFF stress tensor is obtained from (3.81) for $J = 1$ as $P_1 = \sigma_1 / \lambda_1$, this results in

$$P_1 = \sum_{i=1}^{3} \left[\mu_i \lambda_1^{\alpha_i-1} - \lambda_1^{-\frac{1}{2}\alpha_i-1} \right] . \tag{3.142}$$

For a bi-axial loading the stresses can be derived analogous to (3.142). Under the assumption of equal stretches in both directions ($\lambda_1 = \lambda_2$, $\lambda_3 = \lambda^{-2}$ and with the plane stress condition $\sigma_3 = 0$), the component P_1^{bi} of the first PIOLA-KIRCHHOFF stress tensor is given by

$$P_1^{bi} = \sum_{i=1}^{3} \left[\mu_i \lambda_1^{\alpha_i-1} - \lambda_1^{-2\alpha_i-1} \right] . \tag{3.143}$$

An approximation of the experimental data, see Ogden (1972), follows by inserting the constitutive parameters μ_i and α_i in (3.143) according to the given data, see Fig. 3.4. Note that the restriction (3.114)$_2$ is fulfilled by the constitutive parameters. Condition (3.114)$_1$ yields for the given parameters $2\mu = 8.45$. One can furthermore observe that the strain energy function with three terms (3.113) introduced by Ogden (1972) approximates the experimental data of Treloar (1944) for rubber up to strains of 700% very well.

Restriction (3.114)$_1$ must also be met by the NEO-HOOKE material which strain energy function consists only of one term. With $\alpha_1 = 2$ follows $\mu_1 = \mu$. This set of parameters leads with the first PIOLA–KIRCHHOFF stress in case of the uniaxial and bi-axial deformation to

$$P_1 = \mu_1 \left[\lambda_1 - \lambda_1^{-2} \right] \quad \text{and} \quad P_1^{bi} = \mu_1 \left[\lambda_1 - \lambda_1^{-5} \right] . \tag{3.144}$$

Figure 3.4a depicts that the NEO-HOOKE material approximates the experimental results only up to a stretch of $\lambda_1 \approx 1.7$ which denotes strain of $70\,\%$.

For the MOONEY-RIVLIN material, two parameters are chosen according to (3.115) to fit the experimental data. This leads to a better approximation at the beginning of the curve. Again condition (3.114)$_1$ has to be fulfilled. With the choice of $\mu_1 = 2.4$ follows $\mu_2 = -1.825$ and the first PIOLA-KIRCHHOFF stress is given for both cases by

$$\begin{aligned} P_1 &= \mu_1 \left[\lambda_1 - \lambda_1^{-2} \right] + \mu_2 \left[\lambda_1^{-3} - \lambda_1 \right] , \tag{3.145}\\ P_1^{bi} &= \mu_1 \left[\lambda_1 - \lambda_1^{-5} \right] + \mu_2 \left[\lambda_1^{-3} - \lambda_1^{3} \right] . \tag{3.146} \end{aligned}$$

This stress–strain relation approximates the experimental data up to $\lambda_1 \approx 1.7$, as can be seen in Fig. 3.4. However for larger strains, the stresses deviate more from the experimental data than the ones computed with (3.144).

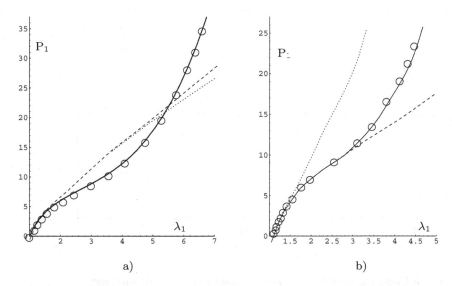

Fig. 3.4 Approximation of experimental data after TRELOAR (○) by different material functions: Neo-Hooke (- -- ---), Mooney-Rivlin (·· ···· ··) and Ogden (————), (**a**) uniaxial and (**b**) biaxial tension

Exercise 3.7: Specify equations (3.120) and (3.121) for the case of a rod which is subjected to a tension force. Compute the CAUCHY stresses under the condition that the contraction of the cross section of the rod is zero. Discuss the results.

Solution: The rod in tension undergoes only a stretch in axial direction, thus the only unknown stretch is λ_1 since $\lambda_2 = \lambda_3 = 1$. Due to that, it holds $J = \lambda_1$ and the remaining component of the deformation gradient $\mathbf{F}$ is $F_{11} = \lambda_1$. With these values, it is possible to evaluate (3.120) directly and it follows with $\gamma = (\Lambda/2 + \mu)$

$$\sigma_1 = \frac{\Lambda}{2\lambda_1}\left(\lambda_1^2 - 1\right) + \frac{\mu}{\lambda_1}\left(\lambda_1^2 - 1\right) = \gamma\left(\lambda_1 - \frac{1}{\lambda_1}\right). \qquad (3.147)$$

This constitutive equation for σ_1 is only zero at $\lambda_1 = 1$ when no deformation is present. One can easily observe that the limit cases are fulfilled: $\lambda_1 \to +\infty \implies \sigma_1 \to +\infty$ and $\lambda_1 \to 0 \implies \sigma_1 \to -\infty$.

The ST. VENANT material law can be stated with $E_{11} = \frac{1}{2}(\lambda_1^2 - 1)$ as

$$S_1 = \frac{\Lambda}{2}\left(\lambda_1^2 - 1\right) + \mu\left(\lambda_1^2 - 1\right) = \gamma\left(\lambda_1^2 - 1\right). \qquad (3.148)$$

This relation will now be transformed to the current configuration. Equation (3.83) yields $\sigma_1 = \lambda_1 S_1$ which leads to

$$\sigma_1 = \gamma\lambda_1\left(\lambda_1^2 - 1\right). \qquad (3.149)$$

This equation fulfils condition $\lambda_1 = 1 \to \sigma_1 = 0$. But for the limiting cases, it yields: $\lambda_1 \to +\infty \implies \sigma_1 \to +\infty$ and $\lambda_1 \to 0 \implies \sigma_1 \to 0$. The last limiting case does not make sense physically, see also Remark 3.4. Figure 3.5 depicts the stress

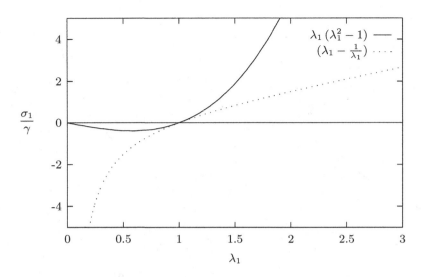

Fig. 3.5 Comparison of the stresses obtained from Neo-Hooke and St. Venant material

response which is completely different when compared to (3.147). Note that both constitutive functions have the same tangent $\Lambda + 2\,\mu$ at $\lambda_1 = 1$. This is consistent with Hooke material law of the linear theory.

3.3.2 Elasto-Plastic Material Laws

Many materials which are widely used in technical applications depict non-linear behaviour even when only small deformations are present. A large class of nonlinear materials can be described by the assumption of elasto-plastic behaviour. Among these are materials like steel, aluminium, concrete but also geo-materials such as rock and soils. Often the associated constitutive models are very complex. In the following sections several equations will be discussed which describe rate-independent constitutive behaviour for the general case of isotropic and kinematic hardening of metals. For a detailed treatment of such constitutive equations and their physical interpretation, see e.g. Hill (1950), Prager (1955), Desai and Siriwardane (1984), Lubliner (1990) and Khan and Huang (1995).

Elasto-Plastic Material Laws of Small Deformations. The phenomenological model of the theory of plasticity has to be based on the fact that plastic flow of a material is an irreversible process. It is described in a solid by an additional strain measure and additional variables, known as plastic strains and hardening parameters.

An example for such constitutive equations is the classical model of elasto-plasticity with isotropic (expansion of flow surface) and kinematic (shift of

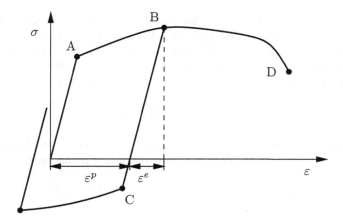

Fig. 3.6 Elasto-plastic constitutive behaviour

origin of flow surface) hardening. Figure 3.6 illustrates the behaviour of that material for the one-dimensional case. Until point A the material reacts elastically that means the loading and unloading path is the same. Plastic deformations occur once the stress reaches the flow stress σ_A in A; after that the stress does not increase as much as in the elastic range. Observe that the flow stress is larger in point B than in point A which is due to hardening. In case of unloading in point B, the stress response follows a straight line which is parallel to the elastic tangent between the origin and point A. Hence unloading is related to elastic behaviour. However, plastic flow occurs again at point C. Here the flow stress has a smaller absolute value than at point B. This observation is related to a movement of the origin of the elastic zone which is called kinematic hardening. A possible softening which is related to a decrease of the flow stress characterizes part BD in the stress–strain diagram.

In the following, a three-dimensional generalization will be given for the phenomenological behaviour described above. In the subsequent development of the mathematical model of elasto-plastic constitutive behaviour, only small deformations occur and the considerations are restricted to metal plasticity. Later a generalized treatment is presented which includes plasticity models, for e.g. concrete or geo-materials, such as sand, clay or rock. For an in-depth treatment of such material laws, see e.g. Desai and Siriwardane (1984), Khan and Huang (1995) and Hofstetter and Mang (1995).

The linear strain tensor (3.17) can be split additively into an elastic and a plastic part when only small strains are present, see Fig. 3.6,

$$\varepsilon = \varepsilon^e + \varepsilon^p . \tag{3.150}$$

The assumption of incompressible plastic deformations is often justified for metals by experimental observations. In such case, it is appropriate to

introduce deviatoric measures. This yields for the total strains with (3.30)
$\mathbf{e} = \boldsymbol{\varepsilon} - \frac{1}{3}\operatorname{tr}\boldsymbol{\varepsilon}\,\mathbf{1}$. (In the following, the deviator is described by $\mathbf{e}$ since the
ALMANSI strain tensor does not occur in this section and thus the notation
can be simplified compared to $\mathbf{e}_D$ in Sect. 3.1.2.) An analogous relation can
be written for the stress tensor which then defines the stress deviator

$$\mathbf{s} = \boldsymbol{\sigma} - \frac{1}{3}\operatorname{tr}\boldsymbol{\sigma}\,\mathbf{1}. \qquad (3.151)$$

The stresses $\boldsymbol{\sigma}$ and the internal variables $\mathbf{q}$ related to the hardening parameters $\boldsymbol{\alpha}$ can be computed by a derivative of the strain energy function ψ

$$\boldsymbol{\sigma} = \rho_0 \frac{\partial\psi(\boldsymbol{\varepsilon}^e, \boldsymbol{\alpha})}{\partial\boldsymbol{\varepsilon}^e}, \qquad \mathbf{q} = -\rho_0 \frac{\partial\psi(\boldsymbol{\varepsilon}^e, \boldsymbol{\alpha})}{\partial\boldsymbol{\alpha}}. \qquad (3.152)$$

For many applications, the assumption of a function ψ is valid which is decoupled in $\boldsymbol{\varepsilon}^e$ and $\boldsymbol{\alpha}$. This leads to

$$\rho_0\,\psi(\boldsymbol{\varepsilon}^e, \boldsymbol{\alpha}) = W_e(\boldsymbol{\varepsilon}^e) + W_v(\boldsymbol{\alpha}), \qquad (3.153)$$

where $W_e(\boldsymbol{\varepsilon}^e)$ is the elastic strain energy function, see e.g. (3.116), and $W_v(\boldsymbol{\alpha})$
denotes a potential function for the hardening variables. The elastic strain
energy function W_e has the explicit form $W_e = \frac{1}{2}\,\boldsymbol{\varepsilon}^e \cdot \mathbb{C}^e[\boldsymbol{\varepsilon}^e]$ for small strains.
The partial derivative of this function with respect to the elastic strains yields
with

$$\boldsymbol{\sigma} = \rho_0 \frac{\partial\psi(\boldsymbol{\varepsilon}^e, \boldsymbol{\alpha})}{\partial\boldsymbol{\varepsilon}^e} = \mathbb{C}^e[\boldsymbol{\varepsilon}^e], \qquad \sigma_{ij} = \mathbb{C}^e_{ijkl}\,\varepsilon^e_{kl} \qquad (3.154)$$

the classical HOOKE'S law of the theory of linear elasticity. Note that (3.154)
can also be written in terms of the deviatoric quantities which leads for
isotropic elastic response to

$$\mathbf{s} = 2\,\mu\mathbf{e}^e \qquad \text{and} \quad p = K\operatorname{div}u. \qquad (3.155)$$

Here μ is the shear modulus and K the bulk modulus (with $K = \lambda + \frac{2}{3}\,\mu$). λ
and μ are also called LAME constants, see also Remark 3.4.

We assume the same structure, given in (3.154), for W_v, see e.g. Lubliner
(1990). Hence $W_v = \frac{1}{2}\,\hat{H}\hat{\alpha}^2 + \frac{1}{3}\,H\,|\boldsymbol{\alpha}|^2$ is defined where $\hat{\alpha}$ are the isotropic
and $\boldsymbol{\alpha}$ kinematic hardening variables. This definition leads with (3.152) to

$$\mathbf{q} = -\frac{2}{3}\,H\,\boldsymbol{\alpha} \quad \text{and} \quad \hat{q} = -\hat{H}\,\hat{\alpha}, \qquad q_{ij} = -\frac{2}{3}\,H\,\alpha_{ij}. \qquad (3.156)$$

The elastic domain of the deformation is restricted by the yield condition.
Mathematically an inequality constraint has to be formulated which depends
upon the stresses and the internal hardening variables. The flow condition
must be able to describe two different phenomena as already pointed out in
Fig. 3.6 for the one-dimensional case. This is the enlargement of the elastic
domain (isotropic hardening) and the shift of the permissible elastic range

(kinematic hardening), depicted schematically in Fig. 3.7a for times t_o and t. The flow condition or yield criterion can be written in general form as

$$f(\boldsymbol{\sigma}, \mathbf{q}, \hat{q}) \leq 0 \qquad (3.157)$$

for isotropic and kinematic hardening. In case of the classical VON MISES plasticity, the flow condition f depends only upon the second invariant of the stress deviator. Hence it can be expressed as

$$f(\mathbf{s}, \mathbf{q}, \hat{q}) = \sqrt{(\mathbf{s} - \mathbf{q}) \cdot (\mathbf{s} - \mathbf{q})} - k(\hat{q}) \leq 0. \qquad (3.158)$$

For linear isotropic and kinematic hardening,

$$f(\mathbf{s}, \mathbf{q}, \hat{q}) = \| \mathbf{s} - \mathbf{q} \| - \sqrt{\frac{2}{3}} \, (Y_0 - \hat{q}) \leq 0 \qquad (3.159)$$

is obtained explicitly. The generalized stress measure $\mathbf{q}$ which is related to kinematic hardening is called back stress. A stress point lies for $f < 0$ in the elastic domain. The stress point is located on the boundary of the flow surface for $f = 0$. This can result in plastic deformations. Values $f > 0$ of the flow condition are not admissible, see Fig. 3.7b.

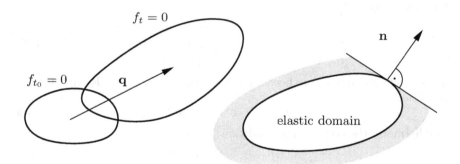

Fig. 3.7a Flow condition   **Fig. 3.7b** Admissible region

Irreversibility of the plastic flow process is expressed by a flow rule. For most metals, an associated flow rule can be used in which the direction of flow is given by the partial derivative of the flow condition with respect to the deviatoric stresses

$$\dot{\mathbf{e}}^p = \lambda \frac{\partial f}{\partial \mathbf{s}}. \qquad (3.160)$$

This equation describes the evolution of the deviatoric plastic strains. The direction of plastic flow is given by $\frac{\partial f}{\partial \mathbf{s}}$ and λ is a scalar which determines the size of the plastic strain increment.

Evolution equations have also to be formulated for the hardening variables. This leads to

$$\dot{\boldsymbol{\alpha}} = \lambda \frac{\partial f}{\partial \mathbf{q}}, \qquad \dot{\hat{\alpha}} = \lambda \frac{\partial f}{\partial \hat{q}}. \tag{3.161}$$

Using the flow condition (3.159), the derivative of f yields

$$\frac{\partial f}{\partial \mathbf{s}} = \frac{\mathbf{s} - \mathbf{q}}{\| \mathbf{s} - \mathbf{q} \|} =: \mathbf{n} \quad \text{and} \quad \frac{\partial f}{\partial \mathbf{q}} = -\mathbf{n} \tag{3.162}$$

which defines the flow direction. Hence the evolution equations

$$\dot{\mathbf{e}}^p = \lambda \, \mathbf{n}, \quad \dot{\boldsymbol{\alpha}} = -\lambda \, \mathbf{n} \quad \text{and} \quad \dot{\hat{\alpha}} = \lambda \sqrt{\frac{2}{3}} \tag{3.163}$$

can be written. Equation (3.163) yields the equivalent plastic strain increment $\dot{\hat{\alpha}} = \sqrt{\frac{2}{3}} \, \| \dot{\mathbf{e}}^p \|$ since $\| \dot{\mathbf{e}}^p \| = \lambda$ is valid. Time integration leads to the equivalent (or effective) strain

$$\hat{\alpha} = \int_0^t \sqrt{\frac{2}{3}} \, \| \dot{\varepsilon}^p \| d\tau, \tag{3.164}$$

which provides a measure for plastic distortion, see e.g. Hill (1950) or Lubliner (1990).

In (3.163), the parameter λ describes the magnitude of plastic flow. Generally, three cases have to be distinguished when a stress point lies on the flow surface $f = 0$:

$$\begin{aligned} \dot{f} < 0 &\implies \lambda = 0 &\quad \text{elasticunloading}, \\ \dot{f} = 0 &\implies \lambda = 0 &\quad \text{neutralloading}, \\ \dot{f} = 0 &\implies \lambda > 0 &\quad \text{plasticflow}. \end{aligned} \tag{3.165}$$

These different cases can be summarized in the so-called KUHN–TUCKER conditions

$$\lambda \geq 0, \qquad f \leq 0, \qquad \lambda f = 0. \tag{3.166}$$

Furthermore, the consistency condition

$$\lambda \dot{f} = 0, \quad \text{if} \quad f = 0 \tag{3.167}$$

is contained in (3.165). With this, all evolution equations for plastic flow and hardening parameters as well as the flow conditions of elasto-plastic flow are known. The incremental form of these equations is derived in Sect. 3.3.4. An algorithm which can be used to integrate the evolution equations is presented in Sect. 6.2.

Generalized Elasto-Plastic Material Equations. Besides the elasto-plastic constitutive equations discussed so far, there exist numerous other models of plasticity which describe the material behaviour and the failure of, e.g. sand, concrete. For a large class of elasto-plastic materials, it is possible to derive a generalized description of the constitutive equations. This will also reflect the fact that there exist many materials of technical interest which can be either described by associative or non-associative flow rules or need multi-surface flow conditions for a proper modelling.

For the general case of an elasto-plastic material with m independent flow surfaces, the subsequent equations and evolution laws are obtained based on the introduced notation. Associative and non-associative plasticity is distinguished in following:

− Stress $\boldsymbol{\sigma}$ and back stress $\mathbf{q}$

$$\begin{aligned} \boldsymbol{\sigma} &= \mathbb{C}\,[\,\boldsymbol{\varepsilon} - \boldsymbol{\varepsilon}^p\,] \\ \mathbf{q} &= -\mathbb{H}\,[\,\boldsymbol{\alpha}\,] \end{aligned} \tag{3.168}$$

− g flow conditions or yield criteria (restrictions for elastic domain)

$$f_g\,(\boldsymbol{\sigma},\mathbf{q}) \le 0 \tag{3.169}$$

− Flow rule and evolution equation for hardening:
  1. associative plasticity

$$\begin{aligned} \dot{\boldsymbol{\varepsilon}}^p &= \sum_{g=1}^{m} \lambda_g \frac{\partial f_g(\boldsymbol{\sigma},\mathbf{q})}{\partial \boldsymbol{\sigma}} \\ \dot{\boldsymbol{\alpha}} &= \sum_{g=1}^{m} \lambda_g \frac{\partial f_g(\boldsymbol{\sigma},\mathbf{q})}{\partial \mathbf{q}} \end{aligned} \tag{3.170}$$

  2. non-associative plasticity

$$\begin{aligned} \dot{\boldsymbol{\varepsilon}}^p &= \sum_{g=1}^{m} \lambda_g\, \mathbf{r}_g(\boldsymbol{\sigma},\mathbf{q}) \\ \dot{\boldsymbol{\alpha}} &= \sum_{g=1}^{m} \lambda_g\, \mathbf{h}_g(\boldsymbol{\sigma},\mathbf{q}) \end{aligned} \tag{3.171}$$

− Loading/unloading conditions in KUHN–TUCKER form

$$\lambda_g \ge 0, \qquad f_g(\boldsymbol{\sigma},\mathbf{q}) \le 0, \qquad \lambda_g\, f_g(\boldsymbol{\sigma},\mathbf{q}) = 0 \tag{3.172}$$

All relations hold for models of plasticity with m flow surfaces. The special case of only one flow surface is naturally included by setting $m = 1$. In (3.168), the two tensors of forth order, $\mathbb{C}$ and $\mathbb{H}$, represent the elastic and the hardening laws. Tensors $\mathbf{r}$ and $\mathbf{h}$ describe the flow direction and the change of hardening for non-associative plasticity in (3.171).

Here the flow conditions depend upon the stress $\boldsymbol{\sigma}$ and not solely on the deviatoric stresses $\mathbf{s}$. The reason for this is that inelastic processes of general non-metallic materials or of metals in which damage has to be considered are pressure sensitive and hence the flow condition has to depend upon the full stress tensor.

Examples for such material behaviour are, e.g. the GURSON model which describes damage in metals due to void nucleation, void growth and ductile fracture in the micro-structure of the metal, see e.g. Gurson (1977). The original model from GURSON was corrected by Tvergaard and Needleman (1984) by introducing an effective porosity

$$
g^*(g) = \begin{cases} g & \text{if} \quad g \le g_c \\ g_c + \frac{1/q_1 - g_c}{g_f - g_c}(g - g_c) & \text{if} \quad g > g_c \end{cases} \tag{3.173}
$$

with the critical volume fraction g_c of the pores at which the voids start to unite. q_1 and g_f are material parameters. With this definition, the flow condition or yield criterion for the GURSON model can be written as

$$
f(\boldsymbol{\sigma}, \sigma_M, f) = \frac{\sigma_e^2}{\sigma_M^2} + 2\, q_1\, g^* \cosh\left(\frac{q_2 \operatorname{tr}\boldsymbol{\sigma}}{2\, \sigma_M}\right) - (q_1\, g^*)^2 - 1, \tag{3.174}
$$

where σ_M is the stress in the matrix material, q_2 another material parameter and $\sigma_e = \sqrt{\frac{3}{2}\mathbf{s}\cdot\mathbf{s}}$ is the VON MISES equivalent stress, see e.g. Hill (1950). The matrix stress σ_M is usually given as a function of the plastic matrix strains ε_M^p

$$
\sigma_M = Y_0 \left(\frac{\varepsilon_M^p}{\varepsilon_0} + 1\right)^{\frac{1}{n}} \tag{3.175}
$$

as a micro-mechanical constitutive relation for the matrix material, see e.g. Tvergaard (1989). Besides these equations, evolution equations for the plastic flow are needed on macro- and micro-mechanical level and for the void growth. These are given for the macro-material by

$$
\dot{\boldsymbol{\varepsilon}}^p = \lambda \frac{\partial f}{\partial \boldsymbol{\sigma}} \tag{3.176}
$$

and for the matrix material on micro-scale by

$$
\dot{\varepsilon}_M^p = \frac{\boldsymbol{\sigma} \cdot \boldsymbol{\varepsilon}^p}{(1 - g)\, \sigma_M}. \tag{3.177}
$$

Void growth is described by an evolution equation which takes into account nucleation f_N and growth f_W of the voids

$$
\dot{g}_l = \dot{g}_N + \dot{g}_W \tag{3.178}
$$

with different evolution laws for each term. For the void growth, the relation

$$
\dot{g}_W = (1 - g)\operatorname{tr}\dot{\boldsymbol{\varepsilon}}^p \tag{3.179}
$$

can be used and for the nucleation of new pores Needleman and Rice (1978) introduced

$$\dot{g}_N = A\left(\varepsilon_M^p\right)\dot{\varepsilon}_M^p,\tag{3.180}$$

where the function $A(\varepsilon_M^p)$ stems from experimental observations. It can be defined by a GAUSSIAN normal distribution

$$A(\varepsilon_M^p) = \frac{g_N}{s_N\sqrt{2\pi}}\,e^{-\frac{1}{2}\left(\frac{\varepsilon_M^p - \varepsilon_N}{s_N}\right)^2}\tag{3.181}$$

with the strain ε_N where nucleation occurs and with the standard deviation s_N. The model has several evolution equations and flow conditions and hence can be treated as a generalized plastic material model.

Remark 3.6: The GURSON model describes softening behaviour of the material due to the growth and nucleation of voids. Such material behaviour, which can also be observed in other elasto-plastic constitutive models used, e.g. in engineering analysis of concrete or soil, depicts localizations and leads in the numerical formulation and simulation to mesh dependent solutions, see Oliver (1995).

Several possibilities can be proposed to overcome this problem by regularization. These are non-local extensions of the model where the evolution of damage in (3.178) is replaced by a non-local description

$$\dot{g}(\mathbf{X}) = \frac{1}{V_r(\mathbf{X})}\int_V \dot{g}_l(\mathbf{X}+\mathbf{S})\,\psi(\mathbf{S})\,dV\tag{3.182}$$

with

$$V_r(\mathbf{X}) = \int_V \psi(\mathbf{S})\,dV \quad \text{and} \quad \psi(\mathbf{S}) = e^{\|\mathbf{S}\|^2 / l_c^2},$$

where $\psi(\mathbf{S})$ is a weighting function and l_c is a characteristic length which defines the size of the influence region of the non-local model. The coordinate $\mathbf{X}$ describes the local point at which the evolution equations have to be evaluated. For a detailed theoretical and numerical treatment of such approach, see e.g. Bazant and Cedolin (1991), Leblond et al. (1994) or Feucht (1999).

Another possible regularization is based on the introduction of gradient dependent evolution equation for the void growth

$$\dot{g} - \frac{l_c^2}{4}\nabla^2\dot{g} = \dot{g}_N + \dot{g}_W .\tag{3.183}$$

This method was proposed in Lasry and Belytschko (1988) for general strains in transient problems and further developed in Sluys (1992), Pamin (1994) and Feucht (1999).

Localizations represent the step from a continuum to a discontinuum in which a part of the body slides along the other at the localization surface. Hence methods which directly introduce the discontinuous behaviour can also be applied to solve such problems. These are based on so-called strong discontinuity approach, see e.g. Simo et al. (1993a), Larsson et al. (1993), Miehe and Schröder (1994) and Oliver (1995). They can be combined with adaptive techniques or special interface elements, see e.g. Leppin and Wriggers (1997).

Another criterion which is often used in soil mechanics is the DRUCKER-PRAGER flow condition, see Drucker and Prager (1952) or Khan and Huang (1995),

$$f(I_\sigma, II_s) = \sqrt{II_s} - \alpha I_\sigma - Y_0 \leq 0 \qquad (3.184)$$

with the material parameters α and Y_0. It depends not only upon the second invariant II_s of the stress deviator s as in the classical VON MISES theory of metal plasticity, see (3.159), but also on a hydrostatic stress term represented here by the first invariant I_σ of the stress tensor σ.

Many more different constitutive models which stem from experimental observations and describe plastic flow can be formulated. These are related to different types of materials, see for an overview, e.g. Desai and Siriwardane (1984), Lubliner (1990), Hofstetter and Mang (1995) or Khan and Huang (1995). Furthermore, different loading conditions like pulsating or dynamic loads can lead to the so-called ratcheting strains, and hence special constitutive descriptions are needed to model such effects, see e.g. Ekh et al. (2000) and Johansson et al. (2005).

Remark 3.7: Inelastic processes result in mechanical dissipation which can only grow when the plastic deformation increases. Based on the local principle of maximum dissipation, some basic properties can be deduced which are needed for the description of plastic flow. The local dissipation is defined by the difference between the stress power and the time derivative of the free energy function. Thus equations (3.66) and (3.153) lead to

$$\mathcal{D} = \sigma \cdot \dot{\varepsilon} - \frac{D}{Dt} \psi(\varepsilon^e, \alpha). \qquad (3.185)$$

The evaluation of the time derivative yields together with (3.150) an expression for the plastic dissipation

$$\mathcal{D} = \sigma \cdot \dot{\varepsilon}^p + q \cdot \dot{\alpha} \geq 0. \qquad (3.186)$$

The principle of maximum plastic dissipation can be formulated as, see e.g. Hill (1950) or Lubliner (1990),

$$\mathcal{D} = \max_{\tau, p} \tau \cdot \dot{\varepsilon}^p + p \cdot \dot{\alpha} \quad \forall \quad \{(\tau, p) \,|\, f(\tau, p) \leq 0\}. \qquad (3.187)$$

It leads to the inequality

$$(\sigma - \tau) \cdot \dot{\varepsilon}^p + (q - p) \cdot \dot{\alpha} \geq 0, \qquad (3.188)$$

where τ and p characterize arbitrary stresses which are contained in the elastic domain. From this variational inequality, it follows that the flow surfaces have to be convex, see e.g. Lubliner (1990). However, this does not exclude flow surfaces with corners which occur frequently in real materials. Such cases require a special mathematical treatment within the framework of non-convex analysis which was substantiated in Moreau (1976), see also Simo (1998).

Based on the variational inequality (3.188) with the constraint $f \leq 0$, a saddle point problem is formulated which yields as result of the flow rule. The actual stress state follows then as extremal value of the functional

$$L(\tau, p, \lambda) = -\tau \cdot \dot{\varepsilon}^p - p \cdot \dot{\alpha} + \lambda f(\tau, p) \longrightarrow EXTREMUM. \qquad (3.189)$$

The constraint $f \leq 0$ is incorporated in this relation by the method of LAGRANGE multipliers. The solution of this saddle point problem at $\tau = \sigma$ and $\mathbf{p} = \mathbf{q}$ leads to

$$\dot{\varepsilon}^p = \lambda \frac{\partial f}{\partial \sigma}, \quad \dot{\alpha} = \lambda \frac{\partial f}{\partial \mathbf{q}}, \quad f(\tau, \mathbf{p}) = 0, \tag{3.190}$$

together with the KUHN–TUCKER conditions, see e.g. Luenberger (1984). Equations (3.190) represent the associative flow rules (3.170), which determine plastic flow for $f = 0$.

Elasto-Plastic Constitutive Equations for Finite Deformations. Models for finite plasticity are essential for engineering analysis, such as metal forming, cutting or pile driving in soils. The formulation of the underlying theoretical background has a long history. However, the possibility to apply such models within numerical simulations has shed new light on some theoretical aspects.

Many authors start from a hypo-elastic constitutive equation for the elastic part of the deformation when finite elasto-plastic deformations have to be considered, see e.g. Khan and Huang (1995). Such material assumption does not represent elasticity in the strict sense, see e.g. Truesdell and Noll (1965) or Simo and Pister (1984). Besides this restriction which can result in unwanted effects, see e.g. Atluri (1984), which are physically meaningless, there exist also problems of numerical nature which will be discussed in the following.

A hypo-plastic constitutive law is presented in a rate form and relates a stress flux, see e.g. (3.97), to the symmetrical spatial velocity gradient $\mathbf{d}$: $\overset{\triangledown}{\tau} = \mathbb{C}[\mathbf{d}^e]$ with an incremental elasticity tensor $\mathbb{C}$. The idea behind this is an additive split of the spatial velocity gradient into an elastic and a plastic part $\mathbf{d} = \mathbf{d}^e + \mathbf{d}^p$ analogous to the assumptions made for the small strain case in the last section. For the correct choice of the stress rates, see e.g. Simo and Hughes (1998) and Khan and Huang (1995). The rate form requires a time integration which results in a costly algorithm. To overcome this disadvantage, algorithms for plasticity were developed based on hyperelastic constitutive equations defined in Sect. 3.3.1. These formulations use a so-called operator-split technique in which the elastic part follows directly via a function evaluation of the hyperelastic material law, see e.g. Simo and Ortiz (1985) or Simo (1988). This circumvents the time integration of the elastic constitutive equation.

Based on the single crystal model for metal plasticity, a multiplicative split of the deformation gradient is introduced instead of the additive split of the strain rates into elastic and plastic parts. For a theoretical foundation, see e.g. Lee and Liu (1967) or Lubliner (1990). This split is defined by

$$\mathbf{F} = \mathbf{F}^e \, \mathbf{F}^p, \tag{3.191}$$

where an intermediate configuration was introduced besides the initial- and spatial configuration, see Fig. 3.8.

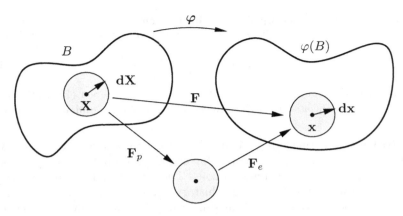

Fig. 3.8 Multiplicative decomposition of the deformation gradient $\mathbf{F}$

Based on this split, the right and left "elastic" CAUCHY–GREEN tensor is introduced by using $\mathbf{F}^e$

$$\tilde{\mathbf{C}}^e = \mathbf{F}^{e^T} \mathbf{F}^e , \qquad \mathbf{b}^e = \mathbf{F}^e \mathbf{F}^{e^T} , \tag{3.192}$$

where the tilde indicates in $(3.192)_1$ that the right CAUCHY–GREEN tensor is referred to the intermediate configuration.

Analogous to the definition of the spatial velocity gradient $\mathbf{l} = \dot{\mathbf{F}} \mathbf{F}^{-1}$, see (3.47), the corresponding elastic and plastic parts are given by

$$\mathbf{l}^e = \dot{\mathbf{F}}^e \mathbf{F}^{e^{-1}} , \qquad \tilde{\mathbf{L}}^p = \dot{\mathbf{F}}^p \mathbf{F}^{p^{-1}} . \tag{3.193}$$

Again tensor $\mathbf{l}^e$ is referred to the spatial configuration while $\tilde{\mathbf{L}}^p$ acts in the intermediate configuration. Since $\mathbf{F}^p$ can be written with (3.191) also as $\mathbf{F}^{e^{-1}} \mathbf{F}$, the following expression is derived with $\dot{\mathbf{F}}^{-1} = -\mathbf{F}^{-1} \dot{\mathbf{F}} \mathbf{F}^{-1}$

$$\tilde{\mathbf{L}}^p = \frac{\partial}{\partial t} \left(\mathbf{F}^{e^{-1}} \mathbf{F} \right) \mathbf{F}^{p^{-1}} = \mathbf{F}^{e^{-1}} \left[\dot{\mathbf{F}} \mathbf{F}^{-1} - \dot{\mathbf{F}}^e \mathbf{F}^{e^{-1}} \right] \mathbf{F}^e , \tag{3.194}$$

which yields with $(3.193)_1$ and (3.47)

$$\mathbf{F}^e \tilde{\mathbf{L}}^p \mathbf{F}^{e^{-1}} = \mathbf{l} - \mathbf{l}^e . \tag{3.195}$$

This equation motivates the following definition of the spatial plastic velocity gradient $\mathbf{l}^p = \mathbf{F}^e \tilde{\mathbf{L}}^p \mathbf{F}^{e^{-1}}$, which results as follows instead of (3.195)

$$\mathbf{l}^p = \mathbf{l} - \mathbf{l}^e . \tag{3.196}$$

The decomposition of the spatial velocity gradient into its symmetric part $\mathbf{d}$ and its antisymmetric part $\mathbf{w}$ yields, with the previous definitions, the additive decomposition of the symmetric velocity gradient

$$\mathbf{d} = \mathbf{d}^e + \mathbf{d}^p .\tag{3.197}$$

This equation is often starting point of rate equations for finite elasto-plastic deformations. Note however that the following definitions are used in (3.195): $\mathbf{d}^p = \mathrm{sym}\,[\,\mathbf{F}^e\,\tilde{\mathbf{L}}^p\,\mathbf{F}^{e^{-1}}\,]$. The additional equation $\mathbf{w} = \mathbf{w}^e + \mathbf{w}^p$ for the anti-symmetric part of the spatial velocity gradient (spin) results from (3.196) and needs further considerations, see e.g. the discussion in Besseling and van der Giessen (1994).

As in the geometrical linear theory of elasto-plastic deformations, the free energy Ψ is written as function of the elastic deformation and of m inner variables α_k $(k = 1, \dots, m)$

$$\Psi(\tilde{\mathbf{C}}^e, \alpha_k) = \tilde{W}(\tilde{\mathbf{C}}^e) + \hat{H}(\alpha_k).\tag{3.198}$$

The corresponding dependence of the strain energy function on $\tilde{\mathbf{C}}^e$ can be found, e.g. in Mandel (1974). By considering the specific stress power $\boldsymbol{\tau} \cdot \mathbf{d}$, the local dissipation $\mathcal{D}$ is stated as

$$\mathcal{D} = \boldsymbol{\tau} \cdot \mathbf{d} - \dot{\Psi}(\tilde{\mathbf{C}}^e, \alpha_k) \geq 0.\tag{3.199}$$

The time derivative of the free energy function yields with $\overset{\bullet}{\tilde{\mathbf{C}}}{}^e = 2\,\mathbf{F}^{e^T}\,\mathbf{d}^e\,\mathbf{F}^e$

$$\mathcal{D} = \left(\boldsymbol{\tau} - 2\,\mathbf{F}^e\,\frac{\partial \tilde{W}}{\partial \tilde{\mathbf{C}}^e}\,\mathbf{F}^{e^T} \right) \cdot \mathbf{d}^e + \boldsymbol{\tau} \cdot \mathbf{d}^p - \sum_{k=1}^m \frac{\partial \tilde{H}}{\partial \alpha_k}\,\dot{\alpha}_k \geq 0.\tag{3.200}$$

From this inequality followed by using the standard arguments of material theory, see e.g. Truesdell and Noll (1965) or Lubliner (1990), the constitutive relations for stresses and hardening variables

$$\boldsymbol{\tau} = 2\,\mathbf{F}^e\,\frac{\partial \tilde{W}}{\partial \tilde{\mathbf{C}}^e}\,\mathbf{F}^{e^T} \quad \text{and} \quad q_k = -\frac{\partial \tilde{H}}{\partial \alpha_k},\tag{3.201}$$

and furthermore the reduced form of the dissipation inequality

$$\mathcal{D} = \boldsymbol{\tau} \cdot \mathbf{d}^p + \sum_{k=1}^m q_k\,\dot{\alpha}_k \geq 0,\tag{3.202}$$

which represents a restriction for the evolution equations of plastic flow. The equations derived above can also be referred to the intermediate configuration which leads to the stress tensor in the intermediate configuration

$$\tilde{\mathbf{S}} = \mathbf{F}^p\,\mathbf{S}\,\mathbf{F}^{p\,T} = \mathbf{F}^{e-1}\,\boldsymbol{\tau}\,\mathbf{F}^{e-T}\tag{3.203}$$

and the constitutive equations

$$\tilde{\mathbf{S}} = 2\,\frac{\partial \tilde{W}}{\partial \tilde{\mathbf{C}}^e} \quad \text{and} \quad q_k = -\frac{\partial \tilde{H}}{\partial \alpha_k}.\tag{3.204}$$

The first term in the reduced dissipation inequality (3.202) can be formulated with respect to the intermediate configuration

$$\boldsymbol{\tau} \cdot \mathbf{d}^p = \tilde{\mathbf{S}} \cdot \left(\tilde{\mathbf{C}}^e \, \tilde{\mathbf{L}}^p \right)^S = \boldsymbol{\Sigma} \cdot \mathbf{L}^p \,.$$

Here $\boldsymbol{\Sigma} = \tilde{\mathbf{C}}^e \, \tilde{\mathbf{S}}$ is the MANDEL stress tensor which can be non-symmetric in the general case. Hence (3.202) can be rewritten as

$$\mathcal{D} = \boldsymbol{\Sigma} \cdot \tilde{\mathbf{L}}^p + \sum_{k=1}^{m} q_k \, \dot{\alpha}_k \geq 0 \,, \qquad (3.205)$$

see Mandel (1974). These relations are valid for general elasto-plastic material behaviour. They will be specified in the following for isotropic materials. For that the assumption of a free energy function is valid which does not depend upon any orientation in the initial configuration. Furthermore, no orientation will enter the constitutive relations in the intermediate configuration which has the consequence of an undetermined plastic spin $\mathbf{w}^p$. In such cases, often the constitutive assumption $\mathbf{w}^p = \mathbf{0}$ is chosen. Further physical interpretations and considerations regarding the plastic spin can be found in Dafalias (1985) or Besseling and van der Giessen (1994).

The elastic domain of a given deformation state is described by a yield condition which is formulated in terms of the KIRCHHOFF stresses and the hardening variables

$$f\left(\boldsymbol{\tau}, \alpha_k \right) \leq 0 \,. \qquad (3.206)$$

The assumption of maximum plastic dissipation at a fixed configuration is valid in case of associative plasticity. This leads together with (3.202) to the evolution equations for the plastic variables

$$\mathbf{d}^p = \lambda \frac{\partial f}{\partial \boldsymbol{\tau}} \,, \qquad \dot{\alpha}_k = \lambda \frac{\partial f}{\partial \alpha_k} \,, \qquad (3.207)$$

see e.g. Simo (1992) or Simo and Miehe (1992). The kinematic part of the first equation in (3.207) can be reformulated using (3.58). This leads with (3.195) and (3.196) to

$$\mathcal{L}_v \mathbf{b}^e = -2 \, \mathbf{F}^e \, \mathrm{sym} \, (\tilde{\mathbf{L}}^p) \, \mathbf{F}^{e^T} = -2 \, \mathrm{sym} \, (\mathbf{l}^p \, \mathbf{b}^e) \,. \qquad (3.208)$$

With the further assumption that the plastic spin is zero ($\mathbf{w}^p = \mathbf{0}$), the relation $\mathbf{d}^p = \mathbf{l}^p$ follows with (3.196) from (3.197). Thus (3.207)$_1$ can be rewritten as

$$-\frac{1}{2} \mathcal{L}_v \mathbf{b}^e = \mathrm{sym} \left(\lambda \frac{\partial f}{\partial \boldsymbol{\tau}} \mathbf{b}^e \right) \,, \qquad (3.209)$$

see Simo and Miehe (1992), where $\mathbf{b}^e$ denotes the LIE derivative defined by

$$\mathcal{L}_v \mathbf{b}^e = -2 \, \mathbf{F}^e \, \frac{1}{2} \left(\tilde{\mathbf{L}}^p + \tilde{\mathbf{L}}^{p^T} \right) \mathbf{F}^{e^T} \,. \qquad (3.210)$$

For the isotropic case, the KIRCHHOFF stresses can be expressed as a function of the left CAUCHY–GREEN tensor $\mathbf{b}^e$, see also (3.111),

$$\boldsymbol{\tau} = 2 \frac{\partial \tilde{W}}{\partial \mathbf{b}^e} \, \mathbf{b}^e \,. \tag{3.211}$$

The left CAUCHY–GREEN tensor follows from

$$\mathbf{b}^e = \mathbf{F}^e \, \mathbf{F}^{eT} = \mathbf{F} \, \mathbf{C}^{p-1} \, \mathbf{F}^T \qquad \text{with} \quad \mathbf{C}^p = \mathbf{F}^{pT} \, \mathbf{F}^p \,. \tag{3.212}$$

This means that $\mathbf{b}^e$ is related to the inverse of the right CAUCHY–GREEN tensor $\mathbf{C}^{p-1}$ which is referred to the initial configuration.

Experiments substantiate that plastic flow does not depend upon the volume change in the body. This fact is synonymous with the assumption $\mathrm{tr}(\mathbf{d}^p) = 0$ or $\det \mathbf{F}^p = 1$ which corresponds to incompressibility of rubber materials, see Sect. 3.3.1. There is a split of the deformation in isochoric and volumetric parts lead to a decomposition of the strain energy function (3.122). Such split can also be defined for the above equation. It yields with (3.28) instead of (3.191) to

$$\mathbf{F} = J^{e \frac{1}{3}} \, \widehat{\mathbf{F}}^e \, \mathbf{F}^p \qquad \text{with} \quad \det \mathbf{F}^p = 1 \,. \tag{3.213}$$

Since it is not obvious that the integration of evolution equation (3.209) preserves this constraint condition, special care is needed in the design of associated numerical algorithms, see Sect. 6.3.

3.3.3 Visco-Elastic and Visco-Plastic Material Behaviour

Many materials exhibit in experiments a behaviour which can only be described by considering real time dependence of deformations and stresses. A frequent application is, e.g. creep of concrete, of metals at high temperatures or of saline rocks, at which the deformations grow under constant stress states with time. Also many polymers depict such behaviour. In the mathematical modelling process of such materials, the associated time dependent (rheological) material behaviour has to be considered. This leads to many different constitutive models which are used in the continuum mechanics description of rheology. Some of them are based on the introduction of springs (elasticity), dampers (viscosity) and friction (plasticity) elements which are combined to match experimental results. For an introduction in the basic theoretical background, see e.g. Findley et al. (1989), which contains also a literature review until 1988. Two simple visco-elastic models will be discussed and their three-dimensional generalization is stated in the framework of finite deformations. Furthermore, visco-plastic material behaviour is described for small and large deformations.

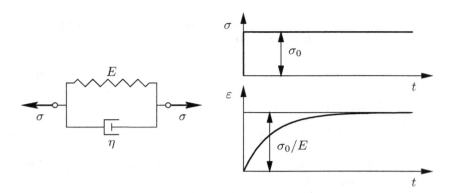

Fig. 3.9a KELVIN-VOIGT material   **Fig. 3.9b** Creep process

Kelvin-Voigt Material. The first model, called KELVIN-VOIGT material, consists of a spring and a damper which act in parallel. The spring describes elastic effect whereas the damper is introduced for the viscous material response, see Fig. 3.9a.

The strain ε due to elongation in damper and spring is equal, see Fig. 3.9a, while the stresses in spring and damper sum up to the total stress ($\sigma = \sigma_E + \sigma_D$). For the one-dimensional case, a constitutive equation, depending on time t, is obtained with the elastic constitutive equation $\sigma_E = E\,\varepsilon$ and the constitutive relation for the damper $\sigma_D = \eta\,\dot{\varepsilon}$

$$\sigma(t) = E\,\varepsilon(t) + \eta\,\dot{\varepsilon}(t) = E\left[\varepsilon(t) + \tau\,\dot{\varepsilon}(t)\right], \qquad (3.214)$$

in which the retardation time is defined by the constant $\tau = \eta/E$. For a force applied instantaneously, see Fig. 3.9b, the strain due to elongation can be computed by integration of (3.214) easily, see e.g. Gross et al. (1999),

$$\varepsilon(t) = \frac{\sigma_0}{E}\left(1 - e^{-(t/\tau)}\right). \qquad (3.215)$$

The solution in Fig. 3.9b depicts that creep occurs which is defined by an increase of strain in time. Within the equilibrium state at time $t \to \infty$, all stress is in the spring ($\varepsilon(\infty) = \sigma_0/E$). Hence, it can be concluded that the KELVIN-VOIGT-body behaves in the first stages like a fluid but in the end of the process like a solid.

The three-dimensional extension of the one-dimensional constitutive model can be found for finite deformations, e.g. in Eringen (1967) or Truesdell and Noll (1965). In the simplest isotropic case, a constitutive equation is obtained for the KIRCHHOFF stress tensor

$$\boldsymbol{\tau} = \alpha_1(I_b\,,II_b\,,III_b)\,\mathbf{1} + \alpha_2(I_b\,,II_b\,,III_b)\,\mathbf{b} + \alpha_3(I_b\,,II_b\,,III_b)\,\mathbf{d}, \quad (3.216)$$

where the first two terms describe the elastic behaviour and the last term the viscous behaviour. The scalar functions α_i depend upon the invariants of the left CAUCHY–GREEN tensor $\mathbf{b}$ (3.25), $\mathbf{d}$ is the rate of deformation tensor, see

(3.49). The constitutive equation (3.216) can be transformed to the initial configuration using (3.49) and (3.84)

$$\mathbf{S} = \mathbf{F}^{-1} \boldsymbol{\tau} \mathbf{F}^{-T} = \alpha_1(I_C, II_C, III_C) \mathbf{C}^{-1} + \alpha_2(I_C, II_C, III_C) \mathbf{1}$$
$$+ \alpha_3(I_C, II_C, III_C) \mathbf{C}^{-1} \dot{\mathbf{E}} \mathbf{C}^{-1} . \qquad (3.217)$$

Here the invariants of the left CAUCHY-GREEN tensor $\mathbf{b}$ are expressed by the invariants of the right CAUCHY-GREEN tensor $\mathbf{C}$, see also (3.105). The first two terms in (3.216) and (3.217) describe the elastic part. These can be expressed by the simple hyper-elastic constitutive equation (3.120). The additional choice of $\alpha_3 = J\eta$ includes viscous effects leading to the following constitutive equation in the current configuration

$$\boldsymbol{\tau} = \frac{\Lambda}{2} (J^2 - 1) \mathbf{1} + \mu (\mathbf{b} - \mathbf{1}) + J\eta \mathbf{d} . \qquad (3.218)$$

This equation can be transformed to the initial configuration, see also (3.119),

$$\mathbf{S} = \frac{\Lambda}{2} (J^2 - 1) \mathbf{C}^{-1} + \mu (\mathbf{1} - \mathbf{C}^{-1}) + J\eta \mathbf{C}^{-1} \dot{\mathbf{E}} \mathbf{C}^{-1} . \qquad (3.219)$$

Often materials which exhibit visco-elastic behaviour (e.g. rubber) are incompressible. In such case, the elastic constitutive relations have to be split in volumetric and deviatoric part according to (3.127) or (3.129).

Neglecting the elastic deformations in the constitutive equation (3.218) yields with (3.84) the following relation for the CAUCHY stress tensor

$$\boldsymbol{\sigma} = \eta \mathbf{d} . \qquad (3.220)$$

This equation describes a linear relation between the rate of deformation tensor and the stresses, and can be used to characterize a viscous compressible fluid.

Maxwell Material. The second model to describe viscous behaviour of materials consists contrary to the KELVIN-VOIGT material of a spring and damping device in series, see Fig. 3.10a.

It is obvious from the one-dimensional model that the stress in damper and spring is equal. The rate of deformations related to spring and damper are split additively ($\dot{\varepsilon} = \dot{\varepsilon}_E + \dot{\varepsilon}_D$) and are obtained from the constitutive relations $\dot{\varepsilon}_E = \dot{\sigma}/E$ and $\dot{\varepsilon}_D = \sigma/\eta$. Hence the total rate of deformation is given by

$$E\dot{\varepsilon}(t) = \frac{1}{\hat{\tau}} \sigma(t) + \dot{\sigma}(t) , \qquad (3.221)$$

with the constant $\hat{\tau} = \eta/E$, also called relaxation time. One can easily show that the response of the material due to a constant stress σ_0 describes a fluid, since the strain increases after an instantaneous jump by σ_0/E linear in time. However, if the strain ($\varepsilon_0 = \sigma_0/E$) is kept constant, then the solution of (3.221) is given by

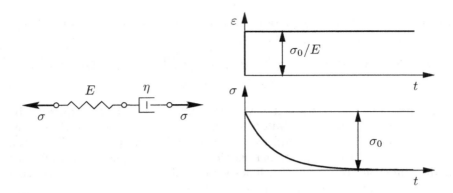

Fig. 3.10a MAXWELL-material   **Fig. 3.10b** Relaxation

$$\sigma(t) = \sigma_0 \, e^{-(t/\hat{\tau})} \tag{3.222}$$

and the stress relaxes from its initial value σ_0 for $t \to \infty$ to zero.

Since this material does not describe a solid body a so-called generalized MAXWELL-model is often defined which consists of a parallel connection of model (3.221) and a spring with constant E^∞. Such a model responds to a constant stress with an instantaneous elastic behaviour, but creeps with time to a limit state. This model is also called linear standard solid. The constitutive response of this model are described for the one-dimensional case by

$$
\begin{aligned}
\sigma(t) &= \sigma_M(t) + \sigma_E(t)\,, \\
E\,\dot{\varepsilon}(t) &= \frac{1}{\hat{\tau}}\,\sigma_M(t) + \dot{\sigma}_M(t)\,, \\
\sigma_E(t) &= E^\infty\,\varepsilon(t)\,.
\end{aligned}
\tag{3.223}
$$

Here stresses σ_M are related to the MAXWELL-model and σ_E related to the parallel connected spring, respectively, see Fig. 3.11.

The three-dimensional version of a linear standard solid is based on the assumption that viscous deformations are only caused by deviatoric stress

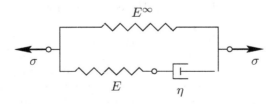

Fig. 3.11 Generalized MAXWELL-model

and strain states, see (3.30) and (3.151). For the MAXWELL material parallel connected to the spring follows

$$\begin{aligned}
\boldsymbol{\sigma}(t) &= \boldsymbol{\sigma}_E(t) + \mathbf{s}_M(t) \\
&= K\,\mathrm{tr}\boldsymbol{\varepsilon}\mathbf{1} + 2\mu\,(\nu^\infty\mathbf{e} + \nu\,\mathbf{q})\,, \\
\dot{\mathbf{e}}(t) &= \frac{1}{\hat{\tau}}\mathbf{q} + \dot{\mathbf{q}}\,,
\end{aligned}\qquad(3.224)$$

where $\nu^\infty + \nu = 1$, which is equivalent to $\nu^\infty = (1 - \nu)$. The value $\mu\nu^\infty$ corresponds to μ^∞, see (3.223)$_3$, since the volumetric parts are purely elastic such that $K = K^\infty$. Instead of the deviatoric stress $\mathbf{s}_M$ the conjugated strain $\mathbf{q}$ is used in (3.224)$_3$, which saves the multiplication by the shear modulus.

The set of equations (3.224) presents a first order differential equation system in time. Furthermore, the constitutive equation can be extended by introducing time dependent parameters to consider aging processes, see e.g. Argyris et al. (1976). It is possible to derive an integral form for the stresses for the assumed linear material behaviour and constant constitutive parameters. Since the volumetric part has no influence on the viscous behaviour, the stress deviator is given by

$$\mathbf{s}(t) = \int_{-\infty}^{t} G(t - \tau)\,\dot{\mathbf{e}}(\tau)\,d\tau\,,\qquad(3.225)$$

where the relaxation function $G(t)$ is defined by

$$G(t) = 2\mu\left[\nu^\infty + \nu\,e^{-(t/\hat{\tau})}\right]\,.\qquad(3.226)$$

A generalization of this model for finite deformation can be found in Simo (1987). For a hyper-elastic constitutive equation, the following relation with respect to the initial configuration can be derived using the 2. PIOLA-KIRCHHOFF stress tensor and (3.126) for pure deviatoric viscous response

$$\begin{aligned}
\mathbf{S} &= \mathbf{S}_M + \mathbf{S}_{ISO}^\infty + \mathbf{S}_{VOL}^\infty\,, \\
\dot{\mathbf{S}}_M + \frac{1}{\hat{\tau}}\mathbf{S}_M &= \frac{d}{dt}\mathbb{P}\left[2\frac{\partial\hat{W}}{\partial\widehat{\mathbf{C}}}\right]\,, \\
\mathbf{S}_{ISO}^\infty &= \mathbb{P}\left[2\frac{\partial\hat{W}^\infty}{\partial\widehat{\mathbf{C}}}\right]\quad\text{and}\quad \mathbf{S}_{VOL}^\infty = J\,\mathbf{C}^{-1}\frac{\partial U^\infty}{\partial J}\,.
\end{aligned}\qquad(3.227)$$

In this equation, the stress $\mathbf{S}^\infty$ is related to the parallel connected spring; this stress was split into a volumetric and deviatoric part, see (3.126). The stress $\mathbf{S}_M$ is associated with the MAXWELL-model; it is determined from the evolution equation (3.227)$_2$. With the assumption that the visco-elastic material consists of identical polymer chains, the strain energy function $\hat{W}$ can be expressed by the strain energy of the elastic part: $\hat{W}_M(\widehat{\mathbf{C}}) = \beta\,\hat{W}^\infty(\widehat{\mathbf{C}})$, with $\beta > 0$, see Govindjee and Simo (1992). Thus the model depends upon the volumetric (U^∞) and deviatoric ($\hat{W}^\infty$) part of the strain energy function. The material behaviour described in (3.227) is called linear visco-elastic behaviour at finite deformations. A generalization for finite visco-elastic material behaviour at finite deformations can be found, e.g. in Reese and Govindjee (1998).

Visco-Plastic Materials. It is well known from experiments that a plasticity model independent on the strain rate, as described in Sect. 3.3.2, is only one possible model for the real physical behaviour. In the area of material theory there exist many papers in which inelastic material models are advocated which do not exhibit a yield surface. Such constitutive models are often employed to describe inelastic behaviour of metals at high temperatures. Examples for such formulations can be found in, e.g. Bodner and Partom (1975), Hart (1976) and Krempl et al. (1986). These constitutive models are often called visco-plastic models in the relevant literature; however, they also could be named nonlinear visco-elastic materials, see e.g. Lubliner (1990). In the following, a visco-plastic material will be defined as a strain rate dependent material with a well-defined yield surface, see Prager (1961) or Perzyna (1966).

The main difference to the rate independent theory of plasticity is related to the fact that the elastic domain, defined by the yield condition $f \leq 0$ in stress space, can be violated. Hence overstresses can occur, see Fig. 3.13, which are outside of the elastic domain defined by $f \leq 0$. In case of pure creep, see e.g. Fig. 3.9b, the elastic domain does not exist; only overstresses occur. Hence the visco-plastic behaviour is described by a coupling of a viscous and a plastic model. The associated rheological model, which is called BINGHAM model, is given by a parallel connection of a damper and a plastic slip element which then is in series with an elastic spring, see Fig. 3.12.

This model reacts elastic as long as the yield stress Y_0 in the plastic slip element is exceeded: $|\sigma| \geq Y_0$. After that plastic flow occurs in combination with the rate dependent response due to the viscus damper. We obtain with the notation, introduced above, and the visco-plastic strain rate $\dot{\varepsilon}_{VP}$

$$\sigma = Y_0 + \eta\,\dot{\varepsilon}_{VP} = E\,\varepsilon_E\,. \tag{3.228}$$

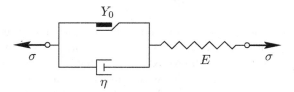

Fig. 3.12 One-dimensional elasto-visco plastic material

Reformulation of this relation yields for $|\sigma| \geq Y_0$

$$\dot{\varepsilon}_{VP} = \frac{1}{\eta}(\sigma - Y_0).\qquad(3.229)$$

Observe that the visco-plastic strain rate depends upon the difference between the total stress (in this case given by the elastic material equation $\sigma = E\varepsilon$) and the yield stress. In general, the projection of the stresses onto the elastic domain $f = 0$ can be performed, see Fig. 3.13 for the multi-dimensional case in which the plastic flow is a function of the stress deviator $\mathbf{s}$. The difference $(\sigma - Y_0)$ is the overstress in the one-dimensional model and $\mathbf{s} - \mathbf{s}^*$ in the multi-dimensional case as depicted in Fig. 3.13. If the yield surface does not exist ($Y_0 = 0$) then the material behaviour reduces to the MAXWELL-model.

Often the so-called FÖPPL symbol is introduced, also known as MACAULEY bracket, to be able to consider the inequality condition $|\sigma| \geq Y_0$ directly within the constitutive equation. From the definition

$$\langle \Phi \rangle = \frac{1}{2}(\Phi + |\Phi|) = \begin{cases} \Phi & \text{for} \quad \Phi > 0 \\ 0 & \text{for} \quad \Phi \leq 0 \end{cases}\qquad(3.230)$$

follows a constitutive equation for a one-dimensional elasto-viscoplastic model, which yields with $\dot{\varepsilon} = \dot{\varepsilon}_E + \dot{\varepsilon}_{VP}$

$$E\dot{\varepsilon} = \dot{\sigma} + \frac{1}{\tau}\langle \sigma - Y_0 \rangle,\qquad(3.231)$$

where τ is defined in the same way as in (3.221).

The three-dimensional generalization of this material model will be presented here for small deformations. In Perzyna (1963), the following ansatz is chosen for the visco-plastic deviatoric strain rates

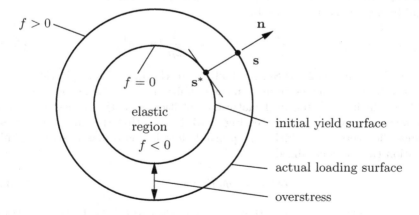

Fig. 3.13 Visco-plastic material behaviour in the stress space

$$\dot{\mathbf{e}}^{vp} = \frac{1}{2\eta} \langle \Phi(\bar{f}) \rangle \, \mathbf{n} \,. \qquad (3.232)$$

Here the definition $\mathbf{n}$ was chosen for $\mathbf{q} = \mathbf{0}$ according to (3.162). The scalar η is a material parameter which describes the visco-plastic behaviour. Observe that this, widely accepted, material model represents a generalization of the constitutive relations given in (3.229). A better approximation of experimental results is provided by an exponential or power function Φ. For the special case of metals under dynamical loading, the following functions can be found in the literature:

$$\text{Power function:} \qquad \Phi = \bar{f}^{\,m} \,,$$
$$\text{Exponential Function:} \qquad \Phi = e^{\bar{f}} - 1 \,.$$

In this relations, the yield function $\bar{f}$ is normalized. For the classical VON MISES plasticity without hardening, the normalized yield criterion is given by: $\bar{f} = \|\mathbf{s}\| / \sqrt{\frac{2}{3}} \, Y_0 - 1$. However, the yield conditions described in Sect. 3.3.2 can be used for $\bar{f}$ in (3.232) also. The simplest choice for a visco–plastic material was stated already in Hohenemser and Prager (1932). It has the form

$$\dot{\mathbf{e}}^{vp} = \frac{1}{2\eta} \langle f \rangle \frac{\partial f}{\partial \mathbf{s}} = \gamma \langle \bar{f} \rangle \, \mathbf{n} \,. \qquad (3.233)$$

In case of elasto-visco-plastic material behaviour, equation (3.232) has to be extended by an elastic constitutive equation for the stresses and the kinematical assumption of an additive split of the strains into elastic and visco-plastic parts

$$\boldsymbol{\sigma} = \mathbb{C} \, [\, \boldsymbol{\varepsilon}^e \,], \quad \boldsymbol{\varepsilon}^e = \boldsymbol{\varepsilon} - \boldsymbol{\varepsilon}^{vp} \,. \qquad (3.234)$$

For the derivation of the constitutive equations for the visco-plastic strain rates (3.232), the concept of overstresses can also be applied. The three-dimensional extension of the one-dimensional model (3.229) for isotropic material behaviour is then given by

$$\dot{\mathbf{e}}^{vp} = \frac{1}{2\eta} \langle \mathbf{s} - \mathbf{s}^* \rangle \,, = \frac{1}{\tau} \frac{1}{2\mu} \langle \mathbf{s} - \mathbf{s}^* \rangle \,, \qquad (3.235)$$

see e.g. Maugin (1992) or Simo (1998). Here the modulus of elasticity E is replaced by the shear modulus μ in the definition of the relaxation time τ. In (3.235), which is written in deviatoric quantities, $\mathbf{s}^*$ is the projection of the stress onto the yield surface, see Fig. 3.13. Due to that the difference $\mathbf{s} - \mathbf{s}^*$ defines the overstress, which is responsible for the visco-plastic strains. This equation can be generalized

$$\dot{\mathbf{e}}^{vp} = \frac{1}{\tau} \langle \Phi(f) \rangle \, (\mathbf{s} - \mathbf{s}^*) \,, \qquad (3.236)$$

see Simo (1998), which essentially corresponds to (3.232). In this relation, the flow vector $\mathbf{n}$, which determines the direction of flow, is replaced by the difference $\mathbf{s} - \mathbf{s}^*$. Based on the assumption of a convex yield surface, both formulations determine the same direction such that (3.236) is equivalent to (3.232), see Simo (1998). Furthermore, (3.233) corresponds to the flow rule (3.160) when instead of the consistency parameter λ the constitutive relation $\gamma \langle \Phi(f) \rangle$ is used. Such re-interpretation of the visco-plastic model can be employed to construct efficient numerical algorithms. The integration of the visco-plastic material equation can then be based on algorithms derived for elasto-plastic problems, see Sect. 6.2.

Remark 3.8: The derivation of constitutive equations for visco-plastic material behaviour can also be derived from the principle of maximum plastic dissipation, see Remark 3.7. In that case, the constraint equation (yield condition) within the optimization problem (3.189) describing the visco-plastic dissipation has to be introduced by a penalty term instead of the LAGRANGE multiplier. This yields for the material model (3.233)

$$\mathcal{D} = -\mathbf{s} \cdot \dot{\mathbf{e}}^{vp} + \frac{1}{2} \gamma \left[f(\mathbf{s}) \right]^2 , \quad \gamma > 0 , \quad \text{for } f(\mathbf{s}) > 0 . \tag{3.237}$$

The scalar γ has to be interpreted as the penalty parameter. For $\gamma \to \infty$, the solution (3.189), describing rate-independent behaviour, follows as limit case. This corresponds to a viscosity which approaches zero since $\gamma \sim \frac{1}{\eta}$. The gradient of $\mathcal{D}$ with respect to the stress deviator yields

$$\dot{\mathbf{e}}^{vp} = \gamma \langle f \rangle \frac{\partial f}{\partial \mathbf{s}} , \tag{3.238}$$

which is equivalent to (3.233). Here the FÖPPL symbol was introduced to reflect condition $f > 0$ in (3.237).

3.3.4 Incremental Form of the Material Equations

Incremental forms of the constitutive equations are discussed in this section. For this purpose, the associated equations are differentiated with respect to time.

Incremental Form of Hyper-Elastic Constitutive Equations. The starting point for the development is the constitutive relation (3.104) which is formulated in terms of the initial configuration. Differentiation with respect to time yields with $W = \rho_0 \psi$

$$\dot{\mathbf{S}} = 2 \frac{\partial^2 W}{\partial \mathbf{C} \, \partial \mathbf{C}} \left[\dot{\mathbf{C}} \right] . \tag{3.239}$$

It provides an incremental relation between the time derivative of the 2. PIOLA-KIRCHHOFF stress tensor $\mathbf{S}$ and the time derivative of the right CAUCHY-GREEN tensor $\mathbf{C}$. With the definition of the incremental constitutive tensor

$$\mathbb{C} = 4 \frac{\partial^2 W}{\partial \mathbf{C} \, \partial \mathbf{C}} , \qquad \mathbb{C}_{ABCD} = 4 \frac{\partial^2 W}{\partial C_{AB} \, \partial C_{CD}} , \tag{3.240}$$

equation (3.239) can be written also as

$$\dot{\mathbf{S}} = \mathbb{C}\left[\frac{1}{2}\dot{\mathbf{C}}\right], \qquad \dot{S}_{AB} = \mathbb{C}_{ABCD}\frac{1}{2}\dot{C}_{CD}, \qquad (3.241)$$

where $\mathbb{C}$ depends upon the actual deformation state.

The transformation of (3.241) to the current configuration can be formulated with the LIE-derivative of the KIRCHHOFF stress tensor (3.96). This derivative is given in index notation as

$$(\mathcal{L}_v \boldsymbol{\tau})_{ik} = F_{iA}\,\dot{S}_{AB}\,F_{kB}\,. \qquad (3.242)$$

With the time derivative of the right CAUCHY-GREEN tensor, see (3.15) and (3.49),

$$\dot{C}_{CD} = 2\,F_{lC}\,d_{lm}\,F_{mD}, \qquad (3.243)$$

the final result can be expressed by

$$(\mathcal{L}_v \boldsymbol{\tau})_{ik} = F_{iA}\,F_{lC}\,F_{mD}\,F_{kB}\,\mathbb{C}_{ABCD}\,d_{lm}\,, \qquad (3.244)$$

where $\mathbf{d}$ is the rate of deformation tensor. Since each basis vector of the incremental constitutive tensor $\mathbb{C}$ in (3.244) is transformed to the current configuration, the spatial incremental constitutive tensor $\mathbf{c}$ can be introduced as

$$\mathbf{c}_{iklm} = F_{iA}\,F_{lC}\,F_{mD}\,F_{kB}\,\mathbb{C}_{ABCD}\,. \qquad (3.245)$$

Hence equation (3.244) can be reformulated with this relation

$$(\mathcal{L}_v \boldsymbol{\tau})_{ik} = \mathbf{c}_{iklm}\,d_{lm}\,, \qquad \mathcal{L}_v \boldsymbol{\tau} = \mathbf{c}\,[\,\mathbf{d}\,]\,. \qquad (3.246)$$

Often the JAUMANN stress rate, defined in (3.97), is used in the literature to describe elasto-plastic material behaviour. Hence the incremental constitutive equation (3.246) will be rewritten for the JAUMANN stress rate. Using (3.98) the relation

$$\begin{aligned}\overset{\nabla}{\boldsymbol{\tau}} &= \mathbf{c}\,[\,\mathbf{d}\,] + \boldsymbol{\tau}\,\mathbf{d} + \mathbf{d}\,\boldsymbol{\tau},\\ \overset{\nabla}{\tau}_{ik} &= \mathbf{c}_{iklm}\,d_{lm} + \tau_{in}\,d_{nk} + d_{in}\,\tau_{nk}\end{aligned} \qquad (3.247)$$

follows which can be abbreviated by placing d_{lm} outside the brackets

$$\overset{\nabla}{\tau}_{ik} = \mathbf{a}_{iklm}\,d_{lm} \quad \text{with} \quad \mathbf{a}_{iklm} = \mathbf{c}_{iklm} + \delta_{il}\,\tau_{km} + \delta_{km}\,\tau_{il}\,. \qquad (3.248)$$

Equation (3.248) is from the physical point of view equivalent to (3.246).

Remark 3.9: Within the constitutive formulations of elasto-plastic material behaviour, it can be assumed in most cases that only small elastic strains occur. Related to this fact the rate equation

$$\overset{\nabla}{\tau}_{ik} = \mathbf{a}_{iklm}^{L}\,d_{lm} \qquad (3.249)$$

is often found in the literature for the elastic part of the deformation. Here $\mathbf{a}^L$ is the classical constant material tensor of the linear theory of elasticity for isotropic material: $\mathbf{a}^L_{iklm} = \Lambda\,\delta_{ik}\,\delta_{lm} + \mu\,(\,\delta_{il}\,\delta_{km} + \delta_{kl}\,\delta_{im}\,)$. Such a material law, however, does not represent a hyperelastic constitutive relation. It is known as hypo-elastic behaviour, see e.g. Truesdell and Noll (1965), and may lead in case of finite plastic deformations to physically wrong responses, see e.g. Atluri (1984).

Hyperelastic Material Equations with Split in Volumetric and Isochoric Parts. Incremental constitutive tensors can be formulated for the constitutive equation (3.125), which is split additively into volumetric and isochoric parts. The derivation of the incremental constitutive relation is based on (3.240) which yields with (3.123) and (3.126) in index notation

$$
\hat{\mathbb{C}}_{ABCD} = 2\frac{\partial S_{AB}}{\partial C_{CD}} \tag{3.250}
$$

$$
2\frac{\partial}{\partial C_{CD}}\left[J\frac{\partial U}{\partial J}\,C^{-1}_{AB} + 2\,J^{-\frac{2}{3}}\left(\mathbb{E}_{ABEF} - \frac{1}{3}\,C^{-1}_{AB}\,C_{EF}\right)\frac{\partial \hat{W}}{\partial \hat{C}_{EF}}\right].
$$

After some analysis, the incremental constitutive tensor is obtained which is related to the volumetric and the isochoric part by using (3.125) or (3.126), respectively,

$$
\hat{\mathbb{C}}_{ABCD} = \hat{\mathbb{C}}^{VOL}_{ABCD} + \hat{\mathbb{C}}^{ISO}_{ABCD} \tag{3.251}
$$

with

$$
\hat{\mathbb{C}}^{VOL}_{ABCD} = \left(J\frac{\partial U}{\partial J} + J^2\frac{\partial^2 U}{\partial J^2}\right)C^{-1}_{AB}\,C^{-1}_{CD} - 2\,J\frac{\partial U}{\partial J}\,\mathbb{E}_{C^{-1}\,ABCD}\,,
$$

$$
\hat{\mathbb{C}}^{ISO}_{ABCD} = -\frac{2}{3}\left[C^{-1}_{CD}\,\mathbb{P}_{ABEF} + C^{-1}_{AB}\,\mathbb{P}_{CDEF} \right. \tag{3.252}
$$

$$
\left. + J^{-\frac{2}{3}}\left(\frac{1}{3}\,C^{-1}_{AB}\,C^{-1}_{DC} - C^{-1}_{AC}\,C^{-1}_{BD}\right)C_{EF}\right]2\frac{\partial \hat{W}}{\partial \hat{C}_{EF}}
$$

$$
+\,\mathbb{P}_{ABEF}\,4\frac{\partial^2 \hat{W}}{\partial \hat{C}_{EF}\partial \hat{C}_{MN}}\,\mathbb{P}_{CDMN}\,.
$$

This can be expressed with (3.126) in direct notation

$$
\hat{\mathbb{C}}_{VOL} = \left(J\frac{\partial U}{\partial J} + J^2\frac{\partial^2 U}{\partial J^2}\right)\mathbf{C}^{-1}\otimes\mathbf{C}^{-1} - 2\,J\frac{\partial U}{\partial J}\,\mathbb{E}_{C^{-1}}\,,
$$

$$
\hat{\mathbb{C}}_{ISO} = -\frac{2}{3}\left[\mathbf{C}^{-1}\otimes\mathbf{S}_{ISO} + \mathbf{S}_{ISO}\otimes\mathbf{C}^{-1}\right] \tag{3.253}
$$

$$
+\frac{2}{3}\,J^{-\frac{2}{3}}\left(\mathbf{C}\cdot 2\frac{\partial \hat{W}}{\partial \hat{\mathbf{C}}}\right)\left[\frac{1}{3}\,\mathbf{C}^{-1}\otimes\mathbf{C}^{-1} - \mathbb{E}_{C^{-1}}\right] + 4\,\mathbb{P}\,\frac{\partial^2 \hat{W}}{\partial \hat{\mathbf{C}}\partial \hat{\mathbf{C}}}\,\mathbb{P}\,.
$$

In these equations, the derivative of the inverse right CAUCHY-GREENtensor $\partial \mathbf{C}^{-1}/\partial \mathbf{C}$ was used which can be expressed in index notation by

$$\frac{\partial C_{AB}^{-1}}{\partial C_{CD}} = -C_{AC}^{-1} C_{BD}^{-1} = -\mathbb{E}_{C^{-1} ABCD} . \qquad (3.254)$$

Now $\partial \mathbf{C}^{-1} / \partial \mathbf{C} = \mathbb{E}_{C^{-1}}$ is introduced to shorten notation. Due the symmetry of $\mathbf{C}$, the tensor $\mathbb{E}_{C^{-1}}$ can be written as

$$\mathbb{E}_{C^{-1} ABCD} = \frac{1}{2} \left(C_{AC}^{-1} C_{BD}^{-1} + C_{AD}^{-1} C_{BC}^{-1} \right) . \qquad (3.255)$$

Hyperelastic Material Equations in Principal Stretches. In the case that the stresses are defined in principal stretches, the formulation of the associated incremental constitutive tensors is more complicated since the changes of the eigenvectors have to be considered. A derivation will be given here for the constitutive equation (3.133). Before the time derivative of the stresses is computed, the time derivative of the right CAUCHY-GREENtensor will be stated. The representation of $\mathbf{C}$ with respect to the principal axes is provided by $\mathbf{C} = \sum_{i=1}^{3} C_{(i)} \mathbf{N}_{(i)} \otimes \mathbf{N}_{(i)}$. Based on this expression $\dot{\mathbf{C}}$ can be formulated, see Ogden (1984),

$$\dot{\mathbf{C}} = \sum_{i=1}^{3} \dot{C}_{(i)} \mathbf{N}_{(i)} \otimes \mathbf{N}_{(i)} + \sum_{i \neq k} \Omega_{(ik)} \left(C_{(k)} - C_{(i)} \right) \mathbf{N}_{(i)} \otimes \mathbf{N}_{(k)} . \qquad (3.256)$$

The last term stems from the time derivative of the eigenvector $\dot{\mathbf{N}}_{(i)} = \mathbf{\Omega} \mathbf{N}_{(i)}$. In this expression, $\mathbf{\Omega}$ represents a skew-symmetric tensor, which is given by $\mathbf{\Omega} = \dot{\mathbf{D}} \mathbf{D}$. The tensor $\mathbf{D}$ describes the transformation between the eigenvectors $\mathbf{N}_{(i)}$ and the cartesian basis $\mathbf{E}_K$, see also (3.135).

An adequate equation to (3.256) follows also for the second PIOLA-KIRCHHOFF-stresses. Since the eigenvalues of the stress tensor, $S_{(i)}$, depend in (3.133) only upon $\lambda_{(i)}$ and hence upon $C_{(i)}$, the second PIOLA-KIRCHHOFF-stresses are given by

$$\dot{\mathbf{S}} = \sum_{i,k=1}^{3} \frac{\partial S_{(i)}}{\partial C_{(k)}} \dot{C}_{(k)} \mathbf{N}_{(i)} \otimes \mathbf{N}_{(i)}$$

$$+ \sum_{i \neq k} \Omega_{(ik)} \left(C_{(k)} - C_{(i)} \right) \left(\frac{S_{(k)} - S_{(i)}}{C_{(k)} - C_{(i)}} \right) \mathbf{N}_{(i)} \otimes \mathbf{N}_{(k)} . \qquad (3.257)$$

By inspection of (3.257), equation (3.241) can be rewritten analogously which leads to an explicit representation of the incremental constitutive tensor

$$\mathbb{C} = \sum_{i,k=1}^{3} \mathbb{L}_{(iikk)} \mathbf{N}_{(i)} \otimes \mathbf{N}_{(i)} \otimes \mathbf{N}_{(k)} \otimes \mathbf{N}_{(k)} \qquad (3.258)$$

$$+ \sum_{i \neq k} \mathbb{L}_{(ikik)} \mathbf{N}_{(i)} \otimes \mathbf{N}_{(k)} \otimes \left(\mathbf{N}_{(i)} \otimes \mathbf{N}_{(k)} + \mathbf{N}_{(k)} \otimes \mathbf{N}_{(i)} \right) .$$

In this equation, the coefficients of the constitutive tensors are

$$\mathbb{L}_{(iikk)} = 2\frac{\partial S_{(i)}}{\partial C_{(k)}},$$

$$\mathbb{L}_{(ikik)} = \left(\frac{S_{(i)} - S_{(k)}}{C_{(i)} - C_{(k)}}\right). \tag{3.259}$$

Since the eigenvalues of $\mathbf{C}$ depend via $C_{(i)} = \lambda_{(i)}^2$ upon the principal stretches, it follows with $\frac{1}{2\lambda_{(i)}}\frac{\partial(\ldots)}{\partial\lambda_{(i)}}$ and $S_{(i)} = \frac{1}{\lambda_{(i)}}\frac{\partial w}{\partial\lambda_{(i)}}$, see (3.134), for the components of the incremental constitutive tensor

$$\mathbb{L}_{(iikk)} = \frac{1}{\lambda_{(k)}}\frac{\partial}{\partial\lambda_{(k)}}\left(\frac{1}{\lambda_i}\frac{\partial w}{\partial\lambda_{(i)}}\right),$$

$$\mathbb{L}_{(ikik)} = \left(\frac{S_{(i)} - S_{(k)}}{\lambda_{(i)}^2 - \lambda_{(k)}^2}\right). \tag{3.260}$$

For practical purposes, it is essential to consider the symmetries of the incremental constitutive tensor $\mathbb{L}$. Due to the symmetry of $\mathbf{S}$ and $\mathbf{C}$,

$$\mathbb{L}_{ijkl} = \mathbb{L}_{jikl} = \mathbb{L}_{ijik} \tag{3.261}$$

follows and additionally for hyperelastic material

$$\mathbb{L}_{ijkl} = \mathbb{L}_{klij} \tag{3.262}$$

can be written.

Observe that principal stretches of the same magnitude yield an undetermined expression in (3.260) for the components $\mathbb{L}_{(ikik)}$. The components $\mathbb{L}_{(ikik)}$ can then be determined by taking the limit, see Chadwick and Ogden (1971). This leads to

$$\lim_{C_i \to C_k} \mathbb{L}_{ikik} = \frac{1}{2}\left(\mathbb{L}_{iiii} - \mathbb{L}_{iikk}\right). \tag{3.263}$$

In some cases, the stress divergence term in the weak form is not given by the second PIOLA-KIRCHHOFF stress tensor. If it is formulated with the first PIOLA-KIRCHHOFF stress tensor, see e.g. (3.289), then the incremental constitutive tensor is computed analogous to (3.240) from

$$\mathbf{A} = \frac{\partial^2 W}{\partial\mathbf{F}\partial\mathbf{F}}. \tag{3.264}$$

An explicit expression for this tensor, derived in Ogden (1984), is given by

$$\mathbb{A}_{iJkL} = F_{iM}\,\mathbb{L}_{MJNL}\,F_{kN} + \delta_{ik}\,S_{JL}, \tag{3.265}$$

in which the components of the incremental constitutive tensor (3.258) are determined from the deformation gradient, the KRONECKER symbol and the second PIOLA-KIRCHHOFF stresses. The incremental constitutive tensor **A** is referred to the initial as well as the current configuration like the first PIOLA-KIRCHHOFF stress tensor **P**. Here the same notation is applied as in the transformation (3.245) in which large indices refer to the initial and small indices to the current configuration.

Remark 3.10: Analogous to the remarks in 3.5(b), it is convenient for computational purposes to bring the fourth order constitutive tensor **L** in matrix form. Note that in this case, due to the time derivative of strain and stresses, components can appear as off diagonal terms. Using the relation $\dot{\mathbf{E}} = \frac{1}{2}\dot{\mathbf{C}}$ yields

$$
\begin{Bmatrix} \dot{S}_{11} \\ \dot{S}_{22} \\ \dot{S}_{33} \\ \dot{S}_{12} \\ \dot{S}_{23} \\ \dot{S}_{31} \end{Bmatrix} = \begin{bmatrix} L_{1111} & L_{1122} & L_{1133} & 0 & 0 & 0 \\ L_{2211} & L_{2222} & L_{2233} & 0 & 0 & 0 \\ L_{3311} & L_{3322} & L_{3333} & 0 & 0 & 0 \\ 0 & 0 & 0 & L_{1212} & 0 & 0 \\ 0 & 0 & 0 & 0 & L_{2323} & 0 \\ 0 & 0 & 0 & 0 & 0 & L_{3131} \end{bmatrix} \begin{Bmatrix} \dot{E}_{11} \\ \dot{E}_{22} \\ \dot{E}_{33} \\ 2\,\dot{E}_{12} \\ 2\,\dot{E}_{23} \\ 2\,\dot{E}_{31} \end{Bmatrix},
$$

$$
\dot{\mathbf{S}} = \mathbf{L}(\dot{\mathbf{E}}). \tag{3.266}
$$

For an implementation within a finite element program all quantities have to be transformed in this equation to the cartesian frame as in (3.139) which is done based on (3.135). This relation can be applied to each base vector in (3.258) which leads to an incremental constitutive tensor $\mathbb{L}_K$ given in terms of the cartesian frame

$$
\mathbb{L}_K = \sum_{i,k=1}^{3} \mathbb{L}_{(iikk)} D_{(i)\,J} D_{(i)\,K} D_{(k)\,L} D_{(k)\,M} \mathbf{E}_J \otimes \mathbf{E}_K \otimes \mathbf{E}_L \otimes \mathbf{E}_M \tag{3.267}
$$

$$
+ \sum_{i \neq k} \mathbb{L}_{(ikik)} D_{(i)\,J} D_{(k)\,K} D_{(i)\,L} D_{(k)\,M} \mathbf{E}_J \otimes \mathbf{E}_K \otimes (\mathbf{E}_L \otimes \mathbf{E}_M + \mathbf{E}_M \otimes \mathbf{E}_L)
$$

For a detailed description of the related matrices, see e.g. Reese (1994) or Reese and Wriggers (1995).

Exercise 3.8: For the constitutive equations (3.119), (a), and (3.127), (b), derive the incremental constitutive tensor which is related to initial and current configuration. Furthermore stated the form of the constitutive tensor for the undeformed initial configuration.

Solution: **(a)** NEO HOOKE **material.** The constitutive equation given in (3.119)

$$
\mathbf{S} = \frac{\Lambda}{2}\,(J^2 - 1)\,\mathbf{C}^{-1} + \mu\,(\mathbf{1} - \mathbf{C}^{-1})
$$

is a function of the strain measure $\mathbf{C}^{-1}$ and the determinant J of the deformation gradient. Their derivatives with respect to the right CAUCHY-GREEN tensor **C** have to be determined for the computation of $\mathbb{C}$. The derivative $\partial J / \mathbf{C}$ is given in (3.124). The derivative of $\mathbf{C}^{-1}$ was presented in (3.254). With these two results

$$
\mathbb{C} = \Lambda J^2\,\mathbf{C}^{-1} \otimes \mathbf{C}^{-1} + [\,2\mu - \Lambda\,(J^2 - 1)\,]\,\mathbb{E}_{C^{-1}},
$$

$$
\mathbb{C}_{ABCD} = \Lambda J^2\,C_{AB}^{-1}\,C_{CD}^{-1} + [\,2\mu - \Lambda\,(J^2 - 1)\,]\,\mathbb{E}_{C^{-1}\,ABCD} \tag{3.268}
$$

follows from the constitutive equation (3.119) in direct and index notation. The transformation of the incremental constitutive tensor $\mathbb{C}$ to the current configuration yields with (3.245) and using

$$C_{AC}^{-1}\, C_{BD}^{-1} = F_{pA}^{-1}\, F_{pC}^{-1}\, F_{qB}^{-1}\, F_{qD}^{-1}\, F_{iA}\, F_{lC}\, F_{mD}\, F_{kB} = \delta_{pi}\, \delta_{pl}\, \delta_{qk}\, \delta_{qm} = \delta_{il}\, \delta_{km},$$

$$
\begin{aligned}
\mathbb{c} &= \Lambda\, J^2\, \mathbf{1} \otimes \mathbf{1} + [\,2\,\mu - \Lambda\,(\,J^2 - 1\,)\,]\, \mathbb{E}, \\
\mathbb{c}_{iklm} &= \Lambda\, J^2\, \delta_{ik}\, \delta_{lm} + [\,2\,\mu - \Lambda\,(\,J^2 - 1\,)\,]\, \mathbb{E}_{iklm}.
\end{aligned}
\tag{3.269}
$$

Here $\mathbf{1}$ is the second order unit tensor and $\mathbb{E}$ is the fourth order unit tensor; both are referred to the current configuration. In index notation $\mathbb{E}$ has the form

$$\mathbb{E}_{iklm} = \frac{1}{2}\left(\delta_{il}\,\delta_{km} + \delta_{im}\,\delta_{kl}\right), \tag{3.270}$$

analogous to $\mathbb{E}_{C^{-1}}$. Within the numerical treatment of elasticity problems using the finite element method, it makes sense to present equation (3.269) in matrix form. For this purpose, the components of the Lie derivative of the Kirchhoff stress tensor, which is, due to (3.95), equal to the Oldroyd stress rate are assembled in a column vector. This procedure is also performed for the components of the rate of deformation tensor, see also (3.266),

$$
\begin{Bmatrix}
\mathcal{L}_v\tau_{11} \\
\mathcal{L}_v\tau_{22} \\
\mathcal{L}_v\tau_{33} \\
\mathcal{L}_v\tau_{12} \\
\mathcal{L}_v\tau_{23} \\
\mathcal{L}_v\tau_{31}
\end{Bmatrix}
=
\begin{bmatrix}
2\mu + \Lambda & \Lambda\, J^2 & \Lambda\, J^2 & 0 & 0 & 0 \\
\Lambda\, J^2 & 2\mu + \Lambda & \Lambda\, J^2 & 0 & 0 & 0 \\
\Lambda\, J^2 & \Lambda\, J^2 & 2\mu + \Lambda & 0 & 0 & 0 \\
0 & 0 & 0 & \alpha & 0 & 0 \\
0 & 0 & 0 & 0 & \alpha & 0 \\
0 & 0 & 0 & 0 & 0 & \alpha
\end{bmatrix}
\begin{Bmatrix}
d_{11} \\
d_{22} \\
d_{33} \\
2\,d_{12} \\
2\,d_{23} \\
2\,d_{31}
\end{Bmatrix},
$$

$$\overset{\Delta}{\tau} = \mathbf{D}\,\mathbf{d} \quad \text{with} \quad \alpha = \mu - \frac{1}{2}\,\Lambda\,(J^2 - 1). \tag{3.271}$$

In the undeformed initial configuration, the deformation gradient is given by $\mathbf{F} = \mathbf{1}$, from which immediately $\mathbf{C}^{-1} = \mathbf{1}$ and $J = 1$ follow. If these quantities are inserted in (3.268), one obtains

$$\mathbb{C}_0 = \Lambda\, \mathbf{1} \otimes \mathbf{1} + 2\,\mu\, \mathbb{E}. \tag{3.272}$$

This equation can also be derived from (3.269), since for $\mathbf{F} = \mathbf{1}$ initial and current configuration are the same. Observe furthermore that the constitutive tensor $\mathbb{C}_0$ is identical, the elasticity tensor of the geometrically linear theory, see e.g. Eschenauer and Schnell (1993). Putting the last result in matrix form yields

$$
\begin{Bmatrix}
\sigma_{11} \\
\sigma_{22} \\
\sigma_{33} \\
\sigma_{12} \\
\sigma_{23} \\
\sigma_{31}
\end{Bmatrix}
=
\begin{bmatrix}
2\mu + \Lambda & \Lambda & \Lambda & 0 & 0 & 0 \\
\Lambda & 2\mu + \Lambda & \Lambda & 0 & 0 & 0 \\
\Lambda & \Lambda & 2\mu + \Lambda & 0 & 0 & 0 \\
0 & 0 & 0 & \mu & 0 & 0 \\
0 & 0 & 0 & 0 & \mu & 0 \\
0 & 0 & 0 & 0 & 0 & \mu
\end{bmatrix}
\begin{Bmatrix}
\epsilon_{11} \\
\epsilon_{22} \\
\epsilon_{33} \\
2\,\epsilon_{12} \\
2\,\epsilon_{23} \\
2\,\epsilon_{31}
\end{Bmatrix},
$$

$$\boldsymbol{\sigma} = \mathbf{D}_0\,\boldsymbol{\epsilon}. \tag{3.273}$$

(b) Neo Hooke material split in volumetric and deviatoric terms. The constitutive equation in (3.240) can be written as

$$\mathbf{S} = \mu\, J^{-\frac{2}{3}} \left[\, \mathbf{1} - \left(\frac{1}{3}\,\mathrm{tr}\,\mathbf{C}\right) \mathbf{C}^{-1} \right] + \frac{K}{2}\,(J^2 - 1)\, \mathbf{C}^{-1} = \mathbf{S}_{ISO} + \mathbf{S}_{VOL}. \tag{3.274}$$

In this equation, the stress depend as in (3.119) only on the right Cauchy-Green tensor and the determinant of the deformation gradient $\mathbf{F}$. Hence the derivative (3.252) can directly be obtained from (3.252) using (3.124) and (3.254). With

$$2\frac{\partial \hat{W}}{\partial \hat{C}_{EF}} = \frac{1}{2}\mu\,\delta_{EF} \quad \text{and} \quad \frac{\partial^2 \hat{W}}{\partial \hat{C}_{EF}\partial \hat{C}_{MN}} = 0$$

and with $U(J) = \frac{K}{4}(J^2 - 1) - \frac{K}{2}\ln J$:

$$\frac{\partial U}{\partial J} = \frac{K}{2}\left(J - \frac{1}{J}\right), \quad \frac{\partial^2 U}{\partial J^2} = \frac{K}{2}\left(J + \frac{1}{J^2}\right)$$

follows

$$
\begin{aligned}
\hat{\mathbb{C}}_{ABCD}^{VOL} &= K\left[J^2\,C_{AB}^{-1}\,C_{CD}^{-1} - (J^2 - 1)\,\mathbb{E}_{C^{-1}\,ABCD}\right], \\
\hat{\mathbb{C}}_{ABCD}^{ISO} &= -\frac{2}{3}\left[C_{CD}^{-1}\,S_{AB}^{ISO} + C_{AB}^{-1}\,S_{CD}^{ISO}\right. \\
&\quad \left. + J^{-\frac{2}{3}}\,C_{EE}\left(\frac{1}{3}\,C_{AB}^{-1}\,C_{DC}^{-1} - \mathbb{E}_{C^{-1}\,ABCD}\right)\right]
\end{aligned}
\tag{3.275}
$$

or in direct notation

$$
\begin{aligned}
\hat{\mathbf{C}}_{VOL} &= K\left[J^2\,\mathbf{C}^{-1}\otimes\mathbf{C}^{-1} - (J^2 - 1)\,\mathbb{E}_{C^{-1}}\right], \\
\hat{\mathbf{C}}_{ISO} &= -\frac{2}{3}\mu\left[\mathbf{C}^{-1}\otimes\mathbf{S}_{ISO} + \mathbf{S}_{ISO}\,\mathbf{C}^{-1}\right. \\
&\quad \left. + J^{-\frac{2}{3}}(\mathrm{tr}\mathbf{C})\left(\frac{1}{3}\,\mathbf{C}^{-1}\otimes\mathbf{C}^{-1} - \mathbb{E}_{C^{-1}}\right)\right].
\end{aligned}
\tag{3.276}
$$

The transformation to the current configuration is performed as in (3.245) and yields

$$
\begin{aligned}
\hat{\mathbf{c}}_{vol} &= K\left[J^2\,\mathbf{1}\otimes\mathbf{1} - (J^2 - 1)\,\mathbb{E}\right], \\
\hat{\mathbf{c}}_{iso} &= -\frac{2}{3}\mu\left[\mathbf{1}\otimes\boldsymbol{\tau}_{iso} + \boldsymbol{\tau}_{iso}\otimes\mathbf{1} + J^{-\frac{2}{3}}(\mathrm{tr}\mathbf{b})\left(\frac{1}{3}\,\mathbf{1}\otimes\mathbf{1} - \mathbb{E}\right)\right]. \tag{3.277}
\end{aligned}
$$

By introducing the following vectors and matrices

$$\mathbf{i} = \begin{Bmatrix} 1 \\ 1 \\ 1 \\ 0 \\ 0 \\ 0 \end{Bmatrix}, \quad \hat{\boldsymbol{\tau}} = \begin{Bmatrix} \tau_{iso\,11} \\ \tau_{iso\,22} \\ \tau_{iso\,33} \\ \tau_{iso\,12} \\ \tau_{iso\,23} \\ \tau_{iso\,31} \end{Bmatrix}, \quad \mathbb{E} = \begin{bmatrix} 1 & & & & & \\ & 1 & & & & \\ & & 1 & & & \\ & & & 1 & & \\ & & & & 1 & \\ & & & & & 1 \end{bmatrix} \tag{3.278}$$

the matrix form of (3.277) is given by

$$
\begin{aligned}
\hat{\mathbf{c}}_{vol} &= K\left[J^2\,\mathbf{i}\,\mathbf{i}^T - (J^2 - 1)\,\mathbb{E}\right], \\
\hat{\mathbf{c}}_{iso} &= -\frac{2}{3}\mu\left[\mathbf{i}\,\hat{\boldsymbol{\tau}}^T + \hat{\boldsymbol{\tau}}\,\mathbf{i}^T + J^{-\frac{2}{3}}(\mathrm{tr}\mathbf{b})\left(\frac{1}{3}\,\mathbf{i}\,\mathbf{i}^T - \mathbb{E}\right)\right].
\end{aligned}
$$

In case of the undeformed configuration, the relations stated above yield with $\mathbf{F} = \mathbf{1}$

$$
\begin{aligned}
\hat{\mathbf{C}}_{vol} &= K\,\mathbf{1}\otimes\mathbf{1}, \\
\hat{\mathbf{C}}_{iso} &= 2\mu\left[\mathbb{E} - \frac{1}{3}\,\mathbf{1}\otimes\mathbf{1}\right], \tag{3.279}
\end{aligned}
$$

which include the split in volumetric and isochoric parts. The tensors in (3.279) are equivalent to the elasticity tensors of the linear theory for the case of an additive split of stresses and strains in volumetric and deviatoric parts according to (3.30). The associated matrix form is given by

$$
\begin{Bmatrix} \sigma_{11} \\ \sigma_{22} \\ \sigma_{33} \\ \sigma_{12} \\ \sigma_{23} \\ \sigma_{31} \end{Bmatrix}
= K \begin{bmatrix} 1 & 1 & 1 & 0 & 0 & 0 \\ 1 & 1 & 1 & 0 & 0 & 0 \\ 1 & 1 & 1 & 0 & 0 & 0 \\ 0 & 0 & 0 & 0 & 0 & 0 \\ 0 & 0 & 0 & 0 & 0 & 0 \\ 0 & 0 & 0 & 0 & 0 & 0 \end{bmatrix}
+ \mu \begin{bmatrix} \frac{4}{3} & -\frac{2}{3} & -\frac{2}{3} & 0 & 0 & 0 \\ -\frac{2}{3} & \frac{4}{3} & -\frac{2}{3} & 0 & 0 & 0 \\ -\frac{2}{3} & -\frac{2}{3} & \frac{4}{3} & 0 & 0 & 0 \\ 0 & 0 & 0 & 1 & 0 & 0 \\ 0 & 0 & 0 & 0 & 1 & 0 \\ 0 & 0 & 0 & 0 & 0 & 1 \end{bmatrix}
\begin{Bmatrix} \epsilon_{11} \\ \epsilon_{22} \\ \epsilon_{33} \\ 2\epsilon_{12} \\ 2\epsilon_{23} \\ 2\epsilon_{31} \end{Bmatrix},
$$

$$
\boldsymbol{\sigma} = (\boldsymbol{D}_{vol} + \boldsymbol{D}_{iso})\,\boldsymbol{\epsilon}. \tag{3.280}
$$

Incremental Form of the Geometric Linear Elasto-Plastic Constitutive Equation. The incremental constitutive equation can be obtained for the case of isotropic and kinematic hardening based on the conditions stated in Sect. 3.3.2. Since the consistency condition (3.167) yields

$$
\dot{f} = \frac{\partial f}{\partial \mathbf{s}} \cdot \dot{\mathbf{s}} + \frac{\partial f}{\partial \mathbf{q}} \cdot \dot{\mathbf{q}} + \frac{\partial f}{\partial \hat{q}} \dot{\hat{q}} = 0, \tag{3.281}
$$

the relation

$$
\begin{aligned}
\dot{f} &= \frac{\partial f}{\partial \mathbf{s}} \cdot \mathbb{C}^e[\dot{\mathbf{e}} - \dot{\mathbf{e}}^p] + \frac{\partial f}{\partial \mathbf{q}} \cdot \dot{\mathbf{q}} + \frac{\partial f}{\partial \hat{q}} \dot{\hat{q}} \\
&= \mathbf{n} \cdot \mathbb{C}^e[\dot{\mathbf{e}}] - \lambda \left(\mathbf{n} \cdot \mathbb{C}^e[\mathbf{n}] + \frac{2}{3} H \mathbf{n} \cdot \mathbf{n} + \frac{2}{3} \hat{H} \right) = 0 \quad (3.282)
\end{aligned}
$$

can be deduced with (3.154), (3.156) and (3.163). By introducing the abbreviation $A = \mathbf{n} \cdot \mathbb{C}^e[\mathbf{n}] + \frac{2}{3} H + \frac{2}{3} \hat{H}$, the latter relation can be solved for λ

$$
\lambda = A^{-1}\,\mathbf{n} \cdot \mathbb{C}^e[\dot{\mathbf{e}}]. \tag{3.283}
$$

Inserting (3.163) into the elastic constitutive equation (3.154) leads finally to

$$
\dot{\mathbf{s}} = \mathbb{C}^e[\dot{\mathbf{e}} - A^{-1}(\mathbf{n} \cdot \mathbb{C}^e[\dot{\mathbf{e}}])\,\mathbf{n}]. \tag{3.284}
$$

Now isotropic material and linear elastic behaviour is assumed. Then $\mathbb{C}^e = K\mathbf{1} \otimes \mathbf{1} + 2\mu\,(\mathbb{E} - \frac{1}{3}\mathbf{1} \otimes \mathbf{1})$ can be written for (3.279) with the bulk modulus K and the shear modulus μ. With these relations, the final expression for the time derivative of the stresses can be completed by adding the compressible part, which is purely elastic in case of VON MISES plasticity,

$$
\dot{p} = \frac{1}{3} \operatorname{tr} \dot{\boldsymbol{\sigma}} = K \operatorname{tr} \dot{\boldsymbol{\varepsilon}}^e, \tag{3.285}
$$

$$
\dot{\mathbf{s}} = 2\mu\dot{\mathbf{e}} - \frac{2\mu}{1 + \frac{H+\hat{H}}{3\mu}}(\dot{\mathbf{e}} \cdot \mathbf{n})\,\mathbf{n}. \tag{3.286}
$$

This equation presents the classical form of the PRANDTL REU constitutive equation for an elasto-plastic material. The combination $\dot{\boldsymbol{\sigma}} = \dot{\mathbf{s}} + \dot{p}\mathbf{1}$ and by placing $\dot{\varepsilon}$ outside the bracket leads to the incremental form of the constitutive equation of the so-called J_2-plasticity: $\dot{\boldsymbol{\sigma}} = \mathbb{C}^{ep}[\dot{\varepsilon}]$. In this relation, the elasto-plastic tangent is explicitly given by

$$\mathbb{C}^{ep} = K\,\mathbf{1} \otimes \mathbf{1} + 2\,\mu\left(\mathbb{E} - \frac{1}{3}\mathbf{1} \otimes \mathbf{1}\right) - 2\,\mu\,\frac{1}{1 + \frac{H+\hat{H}}{3\,\mu}}\,\mathbf{n} \otimes \mathbf{n}. \qquad (3.287)$$

This incremental constitutive tensor is often called elasto-plastic continuum tangent. Until around 1985, this tangent was used within the numerical integration of elasto-plastic processes within the finite element method, leading to an explicit integration scheme for the rate equations, see e.g. Zienkiewicz and Taylor (1991). The incremental tensor given in (3.287) is however not sufficient when an efficient solution algorithm based on an implicit integration of the rate equations shall be constructed, see Simo and Taylor (1985). A detailed discussion of such integration algorithms is presented in Sect. 6.2.2 and can be found also in Simo and Hughes (1998).

3.4 Weak Form of Equilibrium, Variational Principles

For the analysis of nonlinear initial boundary value problems in continuum mechanics, a coupled system of partial differential equations has to be solved which consist of kinematical relations, local balance of momentum and the constitutive equations. The strong form of these equations is presented in the following for hyperelastic solids by two alternative descriptions. These are the description with respect to the initial and current configurations of the bodies. For the description with respect to the initial configuration B, different stress measures can be used like the first PIOLA–KIRCHHOFF stresses or the second PIOLA–KIRCHHOFF stresses which yield two different formulations, see Sect. 3.2,

Kinematics:	$\mathbf{F}$	$\mathbf{E} = \frac{1}{2}(\mathbf{F}^T\,\mathbf{F} - \mathbf{1})$
Equilibrium:	$\operatorname{Div}\mathbf{P} + \rho_0\,\bar{\mathbf{b}} = \rho_0\,\dot{\mathbf{v}}$	$\operatorname{Div}(\mathbf{F}\,\mathbf{S}) + \rho_0\,\bar{\mathbf{b}} = \rho_0\,\dot{\mathbf{v}}$
Constitutive equation:	$\mathbf{P} = \dfrac{\partial W}{\partial \mathbf{F}}$	$\mathbf{S} = \dfrac{\partial W}{\partial \mathbf{E}}$

Additionally the boundary conditions for the displacements have to be prescribed on ∂B_u and boundary conditions for the tractions have to be formulated on ∂B_σ which leads to

$$\mathbf{u} = \bar{\mathbf{u}} \quad \text{on} \quad \partial B_u \qquad \text{and} \qquad \mathbf{P}\,\mathbf{N} = \mathbf{F}\,\mathbf{S}\,\mathbf{N} = \bar{\mathbf{t}} \quad \text{on} \quad \partial B_\sigma.$$

All equations stated above can be transformed to the current configuration $\varphi(B)$ where the constitutive equations are formulated in terms of the CAUCHY stress tensor, $\boldsymbol{\sigma}$, and the KIRCHHOFF stress tensor, $\boldsymbol{\tau}$,

Kinematics: $\qquad\qquad\qquad\qquad \mathbf{b} = \mathbf{F}\,\mathbf{F}^{T}$

Equilibrium: $\qquad\qquad\quad \operatorname{div}\boldsymbol{\sigma} + \rho\,\bar{\mathbf{b}} = \rho\dot{\mathbf{v}} \quad \operatorname{div}\left(\frac{1}{J}\boldsymbol{\tau}\right) + \rho\,\bar{\mathbf{b}} = \rho\dot{\mathbf{v}}$

Constitutive equation: $\qquad \boldsymbol{\sigma} = 2\,\rho\,\mathbf{b}\,\dfrac{\partial\psi}{\partial\mathbf{b}} \qquad\quad \boldsymbol{\tau} = 2\,\mathbf{b}\,\dfrac{\partial W}{\partial\mathbf{b}}$

The displacement boundary conditions are given on $\varphi(\partial B_u)$ as $\mathbf{u} = \bar{\mathbf{u}}$. For the tractions $\boldsymbol{\sigma}\mathbf{n} = \hat{\mathbf{t}}$ on $\varphi(\partial B_\sigma)$ holds.

An analytical solution of these systems of nonlinear partial differential equations is only possible for a selected number of simple initial boundary value problems. Hence approximate methods like the method of finite differences or finite elements have to be applied to solve this set of equations. The use of the finite element method, which is based on a variational formulation of the equations, summarized above, expands the solution range to a broad spectrum of applications. The necessary variational formulation will be described in the following sections based on a referential and spatial description.

Several approaches can be applied to derive the variational formulation which are related to the problem at hand. In case of hyperelastic material responses a functional in the strain energy can be formulated, leading to a variational principle. For arbitrary processes the equations, summarized above, can be fulfilled in a weak sense, which yields a formulation minimizing the error of the finite element approximation for arbitrary test functions, see e.g. Johnson (1987). In the engineering literature, the principle of virtual work is often basis for the derivation of the finite element approximations. It can, however, easily be shown that this formulation is equivalent to using the weak form.

In the following section, several variational formulations are derived which can be applied in the context of finite elements.

3.4.1 Weak Form of Linear Momentum in the Initial Configuration

When an approximation $\mathbf{u}_h$ of the exact solution $\mathbf{u}$ is inserted in the above set of equations, then an error will occur since the approximate solution is usually not equal to the exact solution. Hence the insertion of the approximate solution into the momentum balance equation $\operatorname{Div}\mathbf{P} + \rho_0\,\bar{\mathbf{b}} - \rho_0\,\dot{\mathbf{v}} = 0$ will lead to

$$\operatorname{Div}\mathbf{P}(\mathbf{u}_h) + \rho_0\,\bar{\mathbf{b}} - \rho_0\,\dot{\mathbf{v}}_h = R.$$

The residual R, which denotes the error not fulfilling the momentum balance equation by $\mathbf{u}_h$, will now be reduced to zero in a weak sense by multiplying the residual by a weighting function $\boldsymbol{\eta}$ and by integrating the residual over the whole domain. The vector-valued function $\boldsymbol{\eta} = \{\boldsymbol{\eta}\,|\,\boldsymbol{\eta} = \mathbf{0}$ on $\partial B_u\}$ is often called virtual displacement or test function. This procedure leads to

$$\int_B \text{Div}\, \mathbf{P}(\mathbf{u}_h) \cdot \boldsymbol{\eta}\, dV + \int_B \rho_0\, (\bar{\mathbf{b}} - \dot{\mathbf{v}}_h) \cdot \boldsymbol{\eta}\, dV = 0,$$

which of course also has to hold for exact solution $\mathbf{u}$

$$\int_B \text{Div}\, \mathbf{P} \cdot \boldsymbol{\eta}\, dV + \int_B \rho_0\, (\bar{\mathbf{b}} - \dot{\mathbf{v}}) \cdot \boldsymbol{\eta}\, dV = 0. \tag{3.288}$$

The weak form is also known as principle of virtual work in engineering. Since no further assumptions, like existence of a potential, are made, the weak form is applicable to general problems such as inelastic materials, friction, non-conservative loading, etc.

By partial integration of the first term in (3.288), application of the divergence theorem and introduction of the traction boundary condition, the weak form of linear momentum

$$G\left(\boldsymbol{\varphi}, \boldsymbol{\eta}\right) = \int_B \mathbf{P} \cdot \text{Grad}\, \boldsymbol{\eta}\, dV - \int_B \rho_0\, (\bar{\mathbf{b}} - \dot{\mathbf{v}}) \cdot \boldsymbol{\eta}\, dV - \int_{\partial B_\sigma} \bar{\mathbf{t}} \cdot \boldsymbol{\eta}\, dA = 0 \tag{3.289}$$

is obtained. The gradient of the test function $\boldsymbol{\eta}$ can also be interpreted as the directional derivative of the deformation gradient $D\mathbf{F} \cdot \boldsymbol{\eta}$ also known as variation $\delta\,\mathbf{F}$ of the deformation gradient. In the weak form (3.289), the first PIOLA-KIRCHHOFF stress tensor can be replaced through $\mathbf{P} = \mathbf{F}\,\mathbf{S}$ by the second PIOLA-KIRCHHOFF stress tensor leading to

$$\mathbf{P} \cdot \text{Grad}\, \boldsymbol{\eta} = \mathbf{S} \cdot \mathbf{F}^T\, \text{Grad}\, \boldsymbol{\eta} = \mathbf{S} \cdot \frac{1}{2}\, (\mathbf{F}^T\, \text{Grad}\, \boldsymbol{\eta} + \text{Grad}^T \boldsymbol{\eta}\, \mathbf{F}) = \mathbf{S} \cdot \delta\mathbf{E}, \tag{3.290}$$

where the fact has been used that the scalar product of a symmetrical tensor (here $\mathbf{S}$) with an antisymmetrical part of a tensor is zero. $\delta\,\mathbf{E}$ denotes the variation of the GREEN-LAGRANGE strain tensor which is obtained via the directional derivative

$$
\begin{aligned}
D\mathbf{E} \cdot \boldsymbol{\eta} &= \left. \frac{d}{d\alpha} \frac{1}{2} \left[\mathbf{F}^T\left(\boldsymbol{\varphi} + \alpha\,\boldsymbol{\eta}\right) \mathbf{F}\left(\boldsymbol{\varphi} + \alpha\,\boldsymbol{\eta}\right) - \mathbf{1} \right] \right|_{\alpha=0} \\
&= \left. \frac{d}{d\alpha} \frac{1}{2} \left[\left[\text{Grad}\left(\boldsymbol{\varphi} + \alpha\,\boldsymbol{\eta}\right)\right]^T \text{Grad}\left(\boldsymbol{\varphi} + \alpha\,\boldsymbol{\eta}\right) - \mathbf{1} \right] \right|_{\alpha=0} \\
&= \frac{1}{2} \left[(\text{Grad}\,\boldsymbol{\eta})^T \mathbf{F} + \mathbf{F}^T\, \text{Grad}\,\boldsymbol{\eta} \right] = \delta\mathbf{E}. \tag{3.291}
\end{aligned}
$$

Using (3.290), equation (3.289) can be rewritten as

$$G\left(\boldsymbol{\varphi}, \boldsymbol{\eta}\right) = \int_B \mathbf{S} \cdot \delta\mathbf{E}\, dV - \int_B \rho_0\, (\bar{\mathbf{b}} - \dot{\mathbf{v}}) \cdot \boldsymbol{\eta}\, dV - \int_{\partial B_\sigma} \bar{\mathbf{t}} \cdot \boldsymbol{\eta}\, dA = 0. \tag{3.292}$$

The first term in (3.292) denotes the internal virtual work, also called stress divergence term. The last two terms describe the virtual work of the applied loading and the inertia term.

Exercise 3.9: Within continuum mechanics, there exist several possibilities to describe the internal virtual work in (3.289). If the generalized strain measure (3.18) is evaluated for $\alpha = 1$, then the strain measure $\mathbf{E}^{(1)} = \mathbf{U} - \mathbf{1}$ follows. For this strain measure, the work conjugated stress tensor has to be found.

Solution: Starting from $\mathbf{F} = \mathbf{R}\,\mathbf{U}$, see (3.21), the variation of the strain can be obtained with $\mathbf{R}^{-1} = \mathbf{R}^T$

$$\delta\mathbf{E}^{(1)} = \delta\mathbf{U} = \delta\mathbf{R}^T\,\mathbf{F} + \mathbf{R}^T\,\delta\mathbf{F}.$$

With this result, $\delta\mathbf{F}$ can be replaced in the first term of (3.289)

$$\mathbf{P}\cdot\delta\mathbf{F} = \mathbf{P}\cdot(\,\mathbf{R}\,\delta\mathbf{U} - \mathbf{R}\,\delta\mathbf{R}^T\,\mathbf{F}\,).$$

Based on the trace operation $\mathbf{A}\cdot\mathbf{B} = \mathrm{tr}(\,\mathbf{A}\,\mathbf{B}^T)$ and cyclic exchange, relation

$$\mathbf{P}\cdot\delta\mathbf{F} = \mathbf{R}^T\,\mathbf{P}\cdot\delta\mathbf{U} - \mathbf{P}\,\mathbf{F}^T\cdot\mathbf{R}\,\delta\mathbf{R}^T$$

is deduced. In this relation, the symmetric KIRCHHOFF stress tensor which is defined by $\boldsymbol{\tau} = \mathbf{P}\,\mathbf{F}^T$ does not produce virtual work with the term $\mathbf{R}\,\delta\mathbf{R}^T$ since the latter is skew symmetric $[\delta\,(\mathbf{R}\,\mathbf{R}^T) = \mathbf{R}\,\delta\mathbf{R}^T + \delta\mathbf{R}\,\mathbf{R}^T = \mathbf{0}]$. Hence the work conjugated stress tensor can be assigned to the strain measure $\mathbf{E}^{(1)}$ by $\mathbf{T}_B = \mathbf{R}^T\,\mathbf{P}$ which is known as the symmetric part of the BIOT stress tensor. The associated weak form is then given by

$$G\,(\boldsymbol{\varphi},\boldsymbol{\eta}) = \int_B \mathbf{T}_B\cdot\delta\mathbf{U}\,dV - \int_B \rho_0\,(\bar{\mathbf{b}} - \dot{\mathbf{v}})\cdot\boldsymbol{\eta}\,dV - \int_{\partial B_\sigma} \bar{\mathbf{t}}\cdot\boldsymbol{\eta}\,dA = 0. \qquad (3.293)$$

3.4.2 Weak Form of Linear Momentum in the Current Configuration

The transformation of the weak form (3.289) to the current or spatial configuration is performed by kinematical operations in which the base vectors are *push forward* to the configuration $\varphi(B)$. With the transformation $\boldsymbol{\sigma} = \frac{1}{J}\mathbf{P}\,\mathbf{F}^T$ of the first PIOLA-KIRCHHOFF stress tensor to the CAUCHY stress tensor, see (3.81), equation (3.33) can be rewritten as

$$\mathbf{P}\cdot\mathrm{Grad}\,\boldsymbol{\eta} = J\,\boldsymbol{\sigma}\,\mathbf{F}^{-T}\cdot\mathrm{Grad}\,\boldsymbol{\eta} = J\,\boldsymbol{\sigma}\cdot\mathrm{Grad}\,\boldsymbol{\eta}\,\mathbf{F}^{-1} = J\,\boldsymbol{\sigma}\cdot\mathrm{grad}\,\boldsymbol{\eta}.$$

Furthermore, from (3.12) $dv = J\,dV$ follows which is equivalent to $\rho = \rho_0\,J$. With these relations, the weak form (3.289) can be written in terms of the current configuration

$$g\,(\boldsymbol{\varphi},\boldsymbol{\eta}) = \int_{\varphi(B)} \boldsymbol{\sigma}\cdot\mathrm{grad}\,\boldsymbol{\eta}\,dv - \int_{\varphi(B)} \rho\,(\bar{\mathbf{b}}-\dot{\mathbf{v}})\cdot\boldsymbol{\eta}\,dv - \int_{\varphi(\partial B_\sigma)} \hat{\mathbf{t}}\cdot\boldsymbol{\eta}\,da = 0. \qquad (3.294)$$

In this relations, equation (3.80) has been used to transform the traction vector $\bar{\mathbf{t}}$ to $\varphi(B)$. The symmetry of the CAUCHY stress tensor facilitates the replacement of the spatial gradient of the test function $\boldsymbol{\eta}$ by its symmetric part. Hence with the definition

$$\nabla^S \boldsymbol{\eta} = \frac{1}{2} \left(\operatorname{grad} \boldsymbol{\eta} + \operatorname{grad}^T \boldsymbol{\eta} \right), \qquad (3.295)$$

the weak form follows with respect to the spatial configuration

$$g\left(\boldsymbol{\varphi}, \boldsymbol{\eta}\right) = \int\limits_{\varphi(B)} \boldsymbol{\sigma} \cdot \nabla^S \boldsymbol{\eta} \, dv - \int\limits_{\varphi(B)} \rho \left(\bar{\mathbf{b}} - \dot{\mathbf{v}}\right) \cdot \boldsymbol{\eta} \, dv - \int\limits_{\varphi(\partial B_\sigma)} \hat{\mathbf{t}} \cdot \boldsymbol{\eta} \, da = 0 . \quad (3.296)$$

This relation is, in a formal sense, equivalent to the principle of virtual work of the geometrically linear theory. But here the integral, the stress and virtual strain measures have to be evaluated with respect to the current configuration. Due to this, the nonlinearities do appear, however hidden.

In the further variational formulations presented in this section, the inertia terms $\rho \dot{\mathbf{v}}$ are neglected in order to concentrate on static equilibrium equations.

3.4.3 Variational Functionals

In this section, two variational functionals will be discussed which can alternatively be applied within the discretization process of the finite element method. For a more detailed background, see Washizu (1975).

Principle of Stationary Elastic Potential. In case of a hyper elastic material, there exist a strain energy function W, which describes the elastic energy stored in the solid. Based on this strain energy, the classical principle of the minimum of potential energy can be formulated in the geometrically linear theory. In finite deformation theory, it has in general to be considered that deformations can occur which are non-unique. Hence only a stationary value of the potential can be reached. Under the assumption that the applied loads are conservative, which means path independent (non-conservative loads are described in Exercise 3.12), the functional

$$\Pi\left(\boldsymbol{\varphi}\right) = \int\limits_{B} \left[W(\mathbf{C}) - \rho_0 \,\bar{\mathbf{b}} \cdot \boldsymbol{\varphi} \right] dV - \int\limits_{\partial B_\sigma} \bar{\mathbf{t}} \cdot \boldsymbol{\varphi} \, dA \Longrightarrow STAT \qquad (3.297)$$

can be stated for the static problem. Out of all possible deformations $\boldsymbol{\varphi}$, the ones which make Π stationary fulfil the equilibrium equation. The stationary value of (3.297) can be computed by the variation of Π with respect to the deformation. For this purpose, the directional derivative

$$\delta \Pi = D \, \Pi \left(\boldsymbol{\varphi}\right) \cdot \boldsymbol{\eta} = \frac{d}{d\alpha} \, \Pi \left(\boldsymbol{\varphi} + \alpha \, \boldsymbol{\eta}\right) \bigg|_{\alpha=0} \qquad (3.298)$$

is applied which is also called first variation of Π. The application of this mathematical operation yields

$$D\,\varPi(\varphi) \cdot \boldsymbol{\eta} = \int_B \left[\frac{\partial W}{\partial \mathbf{C}} \cdot D\,\mathbf{C} \cdot \boldsymbol{\eta} - \rho_0\,\bar{\mathbf{b}} \cdot \boldsymbol{\eta} \right] dV - \int_{\partial B_\sigma} \bar{\mathbf{t}} \cdot \boldsymbol{\eta}\,dA = G(\mathbf{u}, \boldsymbol{\eta}) = 0\,.$$

$$(3.299)$$

The directional derivative of the right CAUCHY-GREEN strain tensor can easily be written in terms of the GREEN-LAGRANGE strain tensor, see (3.291)

$$D\,\mathbf{C} \cdot \boldsymbol{\eta} = 2\,D\,\mathbf{E} \cdot \boldsymbol{\eta} \quad \text{or} \quad \delta\mathbf{C} = 2\,\delta\mathbf{E}\,.$$

The partial derivative of W with respect to $\mathbf{C}$ leads to the second PIOLA-KIRCHHOFF stress tensor $\mathbf{S}$, see (3.104): $\mathbf{S} = 2\,\partial W / \partial \mathbf{C}$. Hence equation (3.299) is equivalent to the weak form (3.292) for a hyperelastic material.

The construction of such principle has several advantages. First of all, it can be the basis for mathematical investigations regarding existence and uniqueness of solutions (the latter is however only valid for the linear theory). Secondly, it leads to the development of efficient algorithms for the solution of the resulting non-linear equations on the basis of optimization strategies.

Hu-Washizu Principle. Another variational principle is the HU-WASHIZU principle, see Washizu (1975). It has gained significance early on for the construction of finite elements. This principle can be derived by writing the weak formulation with additional constraint equations which contain kinematics and constitutive equations. Due to this deformations, strains and stresses occur as independent variables. On the contrary, once the principle is constructed its variation with respect to all variables yields the static equilibrium equations, the kinematical relations and the constitutive equation. The formulation of the nonlinear version of the HU-WASHIZU principle can be obtained by using any set of work conjugated variables. Here it will be stated in terms of the deformation gradient $\mathbf{F}$, the first PIOLA-KIRCHHOFF stress tensor $\mathbf{P}$ and the deformation φ

$$\varPi(\varphi, \mathbf{F}, \mathbf{P}) = \int_B [\,W(\mathbf{F}) + \mathbf{P} \cdot (\operatorname{Grad}\varphi - \mathbf{F}\,)\,]\,dV$$

$$- \int_B \varphi \cdot \rho_0\,\bar{\mathbf{b}}\,dV - \int_{\partial B_\sigma} \varphi \cdot \hat{\mathbf{t}}\,dA\,. \qquad (3.300)$$

The variation, according to the definition of the directional derivative, see e.g. (3.291), yields now three independent equations

$$D\varPi(\varphi, \mathbf{F}, \mathbf{P}) \cdot \boldsymbol{\eta} = \int_B (\mathbf{P} \cdot \operatorname{Grad}\boldsymbol{\eta} - \boldsymbol{\eta} \cdot \rho_0\,\bar{\mathbf{b}}\,)\,dV - \int_{\partial B_\sigma} \boldsymbol{\eta} \cdot \hat{\mathbf{t}}\,dA = 0\,,$$

$$D\varPi(\varphi, \mathbf{F}, \mathbf{P}) \cdot \delta\mathbf{P} = \int_B \delta\mathbf{P} \cdot (\operatorname{Grad}\varphi - \mathbf{F}\,)\,dV = 0\,, \qquad (3.301)$$

$$D\varPi(\varphi, \mathbf{F}, \mathbf{P}) \cdot \delta\mathbf{F} = \int_B \delta\mathbf{F} \cdot \left(\frac{\partial W}{\partial \mathbf{F}} - \mathbf{P} \right) dV = 0\,.$$

Observe that they represent the weak form (3.289), the kinematical relation (3.6) and a hyperelastic constitutive equation for $\mathbf{P}$.

A special form of the HU-WASHIZU variational principle can be applied for the construction of finite elements which have to represent nearly incompressible material behaviour. Since incompressibility is associated with a constraint for the volumetric deformation ($J \equiv 1$), the split (3.28) can be used to distinguish volumetric and isochoric parts of the deformation. Based on this idea, Simo et al. (1985a) formulated a three-field functional which is only defined for the volumetric part of the deformation. Hence the independent variable are now the deformation φ, the pressure p and a strain variable θ which is equivalent to J. The last variable has to fulfil the constraint condition $\theta = J$. With the multiplicative split of the deformation gradient (3.28)

$$\bar{\mathbf{F}} = \theta^{\frac{1}{3}} \, \widehat{\mathbf{F}}, \tag{3.302}$$

the split into volumetric and deviatoric parts is achieved. Note that $\widehat{\mathbf{F}} = J^{-\frac{1}{3}} \operatorname{Grad} \varphi$ can be specified in relation (3.302). Furthermore $\bar{\mathbf{C}} = \theta^{\frac{2}{3}} J^{-\frac{2}{3}} \mathbf{C} = \theta^{\frac{2}{3}} \widehat{\mathbf{C}}$ holds with (3.29). In the variational principle also, the strain energy function $W(\mathbf{C})$, see (3.122), has to be defined on the basis of the new variables: $W(\mathbf{C}) = W(\theta^{\frac{2}{3}} \widehat{\mathbf{C}})$. Using the additive split $W = W(\theta) + W(\widehat{\mathbf{C}})$, see (3.122), the following three-field variational functional can be constructed

$$\Pi(\varphi, p, \theta) = \int_B [\, W(\widehat{\mathbf{C}}) + W(\theta) + p(\, J - \theta\,)\,] \, dV$$

$$- \int_B \varphi \cdot \rho_0 \, \bar{\mathbf{b}} \, dV - \int_{\partial B_\sigma} \varphi \cdot \hat{\mathbf{t}} \, dA. \tag{3.303}$$

By considering relations (3.125) and (3.126), the EULER-LAGRANGE equations obtained from this variational principle are

$$D\Pi(\varphi, p, \theta) \cdot \boldsymbol{\eta} = \int_B \left\{ \left(\mathbb{P} \left[2 \frac{\partial W}{\partial \widehat{\mathbf{C}}} \right] + p \, J \, \mathbf{C}^{-1} \right) \cdot \frac{1}{2} \delta \mathbf{C} - \boldsymbol{\eta} \cdot \rho_0 \, \bar{\mathbf{b}} \right\} dV$$

$$- \int_{\partial B_\sigma} \boldsymbol{\eta} \cdot \hat{\mathbf{t}} \, dA = 0,$$

$$D\Pi(\varphi, p, \theta) \, \delta p = \int_B \delta p \, (\, J - \theta\,) \, dV = 0, \tag{3.304}$$

$$D\Pi(\varphi, p, \theta) \, \delta \theta = \int_B \delta \theta \left(\frac{\partial W}{\partial \theta} - p \right) dV = 0.$$

By comparison of relation (3.126) with (3.127), it is clear that the expression $\mathbf{S}_{ISO} + \mathbf{S}_{VOL}$ can be used in $(3.304)_1$ for the first term in the integral. This explains the split into isochoric and volumetric parts.

The terms in $(3.304)_1$ are often written with respect to the spatial configuration since this simplifies the numerical implementation within the finite element method. With the conversion of the variation of the right CAUCHY-GREEN tensor using $(3.33)_1$

$$\delta \mathbf{C} = \mathbf{F}^T \operatorname{Grad} \boldsymbol{\eta} + \operatorname{Grad}^T \boldsymbol{\eta} \, \mathbf{F} = \mathbf{F}^T \left(\operatorname{grad} \boldsymbol{\eta} + \operatorname{grad}^T \boldsymbol{\eta} \right) \mathbf{F}, \qquad (3.305)$$

relation

$$DΠ(\boldsymbol{\varphi}, p, \theta) \cdot \boldsymbol{\eta} = \int_B \left\{ \left(\mathbf{F} \, \mathbb{P} \left[2 \frac{\partial W}{\partial \widehat{\mathbf{C}}} \right] \mathbf{F}^T + p \, J \, \mathbf{1} \right) \cdot \nabla^S \boldsymbol{\eta} - \boldsymbol{\eta} \cdot \rho_0 \, \bar{\mathbf{b}} \right\} dV$$

$$- \int_{\partial B_\sigma} \boldsymbol{\eta} \cdot \hat{\mathbf{t}} \, dA = 0 \qquad (3.306)$$

is deduced with the definition (3.295): $\nabla^S \boldsymbol{\eta} = \frac{1}{2} \left(\operatorname{grad} \boldsymbol{\eta} + \operatorname{grad}^T \boldsymbol{\eta} \right)$. The final result follows with (3.129) and (3.130)

$$DΠ(\boldsymbol{\varphi}, p, \theta) \cdot \boldsymbol{\eta} = \int_B \left\{ \boldsymbol{\tau}_{iso} \cdot \nabla^S \boldsymbol{\eta} + \tau_{vol} \operatorname{div} \boldsymbol{\eta} - \boldsymbol{\eta} \cdot \rho_0 \, \bar{\mathbf{b}} \right\} dV$$

$$- \int_{\partial B_\sigma} \boldsymbol{\eta} \cdot \hat{\mathbf{t}} \, dA = 0,$$

$$DΠ(\boldsymbol{\varphi}, p, \theta) \, \delta p = \int_B \delta p \, (J - \theta) \, dV = 0, \qquad (3.307)$$

$$DΠ(\boldsymbol{\varphi}, p, \theta) \, \delta \theta = \int_B \delta \theta \left(\frac{\partial W}{\partial \theta} - p \right) dV = 0.$$

In this formulations, the integrands are given in terms of the spatial or current configurations whereas the integration still is performed in the initial configuration. The first equation denotes the weak form of the equilibrium (3.296) with the KIRCHHOFF stresses instead of the CAUCHY stresses. The second equation reproduces the constraint condition $J = \theta$ and the third equation yields the constitutive equation for the pressure p, see also $(3.130)_1$.

3.5 Linearizations

Nonlinearities appear in continuum mechanics due to different phenomena. In this respect, geometrical nonlinearities can be mentioned which occur due to the nonlinear strain measures such as the GREEN-LAGRANGE strain tensor introduced in Sect. 3.1.2. Physical nonlinearities stem from nonlinear constitutive behaviour like elasto-plastic or visco-plastic response. Further,

nonlinearities are related to one-sided or unilateral geometrical constraints as
appear in contact problems. These lead to variational inequalities and hence
include nonlinear effects.

Linearizations of the associated models have to be derived for several
reasons when the initial or boundary values are solved. At one hand, the
linearization process can be applied to derive approximate theories which
can still be solved analytically. This is, e.g. the case for the theory of linear
elasticity or for first and second order beam, plate and shell theories. On
the other, hand linearizations are needed within the algorithmic treatment of
the solution process for the nonlinear boundary value problems. This is, e.g.
the case for finite element methods where NEWTON-RAPHSON algorithms are
employed to solve the nonlinear algebraic equation systems, see e.g. Chap. 5.

Due to the different applications, it is desirable to have a general concept
for the linearization process when applied to nonlinear problems. The purpose
of this section is to provide such a background with a unified definition of
linearization and to illustrate the approach using examples. Mathematical
details will be omitted as much as possible.

The idea of a linearization shall be discussed by means of an example. We
assume a scalar valued function f which is defined in $\mathcal{R}$. The function and
its first derivative are required to be continuous (C^1-continuous). With this
assumption, a TAYLOR series expansion of the function f can be developed
at $\bar{x}$

$$f(\bar{x} + u) = \bar{f} + \bar{D}f \cdot u + R. \qquad (3.308)$$

Here, the following notation was used: $\bar{f} = f(\bar{x})$ and $\bar{D}f = Df(\bar{x})$. The operator D denotes the derivative of f with respect to the variable x. The symbol
"$\cdot$" is at this stage a simple multiplication. u is the increment and the remain-
der $R = R(u)$ goes to zero for a small u $\lim_{u \to 0} \frac{R}{|u|} \to 0$. Figure 3.14 depicts
the geometrical interpretation of (3.308). In the case that u can be considered
as being an independent variable at fixed $\bar{x}$ in (3.308) then

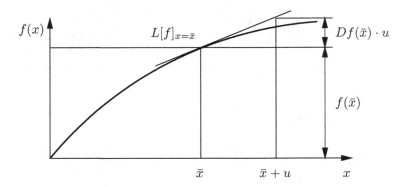

Fig. 3.14 Linearization of function f

$$f(u) = \bar{f} + \bar{D}f \cdot u \qquad (3.309)$$

is the tangent to the curve $f(x)$ at the point $(\bar{x}, \bar{f})$. This result leads to the definition of the linear part of $f(x)$ at $x = \bar{x}$, the *linearization*:

$$L\,[\,f\,]_{x=\bar{x}} \equiv f(u). \qquad (3.310)$$

This one-dimensional result can be easily extended to scalar valued function of points in the three-dimensional space $\mathcal{R}^3$. In that case f is a function of $\mathbf{x}$. The TAYLOR series expansion yields

$$f(\bar{\mathbf{x}} + \mathbf{u}) = \bar{f} + \bar{D}f \cdot \mathbf{u} + R, \qquad (3.311)$$

where $\bar{\mathbf{x}}$ is a point in the EUCLIDIAN space and $\mathbf{u}$ a vector having its origin at point $\bar{\mathbf{x}}$. In more detail

$$\bar{f} = f(\bar{\mathbf{x}}) \qquad \text{and} \qquad \bar{D}f = Df(\bar{\mathbf{x}}) = \left.\frac{\partial f(\mathbf{x})}{\partial \mathbf{x}}\right|_{\mathbf{x}=\bar{\mathbf{x}}} \qquad (3.312)$$

can be written where $\bar{D}f$ is the gradient vector of f at $\bar{\mathbf{x}}$. With this notation, equation (3.311) can be reformulated as

$$f(\bar{\mathbf{x}} + \mathbf{u}) = \bar{f} + \operatorname{Grad} f(\bar{\mathbf{x}}) \cdot \mathbf{u} + R. \qquad (3.313)$$

The symbol "·" in (3.313) denotes here a scalar product between two vectors.

Let us introduce the directional derivative of the function f at $\bar{\mathbf{x}}$ in the direction of the vector $\mathbf{u}$. It is defined by

$$\left.\frac{d}{d\epsilon}\,[\,f(\bar{\mathbf{x}} + \epsilon\,\mathbf{u})\,]\right|_{\epsilon=0},$$

where ϵ is a scalar parameter. Due to the fact that $\bar{\mathbf{x}} + \epsilon\,\mathbf{u}$ describes a straight line in $\mathcal{R}^3$, the directional derivative measures the increment of the function f in the direction of this straight line at point $\bar{\mathbf{x}}$. The directional derivative can be computed using the chain rule

$$\left.\frac{d}{d\epsilon}\,[\,f(\bar{\mathbf{x}} + \epsilon\,\mathbf{u})\,]\right|_{\epsilon=0} = \left[\frac{\partial f(\bar{\mathbf{x}} + \epsilon\,\mathbf{u})}{\partial \mathbf{x}} \cdot \frac{\partial (\bar{\mathbf{x}} + \epsilon\,\mathbf{u})}{\partial \epsilon}\right]_{\epsilon=0} = \frac{\partial f(\mathbf{x})}{\partial \mathbf{x}} \cdot \mathbf{u}.$$

A comparison of the coefficients yields the result

$$\left.\frac{d}{d\epsilon}\,[\,f(\bar{\mathbf{x}} + \epsilon\,\mathbf{u})\,]\right|_{\epsilon=0} = \bar{D}f \cdot \mathbf{u},$$

which coincides with the tangent of f at $\bar{\mathbf{x}}$. Hence the linear part of the function f at $\bar{\mathbf{x}}$ is determined by the value of the function as well as its directional derivative at this point. Note that the directional derivative is a linear operator. Hence all known rules for the differentiation of sums and products can be applied.

The directional derivative can be generalized for functional spaces in a formal way. For this purpose, the C^1-mapping $\mathbf{G} : \mathcal{E} \to \mathcal{F}$ is used, where $\bar{\mathbf{x}}$ and $\mathbf{u}$ are points of the abstract space $\mathcal{E}$. The TAYLOR series expansion yields then

$$\mathbf{G}(\bar{\mathbf{x}} + \mathbf{u}) = \bar{\mathbf{G}} + \bar{D}\,\mathbf{G} \cdot \mathbf{u} + \mathbf{R}, \qquad (3.314)$$

where the symbol "$\cdot$" denotes the inner product associated with the elements of the space. Again

$$\frac{d}{d\epsilon}\left[\mathbf{G}(\bar{\mathbf{x}} + \epsilon\,\mathbf{u})\right]\Bigg|_{\epsilon=0} = \bar{D}\,\mathbf{G} \cdot \mathbf{u} \qquad (3.315)$$

can be written. Due to that result, the linear part of the mapping $\mathbf{G}$ at $\bar{\mathbf{x}}$ is given by

$$\mathbf{L}\,[\,\mathbf{G}\,]_{x=\bar{x}} = \bar{\mathbf{G}} + \bar{D}\,\mathbf{G} \cdot \mathbf{u}\,. \qquad (3.316)$$

Here the elements of the spaces $\mathcal{E}$ and $\mathcal{F}$ can be arbitrary fields, e.g. scalar-, vector- or tensor fields.

To simplify notation the directional derivative is written instead of $\bar{D}\,\mathbf{G}\cdot\mathbf{u}$ in the following in the short form $\Delta\bar{\mathbf{G}}$. Here the bar denotes evaluation at $\bar{\mathbf{x}}$.

3.5.1 Linearization of Kinematical Quantities

The linearization of different kinematical relations is derived in this section, exemplarily. These quantities are selected such that they are formulated with respect to the initial and the current configuration.

Green-Lagrange Strain Tensor. The linear part of the strain measure (3.15) follows with (3.316) as

$$\mathbf{L}\,[\,\mathbf{E}\,]_{\varphi=\bar{\varphi}} = \bar{\mathbf{E}} + \bar{D}\,\mathbf{E} \cdot \mathbf{u} = \bar{\mathbf{E}} + \Delta\bar{\mathbf{E}}\,. \qquad (3.317)$$

In this relation, the directional derivative $\bar{D}\,\mathbf{E} \cdot \mathbf{u} = \Delta\bar{\mathbf{E}}$ has to be computed using (3.315)

$$
\begin{aligned}
\bar{D}\,\mathbf{E} \cdot \mathbf{u} &= \frac{d}{d\epsilon}\left[\frac{1}{2}\mathbf{F}^T(\bar{\varphi} + \epsilon\,\mathbf{u})\,\mathbf{F}(\bar{\varphi} + \epsilon\,\mathbf{u}) - \mathbf{1}\right]\Bigg|_{\epsilon=0}, \\
\Delta\bar{\mathbf{E}} &= \frac{1}{2}\left[\bar{\mathbf{F}}^T\,\mathrm{Grad}\,\mathbf{u} + \mathrm{Grad}^T\mathbf{u}\,\bar{\mathbf{F}}\right].
\end{aligned}
\qquad (3.318)
$$

This result is linear in $\mathbf{u}$ but contains also parts of the deformation at $\bar{\varphi}$, which are represented by $\bar{\mathbf{F}}$. The evaluation of (3.318) at the initial state $\varphi = \mathbf{X}$ yields the strain tensor (3.17) of the linear theory

$$\mathbf{L}\,[\,\mathbf{E}\,]_{\varphi=X} = \mathbf{0} + \frac{1}{2}\,[\,\mathrm{Grad}\,\mathbf{u} + \mathrm{Grad}^T\mathbf{u}\,]\,. \qquad (3.319)$$

Inverse Cauchy-Green Tensor. The inverse CAUCHY-GREEN tensor appears in several constitutive relations which describe finite elastic behaviour. The linearization of inverse arbitrary tensors $\mathbf{T}^{-1}$ can be based on the product $\mathbf{T}\,\mathbf{T}^{-1} = \mathbf{1}$. Starting from this, the linearization of the inverse tensor is computed using the product rule. Thus for the CAUCHY-GREEN tensor

$$D\,(\mathbf{C}\,\mathbf{C}^{-1})\cdot\mathbf{u} = [\,D\,\mathbf{C}\cdot\mathbf{u}\,]\,\mathbf{C}^{-1} + \mathbf{C}\,[\,D\,\mathbf{C}^{-1}\cdot\mathbf{u}\,] = \mathbf{0} \qquad (3.320)$$

is obtained and for the directional derivative of the inverse

$$D\,\mathbf{C}^{-1}\cdot\mathbf{u} = -\mathbf{C}^{-1}\,[\,D\,\mathbf{C}\cdot\mathbf{u}\,]\,\mathbf{C}^{-1} \qquad (3.321)$$

can be written which can be easily computed using the result (3.318), note that $\mathbf{E} = \frac{1}{2}\,(\mathbf{C}-\mathbf{1})$,

$$\Delta\,\mathbf{C}^{-1} = \bar{D}\,\mathbf{C}^{-1}\cdot\mathbf{u} = -\bar{\mathbf{C}}^{-1}\,[\,\bar{\mathbf{F}}^{T}\,\mathrm{Grad}\mathbf{u} + \mathrm{Grad}^{T}\mathbf{u}\,\bar{\mathbf{F}}\,]\,\bar{\mathbf{C}}^{-1}. \qquad (3.322)$$

Based on this expression, the linear part of $\mathbf{C}^{-1}$ is given by

$$\mathbf{L}\,[\,\mathbf{C}^{-1}\,]_{\varphi=\bar{\varphi}} = \bar{\mathbf{C}}^{-1} + \bar{D}\,\mathbf{C}^{-1}\cdot\mathbf{u}. \qquad (3.323)$$

Equation (3.322) can be reformulated by introducing the spatial gradient $\overline{\mathrm{grad}\mathbf{u}} = \partial\mathbf{x}\,/\,\partial\bar{\mathbf{x}}$

$$\bar{D}\,\mathbf{C}^{-1}\cdot\mathbf{u} = -\bar{\mathbf{F}}^{-1}\,[\,\overline{\mathrm{grad}\mathbf{u}} + \overline{\mathrm{grad}}^{T}\mathbf{u}\,]\,\bar{\mathbf{F}}^{-T}. \qquad (3.324)$$

The evaluation of the linear part of the inverse CAUCHY-GREEN tensor with respect to the initial configuration yields with (3.17)

$$\mathbf{L}\,[\,\mathbf{C}^{-1}\,]_{\varphi=X} = \mathbf{1} - 2\,\boldsymbol{\varepsilon} \qquad (3.325)$$

where $\boldsymbol{\varepsilon}$ is the linear strain tensor.

Jacobi Determinant. Another example is provided by the linearization of a scalar quantity, the JACOBI determinant $J = \det\mathbf{F}$. The linear part of this nonlinear function follows from the directional derivative of the determinant

$$\bar{D}\,J\cdot\mathbf{u} = \frac{d}{d\epsilon}\,[\,\det\mathbf{F}(\bar{\varphi}+\epsilon\,\mathbf{u})\,]\Big|_{\epsilon=0}. \qquad (3.326)$$

By applying the chain rule, the result

$$D\,(\det\mathbf{F})\cdot\mathbf{u} = \frac{\partial(\det\mathbf{F})}{\partial\mathbf{F}}\cdot[\,D\,\mathbf{F}\cdot\mathbf{u}\,] \qquad (3.327)$$

follows. For the partial derivative of the determinant of a tensor with respect to the same tensor, the result is

$$\frac{\partial(\det\mathbf{F})}{\partial\mathbf{F}} = J\,\mathbf{F}^{-T}, \qquad (3.328)$$

see also $(3.110)_3$. Since $\mathbf{F} = \operatorname{Grad} \boldsymbol{\varphi}$ is a linear function, the result $\bar{D} J \cdot \mathbf{u} = \Delta \bar{J} = 1 / \bar{J} \bar{\mathbf{F}}^{T} \cdot \operatorname{Grad} \mathbf{u}$ follows and with this also the expression for the linear part

$$L\,[J]_{\varphi=\bar{\varphi}} = \bar{J} + \bar{J}\,\bar{\mathbf{F}}^{-T} \cdot \operatorname{Grad} \mathbf{u}\,. \qquad (3.329)$$

This result can be expressed via $\bar{\mathbf{F}}^{-T} \cdot \operatorname{Grad} \mathbf{u} = \operatorname{tr}(\bar{\mathbf{F}}^{-T}\,\operatorname{Grad}^{T} \mathbf{u}) = \operatorname{tr}(\overline{\operatorname{grad}}^{T} \mathbf{u}) = \overline{\operatorname{div}} \mathbf{u}$ in the form

$$L\,[J]_{\varphi=\bar{\varphi}} = \bar{J} + \bar{J}\,\overline{\operatorname{div}}\,\mathbf{u}\,. \qquad (3.330)$$

Evaluation of (3.329) with respect to the initial configuration yields

$$L\,[J]_{\varphi=X} = 1 + \operatorname{Div} \mathbf{u}\,. \qquad (3.331)$$

Almansi Strain Tensor. The linearization of spatial vectors and tensors is derived by a *pull back* of the spatial objects to the initial configuration. In this configuration, the linearization is performed and the linearized object then is *push forward* to the spatial configuration. This procedure can now be applied to linearize the ALMANSI strain tensor $\mathbf{e} = \frac{1}{2}(\mathbf{1} - \mathbf{b}^{-1})$, see (3.24). With the *pull back* of the strain tensor using (3.35), the linearization

$$
\begin{aligned}
D\,\mathbf{e} \cdot \mathbf{u} &= \bar{\mathbf{F}}^{-T}\{D\,\mathbf{E} \cdot \mathbf{u}\}\bar{\mathbf{F}}^{-1} = \frac{1}{2}\left(\operatorname{Grad} \mathbf{u}\,\bar{\mathbf{F}}^{-1} + \bar{\mathbf{F}}^{-T}\,\operatorname{Grad}^{T} \mathbf{u}\right) \\
&= \frac{1}{2}\left(\overline{\operatorname{grad}}\,\mathbf{u} + \overline{\operatorname{grad}}^{T}\,\mathbf{u}\right) = \nabla_{\bar{x}}^{S}\Delta\mathbf{u} \qquad (3.332)
\end{aligned}
$$

is obtained. By comparing this result to (3.318), it can be observed that

$$\Delta\bar{\mathbf{E}} = \bar{\mathbf{F}}^{T}\nabla_{\bar{x}}^{S}\Delta\mathbf{u}\,\bar{\mathbf{F}} \qquad (3.333)$$

is valid. Furthermore, the linearization of the ALMANSI strain tensor has the same structure as the LIE derivative for differentiation of a spatial object with respect to time, see (3.53).

3.5.2 Linearization of Constitutive Equations

The linearization of constitutive equations can be determined for elastic materials based on the relations stated in Sect. 3.3.1. For inelastic constitutive equations, the linearization for the continuous case can be derived; however in the framework of the finite element method a time integration has to be applied to evaluate the differential evolution equations describing such materials. Due to that, the linearization depends also upon the integration algorithm, see also Remark 3.11. This is also true for rate independent behaviour since in that case a "pseudo time" is introduced to capture the loading history. Hence these linearizations cannot be derived without the knowledge of the integration algorithms. Associated linearizations are presented in Sect. 6.2.

The elastic constitutive equation (3.104) describes the dependence of the 2nd PIOLA-KIRCHHOFF stress tensor on the right CAUCHY-GREEN tensor. The linearization of this constitutive relation follows with (3.316) as

$$\mathbf{L}[\mathbf{S}]_{\varphi=\bar{\varphi}} = \bar{\mathbf{S}} + \bar{D}\mathbf{S} \cdot \mathbf{u} = \bar{\mathbf{S}} + \Delta\bar{\mathbf{S}}$$
$$= \bar{\mathbf{S}} + \frac{\partial\mathbf{S}}{\partial\mathbf{C}}\bigg|_{\varphi=\bar{\varphi}} [\bar{D}\,\mathbf{C}\cdot\mathbf{u}]. \qquad (3.334)$$

Using (3.240) and (3.318), relation

$$\mathbf{L}[\mathbf{S}]_{\varphi=\bar{\varphi}} = \bar{\mathbf{S}} + \bar{\mathbf{C}}[\Delta\bar{\mathbf{E}}] \qquad (3.335)$$

can be written. In comparison with (3.334), this yields

$$\Delta\bar{\mathbf{S}} = \bar{\mathbf{C}}[\Delta\bar{\mathbf{E}}]. \qquad (3.336)$$

This relation has the same structure as the incremental constitutive equation (3.241). The only difference is that the time derivatives have to be exchanged by the directional derivatives. Hence it is not necessary to state the linearizations of the other constitutive equations presented in Sect. 3.3.1. These linearizations are obtained by an evaluation of the incremental constitutive tensors in Sect. 3.3.4 at the deformation state $\bar{\varphi}$.

Exercise 3.10: Derive the linearization of the hyper elastic constitutive equation (3.119)

$$\mathbf{S} = \frac{\Lambda}{2}(J^2 - 1)\mathbf{C}^{-1} + \mu(\mathbf{1} - \mathbf{C}^{-1})$$

with respect to the initial configuration.
Solution: First the kinematical objects J and C^{-1}, appearing in (3.119), have to be linearized. With the results obtained so far, and by using (3.322) and (3.329), the relations

$$L[J]_{\varphi=\bar{\varphi}} = \bar{J} + \frac{1}{\bar{J}}\text{tr}(\bar{\mathbf{F}}^{-T}\,\text{Grad}^T\mathbf{u}),$$
$$\mathbf{L}[\mathbf{C}^{-1}]_{\varphi=\bar{\varphi}} = \bar{\mathbf{C}}^{-1} - \bar{\mathbf{C}}^{-1}(\bar{\mathbf{F}}^T\,\text{Grad}\mathbf{u} + \text{Grad}^T\mathbf{u}\,\bar{\mathbf{F}})\,\bar{\mathbf{C}}^{-1}$$

follow. The evaluation with respect to the initial configuration ($\varphi = \mathbf{X}$) yields with $\bar{\mathbf{C}}^{-1} = \mathbf{1}$, $\bar{\mathbf{F}} = \mathbf{1}$ and $\bar{J} = 1$

$$L[J]_{\varphi=X} = 1 + \text{Div}\mathbf{u},$$
$$\mathbf{L}[\mathbf{C}^{-1}]_{\varphi=X} = \mathbf{1} - 2\boldsymbol{\varepsilon},$$

where in the last equation the linear strain tensor $\boldsymbol{\varepsilon}$ appears, see (3.17). The linearization of the stress tensor $\mathbf{S}$ can now be computed as follows

$$D\mathbf{S}\cdot\mathbf{u}|_{\varphi=X} = \left[\frac{\Lambda}{2}\left\{(2\bar{J}(D\,J\cdot\mathbf{u})\,\bar{\mathbf{C}}^{-1} + (\bar{J}^2 - 1)\,D\,\mathbf{C}^{-1}\cdot\mathbf{u}\right\} - \mu\,D\,\mathbf{C}^{-1}\cdot\mathbf{u}\right]_{\varphi=X}.$$

By inserting the linearizations of the kinematical quantities, the final form

$$\mathbf{L}[\mathbf{S}]_{\varphi=X} = D\mathbf{S}\cdot\mathbf{u}|_{\varphi=X} = \Lambda\,\text{tr}\,\boldsymbol{\varepsilon}\,\mathbf{1} + 2\mu\,\boldsymbol{\varepsilon}$$

is derived at the initial configuration with $\mathrm{Div}\,\mathbf{u} = \mathrm{tr}\,\boldsymbol{\varepsilon}$. This relation represents the classical law of HOOKE used within the linear theory of elasticity. The constitutive parameters Λ and μ are known as LAMÉ constants.

This results could also be obtained by evaluation of the incremental constitutive tensor (3.251) at the initial state $\boldsymbol{\varphi} = \mathbf{X}$, see also Exercise 3.8 (a).

3.5.3 Linearization of the Variational Formulation

The solutions of nonlinear initial boundary value problems in solid mechanics can be obtained in general only by employing approximate solution techniques. Since many of these methods – like the finite element method – rely on a variational formulation of the field equations, the basis for numerical methods are provided by the weak forms of the associated field equations. In solid mechanics, the weak form is also known as principle of virtual work, see (3.289) or (3.292). A discretization of the weak form leads to a set of nonlinear algebraic equations, see Chap. 4.

For the solution of the set of nonlinear equations, many different algorithms are known, for an overview see Chap. 5. Often NEWTON'S method is applied since it possesses the advantage of a quadratic convergence close to the solution point. In case of NEWTON'S method, an improved solution is obtained from the TAYLOR series expansion of the nonlinear equation at the already computed approximate solution. This TAYLOR expansion corresponds in finite element applications to the linearization of the weak form, or in solid mechanics to the linearization of the principle of virtual work, and can be obtained by the directional derivative discussed above. Such linearization will be computed here for solids consisting of hyper elastic materials, further applications can be found in Chaps. 5, 6, 7, 9, 10 and 11.

The linearization will be stated first for the weak form with respect to the initial configuration (3.289). In general, the linearization at a deformation state of the solid is computed which is in equilibrium. This state will be denoted by $\bar{\boldsymbol{\varphi}}$, see Fig. 3.15.

The linear part of the weak form is given by

$$L\,[\,G\,]_{\varphi=\bar\varphi} = G\,(\bar{\boldsymbol{\varphi}}, \boldsymbol{\eta}) + DG\,(\bar{\boldsymbol{\varphi}}, \boldsymbol{\eta}) \cdot \Delta\mathbf{u}\,. \tag{3.337}$$

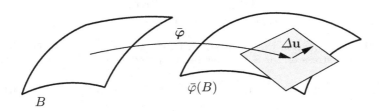

Fig. 3.15 Configuration belonging to the linearization

The operator $G(\bar{\varphi}, \eta)$ corresponds to (3.290), just the deformation $\bar{\varphi}$ is inserted instead of φ. By assuming that the load is conservative the directional derivative of G can be computed in the direction of $\Delta\mathbf{u}$ by only taking the first term in (3.290) into account

$$DG(\bar{\varphi}, \eta) \cdot \Delta\mathbf{u} = \int_B [D\mathbf{P}(\bar{\varphi}) \cdot \Delta\mathbf{u}] \cdot \operatorname{Grad} \eta \, dV, \qquad (3.338)$$

hence all other terms do not depend upon the deformation. The linearization of the first PIOLA-KIRCHHOFF stress tensor yields with $\mathbf{P} = \mathbf{F}\,\mathbf{S}$

$$DG(\bar{\varphi}, \eta) \cdot \Delta\mathbf{u} = \int_B \{ \operatorname{Grad} \Delta\mathbf{u}\,\bar{\mathbf{S}} + \bar{\mathbf{F}}\,[D\mathbf{S}(\bar{\varphi}) \cdot \Delta\mathbf{u}] \} \cdot \operatorname{Grad} \eta \, dV. \quad (3.339)$$

Terms with a bar have to be evaluated at the deformation state $\bar{\varphi}$. The linearization of the second PIOLA-KIRCHHOFF stress tensor can be based on (3.336). It follows

$$D\mathbf{S}(\bar{\varphi}) \cdot \Delta\mathbf{u} = \bar{\mathbb{C}}\,[\Delta\bar{\mathbf{E}}], \qquad (3.340)$$

where the last term is the linearization of the GREEN-LAGRANGE strain tensor $\mathbf{E}$ at $\bar{\varphi}$, see also (3.318). The elasticity tensor $\mathbb{C}$ which is also referred to the initial configuration B is given with (3.240) by

$$\bar{\mathbb{C}} = 4 \left. \frac{\partial^2 W}{\partial \mathbf{C}\,\partial \mathbf{C}} \right|_{\varphi=\bar{\varphi}} \qquad (3.341)$$

at the state $\bar{\varphi}$. The use of (3.341) in (3.339) completes the linearization

$$DG(\bar{\varphi}, \eta) \cdot \Delta\mathbf{u} = \int_B \{ \operatorname{Grad} \Delta\mathbf{u}\,\bar{\mathbf{S}} + \bar{\mathbf{F}}\,\bar{\mathbb{C}}\,[\Delta\bar{\mathbf{E}}] \} \cdot \operatorname{Grad} \eta \, dV. \quad (3.342)$$

Note that $\bar{\mathbb{C}}$ has also to be evaluated at state $\bar{\varphi}$. By applying the trace operation to the second term and by using the symmetry of $\bar{\mathbb{C}}$, a more compact form of (3.342) can be found

$$DG(\bar{\varphi}, \eta) \cdot \Delta\mathbf{u} = \int_B \{ \operatorname{Grad} \Delta\mathbf{u}\,\bar{\mathbf{S}} \cdot \operatorname{Grad} \eta + \delta\bar{\mathbf{E}} \cdot \bar{\mathbb{C}}\,[\Delta\bar{\mathbf{E}}] \} \, dV. \quad (3.343)$$

Here the symmetry in η and $\Delta\mathbf{u}$ can be observed which results from the linearization operation. The first term in (3.343) is often named *initial stress* term since the stresses at the given state appear directly. The second term contains, besides the incremental constitutive tensor $\bar{\mathbb{C}}$, the variation of the GREEN-LAGRANGE strain tensor $\delta\bar{\mathbf{E}} = \frac{1}{2}(\bar{\mathbf{F}}^T \operatorname{Grad} \eta + \operatorname{Grad}^T \eta\,\bar{\mathbf{F}})$ and the increment of the GREEN-LAGRANGE strain tensor $\Delta\bar{\mathbf{E}} = \frac{1}{2}(\bar{\mathbf{F}}^T \operatorname{Grad} \Delta\mathbf{u} + \operatorname{Grad}^T \Delta\mathbf{u}\,\bar{\mathbf{F}})$.

The linearization of the principle of virtual work can be obtained in terms of the current configuration by a *push forward* of the linearization (3.343) to the already computed configuration $\bar{\varphi}$. Using the transformations for the linearization of the GREEN-LAGRANGE strain tensor (3.333), which resulted as *push forward* in $\nabla^S_{\bar{x}}\Delta\mathbf{u}$, it follows for the second term in (3.343)

$$\int_B \nabla^S_{\bar{x}}\boldsymbol{\eta} \cdot \bar{\mathbf{c}}\left[\nabla^S_{\bar{x}}\Delta\mathbf{u}\right] dV .$$

In this equations, the fourth order tensor $\bar{\mathbf{c}}$ can be computed by the transformation (3.240) from $\bar{\mathbf{C}}$.

The first term in (3.343) can be directly recast with $\bar{\boldsymbol{\tau}} = \bar{\mathbf{F}}\,\bar{\mathbf{S}}\,\bar{\mathbf{F}}^T$ as

$$\text{Grad }\Delta\mathbf{u}\,\bar{\mathbf{S}} \cdot \text{Grad }\boldsymbol{\eta} = \bar{\mathbf{F}}\text{Grad }\Delta\mathbf{u}\,\bar{\mathbf{F}}^{-1}\,\bar{\boldsymbol{\tau}}\,\bar{\mathbf{F}}^{-1} \cdot \text{Grad }\boldsymbol{\eta} = \overline{\text{grad}\Delta\mathbf{u}}\,\bar{\boldsymbol{\tau}} \cdot \overline{\text{grad}\boldsymbol{\eta}} . \tag{3.344}$$

This results in the linearization with respect to the known current configuration $\bar{\varphi}$ where all quantities have to be evaluated at $\bar{\varphi}$.

$$Dg(\bar{\varphi},\boldsymbol{\eta}) \cdot \Delta\mathbf{u} = \int_B \left\{ \overline{\text{grad}\Delta\mathbf{u}}\,\bar{\boldsymbol{\tau}} \cdot \overline{\text{grad}\boldsymbol{\eta}} + \nabla^S_{\bar{x}}\boldsymbol{\eta} \cdot \bar{\mathbf{c}}\left[\nabla^S_{\bar{x}}\Delta\mathbf{u}\right] \right\} dV . \tag{3.345}$$

The integral (3.345) can now be referred to the current configuration with the relation $d\bar{v} = \bar{J}dV$. For this purpose, the CAUCHY stress tensor $\bar{\boldsymbol{\sigma}} = \frac{1}{\bar{J}}\bar{\boldsymbol{\tau}}$ is introduced and a further incremental constitutive tensor

$$\bar{\bar{\mathbf{c}}} = \frac{1}{\bar{J}}\,\bar{\mathbf{c}} \tag{3.346}$$

is defined, such that

$$Dg(\bar{\varphi},\boldsymbol{\eta}) \cdot \Delta\mathbf{u} = \int_{\bar{\varphi}(B)} \left\{ \overline{\text{grad}\Delta\mathbf{u}}\,\bar{\boldsymbol{\sigma}} \cdot \overline{\text{grad}\boldsymbol{\eta}} + \nabla^S_{\bar{x}}\boldsymbol{\eta} \cdot \bar{\bar{\mathbf{c}}}\left[\nabla^S_{\bar{x}}\Delta\mathbf{u}\right] \right\} dv \tag{3.347}$$

is obtained.

The deformation state $\bar{\varphi}$ to which the formulation is referred is not known and can only be obtained within the nonlinear solution process by an update of all deformation states in a successive manner. Hence relation (3.347) is known in the literature also as *updated Lagrange* formulation, see e.g. Bathe et al. (1975) or Bathe (1996).

With the above given linearizations all relations needed within the NEWTON method are available as well as for formulations with respect to the initial configuration as with respect to the current configuration. These linearizations are basis for finite element simulations, for more see Chaps. 4 and 5.

Remark 3.11: It should be emphasized that the linearization of the continuous problem as stated in (3.290) does not coincide in all cases with the linearization of the discrete problem which results from a finite element discretization. In case of a finite element formulation of a static problem with continuous interpolations functions for the displacements, the equivalence of the discretization of the linearizations given above and the direct linearization of the discretization of the weak form using finite elements can be shown. In case that the finite element method is applied to problems of elasto-plasticity or other inelastic constitutive equations then the linearization of the discrete form is no longer equivalent to the discretization of the so far given linearizations since the integration algorithm for the evolution equations of the inelastic material response plays a prominent role and hence the linearization depends upon this algorithm, see e.g. Simo et al. (1985a) or for a more general overview Simo and Hughes (1998).

Exercise 3.11: The weak form of equilibrium (3.292) was formulated in Exercise 3.10 depending upon the symmetrical BIOT stress tensor $\mathbf{T}_B$ and the right stretch tensor $\mathbf{U}$. Linearize (3.292) using the constitutive equation (3.119).

Solution: In order to obtain the linearization of the stress divergence term $\mathbf{T}_B \cdot \delta \mathbf{U}$ the linearization of the BIOT stress tenor and the variation of the right stretch tensor have to be linearized on their own. For this purpose, the constitutive equation (3.119) is rewritten such that it represents the BIOT stress tensor in terms of the right stretch tensor. With the relation $\mathbf{T}_B = \mathbf{R}^T \mathbf{P}$, derived in Exercise 3.10, it follows with $\mathbf{P} = \mathbf{F}\,\mathbf{S}$, see (3.82),

$$\mathbf{T}_B = \mathbf{U}\,\mathbf{S}\,.$$

Thus, after some manipulation, the BIOT stress is deduced from the constitutive equation (3.119)

$$\mathbf{T}_B = \frac{\Lambda}{2}\,(J^2 - 1)\mathbf{U}^{-1} + \mu\,(\mathbf{U} - \mathbf{U}^{-1})\,.$$

Based on this result, the linearization of the BIOT stress tensor yields

$$D\,\mathbf{T}_B(\bar{\varphi}) \cdot \varDelta \mathbf{u} = \left. \frac{\partial \mathbf{T}_B}{\partial \mathbf{U}} \right|_{\varphi = \bar{\varphi}} [D\,\mathbf{U} \cdot \varDelta \mathbf{u}] = \bar{\mathbb{C}}_U\,[\varDelta\,\mathbf{U}]\,.$$

Here the incremental constitutive tensor

$$\mathbb{C}_U = \Lambda J^2\,\mathbf{U}^{-1} \otimes \mathbf{U}^{-1} + [\,\mu - \Lambda(J^2 - 1)]\mathbb{E}_{U^{-1}} + \mathbb{E}\mu$$

has the same structure as (3.268). The tensor $\mathbb{E}_{U^{-1}}$ is computed in an analogous way as (3.255).

The linearization of the right stretch tensor can be determined as its variation, see Exercise 3.10,

$$\varDelta \mathbf{U} = \varDelta \mathbf{R}^T \mathbf{F} + \mathbf{R}^T\,\varDelta \mathbf{F}\,.$$

Finally the linearization of the variation of the right stretch tensor $\delta \mathbf{U} = \delta \mathbf{R}^T \mathbf{F} + \mathbf{R}^T \delta \mathbf{F}$ has to be derived. Formally the result

$$\varDelta \delta \mathbf{U} = \varDelta \delta \mathbf{R}^T\,\mathbf{F} + \delta \mathbf{R}^T\,\varDelta \mathbf{F} + \varDelta \mathbf{R}^T\,\delta \mathbf{F}$$

is obtained such that the linearization of the weak form yields

$$D\,G(\bar{\varphi},\boldsymbol{\eta})\cdot\Delta\mathbf{u} = \int\limits_{B} \{\,\delta\mathbf{U}\cdot\bar{\mathbb{C}}_U\,[\Delta\,\mathbf{U}] + \mathbf{T}_B\cdot(\Delta\delta\mathbf{R}^T\,\mathbf{F} + \delta\mathbf{R}^T\,\Delta\mathbf{F} + \Delta\mathbf{R}^T\,\delta\mathbf{F})\,\}\,dV\,.$$

For the final evaluation of this equation, an explicit representation of the linearization of the orthogonal rotation tensor $\mathbf{R}$ is needed. In the two-dimensional case a representation is derived in Exercise 3.2 which is again stated here

$$\mathbf{R} = \begin{bmatrix} \cos\theta & \sin\theta \\ -\sin\theta & \cos\theta \end{bmatrix}, \quad \tan\theta = \frac{F_{12} - F_{21}}{F_{11} + F_{22}}\,.$$

From this form, the variation of $\mathbf{R}$ is computed

$$\delta\mathbf{R} = \frac{\partial\mathbf{R}}{\partial\theta}\,\delta\theta = \mathbf{R}_{,\theta}\,\delta\theta = \begin{bmatrix} -\sin\theta & -\cos\theta \\ \cos\theta & -\sin\theta \end{bmatrix}\delta\theta\,.$$

The variation of the angle θ can be expressed by the variation of the components of the deformation tensor as

$$\delta\theta = \frac{1}{2}\left[(1 + \cos 2\theta)\frac{\delta F_{12} - \delta F_{21}}{F_{11} + F_{22}} - \sin 2\theta\,\frac{\delta F_{11} + \delta F_{22}}{F_{11} + F_{22}}\right]\,.$$

In an analogous way, the linearization $\Delta\mathbf{R}$ follows by exchanging in the last two equations the variation δ by Δ.

Corresponding equations – however, a little bit more complex – are derived for the linearization of the variation of the rotation tensor

$$\Delta\,\delta\mathbf{R} = \mathbf{R}_{,\theta\,\theta}\,\delta\theta\,\Delta\theta + \mathbf{R}_{,\theta}\,\Delta\delta\theta\,,$$

where $\mathbf{R}_{,\theta\,\theta}$ denotes the second derivative of $\mathbf{R}$ with respect to θ. The term $\Delta\delta\theta$ follows from linearization of $\delta\theta$

$$\begin{aligned}\Delta\delta\theta = -\frac{1 + \cos 2\theta}{2\,(\,F_{11} + F_{22})} \quad[\quad &\sin 2\theta\,(\delta F_{12} - \delta F_{21})(\Delta F_{12} - \Delta F_{21}) \\ +\ &\cos 2\theta\,(\delta F_{12} - \delta F_{21})(\Delta F_{11} + \Delta F_{22}) \\ +\ &\cos 2\theta\,(\delta F_{11} + \delta F_{22})(\Delta F_{12} - \Delta F_{21}) \\ -\ &\sin 2\theta\,(\delta F_{11} + \delta F_{22})(\Delta F_{11} + \Delta F_{22})\,]\,.\end{aligned}$$

In the tree-dimensional case, all equations will be even more complex. Due to the fact that the representation using the BIOT stress tensor and the right stretch tensor are equivalent to the weak form in (3.292) always the simpler form (3.292) should be used, as long as no other reasons speak for the formulation based on the BIOT stresses, since the linearization of (3.292) is a lot simpler, see (3.343).

Exercise 3.12: The description of a pressure load resulting from a gas or fluid pressure yields a surface load which depends upon the current state of the deformation. The stress vector is then given by $\bar{\mathbf{t}} = p\,\mathbf{n}$ with the pressure p and the surface normal $\mathbf{n}$. This leads in the weak form (3.283) to the additional term

$$g(\boldsymbol{\varphi},\boldsymbol{\eta}) + g_p(\boldsymbol{\varphi},\boldsymbol{\eta}) = g(\boldsymbol{\varphi},\boldsymbol{\eta}) + \int\limits_{\varphi(\partial B_p)} p\,\mathbf{n}\cdot\boldsymbol{\eta}\,da\,. \tag{3.348}$$

Derive the linearization for this expression.

Solution: It makes sense to write the term (3.348), referred to the current configuration, with respect to the initial configuration when the linearization has to

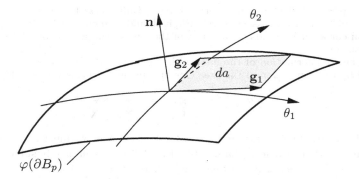

Fig. 3.16 Deformation dependent loads in terms of convective coordinates

be performed. This can be achieved in two ways. The first is based on a transformation of the surface normal $\mathbf{n}\,da$ via the formula of NANSON (3.11) to the initial configuration. This leads to the expression $\int_B p\,J\,\mathbf{F}^{-T}\,\mathbf{N}\cdot\boldsymbol{\eta}\,dA$ which linearization is quite complicated. Simpler is the second approach in which the surface normal is expressed via a cross product in terms of the base vectors, tangent to the surface of the solid. This can be achieved by introducing convective coordinates θ_α on the surface, see Fig. 3.16. With the tangent vectors $\mathbf{g}_\alpha$ ($\alpha = 1,2$), introduced in Figure 3.16, the surface normal is computed

$$\mathbf{n} = \frac{\mathbf{g}_1 \times \mathbf{g}_2}{\|\mathbf{g}_1 \times \mathbf{g}_2\|}.$$

The tangent vectors follow with (3.38) from the deformation by $\mathbf{g}_\alpha = \boldsymbol{\varphi}_{,\alpha}$. Since, furthermore, the area element da is given by $da = \|\mathbf{g}_1 \times \mathbf{g}_2\|d\theta_1\,d\theta_2$ in terms of convective coordinates, the virtual work of the pressure load can be written as

$$g_p(\boldsymbol{\varphi},\boldsymbol{\eta}) = \int_{(\theta_1)}\int_{(\theta_2)} p\,(\boldsymbol{\varphi}_{,1} \times \boldsymbol{\varphi}_{,2})\cdot\boldsymbol{\eta}\,d\theta_1\,d\theta_2. \qquad (3.349)$$

By using these relations, the linearization follows with (3.13) $\boldsymbol{\varphi}_{,\alpha} = (\mathbf{X}+\mathbf{u})_{,\alpha}$ as

$$D\,g_p(\boldsymbol{\varphi},\boldsymbol{\eta})\cdot\Delta\mathbf{u} = \int_{(\theta_1)}\int_{(\theta_2)} p\,(\Delta\mathbf{u}_{,1} \times \boldsymbol{\varphi}_{,2} + \boldsymbol{\varphi}_{,1} \times \Delta\mathbf{u}_{,2})\cdot\boldsymbol{\eta}\,d\theta_1\,d\theta_2. \qquad (3.350)$$

Here it was assumed that only the direction of the pressure load $p\,\mathbf{n}$ but not its magnitude depends upon the deformation. The linearization is derived with respect to the convective coordinates. However, it can be pushed forward to the current deformation state with the relation for the area elements

$$D\,g_p(\boldsymbol{\varphi},\boldsymbol{\eta})\cdot\Delta\mathbf{u} = \int_{\varphi(\partial B_p)} p\,\frac{\Delta\mathbf{u}_{,1} \times \boldsymbol{\varphi}_{,2} + \boldsymbol{\varphi}_{,1} \times \Delta\mathbf{u}_{,2}}{\|\boldsymbol{\varphi}_{,1} \times \boldsymbol{\varphi}_{,2}\|}\cdot\boldsymbol{\eta}\,da. \qquad (3.351)$$

This completes the linearization of the deformation dependent pressure term (3.348).

Further theoretical considerations regarding deformation dependent loads which also concern the conservative or non-conservative character of such loads can be found in Sewell (1967), Schweizerhof (1982), Bufler (1984), Ogden (1984) or Simo et al. (1991). However, as a side remark, it should be mentioned that the appearance of non-conservative loads stems most of the times from a non-complete mechanical model.

The transformation of the integral in (3.351) to the initial configuration is derived by the conversion of the area element, using the basis vectors. This leads to

$$\frac{da}{dA} = \frac{\| \mathbf{g}_1 \times \mathbf{g}_2 \|}{\| \mathbf{G}_1 \times \mathbf{G}_2 \|} = \frac{\| \boldsymbol{\varphi}_{,1} \times \boldsymbol{\varphi}_{,2} \|}{\| \mathbf{X}_{,1} \times \mathbf{X}_{,2} \|}, \qquad (3.352)$$

which can be directly inserted into (3.351). However, within the nonlinear finite element method, the formulation stated in (3.350) is completely sufficient and the most efficient which will be discussed further in Sect. 4.2.5.

4. Spatial Discretization Techniques

Different approximations are made when the method of finite elements is applied to discretize weak forms of the nonlinear problems discussed in the previous chapters. On one hand, the real geometry of a given problem is approximated by finite elements, and on the other hand the fields of the primary variables – displacements, stresses, etc. – are approximated. In the last years, several approaches were developed in order to unify the approximation of geometry and variables. These methods are based on an integration of the finite element analysis software into CAD systems. Successful implementations can be found in Düster et al. (2001) and Hughes et al. (2005) for higher order interpolation methods and in Cirak et al. (2000) for low order techniques. Besides these new integrated general schemes, there is still a need for a general understanding of the underlying theoretical aspects of finite element discretization methods for solids undergoing finite strains. Hence this chapter is focussed on the essential details needed to perform a standard nonlinear analysis using finite elements. Besides the fact that all relevant equations are presented in this chapter, it is assumed that the reader has an understanding of finite element methods for linear problems since basic operations, such as e.g. assembly processes are not discussed. For this, the reader has to consult standard text books for finite elements which are then mentioned in the text.

We approximate the geometry of a body B in the initial configuration by

$$B \approx B^h = \bigcup_{e=1}^{n_e} \Omega_e \,. \tag{4.1}$$

With this the continuous body is subdivided into n_e finite elements. The configuration of one element is described by $\Omega_e \subset B^h$, see Fig. 4.1 for the two-dimensional case. The boundary of the region ∂B^h consist of curves or areas $\partial \Omega_e$ of the elements Ω_e: $\partial B^h = \cup_{e=1}^{n_r} \partial \Omega_e$. This of course is generally an approximation of the real geometry of the boundary ∂B.

An overlapping of finite elements is not allowed; hence the boundaries between finite elements are points, curves or areas. Furthermore, gaps are not allowed in a continuum due to compatibility reasons. Hence the assembled elements have to be continuous in the region B.

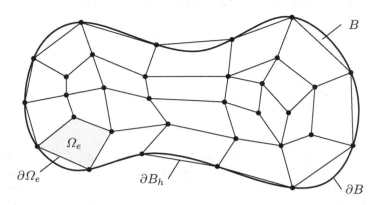

Fig. 4.1 Discretization of body B

4.1 General Isoparametric Concept

Within the finite element methodology, interpolation functions have to be chosen in order to approximate the primary field variables. Hence the exact solution of the mathematical model is approximated within one finite element by

$$\mathbf{u}_{exakt}\left(\mathbf{X}\right) \approx \mathbf{u}_{h}\left(\mathbf{X}\right) = \sum_{I=1}^{n} N_{I}\left(\mathbf{X}\right)\mathbf{u}_{I},\tag{4.2}$$

where the vector $\mathbf{X}$ denotes the position vector with respect to the initial configuration in Ω_{e}, $N_{I}\left(\mathbf{X}\right)$ are the shape functions which are defined in Ω_{e} and the unknown nodal quantities of the primary variable are represented by $\mathbf{u}_{I}$ (these could be, e.g. the nodal displacements $\mathbf{u}_{I} = \{u_{1},\, u_{2},\, u_{3}\}_{I}^{T}$ for a three-dimensional displacement formulation using the weak form (3.292)).

One basic requirement for the choice of the approximation $\mathbf{u}_{h}$ is the convergence of the finite element solution to the true solution of the underlying partial differential equation. Different possibilities exist to construct interpolation or ansatz functions for geometry and variables within the finite element method. For convergence reasons, these functions have to be completed up to the approximation order (e.g. a polynomial including the terms $1\,,x\,,y\,,z\,,x^{2}\,,y^{2}\,,z^{2}\,,xy\,,yz\,,zx$ is complete up to second order), see e.g. Zienkiewicz and Taylor (2000a).

Due to its general applicability, the isoparametric concept is mainly used as interpolation scheme for many engineering problems. Within this concept, geometry and variables are interpolated by the same ansatz functions. Isoparametric elements allow a very good and sufficiently accurate mapping of arbitrary geometries into a finite element mesh. Furthermore, this concept is extremely well suited for nonlinear problems since a discretization of the spatial formulation is easily obtained. This is due to the fact that it makes no difference whether the mapping onto a reference element $\Omega_\square$, needed in the isoparametric formulation is performed from the initial or the spatial configuration. One further advantage stems from the local orthogonal coordinate system at the reference element $\Omega_\square$, which means that neither co- nor contra-variant derivatives have to be computed, even if the bodies under consideration, e.g. shells or beams, depict curved geometries.

As said before, all kinematical variables as well as the geometry (e.g. the geometry in the initial $\mathbf{X}$ or spatial configuration $\mathbf{x}$) are interpolated by the same ansatz functions N_I within the classical isoparametric concept. This can be expressed mathematically within one finite element Ω_e by, see also Fig. 4.2,

$$\mathbf{X}_e = \sum_{I=1}^{n} N_I(\boldsymbol{\xi})\,\mathbf{X}_I , \tag{4.3}$$

$$\mathbf{x}_e = \sum_{I=1}^{n} N_I(\boldsymbol{\xi})\,\mathbf{x}_I . \tag{4.4}$$

Usually a polynomial is chosen for the ansatz function N_I. It is defined in the reference configuration $\Omega_\square$ in order to characterize arbitrary element geometries in the initial or spatial configuration. The ansatz functions within a finite element in the initial configuration Ω_e have been replaced in Eq. (4.3) by the shape functions $N_I(\boldsymbol{\xi})$ defined within the reference element $\Omega_\square$. Thus, for each element Ω_e, a transformation (4.3) has to be performed, which relates

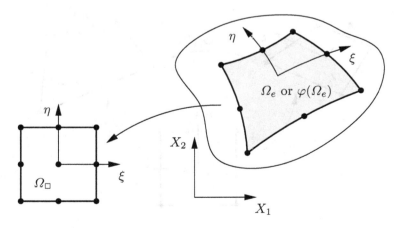

Fig. 4.2 Isoparametric mapping onto the reference configuration

the coordinates $\mathbf{X}_e = \mathbf{X}_e\,(\boldsymbol{\xi})$ to the coordinates $\boldsymbol{\xi}$ of the reference element $\Omega_\square$. This transformation has to fulfil the following requirements:

- For each point within the reference element $\Omega_\square$, there exists one and only one point Ω_e or $\varphi(\Omega_e)$.
- The geometrical nodal points $\mathbf{X}_I$ or $\mathbf{x}_I$ of $\Omega_\square$ are related to points in Ω_e or $\varphi(\Omega_e)$.
- Each part of the boundary on $\Omega_\square$, which is defined by the nodal points of $\mathbf{X}_I$ or $\mathbf{x}_I$ corresponds to the associated part of the boundary of Ω_e or $\varphi(\Omega_e)$.

With these assumptions, such isoparametric transformations preserve the type of element (e.g. a triangle remains a triangle in the initial or deformed finite element configurations). This isoparametric transformation will be used for all coordinate directions in identical manner. The volume or area $\Omega_\square$ of the reference element will basically never be occupied by the real configuration of an element undergoing a physical deformation process. However, the reference configuration $\Omega_\square$ provides an easy way to handle different configuration and can be used for the integration of the element matrices, etc. This especially simplifies formulations related to the current configuration since it is absolutely arbitrary whether the transformation is performed from the reference element to the current or the initial configuration.

   This transformation process is depicted in Fig. 4.3. The mapping of an element from the initial configuration Ω_e to the current configuration $\varphi(\Omega_e)$ is performed using the approximate deformation map $\boldsymbol{\varphi}_h$ which is described by $\boldsymbol{\varphi}_e$ to show its relation to a specific finite element Ω_e. For the mapping, the deformations gradient is needed which is here denoted by $\boldsymbol{F}_e$; hence it is only related to the element Ω_e.

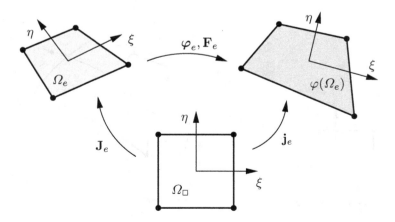

Fig. 4.3 Isoparametric mapping of the deformation of a finite element Ω_e

It is easy to see that the mapping in Fig. 4.3 is the discrete version of the continuum mechanical description of the motion of a body, which is shown in Fig. 3.1. Additionally, the reference configuration $\Omega_\square$ is introduced here in order to be able to describe the isoparametric mapping. From Fig. 4.3, it is easy to deduce the following kinematical relations valid for a finite element

$$\boldsymbol{F}_e = \boldsymbol{j}_e \, \boldsymbol{J}_e^{-1} \quad \text{and} \quad J_e = \det \boldsymbol{F}_e = \frac{\det \boldsymbol{j}_e}{\det \boldsymbol{J}_e} \,. \tag{4.5}$$

These show that the deformation gradient is defined by the isoparametric mapping from $\Omega_\square$ to the initial configuration Ω_e and to the current configuration $\varphi(\Omega_e)$. In this mapping, the gradients $\boldsymbol{j}_e$ and $\boldsymbol{J}_e$ are defined as

$$\boldsymbol{j}_e \;=\; \mathrm{Grad}_\xi \, \mathbf{x}_e = \frac{\partial \mathbf{x}}{\partial \boldsymbol{\xi}} = \sum_{I=1}^{n} N_{I,\xi}(\boldsymbol{\xi}) \, \mathbf{x}_I \otimes \mathbf{E}_\xi \,,$$

$$\boldsymbol{J}_e \;=\; \mathrm{Grad}_\xi \, \mathbf{X}_e = \frac{\partial \mathbf{X}}{\partial \boldsymbol{\xi}} = \sum_{I=1}^{n} N_{I,\xi}(\boldsymbol{\xi}) \, \mathbf{X}_I \otimes \mathbf{E}_\xi \,. \tag{4.6}$$

Since the derivatives $N_{I,\xi}$ are scalar quantities, they can be moved in front of the base vectors $\mathbf{E}_\xi$. Thus the gradients

$$\boldsymbol{j}_e \;=\; \sum_{I=1}^{n} \mathbf{x}_I \otimes N_{I,\xi}(\boldsymbol{\xi}) \, \mathbf{E}_\xi = \sum_{I=1}^{n} \mathbf{x}_I \otimes \nabla_\xi N_I \,,$$

$$\boldsymbol{J}_e \;=\; \sum_{I=1}^{n} \mathbf{X}_I \otimes N_{I,\xi}(\boldsymbol{\xi}) \, \mathbf{E}_\xi = \sum_{I=1}^{n} \mathbf{X}_I \otimes \nabla_\xi N_I \tag{4.7}$$

are obtained explicitly. In these equations, $\nabla_\xi N_I$ is the gradient of the scalar function N_I with respect to the coordinates $\boldsymbol{\xi}$.

With these relations, it is relatively simple to compute gradients related to the initial or current configuration. Exemplarily the gradients of the displacement vector field $\mathbf{u}_h = \mathbf{x}_h - \mathbf{X}$ can be specified within the element Ω_e by

$$\mathrm{Grad}\, \mathbf{u}_e \;=\; \sum_{I=1}^{n} \mathbf{u}_I \otimes \nabla_X N_I \,,$$

$$\mathrm{grad}\, \mathbf{u}_e \;=\; \sum_{I=1}^{n} \mathbf{u}_I \otimes \nabla_x N_I \,. \tag{4.8}$$

In an analogous way, the transformations between the gradients of different configurations, see (3.32), are derived

$$\nabla_\xi N_I = \boldsymbol{J}_e^T \, \nabla_X N_I \quad \text{and} \quad \nabla_\xi N_I = \boldsymbol{j}_e^T \, \nabla_x N_I \,. \tag{4.9}$$

The same holds for the inverse relations

$$\nabla_X N_I = \boldsymbol{J}_e^{-T} \nabla_\xi N_I \quad \text{and} \quad \nabla_x N_I = \boldsymbol{j}_e^{-T} \nabla_\xi N_I; \qquad (4.10)$$

hence the gradients in (4.8) can be written completely in terms of the reference configuration $\Omega_\square$

$$\text{Grad } \mathbf{u}_e = \sum_{I=1}^{n} \boldsymbol{u}_I \otimes \boldsymbol{J}_e^{-T} \nabla_\xi N_I,$$

$$\text{grad } \mathbf{u}_e = \sum_{I=1}^{n} \boldsymbol{u}_I \otimes \boldsymbol{j}_e^{-T} \nabla_\xi N_I. \qquad (4.11)$$

The only difference in the formulation of both gradients in (4.11) consists of the exchange of the gradient $\boldsymbol{j}_e$ by $\boldsymbol{J}_e$, and vice versa. Hence, especially for nonlinear finite element methods, this formulation provides the most flexible approach.

Remark 4.1: The computation of the derivatives related to the coordinates of the initial configuration described above, see Eq. (4.10), is different from the classical approach used in many books describing the finite element method, see e.g. Zienkiewicz and Taylor (2000a) and Bathe (1996). There, the relations

$$\frac{\partial N_I}{\partial X} = \frac{\partial N_I}{\partial \xi} \frac{\partial \xi}{\partial X} + \frac{\partial N_I}{\partial \eta} \frac{\partial \eta}{\partial X}$$

$$\frac{\partial N_I}{\partial Y} = \frac{\partial N_I}{\partial \xi} \frac{\partial \xi}{\partial Y} + \frac{\partial N_I}{\partial \eta} \frac{\partial \eta}{\partial Y}$$

are provided for the two-dimensional case which yield the form

$$\nabla_X N_I = \left\{ \begin{array}{c} \frac{\partial N_I}{\partial X} \\ \frac{\partial N_I}{\partial Y} \end{array} \right\} = \left[\begin{array}{cc} \frac{\partial \xi}{\partial X} & \frac{\partial \eta}{\partial X} \\ \frac{\partial \xi}{\partial Y} & \frac{\partial \eta}{\partial Y} \end{array} \right] \left\{ \begin{array}{c} \frac{\partial N_I}{\partial \xi} \\ \frac{\partial N_I}{\partial \eta} \end{array} \right\} = \bar{\boldsymbol{J}}_e^{-1} \nabla_\xi N_I. \qquad (4.12)$$

Here the matrix $\bar{\boldsymbol{J}}_e$ is the transposed of the matrix $\boldsymbol{J}_e^T$ in (4.10), see also the computation of derivatives for the two-dimensional case in Sect. 4.1.2.

For continuum elements, shell or beam elements for shear elastic formulations, an essential requirement for the ansatz or interpolation functions $N_I(\boldsymbol{\xi})$ is the C^0 continuity. Furthermore, ansatz functions $N_I(\boldsymbol{\xi})$ have to be used which are complete polynomials in the coordinate space X_1, X_2 and X_3 in which the mechanical problem is formulated. Different possibilities exist to construct such interpolation functions.

Ansatz functions which are also well suited for application within the isoparametric concept are provided by the LAGRANGIAN interpolation functions, see e.g. Zienkiewicz and Taylor (2000a). In case of one dimension,

$$N_I(\xi) = \prod_{\substack{J=1 \\ J \neq I}}^{n} \frac{(\xi_J - \xi)}{(\xi_J - \xi_I)} \qquad (4.13)$$

is obtained for a LAGRANGIAN polynomial of order $n - 1$. The index I describes the node at which the polynomial has to assume the value "1" while the ansatz function has the value "0" at all other nodes J of the element Ω_e.

For two- or three-dimensional interpolation, the product form

$$N_J(\xi, \eta) = N_I(\xi)\, N_K(\eta) \quad \text{or} \quad N_J(\xi, \eta, \zeta) = N_I(\xi)\, N_K(\eta)\, N_L(\zeta) \quad (4.14)$$

is used with $J = 1, \ldots, n^{dim}$ and $I, K, L = 1, \ldots, n$ (dim denotes the spatial dimension of the problem). The ansatz functions are defined in the local coordinate system $\boldsymbol{\xi} = \{\, \xi, \eta, \zeta \,\}$. Thus a transformation to the physical coordinates X_1, X_2 or X_3 is necessary, see Figs. 4.2 or 4.3, which defines the problems in the physical space. In the next sections, the isoparametric ansatz functions are specified for one-, two- and three-dimensional applications.

Remark 4.2: For classical beam and shell theories, due to the fact that the mathematical models are of higher order, different interpolation functions are needed which are, e.g. C^1-continuous. The associated formulations and specifications of the interpolation functions will be provided in the sections where the beam and shell theories are described.

4.1.1 One-Dimensional Interpolations

Ansatz functions. Here one-dimensional ansatz or shape functions are discussed which are C^0-continuous. These shape functions are derived from (4.13) by inserting the proper local coordinates into the formula. Since only one coordinate ξ is present, the general ansatz for the coordinates and displacement field in one element is with (4.3) and (4.4) given by

$$X_e = \sum_{I=1}^{n} N_I(\xi)\, X_I\,, \qquad u_e = \sum_{I=1}^{n} N_I(\xi)\, u_I\,, \qquad (4.15)$$

where X_e denotes the coordinate within the element and u_e is the associated displacement field. The value n is related to the number of interpolation functions and $n-1$ determines the order of interpolation, e.g. $n = 3$ is related to a quadratic element. The local coordinate $\xi \in [-1, 1]$ is used within the one-dimensional reference element, see Fig. 4.4.

The shape functions $N_I(\xi)$ follow from (4.13) and are different with respect to the order of the polynomial approximation. The shape functions are stated in the following up to quadratic order. Here the upper index denotes the order of the polynomial, see Fig. 4.4.

– Constant shape function
$$N_1^0(\xi) = 1. \qquad (4.16)$$

– Linear shape functions

$$N_1^1(\xi) = \frac{1}{2}\,(1 - \xi), \qquad N_2^1(\xi) = \frac{1}{2}\,(1 + \xi). \qquad (4.17)$$

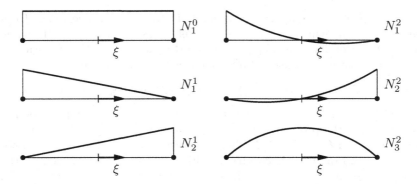

Fig. 4.4 One-dimensional shape functions: constant, linear and quadratic

– Quadratic shape functions

$$N_1^2(\xi) = \frac{1}{2}\xi(\xi - 1), \qquad N_3^2(\xi) = (1 - \xi^2), \qquad N_2^2(\xi) = \frac{1}{2}\xi(1 + \xi).$$
$$(4.18)$$

It can be easily shown that these shape functions fulfil the conditions discussed in the previous section. The isoparametric mapping of a function u_e onto the reference element is obtained by using Eq. $(4.15)_1$.

Computation of Derivatives. To obtain strains, and associated variations or linearizations of the strains, the derivatives of the displacement field have to be computed with respect to the coordinates of the initial or current configuration. Within the isoparametric concept, these derivatives are given within an element Ω_e for a formulation with respect to the initial configurations by

$$\frac{\partial u_e}{\partial X} = \sum_{I=1}^{n} \frac{\partial N_I(\xi)}{\partial X} u_I. \qquad (4.19)$$

Due to the fact that the shape functions depend upon the local coordinates, the chain rule is used to obtain the partial derivatives of N_I with respect to X. This leads for the displacement field u_e to

$$\frac{\partial u_e}{\partial X} = \frac{\partial u_e}{\partial \xi} \frac{\partial \xi}{\partial X} = \left(\sum_{I=1}^{n} \frac{\partial N_I(\xi)}{\partial \xi} u_I \right) \frac{\partial \xi}{\partial X}. \qquad (4.20)$$

The derivative $\frac{\partial \xi}{\partial X}$ can be computed by using the interpolation functions for the coordinates in $(4.15)_1$

$$\frac{\partial \xi}{\partial X} = \left(\frac{\partial X}{\partial \xi} \right)^{-1} = \left(\sum_{I=1}^{n} \frac{\partial N_I(\xi)}{\partial \xi} X_I \right)^{-1} = J_e(\xi)^{-1}. \qquad (4.21)$$

Here the abbreviation J_e was introduced for the derivative $\frac{\partial X_e}{\partial \xi}$.

For the special case of linear shape functions, see (4.17), the following result holds

$$\sum_{I=1}^{n} \frac{\partial N_I(\xi)}{\partial \xi} X_I = \frac{1}{2}(X_2 - X_1) = \frac{1}{2} L_e, \qquad (4.22)$$

with the definition of the element length $L_e = X_2 - X_1$. This leads after insertion in (4.20) to the simple relation

$$\frac{\partial u_e}{\partial X} = \frac{u_2 - u_1}{L_e}, \qquad (4.23)$$

which is constant.

Exercise 4.1: The derivative J_e has to be computed for the quadratic shape functions (4.18). Thus results have to be discussed for the general choice of the position of the middle node $X_3 = (1 - \eta) X_1 + \eta X_2$; here especially the choices $\eta = 1/2, 1/4$ and $3/4$ are of interest.

Solution: With (4.21), the derivatives of the shape functions N_I with respect to the coordinate ξ follow

$$N_{1,\xi} = \xi - \frac{1}{2}, \qquad N_{2,\xi} = \xi + \frac{1}{2}, \qquad N_{3,\xi} = -2\xi.$$

This leads with $L_e = X_2 - X_1$ to the result

$$J_e = \frac{\partial X_e}{\partial \xi} = \left(\xi - \frac{1}{2}\right) X_1 + \left(\xi + \frac{1}{2}\right) X_2 - 2\xi X_3 = \frac{1}{2} L_e + \xi (X_1 + X_2 - 2 X_3).$$

An especially simple form of J_e follows when node 3 is exactly in the middle between nodes 1 and 2 ($\eta = \frac{1}{2}$). Then $X_3 = 1/2 (X_1 + X_2)$ and the second term in J_e disappears, and $J_e = 1/2 L_e$ is constant.

For $\eta = 1/4$ and $\eta = 3/4$, the result

$$J_e = \frac{1}{2}(1 \pm \xi) L_e$$

follows which leads to $J_e = 0$ for $\xi = \pm 1$. Hence the derivative in (4.21) will become singular. This is only in very special cases desired where also the solution of the mechanical problem is singular (e.g. in some fracture mechanics problems). To avoid such singularity, the location of the middle node has to be limited to $\frac{1}{4} < \eta < \frac{3}{4}$. Additionally, the derivative J_e becomes negative for $\eta < \frac{1}{4}$ and $\eta > \frac{3}{4}$ which is equivalent with the fact that isoparametric mapping does not fulfil the condition $J = \det \mathbf{F} > 0$ anymore, since $\det \mathbf{F}_e = \det \mathbf{j}_e / \det \mathbf{J}_e$. In such cases, the requirements stemming from continuum mechanics are not fulfilled; hence these limit the placement of the middle node within the element.

Integration Within the Reference Element. Within the finite element method, the weak form contributions have to be integrated within each finite element Ω_e. Using the isoparametric concept, these integrations are performed within the reference element $\Omega_\square$. Hence the integrals related to the element Ω_e are transformed

$$\int\limits_{(\Omega_e)} g(X)\, dX = \int\limits_{(\Omega_\square)} g(\xi)\, \frac{dX}{d\xi}\, d\xi = \int\limits_{-1}^{+1} g(\xi)\, J_e(\xi)\, d\xi. \qquad (4.24)$$

The integration in the reference element $\Omega_\square$ is defined with respect to the parameter space $[-1 \leq \xi \leq +1]$.

In general, the integration is performed numerically since the product $g(\xi)\, J_e(\xi)$ is usually not polynomial but a rational function. Thus the integral in (4.24) is approximated by

$$\int\limits_{-1}^{+1} g(\xi)\, J_e(\xi)\, \delta\xi \approx \sum_{p=1}^{n_p} g(\xi_p)\, J_e(\xi_p)\, W_p. \qquad (4.25)$$

W_p are the weighting factors and ξ_p are the coordinates of the evaluation points p.

Due to its accuracy and thus efficiency, the GAUSS integration is applied. Table 4.1 provides the weighting factors and the positions of the evaluations points up to order $n_p = 3$. Note that a polynomial of order $p = 2\,n_p - 1$ is integrated exactly by a GAUSS integration with n_p points.

4.1.2 Two-Dimensional Interpolations

Shape Functions. Two-dimensional C^0-continuous shape functions are provided for triangles and quadrilaterals with linear and quadratic order of interpolation. Some results concerning the approximation properties of such interpolations can be found in Chap. 8.

Triangular Elements. The most simple two-dimensional element is a triangle with three nodes. Its linear shape function can be constructed directly or by using the isoparametric formulations. In the latter case, the same shape functions are used for geometry and field variables. Figure 4.5 depicts the

Table 4.1 One-dimensional GAUSS-Integration

n_p	p	ξ_p	W_p	
1	1	0	2	
2	1	$1/\sqrt{3}$	1	
	2	$1/\sqrt{3}$	1	
3	1	$-\sqrt{3/5}$	$5/9$	
	2	0	$8/9$	
	3	$+\sqrt{3/5}$	$5/9$	

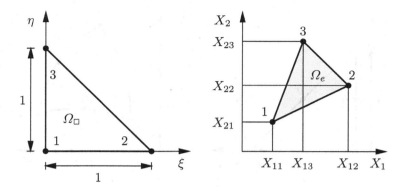

Fig. 4.5 3-Node triangular element

triangular element in its reference configuration $\Omega_\square$, described by the ξ–η-coordinates, and its initial configuration Ω_e using the X_1–X_2 coordinate system.

The shape functions are given by

$$N_1 = 1 - \xi - \eta, \qquad N_2 = \xi, \qquad N_3 = \eta, \tag{4.26}$$

which fulfil the condition of being "1" at their node I and "0" at node J, see (k43). It is easy to verify that the partial derivatives of the shape functions with respect to ξ and η are constant. Hence kinematical quantities, such as the strains, are also constant within the element.

Remark 4.3: A pure displacement triangular element derived on the basis of the linear shape functions has two degrees of freedom per node. Hence the element has in total $2 \times 3 = 6$ unknowns. Of these, three are needed to describe the rigid body motions (two translations and one rotation) and three needed to model the constant strain states.

Besides that the element is very simple and also very robust in nonlinear applications, it does not perform very well since its approximation properties are not too good. Due to that the element is very "stiff", meaning its approximation will yield smaller displacements than an analytical solution. This is especially negative when a structure undergoes bending deformations or when incompressible material has to be considered. In such cases, higher order elements or special elements have to be applied, see also Chap. 10.

A better approximation is provided by the 6-node triangular element, see Fig. 4.6, which is based on quadratic shape functions

$$\begin{aligned}
N_1 &= \lambda\,(\,2\,\lambda - 1\,), & N_4 &= 4\,\xi\,\lambda, \\
N_2 &= \xi\,(\,2\,\xi - 1\,), & N_5 &= 4\,\xi\,\eta, \\
N_3 &= \eta\,(\,2\,\eta - 1\,), & N_6 &= 4\,\eta\,\lambda.
\end{aligned} \tag{4.27}$$

where the abbreviation $\lambda = 1 - \xi - \eta$ has been introduced.

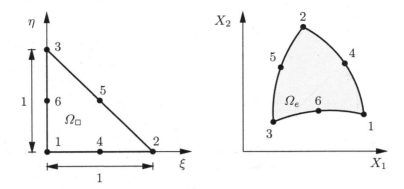

Fig. 4.6 6-node triangular element

Quadrilateral Elements. The simplest quadrilateral element consists of four nodes. The associated interpolation functions for geometry and field variables are bilinear and follow from the product form (4.14) using the one-dimensional shape functions (4.17)

$$N_I(\xi, \eta) = \frac{1}{2}(1 + \xi_I \xi) \frac{1}{2}(1 + \eta_I \eta). \qquad (4.28)$$

where ξ_I and η_I are the corner coordinates defined in Fig. 4.7 for the reference element $\Omega_\square$.

$$\boldsymbol{\xi}_1 = (-1, -1), \quad \boldsymbol{\xi}_2 = (1, -1), \quad \boldsymbol{\xi}_3 = (1, 1), \quad \boldsymbol{\xi}_4 = (-1, 1). \qquad (4.29)$$

In the same way, the shape functions for the 9-node element in Fig. 4.7 follow from the product form (4.14) using quadratic interpolation functions (4.18). The shape function with respect to the reference element $\Omega_\square$ are given for

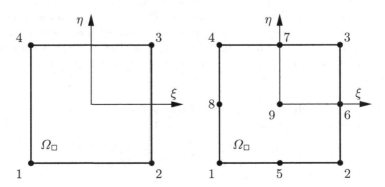

Fig. 4.7 Isoparametric quadrilateral element

– the vertex nodes ($I = 1, 2, 3, 4$):

$$N_I(\xi, \eta) = \frac{1}{4}(\xi^2 + \xi_I\,\xi)(\eta_i^2 + \eta_I\,\eta), \tag{4.30}$$

– the middle edge nodes ($I = 5, 6, 7, 8$):

$$N_I(\xi, \eta) = \frac{1}{2}\xi_I^2(\xi^2 + \xi_I\,\xi)(1 - \eta^2) + \frac{1}{2}\eta_I^2(\eta^2 + \eta_I\,\eta)(1 - \xi^2), \tag{4.31}$$

– and the middle node ($I = 9$):

$$N_9(\xi, \eta) = (1 - \xi^2)(1 - \eta^2). \tag{4.32}$$

This is, however, not the only possibility to formulate the shape functions for the 9-node element. Often a hierarchical formulation – starting from the 4-node element – is applied, see e.g. Zienkiewicz and Taylor (2000a).

Computation of the Derivatives. In order to compute the deformation gradient, strains or the associated variations and linearizations in the weak form of equilibrium (3.292), the derivatives of the deformation or the displacement field are needed. Within the isoparametric concept, these follow from (4.10), in more detail, for a displacement field in Ω_e

$$\frac{\partial \mathbf{u}_e}{\partial X_\alpha} = \sum_{I=1}^{n} \frac{\partial N_I(\xi, \eta)}{\partial X_\alpha} \mathbf{u}_I \qquad (\alpha = 1, 2) \tag{4.33}$$

can be written. The partial derivative of N_I with respect to X_α was given in (4.10). For the two-dimensional case, an explicit expression can be derived

$$\nabla_X N_I = \begin{Bmatrix} N_{I,1} \\ N_{I,2} \end{Bmatrix} = \boldsymbol{J}_e^{-T} \begin{Bmatrix} N_{I,\xi} \\ N_{I,\eta} \end{Bmatrix} \tag{4.34}$$

with the JACOBI matrix $\boldsymbol{J}_e$. The latter describes the transformation between the reference and the initial configuration of the element Ω_e

$$\boldsymbol{J}_e = \sum_{I=1}^{n} \boldsymbol{X}_I \otimes \nabla_\xi N_I = \sum_{I=1}^{n} \begin{Bmatrix} X_{1I} \\ X_{2I} \end{Bmatrix} \begin{Bmatrix} N_{I,\xi} \\ N_{I,\eta} \end{Bmatrix}^T = \begin{bmatrix} X_{1,\xi} & X_{1,\eta} \\ X_{2,\xi} & X_{2,\eta} \end{bmatrix},$$

and $\qquad X_{\alpha,\beta} = \sum_{I=1}^{n} N_{I,\beta}\, X_{\alpha I}.$ $\tag{4.35}$

This explicit form can be used within (4.33) to compute the derivatives with respect to the initial configuration $\mathbf{X}$

$$\begin{Bmatrix} N_{I,1} \\ N_{I,2} \end{Bmatrix} = \frac{1}{\det \boldsymbol{J}_e} \begin{bmatrix} X_{2,\eta} & -X_{2,\xi} \\ -X_{1,\eta} & X_{1,\xi} \end{bmatrix} \begin{Bmatrix} N_{I,\xi} \\ N_{I,\eta} \end{Bmatrix}. \tag{4.36}$$

Exercise 4.2: Determine, for a 3-node triangular element, the deformation gradient $\boldsymbol{F}_e$ within the element Ω_e.

Solution: The deformation gradient can be computed directly from the different isoparametric mappings between the reference configuration and the initial and current configuration using (4.5) and (4.7). The evaluation of these equations is performed for the shape functions (4.26) of the triangular element. With the derivatives

$$N_{1,\xi} = -1, \quad N_{2,\xi} = 1, \quad N_{3,\xi} = 0,$$
$$N_{1,\eta} = -1, \quad N_{2,\eta} = 0, \quad N_{3,\eta} = 1,$$

the JACOBI matrix is obtained, using (4.35) and based on the notation in Fig. 4.5, as

$$\boldsymbol{J}_e = \left\{\begin{matrix} X_{11} \\ X_{21} \end{matrix}\right\}\left\{\begin{matrix} -1 \\ -1 \end{matrix}\right\}^T + \left\{\begin{matrix} X_{12} \\ X_{22} \end{matrix}\right\}\left\{\begin{matrix} 1 \\ 0 \end{matrix}\right\}^T + \left\{\begin{matrix} X_{13} \\ X_{23} \end{matrix}\right\}\left\{\begin{matrix} 0 \\ 1 \end{matrix}\right\}^T$$

$$= \begin{bmatrix} X_{12} - X_{11} & X_{13} - X_{11} \\ X_{22} - X_{21} & X_{23} - X_{21} \end{bmatrix}.$$

The determinant of the transformation matrix $\det \boldsymbol{J}_e$ is then given by

$$\det \boldsymbol{J}_e = (X_{12} - X_{11})(X_{23} - X_{21}) - (X_{13} - X_{11})(X_{22} - X_{21}) = 2\,A_e\,,$$

where A_e is the area of the element. In the same way, the matrix $\boldsymbol{j}_e$ can be computed. Only the coordinates of the initial configuration have to be exchanged by the coordinates of the current configuration

$$\boldsymbol{j}_e = \begin{bmatrix} x_{12} - x_{11} & x_{13} - x_{11} \\ x_{22} - x_{21} & x_{23} - x_{21} \end{bmatrix}.$$

Now the deformation gradient of the 3-node triangular element can be specified explicitly by $\boldsymbol{F}_e = \boldsymbol{j}_e\,\boldsymbol{J}_e^{-1}$

$$\boldsymbol{F}_e = \frac{1}{2\,A_e} \begin{bmatrix} x_{12} - x_{11} & x_{13} - x_{11} \\ x_{22} - x_{21} & x_{23} - x_{21} \end{bmatrix} \begin{bmatrix} X_{23} - X_{21} & X_{11} - X_{13} \\ X_{21} - X_{22} & X_{12} - X_{11} \end{bmatrix}.$$

Since the 3-node triangular element is based on linear shape functions, the deformation gradient is constant within the element Ω_e.

We also can specify the deformation gradient using the displacement variables **u**. With the relation $x_{\alpha I} = X_{\alpha I} + u_{\alpha I}$,

$$\boldsymbol{F}_e = \boldsymbol{1} + \frac{1}{2\,A_e} \begin{bmatrix} u_{12} - u_{11} & u_{13} - u_{11} \\ u_{22} - u_{21} & u_{23} - u_{21} \end{bmatrix} \begin{bmatrix} X_{23} - X_{21} & X_{11} - X_{13} \\ X_{21} - X_{22} & X_{12} - X_{11} \end{bmatrix}$$

is obtained. This result is consistent with the continuum form $\boldsymbol{F} = \boldsymbol{1} + \operatorname{Grad}\boldsymbol{u}$, see (3.14).

Integration Within the Reference Space. For the computation of the weak form (3.292), an integration is necessary over the area of each finite element Ω_e which, in the isoparametric formulation, is performed in the parameter space of the reference element $\Omega_\square$, see Fig. 4.7. Thus the integral has to be transformed from the initial or current configuration to the reference configuration

$$\int\limits_{(\Omega_e)} g(\mathbf{X})\,dA = \int\limits_{(\Omega_\square)} g(\boldsymbol{\xi})\,\det \boldsymbol{J}_e(\boldsymbol{\xi})\,d\square = \int\limits_{-1}^{+1}\int\limits_{-1}^{+1} g(\xi,\eta)\,\det \boldsymbol{J}_e\,d\xi\,d\eta\,. \quad (4.37)$$

The integration over the reference area $\Omega_\Box$ is performed numerically since the product $g(\boldsymbol{\xi}) \det \boldsymbol{J}_e(\xi)$ in general does not represent a polynomial but a rational function. This yields for the integral in (4.37) the approximation

$$\int\limits_{-1}^{+1} \int\limits_{-1}^{+1} g(\xi,\eta)\, \det \boldsymbol{J}_e \, d\xi\, d\eta \approx \sum_{p=1}^{n_p} g(\xi_p,\eta_p)\, \det \boldsymbol{J}_e(\xi_p,\eta_p)\, W_p\,. \qquad (4.38)$$

Normally GAUSS integration is applied due to its efficiency. The weighting factors W_p and the coordinates of the GAUSS or evaluation points ξ_p and η_p are summarized in Table 4.2. Table 4.2 contains GAUSS points and weighting factors up to $n_p = 3 \times 3$. These integration formulae integrate polynomials in $\xi^i\,\eta^k$ exact up to the order $i + k \leq m$. Let us remark that the integration formulae can be derived in the same way as the two-dimensional shape functions by a product form, using one-dimensional GAUSS integration in every coordinate direction. The GAUSS integration has been chosen within the finite element method due to its accuracy and efficiency. Other integration

Table 4.2 Two-dimensional GAUSS integration for quadrilateral elements

m	n_p	p	ξ_p	η_p	W_p	Position of points
1	1	1	0	0	4	
3	4	1	$-1/\sqrt{3}$	$-1/\sqrt{3}$	1	
		2	$+1/\sqrt{3}$	$-1/\sqrt{3}$	1	
		3	$-1/\sqrt{3}$	$+1/\sqrt{3}$	1	
		4	$+1/\sqrt{3}$	$+1/\sqrt{3}$	1	
5	9	1	$-\sqrt{3/5}$	$-\sqrt{3/5}$	$25/81$	
		2	0	$-\sqrt{3/5}$	$40/81$	
		3	$+\sqrt{3/5}$	$-\sqrt{3/5}$	$25/81$	
		4	$-\sqrt{3/5}$	0	$40/81$	
		5	0	0	$64/81$	
		6	$+\sqrt{3/5}$	0	$40/81$	
		7	$-\sqrt{3/5}$	$+\sqrt{3/5}$	$25/81$	
		8	0	$+\sqrt{3/5}$	$40/81$	
		9	$+\sqrt{3/5}$	$+\sqrt{3/5}$	$25/81$	

rules which can be applied for numerical integration of (3.160) can be found in, e.g. Dhatt and Touzot (1985).

The transformation of the integrals from the initial configuration to the reference configuration is different for triangular elements. In general,

$$\int_{(\Omega_e)} g(\mathbf{X})\, dA = \int_0^1 \int_0^{1-\xi} g(\xi,\eta)\, \det \mathbf{J}_e\, d\eta\, d\xi \qquad (4.39)$$

is obtained where the last integral can again be approximated by the numerical integration (4.38). Table 4.3 contains the related evaluation points and weighting functions. These formulae integrate polynomials in the reference space $\xi^k \eta^l$ up to order m (with $m \geq k + l$) exact.

Many more rules for the numerical integration of triangular elements exists which have either different evaluation points and weighting factors or a higher approximation order. Such rules can be found in, e.g Zienkiewicz and Taylor (1989) and Dhatt and Touzot (1985).

Table 4.3 Two-dimensional integration for triangular elements

m	n_p	p	ξ_p	η_p	W_p	position of points
1	1	1	$1/3$	$1/3$	$1/2$	
2	3	1 2 3	$1/2$ 0 $1/2$	$1/2$ $1/2$ 0	$1/6$ $1/6$ $1/6$	
2	3	1 2 3	$1/6$ $2/3$ $1/6$	$1/6$ $1/6$ $2/3$	$1/6$ $1/6$ $1/6$	
3	4	1 2 3 4	$1/3$ $1/5$ $3/5$ $1/5$	$1/3$ $1/5$ $1/5$ $3/5$	$-27/96$ $25/96$ $25/96$ $25/96$	

4.1.3 Three-Dimensional Interpolation

Three-dimensional finite elements can have a variety of different shapes. These can be tetrahedral or hexahedral shapes but also mixtures of both types for special geometries like prisms. In this section, the formulations are restricted to the tetrahedral or hexahedral shaped elements. Shape functions for other element types can be found, e.g. in Dhatt and Touzot (1985). Again the isoparametric formulation is used to be able to generate general shape functions for the discretization of arbitrary three-dimensional geometries.

The shape functions for the three-dimensional hexahedral element are given by

$$N_I(\xi,\eta,\zeta) = \frac{1}{2}\,(1+\xi_I\,\xi)\,\frac{1}{2}\,(1+\eta_I\,\eta)\,\frac{1}{2}\,(1+\zeta_I\,\zeta)\,, \qquad (4.40)$$

which follow from the product form (4.14) together with (4.17). Figure 4.8 describes the associated 8-node hexahedral element in its reference configuration $\Omega_\square$ and its initial configuration Ω_e. The related quadratical shape functions can be derived from (4.14) with (4.18). This yields an interpolation with 27 nodes per element where the functions N_I result from the 27 possible combinations of the local coordinates ξ,η,ζ with their values $-1,0,+1$ at the nodes within the reference element $\Omega_\square$

$$N_I(\xi,\eta,\zeta) = N_I(\xi)\,N_I(\eta)\,N_I(\zeta)\,. \qquad (4.41)$$

The shape functions for tetrahedral elements can be defined analogous to the two-dimensional formulation. This leads to the ansatz functions

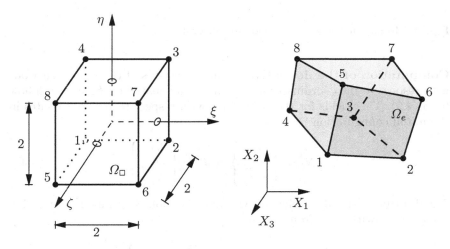

Fig. 4.8 Isoparametric 8-node hexahedral element

– 4-node tetrahedral element (linear shape functions)

$$N_1 = 1 - \xi - \eta - \zeta, \quad N_2 = \xi, \quad N_3 = \eta, \quad N_4 = \zeta. \tag{4.42}$$

– 10-node tetrahedral element (quadratic shape functions)

$$
\begin{aligned}
N_1 &= \lambda\,(2\,\lambda - 1), & N_6 &= 4\,\xi\,\eta, \\
N_2 &= \xi\,(2\,\xi - 1), & N_7 &= 4\,\eta\,\lambda, \\
N_3 &= \eta\,(2\,\eta - 1), & N_8 &= 4\,\zeta\,\lambda, \\
N_4 &= \zeta\,(2\,\zeta - 1), & N_9 &= 4\,\xi\,\zeta, \\
N_5 &= 4\,\xi\,\lambda, & N_{10} &= 4\,\eta\,\zeta,
\end{aligned}
\tag{4.43}
$$

with $\lambda = 1 - \xi - \eta - \zeta$.

The associated node numbers of both elements are depicted in Fig. 4.9.

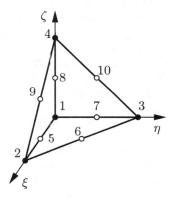

Fig. 4.9 Isoparametric 4 and 10 node tetrahedral

Computation of the derivatives. The derivatives of the shape functions with respect to the coordinates in the initial and spatial configuration follow from (4.9). This yields for the derivatives with respect to the coordinates in the initial configuration

$$\nabla_X N_I = \begin{Bmatrix} N_{I,1} \\ N_{I,2} \\ N_{I,3} \end{Bmatrix} = \boldsymbol{J}_e^{-T} \begin{Bmatrix} N_{I,\xi} \\ N_{I,\eta} \\ N_{I,\zeta} \end{Bmatrix}. \tag{4.44}$$

The JACOBI matrix $\boldsymbol{J}_e$ of element Ω_e which is applied within the derivation is computed with (4.7) from

$$\boldsymbol{J}_e = \sum_{I=1}^{n} \boldsymbol{X}_I \otimes \nabla_\xi N_I = \begin{bmatrix} X_{1,\xi} & X_{1,\eta} & X_{1,\zeta} \\ X_{2,\xi} & X_{2,\eta} & X_{2,\zeta} \\ X_{3,\xi} & X_{3,\eta} & X_{3,\zeta} \end{bmatrix}. \tag{4.45}$$

The components of $\boldsymbol{J}_e$ are given by

$$X_{m,k} = \sum_{I=1}^{n} N_{I,k} \, X_{m\,I} \, ,$$

where the partial derivative denoted by k is the corresponding derivative with respect to ξ, η or ζ, e.g.

$$X_{m,2} = \sum_{I=1}^{n} \frac{\partial N_I}{\partial \eta} \, X_{m\,I} \, .$$

Integration Within the Parameter Space. As in the two-dimensional case, integration of the shape functions and their derivatives over the volume of element Ω_e is carried out in the parameter space of the reference element $\Omega_\square$. This yields

$$\int_{(\Omega_e)} g(\mathbf{X}) \, dA = \int_{(\Omega_\square)} g(\boldsymbol{\xi}) \, \det \boldsymbol{J}_e(\boldsymbol{\xi}) \, d\square = \int_{-1}^{+1} \int_{-1}^{+1} \int_{-1}^{+1} g(\xi, \eta, \zeta) \, \det \boldsymbol{J}_e \, d\xi \, d\eta \, d\zeta \, .$$

$$(4.46)$$

The integration over $\Omega_\square$ is performed numerically. Hence the last integral in (4.46) is approximated by

$$\int_{-1}^{+1} \int_{-1}^{+1} \int_{-1}^{+1} g(\xi, \eta\, \zeta) \, \det \boldsymbol{J}_e \, d\xi \, d\eta \, d\zeta \approx \sum_{p=1}^{n_p} g(\xi_p, \eta_p, \zeta_p) \, \det \boldsymbol{J}_e(\xi_p, \eta_p, \zeta_p) \, W_p.$$

$$(4.47)$$

Again GAUSS point integration is applied. Due to the fact that the coordinates of the GAUSS point $\boldsymbol{\xi}_p$ can easily be derived from the two-dimensional case, see Table 4.2, by expanding these coordinates into the third coordinate direction, they are not explicitly stated here. In case of the tri-linear brick element a $2 \times 2 \times 2$ integration with totally 8 GAUSS points is necessary. Accordingly, for the ansatz with quadratic shape functions, a $3 \times 3 \times 3$ integration with totally 27 GAUSS points has to be used. Since this is rather time consuming, integration rules are stated in Table 4.4 which only need 6 points for an ansatz with tri-linear shape functions and 14 points for an ansatz with quadratic shape functions, see Irons (1971). With this special integration rules, the computational effort is reduced by 25% for linear and by almost 50% for quadratic ansatz functions.

The integration rules for tetrahedral elements are provided in Table 4.5 for linear and quadratic ansatz functions. The coordinates of the sampling points ξ_p, η_p and ζ_p are related to the coordinate system used in Fig. 4.9. Further integration rules for tetrahedral and hexahedral elements can be found in, e.g. Zienkiewicz and Taylor (1989) or Dhatt and Touzot (1985).

Table 4.4 Special integration for hexahedral elements

m	n_p	p	ξ_p	η_p	ζ_p	W_p
2	4	1	0	$\sqrt{2/3}$	$-1/\sqrt{3}$	2
		2	0	$-\sqrt{2/3}$	$-1/\sqrt{3}$	2
		3	$\sqrt{2/3}$	0	$1/\sqrt{3}$	2
		4	$-\sqrt{2/3}$	0	$1/\sqrt{3}$	2
3	6	1	1	0	0	$4/3$
		2	-1	0	0	$4/3$
		3	0	1	0	$4/3$
		4	0	-1	0	$4/3$
		5	0	0	1	$4/3$
		6	0	0	-1	$4/3$
5	14	1	$\sqrt{19/30}$	0	0	$320/361$
		2	$-\sqrt{19/30}$	0	0	$320/361$
		3	0	$\sqrt{19/30}$	0	$320/361$
		4	0	$-\sqrt{19/30}$	0	$320/361$
		5	0	0	$\sqrt{19/30}$	$320/361$
		6	0	0	$-\sqrt{19/30}$	$320/361$
		7	$\sqrt{19/33}$	$\sqrt{19/33}$	$\sqrt{19/33}$	$121/361$
		8	$-\sqrt{19/33}$	$\sqrt{19/33}$	$\sqrt{19/33}$	$121/361$
		9	$\sqrt{19/33}$	$-\sqrt{19/33}$	$\sqrt{19/33}$	$121/361$
		10	$-\sqrt{19/33}$	$-\sqrt{19/33}$	$\sqrt{19/33}$	$121/361$
		11	$\sqrt{19/33}$	$\sqrt{19/33}$	$-\sqrt{19/33}$	$121/361$
		12	$\sqrt{19/33}$	$-\sqrt{19/33}$	$-\sqrt{19/33}$	$121/361$
		13	$-\sqrt{19/33}$	$-\sqrt{19/33}$	$-\sqrt{19/33}$	$121/361$
		14	$-\sqrt{19/33}$	$\sqrt{19/33}$	$-\sqrt{19/33}$	$121/361$

Table 4.5 Three-dimensional integration for tetrahedral elements

m	n_p	p	ξ_p	η_p	ζ_p	W_p
1	1	1	$1/4$	$1/4$	$1/4$	$1/6$
3	5	1	$1/4$	$1/4$	$1/4$	$-2/15$
		2	$1/6$	$1/6$	$1/6$	$3/40$
		3	$1/6$	$1/6$	$1/2$	$3/40$
		4	$1/6$	$1/2$	$1/6$	$3/40$
		5	$1/2$	$1/6$	$1/6$	$3/40$

4.2 Discretization of the Weak Forms

The one-, two- and three-dimensional ansatz functions can now be used to describe the geometry and to discretize the field variables within the kinematical relations and weak forms stemming from continuum mechanics. Here this discretization process will be shown for the variational Eqs. (3.292) and (3.296) derived in Chap. 3.

The region of interest, which is the volume of the solid to be discretized, will be subdivided into n_e finite elements as shown in Fig. 4.1. Within this process, the geometry is approximated using (4.1). An ansatz, as given in (4.3) and (4.4), is then selected for the displacement field $\mathbf{u}$, the coordinates $\mathbf{X}$ and the test function $\boldsymbol{\eta}$ to approximate these quantities within each finite element Ω_e. With these approximations, the integral describing the weak forms is given by

$$\int_B (\dots) \, dV \approx \int_{B_h} (\dots) \, dV_h = \bigcup_{e=1}^{n_e} \int_{\Omega_e} (\dots) \, d\Omega = \bigcup_{e=1}^{n_e} \int_{\Omega_\square} (\dots) \, d\square \qquad (4.48)$$

The operator $\cup$ is chosen here instead of the $\sum$ symbol in order to denote that an assembly process takes place in which all element contributions have not only to be added up but also the kinematical compatibility between the elements has to be fulfilled. The whole process then leads to an algebraic system of nonlinear equations for a given problem, as will be shown later. Hence the assembly process denoted by $\cup$ stands for fulfilment of the inter-element compatibility of the test functions and the displacement field and for the fulfilment of the global displacement boundary conditions. Since the assembly process is the same as the one used within the finite element method for linear problems, it is here not described in detail; detailed descriptions of this process can be found in, e.g. Bathe (1996) and Zienkiewicz and Taylor (2000a).

4.2.1 FE-Formulation of the Weak Form in Initial Configuration

The finite element approximation of the weak form (3.292) is based on the discretization of the stress divergence term (virtual work of the internal forces) $\int_B \mathbf{S} \cdot \delta \mathbf{E} \, dV$, the inertia term $\int_B \rho_0 \dot{\mathbf{v}} \cdot \boldsymbol{\eta} \, dV$ and the prescribed volume forces and surface tractions $\int_B \rho_0 \bar{\mathbf{b}} \cdot \boldsymbol{\eta} \, dV + \int_{\partial B} \bar{\mathbf{t}} \cdot \boldsymbol{\eta} \, dA$.

To formulate the stress divergence, the variation of the GREEN-LAGRANGE strain tensor (4.48) has to be discretized within an element Ω_e. With Eqs. (3.290) and (4.8) relation,

$$\delta \mathbf{E}_e = \frac{1}{2} \sum_{I=1}^{n} \left[\boldsymbol{F}_e^T \left(\boldsymbol{\eta}_I \otimes \nabla_X N_I \right) + \left(\nabla_X N_I \otimes \boldsymbol{\eta}_I \right) \boldsymbol{F}_e \right] \qquad (4.49)$$

is obtained. The finite element approximations of the deformation gradient (3.6) which was applied in (4.49) follows from (4.8) for element Ω_e

$$\mathbf{F}_e = \sum_{K=1}^{n} (\mathbf{x}_K \otimes \nabla_X N_K) . \tag{4.50}$$

To derive the matrix formulation which is convenient for formulating the finite element form, it is advantageous to go back to index notation which yields for (4.49)

$$\delta E_{AB} = \frac{1}{2} \sum_{I=1}^{n} [F_{Ak} N_{I,B} + N_{I,A} F_{kB}] \eta_{k\,I} \tag{4.51}$$

with the components of the deformation gradient: $F_{kB} = \sum_{J=1}^{n} x_{k\,J} N_{J,B}$.

Within the matrix formulation, also called VOIGT notation, the symmetry of the GREEN-LAGRANGE strain tensor and its variation is considered. This leads, in the three-dimensional case, to a vector with six independent components

$$\delta \mathbf{E}_e = \left\{ \begin{array}{c} \delta E_{11} \\ \delta E_{22} \\ \delta E_{33} \\ 2\,\delta E_{12} \\ 2\,\delta E_{23} \\ 2\,\delta E_{13} \end{array} \right\} = \sum_{I=1}^{n} \mathbf{B}_{L\,I}\,\boldsymbol{\eta}_I , \tag{4.52}$$

which is computed by a sum over the nodes of the element using the matrix

$$\mathbf{B}_{L\,I} = \begin{bmatrix} F_{11}\,N_{I,1} & F_{21}\,N_{I,1} & F_{31}\,N_{I,1} \\ F_{12}\,N_{I,2} & F_{22}\,N_{I,2} & F_{32}\,N_{I,2} \\ F_{13}\,N_{I,3} & F_{23}\,N_{I,3} & F_{33}\,N_{I,3} \\ F_{11}\,N_{I,2} + F_{12}\,N_{I,1} & F_{21}\,N_{I,2} + F_{22}\,N_{I,1} & F_{31}\,N_{I,2} + F_{32}\,N_{I,1} \\ F_{12}\,N_{I,3} + F_{13}\,N_{I,2} & F_{22}\,N_{I,3} + F_{23}\,N_{I,2} & F_{32}\,N_{I,3} + F_{33}\,N_{I,2} \\ F_{11}\,N_{I,3} + F_{13}\,N_{I,1} & F_{21}\,N_{I,3} + F_{23}\,N_{I,1} & F_{31}\,N_{I,3} + F_{33}\,N_{I,1} \end{bmatrix} . \tag{4.53}$$

The index L in (4.52) denotes that the matrix $\mathbf{B}_{L\,I}$ depends linearly from the displacements since the deformation gradient can be written as $\mathbf{F} = \mathbf{1} + \text{Grad}\,\mathbf{u}$.

The stresses in the weak form follow from the constitutive equations which have to be employed to model the material behaviour of a given problem. These will be specified in the following chapters in detail. Due to the symmetry of the 2nd PIOLA-KIRCHHOFF stress tensor, see (3.83), only six independent components are used in the vector form of the stresses $\mathbf{S}_e = \{ S_{11}, S_{22}, S_{33}, S_{12}, S_{23}, S_{13} \}^T$. Note that the factor 2 in front of the last three components in (k447d) stem from the fact that the inner product of two symmetric tensors ($S_{IK}\,\delta E_{IK}$ need to be expressed in VOIGT notation by $\delta \mathbf{E}^T \mathbf{S}$. Thus the virtual work of the internal forces can be approximated with finite elements by

$$\int_B \delta \mathbf{E} \cdot \mathbf{S} \, dV \;=\; \bigcup_{e=1}^{n_e} \int_{\Omega_e} \delta \boldsymbol{E}_e^T \, \boldsymbol{S}_e \, d\Omega$$

$$= \; \bigcup_{e=1}^{n_e} \sum_{I=1}^{n} \boldsymbol{\eta}_I^T \int_{\Omega_e} \boldsymbol{B}_{LI}^T \, \boldsymbol{S}_e \, d\Omega \qquad (4.54)$$

$$= \; \bigcup_{e=1}^{n_e} \sum_{I=1}^{n} \boldsymbol{\eta}_I^T \int_{\Omega_\square} \boldsymbol{B}_{LI}^T \, \boldsymbol{S}_e \, \det \boldsymbol{J}_e \, d\square \,.$$

The last equation in (4.54) already contains the evaluation of the integrals with respect to the isoparametric reference element. To shorten the notation, we define the vector

$$\boldsymbol{R}_I \left(\boldsymbol{u}_e \right) = \int_{\Omega_e} \boldsymbol{B}_{LI}^T \, \boldsymbol{S}_e \, d\Omega, \qquad (4.55)$$

which enables us to write the virtual internal work as

$$\int_B \delta \mathbf{E} \cdot \mathbf{S} \, dV = \bigcup_{e=1}^{n_e} \sum_{I=1}^{n} \boldsymbol{\eta}_I^T \, \boldsymbol{R}_I \left(\boldsymbol{u}_e \right) = \boldsymbol{\eta}^T \, \boldsymbol{R} \left(\boldsymbol{u} \right). \qquad (4.56)$$

Within this formulations, $\boldsymbol{\eta}$ is the virtual displacement or test function and $\boldsymbol{R} \left(\boldsymbol{u} \right)$ denotes the force. Their inner product describe the virtual internal work after assembly.

Elements which are purely based on an ansatz for the displacements are generally called T or Q elements, depending upon the selection of shape functions for a triangle or a quadrilateral in two-dimensions. Hence a quadrilateral element with linear shape functions is labeled as Q1 while, e.g. a triangular element with quadratic shape function will be abbreviated by T2. In three dimensions often the abbreviation Q1 is also used for a linear hexahedral element while H1 would also be appropriate. For tetrahedral elements, T is used as well as in the two-dimensional case.

The inertia term is computed by using an ansatz for the velocity in the form $\mathbf{v}_e(\mathbf{X}, t) = \sum_K \mathbf{N}_K(\boldsymbol{\xi}) \, \mathbf{v}_K(t)$ within an element Ω_e. This approach can be viewed as a product form, in which spatial and temporal approximations are split. Thus the accelerations needed in the inertia term are approximated by

$$\dot{\mathbf{v}}_e(\mathbf{X}, t) = \sum_{K=1}^{n} N_K(\boldsymbol{\xi}) \, \dot{\mathbf{v}}_K \,. \qquad (4.57)$$

This yields

$$\int_B \boldsymbol{\eta} \cdot \rho_0 \, \dot{\mathbf{v}} \, dV \;=\; \bigcup_{e=1}^{n_e} \int_{\Omega_e} \boldsymbol{\eta}^T \rho_0 \, \dot{\mathbf{v}}_e \, dV$$

$$= \bigcup_{e=1}^{n_e} \sum_{I=1}^{n} \sum_{K=1}^{n} \boldsymbol{\eta}_I^T \int_{\Omega_e} N_I\, \rho_0\, N_K\, d\Omega\ \dot{\mathbf{v}}_K.$$

With the application of the unit matrix $\boldsymbol{I}$ to the acceleration $\dot{\mathbf{v}}_K = \boldsymbol{I}\,\dot{\mathbf{v}}_K$, the mass matrix

$$\boldsymbol{M}_{IK} = \int_{\Omega_e} \rho_0\, N_I\, N_K\, d\Omega\ \boldsymbol{I} \qquad (4.58)$$

can be introduced and the inertia term takes the form

$$\int_B \rho_0\, \boldsymbol{\eta} \cdot \dot{\mathbf{v}}\, dV = \bigcup_{e=1}^{n_e} \sum_{I=1}^{n} \sum_{K=1}^{n} \boldsymbol{\eta}_I^T\, \boldsymbol{M}_{IK}\, \dot{\mathbf{v}}_K = \boldsymbol{\eta}^T\, \boldsymbol{M}\, \dot{\mathbf{v}}. \qquad (4.59)$$

Here $\boldsymbol{M}$ is the mass matrix and $\dot{\mathbf{v}}$ is the acceleration vector after assembly to the global system. The spatial integration of (4.58) yields a mass matrix which belongs to the finite element Ω_e. It has a similar structure as the first part of the tangent matrix related to the linearization of the weak form, see 4.2.2. This mass matrix is known as consistent mass matrix.

Remark 4.4: In many cases, the consistent mass matrix given in (4.58) is not applied in numerical simulations but a diagonal mass matrix is used instead due to efficiency reasons. There are different ways to determine a diagonal mass matrix which is also called (*lumped mass matrix*). However, all formulations have to obey the condition of mass conservation, see e.g. Hughes (1987) and Bathe (1982). One possibility for the derivation of a diagonal mass matrix is to apply a special quadrature formulae for the integration of (4.58). Since the shape functions N_I (4.14) have the property that they assume at the element node I the value 1 and 0 at all other nodes, a diagonal mass matrix is automatically obtained by using quadrature points which are located at the nodes. This process yields

$$\boldsymbol{M}_{IK} \quad = \quad \int_{\Omega_e} \rho_0\, N_I\, N_K\, d\Omega\ \boldsymbol{I} = \int_{\Omega_\square} \rho_0\, N_I\, N_K\, \det \boldsymbol{J}_e\, d\square\ \boldsymbol{I}$$

$$= \quad \sum_{p=1}^{n_p} \rho_0(\boldsymbol{\xi}_p)\, N_I(\boldsymbol{\xi}_p)\, N_K(\boldsymbol{\xi}_p)\, \det \boldsymbol{J}_e(\boldsymbol{\xi}_p)\, W_p\ \boldsymbol{I},$$

$$\boldsymbol{M}_{IK}^{diag} \quad = \quad \rho_0(\boldsymbol{\xi}_p)\, \det \boldsymbol{J}_e(\boldsymbol{\xi}_p)\, W_p\ \boldsymbol{I} \quad (\text{for } I = K). \qquad (4.60)$$

For $I \neq K$, the numerical integration will be zero, since then at least one shape function is equal to zero. Quadrature formulae which have their evaluation points at element nodes are, e.g. the trapezoidal or SIMPSON rules, see Hughes (1987). When selecting the quadrature rule, mass conservation has to be considered and the right order for the shape functions has to be used. Due to that, the trapezoidal rule can be applied for linear ansatz functions and the SIMPSON rule is useful for quadratic ansatz functions. When performing this type of integrations, it can happen that eventually negative masses are associated with nodes as, e.g. is the case when using an 8-node serendipity element. This is, however, not advisable from the physical point of view. To circumvent such problems, a different diagonalization process has to be constructed.

Another possibility to derive diagonal mass matrices can be found in Hinton et al. (1976). It always leads to positive masses at the nodes. The idea here is to start from a consistent mass matrix and to scale the diagonal terms in such a way that the mass is constant within the element. This procedure yields

$$\boldsymbol{M}_{II}^{diag} = \vartheta_e \, M_{II} \, \boldsymbol{I} \quad \text{with} \quad M_{II} = \int\limits_{\Omega_e} \rho_0 \, N_I^2 \, d\Omega. \tag{4.61}$$

With the scaling factor

$$\vartheta_e = \frac{M_e}{\sum_{I=1}^n M_{II}}, \qquad M_e = \int\limits_{\Omega_e} \rho_0 \, d\Omega.$$

M_e is the total mass of element Ω_e.

The load terms in the weak form are discretized in an analogous way. After inserting the finite element approximation for the virtual displacement or test function $\boldsymbol{\eta}$, it follows

$$\int\limits_{B} \rho_0 \, \boldsymbol{\eta} \cdot \bar{\boldsymbol{b}} \, dV + \int\limits_{\Gamma_\sigma} \boldsymbol{\eta} \cdot \bar{\boldsymbol{t}} \, dA \;=\; \bigcup_{e=1}^{n_e} \sum_{I=1}^{n} \boldsymbol{\eta}_I^T \int\limits_{\Omega_e} \rho_0 \, \bar{\boldsymbol{b}} \, N_I \, d\Omega$$

$$+ \bigcup_{r=1}^{n_r} \sum_{I=1}^{m} \boldsymbol{\eta}_I^T \int\limits_{\Gamma_r} N_I \, \bar{\boldsymbol{t}} \, d\Gamma \,,$$

where n_r denotes the number of element boundaries where loads are applied and Γ_r is the surface of an element subjected to a traction vector $\bar{\boldsymbol{t}}$ which describes the surface loading, see Fig. 4.10a. Note that for the term describing the surface loads shape function have to be defined for $d-1$ dimensions when d is the dimension of the problem. In Fig. 4.10a, the body of a two-dimensional continuum is described which requires the use of a one-dimensional ansatz function within the integral for the surface loads, see Fig. 4.10b. These one-dimensional shape functions are only defined with respect to m nodes describing the surface (in Fig. 4.10b: $m = 2$).

Also here a compact notation will be introduced. With

$$\boldsymbol{P}_I = \int\limits_{\Omega_e} N_I \, \rho \, \bar{\boldsymbol{b}} \, d\Omega \quad \text{and} \quad \boldsymbol{P}_I^\sigma = \int\limits_{\Gamma_r} N_I \, \bar{\boldsymbol{t}} \, d\Gamma, \tag{4.62}$$

the loading terms in the weak form follow

$$\int\limits_{B} \rho \, \boldsymbol{\eta} \cdot \bar{\boldsymbol{b}} \, dV + \int\limits_{\Gamma_\sigma} \boldsymbol{\eta} \cdot \bar{\boldsymbol{t}} \, dA = \bigcup_{e=1}^{n_e} \sum_{I=1}^{n} \boldsymbol{\eta}_I^T \, \boldsymbol{P}_I + \bigcup_{r=1}^{n_r} \sum_{I=1}^{n} \boldsymbol{\eta}_I^T \, \boldsymbol{P}_I^\sigma = \boldsymbol{\eta}^T \, \boldsymbol{P}. \tag{4.63}$$

In this equation, the vector $\boldsymbol{P}$ stands for all applied loads acting on the structure.

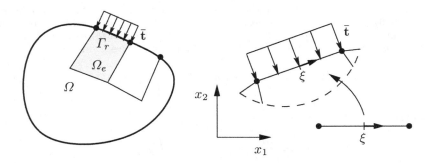

Fig. 4.10 (a) Surface load   (b) Discretization

With the matrix notation introduced in (4.56), (4.59) and (4.63), the weak form (3.292) can be written in its discrete version

$$\boldsymbol{\eta}^T \left[\boldsymbol{M}\dot{\boldsymbol{v}} + \boldsymbol{R}(\boldsymbol{u}) - \boldsymbol{P} \right] = 0 \, . \tag{4.64}$$

Due to the arbitrariness of the virtual displacement $\boldsymbol{\eta}$, the above weak form leads to a nonlinear system of ordinary differential equations

$$\boldsymbol{M}\dot{\boldsymbol{v}} + \boldsymbol{R}(\boldsymbol{u}) - \boldsymbol{P} = \boldsymbol{0} \quad \forall \boldsymbol{u} \in \mathbb{R}^N \, . \tag{4.65}$$

All terms in (4.65) are related to the initial configuration. N denotes the total number of degrees of freedom (dofs) of the system which are combined in the vector of unknowns $\boldsymbol{u}$. $\dot{\boldsymbol{v}}$ is the acceleration vector and $\boldsymbol{M}$ denotes the mass matrix. Often also a damping term of the form $\boldsymbol{C}\boldsymbol{v}$ is introduced in linear and nonlinear simulations to model different physical effects such as material damping or friction in joints or supports. Since these effects result from very different sources, it is very difficult to derive the damping terms in a consistent way using continuum mechanics, see also Sect. 6.1.

In case of vanishing inertia effects and damping ($\boldsymbol{M}\dot{\boldsymbol{v}} = \boldsymbol{C}\boldsymbol{v} = \boldsymbol{0}$), a nonlinear algebraic system of equations

$$\boldsymbol{R}(\boldsymbol{u}) - \boldsymbol{P} = \boldsymbol{0} \quad \forall \boldsymbol{u} \in \mathbb{R}^N \tag{4.66}$$

is obtained instead of the nonlinear system of ordinary differential equations in (4.65). The solution of this algebraic system of equations will be discussed in detail in Chap. 5.

4.2.2 Linearization of the Weak Form in the Initial Configuration

For an efficient numerical treatment of the nonlinear algebraic system of equations, which will either be obtained when a temporal discretization is used in (4.65), see Sect. 6.1, or when (4.66) is used directly, NEWTON method is applied. Since this method needs the derivative of (4.66), the linearization of this algebraic equation system has to be computed. A detailed description of the solution methods is given in the following chapters; hence only the linearization of (4.66) is provided here, the inertia terms are neglected at the moment. As already stated in Remark 3.11, the linearization can be obtained by a discretization of the linearization of the continuum terms provided already in (3.343)

$$DG(\bar{\varphi}, \boldsymbol{\eta}) \cdot \Delta \mathbf{u} = \int_B \left\{ \operatorname{Grad} \Delta \mathbf{u} \, \bar{\mathbf{S}} \cdot \operatorname{Grad} \boldsymbol{\eta} + \delta \bar{\mathbf{E}} \cdot \bar{\mathbb{C}} \left[\Delta \bar{\mathbf{E}} \right] \right\} dV \,. \qquad (4.67)$$

By using

$$\operatorname{Grad} \Delta \mathbf{u}_e = \sum_{K=1}^{n} \Delta \mathbf{u}_K \otimes \nabla_X N_K \,,$$

$$\operatorname{Grad} \boldsymbol{\eta}_e = \sum_{I=1}^{n} \boldsymbol{\eta}_I \otimes \nabla_X N_I, \qquad (4.68)$$

the discretization of the first term is obtained directly

$$\int_B \operatorname{Grad} \Delta \mathbf{u} \, \bar{\mathbf{S}} \cdot \operatorname{Grad} \boldsymbol{\eta} \, dV = \bigcup_{e=1}^{n_e} \sum_{I=1}^{n} \sum_{K=1}^{n} \int_{\Omega_e} (\Delta \mathbf{u}_K \otimes \nabla_X N_K) \, \bar{\mathbf{S}} \cdot (\boldsymbol{\eta}_I \otimes \nabla_X N_I) \, d\Omega \,.$$

This term can be written as, by applying the rules for the dyadic and the scalar product and by considering the matrix form of the scalar product $\Delta \mathbf{u}_K \cdot \boldsymbol{\eta}_I = \boldsymbol{\eta}_I^T \Delta \mathbf{u}_K = \boldsymbol{\eta}_I^T \mathbf{I} \Delta \mathbf{u}_K$,

$$\int_B \operatorname{Grad} \Delta \mathbf{u} \, \bar{\mathbf{S}} \cdot \operatorname{Grad} \boldsymbol{\eta} \, dV = \bigcup_{e=1}^{n_e} \sum_{I=1}^{n} \sum_{K=1}^{n} \boldsymbol{\eta}_I^T \int_{\Omega_e} \bar{G}_{IK} \, \mathbf{I} d\Omega \, \Delta \mathbf{u}_K, \qquad (4.69)$$

where the abbreviation

$$\bar{G}_{IK} = (\nabla_X N_I)^T \, \bar{\mathbf{S}}_e \, \nabla_X N_K \qquad (4.70)$$

was introduced. The matrix form of the scalar product (4.70) can be specified as

$$\bar{G}_{IK} = [\, N_{I,1} \quad N_{I,2} \quad N_{I,3} \,] \begin{bmatrix} \bar{S}_{11} & \bar{S}_{12} & \bar{S}_{13} \\ \bar{S}_{21} & \bar{S}_{22} & \bar{S}_{23} \\ \bar{S}_{31} & \bar{S}_{32} & \bar{S}_{33} \end{bmatrix} \begin{Bmatrix} N_{K,1} \\ N_{K,2} \\ N_{K,3} \end{Bmatrix}, \qquad (4.71)$$

when the gradients are provided in vector form. Relation (4.69) is independent from the constitutive equation, only the stress – computed with respect to the configuration $\bar{\varphi}$ – appears. Hence the matrix defined by (4.69) is often called initial stress matrix.

The seconde term in (3.343)

$$\int_B \delta \bar{\mathbf{E}} \cdot \bar{\mathbb{C}} \left[\varDelta \bar{\mathbf{E}} \right] dV$$

depends upon the constitutive equation via the incremental material tensor $\bar{\mathbb{C}}$ which is computed with respect to the configuration $\bar{\varphi}$. For elastic materials, this tensor is given by, e.g. (3.268). The associated matrix formulation can be found in (3.271). Since $\varDelta \bar{\mathbf{E}}$ has the same structure as $\delta \bar{\mathbf{E}}$, see Remark 3.8, it follows with (4.49)

$$\varDelta \mathbf{E}_e = \frac{1}{2} \sum_{I=1}^{n} \left[\mathbf{F}_e^T \left(\varDelta \mathbf{u}_I \otimes \nabla_X N_I \right) + \left(\nabla_X N_I \otimes \varDelta \mathbf{u}_I \right) \mathbf{F}_e \right]. \qquad (4.72)$$

This yields with (4.53) the matrix form

$$\varDelta \mathbf{E}_e = \sum_{I=1}^{n} \mathbf{B}_{L\,I} \, \varDelta \mathbf{u}_I. \qquad (4.73)$$

Inserting this relation into the second term of (3.343) leads with the incremental constitutive tensor $\bar{\mathbf{D}}$ to

$$\int_B \delta \bar{\mathbf{E}} \cdot \bar{\mathbb{C}} \left[\varDelta \bar{\mathbf{E}} \right] dV = \bigcup_{e=1}^{n_e} \sum_{I=1}^{n} \sum_{K=1}^{n} \boldsymbol{\eta}_I^T \int_{\Omega_e} \bar{\mathbf{B}}_{L\,I}^T \, \bar{\mathbf{D}} \, \bar{\mathbf{B}}_{L\,K} \, d\Omega \, \varDelta \mathbf{u}_K. \qquad (4.74)$$

Adding all terms together, the discretization of the weak form is given by

$$\int_B \left\{ \operatorname{Grad} \varDelta \mathbf{u} \, \bar{\mathbf{S}} \cdot \operatorname{Grad} \boldsymbol{\eta} + \delta \bar{\mathbf{E}} \cdot \bar{\mathbb{C}} \left[\varDelta \bar{\mathbf{E}} \right] \right\} dV = \bigcup_{e=1}^{n_e} \sum_{I=1}^{n} \sum_{K=1}^{n} \boldsymbol{\eta}_I^T \, \bar{\mathbf{K}}_{T\,IK} \varDelta \mathbf{u}_K,$$
$$(4.75)$$

where the matrix $\bar{\mathbf{K}}_{T\,IK}$ is often called tangent matrix

$$\bar{\mathbf{K}}_{T\,IK} = \int_{\Omega_e} \left[(\nabla_X N_I)^T \, \bar{\mathbf{S}}_e \, \nabla_X N_K + \bar{\mathbf{B}}_{L\,I}^T \, \bar{\mathbf{D}} \, \bar{\mathbf{B}}_{L\,K} \right] d\Omega. \qquad (4.76)$$

This matrix is related to the nodal combination I, K within one finite element. When using this notation, the sub-matrix $\bar{\mathbf{K}}_{T\,IK}$ has the size $n_{dof} \times n_{dof}$, where n_{dof} is the number of freedoms of a node (in three-dimensional solid problems $n_{dof} = 3$ and thus $\bar{\mathbf{K}}_{T\,IK}$ has the size 3×3). Indices I and K are the nodes of the element, and hence directly related to the discretization. Thus, for a 10-node tetrahedral element, it follows that $n = 10$; hence the tangent matrix of one element $\bar{\mathbf{K}}_{T_e}$ has the size $(n \cdot n_{dof}) \times (n \cdot n_{dof}) = 30 \times 30$. The element matrices of quadratic hexahedrals are even bigger since these have 27 nodes leading to a tangent matrix $\bar{\mathbf{K}}_{T_e}$ of size 81×81.

Exercise 4.3: Develop the matrix formulation for a two-dimensional finite element with 4-nodes with respect to the initial configuration. Use the ST. VENANT constitutive equation under the assumption of plain strain. Derive as well the residuum as the tangent stiffness matrix.

Solution: For the computation of the residuum, the virtual work expression given in (4.55) has to be specified. For this, the response function for the second PIOLA-KIRCHHOFF stress S and the virtual GREEN-LAGRANGIAN strain δE is needed. Hence, to derive the weak form in terms of the displacements, the ST. VENANT constitutive equation (3.121) has to be presented in matrix form. Equation (3.121) can be rewritten with the fourth order unit tensor $\mathbb{E}$ as

$$\mathbf{S} = (\Lambda \mathbf{1} \otimes \mathbf{1} + 2\mu \mathbb{E})[\mathbf{E}],$$

which immediately leads to a matrix form. In case of the two-dimensional deformation, the constitutive relation for the 2nd PIOLA-KIRCHHOFF tensor is given by

$$\boldsymbol{S} = \boldsymbol{D}\,\boldsymbol{E} = \left\{ \begin{array}{c} S_{11} \\ S_{22} \\ S_{12} \end{array} \right\} = \left[\begin{array}{ccc} \Lambda + 2\mu & \Lambda & 0 \\ \Lambda & \Lambda + 2\mu & 0 \\ 0 & 0 & \mu \end{array} \right] \left\{ \begin{array}{c} E_{11} \\ E_{22} \\ 2\,E_{12} \end{array} \right\}, \tag{4.77}$$

in which the GREEN-LAGRANGIAN strain tensor $\boldsymbol{E}$ has to be specified.

The components of the GREEN-LAGRANGIAN strain tensor (3.15), needed in (4.77), follow with (4.50) in the two-dimensional case for a finite element Ω_e by

$$\boldsymbol{E}_e = \frac{1}{2}(\boldsymbol{F}_e^T \boldsymbol{F}_e - \boldsymbol{I}) \quad \text{with} \quad \boldsymbol{F}_e = \sum_{K=1}^{n} \left[\begin{array}{cc} x_{1K}\,N_{K,1} & x_{1K}\,N_{K,2} \\ x_{2K}\,N_{K,1} & x_{2K}\,N_{K,2} \end{array} \right], \tag{4.78}$$

where the nodal coordinates $x_{\alpha K} = X_{\alpha K} + u_{\alpha K}$ are related to the deformed configuration $\bar{\varphi}$.

The approximation of the virtual GREEN-LAGRANGIAN strains δE given in (4.52) have, by introducing the matrix $\boldsymbol{B}_{L\,I}$, the explicit form in the two-dimensional case

$$\boldsymbol{B}_{L\,I} = \left[\begin{array}{cc} F_{11}\,N_{I,1} & F_{21}\,N_{I,1} \\ F_{12}\,N_{I,2} & F_{22}\,N_{I,2} \\ F_{11}\,N_{I,2} + F_{12}\,N_{I,1} & F_{21}\,N_{I,2} + F_{22}\,N_{I,1} \end{array} \right]. \tag{4.79}$$

Note that it is also possible to use the form $\mathbf{F} = \mathbf{1} + \mathrm{Grad}\,\mathbf{u}$ when deriving the virtual strain expression. Then

$$\delta\,\boldsymbol{E}_e = \sum_{I=1}^{4} [\,\boldsymbol{B}_{0\,I} + \boldsymbol{B}_{V\,I}(\,\boldsymbol{u}\,)\,]\,\boldsymbol{\eta}_I \tag{4.80}$$

is obtained. Here the matrices $\boldsymbol{B}_0$ and $\boldsymbol{B}_L$ have the explicit form

$$\boldsymbol{B}_{0\,I} = \left[\begin{array}{cc} N_{I,1} & 0 \\ 0 & N_{I,2} \\ N_{I,2} & N_{I,1} \end{array} \right]; \quad \boldsymbol{B}_{V\,I} = \left[\begin{array}{cc} u_{1,1}\,N_{I,1} & u_{2,1}\,N_{I,1} \\ u_{1,2}\,N_{I,2} & u_{2,2}\,N_{I,2} \\ u_{1,1}\,N_{I,2} + u_{1,2}\,N_{I,1} & u_{2,2}\,N_{I,1} + u_{2,1}\,N_{I,2} \end{array} \right]. \tag{4.81}$$

The derivative $u_{\alpha,\beta}$ has to be computed at each integration point analogous to the components of the deformation gradient in (4.51). This yields $u_{\alpha,\beta} = \sum_{K=1}^{4} N_{K,\beta}\,u_{\alpha K}$, where the indices α and β assume values of 1 and 2. Note that the matrix $\boldsymbol{B}_{V\,I}$ which describes the nonlinear part in $\mathbf{E}$ disappears for $\mathbf{u} = const$.

Inserting Eq. (4.80) into the virtual work expression for the internal forces (4.55) yields the virtual internal work of one finite element Ω_e

$$R_I\left(u_e\right) = \int_{\Omega_e}\left(B_{0\,I} + B_{V\,I}\right)^T S_e\, d\Omega. \tag{4.82}$$

The load vector can be computed using (4.62). This, however, will not be specified here in detail.

The linearization of (4.82) at $\bar{\varphi}$ yields the tangential stiffness matrix for a finite element. It yields with (4.80) analogous to (4.76)

$$\bar{K}_{T_{IK}} = \int_{\Omega_e}\left[\left(B_{0\,I} + \bar{B}_{V\,I}\right)^T \bar{D}\left(B_{0\,K} + \bar{B}_{V\,K}\right) + \bar{G}_{IK}\,I\right] d\Omega. \tag{4.83}$$

Here all quantities with a bar on top have to be evaluated at $\bar{\varphi}$. A more compact notation of the tangent matrix follows with (4.79)

$$\bar{K}_{T_{IK}} = \int_{\Omega_e}\left[\bar{B}_{L\,I}^T \bar{D}\,\bar{B}_{L\,K} + \bar{G}_{IK}\,I\right] d\Omega. \tag{4.84}$$

For two-dimensional problems, the scalar $\bar{G}_{IK}$ can expressed by the product

$$\bar{G}_{IK} = \left[\, N_{I,1} \quad N_{I,2}\,\right]\begin{bmatrix}\bar{S}_{11} & \bar{S}_{12}\\ \bar{S}_{21} & \bar{S}_{22}\end{bmatrix}\begin{Bmatrix}N_{K,1}\\ N_{K,2}\end{Bmatrix}. \tag{4.85}$$

Both Eqs. (4.83) and (4.85) have to be evaluated at $\bar{\varphi}$. The stresses in (4.85) follow from the constitutive equation (4.77) and the discrete form of the strains (4.78).

The integrals in (4.82) and (4.83) have to be computed by numerical integration. For this, a transformation to the reference configuration, see (4.54) and Fig. 4.11, is advantageous. This yields with (4.55) for the residual a $n_{dof} \times 1 = 2 \times 1$ vector for the node I

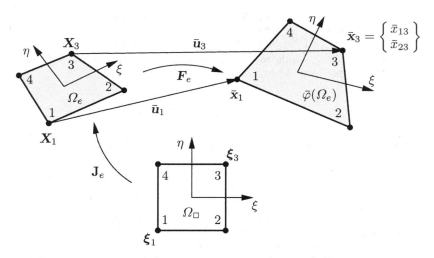

Fig. 4.11 Configurations of the 4-node element

$$\boldsymbol{R}_I\left(\boldsymbol{u}_e\right) = \int\limits_{\Omega_e} (\boldsymbol{B}_{0\,I} + \boldsymbol{B}_{V\,I})^T \boldsymbol{S}_e\, d\Omega \tag{4.86}$$

$$= \int\limits_{\Omega_\square} (\boldsymbol{B}_{0\,I} + \boldsymbol{B}_{V\,I})^T \boldsymbol{S}_e\, \det \boldsymbol{J}_e\, d\square \tag{4.87}$$

$$\approx \sum_{p=1}^{n_p} W_p\, [\,\boldsymbol{B}_{0\,I}(\xi_p\,,\eta_p) + \boldsymbol{B}_{V\,I}(\xi_p\,,\eta_p)\,]^T\, \boldsymbol{S}(\xi_p\,,\eta_p)\, \det \boldsymbol{J}_e(\xi_p\,,\eta_p)\,.$$

For this 4-node element $n_p = 2 = 4$, GAUSS points are sufficient for the numerical integration. The values for the GAUSS point coordinates ξ_p, η_p and the associated weights W_p can be found in Table 4.2. For the computation of the stresses at a GAUSS point $\boldsymbol{S}(\xi_p\,\eta_p)$, the deformation gradient has to be evaluated, see (4.78), in $(\xi_p\,,\eta_p)$. This leads to

$$\boldsymbol{F}_e(\xi_p\,,\eta_p) = \sum_{K=1}^{n} \begin{bmatrix} x_{1K}\, N_{K,1}(\xi_p\,,\eta_p) & x_{1K}\, N_{K,2}(\xi_p\,,\eta_p) \\ x_{2K}\, N_{K,1}(\xi_p\,,\eta_p) & x_{2K}\, N_{K,2}(\xi_p\,,\eta_p) \end{bmatrix}\,. \tag{4.88}$$

The stress at a GAUSS point follows then from (4.77) by using the strains (4.78). Note that in (4.88) a summation (index K) has to be performed overall since all shape functions contribute to the deformation at a GAUSS point within a finite element. In an analogous way, the numerical integration of the tangent matrix (4.83) has to be performed. The sub-matrices for nodes with indices I and K are 2×2 matrices. With Eq. (4.84),

$$\bar{\boldsymbol{K}}_{T_{IK}} = \int\limits_{\Omega_\square} \left[\, \bar{\boldsymbol{B}}_{L\,I}^T\, \bar{\boldsymbol{D}}\, \bar{\boldsymbol{B}}_{L\,K} + \bar{G}_{IK}\, \boldsymbol{I}\,\right]\, \det \boldsymbol{J}_e\, d\square \tag{4.89}$$

$$\approx \sum_{p=1}^{n_p} W_p\, \left[\, \bar{\boldsymbol{B}}_{L\,I}^T(\xi_p,\eta_p) \bar{\boldsymbol{D}}\, \bar{\boldsymbol{B}}_{L\,K}(\xi_p,\eta_p) + \bar{G}_{IK}(\xi_p\,,\eta_p)\, \boldsymbol{I}\,\right]\, \det \boldsymbol{J}_e(\xi_p,\eta_p)$$

is obtained. The vectors $\boldsymbol{R}_I$ and the sub-matrices $\bar{\boldsymbol{K}}_{T_{IK}}$ have to be assembled into a vector and a tangent matrix for the element Ω_e as follows

$$\boldsymbol{R}_e = \left\{ \begin{matrix} \boldsymbol{R}_1 \\ \boldsymbol{R}_2 \\ \boldsymbol{R}_3 \\ \boldsymbol{R}_4 \end{matrix} \right\}_{8\times 1} \qquad \bar{\boldsymbol{K}}_{T_e} = \begin{bmatrix} \bar{\boldsymbol{K}}_{T11} & \bar{\boldsymbol{K}}_{T12} & \bar{\boldsymbol{K}}_{T13} & \bar{\boldsymbol{K}}_{T14} \\ & \bar{\boldsymbol{K}}_{T22} & \bar{\boldsymbol{K}}_{T23} & \bar{\boldsymbol{K}}_{T24} \\ & & \bar{\boldsymbol{K}}_{T33} & \bar{\boldsymbol{K}}_{T34} \\ symm. & & & \bar{\boldsymbol{K}}_{T44} \end{bmatrix}_{8\times 8}\,. \tag{4.90}$$

The total size follows from the number of nodes, here 4, and degree of freedoms per node, here 2. Hence the element residual vector has the size 8×1 and the element tangent is a 8×8 matrix.

With the last equations, all necessary relations, vectors and matrices are known which are needed for a successful implementation of this element into a finite element code. Let us close with the following remarks

- The degree of nonlinearity of (4.82) depends upon the chosen constitutive equation. For the ST. VENANT material, used here, the weak form (4.82) is a cubic polynomial in **u**. This, however, does not hold when, e.g. a compressible Neo-HOOKE material (3.119) is applied.
- For a stress calculation within an actual design, the 2nd PIOLA-KIRCHHOFF stresses have to be transformed to real stresses, e.g. the CAUCHY stresses using (3.83).

– In the case that a linear theory for small strains and displacements is used, then the terms $\bar{G}_{IK}$ and $\bar{B}_{V\,I}.\bar{B}_{V\,K}$ in (4.82) and (4.83) disappear since the evaluation of the residual vector and the tangent matrix has to be evaluated at the state $\bar{\varphi} = 0$. Due to this, the resulting equations become linear in $\mathbf{u}$. Then the resulting vectors and matrices are exactly equivalent to the finite element formulation based on the linear theory of elasticity which yields with the two-dimensional form of (3.273) the linear stiffness matrix

$$\mathbf{K}_{IK} = \int_{\Omega_e} \mathbf{B}_{0\,I}^T \, \mathbf{D}_0 \, \mathbf{B}_{0\,K} \, d\Omega \, .$$

4.2.3 FE-Formulation of the Weak Form in the Current Configuration

The discretization of the weak form with respect to the current configuration (3.294) is derived analogous to the weak form in the initial configuration (3.292). The internal virtual work depends in the current configuration upon the variation of the strain measure $\nabla^S \boldsymbol{\eta}$, see (3.295), which is the *push forward* of the variation of the GREEN-LAGRANGIAN strain. The strain $\nabla^S \boldsymbol{\eta}$ has to be approximated using the finite element shape functions. With $(4.8)_2$, this leads to

$$\nabla^S \boldsymbol{\eta}_e = \frac{1}{2} \sum_{I=1}^{n} \left[(\boldsymbol{\eta}_I \otimes \nabla_x N_I) + (\nabla_x N_I \otimes \boldsymbol{\eta}_I) \right] . \tag{4.91}$$

As in the previous section, it is convenient to use index notation to derive the matrix formulation

$$(\nabla^S \boldsymbol{\eta}_e)_{im} = \frac{1}{2} \sum_{I=1}^{n} \left[\eta_{i\,I} \, N_{I,m} + N_{I,i} \, \eta_{m\,I} \right] . \tag{4.92}$$

Here $N_{I,m} = \partial N_I / \partial x_m$ is the partial derivative of the shape functions with respect to the spatial coordinates x_m. The derivatives can be computed with $(4.10)_2$

$$N_{I,k} = \{j_e^{-1}\}_{1k} \, N_{I,\xi} + \{j_e^{-1}\}_{2k} \, N_{I,\eta} + \{j_e^{-1}\}_{3k} \, N_{I,\zeta} \, , \tag{4.93}$$

where $\{j_e^{-1}\}_{ik}$ are the components of the inverse of the JACOBI-Matrix $\mathbf{j}_e$. Equation (4.92) yields the components of $\nabla^S \boldsymbol{\eta}$, which can be written in vector form

$$(\nabla^S \boldsymbol{\eta})^T = \left[\eta_{1,1} , \eta_{2,2} , \eta_{3,3} , (\eta_{1,2} + \eta_{2,1}) , (\eta_{2,3} + \eta_{3,2}) , (\eta_{3,3} + \eta_{3,1}) \right]$$

when considering symmetry. This vector form can now be approximated by finite elements

$$\nabla^S \boldsymbol{\eta}_e = \sum_{I=1}^{n} \begin{bmatrix} N_{I,1} & 0 & 0 \\ 0 & N_{I,2} & 0 \\ 0 & 0 & N_{I,3} \\ N_{I,2} & N_{I,1} & 0 \\ 0 & N_{I,3} & N_{I,2} \\ N_{I,3} & 0 & N_{I,1} \end{bmatrix} \begin{Bmatrix} \eta_1 \\ \eta_2 \\ \eta_3 \end{Bmatrix}_I = \sum_{I=1}^{n} \mathbf{B}_{0\,I} \, \boldsymbol{\eta}_I \, . \tag{4.94}$$

Note that matrix $\boldsymbol{B}_{0\,I}$ contains no terms which depend directly upon the displacement field; this is here depicted by the index "0".

Remark 4.5: Contrary to matrix $\boldsymbol{B}_{L\,I}$, matrix $\boldsymbol{B}_{0\,I}$ is a sparse matrix. Half of all entries are zero. Thus, when multiplying matrix $\boldsymbol{B}_{0\,\boldsymbol{I}}$ with other matrices or vectors, the zero elements can be neglected. This leads to a faster computation of element matrices. Hence the finite element formulation with respect to the current configuration is more efficient.

The structure of $\boldsymbol{B}_{0\,I}$ coincides exactly with the $\boldsymbol{B}$-matrix of the linear theory, see e.g. Hughes (1987). The only difference – however important – is that the derivatives of the shape functions have to be computed for the geometrically linear element with respect to the coordinates $\mathbf{X}$ of the initial configuration, whereas in the nonlinear case the derivatives have to be computed using (4.92) and (4.93).

With the spatial virtual strains derived above and the vector form of the CAUCHY stress tensor $\boldsymbol{\sigma} = \{\,\sigma_{11}\,,\sigma_{22}\,,\sigma_{33}\,,\sigma_{12}\,,\sigma_{23}\,,\sigma_{13}\,\}^T$, the virtual internal work in (3.294) can be written as

$$
\begin{aligned}
\int_{\varphi(B)} \nabla^S \boldsymbol{\eta} \cdot \boldsymbol{\sigma} \, dv &= \bigcup_{e=1}^{n_e} \int_{\varphi(\Omega_e)} (\nabla^S \boldsymbol{\eta}_e)^T \, \boldsymbol{\sigma}_e \, d\omega \\
&= \bigcup_{e=1}^{n_e} \sum_{I=1}^{n} \boldsymbol{\eta}_I^T \int_{\varphi(\Omega_e)} \boldsymbol{B}_{0\,I}^T \, \boldsymbol{\sigma}_e \, d\omega \qquad (4.95) \\
&= \bigcup_{e=1}^{n_e} \sum_{I=1}^{n} \boldsymbol{\eta}_I^T \int_{\Omega_\square} \boldsymbol{B}_{0\,I}^T \, \boldsymbol{\sigma}_e \, \det \boldsymbol{j}_e \, d\square \,.
\end{aligned}
$$

The last term is already referred to the isoparametric reference element. By comparing this relation with the associated expression in (4.54), it can be noticed that both weak forms differ only in the $\boldsymbol{B}$ matrix, the determinant of the isoparametric map (4.6) and the stress tensor which is now referred to the current configuration. By introducing

$$
\boldsymbol{r}_I\,(\mathbf{u}_e) = \int_{\varphi(\Omega_e)} \boldsymbol{B}_{0\,I}^T \, \boldsymbol{\sigma}_e \, d\omega, \qquad (4.96)
$$

the virtual internal work can be written as

$$
\int_{\varphi(B)} \nabla^S \boldsymbol{\eta} \cdot \boldsymbol{\sigma} \, dv = \bigcup_{e=1}^{n_e} \sum_{I=1}^{n} \boldsymbol{\eta}_I^T \, \boldsymbol{r}_I\,(\mathbf{u}_e) = \boldsymbol{\eta}^T \, \boldsymbol{r}\,(\mathbf{u})\,. \qquad (4.97)
$$

Since the transformation of the volume element is with (3.12) given by $dv = J\,dV$ and, since further, the CAUCHY stress tensor can be expressed by the KIRCHHOFF stress tensor via (3.84) as $\boldsymbol{\tau} = J\,\boldsymbol{\sigma}$, the virtual internal work in (4.96) can be formulated as

$$\int_{\varphi(B)} \nabla^S \boldsymbol{\eta} \cdot \boldsymbol{\sigma} \, dv = \int_B \nabla^S \boldsymbol{\eta} \cdot \boldsymbol{\tau} \, dV. \tag{4.98}$$

Now the integral is transformed from the spatial to the initial configuration. Discretization by finite elements yields

$$\begin{aligned}
\int_B \nabla^S \boldsymbol{\eta} \cdot \boldsymbol{\tau} \, dV &= \bigcup_{e=1}^{n_e} \int_{\Omega_e} (\nabla^S \boldsymbol{\eta}_e)^T \boldsymbol{\tau}_e \, d\Omega \\
&= \bigcup_{e=1}^{n_e} \sum_{I=1}^{n} \boldsymbol{\eta}_I^T \int_{\Omega_e} \boldsymbol{B}_{0I}^T \boldsymbol{\tau}_e \, d\Omega \tag{4.99} \\
&= \bigcup_{e=1}^{n_e} \sum_{I=1}^{n} \boldsymbol{\eta}_I^T \int_{\Omega_\square} \boldsymbol{B}_{0I}^T \boldsymbol{\tau}_e \, \det \boldsymbol{J}_e \, d\square,
\end{aligned}$$

and the vector related to the internal work has the form

$$\boldsymbol{r}_I(\boldsymbol{u}_e) = \int_{\Omega_\square} \boldsymbol{B}_{0I}^T \boldsymbol{\tau}_e \, \det \boldsymbol{J}_e \, d\square. \tag{4.100}$$

The total internal virtual work is then computed by using (4.97).

The approximation of the inertia terms follows the approach leading to (4.59). Similarly, the dead loads are formulated as in (4.63). Thus the discretization of the spatial weak form (3.294) is completed, leading to

$$\boldsymbol{\eta}^T \left[\boldsymbol{M} \dot{\boldsymbol{v}} + \boldsymbol{r}(\boldsymbol{u}) - \boldsymbol{P} \right] = 0. \tag{4.101}$$

This from yields for arbitrary test functions $\boldsymbol{\eta}$ the nonlinear ordinary differential system

$$\boldsymbol{M} \dot{\boldsymbol{v}} + \boldsymbol{r}(\boldsymbol{u}) - \boldsymbol{P} = \boldsymbol{0}. \tag{4.102}$$

For static problems, this system reduces to a set of nonlinear algebraic equations for the unknown nodal displacements $\boldsymbol{u}$

$$\boldsymbol{g}(\boldsymbol{u}) = \boldsymbol{r}(\boldsymbol{u}) - \boldsymbol{P} = \boldsymbol{0}. \tag{4.103}$$

In this formulation, the vector $\boldsymbol{r}(\boldsymbol{u})$ representing the internal virtual work can be computed using either (4.96) or (4.100). Both formulations are equivalent. Relation (4.97) has the same form as the associated relation of the linear theory. Only the gradient of the test function (virtual strains) $\nabla^S \boldsymbol{\eta}$ and the stress tensor $\boldsymbol{\sigma}$ (CAUCHY stress) have to be computed with respect to the current or spatial configuration.

4.2.4 Linearization of the Weak Form in the Spatial Configuration

In the previous section, two weak forms, (4.97) and (4.100), were developed which differ with respect to the configuration, $\varphi(B)$ or B. The associated linearization of the continuum formulation of these weak forms was described in Sect. 3.5.3. Now these forms have to be discretized as in Sect. 4.2.2, where the linearization of the weak form with respect to the initial configuration was described.

The linearization of the spatial weak form (4.97) follows from Eq. (3.347)

$$Dg(\bar{\varphi}, \boldsymbol{\eta}) \cdot \Delta\mathbf{u} = \int\limits_{\bar{\varphi}(B)} \{ \overline{\mathrm{grad}\Delta\mathbf{u}}\,\bar{\boldsymbol{\sigma}} \cdot \overline{\mathrm{grad}\boldsymbol{\eta}} + \nabla_{\bar{x}}^{S}\boldsymbol{\eta} \cdot \bar{\bar{\mathbf{c}}}\,[\nabla_{\bar{x}}^{S}\Delta\mathbf{u}] \} \, dv. \quad (4.104)$$

The first term has exactly the same structure as the associated term of the linearization with respect to the initial configuration. Hence the discretization (4.69) can be adopted directly. Only the derivatives have to be computed with respect to the coordinates $\bar{x}_i$ of the current configuration $\varphi(\bar{B})$. With the discretization of the gradients

$$\overline{\mathrm{grad}\,\Delta\mathbf{u}_e} = \sum_{K=1}^{n} \Delta\mathbf{u}_K \otimes \nabla_{\bar{x}} N_K,$$

$$\overline{\mathrm{grad}\,\boldsymbol{\eta}_e} = \sum_{I=1}^{n} \boldsymbol{\eta}_I \otimes \nabla_{\bar{x}} N_I, \quad (4.105)$$

the first term of the integral (4.104) is given by

$$\int\limits_{\varphi(B)} \overline{\mathrm{grad}\,\Delta\mathbf{u}}\,\bar{\boldsymbol{\sigma}} \cdot \overline{\mathrm{grad}\,\boldsymbol{\eta}}\, dv = \bigcup_{e=1}^{n_e} \sum_{I=1}^{n} \sum_{K=1}^{n} \boldsymbol{\eta}_I^T \int\limits_{\varphi(\Omega_e)} \bar{g}_{IK}\, \mathbf{I} \, d\Omega\, \Delta\mathbf{u}_K. \quad (4.106)$$

Here the abbreviation

$$\bar{g}_{IK} = (\nabla_{\bar{x}} N_I)^T \, \bar{\boldsymbol{\sigma}} \, \nabla_{\bar{x}} N_K \quad (4.107)$$

was utilized. As in (4.71) the matrix form of the scalar product yields

$$\bar{g}_{IK} = [\bar{N}_{I,1} \quad \bar{N}_{I,2} \quad \bar{N}_{I,3}] \begin{bmatrix} \bar{\sigma}_{11} & \bar{\sigma}_{12} & \bar{\sigma}_{13} \\ \bar{\sigma}_{21} & \bar{\sigma}_{22} & \bar{\sigma}_{23} \\ \bar{\sigma}_{31} & \bar{\sigma}_{32} & \bar{\sigma}_{33} \end{bmatrix} \begin{Bmatrix} \bar{N}_{K,1} \\ \bar{N}_{K,2} \\ \bar{N}_{K,3} \end{Bmatrix}. \quad (4.108)$$

This equation is, like (4.69), independent on the constitutive equation, since only the stresses of the spatial configuration $\bar{\varphi}$ enter (4.108).

The second term in (3.347)

$$\int\limits_{\varphi(\bar{B})} \nabla_{\bar{x}}^{S}\boldsymbol{\eta} \cdot \bar{\bar{\mathbf{c}}}\,[\nabla_{\bar{x}}^{S}\Delta\mathbf{u}]\, dv$$

includes the incremental constitutive tensor $\bar{\bar{\mathbf{c}}}$ which has to be computed at configuration $\bar{\varphi}$. Hence this term depends upon the material equation, see e.g. (3.269) in Sect. 3.3.4. With the same arguments as in Sect. 4.2.2 and the relations (3.333) and (4.94), the discretization follows

$$
\int\limits_{\varphi(\bar{B})} \nabla^S_{\bar{x}}\boldsymbol{\eta} \cdot \bar{\bar{\mathfrak{c}}}\, [\nabla^S_{\bar{x}}\Delta\mathbf{u}]\, dv = \bigcup_{e=1}^{n_e} \sum_{I=1}^{n} \sum_{K=1}^{n} \boldsymbol{\eta}^T_I \int\limits_{\varphi(\Omega_e)} \bar{\boldsymbol{B}}^T_{0\,I}\, \bar{\boldsymbol{D}}^M\, \bar{\boldsymbol{B}}_{0\,K}\, d\Omega\, \Delta\mathbf{u}_K\,.
$$

(4.109)

Here the evaluations and derivations of all quantities have to be computed at the state $\bar{\varphi}$. Combining both terms leads to the discretization of the linearization of the spatial weak form

$$
\int\limits_{\bar{\varphi}(B)} \{ \overline{\mathrm{grad}\Delta\mathbf{u}\,\bar{\boldsymbol{\sigma}} \cdot \mathrm{grad}\boldsymbol{\eta}} + \nabla^S_{\bar{x}}\boldsymbol{\eta} \cdot \bar{\bar{\mathfrak{c}}}\, [\nabla^S_{\bar{x}}\Delta\mathbf{u}]\,\}\, dv = \bigcup_{e=1}^{n_e} \sum_{I=1}^{n} \sum_{K=1}^{n} \boldsymbol{\eta}^T_I\, \bar{\boldsymbol{K}}^M_{T_{IK}}\, \Delta\mathbf{u}_K\,.
$$

(4.110)

Here the matrix $\bar{\boldsymbol{K}}^M_{T_{IK}}$

$$
\bar{\boldsymbol{K}}^M_{T_{IK}} = \int\limits_{\varphi(\Omega_e)} \left[\, (\nabla_{\bar{x}} N_I)^T\, \bar{\boldsymbol{\sigma}}_e\, \nabla_{\bar{x}} N_K + \bar{\boldsymbol{B}}^T_{0\,I}\, \bar{\boldsymbol{D}}^M\, \bar{\boldsymbol{B}}_{0\,K} \,\right] d\omega
$$

(4.111)

is the tangent matrix related to the finite element nodes $I\,, K$, see also Sect. 4.2.2.

The discretization of the spatial weak form (4.100) is derived analogously. Thus only the final result is presented

$$
\int\limits_{(B)} \{ \overline{\mathrm{grad}\Delta\mathbf{u}\,\bar{\boldsymbol{\tau}} \cdot \mathrm{grad}\boldsymbol{\eta}} + \nabla^S_{\bar{x}}\boldsymbol{\eta} \cdot \bar{\mathfrak{c}}\, [\nabla^S_{\bar{x}}\Delta\mathbf{u}]\,\}\, dv = \bigcup_{e=1}^{n_e} \sum_{I=1}^{n} \sum_{K=1}^{n} \boldsymbol{\eta}^T_I\, \bar{\boldsymbol{K}}^{MR}_{T_{IK}}\, \Delta\mathbf{u}_K\,.
$$

(4.112)

The matrix $\bar{\boldsymbol{K}}^{MR}_{T_{IK}}$

$$
\bar{\boldsymbol{K}}^{MR}_{T_{IK}} = \int\limits_{\Omega_e} \left[\, (\nabla_{\bar{x}} N_I)^T\, \bar{\boldsymbol{\tau}}_e\, \nabla_{\bar{x}} N_K + \bar{\boldsymbol{B}}^T_{0\,I}\, \bar{\boldsymbol{D}}^{MR}\, \bar{\boldsymbol{B}}_{0\,K} \,\right] d\Omega
$$

(4.113)

is the tangent matrix related to nodes I, K. The matrix form $\bar{\boldsymbol{D}}^{MR}$ of the incremental constitutive tensor $\bar{\mathfrak{c}}$ can be found, e.g. for a NEO-HOOKE material in (3.271). The corresponding relation for $\bar{\boldsymbol{D}}^M$ follows from the transformation using the JACOBI determinant J in accordance with (3.346).

Exercise 4.4: Derive the matrix formulation for an axi-symmetrical finite element undergoing finite elastic deformations with respect to the current or spatial configuration. The constitutive behaviour has to be described by the compressible NEO-HOOKE material given in (3.120). For the approximation of geometry and deformation, either bilinear or bi-quadratic shape functions have to be selected.

Solution: For an axi-symmetrical problem, the geometry of a structure under consideration as well as the loading has to be axi-symmetrical, see Fig. 4.12. It is assumed that the axis of symmetry coincides with coordinate X_2.

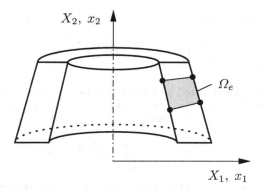

Fig. 4.12 Axi-symmetrical finite 4-node element

Additionally, to the strains in the plane X_1–X_2, hoop strains occur in case of axi-symmetrical deformations. The deformation gradient is then given by

$$\mathbf{F} = \operatorname{Grad}\mathbf{x} = \begin{bmatrix} \frac{\partial x_1}{\partial X_1} & \frac{\partial x_1}{\partial X_2} & 0 \\ \frac{\partial x_2}{\partial X_1} & \frac{\partial x_2}{\partial X_2} & 0 \\ 0 & 0 & \frac{x_1}{X_1} \end{bmatrix}. \tag{4.114}$$

Since the space of the test function is two-dimensional, the scalar product in (3.294) reduces to

$$\boldsymbol{\sigma} \cdot \operatorname{grad}\boldsymbol{\eta} = \sigma_{11}\frac{\partial \eta_1}{\partial x_1} + \sigma_{12}\frac{\partial \eta_1}{\partial x_2} + \sigma_{21}\frac{\partial \eta_2}{\partial x_1} + \sigma_{22}\frac{\partial \eta_2}{\partial x_2} + \sigma_{33}\frac{\eta_1}{x_1}, \tag{4.115}$$

where the components $\eta_{i,k}$ of the spatial gradient have to be computed by partial derivatives with respect to the current coordinates $\mathbf{x}$. For the computation of these spatial derivatives, the isoparametric map is applied, see (4.93).

By using linear or quadratic shape functions for the approximation of the displacement field $\mathbf{u}$, the coordinates $\mathbf{x}$ and the test function $\boldsymbol{\eta}$, the symmetric gradient of the test functions (virtual displacements)

$$(\nabla^S \boldsymbol{\eta})^T = \left[\eta_{1,1}, \eta_{2,2}, \frac{\eta_1}{x_1}, (\eta_{1,2} + \eta_{2,1}) \right]$$

can be discretized, which has the same form as (4.91). This leads to the matrix form

$$\boldsymbol{B}_{0I}^A = \begin{bmatrix} N_{I,1} & 0 \\ 0 & N_{I,2} \\ N_I/x_1 & 0 \\ N_{I,2} & N_{I,1} \end{bmatrix} \tag{4.116}$$

which is analogous to (4.94).

A compact notation for the scalar product introduced in (4.115) will be stated next for one element Ω_e. Introduction of a vector $\boldsymbol{\sigma}^T = [\sigma_{11}, \sigma_{22}, \sigma_{33}, \sigma_{12}]$ which represents the CAUCHY stresses leads to

$$\boldsymbol{\sigma} \cdot \operatorname{grad}\boldsymbol{\eta}\big|_{\Omega_e} = \sum_{I=1}^{n} \boldsymbol{\eta}^T \boldsymbol{B}_{0I}^{A\,T} \boldsymbol{\sigma}.$$

Using this result, the virtual internal work of one element Ω_e is given by

$$\sum_{I=1}^{n} {\boldsymbol{\eta}_I^h}^T 2\pi \int_{\varphi(\Omega_e)} {\boldsymbol{B}_{0I}^A}^T \boldsymbol{\sigma} \, x_1 \, d\omega \, . \tag{4.117}$$

This expression can be inserted in formulation (4.96). Note that the integration has to be performed as well over the coordinates x_1 and x_2 as in circumferential direction. Due to that, the coordinate x_1 appears in (4.117). For the node I of element Ω_e,

$$\boldsymbol{r}_I^A(\boldsymbol{u}_e) = 2\pi \int_{\varphi(\Omega_e)} \boldsymbol{B}_{0I}^T \boldsymbol{\sigma} \, x_1 \, d\omega \tag{4.118}$$

is obtained. In (4.118), the vector of the CAUCHY stress components $\boldsymbol{\sigma}$ has to be expressed by the hyperelastic constitutive equation (3.120). It then depends, via the strain measure directly, on the nodal displacements $\boldsymbol{u}$. This dependency is provided by the left CAUCHY-GREEN tensor $\mathbf{b} = \mathbf{F}\mathbf{F}^T$. Since only the vector $\mathbf{x}$ is known in the current configuration, the deformation gradient $\mathbf{F}$ cannot be specified directly. However it can be computed from its inverse of which is obtained in the spatial configuration $\varphi(B)$ by using the spatial displacement gradient, see (3.34): $\mathbf{F}^{-1} = \mathbf{1} - \operatorname{grad} \boldsymbol{u}$. Within the element Ω_e,

$$\begin{bmatrix} F_{11} & F_{12} \\ F_{21} & F_{22} \end{bmatrix}^{-1} = \begin{bmatrix} 1 & 0 \\ 0 & 1 \end{bmatrix} - \sum_{I=1}^{n} \begin{bmatrix} N_{I,1}\, u_{I1} & N_{I,2}\, u_{I1} \\ N_{I,1}\, u_{I2} & N_{I,2}\, u_{I2} \end{bmatrix} \tag{4.119}$$

is derived for the two-dimensional case. The components F_{11}, F_{12}, F_{21}, F_{22} of the deformation gradient follow then from the inverse of (4.119). In case of axisymmetrical deformations, the component F_{33} has also to be considered. Due to the special form of the deformation gradient (4.114), its value computation can be obtained directly. Since the component of the displacement gradient is given by $u_{3,3} = u_1 / x_1$ in circumferential direction, the component $F_{33}^{-1} = 1 - u_{3,3} = 1 - u_1 / x_1$ of the deformation gradient follows and thus

$$F_{33} = \frac{x_1}{x_1 - u_1} \, .$$

Note that the component F_{33} is expressed solely by terms which are related to the spatial configuration. It is clear that F_{33} could simply be computed from (4.114) which is given in terms of the initial configuration. This, however, does not lead to an efficient finite element code since the isoparametric mapping has then to be carried out twice (for the spatial and the initial configuration).

Using (3.25), the left CAUCHY-GREEN tensor $\mathbf{b} = \mathbf{F}\mathbf{F}^T$ follows by a simple matrix multiplication. The subsequent reordering of $\mathbf{b}$ into a column matrix $\boldsymbol{b}$ leads to a form of $\mathbf{b}$ which can be used in the matrix formulation later on

$$\boldsymbol{b} = \begin{Bmatrix} b_{11} \\ b_{22} \\ b_{33} \\ b_{12} \end{Bmatrix} = \begin{Bmatrix} F_{12}^2 + F_{11}^2 \\ F_{21}^2 + F_{22}^2 \\ F_{33}^2 \\ F_{12}\, F_{22} + F_{21}\, F_{11} \end{Bmatrix} \, . \tag{4.120}$$

Now the dependency of the left CAUCHY-GREEN tensor $\boldsymbol{b}$ on the nodal displacements $\boldsymbol{u}$ is known and the CAUCHY stresses can be computed from (3.120). With the determinant of the deformation gradient $J = \det \mathbf{F}$ and by introducing the unit vector $\boldsymbol{i}^T = [\, 1, 1, 1, 0 \,]$, the hyperelastic constitutive relation yields

$$\sigma = \frac{\Lambda}{2J}\left(J^2 - 1\right)i + \frac{\mu}{J}\left(b - i\right). \tag{4.121}$$

The LAMÉ constants Λ and μ are material parameters, see Sect. 3.3.1. The matrix form of the incremental constitutive tensor belonging to (4.121) can be found in Exercise 3.8 for the three-dimensional case, see (3.271). Here it is specified for the axi-symmetrical case which leads to

$$\boldsymbol{D}^A = \Lambda J^2 \boldsymbol{i}\boldsymbol{i}^T + [\mu - \frac{1}{2}\Lambda\left(J^2 - 1\right)]\boldsymbol{E}, \tag{4.122}$$

where $\boldsymbol{E}$ is a diagonal matrix

$$\boldsymbol{E} = \begin{bmatrix} 2 & 0 & 0 & 0 \\ 0 & 2 & 0 & 0 \\ 0 & 0 & 2 & 0 \\ 0 & 0 & 0 & 1 \end{bmatrix}. \tag{4.123}$$

The linearization of the residual (4.118) yields the tangent matrix which is given as in the three-dimensional case by (4.111). The initial stress matrix follows with (4.106) from

$$\int_{\varphi(B)} \overline{\operatorname{grad}\Delta\mathbf{u}}\,\bar{\sigma}\cdot\overline{\operatorname{grad}\eta}\,dv.$$

Within this expression, the spatial gradients are needed. In case of axi-symmetrical deformations, the following matrix form of the gradient of the test functions $\boldsymbol{\eta}$ is advantageous

$$\overline{\operatorname{grad}\boldsymbol{\eta}_e} = \begin{Bmatrix} \bar{\eta}_{1,1} \\ \bar{\eta}_{1,2} \\ \bar{\eta}_{3,3} \\ \bar{\eta}_{2,1} \\ \bar{\eta}_{2,2} \end{Bmatrix} = \sum_{I=1}^{n} \begin{bmatrix} \bar{N}_{I,1} & 0 \\ \bar{N}_{I,2} & 0 \\ \bar{N}_I/\bar{x}_1 & 0 \\ 0 & \bar{N}_{I,1} \\ 0 & \bar{N}_{I,2} \end{bmatrix} \begin{Bmatrix} \eta_1 \\ \eta_2 \end{Bmatrix}_I = \sum_{I=1}^{n} \bar{\boldsymbol{G}}_I\,\boldsymbol{\eta}_I. \tag{4.124}$$

Using this relation, the initial stress matrix (4.106) is given with

$$\hat{\bar{\sigma}}_e = \begin{bmatrix} \bar{\sigma}_{11} & \bar{\sigma}_{12} & 0 & 0 & 0 \\ \bar{\sigma}_{21} & \bar{\sigma}_{22} & 0 & 0 & 0 \\ 0 & 0 & \bar{\sigma}_{33} & 0 & 0 \\ 0 & 0 & 0 & \bar{\sigma}_{11} & \bar{\sigma}_{12} \\ 0 & 0 & 0 & \bar{\sigma}_{21} & \bar{\sigma}_{22} \end{bmatrix} \tag{4.125}$$

as

$$\int_{\varphi(B)} \overline{\operatorname{grad}\Delta\mathbf{u}}\,\bar{\sigma}\cdot\overline{\operatorname{grad}\eta}\,dv = \bigcup_{e=1}^{n_e}\sum_{I=1}^{n}\sum_{K=1}^{n}\boldsymbol{\eta}_I^T\,2\pi\int_{\varphi(\Omega_e)} \bar{\boldsymbol{G}}_I^T\,\hat{\bar{\sigma}}\,\bar{\boldsymbol{G}}_K\,x_1\,d\omega\,\Delta\mathbf{u}_K. \tag{4.126}$$

By inserting matrices (4.124) and (4.125) in expression (4.126), many unnecessary numerical operations occur when (4.126) is coded, since many components are zero. By hand multiplication of the integrand, these superfluous operations can be avoided and hence the efficiency of the element is increased. With the abbreviation

$$\begin{aligned} \bar{\alpha}_{1\,IK} &= \left(\bar{N}_{I,1}\,\bar{\sigma}_{11} + \bar{N}_{I,2}\,\bar{\sigma}_{21}\right)\bar{N}_{K,1} \\ \bar{\alpha}_{2\,IK} &= \left(\bar{N}_{I,1}\,\bar{\sigma}_{12} + \bar{N}_{I,2}\,\bar{\sigma}_{22}\right)\bar{N}_{K,2} \\ \bar{\alpha}_{3\,IK} &= \frac{\bar{N}_I}{\bar{x}_1}\,\bar{\sigma}_{33}\,\frac{\bar{N}_K}{\bar{x}_1}, \end{aligned}$$

the explicit form

$$\bar{\boldsymbol{G}}_I^T \,\hat{\bar{\boldsymbol{\sigma}}}\, \bar{\boldsymbol{G}}_K = \begin{bmatrix} \bar{\alpha}_{1\,IK} + \bar{\alpha}_{2\,IK} + \bar{\alpha}_{3\,IK} & 0 \\ 0 & \bar{\alpha}_{1\,IK} + \bar{\alpha}_{2\,IK} \end{bmatrix} \qquad (4.127)$$

is obtained.

By considering relations (4.116) and (4.122), the tangent matrix follows using (4.126)

$$\int\limits_{\bar{\varphi}(B)} \{ \overline{\mathrm{grad}\Delta \mathbf{u}}\, \bar{\boldsymbol{\sigma}} \cdot \overline{\mathrm{grad}\eta} + \nabla_{\bar{x}}^S \boldsymbol{\eta} \cdot \hat{\bar{\mathbf{c}}}\, [\nabla_{\bar{x}}^S \Delta \mathbf{u}\,] \} \, dv = \bigcup_{e=1}^{n_e} \sum_{I=1}^{n} \sum_{K=1}^{n} \boldsymbol{\eta}_I^T \, \bar{\boldsymbol{K}}_{T_{IK}}^A \, \Delta \mathbf{u}_K \,. $$

$$(4.128)$$

Here the sub-matrix

$$\bar{\boldsymbol{K}}_{T_{IK}}^A = 2\,\pi \int\limits_{\varphi(\Omega_e)} \left[\bar{\boldsymbol{G}}_I^T \,\hat{\bar{\boldsymbol{\sigma}}}_e\, \bar{\boldsymbol{G}}_K + \bar{\boldsymbol{B}}_{0\,I}^{A\,T}\, \bar{\boldsymbol{D}}^A\, \bar{\boldsymbol{B}}_{0\,K}^A \right] x_1 \, d\omega \qquad (4.129)$$

is referred to the spatial configuration and belongs to the nodal pair I, K of the axi-symmetrical finite element. Again for the second term in $\bar{\boldsymbol{K}}_{T_{IK}}^A$, operations can be saved by hand multiplication since matrices $\bar{\boldsymbol{B}}_{0\,I}^A$ and $\bar{\boldsymbol{D}}^A$ are sparse. Note that all terms with a bar in (4.129) have to be evaluated at the deformation state $\bar{\varphi}$. Inserting the shape functions (4.28) or (4.30) to (4.32) yields then a 4- or 9-node finite element. Of course, all integrals (4.118) and (4.129) have to be computed using numerical integration. This will not be specified here in detail; the approach is similar to the one in Exercise 4.3.

4.2.5 Deformation Dependent Loads

Applied loads can depend upon the deformation in some technical problems. On one hand the load direction can change (in such case also the term *follower loads* is used) and on the other hand the magnitude of the load can decrease or increase. As an example, loading of structures due to fluids or wind can be mentioned. Interaction of structural systems with fluids and gas are special engineering application of high relevance. A related treatment can be found in Hassler and Schweizerhof (2008). An in-depth discussion of the algorithmic treatment of deformation-dependent loads and their discretization can be found in Schweizerhof (1982) and Schweizerhof and Ramm (1984) for the general case and in Simo et al. (1991) and Yosibash et al. (2007) for axi-symmetrical deformations, while the latter paper discusses discretization using high order finite element interpolations.

In this section, only loads which are direction depending are considered. Loads which are always normal to the deformed surface of a solid or structure will be discussed in more detail. Thus the term describing the surface loads $\int_{\Gamma_\sigma} \boldsymbol{\eta} \cdot \bar{\mathbf{t}}\, dA$ will be considered, where $\bar{\mathbf{t}} = \hat{p}\,\mathbf{n}$ was introduced in (4.62). This leads to the surface load term

$$g_p(\boldsymbol{\varphi}, \boldsymbol{\eta}) = \int\limits_{\varphi(\Gamma_\sigma)} \boldsymbol{\eta} \cdot \hat{p}\,\mathbf{n}\, da \,. \qquad (4.130)$$

Here the integration has to be performed with respect to the current configuration. Note that the normal vector **n** depends on the deformation. The discretization of (4.130) can be obtained by introducing convective coordinates with base vectors $\mathbf{g}_\alpha$, see also Exercise 3.12. The coordinates are depicted in Fig. 4.11 for a discretization of the surface of a three-dimensional finite element with quadratic shape functions.

The normal vector of the discretized surface is given in the spatial configuration by

$$\mathbf{n} = \frac{\mathbf{g}_1 \times \mathbf{g}_2}{\| \mathbf{g}_1 \times \mathbf{g}_2 \|} . \tag{4.131}$$

The base vectors are computed from the deformation field by a partial derivative with respect to the convective coordinates ξ and η, $\mathbf{g}_\alpha = \boldsymbol{\varphi},_\alpha$, where $\alpha = 1$ stands for ξ and $\alpha = 2$ for η. Hence the normal vector can be described by

$$\mathbf{n} = \frac{\boldsymbol{\varphi},_\xi \times \boldsymbol{\varphi},_\eta}{\| \boldsymbol{\varphi},_\xi \times \boldsymbol{\varphi},_\eta \|} . \tag{4.132}$$

Since the area element da can be computed with respect to the reference configuration by $da = \| \boldsymbol{\varphi},_\xi \times \boldsymbol{\varphi},_\eta \| \, d\xi \, d\eta$, the deformation-dependent load vector (4.130) can be transformed to the reference configuration of the loaded element surface, see also Eq. (3.349) and the right part of Fig. 4.13,

$$g_p(\boldsymbol{\varphi},\boldsymbol{\eta}) = \int_{\varphi(\Gamma_\sigma)} \boldsymbol{\eta} \cdot \hat{p}\,\mathbf{n}\, da = \int_{\Gamma_{ref}} \boldsymbol{\eta} \cdot \bar{p}\, (\boldsymbol{\varphi},_\xi \times \boldsymbol{\varphi},_\eta)\, d\xi \, d\eta . \tag{4.133}$$

Based on this form, the finite element discretization is straight forward. Using the isoparametric shape functions yields

$$\boldsymbol{\varphi}_e = \mathbf{x}_e = \sum_{A=1}^{m} N_A(\xi,\eta)\, \mathbf{x}_A \tag{4.134}$$

and the derivatives of the components of the position vector x_i

$$x_{i,\alpha} = \sum_{A=1}^{m} N_A(\xi,\eta),_\alpha\, x_{i\,A} \tag{4.135}$$

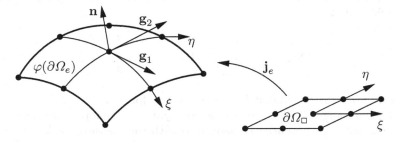

Fig. 4.13 Coordinate systems for deformation dependent loads

for the loaded element surface. Now the cross product in (4.133) can be computed. The result, described in vector form, is

$$\hat{\boldsymbol{n}}_e = \boldsymbol{\varphi}_{e,\xi} \times \boldsymbol{\varphi}_{e,\eta} = \left\{ \begin{array}{l} x_{2,\xi}\, x_{3,\eta} - x_{3,\xi}\, x_{2,\eta} \\ x_{3,\xi}\, x_{1,\eta} - x_{1,\xi}\, x_{3,\eta} \\ x_{1,\xi}\, x_{2,\eta} - x_{2,\xi}\, x_{1,\eta} \end{array} \right\} . \tag{4.136}$$

The discretization of (4.130) follows with these definitions as

$$\int_{\varphi(\Gamma_\sigma)} \boldsymbol{\eta} \cdot \hat{p}\,\mathbf{n}\, da = \bigcup_{r=1}^{n_r} \sum_{A=1}^{m} \boldsymbol{\eta}_A^T\, \boldsymbol{r}_A(\boldsymbol{x}_e)\,, \text{ with } \boldsymbol{r}_A(\boldsymbol{x}_e) = \int_{\partial\Omega_\square} N_A\, \hat{p}\, \hat{\boldsymbol{n}}_e\, d\xi\, d\eta$$
$$\tag{4.137}$$

where $\bigcup_{r=1}^{n_r}$ denotes the assembly of the n_r loaded surfaces and $\partial\Omega_\square$ denotes the surface of the reference element, see right part of Fig. 4.13.

The linearization of the virtual work expression for the deformation-dependent or follower loads (4.137) has to be computed at state $\bar{\boldsymbol{\varphi}}$. This can be derived for constant pressure p using the relations provided in Exercise 3.12. With (4.133), the linearization (3.350) yields

$$D\, g_p(\boldsymbol{\varphi},\boldsymbol{\eta}) \cdot \Delta\mathbf{u} = \int_{\Gamma_{ref}} \boldsymbol{\eta} \cdot \hat{p}\,(\, \Delta\mathbf{u}_{,\xi} \times \bar{\boldsymbol{\varphi}}_{,\eta} + \bar{\boldsymbol{\varphi}}_{,\xi} \times \Delta\mathbf{u}_{,\eta}\,)\, d\xi\, d\eta . \tag{4.138}$$

An explicit evaluation of the cross product leads together with the use of the isoparametric shape functions for the discretization of the spatial coordinates describing the element surface to

$$\int_{\Gamma_{ref}} \boldsymbol{\eta} \cdot \hat{p}\,(\, \Delta\mathbf{u}_{,\xi} \times \bar{\boldsymbol{\varphi}}_{,\eta} + \bar{\boldsymbol{\varphi}}_{,\xi} \times \Delta\mathbf{u}_{,\eta}\,)\, d\xi\, d\eta = \bigcup_{r=1}^{n_r} \sum_{A=1}^{m} \sum_{B=1}^{m} \boldsymbol{\eta}_A^T\, \bar{\boldsymbol{k}}_{AB}\, \Delta\mathbf{u}_B\,,$$
$$\tag{4.139}$$

where the sub-matrix $\bar{\boldsymbol{k}}_{AB}$ has the following form

$$\bar{\boldsymbol{k}}_{AB} = \int_{\partial\Omega_\square} \hat{p}\, N_A\, (\, N_{B,\xi}\, \bar{\boldsymbol{N}}_{,\eta} - N_{B,\eta}\, \bar{\boldsymbol{N}}_{,\xi}\,)\, d\xi\, d\eta . \tag{4.140}$$

By $\bar{\boldsymbol{N}}_{,\alpha}$ (for α the convective coordinate ξ or η has to be used), the skew symmetric matrix

$$\bar{\boldsymbol{N}}_{,\alpha} = \begin{bmatrix} 0 & \bar{x}_{3,\alpha} & -\bar{x}_{2,\alpha} \\ -\bar{x}_{3,\alpha} & 0 & \bar{x}_{1,\alpha} \\ \bar{x}_{2,\alpha} & -\bar{x}_{1,\alpha} & 0 \end{bmatrix} \tag{4.141}$$

is defined. Since the sub-matrix is non-symmetric for nodes A and B also, the total element tangent matrix becomes non-symmetric for pressure loading. Thus, in general, no potential is associated with the pressure load. However under certain boundary conditions, the total assembled tangent matrix can be symmetrical, e.g. for internal pressure in a closed solid. An in-depth discussion can be found in, e.g. Sewell (1967) and Schweizerhof (1982).

For two-dimensional problems, the description of the normal vector $\mathbf{n}$ is a lot simpler, since then the base vector $\mathbf{g}_2$ is expressed by the unit vector $\mathbf{e}_3$ which is perpendicular to the x_1–x_2 plane, see Fig. 4.14. Hence the normal vector

$$\hat{\mathbf{n}}_e = \mathbf{e}_3 \times \boldsymbol{\varphi}_{e,\xi} = \left\{ \begin{array}{c} -x_{2,\xi} \\ x_{1,\xi} \end{array} \right\} \tag{4.142}$$

is obtained. Differentiation of the components of the position vector with respect to the current configuration follows by using the isoparametric shape functions to describe the element surface, see Fig. 4.10 b and Sect. 4.2.1,

$$x_{\alpha,\xi} = \sum_{A=1}^{m} N_A(\xi)_{,\xi}\, x_{\alpha A}\,. \tag{4.143}$$

The expression (4.142) can be inserted directly in (4.137). This leads to

$$\int_{\varphi(\Gamma_\sigma)} \boldsymbol{\eta} \cdot \hat{p}\,\mathbf{n}\, da = \bigcup_{r=1}^{n_r} \sum_{A=1}^{m} \boldsymbol{\eta}_A^T \mathbf{r}_A(\mathbf{x}_e)\,, \quad \text{with} \quad \mathbf{r}_A(\mathbf{x}_e) = \int_{-1}^{+1} \hat{p}\, N_A \left\{ \begin{array}{c} -x_{2,\xi} \\ x_{1,\xi} \end{array} \right\} d\xi\,. \tag{4.144}$$

The tangent matrix can be computed analogous to the three-dimensional derivation. With (3.350) and (4.139), the tangent matrix follows directly from (4.144)

$$\int_{\Gamma_{ref}} \boldsymbol{\eta} \cdot \hat{p}\,(\mathbf{e}_3 \times \Delta \mathbf{u}_{,\xi})\, d\xi = \bigcup_{r=1}^{n_r} \sum_{A=1}^{m} \sum_{B=1}^{m} \boldsymbol{\eta}_A^T\, \bar{\mathbf{k}}_{AB}\, \Delta \mathbf{u}_B, \tag{4.145}$$

where the sub-matrix $\bar{\mathbf{k}}_{AB}$ has the form

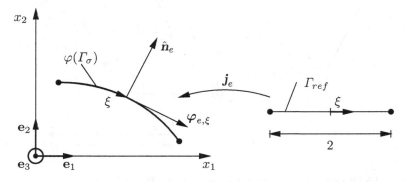

Fig. 4.14 2-D discretization of deformation dependent loads

$$\bar{k}_{AB} = \int\limits_{-1}^{+1} \hat{p}\, N_A\, N_{B,\xi} \begin{bmatrix} 0 & -1 \\ 1 & 0 \end{bmatrix} d\xi. \tag{4.146}$$

Exercise 4.5: A finite element undergoes axi-symmetrical finite deformations. Derive the load vector and tangent matrix for a dependent but constant pressure load. The equations are to be specified for an element with linear isoparametric shape functions for the displacement field.

Solution: In case of axi-symmetric loading, an integration in circumferential direction is necessary to compute the weak form of the load vector. Using (4.144), relation

$$g_p(\boldsymbol{\varphi},\boldsymbol{\eta}) = \int\limits_{\varphi(\Gamma_\sigma)} \boldsymbol{\eta}\cdot\hat{p}\,\mathbf{n}\,da = 2\pi \int\limits_{-1}^{+1} \boldsymbol{\eta}\cdot\hat{p}\,(\mathbf{e}_3 \times \boldsymbol{\varphi}_{,\xi})\,r(\xi)\,d\xi \tag{4.147}$$

is derived. Linear ansatz functions are chosen for the test function (virtual displacement) $\boldsymbol{\eta}$ and the deformation $\boldsymbol{\varphi}$, see (4.17). This yields the discretization depicted in Fig. 4.15. Using such interpolation, the radius r_e can be described within an element

$$r_e = \sum_{B=1}^{2} N_B(\xi)\, r_B \quad \text{with} \quad N_B = \frac{1}{2}\,(1+\xi_A\,\xi). \tag{4.148}$$

The coordinates ξ_A coincide with the nodal coordinates of the reference element $\partial\Omega_\square$, see Fig. 4.15. This is equivalent to specifying $\xi_1 = -1$ and $\xi_2 = 1$. The test function and the deformation are approximated in the same way. Hence, for linear ansatz functions relations,

$$\boldsymbol{\eta} = \sum_{B=1}^{2} N_B(\xi)\,\boldsymbol{\eta}_B\,, \quad \boldsymbol{\varphi}_e = \sum_{B=1}^{2} N_B(\xi)\,\mathbf{x}_B \quad \text{and} \quad \boldsymbol{\varphi}_{e,\xi} = \frac{1}{2}\,(\mathbf{x}_2 - \mathbf{x}_1) \tag{4.149}$$

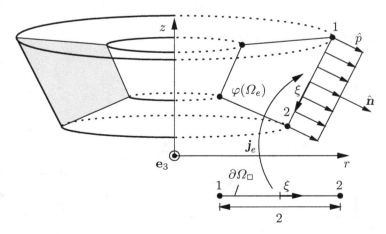

Fig. 4.15 Discretization of an axi symmetrical follower load

follow. The vector $\boldsymbol{\varphi}_{,\xi}$ is tangent to the boundary defined by nodes 1 and 2. Insertion of this discretization into (4.147) yields

$$\int_{\varphi(\Gamma_\sigma)} \boldsymbol{\eta} \cdot \hat{p}\,\mathbf{n}\,da = \bigcup_{r=1}^{n_r} \pi\hat{p}\sum_{A=1}^{2} \boldsymbol{\eta}_A^T \left[\mathbf{e}_3 \times (\mathbf{x}_2 - \mathbf{x}_1)\right] \int_{-1}^{+1} N_A \sum_{B=1}^{2} N_B\, r_B\, d\xi. \quad (4.150)$$

The integral in (4.150) can be solved exactly for linear shape functions

$$\sum_{B=1}^{2} \int_{-1}^{+1} \frac{1}{2}(1 + \xi_A\xi)\frac{1}{2}(1 + \xi_B\xi)\,d\xi = \sum_{B=1}^{2} \left[\frac{1}{2} + \frac{1}{6}\xi_A\xi_B\right] = \sum_{B=1}^{2} \gamma_{AB}.$$

Now the discretized load vector (4.150) can be computed explicitly by using the coordinates defined in Fig. 4.15 and

$$\mathbf{e}_3 \times (\mathbf{x}_2 - \mathbf{x}_1) = \left\{ \begin{array}{c} -(z_2 - z_1) \\ (r_2 - r_1) \end{array} \right\}.$$

This leads to

$$\int_{\varphi(\Gamma_\sigma)} \boldsymbol{\eta} \cdot \hat{p}\,\mathbf{n}\,da = \bigcup_{r=1}^{n_r} \sum_{A=1}^{2} \boldsymbol{\eta}_A^T\, \mathbf{r}_A^R(\mathbf{x}_e). \quad (4.151)$$

The nodal vector $\mathbf{r}_A^R$ depends upon the deformation state

$$\mathbf{r}_A^R = \pi\hat{p} \left\{ \begin{array}{c} -(z_2 - z_1) \\ (r_2 - r_1) \end{array} \right\} (\gamma_{A1}\,r_1 + \gamma_{A2}\,r_2). \quad (4.152)$$

The tangent matrix follows directly from the linearization of (4.147)

$$D\,g_p(\boldsymbol{\varphi},\boldsymbol{\eta}) \cdot \Delta\mathbf{u} = 2\pi \int_{-1}^{+1} \boldsymbol{\eta} \cdot \hat{p}\,[\,(\mathbf{e}_3 \times \Delta\mathbf{u}_{,\xi})\,r(\xi) + (\mathbf{e}_3 \times \boldsymbol{\varphi}_{e,\xi})\,\Delta u_1\,]\,d\xi, \quad (4.153)$$

where the second term considers the change of the radius due to deformation. With the explicit form (4.152) and with the matrix form of the cross product

$$\mathbf{e}_3 \times (\mathbf{x}_2 - \mathbf{x}_1) = \begin{bmatrix} 0 & -1 \\ 1 & 0 \end{bmatrix} (\mathbf{x}_2 - \mathbf{x}_1),$$

the result

$$D\,g_p(\boldsymbol{\varphi},\boldsymbol{\eta}) \cdot \Delta\mathbf{u} = \bigcup_{r=1}^{n_r} \boldsymbol{\eta}_A^T\,\pi\hat{p}\left\{ \begin{bmatrix} 0 & -1 \\ 1 & 0 \end{bmatrix}(\Delta\mathbf{u}_2 - \Delta\mathbf{u}_1)\sum_{C=1}^{2}\gamma_{AC}\,r_C \right.$$
$$\left. + \begin{bmatrix} 0 & -1 \\ 1 & 0 \end{bmatrix}(\mathbf{x}_2 - \mathbf{x}_1)\sum_{B=1}^{2}\gamma_{AB}\Delta u_{1\,B} \right\} \quad (4.154)$$

is obtained. This relation can be rewritten from

$$\bigcup_{r=1}^{n_r} \sum_{A=1}^{2} \sum_{B=1}^{2} \boldsymbol{\eta}_A^T\,\bar{\mathbf{k}}_{AB}^R\,\Delta\mathbf{u}_B \quad (4.155)$$

by using $\Delta \boldsymbol{u}_2 - \Delta \boldsymbol{u}_1 = \sum_{B=1}^{2} \xi_B \, \Delta \boldsymbol{u}_B$. The matrix $\bar{\boldsymbol{k}}_{AB}^{R}$ follows after sum algebraic manipulations

$$\bar{\boldsymbol{k}}_{AB}^{R} = \pi \, \hat{p} \begin{bmatrix} -(z_2 - z_1) \, \gamma_{AB} & -\beta_A \, \xi_B \\ \beta_A \, \xi_B + (r_2 - r_1) \, \gamma_{AB} & 0 \end{bmatrix}, \qquad (4.156)$$

where the abbreviation $\beta_A = \sum_{B=1}^{2} \gamma_{AB} \, r_B$ was introduced. Now all necessary matrices are known which have to be implemented in a finite element program. Note that the analytical integration was only possible since the JACOBI determinant j_e, which performs the isoparametric mapping to the reference element, see e.g. Fig. 4.15, disappears in the integrals (4.151) and (4.153).

5. Solution Methods for Time Independent Problems

The mathematical modelling of technical applications in solid mechanics leads in general to nonlinear partial differential equations, which characterize the associated initial and boundary value problems. Once a problem is spatially discretized by finite elements, a system of ordinary differential equations in time results. If the time dependency can be neglected then the system of non linear ordinary differential equations reduces to a nonlinear algebraic equations system stemming from the finite element discretization

$$G(v) = 0,$$

see (4.65). Here the unknown variables $v \in \mathbb{R}^N$ have to be determined (N is the total number of unknowns). Two different aspects have to be considered when solving the above equation. They are

1. the general solvability of the nonlinear equation systems and
2. the formulation of adequate numerical methods and algorithms.

The first aspect involves the examination of

- existence of solutions in a defined region,
- number of solutions in this region and
- the influence of the change of function G with respect to the solution.

Clarification of these questions needs methods of nonlinear functional analysis which have to be applied to the nonlinear partial differential equations resulting from the modelling procedure. This area cannot be treated in-depth in the framework of this book. However, for applications which fall in the range of the nonlinear theory of elasticity, results can be found in Marsden and Hughes (1983), Ciarlet (1988), Johnson (1987), Brenner and Scott (2002) and Braess (2007). Further, books which contain a general treatment of the above-mentioned questions are provided by Vainberg (1964) or Ortega and Rheinboldt (1970). We will assume, in the following, that the solutions of the equation system $G(v) = 0$ exist in the considered regions and will devote ourself to the numerical methods for the determination of solutions of $G(v) = 0$.

The approximation of solutions of the nonlinear equation system $G(v) = 0$ can be obtained by different methods. Among these are

1. methods for the construction of sets which contain solutions,
2. procedures to find all solutions and
3. methods which just approximate one solution.

The first two methods need, in general, deeper knowledge of the underlying mathematical structure of the associated partial differential equations. Due to that fact only procedures will be considered which yield one approximate solution at a time, but which can be applied to find successively other solutions of $G(v) = 0$. Since a direct solution of $G(v) = 0$ is in general not possible, iterative solution procedures have to be constructed. These methods permit different ways to solve the problem which will be described in detail in the next sections. In general, different procedures can be distinguished:

− methods which base on linearizations,
− minimization schemes or
− reduction methods, which lead to simpler nonlinear equation systems.

When choosing a solution method, the following fundamental questions have to be clarified which constitute the success of an iterative method and its associated algorithm:

− Does the iterative method converge to the solution?
− How fast is the convergence? Does the rate of convergence depend upon the problem size?
− How efficient is the algorithm?
    − How many numerical operations are needed within one iteration step?
    − How many iterations are necessary to converge within a given accuracy?
    − How much memory of the cpu does the iterative method need?

The first question concerns the global convergence characteristics of the iterative method and is essential for the user who needs a robust and reliable iterative method for his/her problem. Also the other questions are of importance. The efficiency depends on several factors which are determined, e.g. by the linear solver within an iterative method, the finite element formulation itself and the convergence properties of the iterative solution method. In case of large problem sizes, a nonlinear finite element equation system with a great number of unknowns is obtained which requires a lot of memory and many numerical operations in the solution process. When the number of operations increases quadratically with respect to the number of unknowns, the method is said to be of order $O(N^2)$. Such method cannot be applied to large problems. Here methods which need only $O(N)$ operations are advantageous and thus present a vivid research area, for an overview, see e.g. Rheinboldt (1984), Hackbusch (1994), Elman et al. (2005), Douglas et al. (2003) and Saad (2003).

In solid mechanics and nonlinear structural mechanics, the range of problems is wide and areas are quit different (geometrical nonlinearity, physical nonlinearity, stability, etc.). Thus there exists up to now no iterative method which can be applied to all different problem areas in an efficient and robust way. Due to that, several methods will be presented which were developed for the solution of the nonlinear equation system $G(v) = 0$.

Finite element approximations with the interpolations described in Chap. 4 lead to a system of nonlinear algebraic equations with N unknowns, see (4.65) in Sect. 4.2.1. This system of equations can be recast for the following considerations in the form

$$G(v, \lambda) = R(v) - \lambda P = 0, \qquad v \in \mathbb{R}^N. \qquad (5.1)$$

The scaling factor λ in front of the load term P is called loading parameter. It is introduced to be able to change the load level with an iterative method. The parameter λ is usually determined by the problem at hand, e.g. as total given load. However, in special iterative methods, it makes sense to consider λ as an unknown variable.

Equation (5.1) will be solved by an iterative method. For the choice of the best method for a special application, the aspects discussed above have to be considered. For nonlinear problems in structural mechanics, there exist a large number of algorithms and solution procedures. The most common ones applied within finite element methods are

- fix-point methods,
- NEWTON-RAPHSON methods,
- quasi-NEWTON methods,
- dynamical relaxation and
- continuation or arc-length methods.

Linear equation solvers provide an essential ingredient for the efficient solution of nonlinear finite element equation systems. This stems from the fact that linearization is used to construct iterative solution schemes leading to very large finite element linear equation systems. The iterative procedures then arrives at the global solution via the solution of several linear sub-problems. While standard elimination methods are often successfully applied to solve the linear equation system of small and middle size finite element discretizations, see e.g. Taylor (2000), one relies on large systems on sparse solvers, see e.g. Duff et al. (1989), Duff (2004) and Schenk and Gärtner (2004). For large systems also iterative equation solvers can be applied successfully. Here the method of pre-conditioned conjugate gradients and multi-grid methods are often employed for symmetrical matrix systems, for mathematical details, see e.g. Ciarlet (1989), Hackbusch (1994), Schwetlick and Kretschmar (1991), Douglas et al. (2003) and Saad (2003). Applications within the method of finite elements in solid mechanics can be found, e.g. in Braess (2007), Kickinger (1996), Korneev et al. (2003). The method of dynamical relaxation makes a "detour" via dynamics to construct a memory saving iterative solver based on an explicit integration method, details can be found in Sect. 6.1.1.

The efficiency of the different methods for the solution of nonlinear equation systems depends also upon the application and its size in terms of number of unknowns. As an example, the method of NEWTON-RAPHSON can be very efficient in combination with direct elimination methods for problems with low number of unknowns. For problems with large number of unknowns, quasi-NEWTON methods or the dynamical relaxation can be more efficient since, even with more iterations, they need less computation time. However, also a combination of the NEWTON-RAPHSON method together with iterative linear solvers can be very efficient and faster than quasi-NEWTON procedures, see e.g. Hackbusch (1994), Meyer (1990), Boersma and Wriggers (1997) and Jung and Langer (2001).

5.1 Solution of Nonlinear Systems of Equations

Three common algorithm which are applied for the solution of nonlinear finite element problems, characterized by Eq. (5.1), will be presented and compared in this section.

5.1.1 Newton-Raphson Method

The most frequently applied scheme for the iterative solution of systems of nonlinear algebraic equation is the NEWTON-RAPHSON algorithm. It is based on a TAYLOR series development of (5.1) at an already known state v_k

$$\boldsymbol{G}(\,\boldsymbol{v}_k + \varDelta\boldsymbol{v}, \,\bar{\lambda}\,) = \boldsymbol{G}(\,\boldsymbol{v}_k, \,\bar{\lambda}\,) + D\,\boldsymbol{G}(\,\boldsymbol{v}_k, \,\bar{\lambda}\,)\,\varDelta\boldsymbol{v} + \boldsymbol{r}(\,\boldsymbol{v}_k, \,\bar{\lambda}\,). \qquad (5.2)$$

The loading parameter $\bar{\lambda}$ denotes the load level for which the solution has to be determined. In (5.2) $D\,\boldsymbol{G} \cdot \varDelta\boldsymbol{v}$ characterizes the directional derivative of $\boldsymbol{G}$ at $\boldsymbol{v}_k$, also referred to as linearization, see Sect. 3.5. The linearization of the vector $\boldsymbol{G}$ yields a matrix, which is also known as HESSE-, JACOBI- or tangent matrix. This matrix will be abbreviated in the following by $\boldsymbol{K}_T$, see Sect. 4.2.2 and 4.2.4. The vector $\boldsymbol{r}$ is the residuum of the TAYLOR series. By neglecting the residuum, the linear equation system $\boldsymbol{G}(\,\boldsymbol{v}_k + \varDelta\boldsymbol{v}, \,\bar{\lambda}\,) = \boldsymbol{0}$ is obtained from (5.2) which is the basis of the following iterative algorithm for the solution of Eq. (5.1).

This algorithm, so far, determines the solution for the load level defined by the load parameter $\bar{\lambda}$. The associated convergence behaviour is depicted in Fig. 5.1 for a one-dimensional problem. For this purpose, the nonlinear equation $G(v, \bar{\lambda}) = R(v) - \bar{\lambda}\,P = 0$ was normalized as: $\hat{R}(v) - \bar{\lambda} = 0$.

The rate of convergence of the NEWTON-RAPHSON scheme is characterized by the inequality $\| \boldsymbol{v}_{k+1} - \boldsymbol{v} \| \leq C \| \boldsymbol{v}_k - \boldsymbol{v} \|^2$, where $\boldsymbol{v}$ is the solution of (5.1), see e.g. Isaacson and Keller (1966, pp. 115) or Schwetlick and Kretschmar (1991, pp. 195). The quadratic convergence of the NEWTON-RAPHSON scheme, which is apparent from the above inequality, has a local

Box 5.1 Algorithm for the NEWTON-RAPHSON scheme

Initial values: $\boldsymbol{v}_0 = \boldsymbol{v}_k$.

Iteration loop $i = 0, 1, \ldots$ *until convergence*

1. Compute $\boldsymbol{G}\,(\boldsymbol{v}_i, \bar{\lambda})$ and $\boldsymbol{K}_T\,(\boldsymbol{v}_i)$
2. Compute the displacement increments: $\boldsymbol{K}_T\,(\boldsymbol{v}_i)\,\Delta\boldsymbol{v}_{i+1} = -\boldsymbol{G}\,(\boldsymbol{v}_i, \bar{\lambda})$
3. Compute new displacement: $\boldsymbol{v}_{i+1} = \boldsymbol{v}_i + \Delta\boldsymbol{v}_{i+1}$
4. Test for convergence

$$\|\,\boldsymbol{G}\,(\boldsymbol{v}_{i+1}, \bar{\lambda})\,\| \quad \begin{cases} \leq \text{TOL} \longrightarrow & \text{Set}: \boldsymbol{v}_{k+1} = \boldsymbol{v}_{i+1}, \quad \text{STOP} \\ > \text{TOL} \longrightarrow & \text{Set } i = i+1 \quad \text{go to 1)} \end{cases}$$

character since it is only valid near the solution point. This convergence behaviour is advantageous since most of the time only a few iterations are needed to obtain the solution of (5.1). A drawback of the NEWTON-RAPHSON scheme stems from the fact that, in every iteration step, the tangent matrix $\boldsymbol{K}_T$ has to be computed and a linear equation system has to be solved, which can be quite time consuming and hence expensive. To shorten the notation, the term NEWTON scheme will be used instead of the historically more correct NEWTON-RAPHSON scheme.

Since it is often very complicated to derive the tangent matrix analytically, different other approaches are possible. One is related to a combination of automatic and symbolic differentiation, see Korelc (1997), and another computed the derivatives by difference quotients. In combination with NEWTON method, the latter approach is called a discrete NEWTON scheme. One possibility is to apply the forward difference quotient. Using its definition, the approximation

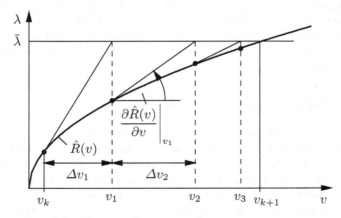

Fig. 5.1 Illustration of the NEWTON-RAPHSON scheme

$$k_m \approx \frac{1}{h_m} \left[\, G(v_i + h_m \, e_m \, , \bar{\lambda}) - G(v_i \, , \bar{\lambda}) \, \right] \qquad (5.3)$$

is obtained for the m-th column of the tangent matrix. In this relation, h_m is the step size and e_m is a vector, which contains zeros everywhere besides at the position m, where it has the value 1. In case of N total unknowns, the tangent matrix can be obtained by N rows k_m as follows

$$K_T = [\, k_1 \quad k_2 \quad \cdots \quad k_m \quad \cdots \quad k_N \,] \, . \qquad (5.4)$$

The step size in (5.3) has to be chosen such that the approximation of the tangent matrix is as good as possible. Optimal is a very small value for h_m. This choice is, however, not possible due to the limited computer accuracy. Suggestions for the practical choice of step size h_m can be found, e.g. in Dennis and Schnabel (1983) or Schwetlick and Kretschmar (1991). In case of a computer accuracy of η, the following estimate is valid

$$h_m = \nu \, (\, | \, (v_m)_i \, | + \tau \,) \qquad \text{with } \nu = 10^{-3} \ldots 10^{-5} < \sqrt{\eta}, \qquad (5.5)$$

where the number τ should be chosen as $\tau = 10^{-3}$ to prevent that h_m becomes zero for $(v_m)_i = 0$. Using such a step size leads even for the discrete NEWTON scheme to quadratic convergence near the solution point.

A disadvantage of this simple scheme is the large number of evaluations of the residual G needed for the approximation of the tangent matrix in (5.4). In detail, when the equation system has N unknowns then also N evaluations are needed which makes the method inefficient for large equation systems. A more efficient way is to use the numerical differenciation procedure on element level; then only n evaluations related to the size of the element residual vector are needed. Furhtermore, this methodology can be helpful during the development of nonlinear finite elements since the analytical derivation of the tangent matrix can be validated with the help of the numerical tangent obtained from (5.4). Also, for complicated constitutive equations, the incremental constitutive tensor can be determined at each GAUSS point of a finite element using the finite difference scheme. Such applications will be discussed in more detail in Sect. 6.2.

5.1.2 Modified Newton Scheme

A simple modification of the NEWTON-RAPHSON algorithm is related to a scheme in which the tangent matrix is not changed in every step of the iteration in Box 5.1. The most simple procedure is to compute and assemble the tangent matrix only in the first step of each load step, see Fig. 5.2. This procedure is known as modified NEWTON method. It has the obvious advantage that the tangent matrix $K_T(v_i)$ has to be inverted or triangulated only once when the equation system $K_T(v_i) \, \Delta v_{i+1} = -G(v_i, \bar{\lambda})$ in Box 5.1 is solved. This leads to considerable savings of computing time since the

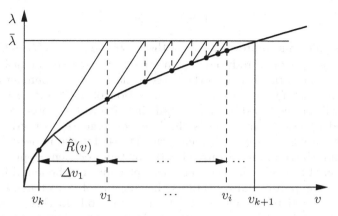

Fig. 5.2 Modified NEWTON-RAPHSON method

inversion or triangulation of $\boldsymbol{K}_T(\boldsymbol{v}_i)$ needs in the worst case $O(N^3)$ operations while the backward substitution needed to compute a new solution with the unchanged tangent matrix only needs $O(N^2)$ operations which is one order of magnitude less.

However these savings may be counterbalanced by the fact that the modified NEWTON method only converges linearly close to the solution point. This behaviour is illustrated in Fig. 5.2 which shows the convergence of the method for a constant tangent matrix during the iterations within one load step. In structural applications, this scheme is also known as method of initial stiffness since $\boldsymbol{K}_T(\boldsymbol{v}_0)$ represents the stiffness of the structure at the initial deformation state $\boldsymbol{v}_0$. Due to its poor convergence behaviour, the modified NEWTON method can only be applied successfully in cases where weak nonlinearities are present.

5.1.3 Quasi-Newton Method

Since the tangential matrix $\boldsymbol{K}_T$ has to be computed, assembled and triangulated at every iteration step within the NEWTON-RAPHSON method, it can be appropriate in case of large dimensional problems to approximate the tangent by a secant. The secant is computed approximately from the known deformation states of the previous iterations. This constitutes the main idea of a quasi-NEWTON method in which the inverse of $\boldsymbol{K}_T(\boldsymbol{v}_i)$, which is needed within the algorithm described in Box 5.1, is approximated with minimal computational effort. Within this process, the equation

$$\boldsymbol{K}_{T\,i}^{QN}\,(\boldsymbol{v}_i - \boldsymbol{v}_{i-1}) = -(\boldsymbol{G}_i - \boldsymbol{G}_{i-1}) \tag{5.6}$$

has to be fulfilled which clearly shows that instead of the tangent matrix $\boldsymbol{K}_T(\boldsymbol{v}_i)$ the secant matrix $\boldsymbol{K}_{T\,i}^{QN}$ is introduced. This relation can also be written for the inverse $\boldsymbol{H}_i^{QN}$ of $\boldsymbol{K}_T^{QN}$ leading to

$$\boldsymbol{H}_i^{QN}\,\boldsymbol{g}_i = \boldsymbol{w}_i\,. \tag{5.7}$$

Here the following abbreviations for the vectors $\boldsymbol{g}_i = -(\boldsymbol{G}_i - \boldsymbol{G}_{i-1})$ and $\boldsymbol{w}_i = \boldsymbol{v}_i - \boldsymbol{v}_{i-1}$ were introduced. The matrix $\boldsymbol{H}_i^{QN}$ denotes a quasi-NEWTON approximation to the inverse of the tangent matrix $\boldsymbol{K}_T$. A number of update algorithms exist for the explicit determination of $\boldsymbol{H}_i^{QN}$. Here the BFGS-method will be discussed in more detail since it has been observed that the associated update formula yields the best convergence properties for solid mechanics problems, see e.g. (Luenberger (1984) for the underlying mathematics and Matthies and Strang (1979)) for finite element applications. An illustration of the associated convergence behaviour can be found in Fig. 5.3, which, for comparison, also contains the tangent of NEWTON method and the slope related to the modified NEWTON method. In this simplified depiction, it can be observed that the evaluation at points $v_k = v_0$ and v_1 leads to a secant which yields a better approximation of the tangent than the modified scheme. However, the secant method does not converge as good as the classical NEWTON method. Mathematically it can be shown that the quasi-NEWTON method converges super linear, see e.g. Luenberger (1984) and Bazaraa et al. (1993).

The BFGS method – named after the originators BROYDEN, FLETCHER, GOLDFARB and SHANNO – was originally developed for nonlinear optimization problems. It belongs to the class of methods which try to approximate the inverse of the tangent matrix in every iteration step of NEWTON method by a secant, see Eq. (5.7). Hence also in case of the application of the BFGS-update, still the algorithm stated in Box 5.1 can be used. Only the equation system under point (2) has to be reformulated by introducing the BFGS-update of the inverse of the secant matrix.

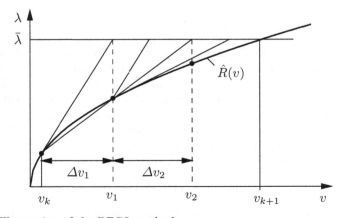

Fig. 5.3 Illustration of the BFGS method

Since a detailed derivation which leads to the BFGS-*updates* goes beyond the scope of this book, only the basic idea behind the update formulae are discussed and the end result are stated. The associate mathematical background and a detailed derivation can be found in, e.g. Luenberger (1984). Applications within the finite element method can be found in Matthies and Strang (1979) or Bathe (1982).

The change of the inverse of the matrix $\boldsymbol{K}_T$ is computed by a rank-two update within the BFGS-method. It is defined as follows

$$\boldsymbol{K}_{T_i}^{-1} \approx \boldsymbol{H}_i^{QN} = (\, \boldsymbol{1} + \boldsymbol{a}_i \, \boldsymbol{b}_i^T \,) \, \boldsymbol{H}_{i-1}^{QN} \, (\, \boldsymbol{1} + \boldsymbol{b}_i \, \boldsymbol{a}_i^T \,) \, . \tag{5.8}$$

Within this equation, the following definitions have been used

$$
\begin{aligned}
\boldsymbol{w}_i &= \boldsymbol{v}_i - \boldsymbol{v}_{i-1}, \\
\boldsymbol{g}_i &= \boldsymbol{G}_{i-1} - \boldsymbol{G}_i, \\
\boldsymbol{a}_i &= \frac{1}{\boldsymbol{g}_i^T \, \boldsymbol{w}_i} \, \boldsymbol{w}_i, \\
\boldsymbol{b}_i &= -\left\{ \boldsymbol{g}_i - \left[\frac{\boldsymbol{w}_i^T \, \boldsymbol{g}_i}{\boldsymbol{w}_i^T \, \boldsymbol{K}_{i-1}^{QN} \, \boldsymbol{w}_i} \right]^{\frac{1}{2}} \boldsymbol{G}_{i-1} \right\} .
\end{aligned}
\tag{5.9}
$$

This *update* scheme preserves the symmetry of $\boldsymbol{K}_{i-1}^{QN}$. Observe that $\boldsymbol{K}_{i-1}^{QN} \, \boldsymbol{w}_i = -\boldsymbol{G}_{i-1}$, and hence this product is already known in (5.9). Due to the form of Eq. (5.8), matrix $\boldsymbol{K}_T$ has to be factorized only once in the beginning of the iteration. This is most time consuming step for an equation system with a large number of unknowns. Furthermore, it can be easily seen that the multiplication of $\boldsymbol{K}_T(\boldsymbol{v}_i)^{-1}$ with $\boldsymbol{G}_{i+1}$, needed for the computation of $\Delta\boldsymbol{v}_{i+1}$ (see point 2 in Box 5.1), involves only scalar products which can be computed in a very fast way. Hence the new approximation of the inverse has not to be computed explicitly. It is sufficient to perform the multiplication with the already known vectors from the iteration steps $0 \le j \le i - 1$. This however requires the storage of all j vectors $\boldsymbol{a}_j$, $\boldsymbol{b}_j$ and $\boldsymbol{g}_j$. In practical application, the number of stored vectors $\boldsymbol{a}_j$, $\boldsymbol{b}_j$ will be limited to a certain number to minimize storage usage. It has been observed that a good value for the number of stored vectors lies in the range $0 < j < 15$. Such procedure is also called partial BFGS-update. Finally, it should be remarked that the simple multiplication of the inverse of $\boldsymbol{K}_T$, according to Eq. (5.8), will destroy the sparse structure of $\boldsymbol{K}_T(\boldsymbol{v})^{-1}$; hence the dyadic products $\boldsymbol{a}_i \, \boldsymbol{b}_i^T$ yield full matrices.

5.1.4 Damped Newton Method, Line-Search

Often it is not possible to apply the full load within one step within a nonlinear solution, even if the problem is purely elastic. This is due to the fact that the NEWTON-RAPHSON method only converges locally, see e.g. Luenberger

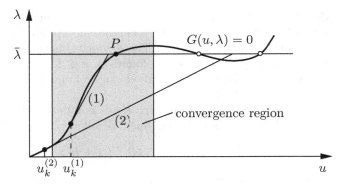

Fig. 5.4 Convergence region of the NEWTON-RAPHSON method

(1984), and that more than one solution is possible in nonlinear problems. In general, convergence to a specific solution can be expected only within a special region. This region is depicted in Fig. 5.4 by the grey region in which the initial starting value of the iteration must be located to obtain a solution related to path (1). Starting values outside the grey region yield tangents which result in an intermediate solution outside the grey region, see path (2) in Fig. 5.4.

Despite these difficulties, there exist several possibilities to compute a solution of the nonlinear problem. At first, the loading can be applied in n_{ink} incremental steps. In that case, the total load, defined by the load parameter $\bar{\lambda}$, will be split as follows: $\bar{\lambda} = \sum_{i=1}^{n_{ink}} \Delta\lambda_i$. Now the algorithm stated in Box 5.1 has to be executed within a loop over all load increments $\Delta\lambda_i$. Hence the nonlinear problem is solved n_{ink}-times which results in a considerable additional expenditure with respect to the original problem. However often in path or history-dependent nonlinear problems (like elasto-plastic or frictional contact problems), it is anyway necessary to apply the load in several steps to capture the correct physical behaviour.

Another possibility is to construct a global method by damping the NEWTON-RAPHSON iteration. For this purpose, the NEWTON-RAPHSON algorithm, see Box 5.1, can be rewritten such that the computation of the new displacement increments, see point (3), is performed by

$$\boldsymbol{v}_{i+1} = \boldsymbol{v}_i + \alpha_i\,\Delta\boldsymbol{v}_{i+1} = \boldsymbol{v}_i - \alpha_i\,\boldsymbol{K}_{T\,i}^{-1}\,\boldsymbol{G}_i\,. \qquad (5.10)$$

With the procedure, a situation is avoided – as depicted by path (2) in Fig. 5.4 – where the solution jumps to another minimum. Such behaviour results usually in a divergence of the NEWTON scheme, especially for large nonlinear equation systems, and hence no solution is obtained at all.

For the damping of the NEWTON method, the damping parameter α_i has to be chosen such that its value is limited by 0 and 1. The selection of a_i could be done heuristically. However, it is more reliable to develop a sound mathematical method for the selection of the damping parameter which of course should not be too costly. Based on the idea of a descent method in which the residuum is reduced in every iteration step

$$\| \, \boldsymbol{G}_{i+1} \, \| = \| \, \boldsymbol{G}(\boldsymbol{v}_i + \alpha_i \Delta \boldsymbol{v}_{i+1}) \, \| < \| \, \boldsymbol{G}_i \, \| \, ,$$

it makes sense to minimize the energy of the system in order to stay in the region of attraction of the solution to point P. Besides the energy, also the function

$$f(\boldsymbol{v}) = \boldsymbol{G}(\boldsymbol{v})^T \, \boldsymbol{G}(\boldsymbol{v}) \tag{5.11}$$

can be minimized since the associated solution is also solution of $\boldsymbol{G}(\boldsymbol{v}) = \boldsymbol{0}$. For the determination of the damping parameter α_i, either the energy of the system or Eq. (5.11) has to be formulated in terms of α_i which yields a scalar relation for the determination of the damping parameter α_i.

As an example, a damped NEWTON scheme for a hyperelastic solid will be discussed. Here the energy $\Pi(\boldsymbol{v})$ can be written, see (3.297), with respect to α_i. Now the requirement for a minimum is $\Pi(\alpha_i) \longrightarrow MIN$. This minimum will be assumed for

$$\frac{\partial \Pi}{\partial \alpha_i} = \frac{\partial \Pi}{\partial \boldsymbol{v}} \frac{\partial \boldsymbol{v}}{\partial \alpha_i} = \boldsymbol{G}(\alpha_i)^T \, \Delta \boldsymbol{v}_{i+1} = 0, \tag{5.12}$$

and hence condition

$$g(\alpha_i) = \Delta \boldsymbol{v}_{i+1}^T \, \boldsymbol{G} \, (\boldsymbol{v}_i + \alpha_i \, \Delta \boldsymbol{v}_{i+1} \, , \bar{\lambda}) = 0 \tag{5.13}$$

can be obtained which the damping parameter α_i has to fulfil. Note that (5.13) represents a nonlinear function with respect to α_i.

A similar result is derived when (5.11) is used

$$g(\alpha_i) = -\boldsymbol{G}_i^T \, \boldsymbol{G} \, (\boldsymbol{v}_i + \alpha_i \, \Delta \boldsymbol{v}_{i+1} \, , \bar{\lambda}) = 0 \, . \tag{5.14}$$

To compute the damping parameter α_i from (5.13) to (5.14), NEWTON method can be applied. Since such a procedure is too costly, other efficient methods to approximately determine α_i have been introduced. These are the so-called inexact *line-search* methods. The solution of (5.13) can be determined by a secant method which is also known as method of *regula falsi*. Within this procedure, only the function $\boldsymbol{G}$ has to be evaluated. The *line-search* is executed only if the function in (5.13) is zero within the interval. With this restriction, which excludes $\alpha_i > 1$, the region of attraction is limited. The solution $g(\alpha_i) = 0$ in the interval $0 \le \alpha_i \le 1$ can be computed iteratively once the sign of the function $g(\alpha_i)$ changes: $g(0) \cdot g(1) < 0$. For this, the values of $g(\alpha_i)$ at 0 and 1 have to be computed. If this condition is fulfilled, then the iteration (with iteration index $k = 1, 2, \ldots$) is started

$$\alpha_i^{k+1} = \alpha_i^k - g(\alpha_i^k) \left[\frac{\alpha_i^k - \alpha_i^{k-1}}{g(\alpha_i^k) - g(\alpha_i^{k-1})} \right] \tag{5.15}$$

to compute α_i for which $g(\alpha_i) = 0$. As already mentioned, it is not necessary to determine the value of α_i for which (5.13) is zero exactly. In most practical applications, the iteration (5.15) can be terminated once the condition

$$| g(\alpha_i^{k+1}) | \le 0.8 \, | g(0) | \tag{5.16}$$

is fulfilled, see e.g. Crisfield (1991). Further criteria for the determination of the iteration (5.15) are provided, e.g. by the ARMIJO rule or the GOLDSTEIN test. These are described in detail in Luenberger (1984) and Bazaraa et al. (1993).

5.1.5 Path-Following or Arc-Length Method

As depicted in Fig. 5.4, the solution path of $G(v, \lambda) = 0$ does not have a unique solution for every load parameter λ. In such case, it is not possible to reach solution paths behind the maximum of the load deflection curve with the methods described so far. In more detail, possible nonlinear solution paths are described in Fig. 5.5. Between points L_1 and L_2, this path even exhibits a decrease of the displacement.

Following the entire solution path of the nonlinear systems of Eq. (5.1) is of practical interest in the case that the overcritical behaviour of a structure has to be known, like in shell buckling. But also material instabilities in softening areas of, e.g. soils or metals have to be determined. Due to this demand, path-following methods were developed which allow to follow arbitrary nonlinear solution paths. The path-following is even possible when singular points are present in which the determinant of the tangent matrix is equal to zero. Methods which allow general path-following are called arc-length or continuation methods. We will describe different variants of these methods in more detail since they present the most general tool to obtain solutions of nonlinear equation systems even on paths in which stable solutions do not exist.

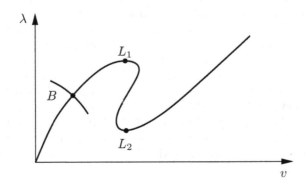

Fig. 5.5 Nonlinear load–deflection diagram

Figure 5.5 characterizes a general nonlinear solution curve of $G(v, \lambda) = 0$. This curve has two limit points L_1 and L_2, which denote a local minimum and maximum of the load level λ. At these points, the determinant of the tangent matrix is zero. This holds also for the bifurcation point B, in which a secondary solution path branches off the primary solution path. For a more precise definition of these points, see Chap. 7, which includes also special algorithms for the determination of these singular points. Here methods will be developed which allow the successive determination of solution points of the nonlinear equation $\boldsymbol{G}(\boldsymbol{v}, \lambda) = \boldsymbol{0}$.

Since path-following methods are well established, a number of different methods are available and documented in the literature. The first work in this field can be found in Riks (1972). Various variants were developed in the following years and investigated with respect to their efficiency and robustness. We refer to Keller (1977), Ramm (1981), Crisfield (1981), Schweizerhof and Wriggers (1986) and Wagner (1991). Overviews can be found in, e.g. Riks (1984), Wagner and Wriggers (1988) and Crisfield and Shi (1991).

The essential idea of a path-following method is to add a constraint condition to the set of nonlinear equations (5.1) from which the unknown load parameter λ can be determined. This extends equation (5.1) and thus the equation system

$$\tilde{\boldsymbol{G}}(\boldsymbol{w}) = \left\{ \begin{array}{c} \boldsymbol{G}(\boldsymbol{v}, \lambda) \\ f(\boldsymbol{v}, \lambda) \end{array} \right\} = \boldsymbol{0}, \quad \boldsymbol{w} = \left\{ \begin{array}{c} \boldsymbol{v} \\ \lambda \end{array} \right\} \tag{5.17}$$

is obtained where the generalized displacement vector $\boldsymbol{w}$ was introduced. Here the general form of the constraint condition is denoted by $f(\boldsymbol{v}, \lambda) = 0$ which is written in terms of the unknown displacement vector and load level. Special techniques can be developed within this framework which include besides different variants of the arc-length methods also load- and displacement control. When the NEWTON-RAPHSON method is be applied to solve Eq. (5.17), a linearization of this set of equations is necessary, see e.g. Schweizerhof and Wriggers (1986). The result of such linearization at a known state $\boldsymbol{w}_i = (\boldsymbol{v}_i, \lambda_i)$ can be stated for the system (5.17) as

$$D\,\tilde{\boldsymbol{G}}(\boldsymbol{w}_i) \cdot \Delta\boldsymbol{w} = \left\{ \begin{array}{c} D\,\boldsymbol{G}(\boldsymbol{v}_i, \lambda_i) \cdot \Delta\mathbf{v} + D\,\boldsymbol{G}(\boldsymbol{v}_i, \lambda_i) \cdot \Delta\lambda \\ D\,f(\boldsymbol{v}_i, \lambda_i) \cdot \Delta\mathbf{v} + D\,f(\boldsymbol{v}_i, \lambda_i) \cdot \Delta\lambda \end{array} \right\} . \tag{5.18}$$

The terms which have to be computed within the linearization process are specified in the following. $D\,\boldsymbol{G}(\boldsymbol{v}, \lambda) = \boldsymbol{K}_T$ represents the tangent matrix which was already introduced in the NEWTON-RAPHSON method, see Sect. 5.1.1. Using Eq. (5.1), the linearization with respect to λ can be determined explicitly: $D\,\boldsymbol{G}(\boldsymbol{v}, \lambda) \cdot \Delta\lambda = -\boldsymbol{P}\,\Delta\lambda$. Furthermore, the definition $D f \cdot \Delta\mathbf{v} = \boldsymbol{f}^T\,\Delta\mathbf{v}$ is introduced, where $\boldsymbol{f}^T = \nabla_v f$ is the gradient of the constraint equation f with respect to the displacement vector $\mathbf{v}$. Finally, the term $D\,f(\boldsymbol{v}, \lambda) \cdot \Delta\lambda = f_{,\lambda}\,\Delta\lambda$ denotes the partial derivative of f with respect to the load parameter λ. With these definitions, the linearization of (5.17) can be stated in matrix form

$$\begin{pmatrix} \boldsymbol{K}_T & -\boldsymbol{P} \\ \boldsymbol{f}^T & f_{,\lambda} \end{pmatrix}_i \begin{Bmatrix} \Delta\boldsymbol{v} \\ \Delta\lambda \end{Bmatrix}_i = - \begin{Bmatrix} \boldsymbol{G} \\ f \end{Bmatrix}_i . \tag{5.19}$$

The matrix of this equation system for the unknown incremental values of displacement vector and load parameter is non-symmetric. One can show that this matrix does not become singular at limit points although the tangent matrix $\boldsymbol{K}_T$ is singular at such points. The matrix in (5.19) is, however, singular at bifurcation points. The non-symmetric equation system (5.19) is usually solved by a partitioning technique to be able to use the symmetric structure of the tangent matrix $\boldsymbol{K}_T$. This procedure leads to an efficient algorithm. However, the property of non-singularity at limit points is lost. This is in practical application of no great significance since the incremental algorithm does not hit, in general, the spot of a limit point directly during path-following.

The partitioning technique – also called block elimination – leads to two equations for the displacement increments $\Delta\boldsymbol{v}_{i+1}$ and the increment of the loading parameter $\Delta\lambda_{i+1}$. By rewriting the first equation of (5.19), the displacement increment is obtained as

$$\Delta\boldsymbol{v}_{i+1} = \Delta\lambda_{i+1}\Delta\boldsymbol{v}_{Pi+1} + \Delta\boldsymbol{v}_{Gi+1} \tag{5.20}$$

with the definitions

$$\Delta\boldsymbol{v}_{Pi+1} = (\boldsymbol{K}_{Ti})^{-1}\,\boldsymbol{P}\,, \quad \Delta\boldsymbol{v}_{Gi+1} = -(\boldsymbol{K}_{Ti})^{-1}\,\boldsymbol{G}_i\,. \tag{5.21}$$

The unknown increment of the loading parameter λ is now determined by the second equation of (5.19). By inserting (5.20), the increment of λ follows as

$$\Delta\lambda_{i+1} = -\frac{f_i + \boldsymbol{f}_i^T \Delta\boldsymbol{v}_{Gi+1}}{(f_{,\lambda})_i + \boldsymbol{f}_i^T \Delta\boldsymbol{v}_{Pi+1}}\,. \tag{5.22}$$

This procedure yields, besides the displacement increment, also the magnitude of the load parameter. The additional effort is not big since the time consuming triangularization of the tangent matrix $\boldsymbol{K}_T$ has only to be performed once. In total, using the block elimination, two scalar products have to be computed for the determination of λ_i with respect to (5.22), and additionally one backward substitution step for the determination of $\Delta\boldsymbol{v}_{Pi+1}$ in $(5.21)_1$.

Since the relations derived so far are linearized in a consistent way, the algorithm based on Eqs. (5.21) to (5.22) will converge quadratically. Contrary to the standard NEWTON-RAPHSON method, the arc-length method needs a predictor step. In this step, the linear equation system $(5.21)_1$ is solved with the right hand side $\boldsymbol{P}$. After that, the load factor $\Delta\lambda_0$ is determined by scaling of the tangent vector which consist of $\Delta\boldsymbol{v}_{P0}$ and $\Delta\lambda_0$. This scaling is based on an increment Δs of the arc-length and yields $\pm\Delta\lambda_0\,\|\Delta\boldsymbol{v}_{P0}\| = \Delta s$, see e.g. Ramm (1981) and Wagner (1991). Figure 5.6 presents a sketch of the

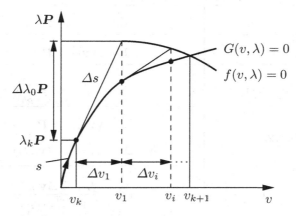

Fig. 5.6 Arc-length method

arc-length method and depicts the fact that the name of the method stems from the parameterization of the solution path by using the arc-length s.

The sign of the incremental loading factor $\Delta\lambda_0$ depends upon the tangent to the solution path and thus will change when the path is followed, see e.g. Fig. 5.5 for the case that the limit point L_1 is bypassed. For positive slopes a positive and for negative slopes a negative sign in front of $\Delta\lambda_0$ has to be chosen. The decision for the correct sign in front of the loading factor can be based on the definiteness of the tangent matrix. However, this leads to difficulties since the tangent matrix is no longer positive definite once a bifurcation point has been bypassed, see e.g. point B in Fig. 5.5. Hence another criterion is needed which is not sensitive to bifurcation points. A simple and easily computable measure is the so-called *current stiffness parameter*, which was introduced in Bergan et al. (1978). This parameter is defined as follows

$$CS_i = \frac{\kappa_i}{\kappa_0} \quad \text{with} \quad \kappa_i = \frac{\boldsymbol{P}^T \Delta \boldsymbol{v}_{i+1}}{\Delta \boldsymbol{v}_{i+1}^T \Delta \boldsymbol{v}_{i+1}} \ . \tag{5.23}$$

Clearly the scalar product $\boldsymbol{P}^T \Delta \boldsymbol{v}_{i+1}$ does not change its sign when a bifurcation point is bypassed, since at that point the load $\boldsymbol{P}$ and the displacement increment $\Delta \boldsymbol{v}_{i+1}$ have the same direction. This is, however, no longer true when a limit point is bypassed. Hence the current stiffness parameter can be used as a measure for the change in direction of the load increment. Further measures which can be defined with respect to a nonlinear solution path can be found in Eriksson (1988).

The complete algorithm for the arc-length method is summarized in Box 5.2. The starting point is given by the displacement vector $\boldsymbol{v}_k$ and its associated load state λ_k, see Fig. 5.6. The notation is conformed to the NEWTON-RAPHSON method. This algorithm differs from the classical

Box 5.2 Algorithm for the arc-length method

1.	Initial values:	$v_0 = v_k$ and Δs
2.	Predictor step	$K_{T0} \Delta v_{P0} = P$
3.	Compute load increment	$\lambda_0 = \lambda_k + \Delta\lambda_0 = \lambda_k \pm \dfrac{\Delta s}{\sqrt{(\Delta v_{P0})^T \Delta v_{P0}}}$
4.	Iteration loop $i = 0, 1, 2, \ldots$	$K_{Ti} \Delta v_{Pi+1} = P$ $K_{Ti} \Delta v_{Gi+1} = -G(v_i, \lambda_i)$
5.	Compute increments	$\Delta\lambda_{i+1} = -\dfrac{f_i + f_i^T \Delta v_{Gi+1}}{f_{,\lambda i} + f_i^T \Delta v_{Pi+1}}$ $\Delta v_{i+1} = \Delta\lambda_{i+1}\Delta v_{Pi+1} + \Delta v_{Gi+1}$
6.	Update	$\lambda_{i+1} = \lambda_i + \Delta\lambda_{i+1}, \quad v_{i+1} = v_i + \Delta v_{i+1}$
7.	Convergence test	$\|G(v_{i+1}, \lambda_{i+1})\| \leq$ TOL $\implies$ Stop otherwise go to 4.

NEWTON-RAPHSON method by the fact that load parameter is computed from the constraint condition $f(v, \lambda) = 0$ for a given arc-length Δs. Optimal is a choice of the constraint in a form which crosses the solution path $G(v, \lambda) = 0$ perpendicularily. This would lead to the most robust method. Unfortunately, the form of the solution path is not known; hence this condition can only be met approximately.

Some selected forms of the constraint condition for the arc-length method are shown in the following table. These are illustrated in Figs. 5.6 to 5.9. Note that the last constraint fulfils the condition of perpendicularity in the best way. Figure 5.7 depicts the classical method of load control which in the frame work of the arc-length method is described by the constraint condition $f = \lambda - \bar{\lambda}$. For a given value $\bar{\lambda}$, this constitutes a straight line parallel to the v-axis. The evaluation of (5.22) for the constraint condition yields $\lambda_k + \Delta\lambda_k = \bar{\lambda}$ or $\Delta\lambda_k = 0$, and hence with (5.20) leads to $\Delta v_k = \Delta v_{Gk}$ which are exactly the equations of the standard NEWTON-RAPHSON method. However, the region behind λ^* cannot be reached by the load control constraint.

This is possible by displacement control which can be observed for the constraint condition $f = v_A - \bar{v}$ in Fig. 5.8. The constraint condition of the displacement control can only be formulated for one component v_A of the displacement field. Hence the user of this method has to select a component of the displacement field which is decisive for the nonlinear process. The specification of Eq. (5.22) leads with $\nabla_v(v_A - \bar{v}) = e_A^T$ (the vector e_A contains

Table 5.1 Examples for constraint conditions

Nr.	Name	Constraint condition
1.	Load control	$f = \lambda - \lambda$
2.	Displacement control BATOZ, DHATT (1979)	$f = v_A - \bar{v}$
3.	Arc-length method RIKS (1972)	$f = (\mathbf{v}_0 - \bar{\mathbf{v}})^T (\mathbf{v} - \mathbf{v}_0)$ $+ (\lambda_0 - \bar{\lambda})(\lambda - \lambda_0)$
4.	Arc-length method CRISFIELD (1981)	$f = \sqrt{(\mathbf{v} - \bar{\mathbf{v}})^T(\mathbf{v} - \bar{\mathbf{v}}) + (\lambda - \bar{\lambda})^2} - \Delta s$

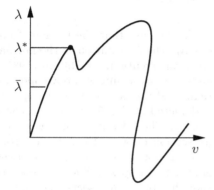

Fig. 5.7 Load control

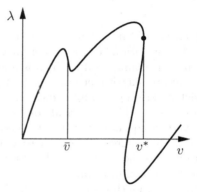

Fig. 5.8 Displacement control

zeros and only at the position A the value 1) to the relation

$$\Delta\lambda_k = -\frac{v_{A_k} - \bar{v} + \mathbf{e}_{Ak}^T \Delta\mathbf{v}_{Gk}}{\mathbf{e}_{Ak}^T \Delta\mathbf{v}_{Pk}} = -\frac{v_{Ak} - \bar{v} + \Delta v_{AGk}}{\Delta v_{APk}}. \tag{5.24}$$

By using a displacement control for the nonlinear solution, the region on the other side of v^* cannot be reached. This is only possible when the arc-length method is applied in conjunction with one of the constraint conditions summarized in Table 5.1. These depend upon the displacement $\mathbf{v}$ as well as on the load parameter λ. The constraint equation introduced by Riks (1972), Table (5.1)$_3$, is linear in $\mathbf{v}$ and λ. It describes a normal plane perpendicular to the tangent $\mathbf{w}_0 - \bar{\mathbf{w}} = (\mathbf{v}_0 - \bar{\mathbf{v}}, \lambda_0 - \bar{\lambda})$ of the curve computed at the last obtained equilibrium state ($\tilde{\mathbf{G}}(\bar{\mathbf{w}}) = \mathbf{0}$). The vector $\mathbf{w}_0$ denotes the solution of the predictor step, see Fig. 5.9. The linearization of the constraint condition, Table (5.1)$_3$, yields

$$\mathbf{f}_k^T = (\mathbf{v}_0 - \bar{\mathbf{v}})^T, \qquad f_{,\lambda k} = \lambda_0 - \bar{\lambda}, \tag{5.25}$$

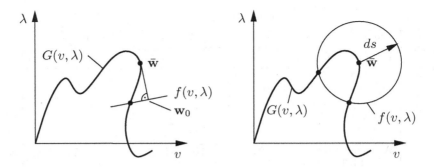

Fig. 5.9 Normal plane   **Fig. 5.10** Spherical surface

which is used in (5.22) to compute $\Delta\lambda_k$. In the case that the tangent from the last iteration, $\boldsymbol{w}_k$, is chosen instead of $\boldsymbol{w}_0$, then a rotation of the normal plane constraint results which is adapted to the solution, see Ramm (1981).

An arc-length method based on a nonlinear constraint was developed by Crisfield (1981) who introduced a sphere around the last converged step, see Table $(5.1)_4$. This spherical constraint is depicted in Fig. 5.10 where the last converged equilibrium state is denoted by $\bar{\boldsymbol{w}}$. The advantage of this condition is that there will be at least one intersection between the spherical constraint and the nonlinear solution path. This is not always the case when the constraint condition introduced by Riks (1972) is applied. The drawback of CRISFIELD methods relates to the fact that the sphere intersects the solution path in most cases at two different points, see Fig. 5.10. Here the "correct" one has to be chosen, see Crisfield (1981). A consistent linearization of the spherical constraint was not provided by Crisfield (1981); it can be found in Schweizerhof and Wriggers (1986). With $g(\boldsymbol{v}, \lambda) = \sqrt{(\boldsymbol{v} - \bar{\boldsymbol{v}})^T(\boldsymbol{v} - \bar{\boldsymbol{v}}) + (\lambda - \bar{\lambda})^2}$, relation

$$\boldsymbol{f}_k^T = (\boldsymbol{v}_k - \bar{\boldsymbol{v}})^T / g(\boldsymbol{v}_k, \lambda_k), \qquad f_{,\lambda k} = (\lambda_k - \bar{\lambda})^T / g(\boldsymbol{v}_k, \lambda_k). \qquad (5.26)$$

is derived for the incremental step k. Further constraint conditions for the arc-length method were discussed in, e.g. Ramm (1981), Fried (1984) and Wagner (1991).

The vector $\boldsymbol{v}$ contains in shell or beam problems besides the nodal displacements also nodal rotations which are, contrary to the displacements, dimensionless. Due to this, it can be advantageous to weight the components of vector $\boldsymbol{v}$ in the constraint conditions, see Box 5.2, by different scaling factors. A discussion related to such weighting of the nodal degrees of freedom (different weights for the displacement and rotations) can be found in Schweizerhof and Wriggers (1986). In the same paper, also the different constraint conditions, summarized in Table 5.1, were compared. It was shown by means of different examples that almost all constraint conditions yield the same robustness of the continuation algorithm with respect to the convergence of the solution. A slightly better performance for large steps can be obtained with the spherical constraint of Crisfield (1981); however this is also the most costly algorithm since always the intersection of two points of the sphere with the solution path has to be checked.

Finally it has to be said that a continuation method, while more robust than the standard undamped NEWTON-RAPHSON scheme, is not a globally convergent algorithm. This means that there exist cases in which algorithms ensuring global convergence have to be applied. One of these algorithm is the *line-search* technique which can be combined with the arc-length method. The procedure corresponds essentially to the algorithm discussed in Sect. 5.1.4. For a detailed description of a combination of arc-length and line-search methods, see Crisfield (1997).

Besides an introduction of line-search algorithms to guarantee global convergence of a continuation methods several methods based on heuristics were developed. These have the goal to adjust the step-length – here the increment of the arc-length Δs – automatically during the solution. This is essential since often regions exist within the nonlinear solution path given by $G(v, \lambda) = 0$ where some solution points can be computed using large step sizes. On the other hand, there can be regions with sharp changes in the curvature (e.g. close to limit points) where a small step size is necessary to obtain a convergent solution. Since the form of the nonlinear solution path is not known a priori, the step size cannot be changed beforehand. A simple but efficient and reliable rule which automatically controls the step size is as follows:

In the case that the number of NEWTON *iterations needed to achieve convergence is larger than 9 steps, the step length is divided into half (note that one has to start from the last computed solution point again). In the case that the number of* NEWTON *iterations is below 5 steps, the step size can be doubled.*

The threshold values 5 and 9 depend upon the solution accuracy TOL, see Box 5.1, and have to be adjusted for different values of TOL. Further algorithms which can control the step size within the arc-length method are discussed in, e.g. Crisfield (1991).

An example in which the arc-length method has to be applied will be considered next. Here the nonlinear response of a relatively simple structure, a star shape dome, is investigated. Its bifurcation behaviour was studied in Wriggers et al. (1988). Here only the primary path of the solution will be computed.

The star shaped dome consists of nonlinear truss elements, see Sect. 9.1, which are modelled by the ST. VENANT elastic constitutive equation with a YOUNG modulus of $E = 1079.6$. The cross sectional area of all trusses is $A = 10$. Top view and front view of the dome are depicted in Fig. 5. 11.

The outer nodes of the finite element mesh are located on a cirlcle with radius $R_o = 50$ while the inner nodes lye on a circle with radius $R_i = 25$. The inner nodes are located at a height of $H_i = 6,216$ and the mid node is

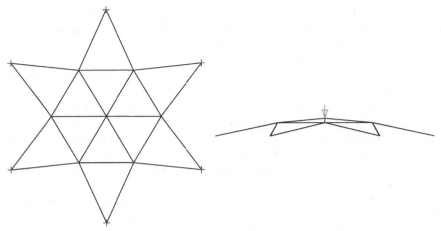

Fig. 5.11 (a) Top view and (b) front view of the star shaped dome

located at a height of $H_i = 8,216$. The structure is simply supported at all outer nodes. A point load is applied at the apex of the dome.

The resulting load displacement curve is plotted in Fig. 5.12. It depicts the complex nonlinear behaviour of this simple structure. It is based on a first snap-through of the inner part of the dome followed by a complex snap-through of the outer part of the dome together with a snap-back of the inner part. Finally, the inner part snaps through again and after that the entire dome depicts again a stable behaviour. To overcome the different limit points in the load displacement curve, the current stiffness parameter was used to

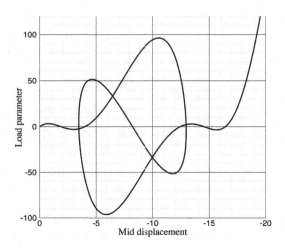

Fig. 5.12 Load displacement curve of the star shaped dome

change load directions. Furthermore, the automatic load stepping scheme, mentioned above, needs to be applied for an efficient solution of this problem. This is due to the fact that the load steps have to be reduced around limit points while they can be increased again once a limit point has been bypassed. In total, 320 load steps were necessary to obtain the load displacement curve in Fig. 5.12.

Exercise 5.1: The system given in Fig. 5.13 is loaded by a prescribed displacement $\bar{v}$ along the boundary. Derive the equations needed to construct an arc-length method under the condition that the boundary loading is enforced via a penalty method.

Solution: Since no external forces are acting on the system, the load vector $\boldsymbol{P}$ in (5.1) is zero. For an enforcement of the boundary displacement, the penalty method is applied. It can be used to include the boundary displacement via the constraint condition

$$\boldsymbol{v} - \lambda\,\bar{\boldsymbol{v}} = \boldsymbol{0}$$

defined at the boundary. This constrains, for a given "load factor" λ, the displacement on the boundary $\boldsymbol{v}$ to be equal to the prescribed displacement $\bar{v}$. This constraint condition can now be considered within the matrix form of the discretized nonlinear equilibrium condition. It yields

$$\boldsymbol{R}(\boldsymbol{v}) + \epsilon\,\mathbb{E}_{\bar{v}}\,(\,\boldsymbol{v} - \lambda\,\bar{\boldsymbol{v}}\,) = \boldsymbol{0}. \tag{5.27}$$

The term $\mathbb{E}_{\bar{v}}$ denotes a diagonal unit matrix which only contains values at nodal points corresponding to the degrees of freedom at which the displacement $\bar{v}$ is prescribed. ϵ is the penalty parameter which has to be chosen sufficiently large such that the constraint is fulfilled accurately enough. This method described in (5.27) for prescribing displacements at the boundary is used since a long time in different finite element programs, see e.g. Bathe (1986). In addition to (5.27), the constraint condition given in (5.17): $f(\boldsymbol{v}, \lambda) = 0$ has now to be formulated for the arc-length method. The linearization yields an equation system for the unknown increments of the nodal displacements $\boldsymbol{v}$ and the load parameter λ

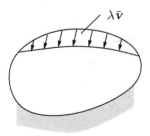

Fig. 5.13 Enforcement of a given displacement within the arc-length method

$$\begin{bmatrix} \boldsymbol{K}_T + \epsilon\,\mathbb{E}_{\bar{v}} & -\epsilon\,\mathbb{E}_{\bar{v}}\,\bar{\boldsymbol{v}} \\ \boldsymbol{f}_{,u}^T & f_{,\lambda} \end{bmatrix} \begin{Bmatrix} \varDelta\boldsymbol{v} \\ \varDelta\lambda \end{Bmatrix} = -\begin{Bmatrix} \boldsymbol{R}(\boldsymbol{v}) + \epsilon\,\mathbb{E}_{\bar{v}}\,(\,\boldsymbol{v} - \lambda\,\bar{\boldsymbol{v}}\,) \\ f(\boldsymbol{v},\lambda) \end{Bmatrix}. \quad (5.28)$$

The solution of this equation system can be obtained analogous to (5.17) leading after some manipulations to

$$\hat{\boldsymbol{K}}_T = \boldsymbol{K}_T + \epsilon\,\mathbb{E}_{\bar{v}} \qquad \text{and} \quad \hat{\boldsymbol{G}} = \boldsymbol{R}(\boldsymbol{v}) + \epsilon\,\mathbb{E}_{\bar{v}}\,(\,\boldsymbol{v} - \lambda\,\bar{\boldsymbol{v}}\,) \qquad (5.29)$$

and

$$\varDelta\boldsymbol{v}_G = -\hat{\boldsymbol{K}}_T^{-1}\,\hat{\boldsymbol{G}} \qquad \text{and} \quad \varDelta\boldsymbol{v}_P = \hat{\boldsymbol{K}}_T^{-1}\,\epsilon\,\mathbb{E}_{\bar{v}}\,\bar{\boldsymbol{v}} \qquad (5.30)$$

with an equation for the displacement increments

$$\varDelta\boldsymbol{v} = \varDelta\boldsymbol{v}_G + \varDelta\lambda\,\varDelta\boldsymbol{v}_P, \qquad (5.31)$$

and for the increment of the load parameter

$$\varDelta\lambda = -\frac{f + \boldsymbol{f}^T\,\varDelta\boldsymbol{v}_G}{(f_{,\lambda}) + \boldsymbol{f}^T\,\varDelta\boldsymbol{v}_P}. \qquad (5.32)$$

These relations are equivalent, besides the definitions in (5.29), to the Eqs. (5.20) and (5.22) of the classical arc-length method.

As an example, the arc depicted in Fig. 5.14 is considered which spans a width related to an angle of $\alpha = 60°$. The inner radius of the arc is $R_i = 100$ and its thickness is $t = 3$. The arc is clamped at both sides. Additionally, a special constraint is introduced which is located at the left half, see Fig. 5.14. It has the width of $b = 2$ and an initial gap of $\delta = 0.1$ with respect to the arc, and hence represents a contact condition. The computation is performed by using isoparametric finite elements with quadratic shape functions to capture the bending behaviour. Three elements are used in thickness direction. In total, 150 elements are applied to discretize the arc. The nonlinear finite element formulation corresponds to the one derived in Exercise 4.3. The constitutive parameters of the ST. VENANT material are the YOUNG modulus $E = 40000$ and the POISSON ratio $\nu = 0.2$. A detailed description of

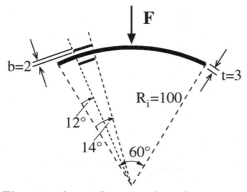

Fig. 5.14 Arc with contact boundary constraints

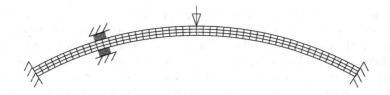

Fig. 5.15 FE-discretization of the arc

the applied contact formulation can be found in Chap. 11. The discretized system is depicted in Fig. 5.15. The computation of the nonlinear solution path is performed by using the arc-length method. Two cases are considered: solution with and without contact. The contact is enforced via the penalty method, see Sect. 11.3. As depicted in Fig. 5.16, the contact constraint leads to a stabilization of the arc structure once the limit point (maximum of the load–displacement curve) is bypassed. This behaviour is related to the acting additional constraint due to contact.

5.2 Solvers for Linear Systems of Equations

As could be observed in the previous sections, the solution of nonlinear problems which lead to systems of the form (5.1) $\boldsymbol{G}(\boldsymbol{v}, \lambda) = \boldsymbol{0}$ require iterative solution methods. Most of the times NEWTON-RAPHSON schemes are applied for the solution. These are based on algorithms as described in Box 5.1 in which a linear system of equations $\boldsymbol{K}_T(\boldsymbol{v}_i)\,\Delta\boldsymbol{v}_{i+1} = -\boldsymbol{G}(\boldsymbol{v}_i)$ has to be solved in each iteration step i.

The equation system stated above is usually symmetrical. However, it can become non-symmetric for special inelastic constitutive models, deformation-

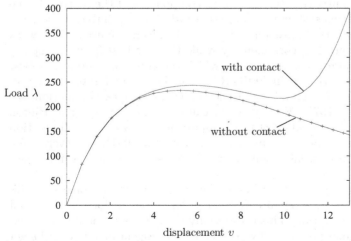

Fig. 5.16 Equilibrium path of the arc with/without contact

dependent loads or beams with finite rotations. A characteristic feature of the tangent matrix $\boldsymbol{K}_T(\boldsymbol{v}_i)$ is its banded structure which results from the locality of the element shape functions. Hence often additional zeros appear, especially in three-dimensional problems, within the band. Thus the notion of sparse matrices is associated with finite element systems.

Since the solution of the above linear equation system requires most of the computing time, it is of great interest to construct fast solvers for such systems, especially since todays, finite element discretizations lead also for nonlinear problems often to systems having wide over 100.000 unknowns. This size actually is one of the factors for the choice of the equation solver, but what "large system" means depends nowadays just on the available computing power. Furthermore, the spatial dimension of the problem has an influence, as will be seen in the following sections which present an overview of some direct and iterative methods for the solution of linear equation systems.

5.2.1 Direct Solvers

Methods which solve linear systems of equations without an iterative process are called direct solvers. There are several algorithms available:

— GAUSS elimination,
— CHOLESKY decomposition,
— frontal solvers,
— sparse solvers and
— block elimination methods.

The associated algorithms will not be described here in detail since these methods can be found in numerous software libraries for mathematical software, see e.g. the references in Golub and van Loan (1989) and Gould et al. (2005). Specially tailored methods for finite elements use the sparsity of the tangent matrix $\boldsymbol{K}_T$. The associated algorithms apply band or profile storage techniques to minimize the storage of zero components of the matrix. In the case that the equation system is so large that it does not fit into core memory, block elimination methods are applied or frontal solvers are used which work on parts of the matrix in core memory while the rest is still on disc space. This leads to a more timely process in which the equation system is solved successively. Codes written in FORTRAN can be found for GAUSS elimination with profile storage in, e.g. Taylor (2000), for block elimination, see Wilson and Dovey (1978). Frontal solver can be found in Owen and Hinton (1980) or Irons and Ahmad (1986). A comparison between block elimination and frontal solvers is provided in, e.g. Taylor et al. (1981). However, these solvers are nowadays outdated since powerful sparse solvers are available, see below.

The advantage of direct solvers lies in their ability to solve even ill-conditioned and negative definite systems of equations as long as round-off does not effect the solution. This is especially interesting for nonlinear applications when singular points are bypassed and the tangent matrix is no longer positive definite. Direct solvers require a lot of operations which is a disadvantage since for large systems these will be of the order $O(N\,b^2)$ for band or profile solvers. Here N is the number of unknowns and b is the width of the band or profile. For really large systems in three-dimensions, this contributes to large solution times. Additionally, so-called fill-in takes place during the elimination phase which changes zero components to non-zero ones. Hence the tangent matrix $\boldsymbol{K}_T$ cannot be stored in compact form without any zero elements. Especially in three-dimensional application, there are many zero components in the tangent matrix $\boldsymbol{K}_T$.

The following table, see Langer (1996), illustrates for a simple case of the LAPLACE equation (only one unknown per node in the discretized system) the increase of the size of the equation system and of the computing times for standard direct solvers. It is assumed in Langer (1996) that the computer has a CPU with 100 MFLOPS which results in the predicted solution times stated in Table 5.2. Especially, for three-dimensional equation systems, it can be observed that a direct solver needs for a discretization with $100^3 = 10^6$ finite elements already 38 GB memory and a computing time of approximately six days. The situation becomes even worse for problems of solid mechanics which have three unknowns per node in the three-dimensional case.

Table 5.2 CPU-time and memory requirement of direct solvers

n	CPU (2D)	Memory (2D)	CPU (3D)	Memory (3D)
20	0.8 ms	31 kB	6.4 s	12.2 MB
50	30 ms	488 kB	65 Min	1192 MB
100	0.5 s	3.6 MB	5.8 days	38.1 GB
200	8 s	30.5 MB	2.1 years	1220 GB
500	5.2 Min	476 MB	—	—

However, as the examples will show, modern sparse solvers have less storage requirements and thus can tackle problems of several million unknowns. In this area, much research was devoted to minimize the disadvantages related to memory requirements and number of operations. In the last ten years, special techniques were developed, based on a compact storage, in which only the elements of the matrix are stored which will obtain a value (fill-in) during the elimination phase. Such solution procedures are subsumed under the term sparse solvers. The associated algorithms and their coding have proved to be efficient for large scale three-dimensional problems. For the mathematical background, see e.g. Duff et al. (1989). A relevant software package is *UMFPACK*, which contains the multi-frontal solver for sparse unsymmetrical

Table 5.3 Memory usage for problems of solid mechanics using direct solvers

n	$M_{Profile}$ (2D)	M_{Sparse} (2D)	$M_{Profile}$ (3D)	M_{Sparse} (3D)
5	2.7 kB	1.7 kB	212 kB	56 kB
10	19 kB	7.3 kB	5.2 MB	480 kB
20	140 kB	29.7 kB	136.8 MB	3.9 MB
40	1.07 MB	120 kB	4.01 GB	31.3 MB

coefficient matrices, see Davis and Duff (1999). Another sparse solver which has a very good performance is *PARDISO*, see Schenk and Gärtner (2004). It works for large sparse symmetrical and unsymmetrical equation systems in a parallel version on shared memory machines. An overview with respect to advantages and disadvantages of different solvers can be found in Gould et al. (2005).

The memory requirement for the two-dimensional plate and the three-dimensional cube depicted in Fig. 5.17 is given in Table 5.3. The edge of plate and cube is subdivided into n finite elements. A point load is applied to one node. Furthermore, the edges of plate and cube are fixed in normal direction at the sides opposite to the load. The memory usage of a profile solver due to Taylor (2000) is compared with a variant of a sparse solver. As already shown in Table 5.2, the extreme increase of memory usage in the three-dimensional case can also be observed in Table 5.3. This is reduced clearly when a direct sparse solver is applied, see e.g. Schenk and Gärtner (2004).

With the growing demand of reliable solutions and the complex problems of today's engineering practice, it is often necessary to apply discretizations

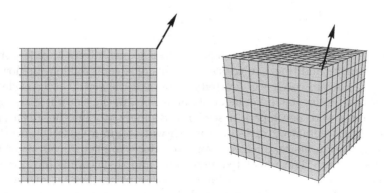

Fig. 5.17 Example for the comparison of memory usage

with over one million finite elements in order to obtain sufficient accurate results. If the problem is, in addition, nonlinear then the equation system has to be solved using NEWTON method. Within this iterative method, a linear equation system of full size has to be solved several times. Hence such large nonlinear engineering problems cannot be solved with direct profile solvers. A way out provide sparse solvers or iterative solution methods since the order with which the size of the system of equation increases cannot be reduced by classical band- or profile solvers.

When band- or profile solvers are used (e.g. in two-dimensional applications), it is of utmost importance to minimize the band or profile of the matrix $\boldsymbol{K}_T$. This reduces as well the memory requirement as the number of operations and with this the total computing time. Several optimization strategies have been developed for this purpose. An overview with regard to problems of structural mechanics can be found in, e.g. Baumann et al. (1990). An essential requirement for band- or profile optimization is that the computing time for the optimization of the band or profile is only a fraction of the total solution time for the equation system. A well known method is the method of Cuthill and McKee (1969). An implementation in FORTRAN can be found, e.g. in Schwarz (1981). A faster technique is described in Hoit and Wilson (1983) who also provided the source code. A detailed discussion of different techniques can be found in Bremer (1986) or Baumann et al. (1990).

5.2.2 Iterative Solution Methods

Finite-element discretizations of nonlinear problems and the associated linearizations lead to large systems of equations with a sparse coefficient matrix within NEWTON method. This is most pronounced in three-dimensional applications within structural mechanics. Besides direct solution methods, also iterative solvers can be applied for the solution of the linear equation systems. Table 5.4, see Langer (1996), depicts the reduction of the memory requirement for the example underlying the results of Table 5.2 when iterative solvers are used. A comparison with Table 5.2 shows clearly that compact storage is a must in three-dimensional problems of large size. Since the compact storage eliminates all zero elements it cannot be applied for direct solvers due to fill-in. Hence compact storage can only be applied within iterative solution procedures.

Table 5.4 Memory requirement for compact storage

n	20	100	200	1000	2000
M (2D)	3.8 kB	81.3 kB	319 kB	7.9 MB	31 MB
M (3D)	83.2 kB	8.3 MB	64.4 MB	7.8 GB	64 GB

Iterative solution strategies are different for symmetric and unsymmetric coefficient matrices. For equation systems with a symmetric coefficient matrix, different methods can be used which are listed below

- method of conjugated gradients (CG-method) for positive definite equation systems,
- LANCZOS algorithm,
- JACOBI- or over-relaxation methods or
- multi-grid methods.

For equation systems with unsymmetric coefficient matrices

- the method of bi-conjugated gradients,
- the GMRES method or
- the CGSTAB algorithm

can be applied.

Iterative solvers are always advantageous when large systems of equations have to be solved since the memory requirement and also the total number of operations is less when compared to direct solvers. However, iterative methods are only efficient when pre-conditioners can be constructed for the problem at hand. These have to be designed such that the number of iterations is low and does not depend upon the number of unknowns. In an optimal case, this would reduce the number of operations to the order $O(N)$. An extensive overview from the mathematical point of view can be found for iterative solvers in Hackbusch (1994). In case of nonlinear problems, an iterative solver will be used within the iterative method to solve the nonlinear system of equations which is, in general, in solid mechanics NEWTON method leading to a sequence of linear problems. The efficient use of iterative solvers within NEWTON type methods needs further considerations since, in the first phases of NEWTON method, the solution can be far away from the correct one. Hence also iterative solvers have not to solve the associated linear equation system with high accuracy which then leads to fewer iterations at this stage.

Method of Pre-conditioned Conjugated Gradients. As an example for iterative solvers, the most popular algorithm applied in solid mechanics is discussed which is the method of pre-conditioned conjugated gradients (PCG-method). It can be applied for the solution of linear equation systems within the i-th iteration step of NEWTON method in Box 5.1: $\boldsymbol{K}_{T\,i}\,\Delta\boldsymbol{v}_{i+1} = -\boldsymbol{G}_i$. To simplify notation, the iteration index is omitted which yields $\boldsymbol{K}_T\,\boldsymbol{v} = \boldsymbol{f}$ with $\boldsymbol{f} = -\boldsymbol{G}$.

The gradient method is based on the fact that if $\bar{\boldsymbol{v}}$ is solution of $\boldsymbol{K}_T\,\boldsymbol{v} = \boldsymbol{f}$ then $\bar{\boldsymbol{v}}$ is also minimum of

$$f(\boldsymbol{v}) = \frac{1}{2}\,\boldsymbol{v}^T\,\boldsymbol{K}_T\,\boldsymbol{v} - \boldsymbol{f}^T\,\boldsymbol{v}. \qquad (5.33)$$

Thus this method includes the assumption of positive definiteness and symmetry of $\boldsymbol{K}_T$ which is not always given in nonlinear analysis, see e.g. descending pathes which have to be solved by arc-length procedures discussed in Sect. 5.1.5. By searching for the zero point of the minimal problem, an iterative method can be constructed. In detail, the minimum of

$$f(\boldsymbol{v}_k + \alpha_k \, \boldsymbol{s}_k) = \min_\alpha \, f(\boldsymbol{v}_k + \alpha \, \boldsymbol{s}_k) \tag{5.34}$$

is determined for a descent direction $\boldsymbol{s}$. This yields an equation for the scalar α_k

$$\alpha_k = \frac{\boldsymbol{r}_k^T \, \boldsymbol{s}_k}{\boldsymbol{s}_k^T \, \boldsymbol{K}_T \, \boldsymbol{s}_k} \qquad \boldsymbol{r}_k = \boldsymbol{f} - \boldsymbol{K}_T \, \boldsymbol{v}_k \,. \tag{5.35}$$

The direction $\boldsymbol{s}_k$ follows from

$$\boldsymbol{s}_k = \boldsymbol{C}^{-1} \, \boldsymbol{r}_k, \tag{5.36}$$

where $\boldsymbol{C}$ is a pre-conditioning matrix (for details see Sect. 5.2.2). With $\nabla_v \, f = -(\boldsymbol{f} - \boldsymbol{K}_T \, \boldsymbol{v})$, the form $\boldsymbol{s}_k = -\boldsymbol{C}^{-1} \, \nabla_v \, f(\boldsymbol{v}_k)$ is obtained from (5.33). Thus the descent direction corresponds to the negative gradient of the function f which explains the name gradient method.

Better convergence properties and thus less number of iterations for the solution of a linear equation provides the method of conjugated gradients. In this method, the descent direction to the minimum is given by the linear combination

$$\boldsymbol{p}_k = \boldsymbol{s}_k + \beta_k \, \boldsymbol{p}_{k-1} \,. \tag{5.37}$$

The parameter β_k is determined from the condition

$$\boldsymbol{p}_k^T \, \boldsymbol{K}_T \, \boldsymbol{p}_{k-1} = (\boldsymbol{s}_k + \beta_k \, \boldsymbol{p}_{k-1})^T \, \boldsymbol{K}_T \, \boldsymbol{p}_{k-1} = 0 \,. \tag{5.38}$$

Hence $\boldsymbol{p}_k$ and $\boldsymbol{p}_{k-1}$ are conjugated with respect to the coefficient matrix $\boldsymbol{K}_T$. This method is stated in Box 5.3 in its algorithmic version where matrix reformulations were introduce to enhance the efficiency of the method.

In the case that $\boldsymbol{K}_T$ is unsymmetric, the method described in Box 5.3 does not work since $\boldsymbol{K}_T$ is no longer positive definite which, however, was presumed in the derivation of the PCG-method. In that case a variant of the PCG-method has to be developed. One possibility is to replenish the unsymmetric equation system with its symmetric part as shown below

$$\begin{bmatrix} \boldsymbol{0} & \boldsymbol{K}_T \\ \boldsymbol{K}_T^T & \boldsymbol{0} \end{bmatrix} \begin{Bmatrix} \boldsymbol{w} \\ \boldsymbol{v} \end{Bmatrix} = - \begin{Bmatrix} \boldsymbol{f} \\ \boldsymbol{0} \end{Bmatrix} \,. \tag{5.39}$$

Box 5.3 Method of pre-conditioned conjugated gradients

PCG $(\boldsymbol{v}, \boldsymbol{f}, \boldsymbol{K})$		
		Choose starting vector: $\boldsymbol{v}_0$
$\boldsymbol{r}_0$	$=$	$\boldsymbol{f} - \boldsymbol{K}_T \, \boldsymbol{v}_0$
FOR		$k = 0, 1, 2, \ldots$
$\boldsymbol{C} \boldsymbol{s}_k$	$=$	$\boldsymbol{r}_k (\text{Pre-conditioning})$
α_k	$=$	$(\boldsymbol{s}_k)^T \boldsymbol{r}_k$
IF $k = 0$		THEN
		$\boldsymbol{p}_0 = \boldsymbol{s}_0$
		ELSE
		$\beta_k = \alpha_k \, / \, \alpha_{k-1}$
		$\boldsymbol{p}_k = \boldsymbol{s}_k + \beta_k \, \boldsymbol{p}_{k-1}$
		END IF
$\boldsymbol{z}_k$	$=$	$\boldsymbol{K}_T \, \boldsymbol{p}_k$
δ_k	$=$	$(\boldsymbol{p}_k)^T \boldsymbol{z}_k$
γ_k	$=$	$\alpha_k \, / \, \delta_k$
$\boldsymbol{r}_{k+1}$	$=$	$\boldsymbol{r}_k - \gamma_k \, \boldsymbol{z}_k$
$\boldsymbol{v}_{k+1}$	$=$	$\boldsymbol{v}_k + \gamma_k \, \boldsymbol{p}_k$
ε_k	$=$	$(\boldsymbol{r}_{k+1})^T \boldsymbol{r}_{k+1}$
UNTIL		CONVERGENCE($\varepsilon_k \leq TOL$)

Having now again a symmetric equation system (5.39), the PCG-method can be applied. Additionally, the special structure of (5.39) can be explored leading to an efficient implementation. In total, this procedure yields an algorithm which is two times as expensive as it is for the symmetric problem, see e.g. Fletcher (1976). More stable and robust methods were constructed after that. These are the GMRES-method, see Saad and Schultz (1986), and the CGSTAB algorithm, see den Vorst (1992). The algorithm for CGSTAB is provided in the following. In this method, the vectors $\boldsymbol{s}$, $\boldsymbol{u}$, $\boldsymbol{w}$, $\boldsymbol{y}$ and $\boldsymbol{z}$ are vectors which are needed for intermediate storage during the computation.

The main cost for the algorithms in Boxes 5.3 and 5.4 is related to the solution of the equation system for pre-conditioning and to the multiplication of matrix $\boldsymbol{K}_T$ with the vectors $\boldsymbol{p}_k$, $\boldsymbol{y}$ and $\boldsymbol{z}$. Often more than 100 iterations are needed to converge to the solution within the PCG-method; hence it is essential to optimize multiplications $\boldsymbol{K}_T \, \boldsymbol{p}_k$ in order to obtain a competitive algorithm. Thus compact storage of $\boldsymbol{K}_T$ is a must such that only non-zero elements of the coefficient matrix are multiplied with the components of the vector. Furthermore, the storage requirement for $\boldsymbol{K}_T$ decreases considerably, see Table 5.3. Another approach is to perform the computation at element level and then to assembly the resulting vector. This method avoids storage of K_T completely but needs more computational effort; however, one may fit the vector/matrix multiplications into the computer cash which then leads to high efficientcy. For a more in-depth treatment of such algorithms, see e.g. Golub and van Loan (1989).

Techniques for Pre-conditioning. As already mentioned, the condition number $\kappa(K_T) = \| K_T \| \| K_T^{-1} \|$ of the large sparse equation system $K_T v - f = 0$ has a considerable influence on the accuracy of direct solvers and on the rate of convergence of iterative solvers. The method of conjugated gradients reduces in the worst case in every step the residual by the factor $(\sqrt{\kappa} - 1)/(\sqrt{\kappa} + 1)$, see Golub and Ortega (1996). The smaller the condition number $\kappa(K_T)$ the better will be the convergence rate. Hence it is desirable to perform a pre-conditioning within the algorithm in Box 5.3. The aim is to lower the condition number, and hence to decrease the number of iterations needed to solve $K_T \Delta v = f$.

The main idea of pre-conditioning is to find a matrix $bisC$ which is similar to K_T but considerablly simpler to invert. By assuming that the pre-conditioning matrix C is symmetric and positive definite and that H is regular, then relation

$$C = H H^T \qquad (5.40)$$

Box 5.4 CGSTAB Algorithm

CGSTAB (v, f, K_T)		
		Choose starting values:
v_0 , r_0	$=$	$f - K_T v_0 , u_0 = p_0 = 0 , \gamma_0 = 0 , \delta_0 = 10^{30}$
FOR		$k = 0, 1, 2, \ldots$
α_{k+1}	$=$	$r_0^T r_k$
β_{k+1}	$=$	$(\alpha_{k+1} \gamma_k) / (\delta_k \alpha_k)$
p_{k+1}	$=$	$r_k + \beta_{k+1}(p_k - \delta_k u_k)$
Cz	$=$	p_{k+1} (Pre – conditioning)
u_{k+1}	$=$	$K_T z$
γ_k	$=$	$\alpha_{k+1} / (u_{k+1}^T r_0)$
w	$=$	$r_k - \gamma_{k+1} p_{k+1}$
Cy	$=$	w (Pre – conditioning)
s	$=$	$K_T y$
δ_{k+1}	$=$	$(s^T r_{k+1}) / (s^T s)$
r_{k+1}	$=$	$w - \alpha_{k+1} s$
v_{k+1}	$=$	$v_k + \gamma_{k+1} z + \delta_{k+1} y$
ε_k	$=$	$(r_{k+1})^T r_{k+1}$
UNTIL		CONVERGENCE($\varepsilon_k \le TOL$)

can be written where H can, for example, be the left upper triangular matrix of a CHOLESKY decomposition of C. The linear equation system $K_T v - f = 0$ can then be reformulated as

$$H^{-1} K_T H^{-T} H^T v - H^{-1} f = \tilde{K}_T \tilde{v} - \tilde{f} = 0 \qquad (5.41)$$

with $\tilde{K}_T = H^{-1} K_T H^{-T}$, $\tilde{v} = H^T v$ and $\tilde{f} = H^{-1} f$. The matrices C or H have to be selected such that the condition number $\kappa(\tilde{K}_T) < \kappa(K_T)$ is reduced. To define the matrix C, consider

$$H^{-T} \tilde{K}_T H^T = H^{-T} H^{-1} K_T H^{-T} H^T = C^{-1} K_T . \qquad (5.42)$$

The choice of $C = K_T$ yields $\tilde{K}_T = I$, and the condition number is optimal ($\kappa(\tilde{K}_T) = 1$). This, however, does not make sense since the original equation system for pre-conditioning has to be solved. However, it can be derived from Eq. (5.42) that C has to be chosen similar to K_T to obtain a reduction of the condition number κ. Since C has to be inverted within each iteration step, the choice of C has to be such that the solution of the equation system in the pre-conditioning phase is as efficient as possible. By decomposing the sparse matrix K_T into $K_T = E + D + F$, the following form for C can be selected after Axelsson (1994)

$$\begin{aligned} C &= (D + \omega E) D^{-1} (D + \omega F) \\ &= D + \omega E + \omega F + \omega^2 E D^{-1} F , \end{aligned} \qquad (5.43)$$

where ω is a relaxation parameter, E denotes the lower triangular matrix, F the upper triangular matrix and D the diagonal of K_T. In the following, four "classical" methods for pre-conditioning will be discussed.

- **Diagonal Scaling:** The most simple possibility to improve the condition number of K_T is diagonal scaling, see Golub and Ortega (1996). For $\omega = 0$, the pre-conditioning in (5.43) reduces to scaling using diagonal elements since $H = D^{1/2}$. In this case, the matrix C is replaced by the diagonal D of K_T. It is obvious that this scaling requires only few operations since the solution of the equation system in the pre-conditioning phase in Box 5.3 is with $C = D$ trivial.

- **JOR Pre-conditioning:** Another possibility to improve the condition number $\kappa(K_T)$ is the application of several JOR relaxation loops, see Schwetlick and Kretschmar (1991). The associated iteration is given by:

$$v_{i+1} = v_i - \omega D^{-1} (K_T v_i - f) . \qquad (5.44)$$

In this case, pre-conditioning in Box 5.3 is carried out implicitly by an algorithm. Comparisons in Boersma and Wriggers (1997) for several solid mechanics applications show that a sufficient pre-conditioning for the CG-method in Box 5.3 is obtained for $\omega = 0.3$ and by using just four JOR relaxation loops. Carrying out more JOR relaxation loops did not lead to a considerable acceleration of convergence of the iterative equation solver in the test cases.

– **Polynomial pre-conditioning:** The inverse of the tangent matrix $\boldsymbol{K}_T^{-1}$ can be approximated by a finite weighted NEUMANN series as a polynomial $P(\boldsymbol{K})$ of low order

$$C = P(\boldsymbol{K}) = \sum_{i=0}^{k} \gamma_i \boldsymbol{K}^i \qquad (5.45)$$

with small k ($k \approx 2$ or 3 is optimal), see Saad (1985). The parameters γ_i have to minimize the polynomial

$$\| \boldsymbol{I} - P(\boldsymbol{K})\boldsymbol{K} \|_2 = \max_{\lambda_i \in \sigma(K)} |1 - \lambda_i P(\lambda_i)|, \qquad (5.46)$$

where $\sigma(K) = \{\lambda\}_{i=1,\cdots,N}$ is the spectrum of $\boldsymbol{K}$. Following Golub and van Loan (1989), this goal can be reached by using CHEBYCHEV-polynomials or by alternatively using a polynomial of least squares with JACOBI-weighting functions.

$$\gamma(\lambda) = \lambda^{\alpha-1} (1 - \lambda)^\beta \qquad (5.47)$$

with $\alpha = 1/2$, $\beta = -1/2$. These polynomials $P(\boldsymbol{K})$ of order $k-1$ minimize

$$\max_{\lambda \in [a,b]} |1 - \lambda P(\lambda)| , \qquad (5.48)$$

where the parameters a and b follow from $a = \lambda_{\min}$ and $b = \lambda_{\max}$. The polynomials of least squares work quite well despite the fact that no eigenvalue estimation is needed as for the CHEBYCHEV-Polynomials. The latter can be advantageous, especially for computers with vector- or parallel architecture.

– **Incomplete Factorization:** The incomplete CHOLESKY-pre-conditioning, see Golub and van Loan (1989), uses an incomplete factorization (IC) of the tangent matrix

$$\boldsymbol{K}_T = \boldsymbol{L}_T \boldsymbol{L}_T^T, \qquad (5.49)$$

where only the entries of $\boldsymbol{K}_T$ in $\boldsymbol{L}_T$ are stored which are non-zero. By this procedure, no additional storage is required in the pre-conditioning phase since the fill-in obtained in a CHOLESKY decomposition is suppressed.

Alternatively, the incomplete decomposition can be enhanced by a diagonal stabilization leading to the modified incomplete pre-conditioning (MIC). In this case, all entries of $\boldsymbol{L}_T$, which are associated with the zero-entries of $\boldsymbol{K}_T$ are added to the diagonal terms of $\boldsymbol{L}_T$. This method leads to a more stable factorization. The pre-conditioning matrix is given for both variants by $\boldsymbol{C} = \boldsymbol{L}_T \boldsymbol{L}_T^T$.

Mainly, direct solvers are applied in classical finite element codes for the solution of sparse equation systems. The reason for this is that iterative solvers cannot be viewed as a general tool ("black box") which are applicable to arbitrary problems in an efficient way. Examples are solids with incompressible materials which are not satisfactorily solvable with the techniques described so far. Here special pre-conditioning schemes are needed, see e.g. Haase et al. (2001). The same problem applies to engineering structures which consist of different structural members, such as solids, shells and beams. Hence iterative solvers need special pre-conditioning techniques for the problem at hand, and are thus "one" pre-conditioning technique cannot be applied in general. This will be shown in Sect. 5.3 by means of examples.

5.2.3 Parallel Equation Solvers

Real engineering problems lead often to discretized systems with a large number of unknowns (10^5–10^7) and hence demand high computing power and capacity for the solution. Examples are complex structures such as drilling rigs, car bodywork, air crafts or shiphulls. Optimization and dynamic analysis of large structures needs extensive computations. Often nonlinear effects have to be considered additionally like inelastic material behaviour or large deformations.

The speed and performance of serial computers grows still very fast; however, at the same time the engineering models become more and more refined. Thus parallel computers are employed for complex simulations. Advantage is a higher computing power and often more essential large main memory. The use of a shared memory parallel computer needs relatively few changes within the finite element code, see e.g. for equation solvers Schenk and Gärtner (2004). On the contrary, the application of distributed memory computers demands new software development with regard to domain decomposition and equation solution. Hence, in the latter case, new algorithms and software structures have to be developed in order to efficiently use parallel computers. One of the most time consuming tasks in a finite element simulation is the solution of the system of equations. Since direct solvers require a lot of communication between the different processors, they can only be efficiently applied in a parallel solver with a small amount of processors. Hence, for massive parallel systems with several hundred nodes, iterative solution strategies have to be used. These need less communication and also less storage. In the last years, a number of different solution algorithms were created with the goal to minimize the number of necessary operations to the order $O(N)$ (with N–number of equations). The relevant algorithms are based on the method of pre-conditioned conjugate gradient methods, hierarchical bases or multigrid methods, see e.g. Hackbusch (1994) and Hackbusch (2003). Practical implementations can be found in, e.g. Balay et al. (2001) and Balay et al. (2004).

The basic ideas which lead to a parallel finite element program will be described in the following, for details, see e.g. Papadrakakis (1993). In this section, the concept of domain decomposition without overlapping domains, see Fig. 5.18, will be described in more detail.

The communication between the domains Ω_s proceeds via the outer nodes of the discretization of Ω_s while the inner nodes are only relevant for the

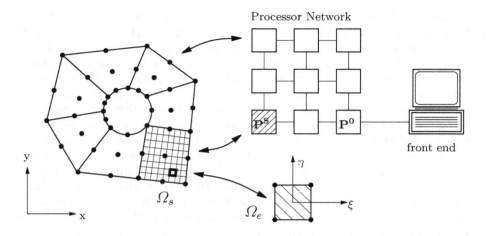

Fig. 5.18 Assignment of a finite element mesh to a processor within domain decomposition

prevailing processors. The domain decomposition is realized by a coarse grid or graph. Each coarse grid element or graph represents a subdomain Ω_s which is associated with exactly one processor $\mathbf{P}^S$, see Fig. 5.18. Hence mesh generation can be performed without communication when non-matching meshes are used, see e.g. Farhat and Roux (1991). For matching meshes, the information of number and spacing of finite elements on the boundaries of the subdomains must be known. Residual vectors and tangent matrices resulting from the elements Ω_e can be assembled without communication. When the division of the mesh in subdomains is performed, care of a good load balancing has to be taken which is needed in order to obtain the same computational load on every processor used for the parallel solution. For this task, there exist different strategies which try to equalize the load on each processor, for an overview see e.g. Axelsson and Barker (2001).

The following steps have to be performed during the solution of a finite element problem using a parallel computer with distributed memory:

1. preparation of input and output for the individual processors,
2. parallel mesh generation,
3. parallel assembly of residuals and tangent matrices,
4. parallel solution of the equation system and
5. parallel postprocessing of the solution.

Here the parallel solution of the finite element equations is the main focuss. For input and output, the specific communication routines have to be considered related to the used parallel computing system. Since mesh generation can be generally performed in parallel on each subdomain, standard tools can be applied, once the spacing of nodes is known on the inter subdomain boundaries. Related software which constructs the sub-domains from a given finite element mesh can be found in Balay et al. (2004). Differences to standard meshing depends upon the load balancing which has to be taken into account during the mesh generation phase, see further Axelsson and Barker (2001). For the communication within the iterative solution of the finite element equation system, a special numbering technique can be applied in which first the corner nodes of the subdomain are numbered, after that nodes C along the interface between the subdomains and finally all inner nodes I are numbered. Using this procedure, a special form of the tangent matrices is obtained which allows for an eventually necessary different treatment of the nodes in a simple way, see Meyer (1990). Since non-overlapping domain decomposition is applied, see Fig. 5.18, neither the computation of the element contributions nor the assembly of the matrices and vectors needs communication; hence these two processes run naturally in parallel.

Conjugated Gradient Method. Within the parallel solution of the finite element equation system, differences to serial versions of the same algorithms are found. The problem is that, due to the coupling of the subdomains, it is not possible to solve for the unknowns without taking the coupling into account. This coupling has to be considered within the solution phase. Here first a parallel version of the pre-conditioned gradient method is discussed, see Sect. 5.2.2.

As in the serial solution of equations, the choice of the adequate preconditioner is important for an efficient solver. Instead of classical preconditioning techniques, e.g. diagonal scaling or incomplete CHOLESKY decomposition, see Sect. 5.2.2, a parallel SCHUR-complement pre-conditioner for the CG-method will be presented which was developed in Meisel and Meyer (1995). This approach employs an iterative solver for the unknowns related to nodes which couple the subdomains. A direct solver will be applied for the unknowns related to the nodes within a domain. The method is described in detail in Meisel and Meyer (1995). Within this approach, the mentioned special numbering of the nodes is used within the subdomains Ω_s which yield the following structure of the stiffness matrix

$$\boldsymbol{K}_s = \left[\begin{array}{cc} \boldsymbol{K}^C & \boldsymbol{K}^{CI} \\ \boldsymbol{K}^{IC} & \boldsymbol{K}^I \end{array} \right]_s . \tag{5.50}$$

Such numbering simplifies the treatment of the coupling nodes C and the internal nodes I since no sorting is necessary. Based on this numbering, the stiffness matrix of one subdomain on processor $\mathbf{P^S}$ has the form

$$\boldsymbol{K} = \left(\begin{array}{cc} \boldsymbol{I} & \boldsymbol{K}^{CI}\boldsymbol{K}^{-I} \\ \boldsymbol{0} & \boldsymbol{I} \end{array} \right) \left(\begin{array}{cc} \boldsymbol{S} & \boldsymbol{0} \\ \boldsymbol{0} & \boldsymbol{K}^I \end{array} \right) \left(\begin{array}{cc} \boldsymbol{I} & \boldsymbol{0} \\ \boldsymbol{K}^{-I}\boldsymbol{K}^{IC} & \boldsymbol{I} \end{array} \right) \tag{5.51}$$

with the SCHUR-complement $\boldsymbol{S} = \boldsymbol{K}^C - \boldsymbol{K}^{CI}\boldsymbol{K}^{-I}\boldsymbol{K}^{IC}$. Matrices with negative superscript denote inverse matrices. When a suitable pre-conditioner $\boldsymbol{V}^C$ is found for the SCHUR-complement S and a pre-conditioner V^I is selected for the inner matrix then a positive definite pre-conditioner for K_T can be constructed as follows

$$C_s = \begin{pmatrix} I & K^{CI} V^{-I} \\ 0 & I \end{pmatrix} \begin{pmatrix} V^C & 0 \\ 0 & V^I \end{pmatrix} \begin{pmatrix} I & 0 \\ V^{-I} K^{IC} & I \end{pmatrix}. \tag{5.52}$$

This pre-conditioner can be applied within the conjugated gradient method described in Box 5.5. The form (5.52) is especially well suited since its inverse form can be determined explicitly, see Axelsson (1994),

$$C_s^{-1} = \begin{pmatrix} I & 0 \\ -V^{-I} K^{IC} & I \end{pmatrix} \begin{pmatrix} V^{-C} & 0 \\ 0 & V^{-I} \end{pmatrix} \begin{pmatrix} I & -K^{CI} V^{-I} \\ 0 & I \end{pmatrix}. \tag{5.53}$$

The degrees of freedom can be condensed out in the inner parts of the subdomains when choosing of $V^I = K^I$ as pre-conditioner. Within the pre-conditioning procedure, it is necessary to further distinguish between edge or vertex and boundary nodes. Vertex nodes belong to more than two boundaries. For the unknowns of these nodes, a simple diagonal pre-conditioning will be applied since communication is minimized that way. The unknowns belonging to the boundary will be pre-conditioned by a linear LAPLACE operator where the resulting equation system is solved by fast FOURIER transform. The parallel version of the PCG-method is depicted in Box 5.5. In this algorithm, two different types of communication between the processors are employed. These are related to the update of a vector at the coupling boundaries between two domains and the summation of a scalar over the entire region Ω which was computed in the domains Ω_s. A local vector in Box 5.5 is denoted by the index (s) while a global vector does not have any index. The subroutine $\mathbf{comm}[\mathbf{x}^{(s)}]$ is used in Box 5.5 to update a vector at the coupling boundaries; an example is $\mathbf{x} = \mathbf{comm}[\mathbf{x}^{(s)}]$. This operation is necessary since the matrix-vector product can only be computed with a global vector due to compatibility of the displacement field. The result yields a local vector, e.g. $\mathbf{y}^{(s)} = K_T^{(s)} \mathbf{x}$, since the matrix $K_T^{(s)}$ is only defined on the processor $\mathbf{P^S}$.

Furthermore, several scalar products have to be computed within the PCG-method. Within the discussed parallel version, a scalar product can be computed only between a local and a global vector. The part $h^{(s)} = \mathbf{x}^T \mathbf{x}^{(s)}$ is computed in processor $\mathbf{P^S}$. The complete scalar product follows then from a global communication $h = \mathbf{sum}[h^{(s)}]$ which is equivalent to $h = \sum_s h^{(s)}$.

Another possibility to pre-condition the PCG-solver can be obtained by using a multi-grid method. These methods are either based on a hierarchical mesh structure or an algebraic decomposition of the stiffness matrix, see e.g. Hackbusch (1994), Boersma and Wriggers (1997), Meynen et al. (1997) and Wriggers and Boersma (1998).

Multi-grid method. Another possibility for the parallel solution of systems of equations is provided by the multi-grid method (MG)), see e.g. Bastian

Box 5.5 Parallel version of the pre-conditioned conjugate gradient method

PCG($\mathbf{v}, \mathbf{f}, \mathbf{K}_T$)

$$\mathbf{r}_0^{(s)} = \mathbf{K}_T^{(s)} \mathbf{v}_0 - \mathbf{f}^{(s)}$$

$$\mathbf{r}_0 = \mathbf{comm}(\mathbf{r}_0^{(s)})$$

FOR $\quad k = 0, 1, 2, \ldots$

$$\mathbf{s}_k = \mathbf{C}^{-1} \mathbf{r}_k \text{ (pre-conditioning using (5.53))}$$

$$\alpha_k^{(s)} = \mathbf{s}_k^T \mathbf{r}_k^{(s)}$$

$$\alpha_k = \mathbf{sum}(\alpha_k^{(s)})$$

IF $k = 0$ THEN

$$\mathbf{p}_0 = \mathbf{s}_0$$

ELSE

$$\beta_k = \alpha_k / \alpha_{k-1}$$

$$\mathbf{p}_k = \mathbf{s}_k + \beta_k \mathbf{p}_{k-1}$$

END IF

$$\mathbf{z}_k^{(s)} = \mathbf{K}_T^{(s)} \mathbf{p}_k$$

$$\delta_k^{(s)} = \mathbf{p}_k^T \mathbf{z}_k^{(s)}$$

$$\delta_k = \mathbf{sum}(\delta_k^{(s)})$$

$$\gamma_k = \alpha_k / \delta_k$$

$$\mathbf{r}_{k+1}^{(s)} = \mathbf{r}_k^{(s)} - \gamma_k \mathbf{z}_k^{(s)}$$

$$\mathbf{v}_{k+1} = \mathbf{v}_k - \gamma_k \mathbf{p}_k$$

$$\mathbf{z}_k = \mathbf{comm}(\mathbf{z}_k^{(s)})$$

$$\mathbf{r}_{k+1} = \mathbf{r}_k - \gamma_k \mathbf{z}_k$$

$$\varepsilon_k^{(s)} = \mathbf{r}_{k+1}^T \mathbf{r}_{k+1}^{(s)}$$

$$\varepsilon_k = \mathbf{sum}(\varepsilon_k^{(s)})$$

UNTIL   CONVERGENCE ($\varepsilon_k < TOL$)

and Wittum (1994) and Hackbusch (2003). A special variant of the multi-grid method is the algebraic multi-grid method (AMG) which can easily be implemented (as black box) into finite element programs, see e.g. Stueben (1983), Ruge (1986), Kočvara and Mande (1987) and Haase et al. (2001). The AMG does not need a special hierarchical mesh structure. It computes the coefficient matrices which belong to different meshes directly from the given tangent matrix of the system, see Brandt et al. (1985), Brandt (1986) and for engineering applications Boersma and Wriggers (1997).

Four points have to be considered within multi-grid methods:

1. a mesh hierarchy with l levels,
2. the determination of transfer operators,
3. an algebraic equation of the form $\mathbf{K}_{Tl} \mathbf{v} = \mathbf{f}$ at each level l and
4. the determination of smoothing operators.

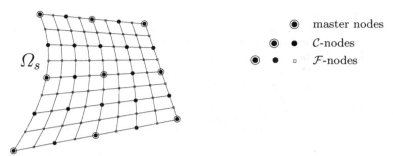

Fig. 5.19 Partitioning of the finite element mesh in coarse and fine grid nodes $\mathcal{C}$- and $\mathcal{F}$

These points are also valid for an algebraic method. However, there are small deviations in the definition of the transfer and smoothing operators when compared to the classical multi-grid method.

The four points are discussed in more detail in the following.

1. Mesh hierarchy: The classical multi-grid method is based on a mesh hierarchy which derives from a coarsening of the fine grid by appropriate geometrical operations. Within the AMG method, the coarse mesh is defined as a subset of the nodes which are associated with the fine mesh without reference to the geometrical mesh structure. Hence there is no geometrical interpretation of the coarse meshes in AMG. The set of all nodal points is subdivided into coarse grid nodes $\mathcal{C}$ and fine grid nodes $\mathcal{F}$, see Fig. 5.19. It illustrates the above definitions by using a typical regular mesh. Additionally, the master nodes are depicted which are introduced to define the shape of the mesh. Before mesh coarsening, it has to be decided how many levels are needed to solve the problem. These levels depend upon the number of nodes (nk) of the fine mesh and the spatial dimension of the problem (ndm). The following relations yield an estimate for the number of levels

$$l_{max} = \left\lfloor \log(nk) \ / \ \log(2^{ndm}) \right\rfloor .$$

For a regular mesh with 2^l nodes in each spatial direction l levels are obtained. This number is usually selected for a multi-grid solution.

The definition of the coarse grid nodes $\mathcal{C}$ is a linear process which has to be performed over all nodes (with numbers $1, \ldots, n_l$). Let $\mathcal{U}\{node\}$ be the set of all neighbouring nodes of the node under consideration ($node$). Then a coarsening follows by executing the algorithm specified in Box 5.6.

Box 5.6 Coarsening procedure

a) Determine the essential coarse grid nodes
b) Define $\mathcal{F} = \emptyset, \mathcal{C} = \{1, \ldots, n_l\}$
c) FOR $node = 1, 2, \ldots, n_l$ DO
    IF $node \in \mathcal{C}$ THEN
        $\mathcal{F} = \mathcal{F} + \mathcal{U}\{node\}/\mathcal{C}$
        $\mathcal{C} = \mathcal{C} - \mathcal{U}\{node\}/\mathcal{C}$

This procedure has to be performed within every mesh related to a level $l = 1, \ldots, l_{max} - 1$. Essential coarse grid points are the ones which have a physical meaning, e.g. for the description of the correct boundary conditions.

2. Transfer-operators: Information between two different levels is transfered by a prolongation, $\boldsymbol{P}$, and a restriction operator $\boldsymbol{R}$. Operator $\boldsymbol{P}$ will be computed directly from the entries in the coefficient matrix in case of the AMG method. In standard multi-grid methods, this task is performed by using the shape functions related to the coarse mesh. The restriction operator $\boldsymbol{R}$ is defined by $\boldsymbol{R} = \boldsymbol{P}^T$. Starting from a row of the linear equation system $\boldsymbol{K}_T \boldsymbol{v} = \boldsymbol{f}$, the transfer operators can be constructed

$$K_{T\,ii}\, v_i = -\sum_j K_{T\,ij}\, v_j + f_i\,. \tag{5.54}$$

The prolongation operator $\boldsymbol{P}$ follows from an approximation of Eq. (5.54). The value v_i in a fine grid node $\mathcal{F}$ is then approximated in terms of coarse grid nodes $\mathcal{C}$ by

$$v_i = \frac{1}{\displaystyle\sum_{k \in \mathcal{C}} |K_{T\,ik}|} \sum_{j \in \mathcal{C}} K_{T\,ij}\, v_j\,. \tag{5.55}$$

This relation guarantees that the sum of all weights for each fine grid node is equal to one. Then an entry in the transfer operator for a degree of freedom i and a neighbouring degree of freedom k is given by

$$P_{ik} = \frac{|K_{T\,ik}|}{\sum_{j \in \mathcal{C}} K_{T\,ij}}\,. \tag{5.56}$$

Details of this procedure can be found, in e.g. Boersma and Wriggers (1997).

3. Equation systems on each level: The coarse grid matrix will be computed algebraically using the transfer operators $\boldsymbol{R}$ and $\boldsymbol{P}$ which yields $\boldsymbol{K}_\mathcal{C} = \boldsymbol{R}\,\boldsymbol{K}_\mathcal{F}\,\boldsymbol{P}$. In the case that $\boldsymbol{R}$ is chosen as $\boldsymbol{P}^T$, then all symmetry properties of the fine grid matrix $\boldsymbol{K}_\mathcal{F}$ are preserved within the coarse grid matrix $\boldsymbol{K}_\mathcal{C}$.

4. Smoothing operator: Within the AMG-method, smoothing follows as in standard multi-grid methods by a GAUSS-SEIDEL iteration or the method of conjugated gradients. The smoothing operation will be described by the operator S.

The parallel version of the AMG-method can then be obtained in two steps. In a starting phase, the coefficient matrix, the transfer operators and the associated coarse grid matrices are determined for all levels. After that the iterative solution of the equation system follows in a second phase.

Starting Phase: Determination of coarse grid points by mapping the distribution of the nodes onto the associated processors which are located on the boundaries of the subdomains. Since these nodes belong to different processors, an exchange of data is necessary. Also the partitioning of inner nodes in fine and coarse grid nodes has to be performed within a processor. For these tasks, first the transfer operators will be computed on the subdomain boundaries (data exchange) then the transfer operators are determined within the subdomain. With this, the coarse grid matrix is computed in parallel. After $\boldsymbol{R}\,\boldsymbol{K}_{\mathcal{F}}\,\boldsymbol{P}$ are computed, a coarse grid matrix $\boldsymbol{K}_{\mathcal{C}}$ is obtained on each subdomain. Note that the matrix product for the determination of the coarse grid matrix $\boldsymbol{R}\,\boldsymbol{K}_{\mathcal{F}}$ will never be executed directly since this destroys the efficiency of the method. A possibility for a fast computation of $\boldsymbol{K}_{\mathcal{C}}$ is described, e.g. in Boersma and Wriggers (1997).

Iteration Phase: A smoothing operation has to be performed within the iterative solution of the equation system. This will be executed by using the parallel smoothing operator $S_{\mathbf{P}}(\mathbf{v}, \boldsymbol{f})$. Within this step, several different methods can be applied, e.g. a parallel GAUSS-SEIDEL algorithm, an incomplete CHOLESKY triangulation or the method of conjugated gradients. Within the iterative solution, vectors have to be transferred between the different computational steps. This is - besides the treatment of coupling nodes - a fully parallel task. Box 5.7 describes the algorithm of the parallel algebraic multigrid

Box 5.7 The parallel algebraic multi-grid method (pAMG)

$$
\begin{array}{ll}
\multicolumn{2}{l}{\underline{\text{pAMG}(\,l, \mathbf{v}, \boldsymbol{f}, \nu\,)}} \\[4pt]
1) & \mathbf{v} \quad \leftarrow \quad S_{\mathbf{P}}(\,\mathbf{v}, \boldsymbol{f}) \\
2) & \boldsymbol{r} \quad = \quad \text{comm}(\boldsymbol{K}_{T\,l}\,\mathbf{v}) - \boldsymbol{f} \\
3) & \boldsymbol{f}_{l+1} \quad = \quad \text{comm}(\boldsymbol{R}\,\boldsymbol{r}) \\
4) & \text{IF } l = l_{max} \text{ THEN} \\
& \qquad \text{Solve } \boldsymbol{K}_{T\,l\,max}\,\mathbf{w} = \boldsymbol{f}_{l\,max} \\
& \text{ELSE} \\
& \qquad \text{Perform } \nu \text{ steps:} \\
& \qquad \text{AMG}(l+1, \mathbf{w}, \boldsymbol{f}_{l+1}, \nu_{l+1}) \\
& \text{END IF} \\
5) & \mathbf{v} \quad \leftarrow \quad \mathbf{v} - \text{comm}(\boldsymbol{P}\,\mathbf{w}) \\
6) & \mathbf{v} \quad \leftarrow \quad S_{\mathbf{P}}(\,\mathbf{v}, \boldsymbol{f})
\end{array}
$$

Fig. 5.20 The different cycles

method (pAMG). For a solution of the linear equation system $\boldsymbol{K}_T\,\boldsymbol{v} = \boldsymbol{f}$, the algorithm is started with the initial conditions $\mathrm{pAMG}(1, \boldsymbol{v}, \boldsymbol{f}, \nu)$. Here the tangent matrix $\boldsymbol{K}_T$ belongs to the finest grid – now denoted by $\boldsymbol{K}_{T\,1}$ - and the vector $\boldsymbol{f}$ describes the loading. The operator $S_{\mathbf{P}}$ is used to perform a smoothing step which is adapted with respect to the parallel computer. The necessary communication routines were already described within the parallel CG-method.

Here the index l is used, which characterizes the grid sequence for all operators applied within the parallel algebraic multi-grid method. Index 1 is related to the finest grid, whereas the index l_{max} is associated with the coarsest grid. Furthermore, the parameter ν has been introduced which indicates how often the multi-grid method has been called. The choice $\nu = 1$ leads to the so-called V-cycle, $\nu = 2$ results in the W-cycle, see Fig. 5.20 ($l = 3$ belongs here to the coarsest grid). The F-cycle is a mixture of the V- and W-cycle. The V-cycle only needs one step at the coarsest grid level, see Fig. 5.20. Contrary, within the W-cycle, there are l_{max}^2 steps at the coarsest grid level necessary. The F-cycle needs l_{max} steps. Hence the V-cycle has a higher parallel efficiency than the W- or F-cycle.

By applying parallel multi-grid methods to solid mechanics problems, it has been observed that the parallel algebraic multi-grid method (pAMG) can act as a robust pre-conditioner for other iterative methods such as the CG-algorithm described in Box 5.5.

5.3 Examples Related to Algorithms and Equation Solvers

The following numerical examples illustrate the behaviour of the iterative solution strategy of some in Chap. 5 mentioned algorithms. The computations, regarding single processor systems, were performed using the finite element program FEAP, see Zienkiewicz and Taylor (2000b), on a PC with an Intel dual core processor and 2.3 GHz clock rate.

5.3.1 Rubber Block

2d Rubber Block. The block shown in Fig. 5.21 is constrained by a rigid plate on the top. This plate is moved downward under the assumption that there is a frictionless interface between plate and block. The initial geometry

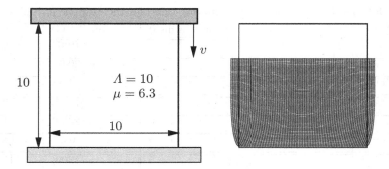

Fig. 5.21 Elastic rubber block   Deformed configuration of
under pressure load   the last load step

is depicted in Fig. 5.21 with the geometrical data of the block and the values of
the constitutive parameters Λ and μ for a Neo-HOOKE material, see Sect. 3.5.

The basis of the finite element formulation is explained in Sect. 4.2.3;
hence details of the implementation will not be discussed, we just note that
a standard displacement formulation is used.

Since the problem is symmetric, only one half of the system is discretized
by 60×120 elements. The total load is applied in different numbers of load
steps. First 10 load steps are used, after that the problem is solved by applying
the load in 5 steps, 2 steps and finally only one step. Since the problem
is elastic, it is path independent, and hence the different numbers of load
steps do not influence the final solution. The deformed configuration which
belongs to the final load step is depicted in Fig. 5.21. It corresponds to a final
compression of 30%.

Four different solution methods were applied to simulate the response of
the structure. These are the NEWTON-RAPHSON method with two direct and
one iterative solver and the BFGS-method. For all methods, the iterations
where terminated when the residual norm was below the tolerance of 10^{-8}.
The NEWTON-RAPHSON method needs the smallest number of equilibrium
iterations when compared with the BFGS-method, as can be seen in Table 5.5.

Furthermore, the number of iterations increases over proportional within
the higher load steps when the BFGS-method is applied. This is related to
the increase in nonlinear deformations. For larger load steps (load applied in
two and one step), the BFGS method does not converge any longer. However,
looking at the total time needed to solve the problem with ten load steps,
the BFGS-method is the most efficient method.

The choice of the termination tolerance has a considerable influence on the
number of iterations within the solution algorithm. The choice of a tolerance
of 10^{-4} leads to a decrease of the number of iterations for all three methods.
However, within the NEWTON method combined with the direct solver, the
number of iterations is not so much reduced since often only one step reduces

Table 5.5 Iteration numbers and computing times for the example in Fig. 5.21

Load step	Newton (direct)	Newton (Pardiso)	Newton (PCG)	BFGS
1	4	4	4 (987)	6
10	6	6	6 (2047)	16
Time	8.6 s	15.1 s	10.9 s	8.5 s
1	4	4	4 (1127)	7
5	6	6	6 (2330)	21
Time	4.5 s	7.9 S	6.3 s	5.4 s
1	5	5	5 (1309)	17
2	7	7	7 (2500)	-
Time	2.3 s	4.1 s	3.2 s	-
1	8	8	8 (2546)	-
Time	1.5 s	2.7 s	2.2 s	-

the residual norm from 10^{-4} to 10^{-8}. This is not true for the BFGS method. Hence it is important for a user to pick the right tolerance for termination so that on one hand the physical problem is solved correctly and on the other hand the solution effort is minimized.

It is also interesting to note that the conjugated iterative solution method with diagonal preconditioning needs many iterations to solve the linear equation system within each load step, see number of all iterations in a load step in brackets in Table 5.5. These iteration numbers have to be compared with the total number of unknowns in the equation system which is here 14.459. Hence the cg-method with diagonal preconditioning is not competitive in this example. Note that the number of cg-iterations increases for higher load steps; this is related to the worsening of the condition number of the tangent matrix K_T with increasing deformation due to the influence of the material tangent and the geometric stiffness terms.

The difference in the computing time of the two direct equation solvers, a standard skyline solver, see Taylor (2000), and the sparse solver *PARDISO*, see Schenk and Gärtner (2004), is related to the overhead needed wthin the sparse solver which slows its computing time down for this small problem.

3d Rubber Block. Next the corresponding three-dimensional problem is investigated. For this purpose, a block made of a compressible hyperelastic material will be discretized and solved by the same methods as in the example above. Again a two direct and one iterative solver are applied in conjunction with the NEWTON-RAPHSON method. The iterative algrithm uses the cg-method, described in Box 5.3, with diagonal preconditioning.

The block, shown in Fig. 5.22, has the dimensions $10 \times 10 \times 5$. The constitutive parameters are chosen such that LAMÉ-constants assume the values

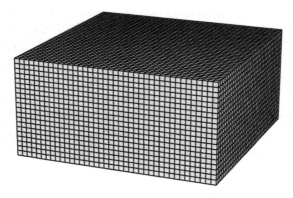

Fig. 5.22 Three-dimensional example for the comparison of different solution algorithms

$\Lambda = 830$ and $\mu = 50$. The block is pressed downwards by a rigid plate; the total compression is 30%. Friction between plate and block is neglected. The load will be applied in 2 and in 3 load steps to compare the different behaviour of the iterative solution methods.

Due to symmetry reasons, only one quarter of the block is discretized by $20 \times 20 \times 20$ eight node brick elements; hence the system has a total of 25620 unknown nodal displacements.

The deformed configuration and the contour of the normal stress σ_{33} in vertical direction is depicted in Fig. 5.23 for a compression of 30%. The number of iterations and the total computing time can be found in Table 5.6. The termination criterion, TOL in Box 5.1, for the residual norm is set to $TOL = 10^{-4}$.

SIGMA_33

-4.79E+02
-1.00E+02
-9.30E+01
-8.60E+01
-7.90E+01
-7.20E+01
-6.50E+01
-5.80E+01
-5.10E+01
-4.40E+01
-3.70E+01
-3.00E+01
-2.06E+01

Fig. 5.23 Deformed block and normal stresses σ_{33} in vertical direction

Table 5.6 Number of iterations and computing times for the example in Fig. 5.22

Load step	Newton (direct)	Newton (Pardiso)	Newton (PCG)	BFGS
1	5	5	5 (773)	16
2	5	5	5 (837)	23
3	6	6	6 (967)	24
Comp. time	348 s	66 s	18 s	105 s
1	5	5	5 (875)	27
2	6	6	6 (1046)	24
Comp. time	241 s	46 s	13 s	79 s

It can be observed that in this example the combination of the NEWTON method with the iterative cg-solver yields the most efficient solution scheme since its total solution time is one order of magnitude less than the times for the other solvers. Only the sparse solver *PARDISO* seems to be competitive since it is "only" four times slower. The numbers in brackets are related to the number of iterations within each load step. The cg-solver needs 160 iterations on average to solve the linear equation system within each NEWTON iteration.

When a larger load step is selected, the number of NEWTON iterations do not increase much. This is different for the BFGS-method. Here the total computing time is increasing with larger load steps.

5.3.2 Solid with an Inclusion

Homogenization is needed for the determination of effective material properties of heterogeneous materials. Within this process, often large three-dimensional representative volume elements (RVE) have to be analysed. In this example, a heterogeneous material is subjected to a constant strain field. The sample consists of a unit cell of a hyperelastic matrix material with YOUNG modulus of $E = 30$ and POISSON ratio of $\nu = 0.3$ and a nearly incompressible particle, YOUNG'S modulus $E = 8$ and POISSON ratio $\nu = 0.499$. The finite element mesh of the unit cell of size $2 \times 2 \times 2$ is shown in Fig. 5.24a. It consists of 56000 elements with 74536 nodes leading in total to 164697 degrees of freedom.

The unit cell is loaded by a prescribed displacement which stems from the constant strain state $\mathbf{u} = \mathbf{H}\,\mathbf{X}$ with

$$\mathbf{H} = \begin{bmatrix} 0.01 & 0.03 & 0.03 \\ 0.03 & 0.01 & 0.03 \\ 0.03 & 0.03 & 0.01 \end{bmatrix}.$$

The finite element simulation is performed using NEWTON method. It yields the deformation of the RVE depicted in Fig. 5.24b. Different load steps were used to obtain the final state. The solution times for the different solvers as

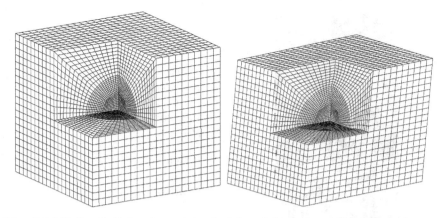

Fig. 5.24 Unit cell, finite element mesh: (**a**) undeformed and (**b**) deformed

well as the number of iterations of the cg-solvers are reported in Table 5.7.
The standard skyline solver in FEAP produced a tangent stiffness matrix
which could not be storered in main memory, and thus this solver could
not be applied for the solution of this problem. The sparse solver *PARDISO*
was still able to generate a solution; however, the solution times are 4 to
5 times higher when compared to the iterative cg-solvers. The latter also
differ depending upon the applied pre-conditioner. The GAUSS-SEIDEL pre-
conditioner needs less iterations than the cg-solver with diagonal scaling. The
total number of iterations is about three times less when the load is applied
in only one step. However, the total computing time is 5% higher since the
GAUSS-SEIDEL pre-conditioning is more time consuming.

The situation changes when either concentrated forces are applied or
bending dominates the response of the solid. Then the iterative solver with
diagonal pre-conditioning will need a lot more iterations, the cg-solver with
GAUSS-SEIDEL pre-conditioning will use more iterations, but then solve the

Table 5.7 Number of iterations and computing times for the example in Fig. 5.24a

Load step	direct (Pardiso)	iter (CG-diag)	iter (cg-Gauss-Seidel)
1	4	4 (5252)	4 (1741)
10	4	4 (5714)	4 (1817)
Comp. time	8452 s	1629 s	1795 s
1	5	5 (7425)	5 (2368)
2	5	5 (7743)	5 (2403)
Comp. time	2114 s	443 s	467 s
1	5	5 (9643)	5 (3033)
Comp. time	1054 s	276 s	290 s

Material data:
$E = 30000$
$\nu = 0.30$
$Y_0 = 30$
$H = 1$

Dimensions:
Height $h = 8$
Width $b = 4$

Fig. 5.25 Plate with hole with elasto-plastic material

problem in shorter time. The sparse solver *PARDISO* is not affected by these differences in loading and thus will use the same time for a linear solution within NEWTON method.

5.3.3 Elasto-Plastic Plate with Hole

Within this example, which exhibits elasto-plastic behaviour, iterative and direct solution methods are compared. The plate with a hole, see Fig. 5.25, is made up of steel and undergoes elasto-plastic deformations during the loading process. To model the hardening behaviour of the steel, a linear isotropic hardening rule, see Sect. 3.3.2, is applied. The derivation of the integration procedures for the evolution equations for the plastic flow can be found in Sect. 6.2.2. This section also contains the associated incremental material tensor, consistent with the integration procedures.

Material data for E, ν, Y_0, H and the dimensions (h, b) of the plate can be found in Fig. 5.25. The simulations are performed by assuming geometrically linear behaviour. Due to symmetry, only a quarter of the plate is modelled by 3750 six node triangular finite elements. This leads to an equation system with 15199 unknown nodal displacements. The plate is loaded by a constant displacement of $v = 0.15$ at its upper side which is employed in 25 load steps.

Table 5.7 summarizes the behaviour of the different algorithms used also in the last examples. The first load step is purely elastic. In the other load steps, the plastic zone evolves until the cross section at the hole is almost completely plastic, see Fig. 5.26.

The termination criterion for the equilibrium iteration was chosen to be $TOL = 10^{-4}$. The convergence behaviour of the different algorithms can be found in Table 5.8. Clearly, the NEWTON method in combination with a

Table 5.8 Number of iterations and computing times for
the example in Fig. 5.25

Load step	Newton (direct)	Newton (Pardiso)	Newton (PCG/Diag)	Newton (PCG/GS)
1	2	2	2 (3030)	2 (825)
5	2	2	2 (3032)	2 (825)
10	6	6	6 (6056)	6 (1622)
15	8	8	8 (10412)	8 (2797)
20	9	9	9 (15458)	9 (4144)
25	10	10	10 (18044)	10 (4961)
Time	72,8 s	68,5 s	187 s	125 s

direct equation solver is the most efficient. The iterative conjugate gradient solver with diagonal scaling needs many iterations, and thus is the slowest method. However, the iterative GAUSS-SEIDEL scheme is also slower than the direct solvers. This is due to the fact that two-dimensional problems have a larger condition number than adequate three-dimensional problems, and thus more iterations are needed for solving the linear equation system by

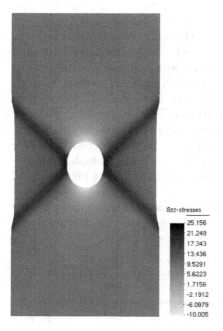

Fig. 5.26 Distribution of stresses σ_z in the 6th load step

the iterative method. Here the GAUSS-SEIDEL scheme is working better and needs one fourth of the iterations of the diagonally scaled method. However, since it needs more computational effort it is only 1.5-times faster. Note that the condition number of the tangent matrix K_T becomes larger when the response of the plate is dominated by plasticity, see Fig. 5.26. This is the case in the last two load steps which again leads to more iterations for the solution of the linear systems within NEWTON method. Since the sparse solver *PARDISO* needs some overhead to compute the sparse structure of the tangent matrix it is only marginally faster for a problem of this size. The plot of the out of plane stresses σ_z is shown in Fig. 5.26 for 100-times magnified deformed configuration.

5.3.4 Problems Solved on Parallel Computers

Iterative solvers are often used on parallel computing platforms, as discussed in the last section. Here examples will show the influence of pre-conditioning and the convergence of algorithms; the results were also discussed in Wriggers and Meynen (1995). While the results are relatively old, they can still be used to depict the general features of the underlying methods and algorithms.

A plate with a hole, see Fig. 5.25, is considered in the first example. It will be solved using a conjugated gradient method (CG). Hence the influence of different pre-conditioners will be of interest and discussed next.

Only a quarter of the plate will be discretized due to symmetry reasons. The load is applied using displacement control. The material behaviour of the plate is assumed to be elasto-plastic with linear isotropic hardening. The constitutive parameters are chosen as $E = 70000 \text{ N/mm}^2$, $H = 2000 \text{ N/mm}^2$, $Y_0 = 243 \text{ N/mm}^2$ and $\nu = 0.2$ which is different from the data reported in Fig. 5.24. The length and width is $36 \times 20 \text{ mm}$. The behaviour of different pre-conditioner can be observed for different numbers of finite elements from Fig. 5.27. The number of iterations decrease for increasing effort for pre-conditioning. The best results are provided by the Schur complement pre-conditioner which does not depend upon the number of finite elements and thus is well suited for parallel computing.

The structure shown in Fig. 5.28 is used in the second example to discuss the efficiency of the parallel algorithms. For this purpose, the two measures

Table 5.9 *speed-up* and *scale-up* for the problem related to Fig. 5.27

speed-up					scale-up				
N_P	n_p	$\sum n$	n_p	$\sum n$	N_P	n_p	$\sum n$	n_p	$\sum n$
16	882		3362		16		14112		53792
32	462	14112	1722	53792	32	882	28224	3362	107584
64	242		882		64		56448		215168

speed-up and *scale-up* will be introduced. The problem in Fig. 5.28 was computed on 16, 32 and 64 nodes (processors) of the parallel system which lead to a number of unknowns from 14112 to 215168, depending on the discretization level. The structure of the mesh was not changed within these computations.

Efficiency of a parallel solution can be measured through two factors, *speed-up* and *scale-up*.

Speed-up monitors the quality of the parallelization

$$Sp(n) = \frac{\text{computing time with 1 processor}}{\text{computing time with } n \text{ processors}} = \frac{T_1}{T_n} \qquad (5.57)$$

Ideally, the speed-up function has a value of 1 for its tangent. This means that a problem can be solved in half time when doubling the number of processors. However, loss due to communication leads in reality to longer computing times when the number of processors is enlarged which is monitored by $Sp(n)$, see (5.57).

The problem depicted in Fig. 5.28 was solved for a constant size of 14112 and 53792 unknowns on different numbers of processors, see Table 5.9. Here N_P denotes the number of processors, n_p the number of unknowns per processor and $\sum n$ the total number of unknowns. A parallel solution with 16 and 64 nodes was compared in Fig. 5.29a. The loss of efficiency can be observed clearly for the larger number of processors since the computing load per node is getting smaller, and then communication times play a mayor role. Thus efficiency can not purely be measured using the definition of speed-up in (5.57).

Based on that observation, another measure for monitoring the efficiency was introduced. It is called *scale-up* and can be used to see how good preconditioning and communication work for increasing numbers of processors.

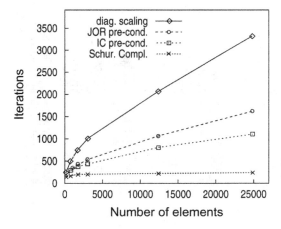

Fig. 5.27 Influence of different pre-conditioners when applied to an elasto-plastic plate problem

$$Sc\left(n\right) = \frac{n \times l \text{ Unknowns on } n \text{ processors}}{l \text{ Unknowns on 1 processor}} = \frac{T_n(l \times n)}{T_1(l)} \tag{5.58}$$

To demonstrate the scale-up measure, a problem with increasing number of unknowns is solved on a system with increasing number of processor such that the computing load on each processor is constant. In an ideal case, the computing time will be constant. Since, however, a cg-solver is not independent upon the element size which reduces within a scale-up simulation, it is possible to observe a decline of the scale-up factor for increasing number of total unknowns.

The problem shown in Fig. 5.28 was again solved by holding the computing load on each processor constant. This was achieved by increasing the number of processor nodes with 882 and 3362 unknowns per processor, see Table 5.9. The total number of unknowns increases within these simulations. All computing times are scaled the simulation time on 16 processors, see Fig. 5.29b.

The hardware for these investigations was at that time a Parsytec Supercluster with 64 nodes each with 8 MB of main memory which is not adequate in standard anymore. However, this system was transputer based and thus had a communication power which was directly adjusted to the computing power. Hence this system was good for general comparison of different solution techniques.

The behaviour of the parallel algebraic multi-grid method (pAMG) is discussed by means of an inelastic shell problem of a pinched cylinder, described in Sect. 9.5.3, see also Meynen et al. (1997). Geometry, loading and material data follow from Fig. 9.25. The cylinder is discretized by the shell elements described in Sect. 9.4.6. Due to symmetry, only one eights of the shell is discretized. Here the influence of the problem size and processor load on the

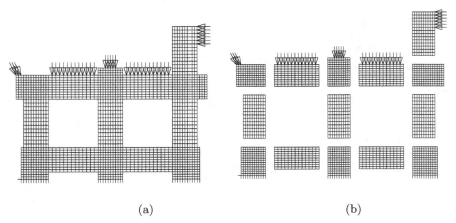

(a)   (b)

Fig. 5.28 (a) Structure and loading, (b) Distribution of the mesh onto 14 processors using subdomains

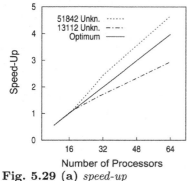
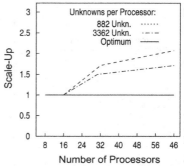

Fig. 5.29 (a) *speed-up*   **Fig. 5.29 (b)** *scale-up*

solution time are investigated. Different finite element discretizations, from 32×32 up to 320×320 elements, were considered which lead to a total number of unknowns from 5445 to 515205 . The finite element discretizations were run on 4, 8 and 16 nodes which consist of a PowerPC processor for computing and a transputer for communication. This system had a fast processor while the transputer for communication was relatively slow.

Figure 5.30 depicts the speed-up for a parallel CG method which employs the pAMG method as pre-conditioner. The results were computed using a V-cycle which has superior behaviour in this case. The simulation with largest load on a node yield the best speed-up. Due to the bad balancing between computing power and communication only a speed-up of 2.23 was reached instead of 4 when the system was solved on 16 nodes instead of 4. More computing nodes did not make sense in this example since the load on a node would be low due to the small coarse grids in the pAMG scheme. Note

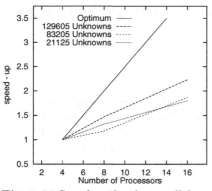
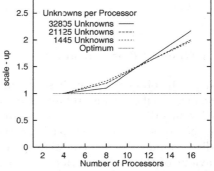

Fig. 5.30 Speed-up for the parallel solution of a shell   **Fig. 5.31** Scale-up for the parallel solution of a shell

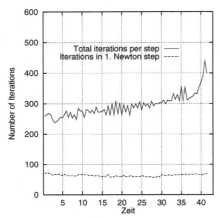

Fig. 5.32 Influence of plasticity on the number of iterations

that it is theoretically possible to achieve a problem independent convergence rate for the multi-grid method but not for the cg-method. Hence, in Fig. 5.31 a deviation from the optimal line can be observed. This is valid for different computing load on the nodes.

The influence of inelastic material behaviour on the number of iteration of the pAMG scheme is depicted in Fig. 5.32. Here the number of iterations within the first NEWTON step and the total number of iterations are plotted for a given load step versus time. The number of NEWTON steps are constant within the first load level. Once plastic flow occurs, the number of NEWTON steps increase within a load level; however, there is no influence of the inelastic behaviour on the iterative solver. The convergence rate is not optimal in this example. It stems from the fact that the shell problem is badly conditioned. Here special pre-conditioners have to be applied, see e.g. Arnold et al. (1997) and Schöberl (1999) in order to enhance the convergence rate. The load deflection curve related to this example is presented in Fig. 9.26; additionally the deformation of the cylinder can be found in Fig. 9.27.

5.3.5 General Observations

As it can be observed from these examples, there is now general concept for the choice of efficient and robust algorithms. This is due to the fact that the solution algorithm, which has to be applied, depends on the spatial dimension, the problem size, the capacity of the main memory of the computer, the condition number of the equation system and last not least on the physical behaviour of the problem. Hence the user of nonlinear finite element methods has to choose the solution algorithms according to his/her knowledge by considering the engineering problem at hand. However, from the examples, some general rules can be extracted for the choice of the right solution procedure:

- The NEWTON-RAPHSON method is advantageous when strongly nonlinear problems are present and high solution accuracy is needed. Often the computation can be performed with large load steps which yields an efficient method. This is especially true for problems related to finite elasticity (like rubber materials) which are path independent.
- Secant methods – like the BFGS method – can be applied efficiently to large two- and three-dimensional problems when weak nonlinearities are present, and the conditioning of the tangent matrix does not allow the use of iterative solvers.
- The combination of iterative solvers and the NEWTON method is efficient for three-dimensional problems, when good pre-conditioning techniques are available for the problem at hand.
- The combination of direct solvers and the NEWTON method is efficient for two- and three-dimensional problems, when sparse solvers like *UMFPACK* or *PARDISO* are applied. This strategy is especially suited for problems which are ill-conditioned, like the analysis of incompressible solids or the numerical simulation of structures which consist of a mixture of beams, shells and solids.
- Parallel solvers are adequate for large three-dimensional problems. Especially when coupled with iterative solvers. Still work has to be done in this area to design efficient solvers for general problems of structural and solid mechanics.

5.3.6 Problems, Which Occur when Running Actual Simulations

In this section, some difficulties are discussed which can materialize when a nonlinear problem is solved. Of course, the possibilities of male function of finite element programs and algorithms are quite wide, but some problems are mentioned which might occur more frequently.

- **The Solution Does not Converge During a Specific Load Step Using Newton's Method.** This means that the load increment is too large for the algorithm being in the convergence region. In such case, use either a method which increments the load automatically, like the arc-length method (Sect. 5.1.5) or the line-search technique (Sect. 5.1.4), or reduce the load step. As a hint, the load steps should be adjusted such that the number of Newton steps n_{steps} is in the range $3 \leq n_{steps} \leq 7$.
- **The Residual Does not Reduce Further in Newton's Method.** Here possibly a problem occurs which is ill-conditioned due to high differences of stiffness in different parts of the discretization. In such cases, the stiffness has to be reduced in order to avoid ill-conditioning or a different discretization has to be used.

− **The Solution Stops with a Zero or Negative Diagonal Element.**
This occurs when a singular point on the solution path is bypassed. In
such case, the arc-length method can be applied to overcome such points
together with an equation solver which is able to compute solutions for
negative definite matrices.

6. Solution Methods for Time Dependent Problems

When dealing with nonlinear partial differential equation which describe the deformation process of solids, then the change of state variables and deformations in time has to be considered. These problems are known as *initial boundary value problems* which additionally depend upon the time. Among engineering applications, related to initial value problems, are vibration analysis of structures or impact problems like car-crash simulations. In such cases, the inertia term in the linear momentum equation, see (4.65) or (4.102), cannot be neglected. Another class of problems is related to inelastic constitutive behaviour, such as elasto-plasticity, visco-plasticity or visco-elasticity. The inelastic response is governed by evolution equations, and thus in general a time dependent process.

Methods and algorithms are discussed in this chapter which can be applied to the above mentioned problem classes with special emphasis to dynamical systems in solid mechanics and nonlinear time dependent constitutive equations. Before starting with the specific application, some general remarks are made concerning the integration of algebraic differential systems or ordinary differential equations systems. These generally appear when the weak forms (3.289) or (3.296) describing a solid are spatially discretized by finite elements which then results to a system of ordinary differential equations in time.

The ordinary differential equation system (4.65) with N unknowns stemming from the spatial discretization can in general be written as

$$\dot{\boldsymbol{L}}(t) = \boldsymbol{P}(t) - \boldsymbol{R}\,[\,\boldsymbol{u}(t)\,]. \qquad (6.1)$$

The quantity $\boldsymbol{L}(t)$ denotes the linear momentum which follows for the discrete system from (4.65) as $\boldsymbol{L}(t) = \boldsymbol{M}\,\boldsymbol{v}(t)$. The vector $\boldsymbol{R}$ represents the vector of internal nodal forces which is equivalent to the stress divergence term. It depends upon the deformation $\boldsymbol{u}$ and the stress states $\boldsymbol{\sigma}(\boldsymbol{u})$, since no constitutive equations has been introduced in (4.65). The vector $\boldsymbol{P}$ contains the prescribed nodal forces due to the applied loading. The type of nonlinearity of (6.1) is determined by the stress divergence term $\boldsymbol{R}$. It depends upon the model which defines the discrete equation of motion (4.65). This can either be a structural model for trusses, beams or shells, see Chap. 9, or a the continuum model, see Chaps. 4 and 10. So far, no constitutive equations are specified in the term $\boldsymbol{R}\,(\boldsymbol{u})$. These could be stated as time independent (elastic) response functions or as time dependent (inelastic) constitutive equations. Hence it is not possible to present here a more specific form for the residual vector $\mathbf{R}$. Detailed information can be found in the chapters describing special applications.

By introducing the velocity $\mathbf{v}(t)$ as a new independent variable, the differential equations system (6.1) can be rewritten as nonlinear ordinary differential system of first order

$$\dot{\mathbf{y}}(t) = \mathbf{g}\,[\,\mathbf{y}(t)\,]\,, \tag{6.2}$$

where the vector $\mathbf{y}^T(t) = \{\,\mathbf{u}(t)\,, \mathbf{v}(t)\,\}$ now contains $2\,N$ unknowns (displacements and velocities).[1]

In general, different algorithms are applied to solve the system (6.1), since the type of the underlying partial differential equation can be different as well as the solution spectrum. Inelastic constitutive behaviour is often of local nature (due to the loading state only parts of a solid will undergo inelastic deformations). Contrary to that the inertia terms in a dynamical system are of global nature since they are present in all parts of a solid. The local behaviour of, e.g. inelastic responses can be considered by devising special algorithms. This will be discussed in detail in Sect. 6.2.

All algorithms, however, rely on an approximation of the time derivatives which have to be chosen within a given time step. As an example, the velocity $\mathbf{v}(t)$ can be approximated by a difference quotient using the displacements at different times

$$\mathbf{v}(t) = \frac{d}{dt}\,\mathbf{u}(t) \approx \frac{1}{\Delta t}\,[\,\alpha\,\mathbf{u}(t_{n-1}) + \beta\,\mathbf{u}(t_n) + \gamma\,\mathbf{u}(t_{n+1})\,]\,. \tag{6.3}$$

The notation used here to describe the time dependent behaviour is depicted in Fig. 6.1 which shows the function of component $u_i(t)$ of the displacement vector $\mathbf{u}(t)$ and its evaluation at different times t_n. A time interval in which the solution of (6.1) is determined is defined as $0 \leq t \leq T$. This time interval is subdivided into m time steps Δt. The time t_n is then given by $(n\,\Delta t)$; in general $t_{n+j} = (n + j)\Delta t$ follows.

It is often advantageous, especially in nonlinear applications, to change the time step size during the numerical integration of the equations of motion since the behaviour of the nonlinear term $\mathbf{R}\,(\,\mathbf{u}\,)$ can change drastically. For that an error analysis is needed. Such analysis is quite complex and will not be discussed here in depth, for an overview and related literature, see e.g. Wood (1990), Sloan et al. (2001) and Ramm et al. (2003).

---

[1] In this chapter, it is assumed that all variables are time dependent. In order to simplify notation this will not be stated explicitly when not needed, e.g. the displacement $\mathbf{u}(t)$ will be written as $\mathbf{u}$ to shorten notation.

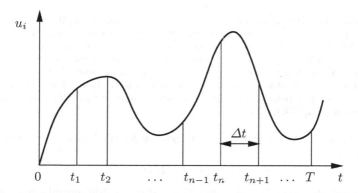

Fig. 6.1 Time dependent behaviour of a displacement component u_i

6.1 Integration of the Equations of Motion

The discrete equations of motions were derived from the weak form of linear momentum in Sect. 4.2. With (6.1), the equation

$$\boldsymbol{M}\,\ddot{\boldsymbol{u}} + \boldsymbol{R}\,(\boldsymbol{u}) = \boldsymbol{P} \tag{6.4}$$

is obtained where $\boldsymbol{M}$ stands for the mass matrix, $\boldsymbol{R}\,(\boldsymbol{u})$ denotes the vector of the internal forces (stress divergence) and $\boldsymbol{P}$ contains the time dependent prescribed loads.

From experimental evidence, it is known that damping effects occur in structures undergoing dynamic motion. These are related, e.g. to viscose effects in the material, internal friction or friction in connections. Such effects occur often in a combined way and are described using the assumption of damping which is proportional to the velocity.[2] Then an additional term has to be introduced in (6.4). For this a damping matrix $\boldsymbol{C}$ is introduced. Often the damping matrix $\boldsymbol{C}$ is assumed to be constant such that the damping force has the form $\boldsymbol{C}\,\dot{\boldsymbol{u}}$. Then it is mostly described by a combination of mass and stiffness matrix ($\boldsymbol{C} = d_1\,\boldsymbol{M} + d_2\,\boldsymbol{K}$). Such approximation is denoted as modal damping. This form has advantages in linear analysis, since a decoupling of the discrete equation system is possible by modal analysis, see e.g. Bathe (1982) or Zienkiewicz and Taylor (1989). Such decomposition can be applied to nonlinear systems as well, however, only in the incremental form of the equations of motion, see also Sect. 6.1.5. Since damping has different origins, as described above, an experimental verification of the damping matrix $\boldsymbol{C}$ or of the parameters d_1 and d_2 is essential in real engineering applications.

---

[2] Other possible damping forces are related to the action of fluids. Then, for a creeping fluid, the damping is also proportional to the velocity; however for turbulent flow the damping force depends quadratically upon the velocitiy.

Remark 6.1: It is also possible to define a nonlinear damping term $C(u, \dot{u})$ instead of the linear approximation $C\dot{u}$. However, such term is not easy to construct and to identify by experiments. Also, a more accurate description of the material damping by a visco-elastic or visco-plastic constitutive model, see Sect. 3.3.3, is possible. This would then effect the vector of internal forces R but not the damping matrix C. The same is true for damping due to frictional forces in connections. These can be modelled correctly by a fine finite element discretization using contact elements, see e.g. Chap. 11.

By including linear damping, the general form of the equations of motions follows as

$$M\ddot{u} + C\dot{u} + R(u) = P. \tag{6.5}$$

This equation can be transformed to a first order differential equation system by introducing the independent variables $\dot{u} = v$ and $\ddot{u} = \dot{v}$

$$\dot{u} = v,$$
$$\dot{v} = M^{-1}\left[P - Cv - R(u)\right]. \tag{6.6}$$

For the description of the algorithms, the letter a is chosen to denote the accelerations $\ddot{u}$ and the letter v denotes the velocities $\dot{u}$. With this notation, the discretized equation of the linear momentum (6.5) has at time t_{n+1} the form

$$M a_{n+1} + C v_{n+1} + R(u_{n+1}) = P_{n+1}. \tag{6.7}$$

The index $(..)_{n+1}$ means that the relevant quantity has to be computed at time t_{n+1}.

Finally, for the definition of an initial value problem, the initial conditions have to be described. These are conditions for the displacements $\bar{u}$ and the velocities $\bar{v}$ at time $t = t_0$ (usually $t_0 = 0$ is selected):

$$u_0 = \bar{u},$$
$$v_0 = \bar{v}. \tag{6.8}$$

The choice of numerical methods for the determination of the time dependent response of the deformation $u(t)$ depends upon the characteristics of the problem. Basically, two options are available for the solution of (6.5) which are known as explicit or implicit integration schemes:

- **Explicit** methods can be easily implemented since the solution at time t_{n+1} depends only upon quantities at time t_n. Explicit methods are very efficient when the mass matrix M in (6.5) is replaced by a *lumped* mass matrix which has diagonal structure, see also Remark 4.4. The disadvantage of explicit methods is the limitation of the time step size due to a stabilization criterion.
- **Implicit** integration schemes replace the time derivatives by quantities which depend as well upon the last time step in (time t_n) as upon the still unknown quantities at time $t_{n+\alpha}$. This requires the solution of a nonlinear

algebraic equation system at every time step. Hence implicit methods have to be combined with methods described in Chap. 5 (e.g. NEWTON method). The advantage of implicit methods is related to the fact that they can be constructed such that they are unconditional stable. Thus the time step size is not limited.

The selection of time step sizes, used either in explicit or implicit algorithms, has to be justified by physics. In case of impact problems (e.g. car-crash analysis) or shock waves moving through a solid, small time steps have to be selected to resolve high frequency parts and travelling waves in order to capture the correct physical behaviour. Hence explicit methods are ideal for such engineering applications. Implicit methods are advantageous for problems where the response of the dynamical system depends mainly upon lower frequencies (e.g. simulation of engine vibrations or vibration of structures). Since both types of physical behaviour occur frequently in engineering applications, explicit and implicit methods will be discussed in the next two sections.

From a mathematical and efficiency point of view, integration algorithms for the solution of the nonlinear equations of motions (6.5) have to be constructed in such a way that they have the same order of accuracy as the spatial finite element approximation, see e.g. Chap. 8, and that they fulfil the essential balance equations. This includes the conservation of linear and angular momentum and – in case of hyper elastic materials – the conservation of mechanical energy. Since classical algorithms used in linear dynamics do not fulfil all of the above requirements when applied to nonlinear problems governed by (6.5), new developments throughout the last years lead to so-called conserving algorithms, see e.g. Simo and Tarnow (1992), Crisfield (1997), Sansour et al. (1997) and Betsch and Steinmann (2000). Two different approaches, derived in Simo and Tarnow (1992) and Sansour et al. (1997), will be discussed in more detail in Sect. 6.1.3. Further conserving integration methods, e.g. starting from the symplectic structure of the equation of motions, will not be discussed, but a literature overview is provided in Sect. 6.1.3.

6.1.1 Explicit Time Integration Methods

When high frequencies (e.g. stemming from impact) or shock waves dominate the solution of a physical problem described by (6.5), then small time steps are required. In such case, the most efficient way to integrate the equations of motions is provided by an explicit method.

The central difference scheme is one of the favourite methods applied to solve the equations of motions in case of solid mechanics or structural problems. Within this scheme, the velocities v and the accelerations a are approximated at time t_n by

$$v_n = \frac{u_{n+1} - u_{n-1}}{2\,\Delta t},$$

$$a_n = \frac{u_{n+1} - 2\,u_n + u_{n-1}}{(\Delta t)^2}. \tag{6.9}$$

By inserting these relations into the discretized form of the linear momentum balance (6.5) at time t_n

$$M(u_{n+1} - 2\,u_n + u_{n-1}) + \frac{\Delta t}{2}\,C(u_{n+1} - u_{n-1}) + (\Delta t)^2 R(u_n) = (\Delta t)^2\,P_n, \tag{6.10}$$

an equation system is obtained from which the unknown displacement u_{n+1} at time t_{n+1} can be computed

$$\left(M + \frac{\Delta t}{2}\,C\right) u_{n+1} = (\Delta t)^2\,[\,P_n - R(u_n)\,] + \frac{\Delta t}{2}\,C\,u_{n-1} + M(2\,u_n - u_{n-1}). \tag{6.11}$$

Here the mass matrix M and the damping matrix C are constant. Hence, for the coefficient matrix $M + \Delta t/2\,C$, a triangular decomposition can be used which leads to an efficient algorithm for the solution of (6.11). Note that the term $R(u_n)$ which contains all nonlinearities appears only on the right hand side.

In the case that M and C are given in diagonal form, (*lumping*), then the inversion of $M + \Delta t/2\,C$ is trivial and only the vectors on the right hand side of (6.11) have to be evaluated.

The initialization of the finite difference scheme needs some special considerations since in (6.11) the values for the displacements u_{-1} have to be determined in order to start the integration process in a consistent way. These displacements can be computed from the initial values at time t_0 by using the initial conditions u_0 and v_0. Based on a second order accurate TAYLOR series expansion for the displacements at time t_{-1}, the relation

$$u_{-1} = u_0 - \Delta t\,v_0 + \frac{(\Delta t)^2}{2}\,a_0 \tag{6.12}$$

is obtained where the accelerations at time t_0 can be computed from the balance of linear momentum (6.7)

$$a_0 = M^{-1}\,[-C\,v_0 - R(u_0) + P_0]. \tag{6.13}$$

A variant of the above stated central difference scheme for the solution of (6.5) can be found in Wood (1990). It is equivalent to the already described method but uses the approximations

$$u_{n+1} = u_n + \Delta t\,v_n + \frac{(\Delta t)^2}{2}\,a_n,$$

$$v_{n+1} = v_n + \frac{1}{2}\,\Delta t\,(a_n + a_{n+1}) \tag{6.14}$$

for displacements and velocities. Equation (6.7) can be applied to determine the accelerations. This yields the algebraic equation system

$$\left(M + \frac{\Delta t}{2}\, C \right) a_{n+1} = P_{n+1} - R \left(u_n + \Delta t\, v_n + \frac{(\Delta t)^2}{2}\, a_n \right) - \frac{\Delta t}{2}\, C\, a_n \, ,$$

(6.15)

which has on the right hand side, besides the loading vector P_{n+1}, only quantities which are measured at time t_n. Thus the initial conditions for displacements and velocities can be incorporated directly in this scheme. Displacements and velocities follow after the solution of (6.15) from (6.14). The coefficient matrix in (6.15) does not change when compared to (6.11); hence the same efficiency as in the first formulation is obtained, especially when lumping procedures are applied.

However, as already mentioned, explicit methods are not unconditionally stable. The critical time step is given for linear problems by

$$\Delta t \le \frac{T_N}{\pi} \, .$$

(6.16)

In this criterion which is named after COURANT, the time T_N denotes the smallest period for a given finite element discretization. It can be estimated based on the element size and the speed of a wave travelling through a solid by, see e.g. Bathe (1996),

$$\Delta t \approx \frac{h}{c_L} \, .$$

(6.17)

h is a characteristic dimension of the smallest element in the FE-mesh and c_L is the velocity of a compression wave in a linear solid ($c_L = 3\, K\, (1-\nu)/\rho\, (1+\nu)$ with the modulus of compression K, the POISSON ration ν and the density ρ).

A critical time step limit can be found for nonlinear problems in Belytschko et al. (1976)

$$\Delta t \le \delta\, \frac{h}{c_L} \, .$$

(6.18)

The constant δ ($0.2 < \delta < 0.9$) is a reduction factor which has to be selected according to the nonlinear properties of the problem under consideration.

Dynamical Relaxation. Finite element programs exist which can only integrate the equations of motions explicitly. Due to this limitation, only dynamical problems can be solved. In order to apply such programs to time independent statical problems, the method of dynamical relaxation can be used. Within this method, the equations of motions are solved by introducing so much damping such that the solution converges quickly to the statical one. A simple possibility is to assume that the damping matrix in (6.10) can be replaced by a multiple of the mass matrix $\vartheta\, M$, see e.g. Skeie et al. (1995). With this assumption, an equation system for the unknown displacements u_{n+1} follows from (6.10)

$$\left(1 + \vartheta \frac{\Delta t}{2}\right) \boldsymbol{M}_{diag} \, \boldsymbol{u}_{n+1} = (\Delta t)^2 \left[\boldsymbol{P}_n - \boldsymbol{R}(\boldsymbol{u}_n)\right] + \left[\left(\vartheta \frac{\Delta t}{2} - 1\right) \boldsymbol{u}_{n-1} + 2\,\boldsymbol{u}_n\right],$$
(6.19)

where the mass matrix is approximated by a diagonal matrix $\boldsymbol{M} = \boldsymbol{M}_{diag}$. The value of ϑ will be selected such that the damping of the system is close to the aperiodic limit case. A value for ϑ can be derived from the ordinary differential equation of a simple one dimensional linear vibrating system $(m\,\ddot{x} + d\,\dot{x} + k\,x = 0)$. The LEHR damping measure can be determined with $d = \vartheta\,m$ as $D = d\,/\,2\,m\,\omega = \vartheta\,/\,2\,\omega$. The aperiodic limit case occurs for $D = 1$ which yields $\vartheta = 2\,\omega$. This of course is only an approximation for a multi-dimensional problem since more than one eigenvalue influences the solution.

The value $\vartheta = 2\,\omega$ is only an approximation for the aperiodic limit case in nonlinear applications since it is assumed that the first natural frequency ω is responsible for the decay of the motion of a dynamical system. However, the first natural frequency will change due to the changing stiffness of a nonlinear system. Thus, it might be necessary to recompute the eigenvalue ω during the dynamic relaxation algorithm.

The first natural frequency is associated with the smallest eigenvalue of the nonlinear finite element system, evaluated at a given state $\bar{\boldsymbol{u}}$. Thus the natural frequency can be computed approximately from the RAYLEIGH quotient

$$\omega^2 = \frac{\boldsymbol{\varphi}^T \boldsymbol{K}_T(\bar{\boldsymbol{u}}) \, \boldsymbol{\varphi}}{\boldsymbol{\varphi}^T \boldsymbol{M} \boldsymbol{\varphi}},$$
(6.20)

when the associated eigenvector $\boldsymbol{\varphi}$ is known. The natural frequency ω is not constant since the tangent stiffness depends upon the deformation state. Hence it is reasonable to determine the parameter ϑ as a function of the deformation. For this task, it is assumed that the eigenvector can be approximated by the displacement increment $\Delta \boldsymbol{u}_{n+1} = \boldsymbol{u}_{n+1} - \boldsymbol{u}_n$. Since the displacement are related to a statical analysis, this assumption is reasonable. With this choice, the following approximation of the first natural frequency and the damping factor ϑ_{n+1} is given by

$$\omega_{n+1}^2 \approx \frac{\Delta \boldsymbol{u}_{n+1}^T \boldsymbol{K}_T \, \Delta \boldsymbol{u}_{n+1}}{\Delta \boldsymbol{u}_{n+1}^T \boldsymbol{M}_{diag} \, \Delta \boldsymbol{u}_{n+1}} = \frac{\Delta \boldsymbol{u}_{n+1}^T \left[\boldsymbol{P}_n - \boldsymbol{R}(\boldsymbol{u}_n)\right]}{\Delta \boldsymbol{u}_{n+1}^T \boldsymbol{M}_{diag} \, \Delta \boldsymbol{u}_{n+1}} \longrightarrow \vartheta_{n+1} = 2\,\omega_{n+1}.$$
(6.21)

In this derivation, the incremental equation system related to NEWTON method $[\boldsymbol{K}_T \, \Delta \boldsymbol{u}_{n+1} = -\boldsymbol{G}(\boldsymbol{u}_n)]$ was used. The computation of ϑ_{n+1} is based on the evaluation of two scalar products, and hence is very efficient. The factor ϑ_{n+1} from (6.21) can now be inserted in (6.19) within each time step.

6.1.2 Implicit Time Integration Methods

Implicit methods can be applied alternatively to solve the nonlinear discrete equations of motion (6.5) for many engineering applications in structural and solid mechanics. The most popular scheme is the NEWMARK method, see Newmark (1959). It is based on the following approximations for displacements and velocities at time t_{n+1}

$$\boldsymbol{u}_{n+1} = \boldsymbol{u}_n + \Delta t \, \boldsymbol{v}_n + \frac{(\Delta t)^2}{2} \, , [\, (1 - 2\,\beta)\, \boldsymbol{a}_n + 2\,\beta \, \boldsymbol{a}_{n+1}\,],$$

$$\boldsymbol{v}_{n+1} = \boldsymbol{v}_n + \Delta t \, [\, (1 - \gamma)\, \boldsymbol{a}_n + \gamma \, \boldsymbol{a}_{n+1}\,] \, . \tag{6.22}$$

These relations not only depend upon quantities at time t_n but also on the accelerations at time t_{n+1}. The parameters β and γ are constants which determine the behaviour of the integration method. They can be selected by the user. A mathematical analysis of NEWMARK method for the linear equations of motions leads to the following inequalities which limit the parameter values: $0 \le \beta \le 0.5$ and $0 \le \gamma \le 1$, see e.g. Bathe (1982) and Zienkiewicz and Taylor (2000a).

The still unknown accelerations $\boldsymbol{a}_{n+1}$ follow from the spatial discretized form of the linear momentum equation (6.7). By using the approximations for $\boldsymbol{u}_{n+1}$ and $\boldsymbol{v}_{n+1}$, the accelerations $\boldsymbol{a}_{n+1}$ have to be determined from the nonlinear algebraic equation system

$$(\, \boldsymbol{M} + \gamma \, \Delta t \, \boldsymbol{C}\,)\, \boldsymbol{a}_{n+1} + \boldsymbol{R}\,(\, \boldsymbol{a}_{n+1}\,,\, \boldsymbol{u}_n\,,\, \boldsymbol{v}_n\,,\, \boldsymbol{a}_n\,) = \boldsymbol{P}_{n+1} - \bar{\boldsymbol{G}}\,(\, \boldsymbol{u}_n\,,\, \boldsymbol{v}_n\,,\, \boldsymbol{a}_n\,) \, . \tag{6.23}$$

All terms which depend linear upon the displacements, velocities and accelerations at time t_n, when (6.22) is inserted in (6.7), are assembled in the vector $\bar{\boldsymbol{G}}$. Equation (6.23) can be solved by using NEWTON method. It yields the accelerations $\boldsymbol{a}_{n+1}$. When these are known, the displacements and velocities follow from (6.22).

Remark 6.2: For the parameter choice $\gamma = 0.5$ and $\beta = 0$, the approximations for displacements and velocities of the explicit central difference methods can be derived from (6.22), see also Eq. (6.14).

Often approximations (6.22) are rewritten in such a way that the velocities and accelerations depend upon the displacements

$$\boldsymbol{a}_{n+1} = \alpha_1 \, (\, \boldsymbol{u}_{n+1} - \boldsymbol{u}_n\,) - \alpha_2 \, \boldsymbol{v}_n - \alpha_3 \, \boldsymbol{a}_n \, ,$$

$$\boldsymbol{v}_{n+1} = \alpha_4 \, (\, \boldsymbol{u}_{n+1} - \boldsymbol{u}_n\,) + \alpha_5 \, \boldsymbol{v}_n + \alpha_6 \, \boldsymbol{a}_n \, . \tag{6.24}$$

Here the following constants were introduced

$$\alpha_1 = \tfrac{1}{\beta\,(\Delta t)^2} \, , \quad \alpha_2 = \tfrac{1}{\beta\,\Delta t} \, , \quad \alpha_3 = \tfrac{1 - 2\,\beta}{2\,\beta} \, ,$$

$$\alpha_4 = \tfrac{\gamma}{\beta\,\Delta t} \, , \quad \alpha_5 = (1 - \tfrac{\gamma}{\beta}) \, , \quad \alpha_6 = (1 - \tfrac{\gamma}{2\,\beta}) \, \Delta t \, .$$

The insertion of relations (6.24) in to the linear momentum equation (6.7) yields now a nonlinear algebraic equation system for the unknown displacements $\boldsymbol{u}_{n+1}$:

$$\begin{aligned}
G\left(u_{n+1}\right) \;=\;& M\left[\alpha_1\left(u_{n+1}-u_n\right)-\alpha_2\,v_n-\alpha_3\,a_n\right] \\
& + C\left[\alpha_4\left(u_{n+1}-u_n\right)+\alpha_5\,v_n+\alpha_6\,a_n\right] \qquad (6.25) \\
& + R\left(u_{n+1}\right)-P_{n+1}=0.
\end{aligned}$$

Again NEWTON method can be applied to determine the unknown displacements at time t_{n+1}. This leads with the tangential stiffness matrix

$$K_T\left(u_{n+1}^i\right) = \left.\frac{\partial R}{\partial u_{n+1}}\right|_{u_{n+1}^i} \qquad (6.26)$$

to the iterative scheme (iteration index i)

$$\begin{aligned}
\left[\alpha_1\,M+\alpha_4\,C+K_T\left(u_{n+1}^i\right)\right]\Delta u_{n+1}^{i+1} \;=\;& -G\left(u_{n+1}^i\right), \\
u_{n+1}^{i+1} \;=\;& u_{n+1}^i+\Delta u_{n+1}^{i+1}, \qquad (6.27)
\end{aligned}$$

which has to be carried out at each time step of the NEWMARK method. The starting value for this iteration is the last converged displacement vector from the last time step : $u_{n+1}^0 = u_n$. The iteration described in (6.27) is terminated when the criterion provided in Box 5.1 is fulfilled.

The accuracy and stability of the NEWMARK method can be analyzed for linear dynamical problems. The results stated in Table 6.1 are due to Wood (1990). It can be observed from this table that the NEWMARK method is optimal for the parameter choice $\gamma = 0.5$. In the case of damping free vibrations, the error in the amplitude is equal to zero. This is equivalent to the conversation of energy. Of equal order of accuracy is the method of central differences ($\gamma = 0$). However, as already mentioned in Sect. 6.1.1, this method is not unconditional stable and has a critical time step of $\Delta t \leq T_N/\pi$, with T_N being the smallest period for a given finite element discretization.

Since the spatial finite element discretization approximates the lower eigenmodes a lot better than the higher one, see Strang and Fix (1973), it is sometimes advantageous to damp the higher modes during the numerical integration process. This often makes in engineering applications sense, since implicit time integration methods are used for problems where the response is governed by the low frequency modes. For the NEWMARK method, the parameter $\gamma > 0.5$ has to be selected. However, following the results in Table 6.1, this leads to a loss of accuracy. Due to that reason, modifications of

Table 6.1 Accuracy and stability of the Newmark- and central difference method

Parameter	Error in amplitude ($C = 0$)	Error in amplitude ($C \neq 0$)	stability
$\gamma = 0.5$	0	$O\left(\Delta t^2\right)$	$\beta \geq 0.25$
$\gamma \neq 0.5$	$O(\Delta t)$	$O\left(\Delta t\right)$	$2\beta \geq \gamma \geq 0.5$
$\gamma = 0$	0	$O\left(\Delta t^2\right)$	$\Delta t \leq \frac{T_N}{\pi}$

the NEWMARK method have been proposed which preserve the order $O(\Delta t^2)$ but damp the high frequencies, see e.g. Hilber et al. (1977) and Wood et al. (1981). The method due to BOSSAK, see Wood et al. (1981), is based on a changed discrete equation of motion (6.7)

$$(1 - \alpha)\, \boldsymbol{M}\, \boldsymbol{a}_{n+1} + \alpha\, \boldsymbol{M}\, \boldsymbol{a}_n + \boldsymbol{C}\, \boldsymbol{v}_{n+1} + \boldsymbol{R}\,(\boldsymbol{u}_{n+1}) - \boldsymbol{P}_{n+1} = \boldsymbol{0}, \qquad (6.28)$$

but still uses the approximations (6.22) for displacements and velocities.

The method developed by Hilber et al. (1977) for problems of linear elastodynamics applies a different approximation to the equations of motion which weighs the displacements. Its nonlinear extension yields, instead of (6.28), to the system of equations

$$\boldsymbol{M}\, \boldsymbol{a}_{n+1} + (1-\alpha)\, [\boldsymbol{C}\, \boldsymbol{v}_{n+1} - \boldsymbol{P}_{n+1}] + \alpha\, [\boldsymbol{C}\, \boldsymbol{v}_n - \boldsymbol{P}_n] + \boldsymbol{R}\, [(1-\alpha)\, \boldsymbol{u}_{n+1} + \alpha\, \boldsymbol{u}_n)] = \boldsymbol{0}.$$
$$(6.29)$$

Also, here the displacements and velocities are computed at time t_{n+1} as in the method of Wood et al. (1981) or NEWMARK by (6.22). The HILBER method damps high frequencies for a parameter choice $0.5 < \alpha < 1$. However, this method needs an evaluation of the vector of internal nodal forces $\boldsymbol{R}$ at time $t_{n+\alpha} = (1 - \alpha)\, t_{n+1} + \alpha\, t_n$, which is not trivial when complex nonlinear constitutive equations dictate the response of a system, and thus history variables have to be considered.

6.1.3 Conserving Algorithms

The implicit time integration methods, which where discussed so far, were essentially developed for linear problems. When these methods are applied to general nonlinear problems in solid mechanics, then they might not preserve all physical quantities, e.g. the balance principles of continuum mechanics. The related quantities are linear momentum, angular momentum and mechanical energy.

When the NEWMARK method is applied to a nonlinear system undergoing finite rotations, the angular momentum is not preserved. Often also the energy is not conserved for elastic systems. It has been shown in Simo and Tarnow (1992) that the NEWMARK method preserves the angular momentum only for the choice of the parameters $\gamma = 1/2$ and $\beta = 0$. The NEWMARK method is then equivalent to the explicit method of central differences, as shown in Remark 6.2. Hence this integration method is even, for the optimal choice of the parameters ($\beta = 1/4$ and $\gamma = 1/2$, no damping of amplitudes), not angular momentum preserving. Contrary to that, the explicit central difference method preserves angular momentum.

The method of Hilber et al. (1977) damps out high frequencies for a choice of the parameter α ($0.5 < \alpha < 1$). But again Simo and Tarnow (1992) showed that this method, even for undamped systems ($\boldsymbol{C} = \boldsymbol{0}$), does not preserve angular momentum.

Hence time integration algorithms which preserve angular momentum, linear momentum and energy, have to be developed knowing that energy conservation can only be achieved for elastic systems. These algorithms have to be especially applied to problems undergoing finite rotations and for long time integration of systems. The latter is essential when dynamical stability of nonlinear systems is investigated.

Starting from the initial work of Simo and Tarnow (1992), different authors developed variants of such schemes. Among these are papers by Kane et al. (1999) who derived variational integrators for conservative mechanical systems that are symplectic and energy and momentum conserving. A scheme based on co-rotational formulations can be found in Crisfield and Shi (1996). Algorithms which include finite deformations of general hyperelastic materials including compressible and incompressibie material response were developed in Gonzalez (2000) independently of the spatial discretization. Another scheme is due to Laursen and Meng (2001) which bases on the extension of a critical stress update formula to encompass generic stored energy functions for the hyperelastic continuum.

A co-rotational energy–momentum scheme which guaranteed conservation was obtained by Kuhl and Crisfield (1999) but additionally was also able to control the decay of total energy by numerical dissipation of unwanted high frequency response. Finally, these schemes were applied within structural elements like beams, see e.g. Doblare (1995), and shells, see e. g. Kuhl and Ramm (1996) and Brank et al. (1998).

The derivation of time integration methods which conserve linear momentum, angular momentum and energy will be based on the following assumptions:

− hyper-elastic constitutive behaviour,
− no action of prescribed loads and
− no prescribed displacements at boundaries such that the body can move freely.

Under these conditions, the conservation of linear and angular momentum as well as of energy can be shown from (3.69), (3.70) and (3.79). These equations lead to the discrete form, written for two time instants at t_n and t_{n+1}

$$\boldsymbol{L}_n = \boldsymbol{L}_{n+1}, \qquad \boldsymbol{J}_n = \boldsymbol{J}_{n+1}, \quad \text{and} \quad E_n = E_{n+1}. \qquad (6.30)$$

In Simo and Tarnow (1992), the following ansatz is chosen for displacements and velocities

$$
\begin{aligned}
\boldsymbol{u}_{n+\alpha} &= \alpha\,\boldsymbol{u}_{n+1} + (1-\alpha)\,\boldsymbol{u}_n\,, \\
\boldsymbol{v}_{n+\alpha} &= \alpha\,\boldsymbol{v}_{n+1} + (1-\alpha)\,\boldsymbol{v}_n
\end{aligned}
\qquad (6.31)
$$

with $0 \le \alpha \le 1$. This approximation is now inserted in the system (6.6) which yields

$$\frac{1}{\Delta t}\left(\boldsymbol{u}_{n+1} - \boldsymbol{u}_n\right) = \boldsymbol{v}_{n+\alpha},$$

$$\frac{1}{\Delta t}\left(\boldsymbol{v}_{n+1} - \boldsymbol{v}_n\right) = \boldsymbol{M}^{-1}\left[\boldsymbol{P}_{n+\alpha} - \boldsymbol{R}(\boldsymbol{u}_{n+\alpha})\right]. \tag{6.32}$$

Damping is neglected due to the assumptions made above. An analysis of the time integration method based on (6.32) shows that linear momentum is preserved for $0 \le \alpha \le 1$. The conservation of angular momentum follows then the special choice $\alpha = 1/2$. Note that the discretized form of the stress tensor is still arbitrary. The proof of these results is based on the weak form of the balance of linear and angular momentum in which the time discretization (6.31) is inserted. Since the test functions, used in the weak form, are arbitrary, it is possible to select a constant value for it, see last assumption. With this choice, linear and angular momentum is preserved for symmetric stress tensors. When also energy has to be preserved then the stresses have to be computed from the constitutive equation such that the energy is constant: $E_{n+1} = E_n$. The kinetic energy K which is part of E is exactly approximated by (6.31) and (6.32), see Simo and Tarnow (1992). Hence the change in the internal energy U has to be considered in more detail. From (3.89) follows for a pure mechanical deformation that $\rho_0\,\dot{u} = \mathbf{S} \cdot \dot{\mathbf{E}} = \mathbf{F}\,\mathbf{S} \cdot \mathrm{Grad}\mathbf{v}$. The integral form of the change of strain energy can be written as

$$\dot{U} = \int_B \mathbf{F}\,\mathbf{S} \cdot \mathrm{Grad}\mathbf{v}\,dV. \tag{6.33}$$

The time integration yields with (6.32) evaluated at $\alpha = \frac{1}{2}$ (conservation of angular momentum)

$$U_{n+1} - U_n = \Delta t \int_B \mathbf{F}_{n+\frac{1}{2}}\,\mathbf{S} \cdot \mathrm{Grad}\mathbf{v}_{n+\frac{1}{2}}dV. \tag{6.34}$$

With $(6.31)_1$, here written for the deformation $\boldsymbol{\varphi}$, the deformation gradient at time $t_{n+\frac{1}{2}}$ is obtained

$$\mathbf{F}_{n+\frac{1}{2}} = \frac{1}{2}\,\mathrm{Grad}\,(\boldsymbol{\varphi}_{n+1} + \boldsymbol{\varphi}_n) = \frac{1}{2}\,(\mathbf{F}_{n+1} + \mathbf{F}_n).$$

The velocity gradient can be approximated with $(6.32)_1$ by

$$\mathrm{Grad}\mathbf{v}_{n+\frac{1}{2}} = \frac{1}{\Delta t}\,\mathrm{Grad}(\boldsymbol{\varphi}_{n+1} - \boldsymbol{\varphi}_n) = \frac{1}{\Delta t}\,(\mathbf{F}_{n+1} - \mathbf{F}_n).$$

After elementary algebraic computations, the change of strain energy can be expressed in terms of the GREEN-LAGRANGE strain tensor (3.15)

$$U_{n+1} - U_n = \int_B \mathbf{S} \cdot (\mathbf{E}_{n+1} - \mathbf{E}_n)dV = 0. \tag{6.35}$$

This constitutes a restriction which has to be fulfilled by the algorithmic evaluation of the constitutive equation. For the ST. VENANT material, it is fulfilled exactly. This can be seen from the strain energy for the ST. VENANT material

$$U = \frac{1}{2} \int_B \mathbf{E} \cdot \mathbb{C}\,[\,\mathbf{E}\,]\,dV\,,$$

such that the 2nd PIOLA-KIRCHHOFF stress tensor which has to fulfil

$$
\begin{aligned}
U_{n+1} - U_n &= \frac{1}{2} \int_B \mathbf{E}_{n+1} \cdot \mathbb{C}\,[\,\mathbf{E}_{n+1}\,] - \mathbf{E}_n \cdot \mathbb{C}\,[\,\mathbf{E}_n\,]\,dV \\
&= \frac{1}{2} \int_B (\mathbf{E}_{n+1} - \mathbf{E}_n) \cdot \mathbb{C}\,[\,\mathbf{E}_{n+1} + \mathbf{E}_n\,]\,dV
\end{aligned}
$$

follows by comparison with (6.35) as

$$\mathbf{S} = \frac{1}{2}\,\mathbb{C}\,[\,\mathbf{E}_{n+1} + \mathbf{E}_n\,]\,. \tag{6.36}$$

The virtual internal work, see (4.56), can be written with respect to the initial configuration as

$$\int_B \delta \mathbf{E}_{n+\frac{1}{2}} \cdot \mathbf{S}\,dV = \boldsymbol{\eta}^T \mathbf{R}\,(\mathbf{u}_{n+\frac{1}{2}}) = \bigcup_{e=1}^{n_e} \sum_{I=1}^{n} \boldsymbol{\eta}_I^T \mathbf{R}_I\,(\mathbf{u}_{n+\frac{1}{2}})\,, \tag{6.37}$$

where the residual vector belonging to element node I is given with (6.36) by

$$\mathbf{R}_I\,(\mathbf{u}_{n+\frac{1}{2}}) = \int_{\Omega_e} \mathbf{B}_{L\,I\,n+\frac{1}{2}}^T \frac{1}{2}\,\mathbb{C}\,[\,\mathbf{E}_{n+1} + \mathbf{E}_n\,]\,d\Omega\,. \tag{6.38}$$

This completes the set of discrete equations needed for the time integration method which preserves linear and angular momentum as well as energy. A disadvantage of this method is related to the fact that the tangent matrix associated with the linearization of (6.38) is non-symmetric. This is due to the fact that the $\mathbf{B}$-Matrix in (6.38) is evaluated at time $t_{n+\frac{1}{2}}$ while the stresses are evaluated at the midpoint of the time step. The linearization can be derived using (4.69), (4.72) and (4.73). Additionally $(6.31)_1$ has to be considered for the computations of the gradients. This yields

$$\mathbf{K}_T = \bigcup_{e=1}^{n_e} \sum_{I=1}^{n} \sum_{K=1}^{n} \boldsymbol{\eta}_I^T \,\bar{\mathbf{K}}_{T\,I\,K}\,\Delta\mathbf{u}_K\,, \tag{6.39}$$

where the second term in matrix $\bar{\mathbf{K}}_{IK}$

$$\bar{\mathbf{K}}_{T\,I\,K} = \frac{1}{2} \int_{\Omega_e} \left[(\nabla_X N_I)^T\,\bar{\mathbf{S}}_e\,\nabla_X N_K + \bar{\mathbf{B}}_{L\,I\,n+\frac{1}{2}}^T\,\bar{\mathbf{D}}\,\bar{\mathbf{B}}_{L\,K\,n+1} \right] d\Omega \tag{6.40}$$

is non-symmetric.

The algorithm, related to this time integration method, can be constructed by inserting $(6.32)_1$ into $(6.32)_2$ which eliminates the velocities. This leads to a nonlinear algebraic equation system for the displacements u_{n+1}

$$G\left(u_{n+1}\right) = \frac{2}{(\Delta t)^2}\, M\left(u_{n+1} - u_n - \Delta t\, v_n\right) - P_{n+\frac{1}{2}} + R(u_{n+\frac{1}{2}}) = 0, \quad (6.41)$$

which can be solved by NEWTON method. Its linearization yields with (6.39) the iterative scheme

$$\left[\frac{2}{(\Delta t)^2}\, M + K_T\right] \Delta u_{n+1}^{i+1} = -G\left(u_{n+1}^i\right),$$

$$u_{n+1}^{i+1} = u_{n+1}^i + \Delta u_{n+1}^{i+1}, \quad (6.42)$$

which has to be executed in each time step of the integration method. As initial value for the displacements, the converged value from the last time has to be used: $u_{n+1}^0 = u_n$. The stopping criterion for the iteration described in (6.42) is the same as the one given in Box 5.1.

The time integration method developed above has the disadvantage that it does not preserve angular momentum for shell problems where rotation fields have to approximated, see Sect. 9.4.6. Additional considerations are necessary to derive preserving algorithms which can be applied for beam or shell problems undergoing finite rotations, see e.g. Crisfield and Shi (1994) and Kuhl and Ramm (1996).

Another development, provided in Sansour et al. (1997) for shells, will be discused below. Here the displacement field u and the rotation field θ are approximated by

$$u_{n+\frac{1}{2}} = \frac{1}{2}\left(u_{n+1} + u_n\right),$$

$$\theta_{n+\frac{1}{2}} = \frac{1}{2}\left(\theta_{n+1} + \theta_n\right). \quad (6.43)$$

Furthermore, it is assumed that the velocities can be computed using

$$v_{n+\frac{1}{2}} = \frac{1}{\Delta t}\left(u_{n+1} - u_n\right),$$

$$\dot{\theta}_{n+\frac{1}{2}} = \frac{1}{\Delta t}\left(\theta_{n+1} - \theta_n\right). \quad (6.44)$$

The membrane and bending strains, given in (9.191) for the general shell and in (9.97) for an axisymmetric shell, are however approximated directly. By using, e.g. (9.97) for membrane- and bending strains relations

$$E_{n+\frac{1}{2}}^m = E_n^m + \frac{1}{2\,\Delta t}\, \dot{E}_{n+\eta}^m,$$

$$E_{n+\frac{1}{2}}^b = E_n^b + \frac{1}{2\,\Delta t}\, \dot{E}_{n+\eta}^b \quad (6.45)$$

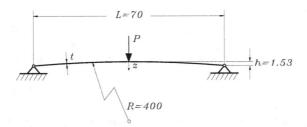

Fig. 6.2 Simply supported shallow arch

are deduced. Sansour et al. (1997) have shown that the energy of a shell is preserved for $\eta = \frac{1}{2}$. For the derivation of the weak form of linear momentum, a specific shell formulation has to be applied. This can be found in Sect. 9.3 for the general shell and in Sect. 9.4 for an axisymmetric shell. The associated discretization in time can then be obtained using the scheme presented above. This however will not be discussed here. The relevant relations and derivations can be found in Sansour et al. (1996).

6.1.4 Numerical Examples

The next two examples are related to the nonlinear dynamical behaviour of beams and shells and their time integration using the methods and algorithms developed in Sect. 6.1.3.

Shallow Arc. In the first example, the shallow arch depicted in Fig. 6.2 is investigated. It has a curvature radius of $R = 400$ and a area of the cross section $A = 1t$. The modulus of elasticity is given by $E = 2 \times 10^7$ and the density by $\rho = 7.5 \times 10^{-5}$. Linear damping with $d = 6 \times 10^{-3}$ is assumed. The applied load $P = F \cos \Omega t$ acts periodically. Its maximum amplitude was selected as $F = 360$ such that it is below but close to the load asociated with the limit point of the arc under statical loading. Under these circumstances,

Fig. 6.3 General load–deflection diagram of the shallow arch

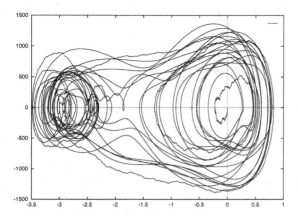

Fig. 6.4 Phase diagram of shallow arch under point load

see Fig. 6.3, an instable saddle point can be expected, and hence chaotic behaviour of the structure will be observed.

The eigenfrequency of the applied point load is selected as $\Omega = 1000$. The time increment was chosen as $\Delta t = 10^{-5}$ and used throughout the entire computation. The arch is discretized by 20 nonlinear finite beam elements. The dynamical structural response is computed with the algorithm based on (6.45). It yields the phase diagram shown in Fig 6.4. Over 10000 time steps were necessary to obtain the phase diagram in Fig. 6.4. Hence a time integration algorithm is needed which does not depict numerical damping, so that the final result only depends upon the physical damping d. It should be noted that the chaotic behaviour of nonlinear dynamical system can depend upon its spatial discretization. Here a spatial discretization is selected which can describe the complete nonlinear behaviour of the solution accurately. The

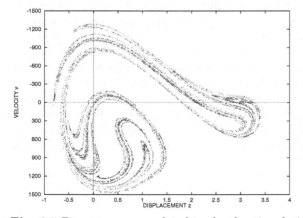

Fig. 6.5 POINCARE map, related to the chaotic solution of the shallow arch

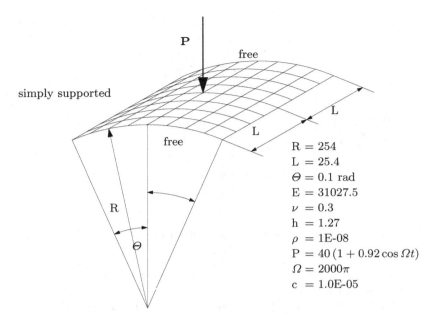

$R = 254$
$L = 25.4$
$\Theta = 0.1$ rad
$E = 31027.5$
$\nu = 0.3$
$h = 1.27$
$\rho = 1E\text{-}08$
$P = 40\,(1 + 0.92\cos\Omega t)$
$\Omega = 2000\pi$
$c = 1.0E\text{-}05$

Fig. 6.6 Cylindrical shell under harmonic loading

dynamical behaviour of the arc under the chosen point load is chaotic as can be seen from the POINCARE map in Fig. 6.5. A discretization with fewer finite elements suppresses eventually the chaotic behaviour, as shown in Sansour et al. (1996).

Cylindrical Shell. The shell shown in Fig. 6.6 is subjected to an applied periodic point load which can lead to chaotic motions.

It is simply supported along the sides in longitudinal direction and is free at its other sides. The point load acts in the middle of the shell segment. Its amplitude and frequency as well as the material and geometry data of the shell can be found in Fig. 6.6. The static load ($P_0 = 40$) is close to the limit load of the shell. Hence the system has an instable saddle point and undergoes a chaotic motion. The nonlinear dynamical behaviour is depicted in the phase diagram in Fig. 6.7. The related POINCARE map is shown in Fig. 6.8. In total, 10^5 time steps were needed to obtain the POINCARE map. The NEWMARK algorithm, described in Sect. 6.1.2, cannot be applied here since only a specific number of time steps can be computed; after that the solution explodes (the displacements and velocities increase beyong limits), see Sansour et al. (1997).

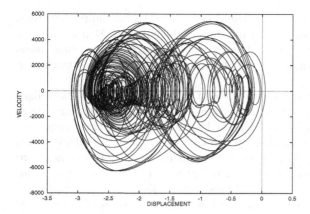

Fig. 6.7 Phase diagram for cylindrical shell

6.1.5 Reduction Techniques for Nonlinear Equations of Motion

Real engineering structures lead to large finite element systems with several hundred thousand unknowns. When the nonlinear dynamical behaviour of such problem has to be investigated using implicit methods, then it is, even for the current available computing power, impossible to perform such computations in an acceptable time frame. Thus methods which reduce the overall computing time have to be found. In linear analysis model, analysis techniques are well established which diagonalize and hence reduce the equation systems, see e.g. Bathe (1996). The basic idea is that most of the energy of the system can be related to a few eigenmodes, and hence the behaviour of the dynamical system is described by these modes. This works well for linear systems; however, for nonlinear systems the modes which contain most

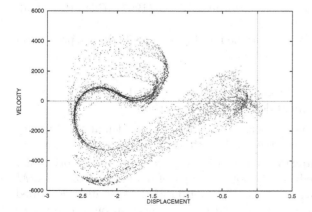

Fig. 6.8 POINCARE map for cylindrical shell

of the energy depend upon the deformation state of the dynamical system. This deformation state changes in time during the solution. Hence the classical modal methods which select a fixed set of eigenvectors as reduced basis for a problem at the beginning of the solution algorithm do not work in a nonlinear setting.

Reduction methods exist for explicit and implicit time integration algorithms. Explicit time integration algorithms are used for shock and wave propagation problems. They are applied basically to short-time computations, and the time step is limited by the COURANT criterion (6.16). Thus the goal for a reduction method within an explicit integration schemes is to allow larger time steps, see Idelsohn and Cardona (1984), and for nonlinear applications, See Bucher (2001). In case of implicit methods, especially for nonlinear applications, the main drawback is the solution of large linear systems within the time and incremental loop, see Sect. 6.1.2. Hence reduction methods decrease the size of the equation system by an appropriate and efficient technique.

Here a number of methods exist which speed up computation time. However, there does not exist a unique way of treating the nonlinear dynamical equations. Methods have been developed for different applications and degrees of nonlinearity, for an extensive review see Noor (1994). Methods based on modal analysis in tangent space can be found in, e.g. Nickell (1976), Leger (1993) and Kirsch et al. (2005). Other methods are the pseudo-force method, see e.g. Bathe and Gracewski (1981). Direct update methods of the eigenvectors which can be found in stability analysis, see e.g. Wriggers et al. (1988), have been employed for dynamic systems in Spiess (2006). The use of modal derivatives was proposed in Idelsohn and Cardona (1985). Since a truncated modal superposition is not very efficient for nonlinear systems, more general bases for reductions were developed such as the RITZ vectors, see e.g. Wilson et al. (1982). Other approaches use substructuring in which a system is subdivided into different smaller substructures. Such technique is related to dynamic condensation, see e.g. Geradin and Rixen (1997), Castanier et al. (2001) and Archer (2001).

Several feasible methods are described in this section. These are projection based methods like

- modal analysis in tangent space,
- methods based on updated eigen- or RITZ vectors and
- the POD (proper orthogonal decomposition) method.

The methods have different properties and hence are optimal for different applications. All methods act in general on the equation system (6.1).

Projection Based Reduction Methods. A projection of the motion of a system onto a subspace reduces the total number of unknowns from N to M in a nonlinear finite element system. The motion of the system is described by a vector $\boldsymbol{u}(t) \in \mathbb{R}^N$, which is replaced by its projection onto a subspace $\boldsymbol{q}(t) \in \mathbb{R}^M$ with

$$u(t) = \boldsymbol{\Psi}\, \boldsymbol{q}(t)\,. \tag{6.46}$$

The projection matrix

$$\boldsymbol{\Psi} = [\,\boldsymbol{\Psi}_1, \ldots, \boldsymbol{\Psi}_M\,] \tag{6.47}$$

contains M base vectors $\boldsymbol{\Psi}_i$ which span the subspace.

The basic formulations starts from the equation of motion (6.4): $\boldsymbol{M}\,\ddot{\boldsymbol{u}} + \boldsymbol{R}\,(\boldsymbol{u}) = \boldsymbol{P}$. By inserting the projection (6.46) and by pre-multiplication with the projection matrix (related to the multiplication of (6.4) by the test function), the reduced matrix equation

$$\boldsymbol{\Psi}^T\,\boldsymbol{M}\boldsymbol{\Psi}\,\ddot{\boldsymbol{q}}(t) + \boldsymbol{\Psi}^T\,\boldsymbol{R}\,(\boldsymbol{\Psi}\,\boldsymbol{q}(t)) = \boldsymbol{\Psi}^T\,\boldsymbol{P}(t) \tag{6.48}$$

is obtained. Here the damping term was neglected. Adding this term is just a technical matter, and thus will not be discussed, for details, see e.g. Spiess (2006). With the definitions $\boldsymbol{M}^* = \boldsymbol{\Psi}^T\,\boldsymbol{M}\boldsymbol{\Psi}$, $\boldsymbol{R}^*(\boldsymbol{q}(t)) = \boldsymbol{\Psi}^T\,\boldsymbol{R}\,(\boldsymbol{\Psi}\,\boldsymbol{q}(t))$ and $\boldsymbol{P}^*(t) = \boldsymbol{\Psi}^T\,\boldsymbol{P}(t)$, the reduced set of equations with M unknowns

$$\boldsymbol{M}^*\,\ddot{\boldsymbol{q}}(t) + \boldsymbol{R}^*(\boldsymbol{q}(t)) = \boldsymbol{P}^*(t) \tag{6.49}$$

is derived.

The problem is now to find an optimal base with a minimum number of base vectors which results in a good approximation of the nonlinear dynamic problem, see Fig. 6.9 for the deviation of the solution resulting from a projection. There are several demands which lead to a good approximation of a nonlinear system:

– The base vectors have to fulfill the essential boundary conditions.
– The base vectors should be orthogonal: $\boldsymbol{\Psi}^T\,\boldsymbol{\Psi} = \boldsymbol{1}$.

It will be necessary to determine the error related to the projection onto the subspace. This is especially needed when the deformation of the nonlinear

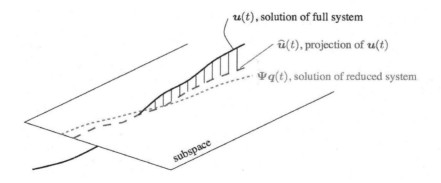

Fig. 6.9 Subspace approximation of a solution

system requires a different selection of the base vectors and thus an update of the base used for the projection. The error related to the reduced projection

$$e_{RP} = \frac{\|\cdot R(\boldsymbol{\Psi}\, q(t)) - \boldsymbol{P}\|}{\|\boldsymbol{P}\| + \|\boldsymbol{M}\boldsymbol{\Psi}\, \ddot{q}(t)\|} \qquad (6.50)$$

can be computed, see e.g. Idelsohn and Cardona (1984) and Spiess (2006). Note that the solution was obtained with (6.49) and the error is computed for the full system.

Besides the use of eigenvectors of the nonlinear system at a specific state $\boldsymbol{u}$, it is possible to use the so-called RITZ vectors, see Wilson et al. (1982). These base vectors are computed from the load case, starting with the first vector

$$\boldsymbol{\psi}_1 = \boldsymbol{K}_T^{-1}\, \boldsymbol{P} \qquad (6.51)$$

which corresponds to the static load. The further RITZ vectors are a KRYLOV sequence of the first vector

$$\boldsymbol{\psi}_i = \boldsymbol{K}_T^{-1}\, \boldsymbol{M}\boldsymbol{\psi}_{i-1}\,, \qquad i = 2,\ldots \qquad (6.52)$$

These vectors have to be orthogonalized, see above, which can be achieved by a GRAM-SCHMIDT or a QR orthonormalization.

As an example, the two-dimensioanl T-beam, depicted in Fig. 6.10, is used to show the performance of the reduction method using RITZ vectors, see also Spiess (2006). The T-beam consists of an elastic ST. VENANT material with $E = 1000$ and $\nu = 0.2$. The data of the geometry are given in Fig. 6.10 as well as the discretization using quadrilateral displacement elements with quadratic shape functions. A time dependent point load of $F(t) = 0.1\,\sin(0.5\,t)$ is applied at the end of the beam. The system is solved using 20 RITZ vectors and compared to a solution of the unreduced system. The resulting vertical displacement at the load point is shown in Fig. 6.11.

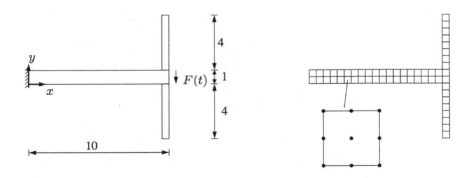

Fig. 6.10 T-beam: geometry and discretization

The full line is the solution which uses updates at given intervals. The dotted line is a RITZ approximation without updates where the RITZ vectors were computed at the beginning of the solution. The dashed, updated reference solution cannot be distinguished from the RITZ reduction with updates and thus underlines, that the base vectors have to be updated in a nonlinear analysis.

Other base vectors, like Lanczos vectors could also be used within the projection based reduction method as well. A discussion can be found in Vukazich et al. (1996).

POD Method. The Proper Orthogonal Decomposition (POD) method is also known as KARHUNEN-LOEVE expansion. It is a method developed for the analysis of data and often used to evaluate experimental data. Its goal is to identify a subspace of the solution space which includes the most relevant parts. The method determines the most energetic modes, the so-called POD vectors, as well as the fraction of energy included in each mode. This enables a better understanding of the dynamics of a system and enhance the analysis of the dynamics when a projection onto some lower dimensional spaces is performed. Although the route of the method is very old and goes back to the beginning of the last century, not many relevant publications can be found applying the method in the field of continuous solid dynamics.

The POD is often used for the analysis of experimental data, in control systems and fluid mechanics. It was first applied to nonlinear dynamics by Kreuzer and Kust (1995); further utilizations are reported in Krysl et al. (2001). Besides these applications, it is also possible to use the vectors obtained by a proper orthogonal decomposition as basis in a projection-based reduction method, see above. These vectors are called POD vectors in the

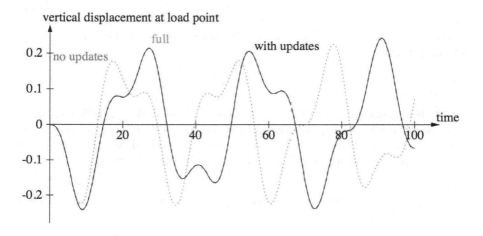

Fig. 6.11 Vertical displacement from solution with Ritz

following. POD- based reduction methods were recently introduced in nonlinear finite element analysis by Krysl et al. (2001), Meyer and Matthies (2003) and also used in Spiess (2006).

To obtain a reduced model within the frame of the POD method from a given high dimenionsal finite element model requires two steps.

1. Extract the basis functions by detailed simulations of high-dimensional systems. This step is also known as Karhunen-Loeve Decomposition (CLD) or Singular Value Decomposition (SVD).
2. Projection of the basis functions to a low-dimensional dynamical model using the Galerkin method.

In the mode extraction step of POD, the displacements $u(t)$ are approximated over some domain of interest as a finite sum in the form of separation of variables:

$$u(t) \approx \sum_{k=1}^{m} a_k(t)\,\psi_k\,. \tag{6.53}$$

This representation is not unique. If the domain bounded, then the vectors ψ_k can be characterized by a FOURIER series, by LEGENDRE polynomials or by CHEVYSHEV polynomials. In the case of the POD method, the vectors ψ_k are chosen from modes, extracted from a singular value decomposition (SVD) analysis.

In the case of discrete data, the displacement vector $u(t)$ is evaluated at P instants of time. In a finite element discretization, this leads to P sets of N finite element unknowns. The motion is regarded as a variation of the displacement vector $u(t)$ around the centre $\tilde{u}$ which defines the time-average of the motion. Thus the motion can be written as

$$u(t) = w(t) + \tilde{u}\,. \tag{6.54}$$

Here $w(t)$ denotes the centred motion. It can be represented in the same way as $u(t)$ in (6.53)

$$w(t) \approx \sum_{i=1}^{n} q_i(t)\,\psi_i \tag{6.55}$$

with the constant basis vectors ψ_i and the time dependent amplitudes $q_i(t)$. The amplitudes are the projections of the centred motion using the associated basis veector: $q_i(t) = \psi_i^T\,w(t)$. By requiring that the basis vectors have to be the best possible representation of $w(t)$, the objective function

$$J(\psi) = \frac{1}{2} \sum_{i=1}^{P} [w_i^T\,\psi]^2 - \lambda\,(\psi^T \psi - 1) \tag{6.56}$$

can be created where normalization of the basis vectors $\psi^T \psi - 1$ was built in as a constraint. Maximizing this function yields the eigenvalue problem

$$(C - \lambda \, \mathbf{1}) \, \psi = \mathbf{0} \tag{6.57}$$

with the so, called covariance matrix

$$C = \sum_{i=1}^{P} \mathbf{w}_i \, \mathbf{w}_i^T . \tag{6.58}$$

The eigenvectors of system (6.57) can be obtained by using a singular value decomposition technique, see e.g. Golub and van Loan (1989). Note that pre-multiplication of the eigenvaue problem by ψ^T yields the relation

$$\lambda = \sum_{i=1}^{P} [\mathbf{w}_i^T \, \psi]^2 \tag{6.59}$$

which is a measure of the participation of a basis vector ψ in the motion of the system.

Since the displacement $\mathbf{u}_i$ can be interpreted as a cloud of points in the N-dimensional space, it is clear that the mean value $\tilde{\mathbf{u}}$ describes the location of the centre of gravity of the cloud. It can be proven that the proper orthogonal decomposition corresponds to a line which points in the main direction of the cloud, i.e. the square sum of the distances of all points to this line are minimal, see e.g. Holmes et al. (1996).

A drawback of the Proper Orthogonal Decomposition is the need to compute the covariance matrix C in (6.58). Its size is $N \times N$ and thus extremely large for finite element problems; additionally C does not have a sparse structure. The matrix size alone can thus render computations of a POD basis impossible. If the number of sample vectors $\mathbf{w}_i$ is smaller than the number of degrees of freedom, $P << N$, the computation of the POD can be obtained with less computational effort by using the method of snapshots, see Holmes et al. (1996). This method yields a $P \times P$ matrix $\tilde{C} = \mathbf{W} \mathbf{W}^T$ with $\mathbf{W} = [\mathbf{w}_1 \ldots \mathbf{w}_P]$ which only has size $P \times P$. The resulting eigenvalue problem is given by

$$(\tilde{C} - \lambda \, \mathbf{1}) \, \mathbf{a} = \mathbf{0}. \tag{6.60}$$

The POD vector is then given by

$$\psi_i = \mathbf{W} \, \mathbf{a}_i . \tag{6.61}$$

These basis vector can now be applied within a reduction method leading to a system as described in (6.49). To show the behaviour of this method, the example of the two-dimensional T-beam, depicted in Fig. 6.10, is considered.

The simulation of the system was performed with a NEWMARK algorithm for the full system. A load of $F(t) = 1.0 \sin(0.5t)$ was applied which has relatively high load level.

This motion was analysed using the proper orthogonal scheme which leads to a set of, POD vectors, depicted in Fig. 6.12. The associated eigenvalues decrease significantly from one POD vector to the next. This fact leads to a faster convergence of the algorithm used to computed the eigenvectors. These POD vectors are then as basis vectors in recomputations of the problem solved by the NEWMARK method. Within this application, it can be shown that the POD method is able to capture the nonlinear motion with a few eigenvectors, and that it converges to the NEWMARK solution of the entire problem, see Fig. 6.13. As can be observed from Fig. 6.12, only the very first vectors are necessary to capture a large portion of the motion. The last two vectors represent deformations which are not directly related to a vibration; however, they are essential to capture the long time response of the motion, see second graph of Fig. 6.13.

A criterion which can be utilized to estimate the completeness of a set of POD vectors can be generated from the fact that the first invariant of matrix C in (6.58) can be computed also by the sum of its eigenvalues; hence $\text{tr}\,C = \sum_{i=1}^{P} \lambda_i$. From this relation, it is possible to find a truncation criterion which determines how many POD vectors should be included into a basis. The part of the motion which is captured by a given basis of R vectors is

$$\varphi = \frac{1}{\text{tr}\,C} \sum_{i=1}^{R} \lambda_i \,. \tag{6.62}$$

Thus the value φ stands for the completeness of the POD vector basis. The latter should be chosen such that $\varphi > 1 - \varepsilon$, with a truncation limit ε that determines the exactness of the basis. The value φ is reported in Fig. 6.12. It can be observed that its value converges very fast against 1 which is in accordance with the very good approximation of the motion in Fig. 6.13.

In the case that similar problems have to be solved or a set of different load cases has to be analysed. Then the POD vector basis computed from one solution can be applied for all other ones. During a computation, the quality of the result can be checked by an error measure, see e.g. (6.50). In the case

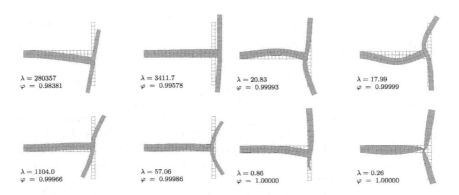

$\lambda = 280357$
$\varphi = 0.98381$

$\lambda = 3411.7$
$\varphi = 0.99578$

$\lambda = 20.83$
$\varphi = 0.99993$

$\lambda = 17.99$
$\varphi = 0.99999$

$\lambda = 1104.0$
$\varphi = 0.99966$

$\lambda = 57.06$
$\varphi = 0.99986$

$\lambda = 0.86$
$\varphi = 1.00000$

$\lambda = 0.26$
$\varphi = 1.00000$

Fig. 6.12 POD vectors

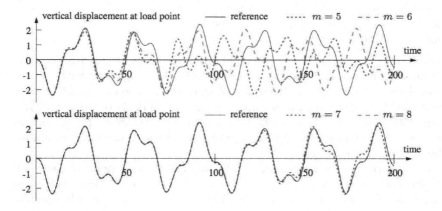

Fig. 6.13 Convergence of the POD method for increasing POD vectors

that the estimated error is beyond a certain threshold a simultation should be recomputed without a reduction technique and the gained data can then be used to improve the set of basis vectors. As the approach is based on resutls from previous simulations, it can only be applied to problems, where the results of the different load cases are not completely different.

6.2 Integration of Inelastic Constitutive Equations for Small Deformations

The constitutive equations discussed in Sect. 3.3.2 describe elasto-plastic, visco-plastic and viscous material behaviour for small deformations (geometrically linear theory). The inelastic behaviour is governed by time dependent evolution equations which can be scalar or vector valued differential equations. These have to be solved by numerical integration algorithms.

The rate equations which have to be integrated are ordinary differential equations which lead in general to an initial value problem of the form

$$
\begin{aligned}
\dot{e}(t) &= f\,[\,e(t)\,], && (6.63) \\
e(0) &= e_0\,.
\end{aligned}
$$

The integration $(I - ALGO)$ is usually be performed by a generalized mid point rule

$$
e_{n+1} = e_n + \Delta t\, f\,(\,e_{n+\theta}\,) \tag{6.64}
$$

with $e_{n+\theta} = (\,1 - \theta\,)\,e_n + \theta\,e_{n+1}\,,\quad 0 \le \theta \le 1$. Here the same notation, as in Sect. 6.1, is chosen with $e_n = e(\,t_n\,)$ and $e_{n+1} = e(\,t_{n+1}\,)$. The approximation (6.64) leads to the following integration algorithms:

− for $\theta = 0$ to the explicit EULER scheme,
− for $\theta = 1$ to the implicit EULER scheme and
− for $\theta = 1/2$ to the *midpoint-rule*.

The scheme (6.64) is for $\theta = \frac{1}{2}$ of second order accuracy. Otherwise, the integration schemes are first order accurate. Further investigations with respect to consistency, stability and accuracy of algorithms based on (6.64) can be found in the mathematical literature, see e.g. Gear (1971) or Stoer and Bulirsch (1990) but also in Simo (1998). Consistency and stability are properties which are essential for establishing convergence of the numerical solution for arbitrary small time steps.

The constitutive equations have to be fulfilled at every point of the solid. It is efficient to preserve this local character within the integration schemes. In case of the spatial discretization, using finite elements, the algorithm will be partitioned in such a way that the integration of the constitutive equations $(I − ALGO)$ is performed locally on element level. Hence the constitutive equations have to be fulfilled at each integration point (GAUSS point) within an element Ω_e. Besides this, the weak form of the equilibrium (6.1) must be obeyed. From these two requirements follows an iteration which has to be performed within each time step of the solution.

Starting with the known displacements $\boldsymbol{u}_n$, the inelastic strains $\boldsymbol{e}_n^{in}$, the stresses $\boldsymbol{\sigma}_n$ and the internal variables $\boldsymbol{\alpha}_n$ at time t_n, an algorithm can be constructed which yields as result the displacements, stresses and inelastic variables at time t_{n+1}

1. Set initial conditions:

$$
\begin{array}{ll}
\boldsymbol{u}_{n+1}^0 = \boldsymbol{u}_n \,, & \boldsymbol{\sigma}_{n+1}^0 = \boldsymbol{\sigma}_n \,, \\
\boldsymbol{e}_{n+1}^{in\,0} = \boldsymbol{e}_n^{in} \,, & \boldsymbol{\alpha}_{n+1}^0 = \boldsymbol{\alpha}_n \,,
\end{array}
$$

2. Iteration loop: $DO\ i = 0, 1, 2, \dots$
   − **Global**: Solve the weak form for $\boldsymbol{u}_{n+1}^{i+1}$

$$
\boldsymbol{G}_{n+1}^{i+1} = \boldsymbol{R}\,(\boldsymbol{u}_{n+1}^{i+1}, \boldsymbol{\sigma}_{n+1}^i) - \boldsymbol{P}_{n+1} = \boldsymbol{0} \,.
$$

   This yields the total strain increment $\Delta \boldsymbol{\varepsilon}_{n+1}^{i+1}$.
   − **Convergence**: $\| \boldsymbol{G}_{n+1}^{i+1} \| < TOL \longrightarrow STOP$
   − **Local** $(I − ALGO)$: Compute with $\Delta \boldsymbol{\varepsilon}_{n+1}^{i+1}$ the stresses $\boldsymbol{\sigma}_{n+1}^{i+1}$, which fulfil the inelastic constitutive equations at the GAUSS point.
   $END\,DO$

A schematic description of the algorithm can be found in Fig. 6.14.

6.2.1 Viscoelastic Material

The basic equations which describe viscous materials are summarized for a the linear standard body in relations (3.224) to (3.226). An integration algorithm for this type of constitutive behaviour is constructed. Additionally, it is assumed that the viscous behaviour is related entirely to the deviatoric strains $\mathbf{e}$. The constitutive relation is given for the linear standard body with Eq. (3.224) and $\nu^\infty = (1 - \nu)$ by

$$\boldsymbol{\sigma}(t) = K \operatorname{tr} \boldsymbol{\varepsilon} \, \mathbf{1} + 2\mu \left[(1 - \nu) \, \mathbf{e} + \nu \mathbf{q} \right],$$

$$\dot{\mathbf{e}}(t) = \frac{1}{\hat{\tau}} \mathbf{q} + \dot{\mathbf{q}}.$$

$\mathbf{q}$ is the strain associated with the deviatoric stress $\mathbf{s}_M$ in $(3.224)_3$. The last relation is a differential equation of first order in time. It can be integrated by using different schemes, see also last section. Here an implicit EULER scheme is employed, which has first order accuracy and is unconditionally stable.

Inserting the integration rule (6.64) with $\theta = 1$ into the rate equation within a time step $\Delta t = t_{n+1} - t_n$ yields

$$\frac{1}{\Delta t} (\mathbf{q}_{n+1} - \mathbf{q}_n) + \frac{1}{\hat{\tau}} \mathbf{q}_{n+1} = \frac{1}{\Delta t} (\mathbf{e}_{n+1} - \mathbf{e}_n). \tag{6.65}$$

This relation can be solved for $\mathbf{q}_{n+1}$

$$\mathbf{q}_{n+1} = \frac{\hat{\tau}}{\hat{\tau} + \Delta t} \mathbf{e}_{n+1} + \frac{\hat{\tau}}{\hat{\tau} + \Delta t} (\mathbf{q}_n - \mathbf{e}_n). \tag{6.66}$$

Inserting (6.66) into the first equation leads with

$$\boldsymbol{\sigma}_{n+1} = K \operatorname{tr} \boldsymbol{\varepsilon}_{n+1} \, \mathbf{1} + 2\mu \left(1 - \nu \frac{\Delta t}{\hat{\tau} + \Delta t} \right) \mathbf{e}_{n+1} + \nu \frac{\hat{\tau}}{\hat{\tau} + \Delta t} (\mathbf{q}_n - \mathbf{e}_n) \tag{6.67}$$

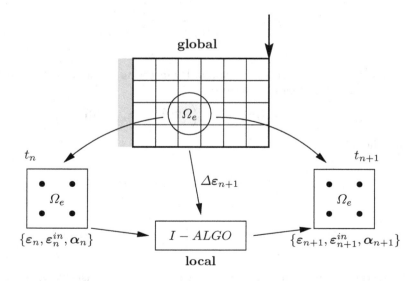

Fig. 6.14. Algorithm for the integration of elasto-plastic material

to an equation for the computation of the stresses at the end of the time step. There is no need for an iteration within the time step since the above relation is linear in the strains. Within the solution procedure, the material tangent which is consistent with (6.67) has to be applied for the computation of the associated tangential stiffness matrix. The material tangent or incremental constitutive tensor follows by considering the definition of the deviatoric strains with (3.279) from $\partial\boldsymbol{\sigma}_{n+1}/\partial\boldsymbol{\varepsilon}_{n+1}$ as

$$\mathbb{C}_{ve} = \frac{\partial\boldsymbol{\sigma}_{n+1}}{\partial\boldsymbol{\varepsilon}_{n+1}} = K\,\mathbf{1}\otimes\mathbf{1} + 2\,\mu\,(1-\nu\,\frac{\Delta t}{\hat{\tau}+\Delta t})[\,\mathbb{E}-\frac{1}{3}\,\mathbf{1}\otimes\mathbf{1}\,]. \qquad (6.68)$$

The material tangent differs from the tangent associated with linear elastic constitutive behaviour (3.280) only by the factor $(1-\nu\,\Delta t/\hat{\tau}+\Delta t)$. Hence the associated finite element code deviates only very little form a code for a linear elastic element. Besides a change in the stress computation, see (6.67), and the necessity to store the strains $\mathbf{e}$ and $\mathbf{q}$ at time t_n, only the above defined factor has to be inserted in the material tangent $\mathbb{C}_{ve}$. For the explicit matrix form of $\mathbb{C}_{ve}$, see (3.280).

Instead of using the implicit EULER scheme for the integration of (6.65), relation (3.225) can be used

$$\mathbf{s}(t) = \int\limits_{-\infty}^{t} G(t-\tau)\,\dot{\mathbf{e}}(\tau)\,d\tau$$

with a direct evaluation of the integral. Under the assumption that a strain acts first at time $t=0$, the integral can be split with $\Delta t = t_{n+1}-t_n$ into

$$\int\limits_{0}^{t}(\bullet)\,d\tau = \int\limits_{0}^{t_n}(\bullet)\,d\tau + \int\limits_{t_n}^{t_n+\Delta t}(\bullet)\,d\tau \qquad (6.69)$$

which constitutes a recursion formula. By inserting relation (3.226), an equation for the stresses can be derived

$$\mathbf{s}_{n+1} = 2\,\mu\,\left[\,(1-\nu)\,\mathbf{e}_{n+1} + \nu\,\left(e^{-(\Delta t/\hat{\tau})}\,\mathbf{h}_n + \Delta\mathbf{h}_{n+1}\right)\right] \qquad (6.70)$$

with the definitions

$$\mathbf{h}_n = e^{-(t_n/\hat{\tau})}\int\limits_{0}^{t_n} e^{(\tau/\hat{\tau})}\,\dot{\mathbf{e}}(\tau)\,d\tau\,,$$

$$\Delta\mathbf{h}_{n+1} = e^{-(t_n+\Delta t/\hat{\tau})}\left(\int\limits_{t_n}^{t_n+\Delta t} e^{(\tau/\hat{\tau})}\,\dot{\mathbf{e}}(\tau)\,d\tau\right). \qquad (6.71)$$

Under the assumption that the strain increment is constant within a time step Δt,

$$\dot{\mathbf{e}}(\tau) \approx \frac{1}{\Delta t} \left(\mathbf{e}_{n+1} - \mathbf{e}_n \right),$$

the integral in $(6.71)_2$ can be solved analytically, see Taylor et al. (1970). After some algebraic manipulations

$$\Delta \mathbf{h}_{n+1} = \frac{\hat{\tau}}{\Delta t} \left(1 - e^{-(\Delta t / \hat{\tau})} \right) \left(\mathbf{e}_{n+1} - \mathbf{e}_n \right) \tag{6.72}$$

is obtained. Since $\mathbf{h}_n$ is already known from the previous steps, all quantities for the computation of the deviatoric stresses are known. The associated incremental constitutive tensor is derived from $\partial \boldsymbol{\sigma}_{n+1} / \partial \boldsymbol{\varepsilon}_{n+1}$

$$\mathbf{C}_{ve} = \frac{\partial \boldsymbol{\sigma}_{n+1}}{\partial \boldsymbol{\varepsilon}_{n+1}} = K \mathbf{1} \otimes \mathbf{1} + 2 \mu \left[(1 - \nu) + \nu \frac{\hat{\tau}}{\Delta t} \left(1 - e^{-(\Delta t / \hat{\tau})} \right) \right] \left[\mathbb{E} - \frac{1}{3} \mathbf{1} \otimes \mathbf{1} \right].$$

$$\tag{6.73}$$

Again, only the factor changes which has to be multiplied by the shear modulus μ when compared to the linear elastic material tangent. Contrary to the implicit EULER scheme, the integration method using (6.70) and (6.72) is of second order accurate, see e.g. Simo and Hughes (1998).

Note that (6.72) has to be divided by Δt. This can lead to numerical difficulties for $\Delta t \to 0$. By using a TAYLOR series expansion of the factor around zero, $\Delta \mathbf{h}_{n+1}$ can be computed also for very small values of Δt in an accurate way. This leads to the following approximation:

$$\frac{\hat{\tau}}{\Delta t} \left(1 - e^{-(\Delta t / \hat{\tau})} \right) \approx 1 - \frac{1}{2} \frac{\Delta t}{\hat{\tau}} + \frac{1}{6} \left(\frac{\Delta t}{\hat{\tau}} \right)^2. \tag{6.74}$$

For the case $\Delta t = 0$, both Eqs. (6.68) and (6.73) lead to the material tensor of the linear elastic constitutive equation (3.280).

6.2.2 Elasto-Plastic Material

Elasto-plasticity is described by the general constitutive equations (3.168) to (3.172) in Sect. 3.3.2. For these equations, a general integration algorithm will be constructed. However, for simplicity, we will restrict ourselves to materials which are described by one flow surface. Algorithms for the more general case are, e.g. described in Simo (1998) or for the special case of non-cohesive soils in Leppin and Wriggers (1997).

Since the differential equations which describe elasto-plastic deformations are stiff in the mathematical sense, see e.g. Nagtegaal (1982), Simo and Taylor (1985) and Simo and Hughes (1998), implicit integration methods, such as the EULER scheme, have to be applied.

This leads, for the case of non-associated plasticity, to a finite difference approximation of Eqs. (3.168–3.172)

$$\boldsymbol{\sigma}_{n+1} = \mathbb{C}\left[\boldsymbol{\varepsilon}_{n+1} - \boldsymbol{\varepsilon}^p_{n+1}\right], \tag{6.75}$$

$$\mathbf{q}_{n+1} = \mathbb{H}\left[\boldsymbol{\alpha}_{n+1}\right], \tag{6.76}$$

$$\frac{1}{\Delta t}\left(\boldsymbol{\varepsilon}^p_{n+1} - \boldsymbol{\varepsilon}^p_n\right) = \dot{\lambda}_{n+1}\,\mathbf{r}(\boldsymbol{\sigma}_{n+1}, \mathbf{q}_{n+1}), \tag{6.77}$$

$$\frac{1}{\Delta t}\left(\boldsymbol{\alpha}_{n+1} - \boldsymbol{\alpha}_n\right) = \dot{\lambda}_{n+1}\,\mathbf{h}(\boldsymbol{\sigma}_{n+1}, \mathbf{q}_{n+1}), \tag{6.78}$$

$$f(\boldsymbol{\sigma}_{n+1}, \mathbf{q}_{n+1}) \le 0. \tag{6.79}$$

This constitutes a set of implicit equations for the unknown stresses and strains. It has to be fulfilled at each GAUSS point of a spatial finite element discretization.

The solution of the equation set (6.75–6.79) is not trivial since the inequality constraint in (6.79) has to be fulfilled. A so-called operator split algorithm has proven to be most efficient for this task. The idea of this procedure is to freeze the plastic variables at the beginning of a time step from t_n to t_{n+1}:

$$\boldsymbol{\varepsilon}^{p\,tr}_{n+1} = \boldsymbol{\varepsilon}^p_n, \qquad \boldsymbol{\alpha}^{tr}_{n+1} = \boldsymbol{\alpha}_n. \tag{6.80}$$

The superscript $()^{tr}$ denotes *trial* quantities which is being fixed for a moment, but may change during the iteration associated with the algorithm. The frozen plastic variables are inserted in Eqs. (6.75) and (6.76)

$$\boldsymbol{\sigma}^{tr}_{n+1} = \mathbb{C}\left[\boldsymbol{\varepsilon}_{n+1} - \boldsymbol{\varepsilon}^{p\,tr}_{n+1}\right],$$

$$\mathbf{q}^{tr}_{n+1} = \mathbb{H}\left[\boldsymbol{\alpha}^{tr}_{n+1}\right]. \tag{6.81}$$

These relations can now be used to check the status (elastic or plastic) within a time step. For this, the trial state is inserted into the yield function

$$f(\boldsymbol{\sigma}^{tr}_{n+1}, \mathbf{q}^{tr}_{n+1})\begin{cases} \le 0 & \Rightarrow \quad \text{elastic}, \\ > 0 & \Rightarrow \quad \text{plastic}. \end{cases} \tag{6.82}$$

In the case that f denotes an elastic state at t_{n+1}, then all plastic variables are updated at the end of the time step by the trial quantities:

$$\boldsymbol{\varepsilon}^p_{n+1} = \boldsymbol{\varepsilon}^{p\,tr}_{n+1}, \qquad \boldsymbol{\alpha}_{n+1} = \boldsymbol{\alpha}^{tr}_{n+1} \tag{6.83}$$

and the local algorithm is terminated for that time step.

In the case that f denotes a plastic state at t_{n+1}, the stress state has to be corrected such that it fulfils the yield condition (6.79). This leads with $\gamma_{n+1} = \dot{\lambda}_{n+1}\,\Delta t$, (6.75) and (6.79) to

$$\mathbf{R}^i_{\sigma} = -\boldsymbol{\varepsilon}_{n+1} + \boldsymbol{\varepsilon}^p_n + \mathbb{C}^{-1}[\boldsymbol{\sigma}^i_{n+1}] + \gamma_{n+1}\,\mathbf{r}(\boldsymbol{\sigma}^i_{n+1}, \mathbf{q}^i_{n+1}) = \mathbf{0},$$

$$\mathbf{R}^i_{q} = \mathbb{H}^{-1}[\mathbf{q}^i_{n+1}] - \boldsymbol{\alpha}_n - \gamma^i_{n+1}\,\mathbf{h}(\boldsymbol{\sigma}^i_{n+1}, \mathbf{q}^i_{n+1}) = \mathbf{0}, \tag{6.84}$$

$$R^i_{f} = f(\boldsymbol{\sigma}^i_{n+1}, \mathbf{q}^i_{n+1}) = 0.$$

These three equations constitute a nonlinear system for the unknowns $\boldsymbol{\sigma}^i_{n+1}$, $\mathbf{q}^i_{n+1}$ and γ^i_{n+1}. For its solution, NEWTON method is applied (iteration index i). This leads to the algorithm

$$\mathbf{A}^i_{n+1} \Delta \mathbf{p}^{i+1}_{n+1} = -\mathbf{R}^i_{n+1}$$
$$\mathbf{p}^{i+1}_{n+1} = \mathbf{p}^i_{n+1} + \Delta \mathbf{p}^{i+1}_{n+1} . \tag{6.85}$$

Matrix $\mathbf{A}$ follows from the linearization of the residuals defined in (6.84). It has the explicit form

$$\mathbf{A}^i_{n+1} = \begin{bmatrix} \mathbb{C}^{-1} + \gamma^i_{n+1} \dfrac{\partial \mathbf{r}^i_{n+1}}{\partial \boldsymbol{\sigma}} & \gamma^i_{n+1} \dfrac{\partial \mathbf{r}^i_{n+1}}{\partial \mathbf{q}} & \dfrac{\partial (\gamma^i_{n+1} \mathbf{r}^i_{n+1})}{\partial \gamma} \\ -\gamma^i_{n+1} \dfrac{\partial \mathbf{h}^i_{n+1}}{\partial \boldsymbol{\sigma}} & \mathbb{H}^{-1} - \gamma^i_{n+1} \dfrac{\partial \mathbf{h}^i_{n+1}}{\partial \mathbf{q}} & -\dfrac{\partial (\gamma^i_{n+1} \mathbf{h}^i_{n+1})}{\partial \gamma} \\ \dfrac{\partial f^i_{n+1}}{\partial \boldsymbol{\sigma}} & \dfrac{\partial f^i_{n+1}}{\partial \mathbf{q}} & \dfrac{\partial f^i_{n+1}}{\partial \gamma} \end{bmatrix} \tag{6.86}$$

with the definitions of the vectors:

$$\mathbf{p}^i_{n+1} = \left\{ \begin{matrix} \boldsymbol{\sigma}^i_{n+1} \\ \mathbf{q}^i_{n+1} \\ \gamma^i_{n+1} \end{matrix} \right\} \quad \text{and} \quad \mathbf{R}^i_{n+1} = \left\{ \begin{matrix} \mathbf{R}^i_\sigma \\ \mathbf{R}^i_q \\ R^i_f \end{matrix} \right\} . \tag{6.87}$$

At the end of this NEWTON iteration, the stresses and plastic variables are known at a single GAUSS point.

Besides the fulfilment of the inequality constraints imposed by plasticity, the global weak form (equilibrium) has to be fulfilled, see also the introductory remarks and Fig. 6.14. For the construction of a global NEWTON method, it is necessary to determine the constitutive tangent $\bar{\boldsymbol{D}}^p_{n+1}$. It occurs in the linearization (4.111) of the weak form (4.97). Note that the geometrical matrix is zero for small strain problems and that all derivatives have to be computed with respect to the initial configuration. Hence the tangent matrix assumes the form

$$\bar{K}^p_{T_{IK}} = \int_{\varphi(\Omega_e)} \bar{\boldsymbol{B}}^T_{0\,I} \, \bar{\boldsymbol{D}}^p_{n+1} \, \bar{\boldsymbol{B}}_{0\,K} \, d\omega . \tag{6.88}$$

The material tangent follows from

$$\bar{\boldsymbol{D}}^p_{n+1} = \frac{\partial \boldsymbol{\sigma}_{n+1}}{\partial \boldsymbol{\varepsilon}_{n+1}} = \frac{\partial \Delta \boldsymbol{\sigma}_{n+1}}{\partial \boldsymbol{\varepsilon}_{n+1}} . \tag{6.89}$$

With the inverse of $\mathbf{A}^i_{n+1}$, which has to be computed in the converged state of the NEWTON iteration given above,

$$(\mathbf{A}^i_{n+1})^{-1} = \begin{bmatrix} \mathbf{A}_{11} & \mathbf{A}_{12} & \mathbf{A}_{13} \\ \mathbf{A}_{21} & \mathbf{A}_{22} & \mathbf{A}_{23} \\ \mathbf{A}_{31} & \mathbf{A}_{32} & \mathbf{A}_{33} \end{bmatrix}^i_{n+1} , \tag{6.90}$$

the derivative is explicitly derived from

$$\frac{\partial \Delta \boldsymbol{\sigma}_{n+1}}{\partial \boldsymbol{\varepsilon}_{n+1}} = \mathbf{A}_{11} . \tag{6.91}$$

This yields the material tangent

$$\bar{D}_{n+1}^p = \mathbf{A}_{11}, \tag{6.92}$$

which is needed within the linearization of the weak form. Hence it is possible to compute the material tangent simply by using matrix $\mathbf{A}$ of the last NEWTON step in (6.85). As can be observed, contrary to elasticity, the material tangent depends upon the chosen integration algorithm. Due to this reason, the material tangent is often called consistent tangent in the literature, since it is consistent with the elasto-plastic algorithm, see e.g. Simo and Taylor (1985) and Simo and Hughes (1998).

The algorithm described above will be discussed now in more depth for a special elasto-plastic material exhibiting nonlinear isotropic and linear kinematic hardening. The plastic flow is determined by the second invariant of the stress deviator II_S (VON MISES material). This material behaviour is known in the literature as J_2-plasticity since the second invariant of the stress deviator is denoted also by J_2. The constitutive equations for such material were described in Sect. 3.3.2 where the internal plastic variables $\mathbf{e}^p, \boldsymbol{\alpha}, \hat{\alpha}$ were introduced together with the generalized stresses $\mathbf{s}, \mathbf{q}, q$. The quantities $\mathbf{s}$ and $\mathbf{e}^p$ denote the deviators of the stress tensor $\boldsymbol{\sigma}$, see (3.151), and the strain tensor $\boldsymbol{\varepsilon}$, respectively. Furthermore, volume change due to plasticity is assumed to be zero (tr $\boldsymbol{\varepsilon}^p = 0$).

The associated material equations can be deduced from (3.150), (3.159), (3.160) and (3.161) in compact form by insertng the movement of the flow surface $\mathbf{q}$ and the equivalent strain $\hat{\alpha}$ directly. Besides the split of the deviatoric strains in elastic and plastic parts

$$\mathbf{e} = \mathbf{e}^e + \mathbf{e}^p, \tag{6.93}$$

the elastic constitutive equation (shear modulus μ, modulus of compression K)

$$\mathbf{s} = 2\,\mu\,\mathbf{e}^e, \qquad p = K\,\mathrm{tr}\,\boldsymbol{\varepsilon}^e \tag{6.94}$$

can be formulated for the deviator stresses $\mathbf{s}$ and the pressure p. Furthermore, the evolution equations for the plastic strains, and the hardening variables are given by

$$\dot{\mathbf{e}}^p = \lambda\,\frac{\partial f}{\partial \mathbf{s}}, \qquad \dot{\mathbf{q}} = -\frac{2}{3}\,H\,\lambda\,\frac{\partial f}{\partial \mathbf{s}}, \qquad \dot{\hat{\alpha}} = \sqrt{\frac{2}{3}}\,\lambda. \tag{6.95}$$

The generalized stresses have to obey the yield condition

$$f(\bar{\mathbf{s}}, \hat{\alpha}) = \|\bar{\mathbf{s}}\| - \sqrt{\frac{2}{3}}\,[Y_0 + \hat{H}(\hat{\alpha})] \le 0 \tag{6.96}$$

where, to shorten notation, $\bar{\mathbf{s}} = \mathbf{s} - \mathbf{q}$ was introduced.

This set of equation, describing the elasto-plastic material behaviour, represents a stiff ordinary differential system. For the integration of such systems which include inequality constraints, operator split methods, based on implicit integration rules are well suited, see e.g. Simo and Hughes (1998) and Simo (1998). They are unconditionally stable. Here a first order scheme, the implicit EULER method, is employed; higher order schemes can be found in e.g. Simo (1998).

Insertion of the implicit EULER rule into the evolution equations for the plastic variables (6.95) yields within the time step $\Delta t = t_{n+1} - t_n$ the plastic strains

$$\frac{1}{\Delta t} \left(\mathbf{e}^p_{n+1} - \mathbf{e}^p_n \right) = \dot{\lambda} \left[\frac{\partial f}{\partial \mathbf{s}} \right]_{n+1} = \frac{1}{\Delta t} \left(\lambda_{n+1} - \lambda_n \right) \mathbf{n}_{n+1} . \tag{6.97}$$

Here the notation $\frac{\partial f}{\partial \mathbf{s}} = \frac{\bar{\mathbf{s}}}{\|\bar{\mathbf{s}}\|} = \mathbf{n}$ is used which was introduced in Sect. 3.3.2. In the same way, relation

$$\frac{1}{\Delta t} \left(\mathbf{q}_{n+1} - \mathbf{q}_n \right) = \frac{1}{\Delta t} \left(\lambda_{n+1} - \lambda_n \right) \frac{2}{3} H \, \mathbf{n}_{n+1} \tag{6.98}$$

is obtained for the kinematic hardening values as well as for the isotropic hardening

$$\frac{1}{\Delta t} \left(\hat{\alpha}_{n+1} - \hat{\alpha}_n \right) = \frac{1}{\Delta t} \left(\lambda_{n+1} - \lambda_n \right) \sqrt{\frac{2}{3}} . \tag{6.99}$$

Equations (6.97–6.99) are now reformulated with $\Delta\gamma_{n+1} = \lambda_{n+1} - \lambda_n$ as

$$\begin{aligned}
\mathbf{e}^p_{n+1} &= \mathbf{e}^p_n + \Delta\gamma_{n+1} \, \mathbf{n}_{n+1}, \\
\mathbf{q}_{n+1} &= \mathbf{q}_n + \Delta\gamma_{n+1} \frac{2}{3} H \, \mathbf{n}_{n+1}, \\
\hat{\alpha}_{n+1} &= \hat{\alpha}_n + \sqrt{\frac{2}{3}} \, \Delta\gamma_{n+1} .
\end{aligned} \tag{6.100}$$

The stresses can be determined at time t_{n+1}

$$\mathbf{s}_{n+1} = 2\mu \left(\mathbf{e}_{n+1} - \mathbf{e}^p_{n+1} \right) = 2\mu \left(\mathbf{e}_{n+1} - \mathbf{e}^p_n \right) - 2\mu \, \Delta\gamma_{n+1} \, \mathbf{n}_{n+1} . \tag{6.101}$$

Observe from (6.101) that the time integration of the elasto-plastic constitutive equations can be written as a predictor–corrector algorithm. Within the predictor step, it is assumed that the plastic variables are "frozen". This yields the generalized *trial* stresses

$$\begin{aligned}
\mathbf{s}^{tr}_{n+1} &= 2\mu \left(\mathbf{e}_{n+1} - \mathbf{e}^p_n \right), \\
\bar{\mathbf{s}}^{tr}_{n+1} &= \mathbf{s}^{tr}_{n+1} - q_n , \\
\hat{\alpha}^{tr}_{n+1} &= \hat{\alpha}_n ,
\end{aligned} \tag{6.102}$$

which represent the first terms in (6.100)$_3$ and (6.101)$_{1,2}$.

Since the plastic variables $\{\, \mathbf{e}_n^p \,,\boldsymbol{\alpha}_n \,, \hat{a}_n \,\}$ are known from the last time step and $\mathbf{e}_{n+1} = \mathbf{e}_n + \Delta\mathbf{e}_{n+1}$ was determined form the global solution of the weak form, the *trial* quantities can be computed directly, see also Fig. 6.14. When the trial quantities fulfil the yield condition (6.96), then the constitutive behaviour is elastic within the time interval $[\, t_n \,, t_{n+1} \,]$, see Fig. 6.15a. If this is the case, the local algorithm for the determination of the plastic variables is terminated.

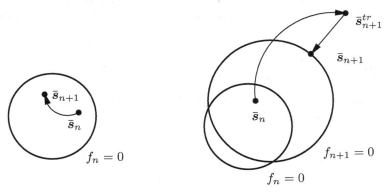

Fig. 6.15 (a) Elastic step   (b) Projection of the stresses

In the case that the trial quantities violate the yield condition at time t_{n+1} – as depicted in Fig. 6.15b – computation of the magnitude of the plastic flow $\Delta\gamma_{n+1}$ as well as its direction $\mathbf{n}_{n+1}$ is needed within the stress computation in (6.101). With Eq. (6.102)$_{1,2}$, relation

$$(\mathbf{s}_{n+1} - \mathbf{q}_{n+1}) = (\bar{\mathbf{s}}_{n+1}^{tr} - [\,2\mu + \frac{2}{3}\,H\,]\Delta\gamma_{n+1}\,\mathbf{n}_{n+1}\,,$$

$$\bar{\mathbf{s}}_{n+1}^{tr} = \bar{\mathbf{s}}_{n+1}\left[\,1 + (\,2\mu + \frac{2}{3}\,H\,)\,\frac{\Delta\gamma_{n+1}}{\|\,\bar{\mathbf{s}}_{n+1}\,\|}\,\right] \qquad (6.103)$$

can be determined from (6.101). It is clear that the term in the square bracket is a scalar. Hence

$$\mathbf{n}_{n+1}^{tr} = \frac{\bar{\mathbf{s}}_{n+1}^{tr}}{\|\,\bar{\mathbf{s}}_{n+1}^{tr}\,\|} = \mathbf{n}_{n+1} \qquad (6.104)$$

is deduced. With this result, the plastic correction step is reduced to the determination of $\Delta\gamma_{n+1}$.

By scalar multiplication of (6.103) with $\mathbf{n}_{n+1}$, relation

$$\|\,\bar{\mathbf{s}}_{n+1}\,\| = \|\,\bar{\mathbf{s}}_{n+1}^{tr}\,\| - (\,2\mu + \frac{2}{3}\,H\,)\Delta\gamma_{n+1} \qquad (6.105)$$

is obtained with (6.104) and $\bar{\mathbf{s}} \cdot \mathbf{n} = \|\bar{\mathbf{s}}\|$. This relation can be inserted into the yield condition (6.96) which has to be fulfilled at time t_{n+1}. This leads with

$$f_{n+1} = \| \bar{s}_{n+1}^{tr} \| - (2\mu + \frac{2}{3} H)\Delta\gamma_{n+1} - \sqrt{\frac{2}{3}} \left[Y_0 + \hat{H}(\hat{\alpha}_n + \sqrt{\frac{2}{3}}\Delta\gamma_{n+1}) \right] = 0$$

(6.106)

to a nonlinear equation for $\Delta\gamma_{n+1}$, which can be solved by NEWTON method. Hence a local iteration has to be performed at each GAUSS point

$$\Delta\Delta\gamma_{n+1}^{j+1} = -\left(\frac{\partial f_{n+1}}{\partial \Delta\gamma_{n+1}} \bigg|_j \right)^{-1} f_{n+1}^j, \qquad (6.107)$$

$$\Delta\gamma_{n+1}^{j+1} = \Delta\gamma_{n+1}^j + \Delta\Delta\gamma_{n+1}^{j+1}. \qquad (6.108)$$

The derivative of f_{n+1} with respect to $\Delta\gamma_{n+1}$ can be stated explicitly

$$\frac{\partial f_{n+1}}{\partial \Delta\gamma_{n+1}} = -2\mu \left(1 + \frac{H + \hat{H}'(\Delta\gamma_{n+1})}{3\mu} \right), \qquad (6.109)$$

where $\hat{H}'$ was introduced for the derivative of $\hat{H}$ with respect to $\Delta\gamma_{n+1}$. Hence the stresses are projected radially back onto the yield surface, see Fig. 6.15b, the integration algorithms is also known in the literature as *radial return* algorithm.

For the special case of linear hardening, relation $\hat{H}(\hat{\alpha}) = \hat{H}\,\hat{\alpha}$ is valid. In this case, the increment $\Delta\gamma_{n+1}$ can be determined explicitly from (6.106)

$$2\mu\,\Delta\gamma_{n+1} = \frac{f_{n+1}^{tr}}{1 + \frac{H+\hat{H}}{3\mu}} \qquad (6.110)$$

with $f_{n+1}^{tr} = \| \bar{s}^{tr} \| - \sqrt{\frac{2}{3}} (Y_0 + \hat{H}\,\hat{\alpha}_n)$.

Once the increment of the plastic flow is known the stresses, the plastic strains and the internal variables can be computed from (6.100). The stresses are then given by

$$\boldsymbol{\sigma}_{n+1} = K\,\mathrm{tr}\,\boldsymbol{\varepsilon}_{n+1} + 2\mu\,(\mathbf{e}_{n+1} - \mathbf{e}_n^p) - 2\mu\,\Delta\gamma_{n+1}\,\mathbf{n}_{n+1}^{tr}. \qquad (6.111)$$

From the stresses at time t_{n+1}, the consistent or algorithmic tangent of the elasto-plastic constitutive equation can be computed using the relations in Sect. 3.3.4. It is needed within the global NEWTON iteration, see Fig. 6.14. From

$$\mathbb{C}_{n+1}^{ep} = \frac{\partial \boldsymbol{\sigma}_{n+1}}{\partial \boldsymbol{\varepsilon}_{n+1}}, \qquad (6.112)$$

the elasto-plastic tangent follows as

$$\mathbb{C}_{n+1}^{ep} = \mathbb{C}^{el} - 2\mu\,\mathbf{n}_{n+1}^{tr} \otimes \frac{\partial \Delta\gamma_{n+1}}{\partial \boldsymbol{\varepsilon}_{n+1}} - 2\mu\,\Delta\gamma_{n+1}\frac{\partial \mathbf{n}_{n+1}^{tr}}{\partial \boldsymbol{\varepsilon}_{n+1}} \qquad (6.113)$$

with $\mathbb{C}^{el} = K\,\mathbf{1}\otimes\mathbf{1} + 2\mu\,(\mathbb{E} - \frac{1}{3}\mathbf{1}\otimes\mathbf{1})$. The partial derivatives with respect to the strains which occur in (6.113) are given by

$$\frac{\partial \Delta \gamma_{n+1}}{\partial \varepsilon_{n+1}} = \frac{\partial \Delta \gamma_{n+1}}{\partial f_{n+1}} \frac{\partial f_{n+1}}{\partial \bar{s}_{n+1}} \frac{\partial \bar{s}_{n+1}}{\partial \varepsilon_{n+1}} = \left(1 + \frac{H + \hat{H}'(\Delta \gamma_{n+1})}{3\mu} \right)^{-1} \mathbf{n}_{n+1}^{tr}$$

(6.114)

and

$$\frac{\partial \mathbf{n}_{n+1}^{tr}}{\partial \varepsilon_{n+1}} = \frac{\partial \mathbf{n}_{n+1}^{tr}}{\partial \mathbf{s}_{n+1}} \frac{\partial \mathbf{s}_{n+1}}{\partial \varepsilon_{n+1}} = -\frac{2\mu}{\|\bar{\mathbf{s}}_{n+1}^{tr}\|} \left[\mathbb{E} - \frac{1}{3} \mathbf{1} \otimes \mathbf{1} + \mathbf{n}_{n+1}^{tr} \otimes \mathbf{n}_{n+1}^{tr} \right]. \quad (6.115)$$

By inserting (6.114) and (6.115) into Eq. (6.113), the algorithmic tangent follows for elasto-plastic material as

$$\mathbb{C}_{n+1}^{ep} = K \mathbf{1} \otimes \mathbf{1} + 2\mu \, a_{n+1} \left(\mathbb{E} - \frac{1}{3} \mathbf{1} \otimes \mathbf{1} \right) - 2\mu \, b_{n+1} \, \mathbf{n}_{n+1}^{tr} \otimes \mathbf{n}_{n+1}^{tr},$$

$$a_{n+1} = 1 - \frac{2\mu \, \Delta \gamma_{n+1}}{\|\bar{\mathbf{s}}_{n+1}^{tr}\|}, \quad (6.116)$$

$$b_{n+1} = \left(1 + \frac{H + \hat{H}'(\Delta \gamma_{n+1})}{3\mu} \right)^{-1} - \frac{2\mu \, \Delta \gamma_{n+1}}{\|\bar{\mathbf{s}}_{n+1}^{tr}\|}.$$

This tangent differs from the incremental elasto-plastic continuum material tensor (3.287) by the scalar factors a_{n+1} and b_{n+1}. For $\Delta \gamma_{n+1} \to 0$, the algorithmic tangent and the continuum tangent become equivalent. This is only the case for $\Delta t \to 0$. This shows the consistency between the integration algorithm and the continuous problem.

The above described integration algorithm is valid for three-dimensional and two-dimensional problems; for the latter plane strain conditions have to be assumed. In case of a plane stress problem, additional considerations are necessary to fulfil the plane stress assumptions within the radial return algorithm. A solution of this problem is provided in Simo and Taylor (1986) where the projection is directly performed in the subspace of the plane stress state. This formulation which is given in similar form in Gruttmann and Stein (1988) will not be described here in detail; related derivations can be found in Simo and Hughes (1998). A projection algorithm for plane stress problems is developed in Sect. 9.4.5 for finite deformations.

It is well known that several constitutive formulations depicts softening behaviour, and hence can lead to localization, see e.g. Remark 3.6. These models can have non-unique solutions and thus are not easily treated by numerical algorithms. Especially, implicit schemes have their problems with such constitutive models. Thus, for geo-materials, explicit schemes using sub-stepping were constructed in, e.g. Sloan (1987b), Sloan et al. (2001) and Sheng and Sloan (2001). Another approach which combines implicit and explicit methods for plasticity and general damage models can be found in Oliver et al. (2006).

6.2.3 Elasto-Viscoplastic Material

Based on the relations in Sect. 3.3.3 for elasto-visco-plastic material, time integration algorithms can be derived for time dependent viscous material behaviour. An essential difference with respect to elasto-plastic materials is dependence of the constitutive relations upon the true rate. Hence the consistency parameter λ follows from a constitutive equation. The constitutive equation from PERZYNA, see (3.233), leads to the relation

$$\lambda = \frac{1}{2\eta} \langle f \rangle . \tag{6.117}$$

With the elastic constitutive equation for the deviatoric quantities, the set of equations describing visco-plastic constitutive behaviour with linear isotropic hardening are obtained

$$\dot{\mathbf{e}}^{vp} = \frac{1}{2\eta} \langle f \rangle \, \mathbf{n} , \qquad \mathbf{s} = 2\mu \, (\mathbf{e} - \mathbf{e}^{vp}) , \qquad \hat{\alpha} = \sqrt{\frac{2}{3}} \frac{1}{2\eta} \langle f \rangle . \tag{6.118}$$

The yield function f is given for linear isotropic hardening by

$$f(\mathbf{s}, \hat{\alpha}) = \|\mathbf{s}\| - \sqrt{\frac{2}{3}} [Y_0 + \hat{H}\,\hat{\alpha}] \leq 0 . \tag{6.119}$$

Integration of (6.117) within the time step $\Delta t = t_{n+1} - t_n$ is performed using an implicit EULER scheme. It leads to

$$
\begin{aligned}
\mathbf{e}_{n+1}^{vp} &= \mathbf{e}_n^{vp} + \frac{\Delta t}{2\eta} \langle f_{n+1} \rangle \, \mathbf{n}_{n+1} , \\
\hat{\alpha}_{n+1} &= \hat{\alpha}_n + \sqrt{\frac{2}{3}} \frac{\Delta t}{2\eta} \langle f_{n+1} \rangle .
\end{aligned}
$$

The constitutive equation can now be written for the deviatoric stresses at time t_{n+1}

$$\mathbf{s}_{n+1} = 2\mu \, (\mathbf{e}_{n+1} - \mathbf{e}_{n+1}^{vp}) = \mathbf{s}_{n+1}^{tr} - 2\mu \frac{\Delta t}{2\eta} \langle f_{n+1} \rangle \, \mathbf{n}_{n+1} . \tag{6.120}$$

Within this relation f_{n+1} is still unknown. As in the previous section, *trial* stresses are defined which indicate the excess of the yield limit

$$\mathbf{s}_{n+1}^{tr} = 2\mu \, (\mathbf{e}_{n+1} - \mathbf{e}_{n+1}^{vp\,tr}) , \tag{6.121}$$

where the visco-plastic strains were set equal to the visco-plastic strains of the last time step: $\mathbf{e}_{n+1}^{vp\,tr} = \mathbf{e}_n^{vp}$. Using the same arguments as in the last section in (6.103), relation

$$\| \mathbf{s}_{n+1} \| = \| \mathbf{s}_{n+1}^{tr} \| - 2\mu \frac{\Delta t}{2\eta} \langle f_{n+1} \rangle \tag{6.122}$$

is obtained from (6.120) to (6.121). Contrary to plasticity, where the consistency parameter follows from the fulfillment of the yield condition at time t_{n+1}, the evaluation of the yield condition

$$f_{n+1} = \| \mathbf{s}_{n+1} \| - \sqrt{\frac{2}{3}} \, [\, Y_0 + \hat{H} \, (\hat{\alpha} + \sqrt{\frac{2}{3}} \frac{\Delta t}{2 \, \eta} \, f_{n+1})] \qquad (6.123)$$

leads to the parameter

$$\Delta \gamma_{n+1} = \frac{\Delta t}{2 \, \eta} \, \langle \, f_{n+1} \, \rangle = \frac{1}{2 \, \mu} \frac{\langle \, f_{n+1}^{tr} \, \rangle}{\frac{2 \eta}{\Delta t} + (\, 1 + \frac{\hat{H}}{3 \, \mu} \,)} \,. \qquad (6.124)$$

With this result, the vicso-plastic strains in (6.120) and hence the stresses in (6.120) are determined, depending on the trial values.

The linearization within the time step Δt is derived as in the last section by differentiation of the stresses with respect to the total strains

$$\mathbb{C}_{n+1}^{vp} = \frac{\partial \boldsymbol{\sigma}_{n+1}}{\partial \boldsymbol{\varepsilon}_{n+1}} \,. \qquad (6.125)$$

From that follows the incremental constitutive tensor at time t_{n+1} by using (6.120) and (6.124)

$$\begin{aligned}
\mathbb{C}_{n+1}^{vp} &= K \, \mathbf{1} \otimes \mathbf{1} + 2 \, \mu \, a_{n+1} \, (\, \mathbb{E} - \frac{1}{3} \, \mathbf{1} \otimes \mathbf{1} \,) - 2 \, \mu \, b_{n+1} \, \mathbf{n}_{n+1}^{tr} \otimes \mathbf{n}_{n+1}^{tr}, \\
a_{n+1} &= 1 - \frac{2 \, \mu \, \Delta \gamma_{n+1}}{\| \mathbf{s}_{n+1}^{tr} \|}, \qquad\qquad\qquad\qquad\qquad\qquad\qquad (6.126) \\
b_{n+1} &= \left(\frac{\eta}{2 \, \mu \, \Delta t} + (\, 1 + \frac{\hat{H}}{3 \, \mu}) \right)^{-1} - \frac{2 \, \mu \, \Delta \gamma_{n+1}}{\| \mathbf{s}_{n+1}^{tr} \|} \,.
\end{aligned}$$

It has the same structure as the elasto-plastic algorithmic tangent (6.116), only the factor and b_{n+1} is different. Hence implementation of this viscoplastic model is very simple when the code for the elaso-plastic model of last section is already available.

6.3 Integration of Constitutive Equations for Finite Deformation Problems

Constitutive equations for finite deformation problems can be formulated in different ways. This was subject - especially in view of numerical algorithms for finite element analysis - of many different research efforts, see e.g. Argyris and Kleiber (1977), Nagtegaal (1982), Argyris et al. (1982), Simo (1988), Nagtegaal et al. (1990), Peric et al. (1992), Simo and Hughes (1998) and Simo

(1998). The choice of the mathematical description depends upon the material at hand but also upon the efficiency of specific solution methods. Two possible formulations and associated integration algorithms are discussed in this section for elasto-plastic material with isotropic hardening which was already described in Sect. 3.3.2. Interesting applications and examples of finite deformation plasticity can be found, e.g. in Peric and Owen (1997).

6.3.1 General Implicit Integration

Implicit EULER schemes were developed for the integration of inelastic constitutive equations of problems undergoing small strains. Such algorithms will now be developed for inelastic constitutive equations at finite strains. These algorithms are developed for a time interval $[t_n, t_{n+1}]$ where it is assumed that the deformation φ and its gradient $\mathbf{F}$ are known at time t_n. This shall also be true for the internal variables $\{\mathbf{F}^e, \boldsymbol{\xi}_\alpha\}$. Hence the set of initial values

$$\mathbf{F}_n = \text{Grad}\,\varphi_n, \qquad \{\mathbf{F}_n^e, \boldsymbol{\xi}_{\alpha_n}\} \tag{6.127}$$

is known. Now the algorithmic approximation of the evolution equations for plastic flow

$$\mathbf{l}^p = \sum_{g=1}^{m} \lambda_g \frac{\partial f_g(\boldsymbol{\tau}, q_\alpha)}{\partial \boldsymbol{\tau}}, \tag{6.128}$$

$$\dot{\xi}_\alpha = \sum_{g=1}^{m} \lambda_g \frac{\partial f_g(\boldsymbol{\tau}, q_\alpha)}{\partial q_\alpha} \tag{6.129}$$

with the KUHN-TUCKER conditions

$$\lambda_g \geq 0, \quad f_g(\boldsymbol{\tau}, q_\alpha) \leq 0, \quad \lambda_g\, f_g(\boldsymbol{\tau}, q_\alpha) = 0 \tag{6.130}$$

has to be constructed. More details related to the derivation of the plastic evolution equations can be found in Sect. 3.3.2. Kinematic hardening is not considered.

The evolution equations in (6.129) include multi-surface plasticity. The summation sign can be neglected in case of only one yield surface.

In Sect. 3.3.2, several relations were derived and definitions were introduced. These can be used to write the spatial plastic rate $\mathbf{l}^p$ as

$$\mathbf{l}^p = \mathbf{F}^e\, \mathbf{L}^p \mathbf{F}^{e-1} \quad \text{with} \quad \dot{\mathbf{F}}^p = \mathbf{L}^p\, \mathbf{F}^p. \tag{6.131}$$

The structure of the last equations leads to an exponential approximations of the evolution of the plastic deformation gradient, see e.g. Simo (1992), Simo and Hughes (1998) and Miehe (1993),

$$\mathbf{F}_{n+1}^p = \exp\left[(t_{n+1} - t_n)\, \mathbf{L}_{n+1}^p\right] \mathbf{F}_n^p. \tag{6.132}$$

Some reformulations – considering the multiplicative split of the deformation gradient $\mathbf{F}_{n+1} = \mathbf{F}_{n+1}^{e}\,\mathbf{F}_{n+1}^{p}$ – yield with (6.132)

$$
\begin{aligned}
\mathbf{F}_{n+1} &= \mathbf{F}_{n+1}^{e}\exp\left[\,(t_{n+1}-t_{n})\,\mathbf{L}_{n+1}^{p}\,\right]\mathbf{F}_{n+1}^{e\,-1}\mathbf{F}_{n+1}^{e}\mathbf{F}_{n}^{p}\\
&= \exp\left[\,(t_{n+1}-t_{n})\,\mathbf{F}_{n+1}^{e}\,\mathbf{L}_{n+1}^{p}\,\mathbf{F}_{n+1}^{e\,-1}\,\right]\mathbf{F}_{n+1}^{e}\mathbf{F}_{n}^{p}\,,
\end{aligned}\qquad(6.133)
$$

where the standard properties of an exponential map where utilized. Equation (6.133) can be resolved with the definition in (6.131)$_1$ and $\Delta t = t_{n+1}-t_n$ and with $\mathbf{F}_{n+1}^{e\,tr} = \mathbf{F}_{n+1}\,\mathbf{F}_{n}^{p\,-1}$ with respect to $\mathbf{F}_{n+1}^{e}$

$$
\mathbf{F}_{n+1}^{e} = \exp\left[\,(\Delta t)\,\mathbf{l}_{n+1}^{p}\,\right]\mathbf{F}_{n+1}^{e\,tr}\,.\qquad(6.134)
$$

The definition of the trial value $\mathbf{F}_{n+1}^{e\,tr}$ is physically motivated since the computation of the elastic part of the deformation gradient is performed for frozen plastic variables $\mathbf{F}_{n+1}^{p} = \mathbf{F}_{n}^{p}$. By inserting (6.129)$_1$ into equation (6.134), relation

$$
\mathbf{F}_{n+1}^{e} = \exp\left[\,-\sum_{g=1}^{m}\lambda_{g}\,\Delta t\,\frac{\partial f_{g}(\boldsymbol{\tau},q_{\alpha})}{\partial\boldsymbol{\tau}}\,\right]\mathbf{F}_{n+1}^{e\,tr}\qquad(6.135)
$$

can be determined. The definitions of the flow increment $\Delta\lambda_{g} = \Delta t\,\lambda_{g}$ and the implicit EULER approximation of (6.129)$_2$ lead to the algorithmic version of the flow rule (6.129)

$$
\mathbf{F}_{n+1}^{e} = \exp\left[\,-\sum_{g=1}^{m}\Delta\lambda_{g}\,\frac{\partial f_{g}(\boldsymbol{\tau},q_{\alpha})}{\partial\boldsymbol{\tau}}\,\right]\mathbf{F}_{n+1}^{e\,tr}\,,\qquad(6.136)
$$

$$
\xi_{\alpha_{n+1}} = \xi_{\alpha_{n}} + \sum_{g=1}^{m}\Delta\lambda_{g}\,\frac{\partial f_{g}(\boldsymbol{\tau},q_{\alpha})}{\partial q_{\alpha}}\qquad(6.137)
$$

and the KUHN-TUCKER conditions

$$
\Delta\lambda_{g} \geq 0\,,\quad f_{g}(\boldsymbol{\tau},q_{\alpha}) \leq 0 \ \text{and}\ \sum_{g=1}^{m}\Delta\lambda_{g}\,f_{g}(\boldsymbol{\tau},q_{\alpha}) = 0\,.\qquad(6.138)
$$

The stresses in these equations is given by

$$
\boldsymbol{\tau}_{n+1} = \mathbf{F}_{n+1}^{e}\left[\frac{\partial W(\mathbf{C}_{n+1}^{e})}{\partial\mathbf{C}^{e}}\right]\mathbf{F}_{n+1}^{e\,T}\qquad(6.139)
$$

and the hardening variables follow from

$$
q_{\alpha_{n+1}} = -\frac{\partial H(\xi_{\alpha_{n+1}})}{\partial\xi_{\alpha}}\,.\qquad(6.140)
$$

The solution of the nonlinear system (6.136) to (6.140) has to be performed locally at each GAUSS. It can be obtained by using NEWTON method.

Note that the constraint condition of incompressibility of the plastic flow, $J^{p} = 1$, is exactly fulfilled by the algorithmic flow rule (6.136), see e.g. Simo (1992).

6.3.2 Implicit Integration with Respect to Principal Axes

In case of isotropic elasto-plastic material behaviour with isotropic hardening, it is possible to simplify the relations derived above. Based on Eq. (3.212), the elastic predictor can be written as

$$\mathbf{b}_{n+1}^{e\,tr} = \mathbf{F}_{n+1}\,\mathbf{C}_n^{p\,-1}\,\mathbf{F}_{n+1}^T\,, \qquad \alpha_{n+1}^{tr} = \alpha_n\,. \tag{6.141}$$

Within these equations, it is assumed that the plastic flow is frozen at time t_n. Hence the inverse plastic right CAUCHY-GREEN tensor $\mathbf{C}^{p\,-1}$ can be used as field to store the loading history as irreversible part of the plastic deformation.

When the predictor step does not leave the admissible region of elastic deformations (this is equivalent with the fulfillment of the yield condition $f(\boldsymbol{\tau}, q) < 0$), then the stresses follow directly from relations (3.201) and the local integration step is terminated

$$\boldsymbol{\tau}_{n+1} = 2\,\left.\frac{\partial \hat{W}}{\partial \mathbf{b}^e}\right|_{\mathbf{b}^e = \mathbf{b}_{n+1}^{e\,tr}} \mathbf{b}_{n+1}^{e\,tr} \quad ; \quad q_{n+1} = -\left.\frac{\partial \hat{H}}{\partial \alpha}\right|_{\alpha = \alpha_{n+1}^{tr}}\,. \tag{6.142}$$

When the yield condition is not fulfilled, then a plastic corrector step has to be employed to fulfil the flow condition $f(\boldsymbol{\tau}, q) = 0$. The correction of stresses and internal variables is computed by a projection scheme as in the previous sections. It projects the stresses onto the yield surface $f(\boldsymbol{\tau}, q) = 0$. The momentary position of the vectors $\mathbf{x} = \mathbf{x}^{tr}$ is fixed within this procedure. This reduces (3.209) to an ordinary differential equation of first order in time

$$\mathcal{L}_v\,\mathbf{b}^e = -2\,\gamma\,\frac{\partial f}{\partial \boldsymbol{\tau}}\,\mathbf{b}^e\,. \tag{6.143}$$

As in the sections above, it will be integrated by an implicit EULER backward scheme, see e.g. Weber and Anand (1990), Cuitino and Ortiz (1992), Peric et al. (1992) and Simo (1992),

$$\mathbf{b}_{n+1}^e = \exp\left[-2\,\underbrace{(t_{n+1} - t_n)\,\gamma}_{\Delta\gamma_{n+1}}\,\left.\frac{\partial f}{\partial \boldsymbol{\tau}}\right|_{n+1}\right]\mathbf{b}_{n+1}^{e\,tr}\,. \tag{6.144}$$

The term $\Delta\gamma\,(\frac{\partial f}{\partial \boldsymbol{\tau}})$ is constant within the time interval $[t_{n+1}, t_n]$.

Since isotropic material was assumed from the beginning, the eigenvector bases of $\mathbf{b}^e$ and $\boldsymbol{\tau}$ are equal. Furthermore, the bases are fixed during the projection onto the yield surface. Thus a spectral decomposition of the elastic strains and the KIRCHHOFF stresses leads to an efficient implementation. For this, the strain and stress tensors are written with respect to their spectral decomposition

$$\mathbf{b}_{n+1}^e = \sum_{A=1}^3 (\lambda_{A\,n+1}^e)^2\,\mathbf{n}_{A\,n+1}^{tr} \otimes \mathbf{n}_{A\,n+1}^{tr}\,,$$

$$\mathbf{b}_e^{tr} = \sum_{A=1}^{3} \lambda_{A\,n+1}^{e\,tr\,2} \, \mathbf{n}_{A\,n+1}^{tr} \otimes \mathbf{n}_{A\,n+1}^{tr} \,, \qquad (6.145)$$

$$\boldsymbol{\tau}_{n+1} = \sum_{A=1}^{3} \tau_{A\,n+1} \, \mathbf{n}_{A\,n+1}^{tr} \otimes \mathbf{n}_{A\,n+1}^{tr} \,.$$

A detailed description of the spectral decomposition of strains and stresses can be found in Eq. (3.23) or (3.131). With such formulation, the algorithmic flow rule (6.144) can be reformulated in principal strains

$$(\lambda_{A\,n+1}^{e})^2 = \exp\left[-2\,\Delta\gamma_{n+1} \left. \frac{\partial f}{\partial \tau_A}\right|_{n+1}\right] \lambda_{A\,n+1}^{e\,tr\,2} \,. \qquad (6.146)$$

By introducing logarithmic strains $\varepsilon_A^e = \ln[\lambda_A^e]$, an even simpler relation can be deduced

$$\varepsilon_{A\,n+1}^{e\,tr} = \varepsilon_{A\,n+1}^{e} + \Delta\gamma_{n+1} \left. \frac{\partial f}{\partial \tau_A}\right|_{n+1} \,. \qquad (6.147)$$

This formulation is equivalent with an additive decomposition of the elastic trial strain in elastic and plastic parts. In Simo (1992), it is shown that the incompressibility of plastic flow is automatically fulfilled by Eq. (6.146) and (6.147) which are presented in principal strains.

For further derivations, it is convenient to write (6.147) in vector form

$$\boldsymbol{\varepsilon}_{n+1}^{e\,tr} = \boldsymbol{\varepsilon}_{n+1}^{e} + \Delta\gamma_{n+1} \left. \frac{\partial f}{\partial \boldsymbol{\tau}}\right|_{n+1} \quad \text{with} \quad \boldsymbol{\varepsilon} = \{\varepsilon_1, \varepsilon_2, \varepsilon_3\} \,. \qquad (6.148)$$

The equivalent plastic strains α are integrated by the implicit EULER method

$$\alpha_{n+1} = \alpha_n + \Delta\gamma_{n+1} \left. \frac{\partial f}{\partial q}\right|_{n+1} \,. \qquad (6.149)$$

Now the following nonlinear system has to be solved for the projection of the stresses onto the yield surface; note that f_{n+1} is zero in case of plastic flow.

$$
\begin{aligned}
\mathbf{r} &= \boldsymbol{\varepsilon}_{n+1}^{e\,tr} - \boldsymbol{\varepsilon}_{n+1}^{e} - \Delta\gamma_{n+1} \left. \frac{\partial f}{\partial \boldsymbol{\tau}}\right|_{n+1} &= \mathbf{0}, \\
r &= \alpha_{n+1} - \alpha_n - \Delta\gamma_{n+1} \left. \frac{\partial f}{\partial q}\right|_{n+1} &= 0, \\
f &= f(\boldsymbol{\tau}_{n+1}, q_{n+1}) &= 0.
\end{aligned}
\qquad (6.150)
$$

As in the previous sections, NEWTON method will be applied. Within this procedure, the values $\boldsymbol{\varepsilon}_{n+1}^{e\,tr}$ and α_n are fixed, see above. This leads to an algorithm in which the plastic variables and projected stresses are determined iteratively.

First equation (6.150) is linearized. This lead for the ith NEWTON step to

$$\mathbf{r}^{i+1} = \mathbf{r}^i - \Delta\varepsilon^{e\,i}_{n+1} - \Delta\Delta\gamma^i_{n+1}\,\mathbf{n}^i_{n+1} - \Delta\gamma^i_{n+1}\,\mathbf{D}^i\,\mathbf{C}^i\,\Delta\varepsilon^{e\,i}_{n+1} = \mathbf{0}\,,$$

$$r^{i+1} = r^i + \Delta\alpha^i_{n+1} - \Delta\Delta\gamma^i_{n+1}n^i - \Delta\gamma^i_{n+1}\,D^i\,C^i\,\Delta\alpha^i_{n+1} = 0\,, \quad (6.151)$$

$$f^{i+1} = f^i + \mathbf{n}^{i\,T}_{n+1}\mathbf{C}^i\,\Delta\varepsilon^{e\,i}_{n+1} + n^i\,C^i\,\Delta\alpha^i_{n+1} = 0\,,$$

where the matrices

$$\mathbf{D}^i = \left.\frac{\partial^2 f}{\partial\boldsymbol{\tau}\partial\boldsymbol{\tau}}\right|_i\,, \qquad \mathbf{C}^i = \left.\frac{\partial\boldsymbol{\tau}}{\partial\boldsymbol{\varepsilon}}\right|_i\,,$$

scalars

$$C^i = \left.\frac{\partial q}{\partial\alpha}\right|_i\,, \qquad D^i = \left.\frac{\partial^2 f}{\partial q\partial q}\right|_i\,, \qquad n^i = \left.\frac{\partial f}{\partial q}\right|_i$$

and the vector

$$\mathbf{n}^i = \left.\frac{\partial f}{\partial\boldsymbol{\tau}}\right|_i$$

were used. Increments of the strain $\Delta\varepsilon^i_{n+1}$ of the hardening variable $\Delta\alpha^i_{n+1}$ and the flow parameter $\Delta\Delta\gamma^i_{n+1}$ can now directly be obtained from the linearized form (6.151). After some algebraic manipulations, these increments follow as

$$\Delta\varepsilon^{e\,i}_{n+1} = (\mathbf{E}^i)^{-1}\,(\mathbf{r}^i - \Delta\Delta\gamma^i_{n+1}\,\mathbf{n}^i_{n+1})\,, \qquad (6.152)$$

$$\Delta\alpha^i_{n+1} = (E^i)^{-1}\,(r^i - \Delta\Delta\gamma^i_{n+1}\,n^i_{n+1})\,, \qquad (6.153)$$

with

$$\mathbf{E}^i = \mathbf{1} + \Delta\gamma^i_{n+1}\,\mathbf{D}^i\,\mathbf{C}^i \quad \text{and} \quad E^i = -1 + \Delta\gamma^i_{n+1}\,D^i\,C^i\,.$$

Here the increment of the flow parameter is given by

$$\Delta\Delta\gamma^i_{n+1} = \frac{f^i + \mathbf{n}^{i\,T}\,\mathbf{H}^i\,\mathbf{r}^i + n^i\,H^i\,r^i}{\mathbf{n}^{i\,T}\,\mathbf{H}^i\,\mathbf{n}^i + n^i\,H^i\,n^i}\,, \qquad (6.154)$$

where the abbreviations

$$\mathbf{H}^i = \mathbf{C}^i(\mathbf{E}^i)^{-1} \quad \text{and} \quad H^i = C^i\,(E^i)^{-1}$$

were introduced. Once the increments are known, the standard update of the variables yields

$$\varepsilon^{e\,i+1}_{n+1} = \varepsilon^{e\,i}_{n+1} + \Delta\varepsilon^{e\,i}_{n+1}\,,$$

$$\alpha^{i+1}_{n+1} = \alpha^i_{n+1} + \Delta\alpha^i_{n+1}\,, \qquad (6.155)$$

$$\Delta\gamma^{i+1}_{n+1} = \Delta\gamma^i_{n+1} + \Delta\Delta\gamma^i_{n+1}\,.$$

The KIRCHHOFF stresses follow by function evaluation from the strain energy function

$$\boldsymbol{\tau}_{n+1}^{i+1} = \left.\frac{\partial W}{\partial \boldsymbol{\varepsilon}_{n+1}^e}\right|_{i+1}. \tag{6.156}$$

The strain energy function, the hardening law and the yield function have to be specified for a given problem. So far - except the assumption of isotropy - no further restrictions are introduced for the choice of a strain energy function. Thus the strain energy function presented in (3.113), see Ogden (1982), can be employed in this algorithmic setting. This strain energy function has to be formulated here in terms of the logarithmic principal strains

$$W = \sum_r \left\{ \frac{\mu_r}{\alpha_r} \left(\exp\left[\varepsilon_1^e \, \alpha_r\right] + \exp\left[\varepsilon_2^e \, \alpha_r\right] + \exp\left[\varepsilon_3^e \, \alpha_r\right] - 3 \right) - \mu_r \ln J \right\}$$
$$+ \frac{\Lambda}{4} \left(J^2 - 1 - 2 \ln J \right)$$

$$\tag{6.157}$$

with $\ln J = \ln(\lambda_1 \lambda_2 \lambda_3) = \varepsilon_1 + \varepsilon_2 + \varepsilon_3$. The material parameters α_r are dimensionless quantities, Λ can be interpreted as LAMÉ constant of the classical HOOKE law, see also Sect. 3.3.1.

Remark 6.2: An alternative strain energy function W_{Lin} was suggested in Simo (1992) for the case of finite plastic deformations but small elastic strains

$$W_{Lin} = \mu \left(\varepsilon_1^{e\,2} + \varepsilon_2^{e\,2} + \varepsilon_3^{e\,2} \right) + \frac{\Lambda}{2} \left(\varepsilon_1^e + \varepsilon_2^e + \varepsilon_3^e \right)^2 \tag{6.158}$$

which often occurs in metal plasticity. Λ and μ are the LAME constants. This function has the advantage that a closed form solution of (6.150) can be derived for linear isotropic hardening

$$q = -\hat{H} \, \alpha \tag{6.159}$$

with the hardening parameter $\hat{H}$. Furthermore, J_2 plasticity after VON MISES is assumed with the yield function

$$f = \|\text{dev}\,\boldsymbol{\tau}\| - \sqrt{\frac{2}{3}} \left(\tau_Y - q \right), \tag{6.160}$$

where τ_Y denotes the yield stress. It follows

$$\Delta\gamma_{n+1} = \frac{f_{n+1}^{tr}}{\mu + \frac{2}{3}\hat{H}},$$

$$\alpha_{n+1} = \alpha_n + \Delta\gamma_{n+1}\sqrt{\frac{2}{3}}, \tag{6.161}$$

$$\boldsymbol{\varepsilon}_{n+1}^e = \boldsymbol{\varepsilon}_{n+1}^{e\,tr} - \Delta\gamma_{n+1}\,\mathbf{n}_{n+1}^{tr}$$

with $\mathbf{n}_{n+1}^{tr} = \text{dev}\,\boldsymbol{\tau}_{n+1}^{tr} / \|\text{dev}\,\boldsymbol{\tau}_{n+1}^{tr}\|$. Here the KIRCHHOFF stress is computed via

$$\boldsymbol{\tau}_{n+1} = \mathbf{C}\,\boldsymbol{\varepsilon}_{n+1}^{e\,tr} - 2\mu\Delta\gamma_{n+1}\,\mathbf{n}_{n+1}^{tr}. \tag{6.162}$$

This relation corresponds to the computation of the stresses performed in the geometrically linear theory for VON MISES plasticity, see (6.101).

Within this formulation, it is necessary to determine the principal values and directions for strain and stress tensors. Within a finite element simulation, all relations given in principal values have to be transformed back to the underlying coordinate system, basically to a CARTESIAN frame. This transformation was already described in (3.135) for the 2nd PIOLA-KIRCHHOFF stresses. The principal directions follow in case of the left CAUCHY-GREEN tensor from the eigenvalue problem

$$(\mathbf{b}_{n+1}^{e\,tr} - \lambda_{i\,n+1}^{e\,tr\,2}\,\mathbf{1})\,\mathbf{n}_{i\,n+1}^{tr} = \mathbf{0}\,. \tag{6.163}$$

The base vector $\mathbf{n}_{i\,n+1}^{tr}$ can be related to CARTESIAN bases $\mathbf{E}_I$ via

$$\mathbf{n}_{i\,n+1}^{tr} = \sum_{J=1}^{3}(\mathbf{E}_J \otimes \mathbf{E}_J) \cdot \mathbf{n}_{i\,n+1}^{tr} = (\mathbf{E}_J \cdot \mathbf{n}_{i\,n+1}^{tr})\,\mathbf{E}_J = D_{iJ\,n+1}\,\mathbf{E}_J\,. \tag{6.164}$$

The principal strains $b_{ij}^{e\,\mathrm{H}}$ of $\mathbf{b}_{n+1}^{e}$ can be written in component form (the index $()_{n+1}$ is neglected in the following transformations to simplify notation)

$$b_{ii}^{e\,\mathrm{H}} = \exp\left[2\,\varepsilon_i^e\right] \quad,\quad b_{ij}^{e\,\mathrm{H}} = 0 \quad \text{for} \quad i \neq j\,. \tag{6.165}$$

The transformation to cartesian bases follows using relation (6.164)

$$\mathbf{b}_{n+1}^{e\,tr} = b_{ij}^{\mathrm{H}}\,\mathbf{n}_i^{tr} \otimes \mathbf{n}_j^{tr} = D_{Ki}\,b_{ij}^{e\,\mathrm{H}}\,D_{Lj}\mathbf{E}_K \otimes \mathbf{E}_L = b_{KL}^e\,\mathbf{E}_K \otimes \mathbf{E}_L\,. \tag{6.166}$$

The components of the stress tensor are transformed in the same way, see (3.137). Writing $\mathbf{b}^e$ and $\boldsymbol{\tau}$ in vector form, see (6.148), yields the modified transformations

$$\bar{b}_A^e = T_{AB}\,\bar{b}_B^{e\,\mathrm{H}}, \quad \bar{\tau}_A = T_{AB}\,\bar{\tau}_B^{\mathrm{H}} \quad \text{with} \quad A, B = 1, 2, \ldots, 6\,, \tag{6.167}$$

where the components of the vectors $\{\bar{b}_A^e\}$ and $\{\bar{\tau}_A\}$ are given by

$$
\begin{aligned}
\{\bar{b}_A^e\}^T &= \{b_{11}^e, b_{22}^e, b_{33}^e, b_{12}^e, b_{13}^e, b_{23}^e\}, \\
\{\bar{\tau}_A\}^T &= \{\tau_{11}, \tau_{22}, \tau_{33}, \tau_{12}, \tau_{13}, \tau_{23}\}.
\end{aligned} \tag{6.168}
$$

The transformation matrix T_{AB} is a 6×6 matrix. It has the same structure as in (3.139) and can be found in explicit form in Reese (1994). Note that only the first three components of the vector $\bar{\tau}_B^{\mathrm{H}}$ are non-zero, since all shear stresses disappear in principal directions, see (3.139) and (6.165).

6.3.3 Consistent Tangent Modulus

When NEWTON method is applied to solve a boundary value problem undergoing finite inelastic deformations, the linearization of the weak form of equilibrium is needed, see e.g. (3.339). Within the linearization of the weak form, it is necessary to compute the material tangent (3.338). As already discussed for small plastic deformations, the projection method derived in the last section will lead to a special form of the material tangent, see algorithmic tangent (6.92), which has to be employed within the NEWTON iteration in order to obtain quadratic convergence.

The derivation of the tangent modulus for elastic constitutive equations, given in principal strains, was already stated in Sect. 3.3.4. Now the material or algorithmic tangent is derived for finite plastic deformations under the restriction of isotropic material behaviour.

First the 2nd PIOLA-KIRCHHOFF stress tensor $\tilde{\mathbf{S}}$ is defined with reference to the intermediate plastic configuration

$$\mathbf{S} = \mathbf{F}^{-1}\,\boldsymbol{\tau}\,\mathbf{F}^{-T} = \mathbf{F}_{pn}^{-1}\,\tilde{\mathbf{S}}\,\mathbf{F}_n^{p\,-T} \qquad \text{with} \qquad \tilde{\mathbf{S}} = \mathbf{F}^{p\,tr\,-1}\,\boldsymbol{\tau}\,\mathbf{F}^{e\,tr\,-T}\,. \tag{6.169}$$

The spectral decomposition of $\tilde{\mathbf{S}}$ assumes the form

$$\tilde{\mathbf{S}} = \sum_{i=1}^{3} \frac{\tau_i}{\lambda_{i\,e}^{tr\,2}}\,\tilde{\mathbf{N}}_i \otimes \tilde{\mathbf{N}}_i = \sum_{i=1}^{3} \tilde{S}_i\,\tilde{\mathbf{N}}_i \otimes \tilde{\mathbf{N}}_i\,. \tag{6.170}$$

The stress increment $\Delta\tilde{\mathbf{S}}$ can be determined based on (6.170)

$$\Delta\tilde{\mathbf{S}} = \frac{\partial\tilde{\mathbf{S}}}{\partial\tilde{\mathbf{C}}^e}\,\Delta\tilde{\mathbf{C}}^e = \sum_{i=1}^{3} \frac{\partial\tilde{\mathbf{S}}}{\partial\varepsilon_i^{e\,tr}} \otimes \frac{\partial\varepsilon_i^{e\,tr}}{\partial\tilde{\mathbf{C}}^e}\,\Delta\tilde{\mathbf{C}}^e \qquad \text{with} \qquad \tilde{\mathbf{C}}^e = \mathbf{F}^{e\,tr\,T}\,\mathbf{F}^{e\,tr}\,. \tag{6.171}$$

The partial derivative of the principal stresses with respect to the logarithmic strains (6.171) yields

$$\frac{\partial\tilde{S}_i}{\partial\varepsilon_j^{e\,tr}} = \frac{1}{\lambda_i^{e\,tr\,4}}\,\big(\underbrace{\frac{\partial\tau_i}{\partial\varepsilon_j^{e\,tr}}}_{C_{ij}^{ALG}}\,\lambda_i^{e\,tr\,2} - \tau_i\,2\,\lambda_i^{e\,tr}\,\underbrace{\frac{\partial\lambda_i^{e\,tr}}{\partial\varepsilon_j^{e\,tr}}}_{\lambda_i^{e\,tr}\,\delta_{ij}} \big)\,, \tag{6.172}$$

where τ_i are KIRCHHOFF stresses. The algorithmic tangent modulus C_{ij}^{ALG} is given for purely elastic behaviour ($f < 0$) by

$$C_{ij}^{ALG} = \frac{\partial^2\hat{W}}{\partial\varepsilon_i^{e\,tr}\,\partial\varepsilon_j^{e\,tr}} \qquad \text{or} \qquad \mathbf{C}^{e\,ALG} = \frac{\partial^2\hat{W}}{\partial\boldsymbol{\varepsilon}^{e\,tr}\,\partial\boldsymbol{\varepsilon}^{e\,tr}}\,. \tag{6.173}$$

In case of plastic flow, a stress point fulfils $f = 0$ and the following incremental form is derived from the relations of the converged NEWTON iteration (6.151)

$$\begin{aligned}
\mathbf{0} &= \Delta\varepsilon_{n+1}^{e\,tr} - \Delta\varepsilon_{n+1}^{e\,i} - \Delta\Delta\gamma_{n+1}^i\,\mathbf{n}_{n+1}^i - \Delta\gamma_{n+1}^i\,\mathbf{D}^i\,\mathbf{C}^i\,\Delta\varepsilon_{n+1}^{e\,i}\,, \\[4pt]
0 &= \Delta\alpha_{n+1}^i - \Delta\Delta\gamma_{n+1}^i n^i - \Delta\gamma_{n+1}^i\,D^i\,C^i\,\Delta\alpha_{n+1}^i = 0\,, \\[4pt]
0 &= \mathbf{n}_{n+1}^{i\,T}\mathbf{C}^i\,\Delta\varepsilon_{n+1}^{e\,i} + n^i\,C^i\,\Delta\alpha_{n+1}^i = 0\,.
\end{aligned} \tag{6.174}$$

These relations are valid for the global NEWTON step at time t_{n+1}. Here the elastic *trial* strains $\varepsilon_{n+1}^{e\,tr}$ are no longer frozen as was the case within the local NEWTON iteration. Equation (6.174) can now be solved for the increment of the flow parameter

$$\Delta\Delta\gamma_{n+1} = \frac{\mathbf{n}_{n+1}^T \mathbf{H}_{n+1}\, \Delta\varepsilon_{n+1}^{e\,tr}}{\mathbf{n}_{n+1}^T \mathbf{H}_{n+1}\,\mathbf{n}_{n+1} + n_{n+1}\, H_{n+1}\, n_{n+1}}\,. \tag{6.175}$$

All quantities in this equation have to be evaluated at the end of the local NEWTON iteration. The algorithmic tangent modulus stemming from the implicit integration scheme follows from (6.152). The incremental form of the KIRCHHOFF stress tensor $\Delta\boldsymbol{\tau}_{n+1} = \mathbf{C}_{n+1}\,\Delta\varepsilon_{n+1}^e$ is given by

$$\Delta\boldsymbol{\tau}_{n+1} = \mathbf{H}_{n+1}\left(\Delta\varepsilon_{n+1}^{e\,tr} - \Delta\Delta\gamma_{n+1}\,\mathbf{n}_{n+1}\right) = \mathbf{C}^{ALG\,p}\,\Delta\varepsilon_{n+1}^{e\,tr}\,. \tag{6.176}$$

The explicit form of the material tangent $\mathbf{C}_p^{ALG}$ for finite plastic deformations can be written as

$$\mathbf{C}^{ALG\,p} = \mathbf{H}_{n+1} - \frac{\mathbf{H}_{n+1}\mathbf{n}_{n+1} \otimes \mathbf{n}_{n+1}^T \mathbf{H}_{n+1}}{\mathbf{n}_{n+1}^T \mathbf{H}_{n+1}\,\mathbf{n}_{n+1} + n_{n+1}\, H_{n+1}\, n_{n+1}}\,. \tag{6.177}$$

Here the tensors and scalars were defined in (6.151).

The second partial derivative in (6.171) is computed as

$$\frac{\partial \tilde{\mathbf{C}}^e}{\partial \varepsilon_j^{e\,tr}} = \sum_{i=1}^{3} \frac{\partial \lambda_i^{e\,tr\,2}}{\partial \varepsilon_j^{e\,tr}}\, \tilde{\mathbf{N}}_i \otimes \tilde{\mathbf{N}}_i = \sum_{i=1}^{3} 2\,\lambda_i^{e\,tr}\, \delta_{ij}\, \lambda_j^{e\,tr}\, \tilde{\mathbf{N}}_i \otimes \tilde{\mathbf{N}}_i = 2\,\lambda_j^{e\,tr\,2}\tilde{\mathbf{N}}_j \otimes \tilde{\mathbf{N}}_j\,. \tag{6.178}$$

This leads to the relation

$$2\,\frac{\partial \varepsilon_i^{e\,tr}}{\partial \tilde{\mathbf{C}}^e} = \frac{1}{\lambda_i^{e\,tr\,2}}\,\tilde{\mathbf{N}}_i \otimes \tilde{\mathbf{N}}_i\,. \tag{6.179}$$

The increment of the stress tensor $\Delta\tilde{\mathbf{S}}$ follows together with (6.172)

$$\Delta\tilde{\mathbf{S}} = \frac{1}{2}\left(\sum_{i=1}^{3}\sum_{j=1}^{3} \frac{C_{ij}^{ALG} - \tau_i\,2\,\delta_{ij}}{\lambda_i^{e\,tr\,2}\lambda_j^{e\,tr\,2}}\,\tilde{\mathbf{N}}_i \otimes \tilde{\mathbf{N}}_i \otimes \tilde{\mathbf{N}}_j \otimes \tilde{\mathbf{N}}_j\right)$$

$$\left[2\sum_{k=1}^{3}(\lambda_k^{e\,tr}\,\Delta\lambda_k^{e\,tr}\,\tilde{\mathbf{N}}_k \otimes \tilde{\mathbf{N}}_k)\right] + \sum_{i=1}^{3}\tilde{S}_i\,\Delta(\tilde{\mathbf{N}}_i \otimes \tilde{\mathbf{N}}_i)\,. \tag{6.180}$$

In this equation, the increment of the eigenvector $\tilde{\mathbf{N}}_i$ is still unknown. It is given by

$$\Delta\tilde{\mathbf{N}}_i = \sum_{i=1}^{3}\underbrace{(\tilde{\mathbf{N}}_j \cdot \Delta\tilde{\mathbf{N}}_i)}_{\Omega_{ji}}\,\tilde{\mathbf{N}}_j\,.$$

The stress and strain increments can be written, based on the already derived relations (3.256) and (3.257), as

$$
\Delta \tilde{\mathbf{S}} \;=\; \sum_{i=1}^{3} \Delta \tilde{S}_i \, \tilde{\mathbf{N}}_i \otimes \tilde{\mathbf{N}}_i + \sum_{i=1}^{3} \sum_{j \neq i=1}^{3} (\tilde{S}_j - \tilde{S}_i)\, \Omega_{ji} \, \tilde{\mathbf{N}}_i \otimes \tilde{\mathbf{N}}_j \,;
$$

$$
\Delta \tilde{\mathbf{C}} \;=\; \sum_{i=1}^{3} 2\, \lambda_i^{e\,tr}\, \Delta \lambda_i^{e\,tr}\, \tilde{\mathbf{N}}_i \otimes \tilde{\mathbf{N}}_i + \sum_{i=1}^{3} \sum_{j \neq i=1}^{3} (\lambda_j^{e\,tr\,2} - \lambda_i^{e\,tr\,2})\, \Omega_{ji}\, \tilde{\mathbf{N}}_i \otimes \tilde{\mathbf{N}}_j \,.
$$

$$(6.181)$$

By comparison of (6.180) with (6.181), the incremental constitutive equation is derived as in (3.241)

$$
\Delta \tilde{\mathbf{S}} = \mathbb{L}\,[\,\tfrac{1}{2}\Delta \tilde{\mathbf{C}}\,]\,.
\tag{6.182}
$$

Here $\mathbb{L}$ is the consistent or algorithmic tangent modulus with reference to the frozen plastic intermediate configuration

$$
\mathbb{L} \;=\; \sum_{i=1}^{3} \sum_{j=1}^{3} \left(\frac{C_{ij}^{ALG} - \tau_i\, 2\, \delta_{ij}}{\lambda_i^{e\,tr\,2}\, \lambda_j^{e\,tr\,2}} \, \tilde{\mathbf{N}}_i \otimes \tilde{\mathbf{N}}_i \otimes \tilde{\mathbf{N}}_j \otimes \tilde{\mathbf{N}}_j \right) +
$$

$$
\frac{1}{2} \sum_{i \neq j} 2\, \frac{\tilde{S}_j - \tilde{S}_i}{\lambda_j^{e\,tr\,2} - \lambda_i^{e\,tr\,2}} \, (\tilde{\mathbf{N}}_i \otimes \tilde{\mathbf{N}}_j \otimes \tilde{\mathbf{N}}_i \otimes \tilde{\mathbf{N}}_j + \tilde{\mathbf{N}}_i \otimes \tilde{\mathbf{N}}_j \otimes \tilde{\mathbf{N}}_j \otimes \tilde{\mathbf{N}}_i)
$$

$$
=\; L_{ijkl}^{H} \, \tilde{\mathbf{N}}_i \otimes \tilde{\mathbf{N}}_j \otimes \tilde{\mathbf{N}}_k \otimes \tilde{\mathbf{N}}_l \,.
$$

$$(6.183)$$

The associated tangent modulus $\mathbf{c}$ is referred to the spatial configuration and follows from the *push forward* of (6.182)

$$
\mathbf{c} \;=\; L_{ijkl}^{H} \, \lambda_i^{e\,tr}\, \lambda_j^{e\,tr}\, \lambda_k^{e\,tr}\, \lambda_l^{e\,tr}\, \mathbf{n}_i^{tr} \otimes \mathbf{n}_j^{tr} \otimes \mathbf{n}_k^{tr} \otimes \mathbf{n}_l^{tr}
$$

$$
=\; c_{ijkl}^{H}\, \mathbf{n}_i^{tr} \otimes \mathbf{n}_j^{tr} \otimes \mathbf{n}_k^{tr} \otimes \mathbf{n}_l^{tr} \,.
$$

$$(6.184)$$

The transformation of the tangent moduli to a CARTESIAN basis can be obtained via relation (6.166). The final result is given with the transformation matrix T_{AB}, see (6.167), and the matrix notation in (6.184) by

$$
\bar{C}_{AB} = T_{AC}\, \bar{C}_{CD}^{H}\, T_{BD} \,.
\tag{6.185}
$$

7. Stability Problems

Real structures respond to external loading often in a nonlinear way. Some nonlinearities are associated with sudden changes of the system behaviour. Among such effects are buckling of cylindrical shells, lateral buckling of beams or snap-through of flat shells but also localization of deformation in inelastic materials.

Points at the load–deflection diagram which are associated with such changes are called instability points since the structural system looses its stability and cannot bear additional loads or has a loss of stiffness and hence can collapse. Simple examples which depict such behaviour are discussed in Sects. 2.1.3 and 2.1.4. The computation of instability points is thus an important part of a structural analysis in which always nonlinear effects have to be considered. Hence it is of interest to develop reliable and efficient methods for the determination and classifications of stability points.

The discussion of stability problems is restricted in this chapter to static and elastic problems. Stability of dynamical systems are discussed from a theoretical point of view, in e.g. Bolotin and Armstrong (1965), Marsden and Hughes (1983) and Simitses (1990). For dynamical stability problems, treated with finite element methods, see e.g. Wriggers and Carstensen (1992) and Briseghella et al. (1998). Stability of inelastic material problems is described, in e.g. Nguyen (2000) and Bazant and Cedolin (2003) theoretically. Applications within the finite element method are treated in, e.g. Petryk and Thermann (1992), Schreyer and Neilsen (1996), Stein et al. (1995) and Steinmann et al. (1997).

7.1 Computation of Stability Points

Instable behaviour of structures is either associated with multiple equilibrium solutions at a certain load level or with the loss of stiffness during loading. Both cases can only occur in nonlinear systems and hence need to be treated on the basis of an adequate theoretical model. A mechanical interpretation for instability was provided, e.g. in Pflüger (1975). It says that a neighbouring equilibrium state N exists, which belongs to the same load level as the given equilibrium state G. Starting from the given equilibrium state, also called ground state,

$$G^{(G)} = R^{(G)} - \lambda^{(G)} P = 0, \tag{7.1}$$

it is possible to compute a neighbouring state by an infinitesimal deviation from the linearized equation which can be computed using the linearization at the ground state

$$K_T^{(G)} \Delta v = -\left(R^{(N)} - \lambda^{(N)} P \right). \tag{7.2}$$

As assumed, the load level is the same for both states such that $\lambda^{(N)} = \lambda^{(G)}$. The neighbouring state is also an equilibrium state which can be expressed by

$$G^{(N)} = R^{(N)} - \lambda^{(N)} P = 0. \tag{7.3}$$

This yields $R^{(N)} = R^{(G)}$, such that the homogeneous equation system occurs

$$K_T^{(G)} \Delta v = 0. \tag{7.4}$$

The ground and neighbouring state for the EULER buckling behaviour of a beam is depicted in Fig. 7.1 together with the associated load–deflection diagram.

From a mathematical point of view the computation of instability points is related to the investigation of tangent stiffness matrix with respect to singularities. The homogeneous equation system (7.4) has non-trivial solutions for det $K_T = 0$. This condition is equivalent to

$$(K_T - \omega_j 1) \phi_j = 0 \tag{7.5}$$

with the eigenvalue ω_j and the eigenvector ϕ_j of the jth eigen solution. An instability point is now denoted by a zero eigenvalue.

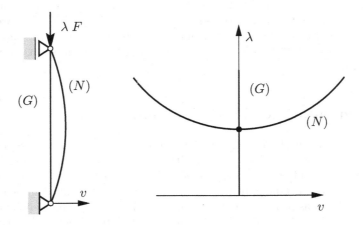

Fig. 7.1 Deformation of a beam and load–deflection diagram

7.1.1 Classical and Linear Buckling Analysis

In the engineering literature and thus also in the engineering codes, linear buckling analysis is mostly applied. This procedure yields the first load level where a system becomes instable. It is based on the formulation of a general eigenvalue problem for the load parameter λ which was introduced in (5.1). The derivation of the eigenvalue problem is provided by the development of the displacements, which occur before reaching the instability point, in a power series relative to the load parameter

$$\boldsymbol{v} = \lambda \boldsymbol{v}^{(1)} + \lambda^2 \boldsymbol{v}^{(2)} + \lambda^3 \boldsymbol{v}^{(3)} + \cdots . \tag{7.6}$$

By considering only linear terms, a linear eigenvalue problem follows. The idea behind this approach is that instability is reached for a stress state which is independent on the displacements close to the instability point. This is often true for structures like beams and shells where the normal forces or membrane states do not depend upon the deformation before the instable behaviour occurs, see also the example in Fig. 7.1.

The linear eigenvalue problem is derived by splitting the tangent stiffness matrix $\boldsymbol{K}_T$, see e.g. (4.76), in the parts

$$\boldsymbol{K}_T = \boldsymbol{K}_L + \boldsymbol{K}_U + \boldsymbol{K}_\sigma, \tag{7.7}$$

where $\boldsymbol{K}_L$ is the linear stiffness matrix, $\boldsymbol{K}_U$ the part of the matrix related to the initial deformations and $\boldsymbol{K}_\sigma$ is the geometric or initial stress matrix. Restriction to the linear part $\boldsymbol{v}_L$ of the displacement yields the "linearized" stiffness matrices

$$\begin{aligned} \hat{\boldsymbol{K}}_U &= \boldsymbol{K}_U(\boldsymbol{v}_L), \\ \hat{\boldsymbol{K}}_\sigma &= \boldsymbol{K}_\sigma(\boldsymbol{v}_L). \end{aligned} \tag{7.8}$$

With this split of the tangential stiffness matrix $\boldsymbol{K}_T$, a linear eigenvalue problem for the load parameter can be deduced, see e.g. Brendel and Ramm (1982),

$$[\boldsymbol{K}_L + \lambda(\hat{\boldsymbol{K}}_U + \hat{\boldsymbol{K}}_\sigma)]\boldsymbol{\phi} = \boldsymbol{0}. \tag{7.9}$$

The associated algorithm can be found in Box 7.1. The influence of the displacement state (initial buckling state) is suppressed when the matrix $\bar{\boldsymbol{K}}_U$ is neglected. Then Eq. (7.9) reduces to the classical eigenvalue problem

$$[\boldsymbol{K}_L + \lambda \hat{\boldsymbol{K}}_\sigma]\boldsymbol{\phi} = \boldsymbol{0}. \tag{7.10}$$

The solution of this eigenvalue problem yields as smallest eigenvalue the critical load factor λ_c and hence the critical load $\boldsymbol{P}_c = \lambda_c \boldsymbol{P}$. The eigenvector $\boldsymbol{\phi}$ belonging to λ_c depicts the form of the failure mode. With relations (7.8) and (7.9), the associated algorithm follows:

Box 7.1 Algorithms for classical linear buckling analysis

1. Solve the linear problem	$K_L\, v_L = P$
2. Solve	$[\,K_L + \lambda_c(\,\hat{K}_U + \hat{K}_\sigma\,)\,]\,\phi = 0$
3. Compute critical values	$P_c = \lambda_c\,P,\qquad v_c = \lambda_c\,v_L$

This method yields generally sufficient accurate results for practical problems, in which only small displacements and deformations occur before an instability point is reached.

7.1.2 General Investigations of Stability

In case of highly nonlinear problems, solutions depend upon the displacement and stress states in the structure. Then the instable points deviate from the ones obtained by the linear buckling analysis. Hence a complete nonlinear computation is necessary. As described in Sect. 5.1.5, an incremental iterative arc-length strategy can be employed to solve the nonlinear problem. Within the arc-length method, the stability behaviour has to be checked by an accompanying investigation.

A simple method for the detection of instability or singular points is the observation of the determinant or equivalently the check of the change of sign of the diagonal elements of the tangential stiffness matrix K_T. A zero element of the diagonal denotes a nontrivial solution of Eq. (7.4). Furthermore, change of sign of the determinant of K_T characterizes the occurence of an instability point. The computation of $\det K_T$ can be obtained nearly without additional effort during the triangularization of K_T. This can be seen from

$$\det K_T = \prod_{i=1}^{ndof} D_{ii} \tag{7.11}$$

with $K_T = L^T D L$, where D_{ii} is an element of the diagonal D. The determinant however is not a well-suited indicator since even for small finite element systems its value is extremely large. Thus it will exceed the limit of the number representation of the computer very fast. By using the logarithm of the determinant, the product formula in (7.11) becomes a sum and thus the numerical value of the determinant will be a lot smaller. However, even this value can be too large for, e.g. a finite element discretization of a shell structure with several thousand unknowns. Thus, in standard simulations, the change of sign of the diagonal elements D_{ii} are used as indicator for the occurrence of a stability point.

Since a finite element discretization leads usually to a positive definite tangent matrix, the first negative diagonal element describes the first instability point. Thus if a computation of a system emplying the arc-length method depicts for a part of the load–deflection curve a negative diagonal element D_{ii} then this part of the path is instable. It can even occur that more than one diagonal elements are negative which of coure is also related to instable paths. Note that an even number of negative diagonal elements lead to a positive determinant of the tangent matrix, but still the system is unstable. Using the sign of the determinant of $\boldsymbol{K}_T$, the following statements regarding the equilibrium state can be made:

Box 7.2 State of equilibrium depending on the sign of the diagonal elements of $\boldsymbol{K}_T$

all $D_{ii} > 0$ $\quad\rightarrow \boldsymbol{K}_T$ pos. def.:		stable equilibrium
at least 1 $D_{ii} = 0$ $\quad\rightarrow \boldsymbol{K}_T$ pos. semidef.:		indifferent equilibrium
at least 1 $D_{ii} < 0$ $\quad\rightarrow \boldsymbol{K}_T$ neg. def.:		instable equilibrium

The condition for instability, $\det \boldsymbol{K}_T = 0$, can also be expressed by the eigenvalue problem $(\boldsymbol{K}_T - \omega \boldsymbol{1}) \, \boldsymbol{\phi} = \boldsymbol{0}$, see (7.5). Another often used approach stems from the linear stability analysis. Then an eigenvalue problem of type

$$
\begin{aligned}
\left[\, \boldsymbol{K}_L + \boldsymbol{K}_U + \lambda_c \boldsymbol{K}_\sigma \right] \boldsymbol{\phi} &= \boldsymbol{0}, \\
\left[\, \boldsymbol{K}_L + \lambda_c (\boldsymbol{K}_U + \boldsymbol{K}_\sigma) \right] \boldsymbol{\phi} &= \boldsymbol{0}
\end{aligned}
$$

is solved, see e.g. Ramm (1976) and Brendel and Ramm (1982). However, this is not very efficient.

Remark 7.1:

1. There is no stringent mathematical reasoning for the split of the tangential stiffness matrix $\boldsymbol{K}_T$ in (7.12) with respect to the load factor λ.
2. The split of the tangential stiffness matrix in three parts does not occur naturally within the finite element formulation. Hence more coding effort is needed, but the computation of the three parts is also more time consuming than computing $\boldsymbol{K}_T$ in one piece.
3. The eigenvalue problem (7.9) is a general one. On the contrary, the eigenvalue problem in (7.5) is a special eigenvalue problem which needs less effort for its solution.
4. The solution of eigenvalue problem (7.12) can only be obtained using special eigenvalue solver since matrices $\boldsymbol{K}_U$ and $\boldsymbol{K}_\sigma$ are generally not positive definite. This can be observed directly from matrix $\boldsymbol{K}_\sigma$ where terms associated with normal stresses appear on the diagonal (which can be plus or minus).

Within the incremental-iterative computation using the arc-length method, see Sect. 5.1.5, the instable points are not determined accurately since only the nonlinear solution path is of concern. However, the appearance of negative diagonals is depicted during the solution with the arc-length method, see Box 7.2. Thus further methods have to be introduced when the singular points need to be computed accurately. A simple method is provided by a bi-section algorithm which can be applied to determine instability points, see e.g. Wagner and Wriggers (1988). This procedure is also propagated in the mathematical literature, see e.g. Keller (1977). The related algorithm needs an arc-length method which can change the loading direction (return on the

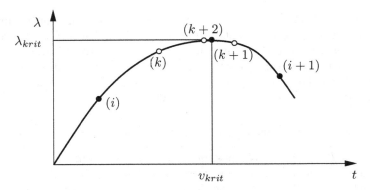

Fig. 7.2 Iteration series for the bi-section method

already computed path) and a possibility to change the arc-length (smaller steps).

Within the bi-section method, the algorithm runs backward on the loading path with half arc-length once a singular point is bypassed from step (i) to $(i+1)$. Starting from the last reached equilibrium state $(k+n)$, this procedure is executed until the eigenvalue ω_j in (7.5) related to the singular point is close to zero within a given tolerance $(\omega_j \leq TOL)$, see Fig. 7.2. In practical application, it is often sufficient to choose a tolerance of $TOL = 10^{-5}$. The series of iterations needed within this procedure is denoted in Fig. 7.2 by $(k), (k+1), \ldots$. In Fig. 7.2, the equilibrium state related to $(k+2)$ is already close to the critical point.

When the bi-section method is applied, it is necessary to change the load direction when the instability point is bypassed during the iteration. Details of such method when applied to structural stability problems can be found in, e.g. Wagner and Wriggers (1988) and Wagner (1991).

The bi-section method can be applied to problems including all possible nonlinearities including inelastic material behaviour. However, in the latter case, the algorithm has to be reformulated since change of loading direction leads in case of plasticity to unloading and thus will not return on the already computed path. Hence the bi-section method has to be formulated in such a way that always loading occurs. This can be done by applying the scheme in such a way that it approaches the stability point from one side, for details see e.g. Wriggers and Simo (1990).

Further possibilities to compute instability points directly are discussed in the next section.

The solution of the eigenvalue problem yields an eigenvector $\boldsymbol{\phi}_j$ related to the critical or instability point with eigenvalue $\omega_j = 0$, see e.g. Spence and Jepson (1984). Using this eigenvector, it is possible to determine the type of the instability at this singular point. This investigation starts from the equilibrium state $(\bar{\mathbf{u}}, \bar{\lambda})$ provided by Eq. (7.1). Its linearization yields

$$\boldsymbol{G}(\boldsymbol{u},\lambda) = \boldsymbol{R}(\bar{\boldsymbol{u}}) + \boldsymbol{K}_T(\bar{\boldsymbol{u}})\,\Delta\boldsymbol{u} - \bar{\lambda}\boldsymbol{P} - \boldsymbol{P}\Delta\lambda = \boldsymbol{0}. \tag{7.12}$$

In this equation, equilibrium demands: $\boldsymbol{R}(\bar{\boldsymbol{u}}) - \bar{\lambda}\boldsymbol{P} = \boldsymbol{0}$. When the remaining part of this relation is multiplied from the left side by the transposed eigenvector $\boldsymbol{\phi}_j^T$ associated with the zero eigenvalue, then the term $\boldsymbol{\phi}_j^T\,\boldsymbol{K}_T(\bar{\boldsymbol{u}})\,\Delta\boldsymbol{u} = 0$. This relation follows from (7.5) since the eigenvalue is zero at the singular point. Hence the only remaining part is $\boldsymbol{\phi}_j^T\,\boldsymbol{P}\Delta\lambda = 0$ which has two possible interpretations. For $\boldsymbol{\phi}_j^T\,\boldsymbol{P} \neq 0$, the load increment has to be zero at the instability point which denotes a snap-through behaviour, see e.g. Fig. 7.2 and the example in Sect. 2.1.4. On the other hand, when $\boldsymbol{\phi}_j^T\,\boldsymbol{P} = 0$, the load increment can be arbitrary and thus instability occurs on the load path which is called bifurcation, see e.g. Fig. 7.1 and the example in Sect. 2.1.3. The result can be summarized as

$$\boldsymbol{\phi}_j^T\,\boldsymbol{P} = \begin{cases} = 0 .. \text{ bifurcation point} \\ \neq 0 .... \text{ limit point} \end{cases}. \tag{7.13}$$

This relation is valid for snap-through and limit points (L) and simple bifurcation points (B), see also Fig. 5.5. The solution branches at a bifurcation point off in two or more paths, as can be seen in Figs. 5.5 and 7.1. The generic loading path (the computation starts along this path) is called primary path while the paths leaving the primary path are called secondary paths. Further classifications of special limit and stability points can be found in, e.g. Jepson and Spence (1985) and Wagner (1991).

When the complete load deflection curve has to be determined for a stability problem, then not only the primary but also the secondary solution paths need to be computed. Thus, at a singular point, the algorithm has to be able to switch to the secondary path. This switch is not always simple since two equilibrium states (on the primary and on the secondary path) are close to each other. A good choice is to use the eigenvectors associated with the zero eigenvalues at the singular point as starting values for an iteration since they provide a tangent to the secondary path, see Jepson and Spence (1985). Relation

$$\boldsymbol{v}_j = \bar{\boldsymbol{v}} + \xi_j\,\frac{\boldsymbol{\phi}_j}{\|\boldsymbol{\phi}_j\|} \tag{7.14}$$

denotes the deformation state which can be employed as starting vector in a combined NEWTON and arc-length method to approach the secondary path. The vector $\bar{\boldsymbol{v}}$ contains the displacement state at the singular point. The magnitude of the factor ξ_j is essential for a successful switch to the secondary path. Related strategies can be found, in e.g. Rheinboldt (1981) and Riks (1984). In the case that the described procedure does not work, Eq. (7.14) can be amended such that instead of the displacement field the coordinates of the structure are updated by the eigenvector belonging to the singularity.

Remark 7.2: When many bifurcation points are close to each other, it is very difficult to change to a secondary path. This is especially true for shell structures where clustering of singular points can occur. Moreover, the main interest in engineering analysis is the path which yields the minimum load-bearing capacity of the structure under investigation. In order to reach this path or to change to another path, inertial effects have to be considered. Examples are snap-through or buckling of shells when a dynamical process is initiated by the physics after passing a singular point. Related algorithms using a static/dynamic approach can be found in, e.g. Riks et al. (1996) and Schweizerhof et al. (2002).

7.2 Direct Computation of Singular Points

Stability points can be detected, computed and classified with the methods described so far. This was performed by monitoring the sign of the diagonals of the tangent stiffness matrix by using the bi-section method and computing eigenvectors at a singular point. However, the bi-section method only leads to linear convergence behaviour; hence for faster (quadratically) convergence different methods have to be applied.

A method which directly determines a singular point and has quadratic convergence properties, see Fig. 7.3, will be developed in the following.

The idea of this method is to use the standard incremental iterative strategy together with the arc-length methods initially. Close to a singular point, the strategy is then switched to an iterative method which is able to compute the singular point directly. Within the following, the questions of designing a direct scheme and of when to switch to a direct method will be addressed.

7.2.1 Formulation of an Extended System

A direct computation of singular points is based on the idea to restrict the set of solution to equilibrium states which are singular. Hence a constraint condition has to be added to the standard finite element weak form which reduces the solution space to singular points.

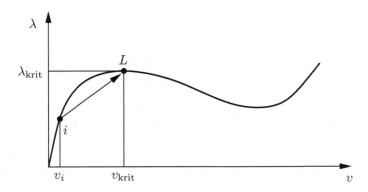

Fig. 7.3 Direct method for the computation of a singular point

One possibility is to add the condition det $\boldsymbol{K}_T = 0$ since this condition is fulfilled at a singular point, but the resulting equations are quite complicated, especially the linearization and its implementation. Furthermore, the determinant of a large finite system is represented by a very large number which can easily be outside the number range of a standard computer. However, the determinant criterion can be applied successfully to stability problems of beams, see Planinc and Saje (1999).

Another formulation uses the eigenvalue problem $(\boldsymbol{K}_T - \omega\,\boldsymbol{1})\,\boldsymbol{\phi} = \boldsymbol{0}$. The eigenvalue ω of the tangent matrix $\boldsymbol{K}_T$ is zero at a singular point. Hence the equation $\boldsymbol{K}_T\boldsymbol{\phi} = \boldsymbol{0}$ can be applied at snap-through and bifurcation to define a singular point. Now the following extended system can be constructed

$$\widehat{\boldsymbol{G}}(\boldsymbol{v},\lambda,\boldsymbol{\phi}) = \left\{ \begin{array}{c} \boldsymbol{G}(\boldsymbol{v},\lambda) \\ \boldsymbol{K}_T(\boldsymbol{v},\lambda)\,\boldsymbol{\phi} \\ l\,(\boldsymbol{\phi}) \end{array} \right\} = \boldsymbol{0}. \tag{7.15}$$

The last equation was introduced to exclude the trivial solution $\boldsymbol{\phi} = \boldsymbol{0}$. This equation can be formulated, e.g. as

$$l\,(\boldsymbol{\phi}) = \|\,\boldsymbol{\phi}\,\| - 1 = 0. \tag{7.16}$$

To compute either snap-through or bifurcation points, Eq. (7.13) can be used. The condition leading to the determination of snap-through points is then given as

$$l\,(\boldsymbol{\phi}) = \boldsymbol{\phi}^T\,\boldsymbol{P} - 1 = 0. \tag{7.17}$$

Due to the choice of the constraint equation, the formulation of the extended system in (7.15) is limited to the computation of simple snap-through and bifurcation points.

The use of extended systems is described in the mathematical literature and applicable to many different problems, see the overview of Mittelmann and Weber (1980). The extended system formulated above was, e.g. applied by Werner and Spence (1984) for the computation of snap-through and symmetrical bifurcation points. In the mathematical literature, these methods are usually only employed for systems with few degrees of freedoms. The extended system was firstly formulated and implemented in the paper of Wriggers et al. (1988) within the finite element method using a consistent linearization, to discuss the stability behaviour of three-dimensional truss systems. Extension to beam and shell problems are reported, e.g. in Wriggers and Simo (1990) and in Fujii and Ramm (1997). Application of the method in optimization can be found in Reitinger and Ramm (1995).

The formulation of a NEWTON procedure for the extended system in (7.15) needs a consistent linearization. This leads to

$$\begin{aligned} \widehat{\boldsymbol{K}}_{T\,i}\,\Delta\boldsymbol{w}_{i+1} &= -\widehat{\boldsymbol{G}}(\boldsymbol{w}_i) \\ \boldsymbol{w}_{i+1} &= \boldsymbol{w}_i + \Delta\boldsymbol{w}_{i+1} \end{aligned} \tag{7.18}$$

with

$$\widehat{\boldsymbol{K}}_{T\,i} = \left.\frac{\partial \widehat{\boldsymbol{G}}}{\partial \boldsymbol{w}}\right|_i \quad \text{and} \quad \boldsymbol{w} = \left\{ \begin{array}{c} \boldsymbol{v} \\ \boldsymbol{\phi} \\ \lambda \end{array} \right\}. \tag{7.19}$$

In order to judge the costs for the solution of the extended system, the Eq. in (7.18) are discussed in detail. In the general case of deformation dependent loads, see Sect. 4.2.5, the tangent matrix not only depends upon the deformations but also on the load parameter λ. For this case, the linearization is given by

$$\begin{bmatrix} \boldsymbol{K}_T & \boldsymbol{0} & -\boldsymbol{P} \\ \nabla_v \left(\boldsymbol{K}_T \, \boldsymbol{\phi} \right) & \boldsymbol{K}_T & \nabla_\lambda \left(\boldsymbol{K}_T \, \boldsymbol{\phi} \right) \\ \boldsymbol{0} & \nabla_\phi \, l \left(\boldsymbol{\phi} \right) & 0 \end{bmatrix} \left\{ \begin{array}{c} \Delta \boldsymbol{v} \\ \Delta \boldsymbol{\phi} \\ \Delta \lambda \end{array} \right\} = - \left\{ \begin{array}{c} \boldsymbol{G}(\boldsymbol{v},\lambda) \\ \boldsymbol{K}_T(\boldsymbol{v},\lambda) \, \boldsymbol{\phi} \\ l \left(\boldsymbol{\phi} \right) \end{array} \right\} \tag{7.20}$$

with

$$\nabla_v \left(\boldsymbol{K}_T \, \boldsymbol{\phi} \right) = \frac{\partial}{\partial \boldsymbol{v}} (\boldsymbol{K}_T \, \boldsymbol{\phi}),$$

$$\nabla_\lambda \left(\boldsymbol{K}_T \, \boldsymbol{\phi} \right) = \frac{\partial}{\partial \lambda} (\boldsymbol{K}_T \, \boldsymbol{\phi}),$$

$$\nabla_\phi \, l \left(\boldsymbol{\phi} \right) = \frac{\partial}{\partial \boldsymbol{\phi}} l \left(\boldsymbol{\phi} \right).$$

The first view on this equation system leads to the impression that the solution of the system (7.20) requires large computational effort. The introduced vector $\boldsymbol{w}$ contains $2\,n+1$ unknowns, the tangent matrix $\widehat{\boldsymbol{K}}_T$ is nonsymmetric and derivatives of the tangential stiffness matrix $\boldsymbol{K}_T$ have to be computed. However, a closer look to $\widehat{\boldsymbol{K}}_T$ reveals that (7.20) can be solved efficiently by a partitioning scheme since the tangent matrix $\boldsymbol{K}_T$ occurs two times on the diagonal of the equation system. The solution algorithms follows the idea of block elimination (equivalent to the scheme used within the arc-length method, see Sect. 5.1.5). Within the iteration step i, the steps described in Box 7.3 have to be executed to solve (7.20).

To start the algorithm, stated in Box 7.3, initial values have to be provided for displacements $\boldsymbol{v}$, load parameter λ and eigenvector $\boldsymbol{\phi}$. While $\boldsymbol{v}$ and λ are known from the last step of the incremental solution procedure, e.g. the arc-length method, the eigenvector is unknown. To fulfil condition (7.16), it cannot be selected as zero vector. However, several possibilities can be followed for the proper selection of a the initial eigenvector $\boldsymbol{\phi}_0$. Below two different choices are described.

1. Unit vector:

$$\boldsymbol{\phi}_0 = \frac{1}{\|\boldsymbol{e}\|} \, \boldsymbol{e} \quad \text{with} \quad \boldsymbol{e} = \{1,1,\dots,1\} \tag{7.21}$$

This starting vector for $\boldsymbol{\phi}$ contains the entire spectrum and hence during the solution procedure the eigenvector $\boldsymbol{\phi}$ can assume the correct form.

2. Eigenvector of $\boldsymbol{K}_T$:

$$\boldsymbol{\phi}_0^0 = \boldsymbol{e} \, / \, \|\boldsymbol{e}\|$$
$$LOOP \ k = 1, \dots, m$$
$$\boldsymbol{\phi}_0^k = \boldsymbol{K}_T^{-1} \, \boldsymbol{\phi}_0^{k-1}$$
$$ENDLOOP$$

m steps of a vector iteration are performed within the algorithm. This leads to a starting vector $\boldsymbol{\phi}_0$ for the extended system which is associated to the current state of $\boldsymbol{K}_T$. Such choice of the starting vector can accelerate the convergence of the computation considerably, especially when the eigenvector belonging to the singular point is close to $\boldsymbol{\phi}_0$. Often only a few iterations (e.g. $m = 3$ to 5) are sufficient to determine $\boldsymbol{\phi}_0$ with sufficient accuracy. Note however that a choice of the starting vector related to $\boldsymbol{K}_T$ can lead to slow convergence when the eigenvector associated with the singular point is orthogonal to $\boldsymbol{\phi}_0$.

It is obvious from the algorithm in Box 7.3 that the matrix $\boldsymbol{K}_T$ has only to be factorized once. This operation is – especially for large systems – essential for computational efficiency, see Sect. 5.2. Hence only a relatively small additional computational effort is needed for the extended system when compared to the standard NEWTON method. Besides the computation of vectors $\boldsymbol{h}_1$ and $\boldsymbol{h}_2$, the equation system has to be solved for three additional right hand sides to compute the vectors $\Delta\boldsymbol{v}_1$, $\Delta\boldsymbol{\phi}_1$ and $\Delta\boldsymbol{\phi}_2$. The non-symmetry of $\widehat{\boldsymbol{K}}_T$ has not to be considered within the algorithm. A judgement concerning the additional effort to compute the derivative of $\boldsymbol{K}_T$ is discussed next.

Box 7.3 Algorithm for the computation of singular points using the extended system

---

1. Solve $\boldsymbol{K}_T \, \Delta \boldsymbol{v}_P = \boldsymbol{P}, \quad \boldsymbol{K}_T \, \Delta \boldsymbol{v}_G = -\boldsymbol{G}$.
2. compute directional derivatives

$$\begin{aligned}
\boldsymbol{h}_1 &= \nabla_v \left(\boldsymbol{K}_T \, \boldsymbol{\phi} \right) \Delta \boldsymbol{v}_P + \nabla_\lambda \left(\boldsymbol{K}_T \, \boldsymbol{\phi} \right), \\
\boldsymbol{h}_2 &= \boldsymbol{K}_T \, \boldsymbol{\phi} + \nabla_v \left(\boldsymbol{K}_T \, \boldsymbol{\phi} \right) \Delta \boldsymbol{v}_G .
\end{aligned}$$

3. Solve $\boldsymbol{K}_T \, \Delta\boldsymbol{\phi}_1 = -\boldsymbol{h}_1, \quad \boldsymbol{K}_T \, \Delta\boldsymbol{\phi}_2 = -\boldsymbol{h}_2$.
4. compute increments

$$\begin{aligned}
\Delta\lambda &= -\frac{\nabla_\phi \, l\,(\boldsymbol{\phi}) \, \Delta\boldsymbol{\phi}_2 + l(\boldsymbol{\phi})}{\nabla_\phi \, l\,(\boldsymbol{\phi}) \, \Delta\boldsymbol{\phi}_1}, \\
\Delta\boldsymbol{v} &= \Delta\lambda \, \Delta\boldsymbol{v}_P + \Delta\boldsymbol{v}_G \quad \Delta\boldsymbol{\phi} = \Delta\lambda \, \Delta\boldsymbol{\phi}_1 + \Delta\boldsymbol{\phi}_2 .
\end{aligned}$$

5. Update displacements, eigenvector and load parameter.

$$\lambda = \lambda + \Delta\lambda, \quad \boldsymbol{v} = \boldsymbol{v} + \Delta\boldsymbol{v}, \quad \boldsymbol{\phi} = \boldsymbol{\phi} + \Delta\boldsymbol{\phi}.$$

7.2.2 Computation of the Directional Derivative of K_T

The formulation of the tangential stiffness matrix follows the rules derived in Sect. 4.2.2

$$K_T = \bigcup_{e=1}^{n_e} \int_{\Omega_e} \{\, \bar{B}_L^T\, \bar{D}\, \bar{B}_L + G^T\, \bar{S}\, G \,\}\, d\Omega \,. \qquad (7.22)$$

The directional derivative of K_T which is needed within the algorithm in Box 7.3 can be described by using the already defined $\bar{B}_L$ matrices, see (4.53). In order to obtain an explicit form of the directional derivative, the $\bar{B}_L$ matrices are split into a constant part and a part which depends linearly upon the displacements v_e

$$\bar{B}_L = B_0 + \bar{B}_{Li}(v_e)\,. \qquad (7.23)$$

The vector h is written as

$$
\begin{aligned}
h &= \nabla_v\,(K_T\phi)\,\Delta v \\
&= \bigcup_{e=1}^{n_e} \eta_e^T \int_{\Omega_e} \{\bar{B}_{Li}^T(\Delta v_e)\,\bar{D}\,\bar{B}_L(v_e) + \bar{B}_L^T(v_e)\,\bar{D}\,\bar{B}_{Li}(\Delta v_e) \quad (7.24) \\
&\quad + G^T\,\Delta\bar{S}\,G\}\,\phi_e\,d\Omega\,.
\end{aligned}
$$

The vector $\Delta\bar{S}$ in (7.22) contains the incremental stresses which follow for a ST. VENANT with constant $\bar{D}$ from

$$\Delta\bar{S} = \bar{D}\,\bar{B}(v_e)\,\Delta v_e\,. \qquad (7.25)$$

A detailed derivation of the above result can be found in Wriggers et al. (1988). It is necessary for the solution of the extended system (7.15) to compute the right hand sides h_1 and h_2, see Box 7.3. The basis for this is provided by (7.24). The first term of h_1 follows with $\Delta v_e = \Delta v_{1e}$ while $\Delta v_e = \Delta v_{2e}$ is necessary for the computation of the second term of h_2. The term $\nabla_\lambda(K_T\,\phi)$ in h_1 is zero for conservative loading.

An estimate for the effort to compute the vectors h_1 and h_2 can be obtained as follows. The $\bar{B}$ matrices are known, see (4.53). The modifications of these matrices needed in (7.24) are trivial. All matrices can be compute element wise. Hence the effort to computed the vectors h_1 and h_2 are basically equivalent to the computation of the residual vector G.

The discussed analytical determination of the derivative of the tangent matrix K_T with respect to the displacements has a relative simple structure since the constitutive equation is linear with respect to GREEN-LAGRANGE strain tensor and the 2. PIOLA-KIRCHHOFF stress. For the class of hyperelastic materials, see Sect. 3.3.1, it is a lot more complex to analytically determine the derivative of K_T. For the class of OGDEN type materials – formulated in principal stretches – the analytical form can be found in Reese (1994) and Reese and Wriggers (1995). The analytical determination of the derivative of K_T for structural elements like shells or beams for finite rotations is even more complex since here the rotations are in $SO(3)$, see Sect. 9.4. Thus it can be advantageous to determine the directional derivative of the tangent K_T numerically. A related procedure was developed in Wriggers and Simo (1990). It starts with the definition of K_T which follows from the directional derivative of the residual G

$$K_T \phi = \nabla_v G(v, \lambda) \phi = \frac{d}{d\epsilon} G(v + \epsilon \phi, \lambda) \Big|_{\epsilon=0}. \qquad (7.26)$$

By using the symmetry of the second derivative of G, the derivative of $K_T \phi$ in the direction of Δv can be written in the following equivalent form

$$
\begin{aligned}
\nabla_v [K_T \phi] \Delta v &= \nabla_v [\nabla_v G(v, \lambda) \phi] \Delta v \\
&= \nabla_v [\nabla_v G(v, \lambda) \Delta v] \phi.
\end{aligned}
\qquad (7.27)
$$

Using these results, the vectors h_1 and h_2, which appear in the algorithm in Box 7.3, can be determined. For this task, only one additional evaluation of the tangent matrix K_T is necessary, leading to

$$\nabla_v [K_T \phi] \Delta v = \frac{d}{d\epsilon} [K_T(v + \epsilon \phi)] \Delta v \Big|_{\epsilon=0}. \qquad (7.28)$$

This relation is now reformulated which yields directly the numerical approximation

$$\nabla_v [K_T \phi] \Delta v = \lim_{\epsilon=0} \frac{1}{\epsilon} [K_T(v + \epsilon \phi) \Delta v - K_T(v) \Delta v]. \qquad (7.29)$$

Using a fixed parameter ϵ, see Remark 7.3, the approximation

$$\nabla_v [K_T \phi] \Delta v \approx \frac{1}{\epsilon} [K_T(v + \epsilon \phi) \Delta v - K_T(v) \Delta v] \qquad (7.30)$$

follows. The application of such approximation, to compute the directional derivative within the algorithm described in Box 7.3, yields expressions for vectors h_1 and h_2:

$$
\begin{aligned}
h_1 &\approx \frac{1}{\epsilon} [(K_T(v + \epsilon \phi) \Delta v_P - P], \\
h_2 &\approx K_T \phi + \frac{1}{\epsilon} [(K_T(v + \epsilon \phi) \Delta v_G + G].
\end{aligned}
\qquad (7.31)
$$

Observe that the load vector P and the residual G are known. Hence the numerical effort to determine vectors h_α ($\alpha = 1, 2$) is limited and does only involve an additional evaluation of the tangent matrix: $K_{T\gamma}(v + \epsilon \phi)$.

Remark 7.3: The following additional considerations are necessary in order to devise an efficient scheme for the computation of the approximate directional derivative:

1. The function evaluation of K_T should only be performed once. All matrix multiplications (7.31) can be carried out at element level such that the matrix $K_T(v + \epsilon\phi)$ does not have to be assembled. The assembly is only needed for the vectors h_α.

2. The choice of the parameter ϵ in (7.31) is an essential ingredient for a successful application of the numerical directional derivative. Its choice depends upon the vector ϕ and the accuracy of the computer. An estimation for ϵ can be found in Dennis and Schnabel (1983). It yields

$$\epsilon = \max_{1 < k < n} \phi_k \eta_{TOL} . \tag{7.32}$$

Here ϕ_k is the k-th component of the vector $\phi \in \mathbb{R}^n$. η_{TOL} is the constant represents the machine accuracy. For double precision, $\eta_{TOL} \approx 10^{-6}$ can be set.

7.2.3 Example: Bifurcation Point of an Arc

The direct method for the computation of singular points, discussed above, will be applied to the arc structure depicted in Fig. 7.4.

The inner radius of the arc is given as $R_i = 100$ while the outer radius has the magnitude $R_a = 103$. Thus the thickness of the arc is $t = 3$. The arc spans an area with an angle of $\alpha = 60°$ and is simply supported on both ends.

The mechanical model is derived with respect to the initial configuration. Its formulation is based on the weak form (3.292) including all nonlinear terms and uses ST. VENANT elastic constitutive equation. The parameters of the ST. VENANT material are $E = 40000$ and $\nu = 0.2$. The finite element formulation is equivalent to the one discussed in Exercise 4.3. The arc is discretized using isoparametric quadrilaterals with quadratic shape functions. Three elements are used in thickness direction; in total 150 elements were applied.

It is now possible to compute the singular point B directly by starting from point S on the load deflection curve, as can be seen in Fig. 7.5. For this, the starting eigenvector was computed using the algorithm described in Remark 7.3.2.

The convergence behaviour of the NEWTON iteration is shown in the following table for the application of the direct scheme to the arc under point

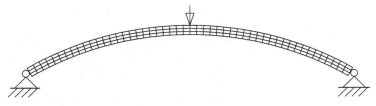

Fig. 7.4 Arc under point load

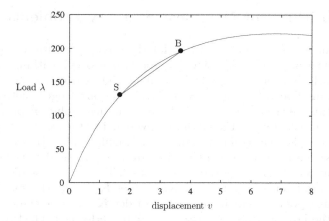

Fig. 7.5 Direct computation of the bifurcation point B

load. It is obvious that the method converges fast and quadratically. Note that the first iterative solution yields already a value for the load parameter λ which only deviates by 2% from the converged value and hence is already very close to the solution.

Residual	λ	v
1.3041823E+02	198.6	−3.0446E+00
9.6354141E+02	202.8	−3.4496E+00
4.9948826E+01	195.8	−3.6558E+00
2.1673679E+01	195.5	−3.6649E+00
3.0736882E−02	195.5	−3.6649E+00
2.6957110E−07	195.5	−3.6649E+00

Once the point is computed, its type of singularity can be evaluated. By using (7.13), a bifurcation point is detected. The eigenform which is associated with the bifurcation point follows also directly from the extended system algorithm in Box 7.3. It is depicted in Fig. 7.6.

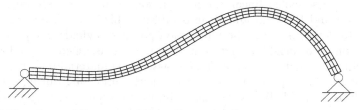

Fig. 7.6 Eigenform of the arc at bifurcation point B

7.3 Algorithms for Nonlinear Stability Problems

The extended systems provide a tool for the direct computation of singular points. Since more than one singular points can occur within a nonlinear solution path, it is advantageous to combine the extended system with the arc-length method described in Sect. 5.1.5. The arc-length method is employed within such strategy to follow the nonlinear solution path. Once a singular point is detected, the method is switched to the extended system for the efficient and direct computation of the singular point. Hence a criterion is needed which indicates when to switch from one method to the other.

Two different heuristic criteria can be used. The first is based on monitoring the determinant of the tangent matrix det K_T along the solution path. Once turning points occur in the course of det K_T then a switch from arc-length method to the extended system has to be performed. The turning point of the determinant is chosen to avoid convergence of the direct scheme to the previous singular point. The second possibility is to use the arc-length method until the number of negative diagonal elements of K_T changes. This depicts that a singular point was bypassed along the nonlinear solution path. After that, the extended system can be applied to exactly determine the singular point. As already noted, for elasto-plastic problems, a one-sided bi-section method has to be executed which avoids unloading. The entire algorithm is summarized in Box 7.4.

Once a stability point is found, it is possible to switch back to the arc-length method to follow the post-critical paths. In case of bifurcation points there are two possibilities, either to follow again the primary or the secondary path. The switch to a secondary path needs a special procedure, see e.g. Wagner and Wriggers (1988) and Eq. (7.14). The related procedure is based on the knowledge of the eigenvector ϕ which belongs to the bifurcation load. This vector is provided either by the solution of (7.5) at the equilibrium point, or it will automatically be computed within the extended system, see Box 7.3. Thus an additional computation of the eigenvector ϕ using (7.5) is not necessary when the extended system is applied. More refined algorithms for the switch from a primary to secondary points can be found in Wagner (1991) and Riks (1984).

Finally, the computation of limit and bifurcation points of elasto-plastic problems is discussed. Under the assumption of an associated flow rule, see Sect. 3.3.2, a so-called linear comparison solid can be introduced following Hill (1958) and Raniecki and Bruhns (1981). This comparison solid can be applied to determine, e.g. the bifurcation points of cylindrical specimen under tension. Two neighbouring solution states are assumed at one load level within an incremental formulation. The associated material tensor is taken to be constant. This approach excludes explicitly unloading during the stability analysis. It yields limits for plastic bifurcation loads, see e.g. Needleman (1972).

Algorithms which can be employed to analyze elasto-plastic finite deformations within a finite element method were described in Sect. 6.3.2. These formulations lead to explicit expressions for the material tangent, consistent with the algorithms, see Sect. 6.3.3. Inserting the material tangent into the relations for the tangent matrix $\boldsymbol{K}_T$ needed within the NEWTON method leads to the same structure as a discretization of the comparison solid. Hence the incremental formulations needed for the NEWTON method can be used to compute elasto-plastic bifurcation loads. For this, it is necessary to keep $\boldsymbol{K}_T$ constant at the equilibrium point to exclude unloading. The eigenvector belonging to the elasto-plastic bifurcation load follows then from (7.5). Related analysis and computations can be found in, e.g. Wriggers and Simo (1990) and Wriggers et al. (1992).

Box 7.4 Combined algorithm for the computation of instability points

1. Computation of equilibrium points at the solution path using arc-length method, see algorithm in Box 5.2

$$\boldsymbol{G}(\boldsymbol{v}, \lambda) = \boldsymbol{R}(\boldsymbol{v}) - \lambda \boldsymbol{P} = \boldsymbol{0}$$

2. Monitoring of diagonal elements or the determinant of $\boldsymbol{K}_T$:
   a) Turning point in the progression of determinant or change of the number of negative diagonal elements: Go to 3
   b) otherwise: Go to 1
3. Computation of a singular point
   a) In case of an elastic problem use extended system:
      i. Compute starting vector for eigenvector ϕ by one or two iterative steps $(i = 0, 1, 2,)$ of an inverse iteration

$$\phi_{i+1} = \boldsymbol{K}_T^{-1} \phi_i \quad \text{with} \quad \phi_0 = \boldsymbol{1}$$

      ii. Compute singular point by extended system, see Box 7.3

$$\widehat{\boldsymbol{G}}(\boldsymbol{v}, \lambda, \phi) = \boldsymbol{0}$$

   b) In case of an elasto-plastic problem use one sided bi-section:
      i. Start with last solution
      ii. Choose half arc-length
      iii. Compute next point on solution path
      iv. If det $\boldsymbol{K}_T < TOL$: Go to 4
      v. When number of neg. diagonal elements is constant: Go to 3(b)ii
      vi. otherwise: Go to 3(b)i
4. Type of stability point

$$\phi^T \boldsymbol{P} = \begin{cases} \neq 0 \dots \text{limit point} & \text{Go to 1} \\ = 0 \dots \text{bifurcation point} & \text{Go to 4(a)(b)} \end{cases}$$

   a) Continue on primary path: Go to 1
   b) Continue on secondary path: Use (7.14) and Go to 1

8. Adaptive Methods

The solution of problems in solid mechanics using finite elements always is an approximation of the analytical solution, besides a few cases, where exact solutions can be obtained for one dimensional linear elastic problems (e.g. trusses or beams). Errors of the finite element approximation consist of discretization errors in space and time and approximations of the real geometry. Each has different influence depending on the type of differential equation to be solved.

Additional error sources are modelling errors which result from the choice of a differential equation to describe real engineering problems. Classically, the choice of a one- or two-dimensional model to describe a three-dimensional engineering problem is an approximation, but also the selection of a constitutive equation does often not represent the physical behaviour adequately. This type of error, which is related to the validation of the mathematical model (question: do we solve the right equations), while very important, is not discussed in the following. Information can be found in, e.g. Stein and Ohnimus (1996), Oden et al. (1996) or Ladeveze and Pelle (2005) and the literature cited in these papers.

8.1 Introduction

To judge the accuracy of finite element solutions, and thus to verify that the equations of the mathematical model are solved correctly, the magnitude of the error of the finite element solutions has to be estimated. Even better would be a method which allows to solve the equations of mathematical model automatically up to a prescribed accuracy by the finite element method. This is, e.g. essential for problems with high local gradients in stress fields which locations are not known a priori. Hence scientists have developed adaptive methods which yield automatically a finite element mesh in which the error is distributed equally over the entire mesh. Different methods were introduced to change mesh or interpolation. These are h-, p- and r-adaption techniques. They are described in Fig. 8.1 where the different refined meshes are depicted assuming that a refinement is necessary in the left lower corner.

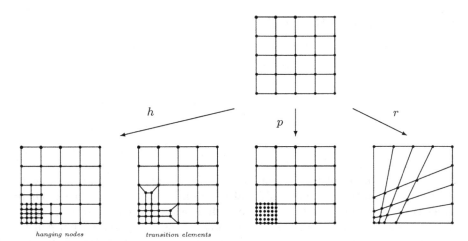

Fig. 8.1 $h-$, $p-$ and $r-$ adaption

- When using h-adaption, the element size is adapted based on computed errors. This technique yields finite element meshes with increasing number of nodes; hence the effort to solve the related finite element problems increases too. Usually three different methods are used to refine a mesh. In the first, the refined part is coupled via constraint equations to the existing mesh. Here so-called *hanging nodes* are allowed which are constraint to the movement of the element edge they are connected to, see first mesh in Fig. 8.1. The second mesh refinement introduces transition elements which lead to a conform mesh, see second mesh in Fig. 8.1. Finally, in the third method, a completely new mesh is generated based on the computed error distribution.
- The p-adaption increases the degree of the interpolation polynomials. This results in a higher order of approximation, but also leads to more unknowns. Furthermore, the bandwidth of the tangent matrix increases and considerably more time is needed to compute the element matrices for a high polynomial degree.
- Finally, the r-adaption method replaces finite element nodes such that the element sizes are optimized within a finite element mesh in order to reduce the overall error of the finite element solution. Within this approach, the structure and size of the finite element equations are preserved. However, negative JACOBIANS can occur at certain GAUSS due to large element distortions. This methods cannot guarantee convergence to the exact solution since the number of elements and the interpolation order are kept constant.

Since the exact solution of a partial differential equation is usually not known, it is not possible to compute the absolute error. Thus methods are needed which can estimate the error of the finite element solution. Hence techniques have to be developed which allow the computation of the finite element error based on the geometry, material data and approximate solution. The related theory is well developed for linear problems and yields error estimation and rules to adapt the element size within a given finite element mesh.

Basically, the approximation of an elliptic variational problem discretized by finite elements converges only if the polynomial degree k of the interpolation functions fulfils the inequality $k + 1 > m$ where the order m of the variational problem is equal to the highest derivative occurring in the problem. This condition can be interpreted in the engineering sense by the requirement that finite element have to be able to reproduce constant strain states exactly.

The order m of a variational problem is equivalent to the order $2\,m$ of the related differential equation of the problem. For linear elasticity or the POISSON equation, $m = 1$ is obtained while the value $m = 2$ is valid for a plate.

The error in displacements between an exact solution $\mathbf{u}$, living in the function space V, and an approximate solution $\mathbf{u}_h$, in the space of ansatz functions V_h of the finite element method, is defined by $\mathbf{e} = \mathbf{u} - \mathbf{u}_h$.

For an elliptical variational problem, the estimate

$$\| \mathbf{u} - \mathbf{u}_h \|_V \le C \, \| \mathbf{u} - \boldsymbol{\eta} \|_V \qquad \forall \boldsymbol{\eta} \in V_h \tag{8.1}$$

is valid, see e.g. Strang and Fix (1973). The norm in (8.1) is defined with respect to the space V in which a unique solution $\mathbf{u}$ of the variational problem exists. This variational problem can be written as bi-linear form

$$a(\mathbf{u}, \boldsymbol{\eta}) = f(\boldsymbol{\eta}). \tag{8.2}$$

The operators a and f of this abstract notation can be specified, e.g. for the case of linear elasticity using (3.17) and (3.273)

$$a(\mathbf{u}, \boldsymbol{\eta}) = \int_{\Omega} \boldsymbol{\varepsilon}(\boldsymbol{\eta}) \cdot \mathbb{C} \, [\, \boldsymbol{\varepsilon}(\mathbf{u}) \,] \, d\Omega \,,$$

$$f(\boldsymbol{\eta}) = \int_{\Omega} \hat{\mathbf{b}} \cdot \boldsymbol{\eta} \, d\Omega + \int_{\Gamma_{\sigma}} \hat{\mathbf{t}} \cdot \boldsymbol{\eta} \, d\Gamma \,. \tag{8.3}$$

Let us remark that the result in (8.1) follows from the error orthogonality which can be deducted by taking the difference of (8.2) and by using $a(\mathbf{u}_h, \boldsymbol{\eta}) = f(\boldsymbol{\eta})$:

$$a(\mathbf{u} - \mathbf{u}_h, \boldsymbol{\eta}) = 0. \tag{8.4}$$

The inequality (8.1) yields another remarkable result: of all functions, $\boldsymbol{\eta}$ in the space V_h, the finite element solution $\mathbf{u}_h$ is the best approximation of the exact solution. That means without knowing the exact solution the best possible approximation is provided by finite elements.

Equation (8.1) can be written with the choice $\boldsymbol{\eta} = \pi_h \, \mathbf{u}$ as

Fig. 8.2 Definition of a characteristic element size h

$$\|\mathbf{u} - \mathbf{u}_h\|_V \leq C \|\mathbf{u} - \pi_h \mathbf{u}\|_V, \tag{8.5}$$

where $\pi_h \mathbf{u} \in V_h$ is a short hand notation for an interpolation of $\mathbf{u}$ with an ansatz function as discussed in Sect. 4.1. The interpolation error on the right hand side $\|\mathbf{u} - \pi_h \mathbf{u}\|_V$ can be estimated, see e.g. Johnson (1987), Brenner and Scott (2002) and Braess (2007). For linear tetrahedral elements, see (4.42), the result is

$$\|\mathbf{u} - \mathbf{u}_h\|_{H^1} \leq C\, h\, |\mathbf{u}|_{H^2}. \tag{8.6}$$

The quantity h describes in this inequality a characteristic element size of the finite element approximation, e.g. the diameter of the element, see Fig. 8.2. The constant C depends upon the chosen mesh, but does not depend upon h. The constant C becomes smaller for regular meshes. In the case that the inner angle of a triangular finite element goes to zero then C goes to infinity and the approximate solution does not converge. The norm $\|\mathbf{u}\|_{H^s}$ in (8.6) is defined as

$$\|\mathbf{u}\|_{H^s} = \left\{ \int_{\Omega} \left[u_i\, u_i + u_{i,j}\, u_{i,j} + \cdots + u_{i,jk\ldots s}\, u_{i,jk\ldots s} \right] d\Omega \right\}^{\frac{1}{2}} < \infty. \tag{8.7}$$

Derivatives of the vector $\mathbf{u}$ occur in this equation up to the degree s. The mathematical spaces related to these norms are called SOBOLEV spaces. A solution $\mathbf{u}$ which fulfils condition (8.7) lies in the SOBOLEV space H^s. The special case

$$\|\mathbf{u}\|_{H^0} = \left\{ \int_{\Omega} \left[u_i\, u_i \right] d\Omega \right\}^{\frac{1}{2}} < \infty \tag{8.8}$$

is also denoted as L^2-norm ($\|\mathbf{u}\|_{H^0} = \|\mathbf{u}\|_{L^2}$).

A natural norm for elasticity problems is the energy norm

$$\|\mathbf{u}\|_E^2 = a(\mathbf{u}, \mathbf{u}). \tag{8.9}$$

This norm is under the condition

$$c\, \|\mathbf{u}\|_V \leq \|\mathbf{u}\|_E \leq C \|\mathbf{u}\|_V$$

with positive constants $c > 0$ and $C > 0$, equivalent to a SOBOLEV norm. The solution of elasticity problems with sufficient smooth boundaries lies in the space H^1. However, since engineers are more often interested in good stress approximations, it might be advantageous to use the L^2-norm of the stresses as an error measure.

Generally, the following asymptotic convergence statement can be made for an elliptical variational problem of order m, which is approximated by a conform finite element method with interpolations of order k, see e.g. Strang and Fix (1973),

$$\| \mathbf{u} - \mathbf{u}_h \|_{H^s} = O(h^{min\,[\,k+1-s\,,2(k+1-m)\,]}) . \tag{8.10}$$

Evaluation of this equation for different interpolations yields different convergence orders. As an example, a linear displacement ansatz $(k = 1)$ is used to discretize a linear elasticity problem $(m = 1)$ and a norm with $(s = 1)$ measures the error. The result is $O(h^{min\,[\,2-1\,,2(2-1)\,]}) = O(h)$; it is equivalent to the statements in (8.6). The general statement (8.10) can be applied to other elliptical problems (e.g. beams, plates and shells) or can be evaluated for interpolation functions of higher polynomial degree. It is obvious that interpolations with higher polynomial degrees yield faster convergence, e.g. a polynomial degree of $k = 3$ yields for the elasticity problem $m = 1$, $s = 1$ the convergence order $O(h^3)$. Numerical results which underline these theoretical statements for different orders of polynomials can be found, e.g. in Ramm et al. (2003), Hughes et al. (2005) and Elguedj et al. (2008).

These results are restricted to problems with smooth boundaries and continuous loading functions. A violation of these assumptions leads eventually to singularities in the solution and the convergence orders stated in (8.10) cannot be obtained. Examples are re-entrant corners in the boundary, like cracks, or point loads. In such cases, higher order polynomials do not enhance the solution since the solution is not smooth enough. Here an h- or r-adaption is preferable.

The mathematical analysis, leading to the results reported so far, is based on a deep knowledge of functional analysis. However, a treatment of this topic is beyond the scope of this book. The interested reader can find the related mathematical background in, e.g. Strang and Fix (1973), Johnson (1987), Verfürth (1996), Brenner and Scott (2002) and Braess (2007).

An adaptive finite element method leads to an approximate solution $\mathbf{u}_h$ which fulfils the inequality for a given error measure, e.g. a condition like

$$\| \mathbf{u} - \mathbf{u}_h \|_{H^1} \leq TOL . \tag{8.11}$$

Here TOL is a prescribed limit. In the case that the error has to be restricted to be, e.g. $\bar{\delta} = 5\,\%$ then instead of (8.11)

$$\delta = \frac{\| \mathbf{u} - \mathbf{u}_h \|_{H^1}}{\| \mathbf{u} \|_{H^1}} \times 100\% \tag{8.12}$$

can be used which is a relative error measure. Here the termination criterion for the adaptive computation is

$$\delta \leq \bar{\delta} \, . \tag{8.13}$$

In engineering applications, often the energy norm is employed for the computation of the relative error leading to

$$\delta = \frac{\| \mathbf{u} - \mathbf{u}_h \|_E}{\| \mathbf{u} \|_E} \times 100\% \leq \bar{\delta} \, . \tag{8.14}$$

The asymptotic results reported so far are not directly applicable for the computation of the relative error. But it is possible to estimate the error on the left side of (8.12) using an approximate solution. The error estimators and indicators needed for this task are discussed in the next sections. These were basically developed by assuming linear elastic behaviour and small deformations, see e.g. Braess (2007).

For nonlinear problems which are of interest within this monograph, only investigations are known which work in the tangent space (linearized problem). However, many demanding engineering tasks can be solved using adaptive schemes by using heuristic extensions of the error measures known from the linear theory which then are called error indicators. The construction of adaptive methods for such problems is the subject of ongoing research, see e.g. Verfürth (1996). In the following, error estimators and indicators will be defined which are based on quantities like deformations or stresses defined in the tangent space. The related quantities for a linear application are then simply given by evaluation of the terms with respect to the initial configuration.

Generally, the following steps have to be executed when applying an adaptive method within finite element techniques, see Box 8.1.

Box 8.1 Adaptive refinement process

1.	Select an appropriate initial mesh, which approximates the geometry of the problem accurately.
2.	Solve the discrete problem.
3.	Compute error estimators or error indicators.
4.	Test, whether the global error lies within the given tolerance.
	If yes, the computation is finished.
	If no, a new mesh has to be constructed.
	Within this process the already computed deformations and internal variables have to projected onto the new mesh.
	Then go to step 2.

The new mesh will usually include areas with refinement and areas with coarsening of the element sizes.

In the case of time dependent problems, distinction is made between a dynamical problem, see Sect. 6.1 and an inelastic application, see Sect. 6.2.

Both problem types require additional considerations which are related to accuracy of the approximation in time and, in case of inelasticity, the error made when history-dependent data are transferred from one mesh to the next, e.g. the plastic variables introduced in Sect. 6.2.2. Hence an adaptive time step control has to be introduced and an adequate projection scheme for the plastic variables has to be constructed, see e.g. Rannacher and Suttmeier (1998) and Johnson and Hansbo (1992).

8.2 Boundary Value Problems and Discretization

The error analysis, natural in numerical mathematics, and the resulting enhancement of the approximate solutions is in standard commercial software only realized in a rudimentary way. However, in practical applications, problems are extremely complex and thus the user of nonlinear finite element software has a hard time to judge the results of such analysis with respect to accuracy of the chosen discretization. With increasing computing power, this problem will be even more pronounced. Thus engineers require that they can compute solutions with controlled and required accuracy. Furthermore, the algorithms have to be reliable and robust which leads to certain demands with respect to the quality of finite element meshes and of nonlinear solvers, see e.g. Chap. 5.

The concept of mesh refinement due to *a posteriori* estimates will be first discussed by means of "linear" elasticity problems which can be formulated with respect to the tangent space of finite elasticity. With this, the methodology error estimation can be also applied to nonlinear problems.

The following methods for estimation and indication of errors can be distinguished in solid mechanics:

— residual error estimators, see e.g. Babuska and Rheinboldt (1978) and Johnson and Hansbo (1992),
— error indicators based on projection methods, see e.g. Zienkiewicz and Zhu (1987), or on the superconvergent patch recovery method, see Zienkiewicz and Zhu (1992),
— error estimation based on equilibrated stresses at element patches, see Ladeveze and Leguillon (1983), Ainsworth and Oden (1992) or Stein and Ohnimus (1996), as well as
— error estimation using dual methods, see e.g. Becker and Rannacher (1996), Rannacher and Suttmeier (1997a) and Ramm and Cirak (1997).

The application of some of these methods will be shown in the context of finite elasticity. Form that, the incremental or linearized boundary value problem of finite elasticity is formulated first.

8.2.1 Boundary Value Problem for Finite Elasticity

The equations which lead to the formulation of boundary value problems in solid mechanics, needed as a basis for the finite element method, were discussed in Chaps. 3 and 4. Especially in detail for the finite element method, the weak form of equilibrium is needed, see (3.292). Here it is formulated using the KIRCHHOFF stresses in the current configuration, see (3.296),

$$\int_\Omega \boldsymbol{\tau} \cdot \nabla_x^S \boldsymbol{\eta} \, dV = \int_\Omega \bar{\mathbf{f}} \cdot \boldsymbol{\eta} \, dV + \int_\Gamma \bar{\mathbf{t}} \cdot \boldsymbol{\eta} \, dA \ . \tag{8.15}$$

The integration is carried out with respect to the initial configuration Ω. The KIRCHHOFF stress tensor $\boldsymbol{\tau} = \mathbf{P}\,\mathbf{F}^T$ and the symmetrical gradient operator $\nabla_x^S(\bullet) = \frac{1}{2}[\nabla_x(\bullet) + \nabla_x^T(\bullet)]$ are computed in the current configuration, see (4.98). Short hand notation for Eq. (8.15) is provided by

$$R(\boldsymbol{\varphi}, \boldsymbol{\eta}) = g(\boldsymbol{\varphi}, \boldsymbol{\eta}) - \lambda f(\boldsymbol{\eta}) = 0 \ . \tag{8.16}$$

$\boldsymbol{\varphi}$ denotes the deformation and $\boldsymbol{\eta}$ denotes the associated variation. The load parameter λ is introduced to be able to scale the magnitude of the applied load. The quantities g and f in (8.16) are defined as

$$g(\boldsymbol{\varphi}, \boldsymbol{\eta}) = \int_\Omega \boldsymbol{\tau} \cdot \nabla_x^S \boldsymbol{\eta} \, dV \ ,$$

$$f(\boldsymbol{\eta}) = \int_\Omega \bar{\mathbf{f}} \cdot \boldsymbol{\eta} \, dV + \int_{\Gamma_\sigma} \bar{\mathbf{t}} \cdot \boldsymbol{\eta} \, dA \ . \tag{8.17}$$

Deformation dependent loads, as discussed in Exercise 3.12, are not considered.

8.2.2 The Linearized Boundary Value Problem

The linearization of the weak from (8.16) is derived with respect to the known deformation state $\bar{\boldsymbol{\varphi}}$

$$R(\bar{\boldsymbol{\varphi}} + \Delta\mathbf{u}, \boldsymbol{\eta}) = R(\bar{\boldsymbol{\varphi}}, \boldsymbol{\eta}) + Dg(\bar{\boldsymbol{\varphi}}, \boldsymbol{\eta}) \cdot \Delta\mathbf{u} + \cdots = 0 \tag{8.18}$$

with $g(\bar{\boldsymbol{\varphi}}, \boldsymbol{\eta})$ from (8.17). The explicit form of the linearization $Dg(\bar{\boldsymbol{\varphi}}, \boldsymbol{\eta})\cdot\Delta\mathbf{u}$ is computed via the directional derivative $g(\bar{\boldsymbol{\varphi}}, \boldsymbol{\eta})$, see Sect. 3.5. With (8.17)

$$Dg(\bar{\boldsymbol{\varphi}}, \boldsymbol{\eta}) \cdot \Delta\mathbf{u} = \int_\Omega \left(\nabla_{\bar{x}}^S \boldsymbol{\eta} \cdot \mathbb{C}_{\bar{x}} [\nabla_{\bar{x}}^S \Delta\mathbf{u}] + \overline{\mathrm{grad}\,\boldsymbol{\eta}} \cdot \overline{\mathrm{grad}\Delta\mathbf{u}\,\bar{\boldsymbol{\tau}}} \right) dV \tag{8.19}$$

can be written, see also (3.345). All derivatives in (8.19) have to be evaluated with respect to the known deformation state $\bar{\boldsymbol{\varphi}}$; this is denoted in (8.19) by

$\bar{x}$. The linearization of the stresses leads to the incremental material tensor $\mathbb{C}_{\bar{x}}$, see Sect. 3.3.4.

The evaluation of the incremental weak form (8.18) with respect to the undeformed initial configuration $\bar{\varphi} = \mathbf{X}$ yields the weak form of the linear theory. This can be written as a bi-linear form, see also (8.2),

$$a(\Delta \mathbf{u}, \boldsymbol{\eta}) = f(\boldsymbol{\eta}) \qquad (8.20)$$

with the definition

$$a(\Delta \mathbf{u}, \boldsymbol{\eta}) = Dg(\mathbf{X}, \boldsymbol{\eta}) \cdot \Delta \mathbf{u} = \int_{\Omega} [\nabla_X^S \boldsymbol{\eta} \cdot \mathbb{C}_X \nabla_X^S \Delta \mathbf{u}] \, dV. \qquad (8.21)$$

The constitutive tensor $\mathbb{C}_X$ is equivalent to the classical HOOKE material tensor when evaluated at the initial configuration, see (3.273). Furthermore, the gradient $\nabla_X^S \Delta \mathbf{u}$ coincides with the strain tensor $\boldsymbol{\epsilon}(\mathbf{u}) = \frac{1}{2}[\operatorname{grad} \mathbf{u} + \operatorname{grad}^T \mathbf{u}]$ of the linear theory of elasticity, where the displacement increment $\Delta \mathbf{u}$ in tangent space is equal to the displacement vector $\mathbf{u}$ of the linear theory.

8.2.3 Discretization

A discretization of the weak form (8.16) is necessary when the boundary value problem of finite elasticity has to be solved by the finite element method. For this, the region $\mathcal{B}$ is subdivided into T non-overlapping finite elements with radius h_T, see Chap. 4. This yields an ansatz space for the finite elements of the form

$$\mathbf{V}_h = \{\boldsymbol{\eta} \in \mathbf{V} \mid \boldsymbol{\eta} \in C(\Omega), \, \boldsymbol{\eta}|_T \in [P(T)]^{ndim}, \, \forall T\}, \qquad (8.22)$$

where the polynomials $P(T)$ of order p_T are defined on T. $ndim$ is the spatial dimension of the problem. Using this ansatz in (8.16) yields the discrete version

$$R(\boldsymbol{\varphi}_h, \boldsymbol{\eta}) = g(\boldsymbol{\varphi}_h, \boldsymbol{\eta}) - \lambda f(\boldsymbol{\eta}) = 0 \quad \forall \boldsymbol{\eta} \in V_h, \qquad (8.23)$$

which represents a nonlinear equation with respect to the deformation state. The solution $\boldsymbol{\varphi}_h \in V_h$ for this equation has to fulfil $R(\boldsymbol{\varphi}_h, \boldsymbol{\eta}) = 0$. With the matrix notation introduced in Sect. 4.2, Eq. (8.17) has the form

$$g(\boldsymbol{\varphi}_h, \boldsymbol{\eta}) = \bigcup_{e=1}^{n_e} \boldsymbol{\eta}_e^T \int_{\Omega_e} \mathbf{B}^T \boldsymbol{\tau}_h \, dV = \boldsymbol{\eta}^T \mathbf{G}(\mathbf{v}),$$

$$f(\boldsymbol{\eta}) = \bigcup_{e=1}^{n_e} \boldsymbol{\eta}_e^T \int_{\Omega_e} \mathbf{N}^T \bar{\mathbf{f}} \, dV + \bigcup_{e=1}^{n_\sigma} \boldsymbol{\eta}_e^T \int_{\Gamma_{\sigma e}} \mathbf{N}^T \bar{\mathbf{t}} \, dA = \boldsymbol{\eta}^T \mathbf{P}.$$

$\mathbf{N}$ contains the ansatz functions and $\mathbf{B}$ contains the associated gradients, which are here related to the current configuration, see also (4.100). For an arbitrary test function $\boldsymbol{\eta} \in V_h$ follows, instead of (8.23), equation

$$\boldsymbol{R}(\boldsymbol{v}) = \boldsymbol{G}(\boldsymbol{v}) - \lambda\,\boldsymbol{P} = \boldsymbol{0}\,, \qquad (8.24)$$

with the load parameter λ, see also (5.1). Equation (8.24) defines a nonlinear algebraic equation system which has to be solved for every load level $\lambda_{n+1} = \lambda_n + \Delta\lambda$. In most cases, NEWTON method is employed for this task, see Box 5.1 in Sect. 5.1.1. This yields an algorithm at each load level λ_{n+1} in which, for $k = 0, 1, \dots$ until convergence, the following set of equations has to be solved:

$$
\begin{aligned}
D\,\boldsymbol{G}(\boldsymbol{v}^k_{n+1})\,\Delta\boldsymbol{v}^k_{n+1} &= -\boldsymbol{R}(\boldsymbol{v}^k_{n+1})\,, \\
\boldsymbol{v}^{k+1}_{n+1} &= \boldsymbol{v}^k_{n+1} + \Delta\boldsymbol{v}^k_{n+1}\,.
\end{aligned}
\qquad (8.25)
$$

The linearization needed in the first equation is given by

$$D\,\boldsymbol{G}(\boldsymbol{v}^k_{n+1}) = \bigcup_{e=1}^{n_e} \int_{\Omega_e} (\,\boldsymbol{B}^T\,\mathbb{C}_{\bar{x}}\,\boldsymbol{B} + \boldsymbol{H}^T\,\bar{\boldsymbol{\tau}}\,\boldsymbol{H}\,)\,dV \qquad (8.26)$$

with the incremental material tensor $\mathbb{C}_{\bar{x}}$ and the discretization of the gradient $\boldsymbol{H}$, see also (4.76). All terms in the integral in (8.26) are related to the already computed displacement states $\boldsymbol{v}^k_{n+1}$.

8.3 Error Estimators and Error Indicators

The application of the GAUSS theorem to the incremental boundary value problem (8.19) yields an incremental operator

$$\boldsymbol{L}_{\bar{x}}(\mathbf{u}) = \mathrm{div}_{\bar{x}}\left(\mathbb{C}_{\bar{x}}\,[\nabla^S_{\bar{x}}\,\mathbf{u}] + \bar{\boldsymbol{\tau}}\,\nabla^S_{\bar{x}}\,\mathbf{u}\right), \qquad (8.27)$$

which belongs to the deformation state $\bar{\varphi}$. Using this operator, equation (8.19) can be formulated as

$$\boldsymbol{L}_{\bar{x}}(\mathbf{u}) = \Delta\lambda\,\mathbf{f}\,. \qquad (8.28)$$

The vector $\mathbf{f}$ denotes the volumetric loads. The term $\mathbf{u}$ represents the exact solution. In the linear case, the operator

$$\boldsymbol{L}_X(\mathbf{u}) = \mathrm{div}\left(\mathbb{C}_X\,[\nabla^S_X\,\mathbf{u}]\right) \qquad (8.29)$$

is obtained by inserting the linear elastic constitutive equation (3.273) in the local form of the momentum balance equation (3.65). This yields the linear boundary value problem

$$\boldsymbol{L}_X(\mathbf{u}) = \mathbf{f}\,. \qquad (8.30)$$

It is assumed that $\mathbf{u}$ is the exact solution of (8.28) or (8.30). $\mathbf{u}_h$ denotes the discrete finite element solution. The difference

$$\mathbf{e}_u = \mathbf{u} - \mathbf{u}_h \qquad (8.31)$$

defines then the error in the displacements. Analogously, an error can be formulated

$$\mathbf{e}_\tau = \boldsymbol{\tau} - \boldsymbol{\tau}_h, \tag{8.32}$$

which is related to the stresses. The task in the next sections is to quantify these errors.

8.3.1 Error Estimation for Nonlinear Problems

The error estimators, known for linear problems, have to be augmented appropriately to estimate errors which occur within the numerical simulation of nonlinear problems. To achieve this, the method of Rheinboldt (1985) is employed who computes the error estimators and indicators for the linearized problem at an equilibrium point, here denoted by $\bar{\varphi}$. The idea of the associated mathematical formulation is sketched in the following.

Let $\mathbf{G}$ be a nonlinear operator, see e.g. (8.16), which maps $\varphi \in V$ from the deformation space V onto the force space V^*

$$\mathbf{G}(\varphi) = \lambda \mathbf{P} . \tag{8.33}$$

λ is the scalar load parameter, see (8.24), and $\mathbf{P} \in V^*$ represents the applied load. In general, the vector valued function $\mathbf{G}$ is obtained from (8.16) by variation. By excluding that the equilibrium point under consideration is a singular point (limit- or bifurcation point), the inverse of the directional derivative of $\mathbf{G}$ exists and is for all φ bounded by $\| \varphi - \bar{\varphi} \| \leq \delta_2$. With these assumptions, the two inequalities hold

$$\| D\,\mathbf{G}^{-1}(\varphi) \| \leq C_1 \quad\text{and}\quad \| D^2\,\mathbf{G}(\varphi) \| \leq C_2 . \tag{8.34}$$

Rheinboldt (1985) derives, by using a series expansion of function $\mathbf{G}(\varphi)$ at the approximate solution φ_h, see Fig. 8.3, with the estimates (8.34), the result

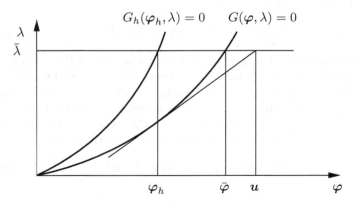

Fig. 8.3 Error estimation for a nonlinear problem

$$\|\mathbf{G}(\bar{\varphi}) - \mathbf{G}(\varphi_h) - D\mathbf{G}(\varphi_h)(\varphi_h - \bar{\varphi})\| \leq \frac{1}{2} C_2 \delta_2^2, \tag{8.35}$$

where $(\bar{\varphi}, \bar{\lambda})$ is the exact equilibrium point which describes the deformation state at which the linearization is computed. After some manipulations, the result

$$\|\bar{\varphi} - \varphi_h + D\mathbf{G}^{-1}(\varphi_h)(\mathbf{G}(\varphi_h) - \bar{\lambda}\mathbf{P})\| \leq \frac{1}{2} C_1 C_2 \delta_2^2 \tag{8.36}$$

follows with $(8.34)_1$. The introduction of

$$\mathbf{w} = \varphi_h - D\mathbf{G}^{-1}(\varphi_h)(\mathbf{G}(\varphi_h) - \bar{\lambda}\mathbf{P}) \tag{8.37}$$

yields the linear problem

$$D\mathbf{G}(\varphi_h)(\mathbf{w} - \varphi_h) = -\mathbf{G}(\varphi_h) + \bar{\lambda}\mathbf{P} \tag{8.38}$$

whose solution is $\mathbf{w}$. From (8.36), the inequality

$$\|\mathbf{w} - \bar{\varphi}\| \leq \frac{1}{2} C_1 C_2 \delta_2^2 \tag{8.39}$$

can be deduced which can be rewritten as

$$\|\bar{\varphi} - \varphi_h\|(1 + c) = \|\bar{\varphi} - \mathbf{u}_h\| \tag{8.40}$$

with $|c| \leq \frac{1}{2} C_1 C_2 \delta_2^2$. Using (8.40), the discretization error $\|\bar{\varphi} - \varphi_h\|$ of the nonlinear problem is expressed by the difference between φ_h and the solution of the linearized problem $\mathbf{u}$, in the case that c is sufficiently small, see also Fig. 8.3. Thus all error estimators and indicators, developed for the linear theory, can be applied for nonlinear problems provided the assumptions contained in (8.34) hold. Here it is necessary that the incremental problem is used which has to be formulated at the computed equilibrium point. The residuals or stresses needed within the error estimation follow from the solution of the linearized problem (8.18) at $(\bar{\varphi}_h, \bar{\lambda})$

$$D\,g(\bar{\varphi}_h, \bar{\lambda}) \cdot \mathbf{u}_h^\epsilon = -R(\bar{\varphi}_h, \bar{\lambda} + \epsilon\lambda). \tag{8.41}$$

$D\,g(\bar{\varphi}_h, \bar{\lambda})$ denotes the linearization which was defined in (8.19). ϵ is a parameter, for which $\epsilon \ll 1$ holds. Hence $\bar{\lambda} + \epsilon\lambda$ describes a perturbed loading state in the neighbourhood of $(\varphi_h, \bar{\lambda})$. The residuals and stresses, needed for the error estimation, are computed from the deformation state $\mathbf{u}_h^\epsilon$ belonging to the perturbed loading state. Instead of (8.18), the solution of the discretized problem (8.25) is used in case of the finite element method. This leads to

$$D\,\mathbf{G}(\bar{\mathbf{v}}, \bar{\lambda})\,\Delta\mathbf{v}_{n+1}^\epsilon = -\mathbf{R}(\bar{\mathbf{v}}, \bar{\lambda} + \epsilon\lambda). \tag{8.42}$$

8.3.2 Residual Based Error Estimator

The error in the displacement field $\mathbf{e}_u = \mathbf{u} - \mathbf{u}_h$ belonging to the incremental problem (8.42) can be expressed in a bi-linear form for the error equivalent to the energy norm

$$\gamma_i(\mathbf{e}_u) = a_i(\mathbf{e}_u, \mathbf{e}_u) = \int_{\Omega_{P_i}} \nabla_x^S(\mathbf{u} - \mathbf{u}_h) \cdot \mathbb{C}_x \left[\nabla_x^S(\mathbf{u} - \mathbf{u}_h) \right] d\,\Omega, \qquad (8.43)$$

where ∇_x^S is the symmetric part of the gradient operator. Ω_{P_i} denotes a patch which consists of a certain number of finite elements Ω_e. With Babuska and Rheinboldt (1978), the inequality

$$C_1 \sum_{i=1}^{M} \|\gamma_i(\mathbf{e}_u)\|_E^2 \leq \|\mathbf{e}_u\|_{H^1}^2 \leq C_2 \sum_{i=1}^{M} \|\gamma_i(\mathbf{e}_u)\|_E^2 \qquad (8.44)$$

holds for the error, when the sum over all patches is computed. For the boundary value problems defined in (8.27) and (8.29), the following computable estimation is valid

$$\|\gamma_i(\mathbf{e}_u)\|_E^2 \leq h_i^2 \int_{\Omega_{P_i}} [\mathbf{L}(\mathbf{u}_h) + \mathbf{f}]^T [\mathbf{L}(\mathbf{u}_h) + \mathbf{f}] \, d\,\Omega + h_i \int_{\partial\Omega_{P_i}} \mathbf{J}(\bar{\boldsymbol{\tau}}_h)^T \mathbf{J}(\bar{\boldsymbol{\tau}}_h) \, ds,$$

$$(8.45)$$

where $\mathbf{J}$ denotes the jumps of the stresses of the approximate solution at the patch boundaries. In the case of linear ansatz functions, the integrals over the region disappear. However, they can often also be neglected within the error estimation when interpolations of higher order are applied.

The methodology used so far is applied in Johnson and Hansbo (1992) differently where the error terms are determined directly from the element and not from the patch contributions. These authors write for linear elastic problems a residual-based error estimator in terms of stresses which consist of different parts related to volume and boundary terms. In more detail, the estimation is given by

$$\| \bar{\boldsymbol{\tau}} - \bar{\boldsymbol{\tau}}_h \|_{E^{-1}}^2 \leq \| h \, C_1 \, R_1(\bar{\boldsymbol{\tau}}_h) \|_{L^2(\Omega)}^2 + \| h \, C_2 \, R_2(\bar{\boldsymbol{\tau}}_h) \|_{L^2(\partial\Omega)}^2 \qquad (8.46)$$

with the following terms defined on a finite element

$$
\begin{aligned}
R_1(\bar{\boldsymbol{\tau}}_h) &= |R_1(\bar{\boldsymbol{\tau}}_h)| = |\operatorname{div} \bar{\boldsymbol{\tau}}_h + \mathbf{f}| && \text{in } T, \\
R_2(\bar{\boldsymbol{\tau}}_h) &= \max_{S \in \partial T} \sup_S \frac{1}{2\,h_T} |[\![\bar{\boldsymbol{\tau}}_h \, \mathbf{n}_S]\!]| && \text{on } \partial T, \\
\text{or } R_2(\bar{\boldsymbol{\tau}}_h) &= \frac{1}{h_T} (\bar{\mathbf{t}} - \bar{\boldsymbol{\tau}}_h \, \mathbf{n}) && \text{on } \partial T \cap \Gamma_\sigma .
\end{aligned} \qquad (8.47)
$$

Ω is the discretized region and $\partial\Omega$ its boundary. Furthermore, h_T is a characteristic element size, see Fig. 8.2. T denotes the finite element volume or area and ∂T its surface or boundary. The jump of the stress vector at two adjacent boundaries is described by the operator $[\![\mathbf{t}]\!] = \mathbf{t}_+ - \mathbf{t}_-$. The L^2 norm was already defined in (8.8); the norm $\|\cdot\|_{E^{-1}}$ in (8.46) is the complimentary energy norm (here written in stress space)

$$\|\bar{\tau} - \bar{\tau}_h\|^2_{E^{-1}} = \int_\Omega (\bar{\tau} - \bar{\tau}_h) \cdot \mathbb{C}^{-1}_{\bar{x}}[\bar{\tau} - \bar{\tau}_h]\,d\Omega \qquad (8.48)$$

with the inverse incremental elasticity tensor $\mathbb{C}^{-1}_{\bar{x}}$.

Equation (8.47) has to be evaluated on element basis within the computation. This leads to

$$\|\bar{\tau} - \bar{\tau}_h\|^2_{E^{-1}} \leq C \sum_T [E_T(h_T, \mathbf{u}_h, \mathbf{f}_T)]^2. \qquad (8.49)$$

The error related to an element E_T is computed for every element as

$$E^2_T = h^2_T \int_T |\,\mathrm{div}\bar{\tau}_h + \mathbf{f}|^2 d\Omega + h_T \int_{\partial T \cap \Omega} \frac{1}{2} |\,[\![\mathbf{t}_h]\!]\,|^2 d\Gamma +$$
$$h_T \int_{\partial T \cap \Gamma_\sigma} |\bar{\mathbf{t}} - \mathbf{t}_h|^2\,d\Gamma. \qquad (8.50)$$

The inequality (8.49) yields an upper bound for the error. It depends upon the element size and the deviation from the discrete solution. The first and third term on the right hand side describe the error in the local equilibrium and in the stress vectors at the boundary. Local equilibrium means that $[\![\mathbf{t}_h]\!] = \mathbf{0}$, see the second term. The jump in the stress vectors at the element boundaries is denoted by $[\![\mathbf{t}_h]\!]$.

A problem related to error estimators which are based on the evaluation of residuals is the determination of the constants C_i. If these constants cannot be estimated accurately then the bounds are not tight enough to control the finite element mesh refinement within an adaptive process. However, the distribution of the error within the finite element mesh can be estimated. When the value for, e.g. the constant C_1 for the volume error is too large compared to the other values C_i, then eventually the errors belonging to these other constants will be neglected while important for the computation of the total error. Often the constants are selected to have the value "1". Furthermore, estimates of the constants can be found in, e.g. Johnson and Hansbo (1992). These estimates are not always precise enough for general problems with arbitrary geometry. However, it is possible to estimate the constants from previous finite element computations within an adaptive mesh refinement procedure as described in Box 8.1 in Sect. 8.1.

8.3.3 Error Indicator Based on the Z^2 Method

Another possibility to determine errors of finite element computations starts directly from the complementary elastic energy (8.48). A simple, but in many case, efficient error indicator can be computed based on a projection method, see Zienkiewicz and Zhu (1987), or based on the so-called superconvergent patch recovery techniques described in Zienkiewicz and Zhu (1992). New investigations have shown that this way of computing finite element errors has a high effectiveness. This means that the error indicated by this methods is very close to the real error, see e.g. Babuska et al. (1994) and Carstensen and Funken (2001). Additionally, it was proven in Carstensen and Bartels (2002) and Bartels and Carstensen (2002) that averaging techniques yield reliable *a posteriori* error control even for unstructured grids.

Within a single finite element of the discretization there exist points at which the stresses have higher order of accuracy. This property is called super convergence, see e.g. Zienkiewicz and Taylor (1989). This fact can be illustrated by means of a simple example. Consider a truss element with linear interpolation. Such ansatz leads to a constant strain, and hence stress in the element which of course is not correct for general loading conditions. Interestingly enough, the stresses at the midpoint of the truss element are very close to the exact solution (and for some special cases actually identical with the exact solution), whereas all other points deviated more or less from the analytical solution. This means that these midpoints can be considered as super convergent points. The same observation holds for quads with bi-linear or triangular elements with linear interpolation. Here also the mid-point has this special feature of super convergence.

The stress values at these superconvergent points can now be used to construct an enhanced stress field which is continuous, see Zienkiewicz and Zhu (1992). This stress field is denoted by $\bar{\tau}^*$. Numerical verification of this method is provided in Babuska et al. (1994).

To formulate the described procedure in an abstract way, a projection operator $\mathbb{P}$ is introduced whose application yields the enhanced stress field $\bar{\tau}^*$

$$\int_{\Omega} \mathbb{P}\left[\bar{\tau}^* - \bar{\tau}_h\right] d\Omega = \mathbf{0} \,. \tag{8.51}$$

An efficient technique to compute this projection is based on a least squares minimum functional, see Zienkiewicz and Taylor (1989). Within this method, the enhanced continuous stresses $\bar{\tau}^*$ are computed from the stresses at the superconvergent points $\bar{\tau}_h$ via the least squares integral which minimizes the error

$$\int_{\Omega} \left[\bar{\tau}^* - \bar{\tau}_h\right]^2 d\Omega \to MIN \,. \tag{8.52}$$

Within this functional, an ansatz for the continuous stresses is made

$$\bar{\tau}^* = \sum_{I=1}^{n} N_I \,\hat{\tau}_I, \tag{8.53}$$

which then leads to an equation system for the computation of the nodal stresses $\hat{\tau}_I$. The minimum of (8.52) is obtained for

$$M_p \, \hat{\tau} = t_p \qquad (8.54)$$

with the matrices and vectors being defined as

$$M_p = \bigcup_{e=1}^{n_e} \sum_{I=1}^{n} \sum_{K=1}^{n} \int_{\Omega_e} N_I \, N_K \, I \, d\Omega,$$

$$t_p = \bigcup_{e=1}^{n_e} \sum_{I=1}^{n} \int_{\Omega_e} N_I \, \hat{\tau}_h \, d\Omega. \qquad (8.55)$$

The matrix M_p is, besides the missing density ρ_0, equivalent to the mass matrix defined in (4.58). The most efficient solution of this Eq. (8.54) is obtained by using a lumped matrix instead of the structure of M_p obtained by using the interpolation (8.53) since a lumped matrix leads to a diagonal form of M_p, see also Remark 4.2. However, this yields a different and thus only approximate result for the nodal stresses t_p.

There exist many different possibilities to compute the enhanced continuous stress field. One method which is based on the introduction of patches can be found in, e.g. Zienkiewicz and Zhu (1992).

The straightforward application of these methods does not yield the best results at boundaries or at interfaces at which material properties change. Enhanced methods which take boundaries into account were proposed, e.g. by Wiberg et al. (1994).

Based on the projected enhanced stress field (8.53) computed in (8.54), it is possible to indicate the finite element error with (8.48) by

$$\| \, \bar{\tau} - \bar{\tau}_h \, \|_{E^{-1}}^2 \leq \int_{\Omega} (\, \bar{\tau}^* - \bar{\tau}_h \,) \cdot \mathbb{C}^{-1} [\, \bar{\tau}^* - \bar{\tau}_h \,] \, d\Omega. \qquad (8.56)$$

The total error in (8.56) is now computed from the sum of the errors related to the elements T:

$$\| \, \bar{\tau} - \bar{\tau}_h \, \|_{E^{-1}}^2 = \| \, e_\tau \, \|_{E^{-1}}^2 \leq \sum_T \| \, e_\tau \, \|_T^2 \qquad (8.57)$$

with

$$\| \, e_\tau \, \|_T^2 = \int_T (\, \bar{\tau}^* - \bar{\tau}_h \,) \cdot \mathbb{C}^{-1} [\, \bar{\tau}^* - \bar{\tau}_h \,] \, d\Omega. \qquad (8.58)$$

It is obvious that a great advantage of this method lies in the fact that no constants have to be computed.

8.3.4 Error Estimators Based on Dual Methods

Strategies for the indication and estimation of errors within finite element solutions, so far, are based on global quantities like the L^2-norm of stresses or the energy norm. These lead to errors for the stress fields as discussed in the last two sections. In practical applications, however, the engineer wants eventually not to control global quantities but local quantities or other quantities of interest like a specific displacement or stress or more global but specific quantities like the J-integral in fracture mechanics or a drag coefficient in fluid mechanics. Based on this demand, methods which employ local error functionals were developed. Related mathematical analysis can be found in Becker and Rannacher (1996) and Rannacher and Suttmeier (1997b). Engineering applications with respect to shell problems were considered in Ramm and Cirak (1997) and for problems including contact in Wriggers et al. (2000), for an overview see also Ramm et al. (2003).

The error estimation is computed from another evaluation of the equation system related to the finite element problem and by application of known error estimators and indicators. In total, the combination of discretization error for the given problem and a related dual problem yields the desired local error quantity. The formal approach is discussed in the following sections.

Error Control of the Displacements. Starting point for the derivation of the local error estimation is, as for the computation of the residual error estimator, the differential equation (8.29) for the discretization error $\mathbf{e}_u = \mathbf{u} - \mathbf{u}_h$

$$\mathbf{L}_{\bar{x}}(\mathbf{u} - \mathbf{u}_h) = \mathbf{L}(\mathbf{e}_u) = \mathbf{f} - \mathbf{L}_{\bar{x}}(\mathbf{u}_h) = \mathbf{R}_1 . \tag{8.59}$$

$\mathbf{L}_{\bar{x}}$ denotes the differential operator of the incremental or linear problem. $\mathbf{R}_1$ is the residuum associated with the internal energy of the element. With the test function $\boldsymbol{\eta}$, the weak form of equilibrium is given by the bi-linear form

$$a(\mathbf{e}_u, \boldsymbol{\eta}) = \sum_T \left[\int_T (\operatorname{div} \bar{\boldsymbol{\tau}}_h + \mathbf{f}) \cdot \boldsymbol{\eta} \, d\Omega \right.$$
$$\left. + \int_{\partial T \cap \Gamma_\sigma} (\hat{\mathbf{t}} - \bar{\boldsymbol{\tau}}_h \mathbf{n}) \cdot \boldsymbol{\eta} \, d\Gamma + \int_{\partial T \cap \Omega} \frac{1}{2} [\![\mathbf{t}_h]\!] \cdot \boldsymbol{\eta} \, d\Gamma \right] . \tag{8.60}$$

The first integral represents the virtual work, which stresses and body forces perform along the virtual displacement. The second integral contains the virtual work of the jump related to the stress vectors $\bar{\boldsymbol{\tau}}_h \mathbf{n}$ along the boundaries at which stress vectors $\hat{\mathbf{t}}$ are inscribed. The third integral is associated with the virtual work of the stress vector jumps between adjacent finite elements within the mesh. The factor $1/2$ results from the fact that two finite elements have always only one common edge (2d) or surface (3d) within the mesh.

To simplify notation, all element related quantities are summarized in $\mathbf{R}_1$ while $\mathbf{R}_2$ contains all jump terms

$$a(\mathbf{e}_u, \boldsymbol{\eta}) = \sum_T \{(\mathbf{R}_1, \boldsymbol{\eta})_T + (\mathbf{R}_2, \boldsymbol{\eta})_{\Gamma_T}\} . \tag{8.61}$$

An additional dual problem is formulated to estimate the error of a certain displacement (e.g. the displacement component i at point $\mathbf{x} = \hat{\mathbf{x}}$)

$$\operatorname{div} \boldsymbol{\tau}(\mathbf{G}) + \boldsymbol{\delta}_i(\hat{\mathbf{x}}) = \mathbf{0} \,. \tag{8.62}$$

Its weak form is given by

$$a(\mathbf{G}, \boldsymbol{\eta}) = (\boldsymbol{\delta}_i, \boldsymbol{\eta}) \,. \tag{8.63}$$

$\boldsymbol{\delta}_i$ is the DIRAC delta vector in this dual problem pointing at $\hat{\mathbf{x}}$ in direction i. This is equivalent to applying a point load. $\mathbf{G}$ denotes the GREEN function associated to a point load at $\hat{\mathbf{x}}$ in direction i.

From the linear theory of elasticity, the theorem of BETTI-MAXWELL is known. Its application to the bi-linear form of the error (8.61) and to the dual problem (8.63) yields the relation

$$(\mathbf{e}_u, \boldsymbol{\delta}_i) = \sum_T \{(\mathbf{R}_1, \mathbf{G})_T + (\mathbf{R}_2, \mathbf{G})_{\Gamma_T}\} \,. \tag{8.64}$$

The term on the left hand side denotes the work of a point load along the error $\mathbf{e}_u$. Thus it is equal to the local error $e_i(\hat{\mathbf{x}})$ of the ith component of the displacement at point $\hat{\mathbf{x}}$. By inserting $\mathbf{G}$ in Eq. (8.61) instead of the test function $\boldsymbol{\eta}$, the local error is expressed by the bi-linear form

$$e_i(\hat{\mathbf{x}}) = a(\mathbf{e}_u, \mathbf{G}) \,. \tag{8.65}$$

The solution of the dual problem is not known. However, it can be determined numerically. For this, the same discretization and tangent matrix is used, however, with a different right hand side which is given by a point load in the direction of the ith displacement component. Hence only a further load case has to be computed using the same tangent matrix.

Considering further GALERKIN orthogonality (the error is orthogonal to the ansatz space: $a(\mathbf{e}_u, \mathbf{G}_h) = 0$, see e.g. (8.4) or Johnson 1987), the local error can be written, by introducing the finite element approximation of the dual problem $\mathbf{G}_h$, as

$$e_i(\hat{\mathbf{x}}) = a(\mathbf{e}_u, \mathbf{G} - \mathbf{G}_h) \,. \tag{8.66}$$

Application of CAUCHY-SCHWARZ inequality ($|(\mathbf{u}, \mathbf{v})| \leq \|\mathbf{u}\| \|\mathbf{v}\|$ with $(\mathbf{u}, \mathbf{v}) = \int_\Omega \mathbf{u} \cdot \mathbf{v} \, d\Omega$ and $\|\mathbf{u}\|^2 = \int_\Omega \mathbf{u} \cdot \mathbf{u} \, d\Omega$) yields an estimation for the local error. It follows from weighting of the error energy of the primal problem $a(\mathbf{e}_u, \mathbf{e}_u)$ (8.61) and the error energy of the dual problem $a(\mathbf{G} - \mathbf{G}_h, \mathbf{G} - \mathbf{G}_h)$ (8.63)

$$e_i^2(\hat{\mathbf{x}}) \leq a(\mathbf{e}_u, \mathbf{e}_u) \, a(\mathbf{G} - \mathbf{G}_h, \mathbf{G} - \mathbf{G}_h) \,. \tag{8.67}$$

The second term acts like a weighting function which filters the influence of the total error distribution related to the local displacement error. Inequality (8.67) can be computed element wise

$$e_i^2(\hat{\mathbf{x}}) \leq \sum_T a(\mathbf{e}_u, \mathbf{e}_u)_T \, a(\mathbf{G} - \mathbf{G}_h, \mathbf{G} - \mathbf{G}_h)_T \,. \tag{8.68}$$

The single terms can now be estimated using the methods of the last section. By employing the error indicator derived by Zienkiewicz and Zhu (1987), the terms in (8.68) can be approximately computed from

$$a(\mathbf{e}, \mathbf{e})_T = \int_{\Omega_T} (\boldsymbol{\tau}^*(\mathbf{u}_h) - \boldsymbol{\tau}(\mathbf{u}_h)) \cdot \mathbb{C}^{-1}[\boldsymbol{\tau}^*(\mathbf{u}_h) - \boldsymbol{\tau}(\mathbf{u}_h)] \, d\Omega \tag{8.69}$$

and

$$a(\mathbf{G} - \mathbf{G}_h, \mathbf{G} - \mathbf{G}_h)_T = \int_{\Omega_T} (\boldsymbol{\tau}^*(\mathbf{G}_h) - \boldsymbol{\tau}(\mathbf{G}_h)) \cdot \mathbb{C}^{-1}[\boldsymbol{\tau}^*(\mathbf{G}_h) - \boldsymbol{\tau}(\mathbf{G}_h)] \, d\Omega. \tag{8.70}$$

The stresses, necessary to evaluate (8.69) and (8.70), are computed from the associated linearized equation system (8.42) with the same argument as in the previous sections. This is mathematically not completely consistent but leads usually to good estimations. For a mathematically correct approach, see Rannacher and Suttmeier (1997b)). The stresses $\bar{\boldsymbol{\tau}}(\boldsymbol{\varphi}_h)$ and the associated enhanced stresses $\bar{\boldsymbol{\tau}}^*(\boldsymbol{\varphi}_h)$ follow from the perturbed loading state $\bar{\lambda} + \epsilon\lambda$, which yields a solution close to the equilibrium point $(\bar{\boldsymbol{\varphi}}_h, \bar{\lambda})$ and thus a solution of the linearized problem. The values $\boldsymbol{\tau}(\mathbf{G}_h)$ and $\boldsymbol{\tau}^*(\mathbf{G}_h)$ are determined in the same way. These follow from an increase of the displacement due to an additional point load $\boldsymbol{\delta}_i$.

Analogous to Eq. (8.12), the absolute error in (8.68) can be replaced by the relative error measure

$$\eta = \sqrt{\frac{e^2(\hat{\mathbf{x}})}{e^2(\hat{\mathbf{x}}) + u_h^2(\hat{\mathbf{x}})}} \,. \tag{8.71}$$

Computation of the Stress Error. Local error of the stresses at a given point $\hat{\mathbf{x}}$ can be estimated using the same approach as for the displacement at $\hat{\mathbf{x}}$. In this case, however, a discontinuity has to be prescribed for the related displacements of the dual problem. This leads to

$$\operatorname{div} \boldsymbol{\tau}(\mathbf{z}) + \frac{\partial}{\partial x_j} \boldsymbol{\delta}_i(\hat{\mathbf{x}}) = \mathbf{0} \,. \tag{8.72}$$

Again the application of the BETTI-MAXWELL theorem and the GALERKIN orthogonality together with the CAUCHY-SCHWARZ inequality yields the error in the displacement gradient or in the associated stress value

$$(\mathbf{e}_u, \frac{\partial}{\partial x_j} \boldsymbol{\delta}_i) = \frac{\partial e_i(\hat{\mathbf{x}})}{\partial x_j} = \sum_T \{(\mathbf{R}_1, \mathbf{z})_{\Omega_T} + (\mathbf{R}_2, \mathbf{z})_{\Gamma_T}\} = a(\mathbf{e}_u, \mathbf{z} - \mathbf{z}_h) \,. \tag{8.73}$$

The discontinuity of a displacement cannot be applied directly within a two- or three-dimensional problem. Here a regularization is necessary. The simplest approach is to replace the jump in the displacement component by a group of point loads being in equilibrium. This means in practice that two point loads of equal magnitude are attached to two neighbouring nodes in the finite element mesh which act in opposite direction. This pair of forces has to be near the point $\hat{\mathbf{x}}$ of displacement discontinuity.

8.4 Error Estimation for Plasticity

In the case of inelastic problems, additionally, an error in time has to be considered besides the discretization error in space. The error in time depends upon the selected integration algorithm, see Sect. 6.2. Error estimation in time is not well developed for inelastic constitutive equations; however some recent results can be found in, e.g. Rannacher and Suttmeier (1997b), Ladeveze (1998), Rannacher and Suttmeier (1999) and Ladeveze and Pelle (2005).

Here only the spatial discretization error is considered within a time step $\Delta t = [t_n , t_{n+1}]$. The applied methodology was proposed in Wriggers and Scherf (1995). Further error estimations for elasto-plastic problems are discussed in Bass and Oden (1987), Johnson and Hansbo (1992), Peric and Owen (1994), Fourment and Chenot (1995), Rannacher and Suttmeier (1998) and Perić et al. (1999).

To simplify the formulation and derivation of an error indicator for an elasto-plastic material, linear hardening is considered, see Sect. 6.2.2. There however is no limitation for the application of this error indicator to more complex elasto-plastic constitutive equations also.

In case of J^2 VON MISES plasticity with linear isotropic hardening, the plastic strains ε_n^p and the hardening variable $\hat{\alpha}_n$ are known at time t_n. The radial return or projection method described in 6.2.2 for the integration of the elasto-plastic material equations starts from the trial state of the deviator stresses and the hardening variable

$$
\begin{aligned}
\mathbf{s}_{n+1}^{tr} &= 2\,\mu\,(\,\mathbf{e}_{n+1} - \mathbf{e}_n^p\,)\,, \\
\hat{\alpha}_{n+1}^{tr} &= \hat{\alpha}_n\,.
\end{aligned}
$$

The projection onto the yield surface produces with $(6.100)_3$ and (6.101) the stresses and hardening variable at time t_{n+1}

$$
\begin{aligned}
\mathbf{s}_{n+1} &= \mathbf{s}_{n+1}^{tr} - 2\,\mu\,\Delta\gamma_{n+1}\,\mathbf{n}_{n+1}\,, \\
\hat{\alpha}_{n+1} &= \hat{\alpha}_{n+1}^{tr} + \sqrt{\tfrac{2}{3}}\,\Delta\gamma_{n+1}\,.
\end{aligned}
$$

The consistency parameter $\Delta\gamma_{n+1}$ can be stated for linear isotropic hardening explicitly, see also (6.110),

$$\Delta \gamma_{n+1} = \frac{f_{n+1}^{tr}}{2\,\mu + \frac{2}{3}\,\hat{H}}\,, \tag{8.74}$$

where f_{n+1}^{tr} is the yield condition evaluated using the trial quantities

$$f_{n+1}^{tr} = \|\,\mathbf{s}_{n+1}^{tr}\,\| - \sqrt{\frac{2}{3}}\,(Y_0 + \hat{H}\,\alpha_{n+1}^{tr})\,. \tag{8.75}$$

These relations can be solved for the total strains ε_{n+1} at time t_{n+1} by considering the elastic constitutive equations for isotropic material

$$\varepsilon_{n+1} - \mathbf{e}_n^p = \frac{1}{2\mu}\mathbf{s}_{n+1} + \frac{1}{9K}\mathrm{tr}\,\bar{\tau}_{n+1}\mathbf{1} + \frac{3}{2\hat{H}}\,[\mathbf{s}_{n+1} - \Pi(\mathbf{s}_{n+1})]\,. \tag{8.76}$$

K is the modulus of compression and μ the shear modulus of the elastic constitutive equation. Here

$$\Pi(\mathbf{s}_{n+1}) = \begin{cases} \mathbf{s}_{n+1} : & \text{for an elastic step} \\[2mm] \dfrac{\|\,\mathbf{s}_n\,\|}{\|\,\mathbf{s}_{n+1}\,\|}\,\mathbf{s}_{n+1} : & \text{for a plastic step} \end{cases}$$

is a projection which describes the increase of the plastic strains in (8.76). With these relations, the error in the strain field can be determined within a time step Δt

$$\begin{aligned} (\varepsilon - \mathbf{e}_n^p) - (\varepsilon_h - \mathbf{e}_{h\,n}^p) &= \frac{1}{2\mu}\,(\mathbf{s} - \mathbf{s}_h) + \frac{1}{9K}\,\mathrm{tr}\,(\bar{\tau} - \bar{\tau}_h)\,\mathbf{1} \\ &\quad + \frac{3}{2\hat{H}}\,[\mathbf{s} - \Pi(\mathbf{s}) - (\mathbf{s}_h - \Pi(\mathbf{s}_h))]\,. \end{aligned} \tag{8.77}$$

The multiplication by $(\bar{\tau} - \bar{\tau}_h)$ and the integration over the region Ω yields by using the monotony relation $[\Pi(\mathbf{q}) - \Pi(\mathbf{p})]\cdot(\mathbf{q} - \mathbf{p}) \geq 0$ (which relates to the dissipation inequality (3.188))

$$\|\,\bar{\tau} - \bar{\tau}_h\,\|_{E^{-1}}^2 \leq \int_\Omega [(\varepsilon - \varepsilon_h) - (\mathbf{e}_n^p - \mathbf{e}_{h\,n}^p)]\cdot(\bar{\tau} - \bar{\tau}_h)\,d\Omega\,. \tag{8.78}$$

This relation can be employed to estimate the error within a time step. By splitting the strains in (8.78) into an elastic and incremental plastic part,

$$\varepsilon_{n+1} = \varepsilon_{n+1}^e + \varepsilon_{n+1}^p \implies \varepsilon_{n+1} - \mathbf{e}_n^p = \varepsilon_{n+1}^e + \Delta\mathbf{e}_{n+1}^p \tag{8.79}$$

can be deduced. The error is then

$$\begin{aligned} \|\,\bar{\tau} - \bar{\tau}_h\,\|_{E^{-1}}^2 &\leq \int_\Omega (\Delta\mathbf{e}^p - \Delta\mathbf{e}_h^p)\cdot(\bar{\tau} - \bar{\tau}_h)\,d\Omega \\ &\quad + \int_\Omega (\varepsilon^e - \varepsilon_h^e)\cdot(\bar{\tau} - \bar{\tau}_h)]\,d\Omega\,. \end{aligned} \tag{8.80}$$

Improved strains and stresses can be computed based on the methods developed in Sect. 8.3.3. By inserting these values in (8.80), the error can be determined for every finite element

$$
\begin{aligned}
(E_T^{ep})^2 = (\|\,\mathbf{e}_\tau\,\|_T^{ep})^2 \;\approx& \int_T (\,\Delta\,\mathbf{e}^{*\,p} - \Delta\,\mathbf{e}_h^p\,)\cdot(\bar{\boldsymbol{\tau}}^* - \bar{\boldsymbol{\tau}}_h)\,\mathrm{d}\Omega \\
&+ \int_T (\,\boldsymbol{\varepsilon}^{*\,e} - \boldsymbol{\varepsilon}_h^e\,)\cdot(\bar{\boldsymbol{\tau}}^* - \bar{\boldsymbol{\tau}}_h)\,\mathrm{d}\Omega.
\end{aligned}
\tag{8.81}
$$

Hence this method is an augmentation of the methods described in Sect. 8.3.3 to elasto-plastic problems. It can easily be implemented and yields good results in practical applications, see the example in Sect. 8.7.2 or in Han (1999).

8.5 Mesh Refinement

The adaptive refinement of a mesh defines mathematically an optimization problem since the goal is to find a mesh which yields the best finite element approximation. Mathematically the problem can be stated as: construct a finite element mesh such that the solution of the inequality

$$
\|\,\boldsymbol{\tau} - \boldsymbol{\tau}_h\,\|_{E^{-1}} \le \sum_T [E_T(h_T, \mathbf{u}_h, \bar{\mathbf{f}}_T)]^2 \le TOL
\tag{8.82}
$$

is fulfilled. Here TOL is a given tolerance. The given constraint is that the costs to compute $\mathbf{u}_h$ or $\boldsymbol{\tau}_h$ which fulfil (8.82) are minimal. The element error E_T in (8.82) can be computed either from

$$
\begin{aligned}
E_{T1}^2 &= E_T^2 \;,\; \text{see Eq. (8.50), or from} \\[4pt]
E_{T2}^2 &= \|\,\mathbf{e}_\tau\,\|_T^2 \;,\; \text{see Eq. (8.58), or from} \\[4pt]
E_{T3}^2 &= e_i^2 \;,\; \text{see Eq. (8.68), or from} \\[4pt]
E_{T4}^2 &= (E_T^{ep})^2 \;,\; \text{see Eq. (8.81)}.
\end{aligned}
\tag{8.83}
$$

One measure for the numerical effort needed to solve a finite element problem is the maximum number of unknown, since it is related directly to the computing time. Of course other measures would be possible too, but here the first criterion will be employed.

Since the exact solution of the partial differential equation is not known, it is additionally required that the error should be equally distributed over the mesh and hence be equal in all the elements.

With (8.82), the inequality

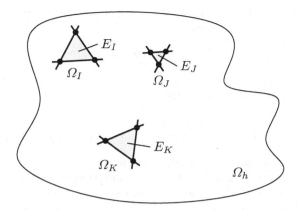

Fig. 8.4 Equally distributed error with a finite element mesh

$$\sum_T E_T^2 \leq TOL \,, \qquad (8.84)$$

has to be fulfilled. This inequality provides also a criterion for the termination of the adaptive algorithm.

With the requirement that the error is equal within all finite elements of a mesh and fulfils (8.84), an optimal mesh is created. For the finite elements I, J and K, the relation

$$E_I = E_J = E_K \qquad (8.85)$$

is valid, see Fig. 8.4. The total error can be written under this assumption as

$$\sum_T E_T^2 = n_e \, E_T^2 \,. \qquad (8.86)$$

Here n_e is the number of finite elements in the mesh. Equation (8.84) leads together with the last equation to a criterion which indicates when a finite element has to be refined

$$E_T^2 \leq \frac{TOL}{n_e} \,. \qquad (8.87)$$

Another possibility to determine which elements have to be refined can be found in Zienkiewicz and Taylor (1989). The authors start from a relative error δ, see (8.14), with the goal that it has to be less than a given tolerance $\bar{\delta}$. As in (8.87), the error will be distributed equally over the mesh. With the notation $\| \mathbf{e} \|_e$ for the error within a single finite element condition,

$$\| \mathbf{e} \|_e \leq \bar{\delta} \sqrt{\frac{\| \mathbf{e}_\tau \|^2 + \| \boldsymbol{\tau}_h \|^2}{n_e}} = \bar{e}_{n_e} \qquad (8.88)$$

is obtained. $\bar{e}_{n_e}$ denotes the maximum error related to a finite element Ω_e. The error in inequality (8.88) is computed from the stresses. However, this inequality can also be formulated with respect to displacements and strains. By defining the dimensionless factor

$$\beta_e = \frac{\| \mathbf{e}_\tau \|_e}{\bar{e}_{n_e}}, \tag{8.89}$$

the refinement criterion can be written as

$$\beta_e \begin{cases} > 1 & \text{refinement,} \\ \leq 1 & \text{no refinement or mesh coarsening.} \end{cases} \tag{8.90}$$

Different strategies can be employed to refine or coarsen a finite element mesh. One possibility is to subdivide an element into two, once the error measure β_e indicates refinement. This leads to a so-called hierarchical refinement. This procedure, however, can lead to a large number of refinement steps in order to obtain a solution which fulfils the tolerance $\bar{\delta}$. A positive feature of this method is related to the fact that the hierarchical structure can be directly used within a pre-conditioning method for an iterative equation solver, see also Sect. 5.2.2.

A method which often leads to faster convergence with less adaptive remeshing is based on the introduction of a density function. This function is used to construct a new FE-mesh Ω^h_{n+1}. The general idea is to start from Eq. (8.10) which states that the error within a finite element is proportional to $O(h^{k+1-s})$. This proportionality can be used to specify the new element size of the adapted mesh as

$$h_{e\,n-1} = \beta_e^{-\frac{1}{k+1-s}}\, h_{e\,n}. \tag{8.91}$$

In this expression, value k denotes the complete polynomial order of the finite element ansatz and s is the norm which measures the error. The error norm for the stresses in, e.g. (8.88) is then related to $s = 1$.

Based on the above considerations, the general algorithm can be stated for an h-adaptive method. Based on Eq. (8.87), the following steps have to be performed.

Generate the starting mesh: $\mathcal{M}_i$, set $i = 0$
1. Loop over all load steps: $t = k\,\Delta t$, $k = 1, \ldots$
2. Iteration for the solution of the problem using (8.25)
3. Optimization of the mesh
   − Compute E_T^2
   − IF $\sum E_T^2 < TOL \Longrightarrow k = k + 1$, GO TO 1
   − IF $E_T^2 > TOL\,/\,N \Longrightarrow$ refine element T
   − Set $i = i + 1$
   − Generate new mesh $\mathcal{M}_i$
     • Transfer the history data, if necessary
     • Mesh smoothing, if necessary
   − GO TO 2

Within the algorithm an arbitrary method can be applied to generate the new mesh. This is also valid for the transfer of the history data which has to be performed when inelastic material behaviour has to be considered. The additional error which stems from this transfer will be discussed in Sect. 8.6.2. Mesh smoothing is always used when the geometry of the finite elements deteriorates within the refinement process, e.g. when the inner angles of elements become too small which would lead to a degradation of the convergence rates. Details of related implementations can be found in Scherf (1997) and Han (1999). The algorithms can be applied in a similar way when the relative error measure (8.89) is used.

8.6 Adaptive Mesh Generation

A large number of algorithms for automatic mesh generation was developed within the last years. Hence different approaches exist for two- and three-dimensional meshes consisting of triangles, quadrilaterals, tetrahedra or hexahedra. Here some of the most popular methods are described.

Within adaptive procedures refined and coarsened meshes have to be constructed automatically; hence robust meshing tools are needed which yield new meshes with good sized elements. Furthermore the case, that history data have to be transferred from one mesh to the other, has to be considered. The same holds for deformation states when finite deformation processes are simulated by an adaptive scheme.

8.6.1 Mesh Generation

Adaptive finite element simulations require automatic schemes for the generation of new finite element meshes. The region under consideration is discretized by triangles or quadrilateral in a two-dimensional problem. In the three-dimensional case, the elements are either tetrahedra or hexahedra. Additionally, a two-dimensional surface has to be discretized in three-dimensional space when a shell problem is investigated. The form and distribution of finite elements is ideally generated automatically by the algorithm. These algorithms base on a geometry description which is mostly provided by a CAD model, e.g. by BEZIER, *NURBS* or other smooth functions. In order to obtain a convergent solution for such a geometry model, the mesh refinement algorithm has to use the geometry model defined by the CAD system, for special issues related to this problem, see e.g. Yagawa et al. (1995) and Ribó et al. (2002).

Basically, algorithms for the generation of structured and unstructured mesh have to distinguished. Within adaptive finite element refinements, the latter algorithms can be applied in a more flexible way, and hence these algorithms will be discussed in more detail. Meshing algorithms for triangles and tetrahedra are different from the ones needed to generate quadrilateral or hexahedral element meshes. Hence both type of algorithms will be discussed separately.

Mesh algorithms for triangles and tetrahedrons are constructed by using different methods. Among them are:

- The *octree* technique places a mesh of cells on a surface or in a volume of a body. At the boundaries, the mesh has to be adapted. This is done by a recursive subdivision of the cells until the boundary of the geometrical is sufficiently approximated, see e.g. Shepard and Georges (1991).
- The DELAUNAY method relies on a good placement of coordinate points within the area to be meshed. These control the density of the mesh. Several algorithms are available to generate these points. The DELAUNAY method is then applied to create a triangularization by triangles or tetrahedrons. Algorithms for two-dimensional meshing can be found, e.g. in Sloan (1987a) and Sloan (1993).
- The *advancing front* method starts the triangularization at the boundary of the area to be meshed and introduces there a layer of finite elements which defines the boundary for the next layer of elements. Here overlapping has to be avoided when the gap between the fronts closes. Related algorithms are described, e.g. in Löhner (1996).
- The algorithm of *recursive region splitting* places nodes on the boundary of the region to be meshed by considering a certain density distribution. After that, starting from the boundary, the area to be meshed is subdivided into smaller areas recursively. This yields a region which is in the end subdivided in triangles and quadrilaterals. To obtain a mesh consisting purely out of triangles, the quadrilaterals are then simply divided into two triangles using the shortest diagonal, see e.g. Bank (1990).

Meshing algorithms for quadrilaterals or hexahedrals base either on indirect methods which start from a triangularization using triangles or tetrahedra or methods which generate quadrilateral or hexahedral meshes directly.

- An indirect method is based on the fact that four quadrilaterals can be generated from two triangles. Furthermore, single triangles can be subdivided into three quadrilaterals. The combination of both leads to the construction of a quadrilateral mesh, see e.g. Rank et al. (1993). In the same way, one can proceed in three dimensions. However, so far these methods do not lead to satisfactory results.
- Direct methods use, in the two-dimensional case, either the *advancing front* technique, see e.g. Zhu et al. (1991), or rely on algorithms, which subdivide the area into simpler regions which can easily be meshed by quadrilaterals, see e.g. Joe (1995).
- Other algorithms, for the generation of three-dimensional hexahedral meshes, use medial surfaces together with a so called plastering or directly adjust meshes. An overview can be found in Owen (1999) for related techniques.

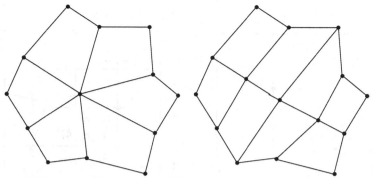

Fig. 8.5 Mesh smoothing for quadrilaterals

Shell problems with arbitrary free form surfaces are located between two- and three-dimensional mesh generation since a mesh has to be constructed on a two-dimensional surface in three-dimensional space. Associated algorithms are often based on a two-dimensional generation which then is mapped from a plane surface to the free form surface in space. However, often this mapping is not valid globally and hence subregions have to be mapped and tied together. Furthermore, the elements generated on a plane surface which are then projected onto the free form surface can be distorted by the mapping. Possible solutions are discussed in, e.g. Rehle (1996).

The algorithms described above are first applied to generate the starting mesh within an adaptive calculation. During the adaptive mesh refinement steps, the mesh is altered either by computing a density function which leads to a complete remeshing of the problem or a subdivision of existing elements related to the error measures. A mesh smoothing procedure has to be applied often in the last case in order to avoid elements which are highly distorted, which then effects the quality of the solution, see also Sect. 8.6. Mesh smoothing can be performed using different algorithms. A very simple one minimizes the number of elements which are connected to a node, see e.g. Fig. 8.5. Within this procedure, new nodes are inserted, see e.g. Han (1999). This procedure is only applied to elements which are too distorted; it results to slightly more elements. Based on the better aspect ratios of the elements, the constant in the error measure (8.6) is reduced, and hence better results are obtained. Another possibility to enhance meshes is related to an elimination of nodes, associated strategies can be found, in e.g. Zhu et al. (1991). Algorithms which smooth meshes for shell discretizations were developed in Riccius et al. (1997).

8.6.2 Transfer of History Variables

Evolution equations which occur in the modelling of inelastic problems include often history variables, see e.g. Sect. 3.3.2. These history variables have

Mesh i   *Close up: Mesh i+1*

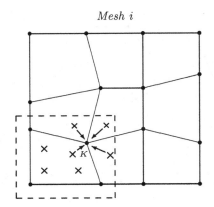 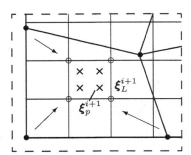

Fig. 8.6 Transfer of history variables

to be transferred from one mesh to the next within an adaptive refinement scheme. Different algorithms can be applied for the transfer of these variables, see e.g. Ortiz and Quigley (1991) and Peric et al. (1996). The following algorithm is often applied. It is presented next and partly depicted in Fig. 8.6.

1. L^2-projection of history variables $\boldsymbol{\alpha}$, which are related to GAUSS-points, onto the nodes within mesh $i \Rightarrow \boldsymbol{\alpha}_K^i$. This is performed by the methods described in Sect. 8.3.3, see left part of Fig. 8.6.
2. Interpolation of the data within mesh Ω_e^i by isoparametric shape functions:

$$\boldsymbol{\alpha}^i = \sum_{K=1}^{n} N_K(\boldsymbol{\xi}^i)\, \boldsymbol{\alpha}_K^i\,.$$

3. Search for point $\boldsymbol{\xi}_L^{i+1}(\Omega_e^{i+1})$ in the previous mesh Ω_e^i. For this, the closest point and its associated elements have to be located in the previous mesh. This procedure leads to a local nonlinear equation system since the inverse of the nonlinear isoparametric mapping is needed.
4. Evaluation of the interpolation at the nodes $\boldsymbol{\xi}_L^{i+1}$ of the new mesh Ω_e^{i+1}:

$$\boldsymbol{\alpha}_L^{i+1} = \sum_{K=1}^{n} N_K(\boldsymbol{\xi}_L^{i+1})\, \boldsymbol{\alpha}_K^i\,.$$

5. The isoparametric mapping defines the history variables at the GAUSS points $\boldsymbol{\xi}_p^{i+1}$ of the new mesh:

$$\boldsymbol{\alpha}_p^{i+1} = \sum_{L=1}^{n} N_K(\boldsymbol{\xi}_p^{i+1})\, \boldsymbol{\alpha}_L^{i+1}\,.$$

After executing these steps, all history data are transferred to the new mesh. Using these data, equilibrium has to be fulfilled before the next load increment can be applied since the global equilibrium is violated by the above procedure. The associated error in the equilibrium is provided with (4.54) by

$$\boldsymbol{G}^{i+1} = \bigcup_{e=1}^{n_e} \sum_{I=1}^{n} \boldsymbol{\eta}_I^T \int\limits_{(\Omega_e)} [\, \mathbf{B}_I^T \, \mathbf{S}_e(\boldsymbol{\alpha}^{i+1}) - N_I \, \mathbf{p} \,] \, d\Omega \neq 0 \,. \tag{8.92}$$

The aberration from equilibrium can be eliminated by an iteration before the next load step is executed. This iteration can be computationally intensive, see e.g. Han (1999). The underlying strategy is named strategy I throughout the remaining part of this chapter. An alternative method includes the residuum $\boldsymbol{G}^{i+1}$ after the transfer of the history data directly within the next load step. This however yields often a large deviation from the equilibrium such that NEWTON method will often not converge in the next load step, even when a line search is applied, see Sect. 5.1.4).

There exists no explicit error measure for the transfer error such that it is not clear beforehand how big this error is. Exemplary simulations show however that the transfer error is not negligible, see e.g. Habraken and Cescotto (1990). Since variables have to be transferred which represent the history of the material at a GAUSS point and hence are necessary for a successful analysis, this error has to be controlled.

A simple way to control the error is to avoid the transfer of history variable completely; however also this strategies has disadvantages. It will be named strategy II throughout the remaining chapter. The idea is to compute, with a given starting mesh, the complete nonlinear response of a system. Within the individual load step, the discretization error is computed and its distribution is used to determine the distribution of the element sizes for that load step, based on the methods described before. At the end of the nonlinear computation, the information gained within all different load steps will be added up and used to construct a new mesh. Technically, the density functions of all load steps are overlaid in order to determine the location and size of the refinement. After that, the computation will be repeated for all load steps starting from the very first load step. This adaptive simulation is terminated once the prescribed tolerance ($\delta \leq \bar{\delta}$) is fulfilled. Strategy II has following advantages:

- No transfer of history variables is needed. Thus the associated error is eliminated.
- Singular points are naturally included in this type of adaptive analysis. This is not the case by the first strategy since a singular point can be distinctly different for refined meshes.
- The code development is a lot easier since the necessary search processes needed for the transfer of the history variables are omitted.

– Complex element formulations as the *enhanced strain* elements need additional transfer of internal variables. This is not required when the nonlinear computation is repeated using a new mesh.

Naturally strategy II also has disadvantages:

– Due to repeated computations of the nonlinear problem using different meshes, a high computational effort is required.
– The mesh generated by this adaptive process does not lead to optimal meshes for all load steps.
– Finite strain problems with large distortion of the elements cannot be handled by this strategy. Here a remeshing of the deformed geometry is necessary at certain load steps.

This discussion shows that the best strategy depends upon the particular problem. As example forming processes have to be simulated using strategy I with remeshing, see e.g. Ortiz and Quigley (1991) and Fourment and Chenot (1995). However, strategy II proves to be successful for the elasto-plastic analysis of shell structures, see Han (1999).

8.7 Examples

The different error estimators and indicators are compared in this section by means of two examples depicting different nonlinearities. All simulations were performed using an extended version of the finite element program FEAP, see Zienkiewicz and Taylor (1989). The finite element meshes were generated based on algorithms developed in Bank (1990) and Rank et al. (1993). The mesh generation for triangular elements is carried out by using the algorithm provided in Sloan (1987a) and Sloan (1993). Quadrilateral element meshes were created by a complete remeshing based on density functions as discussed in Rank et al. (1993).

8.7.1 Hertzian Contact Problem

The first example is a HERTZ contact problem of a cylinder and a rigid surface. The elastic cylinder has a YOUNG modulus of $E = 7000$ and a POISSON ratio of $\nu = 0.3$. The rigid surface is modelled by an elastic material with high stiffness ($E = 100000$ and $\nu = 0.45$). The cylinder has a radius of $r = 1$ and is loaded at its upper part by a distributed load with a resulting force of $F = 100$. Due to symmetry, only one half of the problem is discretized.

For the HERTZ contact problem, there exist analytical solutions. Hence the computations of the adaptive methods can directly be compared with the analytical solution. The analytical solution, however, contains approximations regarding the geometry, see e.g. Szabó (1977), but these do not influence the results for small deformations. The maximum contact pressure between the cylinder of radius r and a rigid plate is given for plane strain conditions by

$$p_{max} = \sqrt{\frac{F}{\pi\,r}\frac{E}{(1+\nu)(1-\nu)}}\,.$$

This value will be used to discuss the quality of the different error measures applied within a finite element analysis.

The three different error measures are used in this example to control the adaptive computation and to obtain a converged result. One measure is the residual based error estimator of Johnson and Hansbo (1992) which was enhanced for contact in Carstensen et al. (1999). The Z^2 error indicator presented in Zienkiewicz and Zhu (1987) is also applied, as well as the dual error estimator, for local quantities derived by Rannacher and Suttmeier (1997b).

All error estimators and indicators were described in this chapter. They have to be adapted to contact problems, see Wriggers et al. (2000), which however is not essential for the general behaviour of the different error measures and the discussion of the results. The maximum contact pressure was selected at the contact interface as a goal quantity for the local error estimation. Then, with respect to (8.73), a group of two forces, being in equilibrium, was applied at the contact interface. These yield the appropriate displacement jump needed for the computation of the error within the dual method.

The starting mesh which only consist of 258 finite elements is depicted in Fig. 8.7.

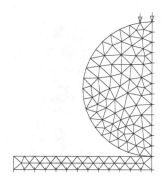

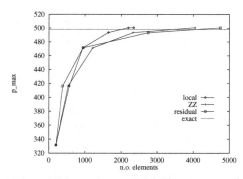

Fig. 8.7 Initial mesh: 258 el   **Fig. 8.8** Convergence behaviour

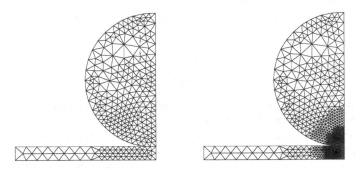

Fig. 8.9 Mesh refinement: dual error estimator

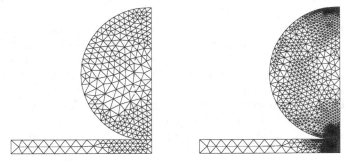

Fig. 8.10 Mesh refinement: Z^2 error indicator

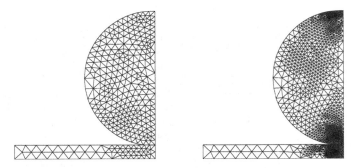

Fig. 8.11 Mesh refinement: residual based error estimator

Figures 8.9–8.11 show meshes which were generated by the adaptive method based on the different error measures. The last mesh, on the right hand side, is related to the converged solution.

The maximum value of the contact pressure is compared in Figure 8.8 with the analytical solution $p_{max} = 494,83$. It can be seen clearly that the dual error estimator, whose target was to produce the best mesh for the contact pressure, does only need half of the number of finite elements to converge. The associated mesh depicts that the mesh is only refined within the area of interest close to the contact surface, stemming from the superposition of primal and dual solution.

Contrary to that, the residual method and the Z^2 error indicator refine the mesh where gradients in the stress field occur. Hence also the area beneath the applied load is refined as well. Both methods control the error within the entire mesh.

Hence the dual or local estimation is more efficient. This, however, this is only true when only one quantity is of interest for the design engineer. This could be a displacement, a stress as in this example, but also an integral measure like the J-integral or a total load. Within complex structures, however, it is not always clear from the beginning where maximal values, like e.g. stresses, occur. These values, however, are often essential for the design of a structure. Thus an error measure - as provided by the residual method or the Z^2 indicator - has to be applied which includes all quantities and then leads to refined meshes where the error is globally limited by the prescribed tolerance $\bar{\delta}$.

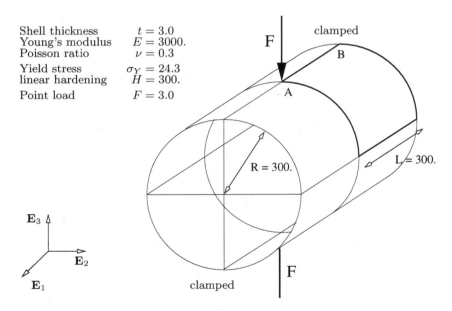

Shell thickness	$t = 3.0$
Young's modulus	$E = 3000.$
Poisson ratio	$\nu = 0.3$
Yield stress	$\sigma_Y = 24.3$
linear hardening	$H = 300.$
Point load	$F = 3.0$

Fig. 8.12 Cylinder under point load

8.7.2 Elasto-Plastic Deformation of a Cylindrical Shell

The second example describes the application of adaptive techniques to inelastic deformations. A cylindrical shell will be subjected to a point load.
It consists of an elasto-plastic material and will undergo large deflections
and rotations. This example is selected to depict the difference between the
approaches for the transfer of history data as described in Sect. 8.6.2.

Geometry, boundary conditions and constitutive data can be found in
Fig. 8.12, see also Eberlein (1997) and Wriggers et al. (1996). The shell theory
applied in this example is discussed in Sect. 9.4.

The computation is performed using a displacement driven approach.
Hence a displacement is prescribed at point A which relates to the location of the point load, see Fig. 8.12. Symmetry of the shell geometry and
loading allows to only discretize one eighth of the cylindrical shell.

The prescribed displacement at point A is incrementally increased up to
the value of 120 which is 40% of the cylinder radius and thus relates to a
large deflection of the shell. Due to the nonlinearity of the problem, the step
size has to be changed throughout the computation. Thus, for the first 20

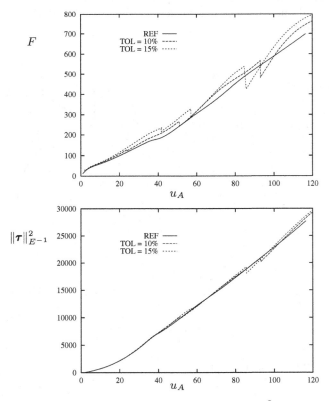

Fig. 8.13 Strategy I: Load–deflection curve and energy norm $\|\boldsymbol{\tau}\|_{E-1}^{2}$.

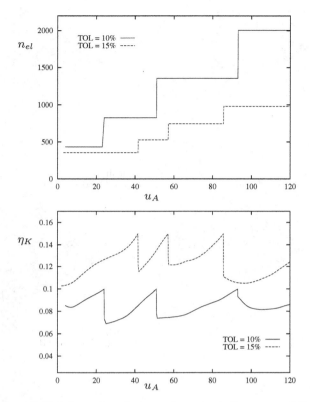

Fig. 8.14 Strategy I: Number of used elements and relative error of the adaptive computation

steps, an increment of $\Delta u_A = 1$ was used. For the next 60 steps, the step size amounted to $\Delta u_A = 0.5$ and for the last steps $\Delta u_A = 0.25$ was selected. Note that no adaptive load incremental procedure is employed what usually should be done in such simulations. However, in this way, the adaptive computation affects only the spatial discretization.

By means of test computations, it was secured that the chosen load steps were small enough to correctly reproduce the load history.

A further necessity is to limit the minimum element size to $1/8$ of the shell thickness, leading to $h_{min} = 3/8$. This avoids the singularity which would occur due to the point load (here prescribed displacement) at point A. It is equivalent to the distribution of the load over an area which is related to the minimum element size and from the practical point sufficient.

The adaptive solutions are computed using relative tolerances of 15 and 10% with respect to the relative error η_K in the energy norm written in stress space, see (8.48). However, here the relative error could also be measured using the L^2-norm for the stresses. The results are compared with a reference solution which was obtained by a simulation with a regular mesh of 5000

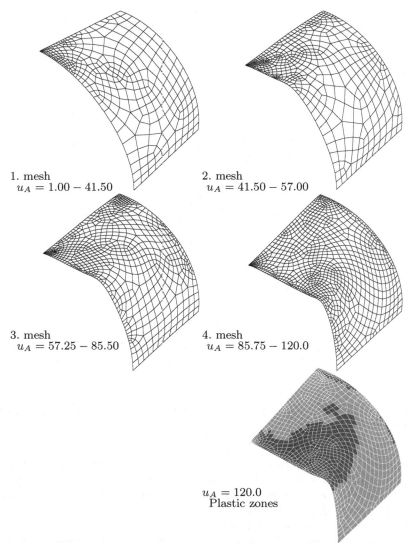

1. mesh
$u_A = 1.00 - 41.50$

2. mesh
$u_A = 41.50 - 57.00$

3. mesh
$u_A = 57.25 - 85.50$

4. mesh
$u_A = 85.75 - 120.0$

$u_A = 120.0$
Plastic zones

Fig. 8.15 Strategy I: Adaptive meshing in the deformed configuration of an adaptive computation with a tolerance of 15%

elements. Strategies I and II described in Sect. 8.6.2 – with and without transfer of history variables – were applied.

The load–displacement curves are depicted in the first graph of Fig. 8.13 for strategy I. The results of the adaptive computation are in good agreement with the reference solution for a displacement up to $u_A = 80.0$. This is especially true after a remeshing was performed. In that case, related to

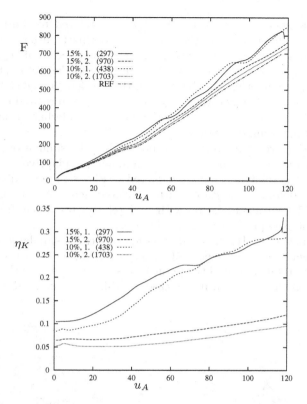

Fig. 8.16 Strategy II: Load–displacement curve, energy norm $\|\boldsymbol{\tau}\|_K$ and relative error

the increase of the number of elements, the discretized shell becomes more flexible and thus the reaction force at point A is reduced.

Between $u_A = 80$ and $u_A = 100$, both adaptive solutions depict an over proportional, non-physical decrease in the load which is linked to the equilibrium state after a remeshing step where a transfer of all variables took place. Cause of this strong decrease are the "soft" response of the shell with finer discretization and the error due to the transfer of the plastic part of the right CAUCHY-GREEN tensor $\mathbf{C}^p$ and the hardening variable α, see Sect. 9.4. An additional error occurs within the transfer process since a plane stress state is assumed locally for the projection of $\mathbf{C}^p$ and due to the curvature of the shell; then a transformation to the global coordinate system is necessary.

Further error originates from the plastic bend which moves with increasing load over the shell surface and is characteristic for this problem. Since this is a localization phenomenon, the results are strongly dependent on the element size in this area, and hence also the refined mesh has to move with the plastic bend. The shell exhibits high bending strains in this area. Thus the results are sensible with respect to errors which are related to the transfer of $\mathbf{C}^p$ and α. With finer meshes, the error due to this transfer decreases. Thus an adequately refined mesh, using smaller tolerances, will not lead to the sudden decrease of the load F, see first diagram in Fig. 8.13. The error norm $\| \bar{\tau} - \bar{\tau}_h \|_{E^{-1}}$ smoothes this behaviour; the sudden change can be observed, but the value of the norm stays very close to the norm related to the reference solution, see second diagram in Fig. 8.13. This underlines that local effects originate the strong decrease in F.

Four adaptive steps were performed within the simulations with the relative tolerance of 10 and 15%. The number of adaptively generated elements and the relative error are depicted in Fig. 8.14 depending on the load steps. The deformation states with overlayed generated meshes are shown in Fig. 8.15 for the simulation with 15% relative tolerance; furthermore the plastic zone is depicted for the final state.

Within strategy II, the entire loading process is simulated using a given finite element mesh, see Sect. 8.6.2. The starting mesh of the computation was already refined around the point load since here, due to the singularity, a mesh refinement will occur in any case. By this pre-refining, using engineering

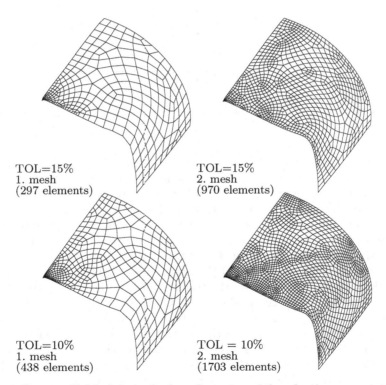

TOL=15%
1. mesh
(297 elements)

TOL=15%
2. mesh
(970 elements)

TOL=10%
1. mesh
(438 elements)

TOL = 10%
2. mesh
(1703 elements)

Fig. 8.17 Strategy II: Mesh in its final configuration of the adaptive computations

knowledge of the solution behaviour, unnecessary simulations over all load steps are avoided. Within each simulation of the loading process, the error is determined and stored.

Due to the above described choice of a starting mesh, only one additional mesh had to be constructed within the adaptive procedure to keep the error below the prescribed tolerances throughout all load steps. This can be observed from the diagram depicted in Fig. 8.16. Again η_K is related to the norm $\| \bar{\tau} - \bar{\tau}_h \|_{E^{-1}}$, which has a dimensionless form, see (8.89).

The first diagram in Fig. 8.16 represents the load–displacement curve computed within the adaptive simulation. Contrary to strategy I, no jumps occur and thus this strategy is much more robust. It can also be observed that the solution converges rapidly to the reference solution, even for relative large tolerances. This however is only true for the global values like the load–displacement curve. Once local stresses or strains are of interest, smaller tolerances have to be selected which then leads to further refinement. However, when local stresses are of interest, the better strategy is to apply the dual method in order to zoom in on specific stress values, see also the last example. Fig. 8.17 depicts the deformed shell in its final configuration. It is interesting to note that strategy I and II yield basically the same number of finite elements; also the density distribution of the finite element sizes is equal for both strategies.

9. Special Structural Elements

Trusses, beams and shells belong to the most important structural elements in engineering practise. Many structures in civil engineering – like masts, domes, frames or cooling towers – consist of such structural elements. But also in mechanical engineering – car bodies, robots or general machines – can be modelled by beams and shells. The reliable mechanical and mathematical description of trusses, beams and shells is of great significance. Hence it is, since a long time, under investigation and assoicated with great names like GALILEO, LEIBNIZ, MARIOTTE, BERNOULLI, EULER and KIRCHHOFF. Linear and approximate nonlinear theories are known for a long time and have been introduced to the engineering codes. Especially stability problems were solved by different approximate theories and associated numerical methods, see also the introduction, Sect. 2.1, and Chap. 7. However, due to the development of inexpensive computer hardware, it is possible today to perform numerical simulations based on completely nonlinear theories. Thus the general description of finite deformation states of such structural members has found its way into modern numerical simulation tools like the finite element method. Due to this development, it is not necessary to discuss the validity of approximate theories since no restrictions with respect to deflections and rotations are made in this approach.

The structural elements are modeled in case of trusses and beams by one-dimensional models which, however, are imbedded in three-dimensional space. The same holds for two-dimensional shell models. All models are characterized by a description of the geometry as a curve or surface in space. Formulations which can be applied to describe the spatial curves or surfaces are provided by introduction of arc-length of a curve, convective coordinates for surfaces or simply reference to a cartesian coordinate system using an approximation of the initial geometry by polynomial patches. The last is often chosen within finite element approximations since it naturally fits into the isoparametric concept, see Sect. 4.1.

The latter leads often to a discretization of curved spatial beams by a number of straight finite elements with linear interpolation. Such approximation of the geometry, however, will create additional errors besides the discretization error of the deformation field. Often this error, due to the approximation of the geometry, vanishes with increasing number of finite elements. In that case, it is essential that the additional element coordinates are always related to the exact geometry. These considerations also hold for shells. However, they react more sensitive to errors in geometry since the surface properties can easily be changed locally. A bi-linear isoparametric element, for example, approximates in general hyperbolic surfaces for unstructured meshes, even if the global surface is a sphere or a cylinder. These errors diminish for higher order finite element approximations. Additionally, in the last few years, different approaches have been proposed which link the geometry directly to the finite element approximation by using the same ansatz, like *NURBS*, for geometry and finite element interpolation.

In the following, first truss, beam and axisymmetrical shell elements will be discussed and then general shell elements for arbitrary three-dimensional surfaces will be derived.

9.1 Nonlinear Truss Element

This section is concerned with the derivation of a three-dimensional truss element. Truss structures are assumed to consist of straight members which are connected by hinges and are loaded only at their connection points. Thus trusses will only endure tension and compression forces but no bending and torsional moments. There are no kinematical restrictions related to the magnitude of the deformation; hence this formulation will be geometrically exact. The constitutive equations are first formulated for purely elastic behaviour. Especially the material relation small strain of ST. VENANT, see (3.121), and the hyperelastic OGDEN material for finite strain, see (3.113), are applied. Finally, elasto-plastic constitutive relations are introduced which can describe small elastic but finite plastic deformations under the assumption of isotropic hardening.

9.1.1 Kinematics and Strains

For trusses, the three-dimensional kinematical relations have to be specialized for the one-dimensional case, see Sect. 3.1. Such formulation is sufficient for trusses since these are only loaded along their axis.

The formulation will be presented with respect to the initial configuration Ω. The deformed configuration of the truss $\varphi(\Omega)$ can be described by, see also Fig. 9.1,

$$\varphi(\mathbf{X}) = (X + u(X))\mathbf{e}_1 + (Y + v(X))\mathbf{e}_2 + (Z + w(X))\mathbf{e}_3. \qquad (9.1)$$

Here X, Y and Z are the coordinates with respect to a cartesian basis in the initial configuration. The associated displacements are denoted by u, v, w.

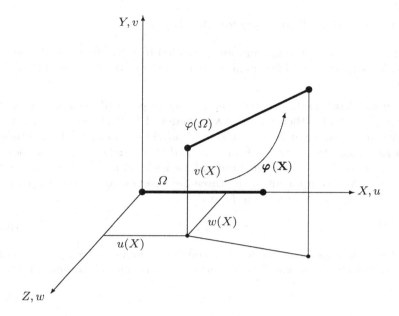

Fig. 9.1 Truss: deformed and undeformed configuration

Remark 9.1: Note that a special location of the truss element is assumed with the local axis of the truss being the X-axis. A general initial position of the truss can be described by appropriate transformations in which the local truss axis (here X-axis) is related to the global coordinate system in which a truss structure is described. The transformation of the local initial configuration to the global initial configurations is the same as in the linear theory, see e.g. Crisfield (1991). Between the local cartesian frame with base vectors $\{\mathbf{E}_i^l\}$ and the global cartesian frame with base vectors $\{\mathbf{E}_k^g\}$, the relation

$$\mathbf{E}_i^l = (\,\mathbf{E}_i^l \cdot \mathbf{E}_k^g\,)\,\mathbf{E}_k^g \tag{9.2}$$

is obtained. Here the scalar product $\mathbf{E}_i^l \cdot \mathbf{E}_k^g$ denotes the directional cosine between the local i and the global coordinate axis k.

Using (9.1), the deformation gradient (3.6) can be specified

$$\mathbf{F} = \text{Grad } \mathbf{x} = \begin{bmatrix} 1 + u,_X & 0 & 0 \\ v,_X & 1 & 0 \\ w,_X & 0 & 1 \end{bmatrix} \tag{9.3}$$

with the JACOBI-determinant $J = \det \mathbf{F} = 1 + u,_X$. Inserting $\mathbf{F}$ in the GREEN-LAGRANGE strain $\mathbf{E} = \frac{1}{2}(\mathbf{F}^T\mathbf{F} - \mathbf{1})$, (3.15) yields the relevant strain component related to the X-coordinate for a truss

$$E_X = u,_X + \frac{1}{2}(u,_X^2 + v,_X^2 + w,_X^2). \tag{9.4}$$

9.1.2 Constitutive Equations for the Truss

Two elastic constitutive equations are presented here for the truss. These are the ST. VENANT material for small strains and the OGDEN material for finite strains.

St. Venant Material. If a truss undergoes large displacements but only endures small strains then ST. VENANT material (3.121), see Sect. 3.3.1, is adequate for the constitutive description. It provides a linear relation between the GREEN-LAGRANGE strains $\mathbf{E}$ and the 2nd PIOLA-KIRCHHOFF stresses $\mathbf{S}$. Since the truss element is loaded only along its local axis, it is sufficient to consider the first component of the three-dimensional constitutive equation. This leads with YOUNG modulus E to

$$S_X = E\,E_X\,. \tag{9.5}$$

For completeness also strains induced by a change of temperature are introduced in this equation. These thermal strains can be computed from

$$E_\Theta = \alpha_T\,(\Theta - \Theta_A) \tag{9.6}$$

with the thermal expansion coefficient α_T. Θ_A is a given reference temperature. Relation (9.6) is only valid for small strains since the GREEN-LAGRANGE strains are work conjugate to the 2nd PIOLA-KIRCHHOFF stress which is not real physical stress, see Sect. 3.2.4. However, for small strains, this stress deviates only very little from the KIRCHHOFF stresses. With (9.4) and (9.6), the thermo-elastic ST. VENANT constitutive equation for the truss

$$S_X = E\,[\,E_X - E_\Theta\,] = E\left[u,_X + \frac{1}{2}\,(u,_X^2 + v,_X^2 + w,_X^2\,) - \alpha_T\,(\Theta - \Theta_A)\right] \tag{9.7}$$

is obtained.

Ogden Material. Finite elastic deformations can be described by the material equation of OGDEN (3.113), see Sect. 3.3.1. The strain energy is given in terms of the principal stretches

$$W(\lambda_i) = \sum_r \frac{\mu_r}{\alpha_r}\,[\,\lambda_1^{\alpha_r} + \lambda_2^{\alpha_r} + \lambda_3^{\alpha_r} - 3\,]\,. \tag{9.8}$$

The constants μ_r and α_r are material parameters. λ_i denote the principal stretches which follow in general from the spectral decomposition of the strain tensor. Due to the one-dimensional loading of a truss element, only the principal stretch λ_1 is relevant. It can be directly computed from (3.23) and (9.4). Hence $E_X = \frac{1}{2}\,(\lambda_1^2 - 1)$ is obtained which leads to

$$\lambda_1 = \sqrt{2\,E_X + 1}\,. \tag{9.9}$$

The constitutive relations of OGDEN are often applied to rubberlike materials. In that case, incompressible material behaviour has to be considered additionally. The CAUCHY stresses are then given with (3.134) by

$$\sigma_i = \lambda_i \frac{\partial W}{\partial \lambda_i} + p \,, \tag{9.10}$$

where p is the pressure. It can be determined from the condition of incompressibility $\mathbf{F} = 1$. The stresses follow from (3.142), as described in Exercise 3.6,

$$\sigma_1 = \sum_r \mu_r \left[\lambda_1^{\alpha_r} - \lambda_1^{-\frac{1}{2}\alpha_r} \right]. \tag{9.11}$$

Since the truss element will be formulated with respect to the initial configuration, the CAUCHY stress σ_1 has to be transformed (pulled back) to the 2nd PIOLA-KIRCHHOFF stress. The relation $\mathbf{S} = J \mathbf{F}^{-1} \boldsymbol{\sigma} \mathbf{F}^{-T}$, see also (3.82), yields – by considering incompressibility ($J = 1$) – for the stress component in direction of the local truss axis

$$S_X = \frac{1}{\lambda_1^2} \sigma_1 \,. \tag{9.12}$$

Thus the OGDEN material can be formulated for the truss element in terms of the 2nd PIOLA-KIRCHHOFF stresses and the GREEN-LAGRANGE strains by

$$S_X = \sum_r \mu_r \left[\lambda_1^{\alpha_r - 2} - \lambda_1^{-\frac{1}{2}\alpha_r - 2} \right]. \tag{9.13}$$

Here the stretch λ_1 can be expressed by E_X when using (9.9).

Since both constitutive equations, the OGDEN and the ST. VENANT material, are related to the initial configuration and expressed in terms of the 2nd PIOLA-KIRCHHOFF stresses and the GREEN-LAGRANGE strains it is possible to derive a uniform variational formulation for both materials.

9.1.3 Variational Formulation and Linearization

The finite element formulation of the nonlinear truss element is based on the weak form of equilibrium. With respect to Sect. 3.4.1, the following one-dimensional version of the weak form can be stated for a truss element

$$G(\mathbf{u}) = \int_{(X)} \delta E_X S_X A \, dX - \int_{(X)} \eta_X b_X A \, dX - \sum_k \eta_{Xk} P_k = 0 \,, \tag{9.14}$$

where S_X is provided by one of the constitutive Eqs. (9.5) or (9.13). b_X are the body forces, P_k denote applied nodal forces and A is the cross sectional area of the initial configuration. η_X is the X-component of the test function. The variation of the GREEN-LAGRANGE strains δE_X is given with (9.4) by

$$\delta E_X = (1 + u,_X)\,\eta_{X,X} + v,_X\ \eta_{Y,X} + w,_X\ \eta_{Z,X}, \qquad (9.15)$$

where $\eta_{X,X}$, $\eta_{Y,X}$ and $\eta_{Z,X}$ are the derivatives of the components of the test function. The constitutive equations of the last section for S_X yield a strongly nonlinear weak form (9.14).

NEWTON method is usually applied for the solution of (9.14), see Sect. 5.1.1. In that case, the linearization of (9.14) is needed which leads to the tangent stiffness of a truss element. By using the concept of directional derivatives, see Sect. 3.5, the tangent stiffness is derived

$$DG(\mathbf{u}) = \int_{(X)} \delta E_X \Delta S_X\, A\, dX + \int_{(X)} \Delta \delta E_X\, S_X\, A\, dX. \qquad (9.16)$$

Here the linearization of the strains is given by (9.4)

$$\Delta E_X = \Delta u,_X\ (1 + u,_X) + \Delta v,_X\ v,_X + \Delta w,_X\ w,_X \qquad (9.17)$$

as well as the linearization of the variation of the strains (9.15)

$$\Delta \delta E_X = \Delta u,_X\ \eta_{X,X} + \Delta v,_X\ \eta_{Y,X} + \Delta w,_X\ \eta_{z,X}. \qquad (9.18)$$

The linearization of the constitutive Eqs. (9.5) yields the increment of stresses

$$\Delta S_X = E\,\Delta E_X = E\,[\,\Delta u,_X\ (1 + u,_X) + \Delta v,_X\ v,_X + \Delta w,_X\ w,_X\,]. \qquad (9.19)$$

Analogously for the OGDEN material, see (9.13), the linearization

$$\Delta S_X = C(\lambda_1)\,\Delta E_X \qquad (9.20)$$

is derived. The relation (9.9) leads to $\Delta\lambda_1 = \lambda_1^{-1}\,\Delta E_X$, and thus the incremental material tensor is given by

$$C(\lambda_1) = \sum_r \mu_r \left[(\alpha_r - 2)\lambda_1^{(\alpha_r - 4)} + (\frac{1}{2}\alpha_r + 2)\,\lambda_1^{-(\frac{1}{2}\alpha_r + 4)} \right]. \qquad (9.21)$$

9.1.4 Finite-Element Model

The discretization of Eqs. (9.14) and (9.16) is obtained using finite elements. For that, the displacements u, v, w and the test functions η_X, η_Y, η_Z are approximated by linear shape functions. Of course, also quadratic or higher order interpolations could be introduced, but for most applications linear elements are sufficient since the linear shape functions are the solution of the homogeneous differential equations of the truss in the geometrically linear case when linear elastic constitutive behaviour is assumed. Thus the interpolation

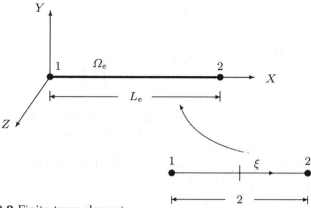

Fig. 9.2 Finite truss element

$$u_e = \sum_{K=1}^{2} N_K(\xi)\, u_K, \quad v_e = \sum_{K=1}^{2} N_K(\xi)\, v_K \ \text{ and } \ w_e = \sum_{K=1}^{2} N_K(\xi)\, w_K,$$
(9.22)

is used. Here $N_K(\xi)$ are given in terms of the linear shape functions (4.17). The quantities u_K, v_K and w_K represent the nodal degrees of freedom which are related to the truss element Ω_e which has the length L_e, see Fig. 9.2.

Using the shape functions (9.22), the displacement gradients follow with (4.23) by

$$u_{e,X} = \frac{u_2 - u_1}{L_e}, \quad v_{e,X} = \frac{v_2 - v_1}{L_e} \ \text{ and } \ w_{e,X} = \frac{w_2 - w_1}{L_e}.$$
(9.23)

By inserting (9.22) and (9.23) into the variational formulation (9.14), the matrix form

$$G_h(\mathbf{u}_h, \boldsymbol{\eta}) = \boldsymbol{\eta}^T \mathbf{G}(\mathbf{v}) = \boldsymbol{\eta}^T \bigcup_{e=1}^{n_{el}} \mathbf{G}_e(\mathbf{v}) = \boldsymbol{\eta}^T \bigcup_{e=1}^{n_{el}} (\mathbf{R}_e - \mathbf{P}_e)$$
(9.24)

is obtained. Here $\mathbf{v}$ is the vector of the nodal displacements related to the element Ω_e

$$\mathbf{v}^T = \{\, u_1, v_1, w_1, u_2, v_2, w_2 \,\}$$
(9.25)

and $\boldsymbol{\eta}$ denotes the vector of the nodal testfunctions

$$\boldsymbol{\eta}^T = \{\, \eta_{X1}, \eta_{Y1}, \eta_{Z1}, \eta_{X2}, \eta_{Y2}, \eta_{Z2} \,\}.$$
(9.26)

The residual $\mathbf{R}_e$ describes the stress divergence

$$\boldsymbol{R}_e(\boldsymbol{v}) = A \begin{bmatrix} (1 + u_{e,X})\, S_X \\ v_{e,X}\, S_X \\ w_{e,X}\, S_X \\ -(1 + u_{e,X})\, S_X \\ -v_{e,X}\, S_X \\ -w_{e,X}\, S_X \end{bmatrix}. \tag{9.27}$$

This formulation is valid as well for the material model of St. Venant as for the model of Ogden. One only has to insert the associated constitutive equation for S_X given by (9.5) or (9.13).

The matrix form of the linearization follows from (9.16) and (9.24). It leads to the tangential stiffness matrix $\boldsymbol{K}_T$

$$\boldsymbol{K}_T = \bigcup_{e=1}^{n_{el}} D\, \boldsymbol{G}_e(\boldsymbol{v}). \tag{9.28}$$

$\boldsymbol{K}_T$ represents the global tangent operator, which results from the assembly of the (6×6) element stiffness matrices. The element stiffness $\boldsymbol{K}_T^e$ can be written explicitly as

$$\boldsymbol{K}_T^e = \begin{bmatrix} (\boldsymbol{A}_1 + \boldsymbol{A}_2) & -(\boldsymbol{A}_1 + \boldsymbol{A}_2) \\ -(\boldsymbol{A}_1 + \boldsymbol{A}_2) & (\boldsymbol{A}_1 + \boldsymbol{A}_2) \end{bmatrix}. \tag{9.29}$$

The first term of (9.29) is given by

$$\boldsymbol{A}_1 = \frac{HA}{L_e} \begin{bmatrix} (1 + u_{e,X})^2 & (1 + u_{e,X})v_{e,X} & (1 + u_{e,X})w_{e,X} \\ (1 + u_{e,X})v_{e,X} & v_{e,X}^2 & v_{e,X}w_{e,X} \\ (1 + u_{e,X})w_{e,X} & v_{e,X}w_{e,X} & w_{e,X}^2 \end{bmatrix}. \tag{9.30}$$

H describes the tangent modulus of the material model. For the St. Venant material, it is given by $H = E$, while for the Ogden material it has the form $H = C(\lambda_1)$. The second term of (9.29)

$$\boldsymbol{A}_2 = \frac{S_X A}{L_e} \begin{bmatrix} 1 & 0 & 0 \\ 0 & 1 & 0 \\ 0 & 0 & 1 \end{bmatrix} \tag{9.31}$$

is often denoted as initial stiffness matrix. The associated equation for the stresses, St. Venant or Ogden material, has to be used in (9.31). The index h, often used to indicate the approximation of the field quantities, was suppressed to simplify the notation. The finite element formulation for the truss can now be implemented in a finite element program for the discussed elastic constitutive equations based on this matrix form.

Exercise 9.1: Consider a truss element undergoing elasto-plastic deformations. For the assumption of von Mises plasticity, the algorithm for the stress computation as well as the associated linearization have to be developed using the relations derived in Sect. 6.3.2. Two different material models should be considered:

(a) Formulate the constitutive relations for the truss by assuming small elastic strains using (6.158). Here the finite plastic strains can be modelled for isotropic hardening, see (6.159).

(b) Formulate a constitutive equation for finite incompressible elastic deformations using OGDEN material and large plastic deformations with the nonlinear hardening $q = -(H^l \alpha + H^{nl} \alpha^\delta)$.

Solution: The three-dimensional continuum equation are provided in Sect. 3.3.2. All quantities can be related to principal axes since a one-dimensional structure is considered. Hence the algorithms stated in Sect. 6.3.2 can be applied. The stretches are split, based on (3.191), by

$$\lambda = \lambda^e \lambda^p .$$

To derive a truss element which can describe the elasto-plastic deformations, all relations which follow from the integration of the constitutive equations have to be inserted in Eqs. (9.27) to (9.31). Hence the 2nd PIOLA-KIRCHHOFF stress S_X and the tangent modulus H have to be determined within a time step $[t_n, t_{n+1}]$. The equations from Sect. 6.3.2 simplify considerably for the one-dimensional case. Here the *trial* stretch

$$\lambda_{n+1}^{e\,tr} = \frac{\lambda_{n+1}}{\lambda_n^p}$$

is obtained together with its associated logarithmic strain

$$\varepsilon_{n+1}^{e\,tr} = \ln [\lambda_{n+1}^{e\,tr}] .$$

In case (a), the one-dimensional flow condition follows from (3.159) as

$$f(\tau) = |\tau| - (Y_0 + \hat{H} \alpha) \leq 0$$

and from (3.210) the flow rule. The elastic constitutive equation relates the KIRCHHOFF stress τ via

$$\tau = E \varepsilon^e$$

to the logarithmic strains $\varepsilon^e = \ln \lambda^e$, see Sect. 6.3.2. The only material parameter E is YOUNG modulus which results from the strain energy function defined in *Remark 6.2* for small strains. E does not represent the modulus of elasticity introduced in (9.5) since it relates different strains and stresses. The *trial* KIRCHHOFF stress is then given by

$$\tau_{n+1}^{tr} = E \varepsilon_{n+1}^{e\,tr} .$$

This stress has to be inserted in the flow condition

$$f_{n+1}^{tr} = |\tau_{n+1}^{tr}| - (Y_0 + \hat{H} \alpha_{n+1}^{tr}) ,$$

where the trial value of the hardening parameter $\alpha_{n+1}^{tr} = \alpha_n$ is determined from the last time step. With f_{n+1}^{tr}, it can be distinguished whether a truss element behaves elastically or plastically. The relations presented in Sect. 6.3.2 yield explicitly:

$-\ f_{n+1}^{tr} < 0 \Longrightarrow$ elastic:

   1. 2nd PIOLA-KIRCHHOFF stress: $S_{X\,n+1} = \dfrac{\tau_{n+1}^{tr}}{(\lambda_{n+1}^{e\,tr})^2}$

   2. Tangent modulus (see Sect. 6.3.3): $H_{n+1} = \dfrac{E - 2\,\tau_{n+1}^{tr}}{(\lambda_{n+1}^{e\,tr})^4}$

$-\ f_{n+1}^{tr} \geq 0 \Longrightarrow$ plastic:

1. Increment of the consistency parameter: $\Delta\gamma_{n+1} = \dfrac{f_{n+1}^{tr}}{E + \hat{H}}$.

2. Elastic strain: $\varepsilon_{n+1}^{e} = \varepsilon_{n+1}^{e\,tr} - \Delta\gamma_{n+1} \dfrac{\tau_{n+1}^{tr}}{|\tau_{n+1}^{tr}|}$.

3. Hardening parameter: $\alpha_{n+1} = \alpha_n + \Delta\gamma_{n+1}$.

4. KIRCHHOFF stress: $\tau_{n+1} = E\,\varepsilon_{n+1}^{e}$.

5. Algorithmic tangent modulus: $C_p^{ALG} = \dfrac{E\,\hat{H}}{E + \hat{H}}$.

6. Update of the plastic variables: $\lambda_{n+1}^{p} = \dfrac{\lambda_{n+1}}{\exp[\varepsilon_{n+1}^{e}]}$.

7. 2nd PIOLA-KIRCHHOFF stress: $S_{X\,n+1} = \dfrac{\tau_{n+1}}{(\lambda_{n+1}^{e\,tr})^{2}}$.

8. Tangent modulus (see Sect. 6.3.3): $H_{n+1} = \dfrac{C_p^{ALG} - 2\,\tau_{n+1}}{(\lambda_{n+1}^{e\,tr})^{4}}$.

With the above relations, all equations needed for the elasto-plastic analysis of a truss element are known. They just have to be inserted into the finite element formulation presented in the previous section.

The truss structure depicted in Fig. 9.3a in its initial and deformed configuration is computed using the model described above. The material parameters are $E = 21000$, $A = 10$, $Y_0 = 24$ and $\hat{H} = 10$. Under loading, the system first deforms elastically. From a specific load level on, see load–displacement curve in Fig. 9.3b, the trusses close to the support undergo plastic deformations. Thus a type of plastic hinge develops within the first segment at the support and the stiffness of the system reduces drastically which is expressed in the graph by the almost horizontal part of the load deflection curve.

In case (b), the flow condition

$$f(\tau) = |\tau| - \left(Y_0 + \hat{H}^l\,\alpha + \hat{H}^{nl}\alpha^\delta\right) \leq 0$$

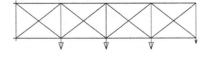

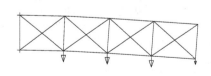

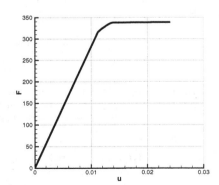

Fig. 9.3a Truss structure: system and deformation

Fig. 9.3b Load–displacement curve of the truss structure

is used which exhibits nonlinear hardening. Additionally, a one-dimensional incompressible OGDEN material, see (9.11), is applied which is formulated here in terms of the KIRCHHOFF stress

$$\tau = \sum_i \mu_i \left(\lambda^{\beta_i} - \lambda^{-\beta_i/2} \right) = \sum_i \mu_i \left(\exp(\beta_i\,\varepsilon) - \exp(-\beta_i\,\varepsilon/2) \right).$$

For the algorithmic treatment, the trial stresses are needed. They follow from the elastic strains by assuming frozen plastic variables ($\lambda^{e\ tr}_{n+1} = \lambda_{n+1}\,/\,\lambda^p_n$)

$$\tau^{tr}_{n+1} = \sum_i \mu_i \left(\exp(\beta_i\,\varepsilon^{e\ tr}_{n+1}) - \exp(-\beta_i\,\varepsilon^{e\ tr}_{n+1}/2) \right).$$

The elastic tangent modulus is deriveed from

$$\tilde{E}_{n+1} = \frac{\partial \tau^{tr}_{n+1}}{\partial \varepsilon^{e\ tr}_{n+1}} = \sum_i \mu_i\,\beta_i \left[(\lambda^{e\ tr}_{n+1})^{\beta_i} + 0.5\,(\lambda^{e\ tr}_{n+1})^{-\beta_i/2} \right].$$

The trial stresses are inserted in the flow condition to check whether elastic or plastic behaviour is related to the stress state

$$f^{tr}_{n+1} = |\tau^{tr}_{n+1}| - \left[Y_0 + \hat{H}^l\,\alpha^{tr}_{n+1} + \hat{H}^{nl}(\alpha^{tr}_{n+1})^\delta \right].$$

The following cases can be distinguished

− $f^{tr}_{n+1} < 0$ elastic leading to:

1. 2nd PIOLA-KIRCHHOFF stresses: $S_{X\,n+1} = \dfrac{\tau^{tr}_{n+1}}{(\lambda^{e\ tr}_{n+1})^2}$.

2. Tangent modulus: $H_{n+1} = \dfrac{\tilde{E}_{n+1} - 2\,\tau^{tr}_{n+1}}{(\lambda^{e\ tr}_{n+1})^4}$.

− $f^{tr}_{n+1} > 0$ plastic. Here the following steps are needed to determine stresses and associated tangent modulus:

1. First, the increment of the consistency parameter $\Delta\gamma_{n+1}$ is computed from $f_{n+1} = 0$. Here f_{n+1} has the form

$$f_{n+1} = \tau(\varepsilon^{e\ tr}_{n+1} - \Delta\gamma_{n+1}) - \left[Y_0 + H^l\,(\alpha_n + \Delta\gamma_{n+1}) + H^{nl}\,(\alpha_n + \Delta\gamma_{n+1})^\delta \right].$$

$\tau(x)$ means that the argument x has to be inserted into the constitutive equation of OGDEN. Contrary to the last example, the flow condition is a nonlinear function in $\Delta\gamma_{n+1}$. This requires a local iteration to find the roots. After that the remaining quantities can be computed.

2. Elastic strains $\varepsilon^e_{n+1} = \varepsilon^{e\ tr}_{n+1} - \Delta\gamma_{n+1}\dfrac{\tau^{tr}_{n+1}}{|\tau^{tr}_{n+1}|}$.

3. Hardening parameter $\alpha_{n+1} = \alpha_n + \Delta\gamma_{n+1}$.

4. KIRCHHOFF stresses (from the OGDEN material)

$$\tau_{n+1} = \sum_i \mu_i \left(\exp(\beta_i\,\varepsilon^e_{n+1}) - \exp(-\beta_i\,\varepsilon^e_{n+1}/2) \right).$$

5. The algorithmic tangent modulus, needed within the FE-formulation, follows
   with respect to Sect. 6.3.2

$$C_p^{\text{ALG}} = \frac{\partial \tau(\varepsilon_{n+1}^e)}{\partial \varepsilon_{n+1}^{e\ \text{tr}}} = \frac{\partial \tau(\varepsilon_{n+1}^e)}{\partial \varepsilon_{n+1}^e} \cdot \frac{\partial \varepsilon_{n+1}^e}{\partial \varepsilon_{n+1}^{e\ \text{tr}}}.$$

The first term on the right hand side is equivalent to the elastic tangent modulus $\tilde{E}_{n+1}$ stated above. The second term is computed from

$$\frac{\partial}{\partial \varepsilon^{e\ \text{tr}}} \left(\varepsilon_{n+1}^e \right) = \frac{\partial}{\partial \varepsilon^{e\ \text{tr}}} \left(\varepsilon_{n+1}^{e\ \text{tr}} - \Delta\gamma_{n+1} \right) = \left(1 - \frac{\partial \Delta\gamma_{n+1}}{\partial \varepsilon^{e\ \text{tr}}} \right),$$

where the derivative of the consistency parameter can be determined by using
the flow condition

$$\frac{\partial f}{\partial \varepsilon_{n+1}^{e\ \text{tr}}} \cdot \text{d}\varepsilon_{n+1}^{e\ \text{tr}} + \frac{\partial f}{\partial \Delta\gamma_{n+1}} \cdot \text{d}\Delta\gamma_{n+1} \equiv 0.$$

This yields

$$\frac{\text{d}\Delta\gamma_{n+1}}{\text{d}\varepsilon_{n+1}^{e\ \text{tr}}} = -\frac{\left(\dfrac{\partial f}{\partial \varepsilon_{n+1}^{e\ \text{tr}}} \right)}{\left(\dfrac{\partial f}{\partial \Delta\gamma_{n+1}} \right)} = \frac{\tilde{E}_{n+1}}{\tilde{E}_{n+1} + \tilde{H}_{n+1}}$$

with

$$\tilde{H}_{n+1} = H^l + \delta \left(\alpha_n + \Delta\gamma_{n+1} \right)^{\delta-1} H^{nl}.$$

Hence C_p^{ALG} has the form

$$C_p^{ALG} = \frac{\tilde{H}_{n+1} \tilde{E}_{n+1}}{\tilde{H}_{n+1} + \tilde{E}_{n+1}}.$$

6. The update of the plastic variables yields a plastic stretch at the end of the
   time step:

$$\lambda_{n+1}^p = \frac{\lambda_{n+1}}{\exp[\varepsilon_{n+1}^e]}.$$

7. The 2nd PIOLA-KIRCHHOFF stress which has to be inserted in (9.27) is computed from the KIRCHHOFF stress:

$$S_{X\ n+1} = \frac{\tau_{n+1}}{(\lambda_{n+1}^{e\ \text{tr}})^2}.$$

8. The incremental tangent modulus, needed in (9.29), is given by:

$$H_{n+1} = \frac{C_p^{ALG} - 2\,\tau_{n+1}}{(\lambda_{n+1}^{e\ \text{tr}})^4}.$$

The constitutive equations discussed in (b) exhibit nonlinear isotropic hardening and nonlinear elastic behaviour. They can, e.g. be used to model the material response of polymer strings, see e.g. Bidmon (1989), which occur in composite materials.

A one dimensional test model of such string is shown on the left hand side of Fig. 9.4 where a point load is applied at the end of the string. Adaption of the material parameters of this model to experimental results, provided in Bidmon (1989), leads to the elastic constants μ and α, the yield limit σ_{Y0} and the hardening parameters H_{lin}, H_{nl} and k as depicted in Fig. 9.4.

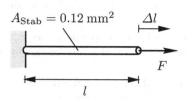

Material parameters		
μ	=	195.95 N/mm^2
α	=	30.37
σ_{Y0}	=	111.5 N/mm^2
H_{lin}	=	5969.49 N/mm^2
H_{nl}	=	43977191 N/mm^2
k	=	4.37

Fig. 9.4 One-dimensional elasto-plastic truss model for polymer strings

The load–displacement curve of the truss shown in Fig. 9.4 is now computed using the algorithm stated above by increasing the point load incrementally. This curve, see Fig. 9.5, is in good agreement with the experimental results in Bidmon (1989). Note that the nonlinear behaviour is matched as well for the loading as for the unloading phase by the chosen constitutive equation.

9.2 Two-dimensional Geometrically Exact Beam Element

Nonlinear theories for beams were developed in the last three decades. They all can be applied for finite element discretization. Generally, three approaches have to be distinguished.

— The first is based on the assumption of small strains. It introduces a frame undergoing finite rigid rotations and formulates the strains and stresses relative to the rotations of the frame and is known as *co-rotational* formulation. Thus the strains have to be small but large deflections and rotations can be investigated. Finite element schemes which is based on such formulations can be found in z.B. in Oran and Kassimali (1976), Wempner (1969), Rankin and Brogan (1984), Lumpe (1982), Crisfield (1991) and Crisfield (1997).

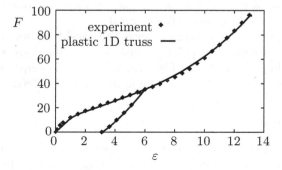

Fig. 9.5 Load-displacement of a polymer strings

- The second approach uses the continuum equations and introduces the beam kinematics by special isoparametric finite element interpolations. This approach is known as degenerated continuum approach, see e.g. Bathe and Bolourchi (1979), Dvorkin et al. (1988) or the textbooks by Bathe (1996) and Crisfield (1997).
- The third approach is based on the formulation of nonlinear rod and beam theories which have, as only restriction, the classical assumption of "plane cross sections remain plane". Here no other approximations are made; hence the strains, deflections and rotations can be finite and these theories are called *geometrically exact*. The development of these beam theories goes back to the work by Reissner (1972). A generalization for the three-dimensional case can be found in Simo (1985). Based on this theoretical background, several authors developed associated finite element formulations, see Simo and Vu-Quoc (1986), Pimenta and Yojo (1993), Jelenic and Saje (1995), Gruttmann et al. (1998) and Mäkinen (2007). For a nonlinear formulation of curved beam elements, see e.g. Ibrahimbegovic (1995). In Gruttmann et al. (2000) elasto-plastic material and in Romero and Armero (2002) dynamics were considered within the geometrically exact framework.

The latter beam theories include also arbitrary loading of truss and cable structures and thus can be applied in a general way to one-dimensional construction elements.

Besides very limited and simplified examples these nonlinear beam theories cannot be solved analytically. However, within the framework of finite element methods, there is no difficulty to apply the nonlinear beam formulations. Thus many complex engineering problems like the starting of a rotor blade or the opening of an antenna structure in space can be solved.

However even now many software tools for civil engineering still use the so-called second order theories as basis for the finite element implementation. These theories include nonlinear effects but are restricted to small rotations. They can be applied for limit load computations and stability investigations. These theories stem from the times where analytical solutions where needed to solve such problems; however when using modern computers the geometrically exact theories can be applied instead.

In this chapter different beam theories are considered for the two-dimensional case. Additionally, associated numerical formulations are derived for finite element implementations. The theories are then compared by means of an example which depicts the limits in application of different approaches.

9.2.1 Kinematics

The two-dimensional theory for the geometrically exact beam element is based on the assumptions of plane cross sections. It was derived in Reissner (1972) and is valid for finite deflections, rotations and strains.

Often it is sufficient to consider only small strains in beams which leads to the use of the St. Venant constitutive equations. Of course, also constitutive equations describing finite elastic strains or elasto-plastic deformations can be included, see e.g. Kahn (1987) and Ehrlich and Armero (2005), where a plastic hinge theory within the framework of geometrically exact beams was developed.

The nonlinear strain measures for the shear elastic beam can be found in Reissner (1972). They are based on the kinematical assumption for the beam deformation

$$\varphi = \left\{ \begin{array}{c} X_1 + u(X_1) \\ w(X_1) \end{array} \right\} + X_2 \left\{ \begin{array}{c} -\sin\psi(X_1) \\ \cos\psi(X_1) \end{array} \right\} = \varphi|_{X_2=0} + X_2\,\mathbf{t}\,, \qquad (9.32)$$

see also Fig. 9.6. Here the initial configuration of the beam is straight and the local axis coincides with the global axis. When the beam is arbitrarily located in space then an additional transformation has to be applied, see Remark 9.1.

The associated strain–deflection relations were derived in Reissner (1972) by using the principle of virtual work. This leads to strain measures for the axial strain ϵ, the shear strain γ and the curvature κ:

$$\begin{aligned} \epsilon &= (1+u')\cos\psi + w'\sin\psi - 1\,, \\ \gamma &= w'\cos\psi - (1+u')\sin\psi\,, \\ \kappa &= \psi'\,, \end{aligned} \qquad (9.33)$$

where u is the displacement in axial direction, w the deflections and ψ is the rotation, see Fig. 9.6. By $()'$ the derivative with respect to the coordinate X_1

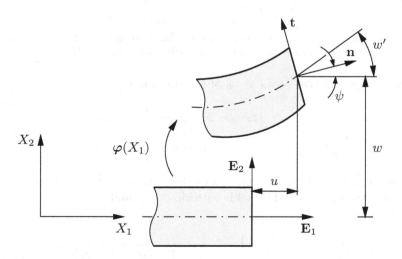

Fig. 9.6 Beam kinematics

are denoted. Note that the strain $(9.33)_3$ for the curvature is linear in ψ. The strain measures can also be formulated in matrix notation

$$\epsilon = \mathbf{T}(\psi)\,\mathbf{u}' - \mathbf{N}, \qquad (9.34)$$

where the matrices

$$\epsilon = \left\{\begin{matrix} \epsilon \\ \gamma \\ \kappa \end{matrix}\right\}, \quad \mathbf{T}(\psi) = \begin{bmatrix} \cos\psi & \sin\psi & 0 \\ -\sin\psi & \cos\psi & 0 \\ 0 & 0 & 1 \end{bmatrix}, \quad \mathbf{u}' = \left\{\begin{matrix} 1 + u' \\ v' \\ \psi' \end{matrix}\right\}, \quad \mathbf{N} = \left\{\begin{matrix} 1 \\ 0 \\ 0 \end{matrix}\right\}$$

are introduced. Relation (9.34) shows the simple structure of these nonlinear strain measures. The nonlinearity is only associated with the sin and cos functions in which the rotation angle ψ occurs. These act on u' and w' by the rotation matrix $\mathbf{T}$. Note that the rotation matrix $\mathbf{T}$ describes the rotation of the basis $(\mathbf{E}_1, \mathbf{E}_2)$ to $(\mathbf{n}, \mathbf{t})$, see Fig. 9.6.

Remark 9.2:

1. The vector $\mathbf{u}'$ in (9.34) is equivalent to the components of the deformation gradient (9.3). Since the right stretch tensor $\mathbf{U} = \mathbf{R}^T\mathbf{F}$ follows from (3.21), the strain ϵ with $\mathbf{T} = \mathbf{R}^T$ can be viewed as a strain measure equivalent to the right stretch tensor. This relates ϵ to the generalized strains in (3.18) for $\alpha = 1$.
2. Based on (9.32), the GREEN-LAGRANGE strains can be determined for the beam model. These follow with (3.15) and (3.41) from

$$\mathbf{E} = \left\{\begin{matrix} E_{11} \\ 2\,E_{12} \end{matrix}\right\} = \left\{\begin{matrix} \tfrac{1}{2}\,(g_{11} - G_{11}) \\ g_{12} - G_{12} \end{matrix}\right\}$$

with

$$\begin{matrix} g_{11} = \boldsymbol{\varphi}_{,1} \cdot \boldsymbol{\varphi}_{,1} & G_{11} = \mathbf{E}_1 \cdot \mathbf{E}_1 = 1 \\ g_{12} = \boldsymbol{\varphi}_{,1} \cdot \boldsymbol{\varphi}_{,2} & G_{12} = \mathbf{E}_1 \cdot \mathbf{E}_2 = 0 \end{matrix}.$$

These strains can be used to formulate the kinematics of the beam instead of (9.33). Such approach was used in, e. g. Gruttmann et al. (2000).

Based on a TAYLOR series at ψ_0 of the sin- and cos function

$$\begin{aligned} \sin(\psi_0 + \psi) &= \sin\psi_0 + \cos\psi_0\,\psi + \frac{1}{2}\sin\psi_0\psi^2 + \cdots \\ \cos(\psi_0 + \psi) &= \cos\psi_0 - \sin\psi_0\,\psi - \frac{1}{2}\cos\psi_0\psi^2 + \cdots \end{aligned}$$

a consistent linearization of (9.33) can be derived, see Sect. 3.5. Including all terms up to second order this yields with $\sin\psi \approx \psi$ and $\cos\psi \approx 1 - \frac{1}{2}\,\psi^2$ for $\psi_0 = 0$

$$\left.\begin{matrix} \epsilon &=& u' + w'\,\psi - \frac{1}{2}\,\psi^2 \\ \gamma &=& w' - (1 + u')\,\psi \\ \kappa &=& \psi' \end{matrix}\right\} \longrightarrow \boxed{\epsilon = \bar{\mathbf{T}}(\psi)\,\mathbf{u}' - \bar{\psi} - \mathbf{N}}. \qquad (9.35)$$

The skew symmetric matrix $\bar{\mathbf{T}}$ is the linearization of $\mathbf{T}|_{\psi_0=0}$. It is, together with the vector $\bar{\psi}$, defined as

$$\bar{\mathbf{T}}(\psi) = \begin{bmatrix} 1 & \psi & 0 \\ -\psi & 1 & 0 \\ 0 & 0 & 1 \end{bmatrix} \quad \text{and} \quad \bar{\psi} = \left\{ \begin{array}{c} \frac{1}{2}\psi^2 \\ 0 \\ 0 \end{array} \right\}.$$

The approximation (9.35) contains all quadratic terms. The strain measures are not much simpler than the exact ones provided in (9.33); only the trigonometric functions of the angle ψ disappear.

A further approximation can be obtained by neglecting the strain part u' in axial direction when the shear strains in (9.35) are derived. By additionally enforcing the BERNOULLI assumption ($w' = \psi$), the strains for a theory of moderate rotations can be formulated based on the BERNOULLI kinematics:

$$\epsilon = u' + \frac{1}{2}w'^2, \qquad \kappa = w''. \tag{9.36}$$

Here the only nonlinear term is w'^2.

Exercise 9.2: Derive the strains related to the BERNOULLI beam assumption for a geometrically nonlinear kinematic by using the strain measures (9.33).

Solution: The BERNOULLI assumption prevents the occurrence of shear strains. Hence γ is always zero. With this constraint condition, the strain measures of the BERNOULLI theory can be derived. The constraint $\gamma = 0$ yields

$$w'\cos\psi = (1+u')\sin\psi.$$

In order to eliminate the angle ψ, the first equation in (9.33) is squared which leads to

$$(\epsilon + 1)^2 = (1+u')^2\cos^2\psi + w'^2\sin^2\psi + 2(1+u')\,w'\sin\psi\cos\psi.$$

By inserting the shear constraint, the strain in axial direction is obtained depending upon u and the deflection w

$$\epsilon = \sqrt{(1+u')^2 + w'^2} - 1. \tag{9.37}$$

In the same way, a relation for the curvature κ can be derived. The derivative of the shear constraint yields with $(9.33)_1$

$$w''\cos\psi - u''\sin\psi = (\epsilon + 1)\,\psi'.$$

The multiplication of this equation by $(9.33)_1$ leads after some algebra to

$$\kappa = \frac{w''(1+u') - u''w'}{(\epsilon + 1)^2} = \frac{w''(1+u') - u''w'}{(1+u')^2 + w'^2}, \tag{9.38}$$

which can already be found in, e.g. Kappus (1939). It is obvious that the strain–displacement relation of the shear elastic beam has a simpler form.

9.2.2 Weak Form of Equilibrium

The weak form of equilibrium which is equivalent to the principle of virtual work can be stated for shear elastic beams as

$$G(\mathbf{u}, \boldsymbol{\eta}) = \int_0^l (N\, \delta\epsilon + Q\, \delta\gamma + M\, \delta\kappa)\, dx - \int_0^l (n\, \delta u + q\, \delta w)\, dx = 0\,. \quad (9.39)$$

N, Q, M are the stress resultants and n and q, respectively, are the loading in axial and perpendicular direction related to the beam axis. The definition of the stress resultants can be found in the next section. By introducing the vector containing the stress resultants $\mathbf{S}^T = \{\, N, Q, M\,\}$, Eq. (9.39) can be written in the compact form

$$G(\mathbf{u}, \boldsymbol{\eta}) = \int_0^l \delta\boldsymbol{\epsilon}^T \mathbf{S}\, dx - \int_0^l \boldsymbol{\eta}^T \mathbf{q}\, dx = 0\,, \quad (9.40)$$

where the loads are combined in $\mathbf{q} = \{\, n, q, 0\,\}^T$ and the variations of the deformations are given by $\boldsymbol{\eta} = \{\, \delta u, \delta w, \delta\psi\}^T$. The strains $\boldsymbol{\epsilon}$ stem from the definition given in (9.34). For the geometrically exact model, the variation of the strains in (9.34) yields

$$\delta\boldsymbol{\epsilon} = \mathbf{T}(\psi)\, \boldsymbol{\eta}' + \frac{\partial \mathbf{T}(\psi)}{\partial \psi}\, \mathbf{u}'\, \delta\psi\,. \quad (9.41)$$

By inserting this relation in the weak form (9.40), the first term can be specified as

$$\int_0^l \delta\boldsymbol{\epsilon}^T \mathbf{S}\, dx = \int_0^l \left[\boldsymbol{\eta}'^T \mathbf{T}(\psi)^T + \delta\psi\, \mathbf{u}'^T \left(\frac{\partial \mathbf{T}(\psi)}{\partial \psi} \right)^T \right] \mathbf{S}\, dx\,. \quad (9.42)$$

This is the stress divergence term of the geometrically exact beam model.

This weak form represents a nonlinear functional with respect to the displacements and rotations. Since analytical solutions are only available for special cases, the finite element method will be applied to solve (9.40).

Before the finite element discretization is formulated, the weak forms associated with the strain measures (9.35) and (9.36) are stated. Since the weak form (9.39) for the beam does not change in general, only the related variation of the strains has to be inserted. The variation of (9.35) leads to

$$\delta\boldsymbol{\epsilon} = \bar{\mathbf{T}}(\psi)\, \boldsymbol{\eta}' + \left[\frac{\partial \bar{\mathbf{T}}(\psi)}{\partial \psi}\, \mathbf{u}' - \frac{\partial \bar{\psi}}{\partial \psi} \right] \delta\psi\,, \quad (9.43)$$

which can be inserted in (9.40). In the same way, the variation of (9.36) is obtained

$$\delta\epsilon = \delta u' + w'\,\delta w',$$
$$\delta\kappa = \delta w'',\qquad (9.44)$$

which leads, neglecting the shear term in (9.39), to the weak form related to the theory of moderate rotations

$$G(u,w,\delta u,\delta w) = \int_0^l \left((\delta u' + w'\,\delta w')\,N + \delta w''\,M \right) dx - \int_0^l \left(n\,\delta u + q\,\delta w \right) dx = 0.$$

$$(9.45)$$

9.2.3 Constitutive Equations

Within the simulation of beam structures, it can be assumed for most applications that the strain are small, even for large deflections and rotations. Hence it is possible to describe elastic material behaviour by the classical HOOKE law of the linear theory. It relates within the geometrically exact theory the 1st PIOLA-KIRCHHOFF stresses which are back-rotated using matrix $\mathbf{T}_B = \mathbf{T}^T\,\mathbf{P}$ with the strains in (9.33). These stresses can be interpreted as BIOT stresses, see also (3.293). However, the stress $\mathbf{T}_B$ does not follow from the polar decomposition of the deformation gradient. With the engineering strains $\mathbf{E}^{(1)}$, which are equivalent to ϵ, see Remark 9.2$_1$, the stresses

$$\left\{\begin{matrix} T_{11} \\ T_{12} \end{matrix}\right\} = \begin{bmatrix} E & 0 \\ 0 & G \end{bmatrix} \left\{\begin{matrix} \epsilon + X_2\,\kappa \\ \gamma \end{matrix}\right\} \quad\text{or}\quad \mathbf{T}_B = \mathbf{C}\,\mathbf{E}_B \qquad (9.46)$$

are obtained as in the linear theory based on (3.273). The engineering strains are described by $\mathbf{E}_B$.

The integration of the stresses over the cross sectional area (width b and height h) yields the stress resultants

$$N = \int_{(h)} T_{11}\,b\,dX_2, \quad Q = \int_{(h)} T_{12}\,b\,dX_2 \quad\text{and}\quad M = \int_{(h)} T_{11}\,X_2\,b\,dX_2 \;(9.47)$$

leading to the compact form

$$\mathbf{S} = \mathbf{D}\,\epsilon \quad\text{with}\quad \mathbf{D} = \begin{bmatrix} EA & 0 & 0 \\ 0 & G\hat{A} & 0 \\ 0 & 0 & EI \end{bmatrix}, \qquad (9.48)$$

with YOUNG modulus E, the shear modulus G, the cross sectional area A and the moment of inertia I. The introduction of the shear area $\hat{A}$ is related to a shear correction term which is needed within to correct the violation of the boundary condition for the shear stresses ($T_{12} = 0$) at $X_2 = \pm h/2$ due to the beam model assumption of "plane sections remain plane".

For inelastic constitutive relations, an evolution equation for the inelastic part of the strains has to be formulated. This depends on the material at hand, see Sects. 3.3.2 and 3.3.3. Two possibilities exist within the beam theory to consider inelastic deformations. One is based on a two- or three-dimensional formulations of the inelastic constitutive equations using the stresses. The other possibility makes use of so-called integrated constitutive equations which are directly formulated in terms of stress resultants. Both variants will be described in the following for elasto-plastic material behaviour.

Stress Formulation. When the inelastic response of a beam is modelled using directly the stress then the constitutive equations of the two- or three-dimensional continuum can be adopted easily. Since the stress field over the cross sectional area of the beam is then no longer bi-linear, these stresses cannot be integrated analytically in order to get the stress resultants. Thus a numerical integration over the thickness has to be applied to obtain the stress resultants which have to be inserted in the weak form (9.39), see Fig. 9.7. The numerical integration is performed by GAUSS integration in the left part of the figure while in the right part the cross section is subdivided into layers which can be integrated separately.

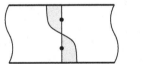

Fig. 9.7 Integration over the cross sectional area

The strains in (9.46) are additively split into an elastic and a plastic part

$$\boldsymbol{E}_B = \boldsymbol{E}_B^{el} + \boldsymbol{E}_B^{pl}. \tag{9.49}$$

The constitutive behaviour of the elastic part is described by (9.46). For the plastic part, a flow condition has to be formulated, the plastic evolution equation and a hardening law, see Sect. 3.3.2. Here these constitutive equations are selected analogous to Eqs. (3.160), (3.159) and (3.156). The evaluation of the flow condition (3.159) yields with linear hardening $Y(\hat{\alpha}) = Y_0 + \hat{H}\,\hat{\alpha}$ for the presented beam formulation

$$f_B(\boldsymbol{T}_B\,,\hat{\alpha}) = \sqrt{\boldsymbol{T}_B^T\,\boldsymbol{P}\,\boldsymbol{T}_B} - Y(\alpha) \le 0 \quad \text{with} \quad \boldsymbol{P} = \begin{bmatrix} 1 & 0 \\ 0 & 3 \end{bmatrix}. \tag{9.50}$$

The evolution of the plastic deformation is determined by the flow rule. With (3.160), the rate of the plastic strain is given by

$$\dot{\boldsymbol{E}}_B^{pl} = \lambda \frac{\partial f_B}{\partial \boldsymbol{T}_B} = \lambda \frac{\boldsymbol{P}\,\boldsymbol{T}_B}{\sqrt{\boldsymbol{T}_B^T\,\boldsymbol{P}\,\boldsymbol{T}_B}} := \lambda\,\boldsymbol{N}_B \qquad \dot{\hat{\alpha}} = \lambda\,. \tag{9.51}$$

Finally, the loading/unloading conditions (3.166) have to be considered.

These relations have be integrated in time. The general algorithms for this purpose are provided in Sect. 6.2. Within a time step $\Delta t = t_{n+1} - t_n$, an implicit EULER method is applied. This leads for the evolution equations to

$$\boldsymbol{E}_{B\,n+1}^{pl} = \boldsymbol{E}_{B\,n}^{pl} + \zeta_{n+1}\,\boldsymbol{N}_{B\,n+1} \quad \text{and} \quad \hat{\alpha}_{n+1} = \hat{\alpha}_n + \zeta_{n+1} \tag{9.52}$$

with $\zeta_{n+1} = \lambda\,\Delta t$. Making use of the predictor–corrector method, see equation (6.102) in Sect. 6.2.2, the relations

$$\begin{aligned}
\boldsymbol{E}_{B\,n+1}^{tr} &= \boldsymbol{E}_{B\,n+1} - \boldsymbol{E}_{B\,n}^{pl}, \\
\boldsymbol{T}_{B\,n+1} &= \bar{\boldsymbol{C}}(\zeta_{n+1})\,\boldsymbol{E}_{B\,n+1}^{tr}
\end{aligned} \tag{9.53}$$

are obtained. Matrix $\bar{\boldsymbol{C}}$ follows after some algebraic manipulation and by using $f_B = 0$ for plastic flow can be stated explicitly

$$\bar{\boldsymbol{C}}(\zeta) = [\,\boldsymbol{C}^{-1} + \beta\,\boldsymbol{P}\,]^{-1} = \begin{bmatrix} \frac{E}{1+E\,\beta} & 0 \\ 0 & \frac{G}{1+3\,G\,\beta} \end{bmatrix} \quad \text{with} \quad \beta = \frac{\zeta_{n+1}}{Y_{n+1}}\,. \tag{9.54}$$

The consistency parameter ζ_{n+1} is still unknown in this relation. It follows from the fulfilment of the flow condition (9.50): $f_B(\boldsymbol{T}_{B\,n+1}, \hat{\alpha}_{n+1}) = 0$ at time t_{n+1}. Since the relation is nonlinear in ζ_{n+1}, NEWTON method has to be applied for the solution.

Based on these considerations, the stresses can be computed at each integration point within the cross section of the beam

1. Predictor step: $\boldsymbol{T}_{B\,n+1}^{tr} = \boldsymbol{C}\,\boldsymbol{E}_{B\,n+1}^{tr}$.
2. Insertion into the flow condition:
   a) For $f_B(\boldsymbol{T}_{B\,n+1}^{tr}, \hat{a}_n) \leq 0$ the stress point is in the elastic range, which yields

$$\boldsymbol{T}_{B\,n+1} = \boldsymbol{T}_{B\,n+1}^{tr} \quad \text{and} \quad \Delta\boldsymbol{T}_{B\,n+1} = \boldsymbol{C}\,\Delta\boldsymbol{E}_{B\,n+1}\,. \tag{9.55}$$

   b) For $f_B(\boldsymbol{T}_{B\,n+1}^{tr}, \hat{a}_n) > 0$, a plastic corrector step has to be performed leading to

$$\begin{aligned}
\boldsymbol{T}_{B\,n+1} &= \bar{\boldsymbol{C}}\,\boldsymbol{E}_{B\,n+1}^{tr} \quad \text{and} \\
\Delta\boldsymbol{T}_{B\,n+1} &= \left[\, \bar{\boldsymbol{C}} - \frac{\bar{\boldsymbol{C}}\,\boldsymbol{N}_B\,\boldsymbol{N}_B^T\,\bar{\boldsymbol{C}}}{\boldsymbol{N}_B^T\,\bar{\boldsymbol{C}}\,\boldsymbol{N}_B + \hat{H}/(1-\beta\,\hat{H})} \,\right]\,\Delta\boldsymbol{E}_{B\,n+1}\,.
\end{aligned} \tag{9.56}$$

The incremental form of the constitutive equation is denoted here by $\Delta\boldsymbol{T}_{B\,n+1}$ which is needed in the linearization of the weak form.

The stress resultants are now obtained by integration over the cross section of the beam using (9.47). The stresses which have to be inserted in this equation follow either from (9.55) or from (9.56).

This procedure models the distribution of the stresses over the cross section within the assumptions of the beam theory in a realistic way. The computation, however, is time-consuming since the stresses have to be determined at each integration point within the cross section in order to capture the nonlinear distribution of the stresses. Furthermore, the plastic variables $\boldsymbol{E}_B^{pl}$ and $\hat{\alpha}$ have to be stored at each integration point as history variables. The integration is depicted in the left part of Fig. 9.7 for the GAUSS quadrature. Examples and finite element formulations based on this approach can be found, e.g. in Vogel (1965), Becker (1985) and for three-dimensional beam structures, see Gruttmann et al. (2000).

Stress Resultant Formulation. A more efficient scheme which can be used to analyze the elasto-plastic response of beams is provided by the so-called method of plastic hinges. It is based directly on the stress resultants for which now the nonlinear constitutive equations have to be formulated. Different approaches for steel are discussed in Windels (1970). They lead to nonlinear interaction relations between the different stress resultants. Hence the flow condition will now be formulated in terms of the stress resultants. Simplification can be found in civil engineering codes, like DIN 18800, which have the uniform structure

$$\Phi(N, Q, M) = \alpha_1 \frac{Q}{Q_{pl}} + \alpha_2 \frac{N}{N_{pl}} + \frac{M}{M_{pl}} - \alpha_3 \leq 0. \qquad (9.57)$$

Such formulations are limited in their range of application, but good results can be obtained for many practical problems, see e.g. Henning (1975), Vogel (1985) and Becker (1985). Explicit formulations can be found in the engineering literature which is quite extensive.

Formulations of the inelastic constitutive behaviour of beams undergoing large deflections and rotations can be found together with the associated finite element formulation in, e.g. Kahn (1987), Simo et al. (1984) and Ehrlich and Armero (2005).

By inserting the constitutive relations in the principle of virtual work or weak form (9.39), the nonlinear beam model is complete. This model and its approximations are now basis for the finite element discretization procedures discussed in the next section.

9.2.4 Finite Element Formulation

For the finite element discretization of the shear elastic geometrically exact beam model linear shape functions can be selected since the weak form needs only ansatz functions which are C^0-continuous. This is equivalent to the linear case, see e.g. Hughes (1987).

Hence polynomials (4.17) or (4.18) can be applied as finite element interpolations for the axial displacements u, the deflection w and the rotation ψ. Computations show that elements with quadratic shape functions yield

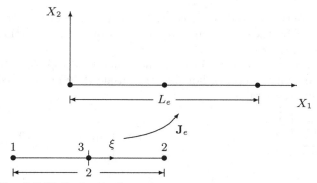

Fig. 9.8 Finite beam element

better approximations of the solutions than linear ones. Mathematically this leads to a higher convergence order. This is supported by convergence analysis of the linear case, see e.g. Chap. 8. Quadratic interpolations are based on elements with three nodes, see Fig. 9.8. The associated interpolation functions can be found in (4.18). In general, the finite element approximation for a finite element is given by

$$u_e = \sum_{I=1}^{n} N_I(X_1)\, u_I, \quad w_e = \sum_{I=1}^{n} N_I(X_1)\, w_I \quad \text{and} \quad \psi_e = \sum_{I=1}^{n} N_I(X_1)\, \psi_I,$$

(9.58)

for u, w and ψ. n is the number of nodes defining an element (linear ansatz: $n = 2$, quadratic ansatz: $n = 3$). The nodal values u_I, w_I and ψ_I will be determined by inserting these interpolations into the weak form. Inserting (9.58) into (9.34) yields the element strains. By combining the nodal values in the vector $\boldsymbol{u}_I = \{u_I, w_I, \psi_I\}^T$, the explicit form of the strains is obtained

$$\boldsymbol{\epsilon}_e = \sum_{I=1}^{n} \boldsymbol{T}(\psi_e)\, \boldsymbol{B}_{0I}\, \boldsymbol{u}_I - \boldsymbol{N} \quad \text{with} \quad \boldsymbol{B}_{0I} = \begin{bmatrix} N'_I & 0 & 0 \\ 0 & N'_I & 0 \\ 0 & 0 & N'_I \end{bmatrix}. \quad (9.59)$$

The term ()' denotes the derivative with respect to the coordinate X_1. The angle ψ_e in the rotation matrix $\boldsymbol{T}$ is computed using (9.58)$_3$.

The variation of the strains (9.41) can now be expressed by the shape functions. This leads to

$$\delta\boldsymbol{\epsilon}_e = \sum_{I=1}^{n} \boldsymbol{B}_I\, \boldsymbol{\eta}_I \quad \text{with} \quad \boldsymbol{B}_I = \begin{bmatrix} N'_I \cos\psi_e & N'_I \sin\psi_e & \alpha_e\, N_I \\ -N'_I \sin\psi_e & N'_I \cos\psi_e & \beta_e\, N_I \\ 0 & 0 & N'_I \end{bmatrix},$$

(9.60)

where the abbreviations α_e and β_e are defined as

$$\begin{aligned}\alpha_e &= -(1+u'_e)\sin\psi_e + w'_e\cos\psi_e, \\ \beta_e &= -(1+u'_e)\cos\psi_e - w'_e\sin\psi_e.\end{aligned} \tag{9.61}$$

By inserting this result in the stress divergence term, the weak form (9.42) can be completely expressed by the displacements and rotations after inserting the constitutive equation (9.48).

In an analogous way, the discrete form of the other two approximate nonlinear beam theories can be derived. For this, the ansatz (9.58) is inserted into (9.35) to determine the strains

$$\epsilon_e = \sum_{I=1}^{n} \begin{bmatrix} N'_I & N'_I\psi_e & -\tfrac{1}{2}\psi_e N_I \\ 0 & N'_I & -(1+u'_e)N_I \\ 0 & 0 & N'_I \end{bmatrix} u_I. \tag{9.62}$$

The variation of these strains is given by

$$\delta\epsilon_e = \sum_{I=1}^{n} B_I^S \eta_I \quad \text{with} \quad B_I^S = \begin{bmatrix} N'_I & N'_I\psi_e & (w'_e - \psi_e)N_I \\ -N'_I\psi_e & N'_I & -(1+u'_e)N_I \\ 0 & 0 & N'_I \end{bmatrix}, \tag{9.63}$$

which then can be inserted in (9.42). Again the constitutive equation (9.48) is applied.

This completes the formulation of the different finite element models for nonlinear beams since the stress resultants in (9.39) can be obtained via the material equation (9.48) in terms of the nodal quantities u_I. Generally, this leads to the nonlinear equation system

$$G(u,\eta) = \bigcup_{j=1}^{n_e} \sum_{i=I}^{n} \eta_I^T [R_I(u_I) - P_I] = 0 \tag{9.64}$$

with the vectors

$$R_I(u_I) = \int_0^{L_e} B_I^T(u_I) S_e(u_I)\,dx,$$

$$P_I = \int_0^{L_e} N_I q_e\,dx. \tag{9.65}$$

The symbol $\cup$ in (9.64) describes the assembly of all elements including the enforcement of continuity for the displacements and rotations along the element boundaries. L_e is the length of one finite element, see Fig. 9.4. Note that the integrals in (9.65) cannot be computed analytically in general. Thus GAUSS quadrature is applied, see Table 4.1. By transforming the integral to the local finite element coordinate ξ, see Fig. 9.8, relation

$$\boldsymbol{R}_I(\boldsymbol{u}_I) = \int\limits_{-1}^{+1} \boldsymbol{B}_I^T(\xi)\, \boldsymbol{S}_e(\xi)\, \frac{L_e}{2}\, d\xi \approx \sum_{p=1}^{n_p} \boldsymbol{B}_I^T(\xi_p)\, \boldsymbol{S}_e(\xi_p)\, \frac{L_e}{2}\, W_p \qquad (9.66)$$

is obtained using (4.25). For a linear two-node element, one-point integration is sufficient to exactly integrate the bending part. In that case, the shear term is underintegrated. The latter, however, is advantageous since it avoids shear locking. This effect is well known from the linear theory and will not be studied in detail. A complete discussion of this phenomenon can be found in, e.g. in Hinton and Owen (1979) or Hughes (1987). For a quadratic element with three nodal points, two GAUSS points for the integration of (9.66) have to be selected.

When an elasto-plastic material, see (9.55) and (9.56), is considered only the stresses and not the stress resultants are given. With the strains (9.46), the weak form can be reformulated. For that the projection tensor $\boldsymbol{P}_B$ is introduced which can be applied to compute from the strains $\boldsymbol{E}_B$ the strains $\boldsymbol{\epsilon}$ using (9.34). Furthermore, $\boldsymbol{P}_B$ transforms the stresses $\boldsymbol{T}_B$ to the stress resultants $\boldsymbol{S}$. With

$$\boldsymbol{P}_B = \begin{bmatrix} 1 & 0 & X_2 \\ 0 & 1 & 0 \end{bmatrix}, \qquad (9.67)$$

the strains and stress resultants are provided

$$\boldsymbol{E}_B = \boldsymbol{P}_B\, \boldsymbol{\epsilon} \quad \text{and} \quad \boldsymbol{S} = \int\limits_{(A)} \boldsymbol{P}_B^T\, \boldsymbol{T}_B\, dA, \qquad (9.68)$$

where the integral has to be evaluated with respect to the cross sectional area. Such integration takes into account the nonlinear material behaviour within the cross section. Here numerical integration is applied using either GAUSS quadrature, or even more simple, trapezoidal or mid-point rule. From

$$\int\limits_{(x)} \delta\boldsymbol{\epsilon}^T\, \boldsymbol{S}\, dx = \int\limits_{(x)} \int\limits_{(A)} \delta\boldsymbol{E}_B^T\, \boldsymbol{T}_B\, dA\, dx, \qquad (9.69)$$

the stress divergence term of the weak form follows with (9.68)

$$\boldsymbol{R}_I(\boldsymbol{u}_I) = \int\limits_0^{L_e} \boldsymbol{B}_I^T(\boldsymbol{u}_I) \left[\int\limits_{(A)} \boldsymbol{P}_B^T\, \boldsymbol{T}_B\, dA \right] dx. \qquad (9.70)$$

The numerical integration will be provided for a beam having a cross section with changing width $b(X_2)$. Based on this assumption, also I-beams can be considered. The height h of the cross section is parameterized by the coordinate χ. The application of GAUSS integration with n_q points over the cross sectional height leads analogous to (9.66) to

$$\boldsymbol{R}_I(\boldsymbol{u}_I) \quad = \quad \int\limits_{-1}^{+1} \boldsymbol{B}_I^T(\xi) \left[\int\limits_{-1}^{+1} \boldsymbol{P}_B^T(\chi)\, \boldsymbol{T}_B(\xi,\chi)\, b(\chi)\, \frac{h}{2}\, d\chi \right] \frac{L_e}{2}\, d\xi \qquad (9.71)$$

$$\approx \quad \sum_{p=1}^{n_p} \sum_{q=1}^{n_q} \boldsymbol{B}_I^T(\xi_p)\, \boldsymbol{P}_B^T(\chi_q)\, \boldsymbol{T}_B(\xi_p,\chi_q)\, b(\chi_q)\, \frac{h}{2}\, \frac{L_e}{2}\, W_p\, W_q\,.$$

The nonlinear weak form is now completed for a beam element based on the geometrically exact theory. For the approximate theories, the associated $\boldsymbol{B}$ matrices and the stresses have to be inserted; the latter follow from the approximate strains via the constitutive equation. The solution of (9.64) will be discussed next after the exercise.

Exercise 9.3: Derive the discretized weak form of (9.45) using the strains measures (9.36) of the BERNOULLI kinematics.

Solution: The weak form is provided by (9.45). The strains (9.36) based on BERNOULLI kinematics need shape for the deflection which are C^1-continuous since second order derivatives of the deflection w with respect to the coordinate X_1 occurs in the strain measure. For the axial displacement u, a linear interpolation function N_I is selected due to (4.17). For the deflection w cubical HERMITE functions H_I are applied, which are also used to discretize beam elements for the linear theory, see e.g. Hughes (1987) and Gross et al. (1999). Based on these functions, the following ansatz can be introduced for the deflection

$$w_e = H_1\, w_1 + \bar{H}_1\, w_1' + H_2\, w_2 + \bar{H}_2\, w_2'\,.$$

H_α and $\bar{H}_\alpha$ are cubical polynomials. By using the reference coordinate defined in Fig. 9.8, $-1 \le \xi \le 1$, the polynomial have the explicit form

$$H_1 \quad = \quad \frac{1}{4}\left(2 - 3\xi + \xi^3\right), \quad \bar{H}_1 = \frac{1}{4}\left(1 - \xi - \xi^2 + \xi^3\right),$$

$$H_2 \quad = \quad \frac{1}{4}\left(2 + 3\xi - \xi^3\right), \quad \bar{H}_2 = \frac{1}{4}\left(-1 - \xi + \xi^2 + \xi^3\right). \qquad (9.72)$$

As for the isoparametric C^0 continuous ansatz the following transformation of coordinate system holds $X_1 = \frac{L_e}{2}(\xi + 1)$ with the length of a beam element L_e, see Fig. 9.4. The polynomials (9.72) have the property

$$\begin{aligned} H_1(-1) &= 1, & \bar{H}_1(-1) = H_2(-1) = \bar{H}_2(-1) = 0 \\ \bar{H}_1'(-1) &= 1, & H_1'(-1) = H_2'(-1) = \bar{H}_2'(-1) = 0, \quad \text{etc.} \end{aligned}$$

This leads to the interpolation within an element e in terms of ξ

$$w_e(\xi) = H_1(\xi)\, w_1 + \bar{H}_1(\xi)\, \frac{dw_1}{d\xi} + H_2(\xi)\, w_2 + \bar{H}_2(\xi)\, \frac{dw_2}{d\xi}\,.$$

In this expression, the derivative of the deflection has to be transformed to the coordinate X_1. With $dX_1 = (L_e/2)\, d\xi$, the relation $dw/d\xi = dw/dX_1\,(dX_1/d\xi) = w'\,(L_e/2)$ is derived and hence

$$w_e(\xi) = \sum_{I=1}^{2} \left[H_I(\xi)\, w_I + \bar{H}_I(\xi)\, \frac{L_e}{2}\, w_I' \right]\,. \qquad (9.73)$$

The strains $\epsilon_e^B = \{\epsilon, \kappa\}^T$ can be written as

$$\epsilon_e^B = \sum_{I=1}^{2} \begin{bmatrix} N'_I & \frac{1}{2}w'_e H'_I & \frac{1}{2}w'_e \bar{H}'_I \frac{L_e}{2} \\ 0 & H''_I & \bar{H}''_I \frac{L_e}{2} \end{bmatrix} \mathbf{u}_I^B, \tag{9.74}$$

where $\mathbf{u}_I^B = \{u_I, w_I, w'_I\}^T$. The variation of ϵ_e^B yields

$$\delta\epsilon_e^B = \sum_{I=1}^{2} \mathbf{B}_I^B \, \eta_I \quad \text{with} \quad \mathbf{B}_I^B = \begin{bmatrix} N'_I & w'_e H'_I & w'_e \bar{H}'_I \frac{L_e}{2} \\ 0 & H''_I & \bar{H}''_I \frac{L_e}{2} \end{bmatrix}. \tag{9.75}$$

This relation can be inserted in (9.45) which results in the matrix formulation of the weak form

$$G^B(\mathbf{u}, \eta) = \bigcup_{j=1}^{n_e} \sum_{I=1}^{2} \delta\mathbf{u}_I^B \left[\int_{-1}^{1} \mathbf{B}_I^{B\,T}(\mathbf{u}_e)\, \mathbf{S}_e(\mathbf{u}_e) \frac{L_e}{2}\, d\xi \right.$$

$$\left. - \int_{-1}^{1} [\, N_I\, \mathbf{e}_1\, n_e + (\, H_I\, \mathbf{e}_2 + \bar{H}_I(\xi)\, \mathbf{e}_3\,)\, q_e\,]\, d\xi \right] = 0 \tag{9.76}$$

with the vectors $\mathbf{e}_1 = \{1, 0, 0\}^T$, $\mathbf{e}_2 = \{0, 1, 0\}^T$ and $\mathbf{e}_3 = \{0, 0, \frac{L_e}{2}\}^T$.

Discretization of the Linearization of the Weak Form. Based on the linearization of (9.64) or (9.76), the methods described in Chap. 5 can be applied to solve (9.64) or (9.76), like the NEWTON or the arc-length method. Within this linearization process, it is essential that no terms are neglected in order to obtain the quadratic convergence properties of NEWTON method.

The complete linearization of the weak form (9.39) can be derived once a constitutive equation is selected. Here (9.48) is used and the weak form

$$G(\mathbf{u}, \eta) = \int_0^l (\,\delta\epsilon\, EA\,\epsilon + \delta\gamma G\hat{A}\,\gamma + \delta\kappa\, EI\,\kappa\,)\, dx - \int_0^l (\, n\,\delta u + q\,\delta w\,)\, dx = 0 \tag{9.77}$$

is obtained. The linearization of G is a formal procedure, see Sect. 3.5.3. It yields

$$DG(\mathbf{u}, \eta) \cdot \Delta\mathbf{u} = \int_0^l (\,\delta\epsilon\, EA\, \Delta\epsilon + \delta\gamma G\hat{A}\, \Delta\gamma + \delta\kappa\, EI\, \Delta\kappa\,)\, dx +$$

$$+ \int_0^l (\, \Delta\delta\epsilon\, N + \Delta\delta\gamma Q + \Delta\delta\kappa M\,)\, dx. \tag{9.78}$$

When elasto-plastic constitutive equations like (9.55) and (9.56) are used, then again the integration over the cross sectional area has to be performed.

Since the mathematical rules which are applied to compute variations and linearization do not differ, see Sect. 3.5, the variation $\delta\epsilon$ and linearization $\Delta\epsilon$ of the strains have the same structure. Only $\boldsymbol{\eta}$ has to be exchanged by $\Delta\mathbf{u}$. With this observation, the discretization of the first integral can be stated directly without further derivations. The discretization of the second integral has still to be computed since it contains linearizations of the virtual strains $\delta\epsilon$. With (9.33), relation

$$
\begin{aligned}
\Delta\delta\epsilon &= [-\delta u' \sin\psi + \delta w' \cos\psi]\Delta\psi + [-\Delta u' \sin\psi + \Delta w' \cos\psi]\delta\psi + \\
&+ \delta\psi[-(1 + u') \cos\psi - w' \sin\psi]\Delta\psi\,, \\
\Delta\delta\gamma &= [-\delta u' \cos\psi - \delta w' \sin\psi]\Delta\psi + [-\Delta u' \cos\psi - \Delta w' \sin\psi]\delta\psi + \\
&+ \delta\psi[(1 + u') \sin\psi - w' \cos\psi]\Delta\psi\,, \\
\Delta\delta\kappa &= 0
\end{aligned}
\tag{9.79}
$$

is obtained for the geometrically exact model. By introducing the finite element interpolations (9.58), the tangent stiffness matrix $\boldsymbol{K}_T$ is formulated based on the linearization (9.78). Its explicit form for the geometrically exact model is given by

$$
\boldsymbol{K}_T = \bigcup_{j=1}^{n_e} \sum_{I=1}^{n} \sum_{K=1}^{n} \boldsymbol{K}_{T\,IK}
\tag{9.80}
$$

with

$$
\boldsymbol{K}_{T\,IK} = \int_0^{L_e} \boldsymbol{B}_I^T \, \boldsymbol{D} \, \boldsymbol{B}_K \, dx + \int_0^{L_e} (N \, \boldsymbol{G}_{IK}^N + Q \, \boldsymbol{G}_{IK}^Q) \, dx\,.
\tag{9.81}
$$

The integration is performed numerically as for the residual in (9.66). This leads finally to the explicit form of the tangent matrix

$$
\begin{aligned}
\boldsymbol{K}_{T\,IK} &= \int_{-1}^{+1} [\,\boldsymbol{B}_I^T \, \boldsymbol{D} \, \boldsymbol{B}_K + N \, \boldsymbol{G}_{IK}^N + Q \, \boldsymbol{G}_{IK}^Q\,] \frac{L_e}{2}\, d\xi \\
&\approx \sum_{p=1}^{n_p} \left[\boldsymbol{B}_I^T(\xi_p) \boldsymbol{D} \boldsymbol{B}_K(\xi_p) + N(\xi_p)\, \boldsymbol{G}_{IK}^N(\xi_p) + Q(\xi_p)\, \boldsymbol{G}_{IK}^Q(\xi_p) \right] \frac{L_e}{2} W_p\,.
\end{aligned}
$$

In the case of an elasto-plastic material, integration over the cross section has to be performed. For the materials described in (9.55) and (9.56), the tangent matrix can be determined using the projection operator (9.51), the variable width $b(X_2)$ and the parametrization of the coordinate X_2 in thickness direction by χ. The application of a GAUSS quadrature with n_q points in thickness direction yields the sub-tangent matrix for the nodal combination $I\,,K$

$$
\boldsymbol{K}_{T\,IK} = \int_{-1}^{+1}\int_{-1}^{+1} [\,\boldsymbol{B}_I^T \, \boldsymbol{P}_B^T \, \hat{\boldsymbol{C}} \boldsymbol{P}_B \, \boldsymbol{B}_K + T_{B\,11} \, \boldsymbol{G}_{IK}^N + T_{B\,12} \, \boldsymbol{G}_{IK}^Q\,] \, b \, \frac{h}{2} \, \frac{L_e}{2} \, d\chi \, d\xi
$$

$$\approx \sum_{p=1}^{n_p} \sum_{q=1}^{n_q} \left[\boldsymbol{B}_I^T(\xi_p) \boldsymbol{P}_B^T(\chi_p) \, \hat{\boldsymbol{C}}(\xi_p, \chi_q) \, \boldsymbol{P}_B(\chi_q) \boldsymbol{B}_K(\xi_p) \right.$$

$$\left. + \quad T_{B\,11}(\xi_p, \chi_q) \, \boldsymbol{G}_{IK}^N(\xi_p) + T_{B\,12}(\xi_p, \chi_q) \, \boldsymbol{G}_{IK}^Q(\xi_p) \right] b(\chi_q) \frac{L_e}{2} W_p \, W_q \, .$$

In this equation, either the elastic modulus (9.55) or the elasto-plastic modulus (9.56) has to be inserted for the material tangent $\hat{\boldsymbol{C}}$.

The matrices which are used in (9.81) are given by

$$\boldsymbol{G}_{IK}^N = \begin{bmatrix} 0 & 0 & -N'_I\,N_K\,\sin\psi_e \\ 0 & 0 & N'_I\,N_K\,\cos\psi_e \\ -N_I\,N'_K\,\sin\psi_e & N_I\,N'_K\,\cos\psi_e & \alpha_3\,N_I N_K \end{bmatrix}$$

$$\boldsymbol{G}_{IK}^Q = \begin{bmatrix} 0 & 0 & -N'_I\,N_K\,\cos\psi_e \\ 0 & 0 & -N'_I\,N_K\,\sin\psi_e \\ -N_I\,N'_K\,\cos\psi_e & -N_I\,N'_K\,\sin\psi_e & \alpha_4\,N_I N_K \end{bmatrix} .$$

They stem from the linearization of the virtual strains. The abbreviations are defined as

$$\begin{aligned} \alpha_3 &= -(1 + u'_e)\cos\psi_e - w'_e\,\sin\psi_e \, , \\ \alpha_4 &= (1 + u'_e)\sin\psi_e - w'_e\,\cos\psi_e \, . \end{aligned} \tag{9.82}$$

In the same way, the tangential stiffness matrix is derived for the approximate theories. This leads for the formulation based on (9.35) with matrix $\boldsymbol{B}_I^S$ from (9.63) to

$$\boldsymbol{K}_T^S = \bigcup_{j=1}^{n_e} \sum_{I=1}^{n} \sum_{K=1}^{n} \left[\int_0^{L_e} \boldsymbol{B}_I^{S^T} \boldsymbol{D}\boldsymbol{B}_K^S \, dx + \int_0^{L_e} (N\,\boldsymbol{G}_{IK}^{SN} + Q\,\boldsymbol{G}_{IK}^{SQ})\, dx \right] \tag{9.83}$$

with

$$\boldsymbol{G}_{IK}^{SN} = \begin{bmatrix} 0 & 0 & 0 \\ 0 & 0 & N'_I\,N_K \\ 0 & N_I\,N'_K & -N_I N_K \end{bmatrix} , \quad \boldsymbol{G}_{IK}^{SQ} = \begin{bmatrix} 0 & 0 & -N'_I\,N_K \\ 0 & 0 & 0 \\ -N_I\,N'_K & 0 & 0 \end{bmatrix} .$$

This result can also be derived directly from the above relations of the geometrically exact theory. In that case, the limits $\sin\psi \to \psi$ and $\cos\psi \to 1$ have to be computed for $\psi \to 0$.

The previous set of equations is sufficient to establish the tangential stiffness matrices and residuals needed with a solution method like the NEWTON scheme. These equations are related to the local coordinate system of the straight beam axis. Since beam members are used in most cases within the construction of complex structures, like multi-storey frames in which the beams are located in different positions, the matrices and vectors have to be transformed to a global coordinate system. This can be performed in the same way as in the linear theory since all equations are referred to the initial configuration. The transformation and associated matrices can be found, e.g. in Crisfield (1991), Zienkiewicz and Taylor (2000a), Hughes (1987) and Bathe (1996), see also Remark 9.1. By such transformation, the local nodal displacements and rotations $\boldsymbol{u}_I^l$ at node I, see (9.58), can be expressed in terms of the global deformations $\boldsymbol{u}_I^g$ via

$$\boldsymbol{u}_I^l = \bar{\boldsymbol{T}}_I \, \boldsymbol{u}_I^g \,. \tag{9.84}$$

The explicit form is given in the two-dimensional case by

$$\left\{ \begin{matrix} u_I \\ w_I \\ \psi_I \end{matrix} \right\}^l = \begin{bmatrix} \cos\alpha & \sin\alpha & 0 \\ -\sin\alpha & \cos\alpha & 0 \\ 0 & 0 & 1 \end{bmatrix} \left\{ \begin{matrix} u_I \\ w_I \\ \psi_I \end{matrix} \right\}^g \,. \tag{9.85}$$

The angle α refers to the angle between the local and global coordinate axis X_1. This transformation can be written in the same way for the virtual nodal displacements $\boldsymbol{\eta}_I$. Using this transformation, the local form (9.64) of the residual vector $\boldsymbol{R}_I^l$ and the local form of (9.81) of the tangent matrix $\boldsymbol{K}_{T\,IK}^l$ can be expressed in terms of the global coordinates

$$\boldsymbol{R}_I^g = \bar{\boldsymbol{T}}_I^T \, \boldsymbol{R}_I^l \quad \text{and} \quad \boldsymbol{K}_{T\,IK}^g = \bar{\boldsymbol{T}}_I^T \, \boldsymbol{K}_{T\,IK}^l \, \bar{\boldsymbol{T}}_I \,. \tag{9.86}$$

Exercise 9.4: Derive the linearization for the strain measures and weak form of Exercise 9.3. The result has to be compared with the equations stemming from the second order theory, often used in practical engineering applications.

Solution: Using equation (9.36), the variation and linearization of the strain measures can be computed

$$\begin{aligned} \delta\epsilon &= \delta u' + w' \, \delta w' \,, & \delta\kappa &= \delta\psi' \,, \\ \Delta\epsilon &= \Delta u' + w' \, \Delta w' \,, & \Delta\kappa &= \Delta\psi' \,. \end{aligned}$$

They have the same structure and thus can both be discretized using (9.75). The linearization of the virtual strains yields furthermore

$$\Delta\delta\epsilon = \Delta w' \, \delta w' \,, \quad \Delta\delta\kappa = 0 \,,$$

which leads, as for the geometrically exact model, to the tangent matrix

$$\boldsymbol{K}_T^B = \bigcup_{j=1}^{n_e} \sum_{I=1}^{n} \sum_{K=1}^{n} \left[\int_0^{L_e} \boldsymbol{B}_I^{B\,T} \, \boldsymbol{D}^B \boldsymbol{B}_K^B \, dx + \int_0^{L_e} N \, \boldsymbol{G}_{IK}^B \, dx \right] \,. \tag{9.87}$$

The matrix $\boldsymbol{G}_{IK}^B$ has the form

$$\boldsymbol{G}_{IK}^B = \begin{bmatrix} 0 & 0 & 0 \\ 0 & H'_I \, H'_K & H'_I \, \bar{H}'_K \, \frac{L_e}{2} \\ 0 & \bar{H}'_I \, H'_K \, \frac{L_e}{2} & \bar{H}'_I \, \bar{H}'_K \, \frac{L_e^2}{4} \end{bmatrix} \,. \tag{9.88}$$

In the case of the second order theory only the influence of the normal force is considered at the deformed system, see e.g. Petersen (1980). Hence the nonlinear part can be neglected in the strain measure (9.36) which yields the linear strain

$$\epsilon^{II} = \sum_{I=1}^{2} \begin{bmatrix} N'_I & 0 & 0 \\ 0 & H''_I & \bar{H}''_I \frac{L_e}{2} \end{bmatrix} \mathbf{u}_I \,. \tag{9.89}$$

and its variation

$$\delta\epsilon^{II} = \sum_{I=1}^{2} \boldsymbol{B}_I^{II} \, \boldsymbol{\eta}_I \quad \text{with} \quad \boldsymbol{B}_I^{II} = \begin{bmatrix} N'_I & 0 & 0 \\ 0 & H''_I & \bar{H}''_I \frac{L_e}{2} \end{bmatrix} \,. \tag{9.90}$$

Instead the normal force is split into a linear part $EA\,u'$ and a nonlinear one $N\,w'\delta w'$ within the principle of virtual work (9.45), leading to

$$G(\mathbf{u}, \boldsymbol{\eta}) = \int_0^l (\delta u' \, EA\, u' + N\, w' \, \delta w' + \delta w'' \, EI\, w'') \, dx - \int_0^l \boldsymbol{\eta}^T \mathbf{q}\, dx = 0 \,. \tag{9.91}$$

Here it is assumed that the normal force is constant within each load step; hence it can be viewed as constant within the linearization. This yields the tangential stiffness matrix

$$\boldsymbol{K}_T^{II} = \bigcup_{j=1}^{n_e} \sum_{I=1}^{n} \sum_{K=1}^{n} \left[\int_0^{L_e} \boldsymbol{B}_I^{II\,T} \boldsymbol{D}^B \boldsymbol{B}_K^{II} \, dx + \int_0^{L_e} N\, \boldsymbol{G}_{IK}^B \, dx \right] , \tag{9.92}$$

where $\boldsymbol{G}_{IK}^B$ is equivalent to the term stated in (9.87). Thus the tangent matrix has the same form as the tangent matrix of the theory of moderate rotations. Only the influence of the nonlinear term w'^2 is neglected in the constitutive equation for the computation of the normal force.

9.2.5 Example

A geometrically exact theory and different consistent approximate theories with the strain measures (9.35) were derived for a two-dimensional beam in the previous sections. Besides this, the second order theory as most simple theory to include nonlinear behaviour within beam structures was presented which relies on further simplifications. It is now of certain value to discuss the applicability of the different theories. While the geometrically exact theory can be applied in all cases, the approximate theory based on (9.35) is restricted to relative small displacements and rotations. It can, however, be applied to solve beam structures with vanishing bending stiffness like ropes or chains. The classical second order theory is not applicable when the normal forces to be considered stem from the nonlinear tern w'^2 in the constitutive equations. In such cases, the theory of moderate rotations based on (9.36) has to be used. This is, e.g. the case when the stiffening effect of boundary conditions has to be considered, like e.g. a beam between to spatially fixed supports.

A comparison of the different theories is now presented by means of an example. The selected frame structure is depicted in Fig. 9.9 which also shows geometrical and material data. The frame is loaded by a point force $F = \lambda \cdot 1$

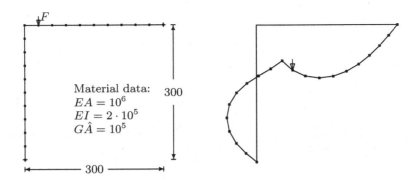

Fig. 9.9 Frame structure   **Fig. 9.10** Deformed structure

acting in vertical direction. Twenty one finite beam elements are applied to discretize the structure. The computation of the load deflection curve is performed by using an arc-length method discussed in Sect. 5.1.5. Figure 9.10 shows the deformed configuration for a load factor $\lambda = 45$, which was computed with the geometrically exact beam model.

Figure 9.11 depicts the load deflection curves for the geometrically exact model, the other two approximate theories and the theory of second order. In this figure, the load is plotted versus the vertical displacement under the point load. Differences are clearly visible. The classical second order theory follows the results of the geometrically exact theory only for small displacements and hence can only be applied for problems with small displacements and rotations. Hence the second order theory cannot be applied to model post-critical states of structures. However, since many beam structures only

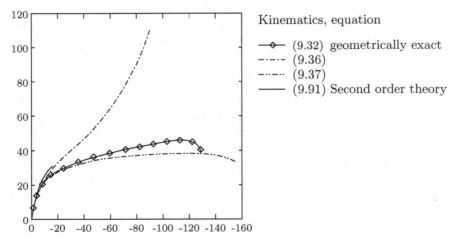

Fig. 9.11 Load deflection curve of the frame

undergo small displacements and rotations in practical applications, the second order theory has its eligibility.

The theory of moderate rotations based on equation (9.35) is closer to the exact model, but deviates from the result of the geometrically exact theory when the load factor is larger than $\lambda = 28$. It is interesting to note that this simple model recovers the correct tendency of the solution and hence can be used to estimate the behaviour in post-critical states.

9.2.6 Summary

This section presented a short overview with respect to different beam theories spanning the arc from geometrically exact models to second order theories. For all theoretical models, weak forms were developed and the discretization leading to residuals and tangent matrices for the nonlinear beam theories were derived. In summary, the following statements can be made:

- The effort for the computation of residuals vectors and tangent matrices is almost identical for the geometrically exact model and the approximate theories.
- One can always use the best (geometrically exact) model to discretize beam structures. Such approach ensures that, from the model point of view, the finite element analysis will converge to the correct theoretical solution.
- Exact analytical solution of approximate theories (these can be developed for the second order theory) are not exact when it comes to modelling a nonlinear beam problem.
- From the results of the example in Fig. 9.9, it can be deduced that approximate theories are useful for a wide range of problems in which the effects of the nonlinear behaviour yield only small displacements and rotations.

9.3 Axisymmetric Shell Element

Many engineering shell problems can be described by axisymmetric deformation states since different structures like cylindrical or spherical shells are axisymmetric. Problems which belong to this class are air-filled springs made of rubber or forming processes including inelastic deformations. Also deformations related to biomechanical systems, like e.g. arteries, can be modeled by axisymmetric shell elements when geometry and loading are adequate. Axisymmetric formulations can be used for the general problem class of shells of revolution. These are very efficient since they basically reduce the dimension by one and hence lead to equations systems which are very sparse and can be solved in a fraction of the computing time of the associated three-dimensional shell computations.

Shell theories base usually on the kinematic assumption either of shear elastic or of shear rigid models. A special case is related to pure membrane behaviour which can be applied to describe certain phenomena and structures in engineering but also in biomechanics. Technical applications are deep drawing processes of metal sheets, rubber balloons under internal pressure or the deformation of blood cells which depict no or negligible bending energy. Hence besides the shell equations the equations for pure membrane states will also be presented in the following.

All mentions problem areas include, besides finite deformations, nonlinear material behaviour. Finite deformations can be formulated, as in the previous section, with respect to the initial configuration. For the constitutive model, an hyperelastic *Ogden* material will be applied, see Sect. 3.3.1, which can be used to model a large range of rubberlike materials. Furthermore, finite inelastic strains will be considered for metals and in biomechanical applications, see e.g. Wriggers et al. (1995), Holzapfel et al. (1996a) and Holzapfel et al. (1996b).

The finite element formulation for the axisymmetrical shell follows to a large extend the derivations in Wriggers et al. (1995). It bases on the assumption of a straight element describing the geometry of a cone which, however, can be used to model arbitrary geometries approximatively. The special case of such finite element shell formulation is a membrane which was explicitly derived in Wriggers and Taylor (1990).

9.3.1 Kinematics and Strains of the Axisymmetrical Shell

Again, as for beams, a geometrically exact model is introduced for the shell which has, besides the kinematical assumption of the shell theory, no further restrictions with respect to displacements, rotations and strains.

The relevant nonlinear strains measures will not be derived in detail; they can be found in e.g. Wagner (1990) for shear elastic rotational shells. Due to the formulation, which takes shear deformation into account, simple C^0-elements can be applied to discretize the nonlinear shell equations. In case of thin shells, the shear elastic theory is not necessary since the classical KIRCHHOFF–LOVE theory is sufficient to model these problems in engineering applications. Such model however would need a discretization on the basis of C^1-elements since the ansatz functions have to be continuous for the displacements and for the derivatives. Thus the resulting finite element formulation is more complex, for the linear case see e.g. Zienkiewicz and Taylor (1989).

It is well known that locking occurs in the thin shell limit within a shear elastic finite element model. In that case, the shear deformations tend to zero and this constraint is responsible for locking. In case of the axisymmetrical shell formulation, locking can be circumvented by reduced integration. In case of a two-node linear element, reduced integration does not lead to a rank deficient tangent matrix; the associated linear case was discussed in

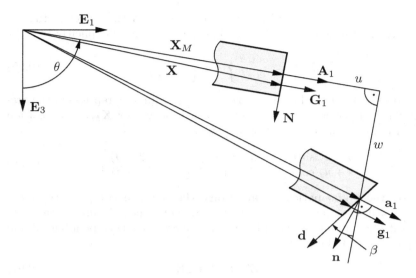

Fig. 9.12 Kinematics for shells of revolution

Zienkiewicz et al. (1977). Due to this fact, it is possible to apply the shear elastic shell theory without any complications.

Here a finite element formulation for thin shells will be developed. In that case, the shear deformation is almost zero and hence this part of the strain energy can be modeled by a penalty method in order to enforce zero shear in the finite element model. Such model is called quasi-KIRCHHOFF model and is only applied in order to be able to use simple C^0-elements for the discretization of the shells. More complex models including real shear elastic behaviour and thickness change can be found in e.g. Eberlein (1997).

The deformation of the shell continuum - based on these preliminary remarks - is described in Fig. 9.12. A point in shell space is given by the position vector $\mathbf{X}$ which can be expressed by the position vector $\mathbf{X}_M$ to the shell midsurface, the normal vector, $\mathbf{N}$, and the local coordinate in thickness direction, ξ,

$$\mathbf{X} = \mathbf{X}_M + \xi \, \mathbf{N} \, . \tag{9.93}$$

All quantities describe the initial configuration of the shell. The coordinate ξ is restricted to $-\frac{t_0}{2} \leq \xi \leq +\frac{t_0}{2}$ with t_0 being the shell thickness in the initial configuration.

In the same way, a point in the deformed shell continuum is given by

$$\mathbf{x} = \mathbf{x}_M + \xi \, \mathbf{d} \, . \tag{9.94}$$

Again $\mathbf{x}_M$ is related to the deformed shell midsurface. The director vector $\mathbf{d}$ describes the rotation of the cross section relative to the midsurface. Such geometrical assumption includes shear deformation. With respect to the notation presented in Fig. 9.12, the position vector is stated explicitly by

$$\mathbf{X} = \left\{ \begin{array}{c} s \sin\theta \\ s \cos\theta \end{array} \right\} + \xi \left\{ \begin{array}{c} -\cos\theta \\ \sin\theta \end{array} \right\} \tag{9.95}$$

with respect to the initial configuration. The coordinate s represents the arclength which describes the length of the position vector $\mathbf{X}_M$. The position vector in the current configuration is now formulated as

$$\mathbf{x} = \left\{ \begin{array}{c} (s+u)\sin\theta - w\cos\theta \\ (s+u)\cos\theta + w\sin\theta \end{array} \right\} + \xi \left\{ \begin{array}{c} -\cos(\theta-\beta) \\ \sin(\theta-\beta) \end{array} \right\} . \tag{9.96}$$

Based on these kinematical assumptions, the physical components of the GREEN–LAGRANGE strain tensor can be derived for the axisymmetrical shell. In general, the strain tensor can be split in a membrane (m), bending (b) and shear part (s) as follows

$$\mathbf{E} = \mathbf{E}^m + \mathbf{E}^s + \xi \mathbf{E}^b . \tag{9.97}$$

Again ξ is the local coordinate in thickness direction. Terms which are multiplied by ξ^2 have been neglected in this derivation since only thin shells are considered and thus this term can be neglected due to its smallness.

The strains in meridian direction, E_1, and the hoop strains, E_2, follow as well as the shear strains E_{13} from the kinematical relation (9.96)

$$E_1^m = u_{,s} + \tfrac{1}{2}(u_{,s}^2 + w_{,s}^2),$$

$$E_2^m = e_\theta + \tfrac{1}{2}e_\theta^2 \text{ with } e_\theta = \tfrac{1}{r}(u\sin\theta - w\cos\theta), \quad r = s\sin\theta,$$

$$E_1^b = -[(1+u_{,s})\cos\beta + w_{,s}\sin\beta]\beta_{,s},$$

$$E_2^b = \frac{\cos\theta}{r} - r\,c_2(1+e_\theta) \text{ with } c_2 = \frac{1}{r^2}(\sin\theta\sin\beta + \cos\theta\cos\beta),$$

$$E_{13}^s = -(1+u_{,s})\sin\beta + w_{,s}\cos\beta . \tag{9.98}$$

In these, equation u and w are displacements with respect to the local coordinate system. β is the angle of rotation which describes a plane rotation of the director vector $\mathbf{d}$, see Fig. 9.12. These strain measures include no further approximations than the ones due to the kinematical model (9.96); hence they are valid for finite strains and rotations of shells of revolution. The strains E_γ can also be expressed by the principal strains λ_γ, see (3.23). For shells of revolution with axisymmetrical loading, this is quite simple, since only the strains

$$E_1 = E_1^m + \xi E_1^b \quad \text{and} \quad E_2 = E_2^m + \xi E_2^b \tag{9.99}$$

occur which are perpendicular to each other. Hence these are principal strains by definition. From relation $E_\gamma = 1/2\,(\lambda_\gamma^2 - 1)$, the principal strains follow with (9.98) in explicit form

$$\lambda_1 = \sqrt{(1 + u_{,s})^2 + w_{,s}^2 - 2\,\xi\,[(1 + u_{,s})\cos\beta + w_{,s}\,\sin\beta]\,\beta_{,s}}\,,$$

$$\tag{9.100}$$

$$\lambda_2 = \sqrt{(1 + e_\theta)^2 + 2\,\xi\,[\,\tfrac{\cos\theta}{r} - r\,c_2\,(1 + e_\theta)\,]}\,.$$

The principal strain λ_3 can be obtained from the constitutive equation by assuming plane stress which is a valid assumption for thin shells.

For further reference, the determinant J of the deformation gradient $\mathbf{F}$ is stated which is simply the product of the principal strains

$$J = \lambda_1\,\lambda_2\,\lambda_3\,. \tag{9.101}$$

Note that incompressible material response ($J = 1$) can be included exactly within this model. Since the direction of principal strain λ_3 is normal to the shell midsurface for thin shells, it is simply given by the thickness change $\lambda_3 = t/t_0$. Thus the current shell thickness is given for incompressibility by

$$t = \frac{t_0}{\lambda_1\,\lambda_2} \tag{9.102}$$

with the thickness t_0 of the shell in the initial configuration.

9.3.2 Variational Formulation

The principel of virtual work or weak, for the general form see (3.292), can be stated under the assumption of axisymmetrical geometry and loading with respect to the initial configuration as

$$G(\mathbf{u}, \boldsymbol{\eta}) = 2\pi \left[\int_{(C)} \int_\xi S_\gamma\, \delta E_\gamma\, r\, d\xi\, dS + \epsilon \int_{(C)} \int_\xi E_{13}^s\, \delta\, E_{13}^s\, r\, d\xi\, dS \right]$$

$$-2\pi \int_{(C)} \int_\xi \hat{t}_\gamma\, \eta_\gamma\, r\, d\xi\, dS = 0\,. \tag{9.103}$$

In this equation, the sum convention is used where the Greek indices assume values 1 and 2 ($\gamma = 1, 2$). The curve describing the shell of revolution is denoted by C in (9.103). The radius of the shell is described by r; it is measured in the initial configuration. The virtual strains δE_γ result from (9.99) and (9.98). η_γ denotes the variation of u_γ, which have to fulfil the essential boundary conditions: $\{\eta_\gamma \mid \eta_\gamma = 0 \quad \text{auf } \partial C_u\}$, as already discussed in Sect. 3.4.1.

The stress vector $\hat{t}_\gamma$ in (9.103) describes surface loads acting on the shell. For a deformation dependent pressure load p, the relation $\hat{t}_\gamma = p\,n_\gamma$ is obtained, where n_γ are the components of the normal vector in the current configuration. The explicit finite element formulation for this loading case can be found in Sect. 4.2.5, see Exercise 4.5.

The work conjugated 2nd PIOLA–KIRCHHOFF stresses S_γ are computed from the CAUCHY stresses by $\mathbf{S} = J\,\mathbf{F}^{-1}\,\boldsymbol{\sigma}\,\mathbf{F}^{-T}$, see (3.83). For the principal stresses, this simplifies to

$$S_1 = J\,\lambda_1^{-2}\,\sigma_1, \qquad S_2 = J\,\lambda_2^{-2}\,\sigma_2. \tag{9.104}$$

With $J = 1$, relations

$$S_1 = \lambda_1^{-2}\,\sigma_1, \qquad S_2 = \lambda_2^{-2}\,\sigma_2 \tag{9.105}$$

follow for the special case of incompressibility.

9.3.3 Constitutive Equations

Three different constitutive equations are discussed in the following and applied to finite deformation states of shells of revolution. The materials belong to different engineering problem classes and include constitutive relations for rubberlike materials, metals and biomechanical materials.

Rubberlike Materials. Rubberlike materials can best be described by the constitutive equation (3.113) of Ogden (1972) which was developed for incompressible materials. The constitutive relation is based upon a strain energy function formulated in terms of the principal stretches λ

$$W(\lambda_i) = \sum_r \frac{\mu_r}{\alpha_r}\,[\,\lambda_1^{\alpha_r} + \lambda_2^{\alpha_r} + \lambda_3^{\alpha_r} - 3\,]. \tag{9.106}$$

μ_r and α_r are constitutive constants which have to be deducted by experiments. The CAUCHY stresses for an incompressible material follow with (3.134) analogous to (9.10)

$$\sigma_i = \lambda_i\,\frac{\partial W}{\partial \lambda_i} + p, \tag{9.107}$$

where p is the pressure which is related to the constraint $J = 1$. In case of a plane stress state, the pressure can be determined from the equation $\sigma_3 \equiv 0$. With the incompressibility condition $\lambda_3 = (\lambda_1\,\lambda_2)^{-1}$, the result

$$\sigma_\gamma = \sum_r \mu_r\,[\,\lambda_\gamma^{\alpha_r} - (\lambda_1\,\lambda_2)^{-\alpha_r}\,] \tag{9.108}$$

is derived. By inserting this relation in (9.105), the 2nd PIOLA–KIRCHHOFF stresses are obtained which are used in the weak form (9.103).

The incremental constitutive tangent with reference to the initial configuration is derived from

$$L_{\gamma\delta} = \frac{\partial S_\gamma}{\partial E_\delta} = \frac{\partial S_\gamma}{\partial \lambda_\beta}\frac{\partial \lambda_\beta}{\partial E_\delta}, \tag{9.109}$$

see also Sect. 3.3.4. The components of the material tangent can be stated explicitly as

$$L_{\gamma\gamma} = \frac{1}{\lambda_\gamma^4}\sum_r \mu_r \left[(\alpha_r - 2)\,\lambda_\gamma^{\alpha_r} + (\alpha_r + 2)(\lambda_1\,\lambda_2)^{-\alpha_r}\right] \quad (\gamma = \delta),$$

$$L_{\gamma\delta} = \frac{1}{\lambda_\gamma^2\,\lambda_\delta^2}\sum_r \mu_r\,\alpha_r\,(\lambda_1\,\lambda_2)^{-\alpha_r} \quad (\gamma \neq \delta). \tag{9.110}$$

Note that the constitutive parameters μ_r and α_r have to be selected such that the conditions presented in (3.114) are met. Since the incremental material tensor has to coincide with the material tensor of a linear elastic material (HOOKE'S law) for small strains, condition $2\mu = \sum_r \mu_r\alpha_r$ has to be fulfilled additionally. This can be shown by an evaluation of the material tangent given in (9.110) at the undeformed initial configuration ($\lambda_\gamma = 1$):

$$\mathbf{L}\big|_{\lambda_\gamma=1} = \sum_r \mu_r\alpha_r \begin{bmatrix} 2 & 1 \\ 1 & 2 \end{bmatrix}. \tag{9.111}$$

The linear law of HOOKE can be written for a membrane as

$$\mathbf{C} = \frac{E}{1 - \nu^2}\begin{bmatrix} 1 & \nu \\ \nu & 1 \end{bmatrix}. \tag{9.112}$$

With $2\mu(1 + \nu) = E$, matrix $\mathbf{C}$ is equivalent to $\mathbf{L}$ in (9.110) for the case of incompressibility ($\nu = 0.5$).

Metal Plasticity. The general constitutive equations for the description of metals undergoing finite elasto-plastic deformations were formulated in Sect. 3.3.2. The kinematics of such deformations is based on the multiplicative decomposition of the deformation gradient, see (3.191),

$$\mathbf{F} = \mathbf{F}^e\,\mathbf{F}^p. \tag{9.113}$$

The incompatible intermediate configuration is described by the tensor $\mathbf{F}^e$. This configuration is assumed to be stress-free. The multiplicative decomposition can be formulated in principal stretches in case of axisymmetrical deformations

$$\lambda_i = \lambda_i^e\,\lambda_i^p. \tag{9.114}$$

By introducing the logarithmic strains (6.147) which denote the eigenvalues of the HENCKY tensor $\mathbf{E}^{(0)} = \ln \mathbf{U}$, see (3.19),

$$\varepsilon_i = \ln \lambda_i, \quad \varepsilon_i^e = \ln \lambda_i^e, \quad \varepsilon_i^p = \ln \lambda_i^p,$$
$$e = \ln J = \varepsilon_1 + \varepsilon_2 + \varepsilon_3, \tag{9.115}$$

the multiplicative decomposition (9.114) can be reformulated as an additive decomposition

$$\varepsilon_i = \varepsilon_i^e + \varepsilon_i^p. \tag{9.116}$$

The plastic flow is modeled as isochoric process which leads to the constraint

$$\det \mathbf{F}^p = J^p = \lambda_1^p \lambda_2^p \lambda_3^p = 1 \quad \text{or} \quad e^p = \varepsilon_1^p + \varepsilon_2^p + \varepsilon_3^p = 0. \tag{9.117}$$

Furthermore, the elastic strains are split into a volumetric and deviatoric part

$$\bar{\lambda}_i^e = J^{-\frac{1}{3}} \lambda_i^e \quad \text{with} \quad \bar{\lambda}_1^e \bar{\lambda}_2^e \bar{\lambda}_3^e = 1. \tag{9.118}$$

The associated additive split of the elastic strains follows with (9.115)

$$\bar{\varepsilon}_i^e = \varepsilon_i^e - \tfrac{1}{3}e \quad \text{with} \quad \bar{\varepsilon}_1^e + \bar{\varepsilon}_2^e + \bar{\varepsilon}_3^e = 0. \tag{9.119}$$

Equations (9.116) and (9.119) show that the strains can be split additively, as in the small strain theory, when logarithmic strains are introduced. Hence the finite elasto-plastic strains can be modeled as in the geometrically linear theory, see also Sect. 6.2.

For isotropic material behaviour, it is consistent to introduce a strain energy function for small elastic strains, see Remark 6.2 (6.158), which depends only on the elastic strains

$$W_{Lin}(\varepsilon_i^e) = \frac{\Lambda}{2} e^2 + \mu \left[(\varepsilon_1^e)^2 + (\varepsilon_2^e)^2 + (\varepsilon_3^e)^2 \right]. \tag{9.120}$$

Λ and μ are the LAMÉ constants, see (3.119). The strain energy function (9.120) is equivalent to the strain energy function for linear elastic material, see e.g. Malvern (1969). The restriction to small elastic strains does not pose a problem for metal plasticity since only small elastic strains occur on the onset of plastic deformations.

The principal stresses τ_i of the KIRCHHOFF stress tensor are work conjugate to the logarithmic strains ε_i^e as shown in Hill (1970) and Hoger (1987). With this result and by using the chain rule, the principal stresses follow from the strain energy function

$$\tau_i = \frac{\partial W_{Lin}(\varepsilon_i^e)}{\partial \varepsilon_i^e} = \Lambda e + 2\mu \varepsilon_i^e. \tag{9.121}$$

This relation can be reformulated with (9.119), leading to a decoupled representation for the volumetric and deviatoric part

$$\tau_i = K e + 2\mu \bar{\varepsilon}_i^e. \tag{9.122}$$

The strains in thickness direction are eliminated by the assumption of a plane stress state

$$\tau_3 = K\,e + 2\mu\bar{\varepsilon}_3^e = 0\,. \tag{9.123}$$

The strain $\bar{\varepsilon}_3^e$ can be substituted using $(9.119)_2$ which yields an equation for the volumetric strain

$$e = \frac{2\mu}{K + \frac{4}{3}\mu}\,(\varepsilon_1^e + \varepsilon_2^e)\,.$$

The stresses τ_α follow with $K = \frac{2\mu\,(1+\nu)}{3(1-2\nu)}$ and $\mu = \frac{E}{2(1+\nu)}$ and some algebraic manipulations

$$\tau_\alpha = \frac{E}{1-\nu^2}\left[\,(1-\nu)\,\varepsilon_\alpha^e + \nu\,(\varepsilon_1^e + \varepsilon_2^e)\,\right]. \tag{9.124}$$

Note that the relation (9.124) fulfils the plane stress state in an exact manner.

The flow rule can be derived from the principle of maximum plastic dissipation, which was already described in Remark 3.6. For axisymmetrical shells, the dissipation is given by

$$D^p = \tau_i\,\dot{\varepsilon}_i^p \longrightarrow \max\,. \tag{9.125}$$

To fulfil the convex flow condition $f(\tau_i) = 0$, a LAGRANGE function

$$L^p(\tau_i, \dot{\gamma}) = -\tau_i\,\dot{\varepsilon}_i^p + \dot{\gamma}\,f(\tau_i) \tag{9.126}$$

is introduced in which $\dot{\gamma}$ represents the LAGRANGE multiplier. The solution of the saddle point problem (9.126) has to fulfil the constraint conditions

$$\frac{\partial L^p}{\partial \tau_i} = -\dot{\varepsilon}_i^p + \dot{\gamma}\,\frac{\partial f}{\partial \tau_i} = 0 \tag{9.127}$$

and yields the associated flow rule

$$\dot{\varepsilon}_i^p = \dot{\gamma}\,\frac{\partial f}{\partial \tau_i}\,. \tag{9.128}$$

For metal plasticity, the classical VON MISES flow condition or yield function can be formulated in terms of the deviatoric KIRCHHOFF stresses $\mathbf{s}$

$$f(\boldsymbol{\tau}, \alpha) = \frac{3}{2}tr(\mathbf{s}^2) - Y^2(\alpha) \le 0 \qquad \mathbf{s} = \mathrm{dev}\,\boldsymbol{\tau}\,, \tag{9.129}$$

where linear isotropic hardening is assumed

$$Y(\alpha) = Y_0 + \hat{H}\,\alpha\,. \tag{9.130}$$

The magnitude of hardening depends on the equivalent plastic strains α. Y_0 is the initial yield stress and $\hat{H}$ defines the isotropic hardening coefficient.

The principal values of the deviatoric stress tensor $\mathbf{s}$ can be presented for the plane stress case in matrix form

$$\mathbf{s} = \mathbf{A}\boldsymbol{\tau} \quad \text{with} \quad \mathbf{s} = \left\{ \begin{array}{c} s_1 \\ s_2 \end{array} \right\}, \quad \boldsymbol{\tau} = \left\{ \begin{array}{c} \tau_1 \\ \tau_2 \end{array} \right\} \text{ and } \mathbf{A} = \left[\begin{array}{cc} 2 & -1 \\ -1 & 2 \end{array} \right].$$
(9.131)

With this matrix formulation, the flow condition (9.129) can be reformulated as

$$f(\boldsymbol{\tau}, \alpha) = g^2(\boldsymbol{\tau}) - Y^2(\alpha) \le 0 \quad \text{with} \quad g^2(\boldsymbol{\tau}) = \frac{1}{2} \boldsymbol{\tau}^T \mathbf{A} \boldsymbol{\tau}.$$
(9.132)

The plastic strains $\boldsymbol{\varepsilon}^p = \{\varepsilon_1^p, \varepsilon_2^p\}$ are obtained by an implicit EULER integration of the flow rule, as already described for the three-dimensional case in Sect. 6.2.2. This leads for the flow rule (9.128) with (9.132) to

$$\boldsymbol{\varepsilon}_{n+1}^p = \boldsymbol{\varepsilon}_n^p + \int_{t_n}^{t_{n+1}} \dot{\gamma} \frac{\partial f}{\partial \boldsymbol{\tau}} d\bar{t} = \boldsymbol{\varepsilon}_n^p + \gamma \mathbf{A} \boldsymbol{\tau}_{n+1}.$$
(9.133)

$\gamma = \int_{t_n}^{t_{n+1}} \dot{\gamma} \, d\bar{t}$ is the increment of the consistency parameter within the time step t_{n+1}. The plastic incompressibility, expressed by (9.117)$_2$, is fulfilled automatically since $\mathbf{s}$ is purely deviatoric.

In the next step, the evolution equation for the hardening parameter α is derived from the flow condition (9.129). With (3.161), relation

$$\dot{\alpha} = 2 \dot{\gamma} Y(\alpha)$$
(9.134)

follows. Using now an implicit EULER integration for the rate of the hardening variable within time step t_{n+1} yields

$$\alpha_{n+1} = \alpha_n + 2\gamma Y(\alpha_{n+1}).$$
(9.135)

This leads to an update formula for the hardening variable

$$Y(\alpha_{n+1}) = Y_0 + \hat{H} \left[\alpha_n + 2\gamma Y(\alpha_{n+1}) \right] \frac{Y_n}{1 - 2\gamma \hat{H}}$$
(9.136)

with $Y_n = Y_0 + \hat{H} \alpha_n$. Due to the implicit EULER method, the plastic strains are implicit functions of the consistency parameter such that an iterative algorithm has to be applied for its solution, see also Sect. 6.2.2. This iteration is performed within the framework of the predictor–corrector schemes described in Sect. 6.2.2.

In matrix notation, the principal values of the elastic KIRCHHOFF stresses, given by (9.124), can be written at the beginning of a new time step as *trial* stresses

$$\boldsymbol{\tau}^{tr} = \mathbf{C} \boldsymbol{\varepsilon}^e = \mathbf{C} (\boldsymbol{\varepsilon}_{n+1} - \boldsymbol{\varepsilon}_n^p),$$
(9.137)

where in addition to (9.131) the following definitions have been introduced

$$\mathbf{C} = \frac{E}{1-\nu^2} \begin{bmatrix} 1 & \nu \\ \nu & 1 \end{bmatrix}, \ \boldsymbol{\varepsilon} = \left\{ \begin{matrix} \varepsilon_1 \\ \varepsilon_2 \end{matrix} \right\}, \ \boldsymbol{\varepsilon}^e = \left\{ \begin{matrix} \varepsilon_1^e \\ \varepsilon_2^e \end{matrix} \right\}, \ \boldsymbol{\varepsilon}^p = \left\{ \begin{matrix} \varepsilon_1^p \\ \varepsilon_2^p \end{matrix} \right\}. \quad (9.138)$$

Now the predictor–corrector method, described in Sect. 6.2.2, is applied. In case that the *trial* stresses fulfil the flow condition ($f(\boldsymbol{\tau}^{tr}, \alpha^{tr}) \leq 0$), then the material point undergoes purely elastic deformation within the time step. When the *trial* stresses do not fulfil the flow condition

$$f(\boldsymbol{\tau}^{tr}) = \frac{1}{2}\boldsymbol{\tau}^{tr\,T} \mathbf{A} \boldsymbol{\tau}^{tr} - Y(\alpha_n)^2 > 0, \quad (9.139)$$

then the time step includes elasto-plastic deformations. In that case, the elasto-plastic stresses follow from (9.116) and (9.133) in the time step t_{n+1}

$$\boldsymbol{\varepsilon}_{n+1} \ = \ \boldsymbol{\varepsilon}_{n+1}^e + \boldsymbol{\varepsilon}_{n+1}^p \ = \ \mathbf{C}^{-1}\boldsymbol{\tau}_{n+1} + \boldsymbol{\varepsilon}_n^p + \gamma\mathbf{A}\boldsymbol{\tau}_{n+1}. \quad (9.140)$$

A reformulation of (9.140) yields

$$\boldsymbol{\tau}(\gamma)_{n+1} = (\mathbf{C}^{-1} + \gamma\mathbf{A})^{-1}\,(\boldsymbol{\varepsilon}_{n+1} - \boldsymbol{\varepsilon}_n^p) = \bar{\mathbf{C}}(\gamma)\,(\boldsymbol{\varepsilon}_{n+1} - \boldsymbol{\varepsilon}_n^p) \quad (9.141)$$

with

$$\bar{\mathbf{C}}(\gamma) = [(\frac{1}{E} + 2\gamma)^2 + (\frac{\nu}{E} + \gamma)^2]^{-1} \begin{bmatrix} \dfrac{1}{E} + 2\gamma & \dfrac{-\nu}{E} - \gamma \\ \dfrac{-\nu}{E} - \gamma & \dfrac{1}{E} + 2\gamma \end{bmatrix}. \quad (9.142)$$

The stresses $\boldsymbol{\tau}$ are functions of the consistency parameter γ, which follows from the flow condition $f(\gamma) = 0$. The associated nonlinear algebraic equation for γ is solved using NEWTON'S method, see also Sect. 6.2.2. Analogous to (6.108), the iteration scheme

$$\gamma_{i+1} = \gamma_i - f(\gamma_i)/f'(\gamma_i) \quad (9.143)$$

is obtained with

$$\begin{aligned} f(\gamma_i) &= g^2(\gamma_i) - Y^2(\gamma_i), \\ f'(\gamma_i) &= -\mathbf{s}^T \bar{\mathbf{C}} \mathbf{s} - 4Y^2(\gamma_i)\bar{H}, \\ \bar{H} &= \hat{H}(1 - 2\gamma_i\,\hat{H})^{-1}. \end{aligned} \quad (9.144)$$

In case of incompressible behaviour ($\nu = 0.5$), a closed form solution can be derived for the consistency parameter

$$\gamma^{inc} = \frac{1-\kappa}{6\mu\kappa} \quad \text{with} \quad \kappa = \frac{Y_n + \delta\,g^{tr}}{g^{tr}\,(1+\delta)} \quad \text{and} \quad \delta = \frac{\hat{H}}{3\mu}. \quad (9.145)$$

The solution (9.145) can be applied as starting value $\gamma_0 = \gamma^{inc}$ for the more general case in (9.143).

Finally the *update* of the intermediate configuration can be obtained using (9.133). With these relations, all stresses which enter the weak form (9.103) are known from (9.141) with the solution of (9.143). The material matrix needed in the linearization of the shell element is given by (9.142) where the solution of (9.143) has to be inserted.

Biomechanical Material Behaviour. Arteries consist of biological material which can be described in general by viscoelastic constitutive response. Furthermore, these biological structures undergo eventually finite deformations. Hence they are considered here as an example for a biomechanical constitutive materials. For reasons of simplicity, the following formulation is restricted to the description of the elastic part of the more general constitutive behaviour which is e.g. valid to model the response of an aorta under certain loading conditions.

Arteries change their material response in the whole range from small to finite strains. Based on experimental evidence, a strain energy function can be introduced which describes isotropic response for small strains but anisotropic constitutive behaviour at large strains. When the isotropic behaviour is represented by a simple NEO–HOOKE model, see (3.116), and when a formulation due to Chuong and Fung (1983) is selected for the anisotropic part, then the strain energy

$$W(\mathbf{E}) = c_1 \left(I_E - 3 \right) + c_2 \, e^{Q-1} \tag{9.146}$$

follows. Here $\mathbf{E}$ is the GREEN–LAGRANGE strain tensor and I_E its first invariant. c_1 and c_2 are constitutive constants and Q is a function of the strains E_{AA} in circumferential, axial and radial direction of the artery

$$Q = a_1 \, E_{11}^2 + a_2 \, E_{22}^2 + a_3 \, E_{33}^2 + +2 \, a_4 \, E_{11} \, E_{22} + 2 \, a_5 \, E_{22} \, E_{33} + 2 \, a_6 \, E_{11} \, E_{33} \, . \tag{9.147}$$

The components E_{AB} are assumed to be zero for $(A \neq B)$. In Q occur six more material constants which have to be determined by experiments as well as c_1 and c_2. Furthermore, experimental tests have shown that arteries depict incompressible material behaviour; hence $J = \lambda_1 \lambda_2 \lambda_3 = 1$ can be introduced.

Within the constitutive model for axisymmetrical shells, it is further assumed that the stresses in thickness direction vanish. Thus the 2nd PIOLA–KIRCHHOFF stresses are present only in axial and circumferential direction (S_1, S_2). Analogous to the derivation of (9.108), for incompressible rubberlike material, the principal stresses

$$
\begin{aligned}
S_1 &= 2 \, c_1 \left\{ 1 - [(2 \, E_2 + 1)(2 \, E_1 + 1)^2]^{-1} \right\} \\
&\quad + 2 \, c_2 \left(a_2 \, E_1 + a_4 \, E_2 \right) e^Q \, , \\
S_2 &= 2 \, c_1 \left\{ 1 - [(2 \, E_2 + 1)^2 (2 \, E_1 + 1)]^{-1} \right\} \\
&\quad + 2 \, c_2 \left(a_2 \, E_1 + a_4 \, E_2 \right) e^Q
\end{aligned}
\tag{9.148}
$$

are obtained from (9.146) and (9.147) with $E_\gamma = \frac{1}{2}(\lambda_\gamma^2 - 1)$, see Holzapfel et al. (1996b). The constants a_3, a_5 and a_6 in (9.147) are zero.

The incremental material tangent $\mathbf{L}$ follows from the stresses in (9.148) with respect to the initial configuration

$$L_{11} = 8c_1\left[(2E_2+1)(2E_1+1)^3\right]^{-1} + 2c_2\left[a_2 + 2(a_2 E_1 + a_4 E_2)\right]e^Q ,$$

$$L_{12} = 4c_1\left[(2E_2+1)^2(2E_1+1)^2\right]^{-1} \qquad (9.149)$$

$$+2c_2\left[a_4 + 2(a_2 E_1 + a_4 E_2)(a_4 E_1 + a_1 E_2)\right]e^Q ,$$

$$L_{22} = 8c_1\left[(2E_2+1)^3(2E_1+1)\right]^{-1} + 2c_2\left[a_1 + 2(a_4 E_1 + a_1 E_2)\right]e^Q ,$$

with $L_{12} = L_{21}$. For this constitutive description of arteries, a parameter identification was performed in Holzapfel et al. (1996b) to obtain the constitutive parameters from experimental data and then applied to simulate the dilatation of arteries by balloons.

9.3.4 Finite Element Formulation

The kinematical relations (9.98) and the weak form of equilibrium (9.103) define together with the constitutive equation the boundary value problem of the axisymmetrical shell. The solution of this set of nonlinear equations by the finite element method requires a discretization for which interpolation functions have to be selected. Here C^0 interpolations are sufficient for the presented 3-parametertheory. Hence the linear shape functions given in (4.17) can be used. These are applied to discretize displacements and rotations as well as the associated test functions which all are related to the local shell midsurface, see Fig. 9.13. The general interpolation

$$\mathbf{u}_e = \sum_{I=1}^{2} N_I(\zeta)\,\mathbf{u}_I , \qquad (9.150)$$

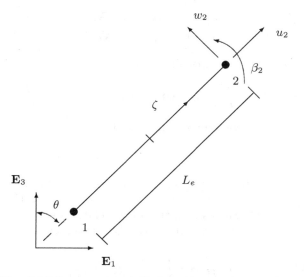

Fig. 9.13 3-parameter-shell element

includes the displacements u and w as well as the rotation β. They are assembled in the vector $\mathbf{u}_e = \{\, u_e\,, w_e\,, \beta_e\,\}^T$. In the same way, the test function (or virtual displacements and rotations) are given by

$$\boldsymbol{\eta}_e = \sum_{I=1}^{2} N_I(\zeta)\,\boldsymbol{\eta}_I\,, \tag{9.151}$$

with $\boldsymbol{\eta}_e = \{\, \delta u\,, \delta w\,, \delta \beta\,\}^T$.

By inserting these interpolation functions into (9.98), the discretized strains of the axisymmetrical shell are obtained. The virtual strains for membrane, bending and shear are part of the weak form (9.103). The shear strains are necessary for the fulfillment of the constraint condition $E_{13} = 0$ when using a penalty method within the quasi-KIRCHHOFF formulation. With (9.151) and (9.98), the variation of the strains is given by

$$\left\{ \begin{matrix} \delta E_1 \\ \delta E_2 \end{matrix} \right\} = \sum_{I=1}^{2} \boldsymbol{B}_I^{mb}\,\boldsymbol{\eta}_I, \qquad \delta E_{13} = \sum_{I=1}^{2} \boldsymbol{B}_I^{pen}\,\boldsymbol{\eta}_I\,. \tag{9.152}$$

The $\boldsymbol{B}$ matrices are defined as follows

$$\boldsymbol{B}_I^{mb} = \begin{bmatrix} B_{11} & B_{12} & B_{13} \\ B_{21} & B_{22} & B_{23} \end{bmatrix} \tag{9.153}$$

with

$$
\begin{aligned}
B_{11} &= (\,1 + u_{,\varsigma} - \xi\,\beta_{,\varsigma}\cos\beta\,)\,N_{I,\varsigma} \\
B_{12} &= (\,w_{,\varsigma} - \xi\,\beta_{,\varsigma}\sin\beta\,)\,N_{I,\varsigma} \\
B_{13} &= \xi\,\{[\,(1 + u_{,\varsigma})\sin\beta - w_{,\varsigma}\cos\beta\,]\,\beta_{,\varsigma}\,N_I \\
&\quad - [\,(1 + u_{,\varsigma})\cos\beta + w_{,\varsigma}\sin\beta\,]\,N_{I,\varsigma}\,\} \\
B_{21} &= [\,(1 + e_\theta)\,\frac{\sin\theta}{r} - \xi\,c_2\,\sin\theta\,]\,N_I \\
B_{22} &= [\,-(1 + e_\theta)\,\frac{\cos\theta}{r} + \xi\,c_2\,\cos\theta\,]\,N_I \\
B_{23} &= \frac{\xi}{r}\,(1 + e_\theta)(\,\cos\theta\,\sin\beta - \sin\theta\,\cos\beta\,]\,N_I
\end{aligned}
$$

and

$$\boldsymbol{B}_I^{pen} = [\,-\sin\beta\,N_{I,\xi}\,,\cos\beta\,N_{I,\xi}\,,-[(1 + u_{,\varsigma})\cos\beta + w_{,\varsigma}\sin\beta]N_I\,]\,. \tag{9.154}$$

The notation introduced in (9.98) was used in the above relations also.

Now the discrete weak form, based on (9.103), can be written as

$$G(\mathbf{u}\,,\boldsymbol{\eta}) = \bigcup_{e=1}^{n_e} 2\,\pi \sum_{I=1}^{2} \boldsymbol{\eta}_I^T \left[\int_{-1}^{+1}\!\!\int_{-1}^{+1} (\boldsymbol{B}_I^{mb})^T \left\{ \begin{matrix} S_1 \\ S_2 \end{matrix} \right\} r\,\frac{L_e}{2}\,\frac{t_0}{2}\,d\hat{\xi}\,d\zeta + \right.$$

$$\epsilon\, t_0 \int_{-1}^{+1} (\boldsymbol{B}_I^{pen})^T\, E_{13}\, r\, \frac{L_e}{2} d\zeta - \int_{-1}^{+1} N_I \left\{ \begin{matrix} \hat{t}_1 \\ \hat{t}_2 \end{matrix} \right\} r\, \frac{L_e}{2} d\zeta \right] . \quad (9.155)$$

L_e denotes the length of the finite element, see Fig. 9.13. Since a nonlinear constitutive equation (e.g (9.107) or (9.142)) is used, it is not possible to analytically integrate over the shell thickness. Hence numerical integration has to be applied. Here GAUSS-Integration is selected, for the location of the GAUSS points see Table 4.1. Within this approach, a transformation of the thickness variable to the length "2" has to be performed such that a new variable $\hat{\xi} = \frac{t_0}{2}\xi$ appears in (9.155).

The second term in (9.155) can be analytically integrated over the thickness since a linear "constitutive equation" with the penalty parameter ϵ is used within the quasi-KIRCHHOFF model.

The linearization of the continuous form can be derived from (9.103) which then yields the tangent operator necessary for the NEWTON scheme

$$DG(\mathbf{u},\boldsymbol{\eta}) \cdot \Delta\mathbf{u} \;=\; 2\pi \left[\int_{(C)} \int_{\xi} (L_{\gamma\nu}\,\delta E_\gamma\,\Delta E_\nu + S_\gamma \Delta\delta E_\gamma)\, r\, d\xi\, dS \right.$$

$$\left. +\, \epsilon\, t_0 \int_{(C)} (\,\Delta E_{13}\,\delta\, E_{13} + E_{13}\,\Delta\delta E_{13}\,)\, r\, dS \right] . \quad (9.156)$$

The matrix form of the variations of E_γ and E_{13} is known from the discretization (9.152). The linearization of the strains has the same structure as the variation since variation are based on the directional derivative. Hence

$$\left\{ \begin{matrix} \Delta E_1 \\ \Delta E_2 \end{matrix} \right\} = \sum_{I=1}^{2} \boldsymbol{B}_I^{mb}\,\Delta\mathbf{u}_I, \qquad \Delta E_{13} = \sum_{I=1}^{2} \boldsymbol{B}_I^{pen}\,\Delta\mathbf{u}_I \qquad (9.157)$$

can be written by directly using (9.152) to (9.154). The linearization of the variation yields for a finite element

$$\Delta\delta E_\gamma \;=\; \sum_{I=1}^{2} \sum_{K=1}^{2} \boldsymbol{\eta}_I^T\, \boldsymbol{G}_{\gamma IK}^{mb}\,\Delta\mathbf{u}_K \qquad \gamma = 1,2, \qquad (9.158)$$

$$\Delta\delta E_{13} \;=\; \sum_{I=1}^{2} \sum_{K=1}^{2} \boldsymbol{\eta}_I^T\, \boldsymbol{G}_{IK}^{pen}\,\Delta\mathbf{u}_K . \qquad (9.159)$$

Finally,

$$G_h(\mathbf{u},\boldsymbol{\eta}) \cdot \Delta\mathbf{u} = \bigcup_{e=1}^{n_e} \sum_{I=1}^{2} \sum_{K=1}^{2} \boldsymbol{\eta}_I^T\, 2\pi\, \boldsymbol{K}_{IK}\,\Delta\mathbf{u}_K \qquad (9.160)$$

is obtained, where $\boldsymbol{K}_{IK}$ is the tangential stiffness matrix related to node I and K.

$$
\boldsymbol{K}_{IK} = \int\limits_{-1}^{+1} \int\limits_{-1}^{+1} [(\boldsymbol{B}_I^{mb})^T \boldsymbol{L} \boldsymbol{B}_K^{mb} + S_\gamma \, \boldsymbol{G}_{\gamma IK}^{mb}] \, r \, \frac{L_e}{2} \, \frac{t_0}{2} \, d\hat{\xi} \, d\zeta \; +
$$

$$
\epsilon \, t_0 \int\limits_{-1}^{+1} [(\boldsymbol{B}_I^{pen})^T \boldsymbol{B}_K^{pen} + E_{13} \, \boldsymbol{G}_{IK}^{pen}] \, r \, \frac{L_e}{2} \, d\zeta \,. \tag{9.161}
$$

The analytical expression of the matrices $\boldsymbol{G}_{IK}^{mb}$ and $\boldsymbol{G}_{IK}^{pen}$ is rather complex. Hence the explicit forms will not be provided here. They can be found in Eberlein (1997) and Eberlein et al. (1993).

All matrices are referred to the local coordinate system defined in Fig. 9.13. A transformation of these matrices to the global coordinate system with the base vectors $\boldsymbol{E}_i$ has to be performed in order to be able to assemble elements to model an arbitrarily shaped axisymmetrical shell. This transformation, however, is the same as for beam elements and hence is standard, see Remark 9.1 for details. Again the transformation, valid for the linear theory, can be applied since the shell element is formulated with respect to the initial configuration.

Remark 9.3: Nonlinear shell problems depict often stability behaviour and snapthrough or bifurcation can occur as discussed in Chap. 7. Hence it is often necessary to use, within the computation of the nonlinear response of shell structures, a continuation method which is able to follow arbitrary solution paths. Commonly arc-length methods are employed, see Sect. 5.1.5, which provide a stable solution scheme for such problems. Additionally, the extended system could be applied to locate stability or singular points, see Sect. 7.2.

Exercise 9.5: Specialize the equations of the axi-symmetrical shell for a pure membrane state. Assume rubber elastic material and derive the residual and linearization of the membrane element explicitly.

Solution: Shear strains as well as strains resulting from curvature can be neglected in a pure membrane states. This leads to a simplification of equations (9.98)

$$
\epsilon_1 = u_{,s} + \frac{1}{2} u_{,s}^2 + \frac{1}{2} w_{,s}^2 \tag{9.162}
$$

$$
\epsilon_2 = e_\theta + \frac{1}{2} e_\theta^2 \qquad \text{with} \quad e_\theta = (\cos\theta \, u + \sin\theta \, w) \, / \, R \,, \tag{9.163}
$$

where ϵ_1 and ϵ_2 are the membrane strains in meridional and circumferential directions. u and w are the displacements with respect to the local coordinate system, see Fig. 9.12. The principal stretches follow from the strains in (9.103) identical to the derivation for (9.100)

$$
\lambda_1 = \sqrt{(1 + u_{,s})^2 + w_{,s}^2} \,, \tag{9.164}
$$

$$
\lambda_2 = 1 + e_\theta \,. \tag{9.165}
$$

The principle of virtual displacements or the weak form of equilibrium for the membrane can be deduced from (9.103)

$$G(\mathbf{u}, \boldsymbol{\eta}) = 2\pi t_0 \left[\int_{(C)} S_\gamma \, \delta\epsilon_\gamma \, r \, dS - \int_{(C)} \hat{\mathbf{t}} \cdot \boldsymbol{\eta} \, dS \right] = 0 \quad (\gamma = 1, 2), \qquad (9.166)$$

where the variations of the membrane strains (9.163) have to be inserted. Since the stresses in a membrane are constant over the thickness, an analytical integration over the thickness can be performed. The variations of (9.163) are given by

$$\begin{aligned} \delta\epsilon_1 &= (1 + u_{,s})\delta u_{,s} + w_{,s}\,\delta w_{,s}\,, \\ \delta\epsilon_2 &= (1 + e_\theta)\,\delta e_\theta \qquad \text{with} \quad \delta e_\theta = (\cos\theta\,\delta u + \sin\theta\,\delta w)\,/\,r\,. \end{aligned} \qquad (9.167)$$

For the linearization of (9.166), all field quantities which depend upon the displacements have to be considered. This yields ΔS_γ, $\Delta\epsilon_\gamma$ and $\Delta\delta\epsilon_\gamma$. Using the product rule,

$$D\,G(\mathbf{u}, \boldsymbol{\eta}) \cdot \Delta\mathbf{u} = \int \left[\delta\epsilon_\gamma \frac{\partial S_\gamma}{\partial \epsilon_\beta} \Delta\epsilon_\beta + S_\gamma\,\Delta\delta\epsilon_\gamma \right] dS \qquad (9.168)$$

is obtained. All derivatives of S_γ with respect to ϵ_β follow from (9.110). Hence the incremental stresses can be written in matrix form $\Delta\mathbf{S} = \mathbf{L}\,\Delta\boldsymbol{\epsilon}$ or explicitly as

$$\left\{ \begin{array}{c} \Delta S_1 \\ \Delta S_2 \end{array} \right\} = \left[\begin{array}{cc} L_{11} & L_{12} \\ L_{21} & L_{22} \end{array} \right] \left\{ \begin{array}{c} \Delta\epsilon_1 \\ \Delta\epsilon_2 \end{array} \right\}. \qquad (9.169)$$

The linearization of the strains $\Delta\epsilon_\gamma$ and of the virtual strains $\Delta\delta\epsilon_\gamma$ can then be deduced, leading to

$$\begin{aligned} \Delta\epsilon_1 &= (1 + u_{,s})\,\Delta u_{,s} + w_{,s}\,\Delta w_{,s}\,, & (9.170) \\ \Delta\epsilon_2 &= (1 + e_\theta)\,(\cos\theta\,\Delta u + \sin\theta\,\Delta w\,)/r\,, & (9.171) \\ \Delta\delta\epsilon_1 &= \delta u_{,s}\,\Delta u_{,s} + \delta w_{,s}\,\Delta w_{,s}\,, & (9.172) \\ \Delta\delta\epsilon_2 &= (\cos\theta\,\delta u + \sin\theta\,\delta w\,)(\cos\theta\,\Delta u + \sin\theta\,\Delta w\,)/r^2\,. & (9.173) \end{aligned}$$

A simple isoparametric finite element with two nodes can be formulated based on the equations for the membrane derived above. The displacements are approximated by a linear interpolation

$$u_\gamma = \sum_{I=1}^{2} N_I(\zeta)\,u_{I\,\gamma}\,. \qquad (9.174)$$

Using the same interpolation for the virtual displacements or test function $\eta_\gamma = \sum_{I=1}^{2} N_I(\xi)\,\eta_{I\,\gamma}$ leads to the discretized version of variations (9.167) and its linearizations (9.173)

$$\left\{ \begin{array}{c} \delta\epsilon_1 \\ \delta\epsilon_2 \end{array} \right\} = \sum_{I=1}^{2} \mathbf{B}_I \left\{ \begin{array}{c} \eta_1 \\ \eta_2 \end{array} \right\}_I\,, \qquad \left\{ \begin{array}{c} \Delta\epsilon_1 \\ \Delta\epsilon_2 \end{array} \right\} = \sum_{I=1}^{2} \mathbf{B}_I \left\{ \begin{array}{c} \Delta u_1 \\ \Delta u_2 \end{array} \right\}_I\,, \qquad (9.175)$$

where the $\mathbf{B}$-matrix is given by

$$\mathbf{B}_I = \sum_{I=1}^{2} \left[\begin{array}{cc} (1 + u_{,\zeta})\,N_{I,\zeta} & w_{,\zeta}\,N_{I,\zeta} \\ (1 + e_\theta)\,\frac{\cos\theta}{r}\,N_I & (1 + e_\theta)\,\frac{\sin\theta}{r}\,N_I \end{array} \right]. \qquad (9.176)$$

Furthermore, the linearization of the virtual strains (9.173) have to be approximated. This yields

$$\Delta\delta\epsilon_\gamma = \sum_{I=1}^{2} \sum_{k=1}^{2} \langle \eta_1 , \eta_2 \rangle_I \, \boldsymbol{G}_{\gamma I} \, \boldsymbol{G}_{\gamma K}^T \left\{ \begin{array}{c} \Delta u_1 \\ \Delta u_2 \end{array} \right\}_J . \tag{9.177}$$

The operator matrices $\boldsymbol{G}_\gamma$ are given by

$$\boldsymbol{G}_{1 I} = \left\{ \begin{array}{c} N_{I,\varsigma} \\ N_{I,\varsigma} \end{array} \right\} \qquad \boldsymbol{G}_{2 I} = \left\{ \begin{array}{c} \frac{\cos\theta}{R} N_I \\ \frac{\sin\theta}{R} N_I \end{array} \right\} . \tag{9.178}$$

Based on these results, the weak form of the equilibrium can be formulated for the membrane element. It is obtained by inserting (9.175) into (9.166)

$$G_e(\mathbf{u},\boldsymbol{\eta}) = 2\pi t_0 \sum_{I=1}^{2} \eta_I \left[\int_{-1}^{1} \boldsymbol{B}_I^T \left\{ \begin{array}{c} S_1 \\ S_2 \end{array} \right\} r \frac{L_e}{2} \, d\zeta - \int_{-1}^{1} N_I \left\{ \begin{array}{c} \hat{t}_1 \\ \hat{t}_2 \end{array} \right\} r \frac{L_e}{2} \, d\zeta \right] . \tag{9.179}$$

L_e denotes the length of the finite element. The 2nd PIOLA-KIRCHHOFF stress tensor S_γ in (9.179) has to be computed from the nonlinear constitutive equation (9.108) using (9.105).

The finite element approximation of the linearization (9.168) yields, for an element e analogous to (9.160),

$$D G^e(\mathbf{u},\boldsymbol{\eta}) \, \Delta\mathbf{u} = \sum_{I=1}^{2} \sum_{K=1}^{2} \eta_I \, \boldsymbol{K}_{TIK} \, \Delta\mathbf{u}_K . \tag{9.180}$$

Here the tangential stiffness matrix related to node I and K is given by

$$\boldsymbol{K}_{TIK} = 2\pi t_0 \int_{-1}^{1} (\boldsymbol{B}_I^T \, \boldsymbol{L} \, \boldsymbol{B}_K + S_1 \, \boldsymbol{G}_{1 I} \boldsymbol{G}_{1 K}^T + S_2 \, \boldsymbol{G}_{2 I} \boldsymbol{G}_{2 K}^T) \, r \frac{L_e}{2} \, d\zeta , \tag{9.181}$$

where the incremental constitutive matrix $\boldsymbol{L}$ is defined by (9.109) and (9.169). Note that a one point GAUSS-quadrature is sufficient to integrate (9.179) and (9.181).

9.4 General Shell Elements

Axi-symmetric shell elements were discussed in the previous section for applications with finite elastic and inelastic deformations. The generalization of these formulations to three-dimensional general shell elements will be derived in this chapter.

Basically any shell could be discretized using three-dimensional solid elements as discussed in Chap. 4. This approach is depicted on the left side of Fig. 9.14. A linear interpolation through the thickness is then in close accordance with the assumption that plane cross sections remain plane during the deformation.[1] Additionally, change of thickness is also taken into account in this model. It is well known, see e.g. Zienkiewicz et al. (1971) for the linear

---

[1] Note that a plane isoparametric surface assumes, in general, a hypar surface after deformation.

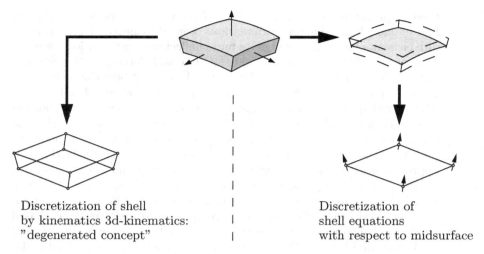

Discretization of shell		Discretization of
by kinematics 3d-kinematics:		shell equations
"degenerated concept"		with respect to midsurface

Fig. 9.14 Discretization of shells

case, that a pure displacement formulation, as defined in Sects. 4.1.3 and 4.2, leads to *locking* in the thin shell limit. Hence special interpolations have to be employed when three-dimensional solid formulations are used to discretize shells in order to eliminate locking, see e.g. Chap. 10.

Classically the approach described on the right side of Fig. 9.14 is followed for the development of nonlinear shell elements. However, locking can also occur for specific shell formulations which are derived in that way.

9.4.1 Introductory Remarks

The description of the shell continuum, the kinematics, the weak form of linear momentum and the constitutive equations can be stated in various ways. Some basic concepts will be discussed briefly in this introduction.

Shell continuum and shell kinematics. Several possible formulations can be applied for the description of a shell continuum when finite element analysis is concerned. These are depicted in Fig. 9.14 and discussed in the following.

− Classically, the description of shells is based on the definition of a middle surface, see the right path in Fig. 9.14. Using such parametrization, the kinematics, weak form and constitutive equations can be developed from the three-dimensional continuum equations. By using this approach, many different approximations for the kinematical description of the shell model in thickness direction can be developed. Different assumptions lead to equations for thin and thick shells which additionally can model deformations in thickness direction. Depending on the number of kinematical variables, these approaches are denoted 5-, 6- or 7-parameter theories. Within this line of modeling, different formulations were developed to construct finite elements for shells undergoing finite deflections and rotations, see e.g. Simo et al. (1990), Simo et al. (1990B), Onate and Cervera (1993), Sansour (1995), Eberlein and Wriggers (1999), Cirak et al. (2000), Campello et al. (2003), Pimenta et al. (2004) and Gruttmann and Wagner (2005).

– The second approach – called degenerated concept – uses the equations of a three-dimensional solid and introduces the shell kinematics at the discretization level, see left path of Fig. 9.14. As in shell theory a reference midsurface is chosen, see e.g. Ramm (1976) and Bathe (1982). Within this approach, no shell theory – besides the kinematical assumption – is needed for the discretization of a shell continuum. Hence this approach is conceptional simple. However the introduction of stress resultants is not naturally included in this formulation. A comparison of finite elements based on classical shell theory and on the degenerated concept was investigated in Büchter and Ramm (1992), who found that both approaches lead under certain circumstances to the same finite element formulations.

– A third approach starts directly from the continuum elements discussed in Sect. 4.2. No shell midsurface is introduced explicitly but the node of the continuum elements are located at the upper and lower side of the shell continuum for discretizations using low order element see e.g. Hughes and Liu (1981), Schoop (1986) Kühborn and Schoop (1992), Seifert (1996), Miehe (1998) and Hauptmann and Schweizerhof (1998). Higher order element formulations were discussed in Düster et al. (2001).

In all mentioned formulations, special measures have to be taken to avoid locking effects.

In most approaches, it is assumed that plane cross sections remain plane during the deformation of the shell continuum. This yields theories which include shear deformations which requires only C^0-continuous discretizations within the finite element method. The classical KIRCHHOFF–LOVE hypothesis would, of course, be a natural assumption for kinematics. of thin shells too. However, this additional assumption requires C^1-continuous interpolation functions which cannot be constructed by interpolations using only primary variables for triangular and quadrilateral finite elements. In this context, a new approach which combines interpolations of the deformations with the CAD description of the shell surfaces are of interest. These discretization employ BEZIER or other C^1-continuous polynomials, see e.g. Cirak et al. (2000), Onate and Cervera (1993) and lately Hughes et al. (2005) who introduced the notion of *isogeometric* analysis. Such formulations deviate from the classical finite element concept since the C^1-continuity is not fulfilled on element but on patch level by using a patch of elements to define the interpolation functions. Such formulations have an advantage because shells are highly sensitive to geometry changes and BEZIER or *NURBS* surfaces ideally map complex geometries into the shell model.

When no further assumptions besides "plane cross sections remain plane" are introduced within the derivation of the shell equations from the nonlinear continuum equations, this shell theory is called "geometrically exact". Geometrically exact shell theories were developed during the last two decades since growing computer power enabled the engineer to perform numerical simulations of complex nonlinear shell problems without assuming any approximation with respect to the size of rotations and deflections.

First investigations using geometrically exact shell theories can be found in Simo et al. (1989) and Wriggers and Gruttmann (1989). New in this work is the formulation of a singularity free parametrization of the rotations and the use of the isoparametric formulation for approximation of the shell geometry. The latter eliminates the appearance of co- and contra-variant derivatives which are replaced by standard partial derivatives. This direct approach was basis for different theoretical formulations leading to new finite elements for nonlinear shell problems, see e.g. Simo et al. (1990), Basar and Ding (1990), Wriggers and Gruttmann (1993), Wagner and Gruttmann (1994), Basar and Ding (1996) and Bischoff and Ramm (1997); a further in depth discussion can be found in Bischoff et al. (2004).

As for beams, the deformations of shells can be classified and thus related descriptions and reduced models for the kinematics can be defined. A fundamental investigation of Pietraszkiewicz (1978) differentiates – based on a separate consideration of deflections and rotations – the following types of rotations: small, moderate, large and finite. These distinctions can be made as well for theoretical considerations as for the validation of approximate shell theories, which are basis for analytical solutions. Today many nonlinear shell theories are still based on the assumption of moderate rotations, since this yields a simple description of the rotation field. With this mathematical model relatively large rotations – up to 8 degrees – can be described which are sufficient for many engineering applications. The related equations will not be provided here; they can be found in e.g. Naghdi (1972). However, with today's computing power, engineers tend to apply geometrically exact models to solve nonlinear shell problems.

Constitutive Equations. Not only the shell kinematics require special attention but also the formulation of constitutive equations for shells. Here different requirements due to the use of a nonlinear shell element have to be fulfilled. For example in case of a stability problem like shell buckling an elastic constitutive equation for small strains can be sufficient when it can capture the rigid body rotations due to finite deformations. For this purpose, the ST. VENANT constitutive equation, see (3.121), is sufficient. Biomedical applications – such as the analysis of skin or muscles – need the formulation of anisotropic elastic constitutive equations for large strains, see Sect. 9.3.3. Pneumatic springs made from rubber require hyper elastic constitutive equations. Here the relations of OGDEN type (3.113) can be mentioned. The simulation of metal forming processes, e.g. deep drawing of sheets, needs the formulation of constitutive equations for isotropic or anisotropic finite elasto-plastic deformations. These were derived for isotropic materials in Sect. 3.3.2.

In general, the shell equations can be derived in stress resultant form or obtained by directly using the stresses of the three-dimensional theory. This has some implications regarding the formulation of constitutive equations for shells.

- Within the classical development of shell theories, the constitutive equations are mapped to the midsurface of the shell continuum by introduction of stress resultants. This is very efficient in case of elastic ST. VENANT materials. However, this process implies simplifications even for hyperelastic deformations, see e.g. Libai and Simmonds (1992). But as shown in Campello et al. (2003), these approximations can be circumvented when an adequate nonlinear constitutive relation is introduced. Shell elements for large deformations, which are based on the introduction of stress resultants, were developed in e.g. Simo et al. (1990B), Wriggers and Gruttmann (1993), Krätzig (1993), Sansour (1995) and Campello et al. (2003).
- An alternative derivation of shell theories starts directly with the stresses of the continuum. There the three-dimensional constitutive equations can be applied directly. The resulting shell equations (weak from) have then to be integrated over the shell thickness, see e.g. Betsch et al. (1996) and Eberlein and Wriggers (1999) and the schematic description in Fig. 9.7.
- Additionally different layers across the shell thickness can be introduced. Then for each layer a different constitutive equation can be formulated, as e.g. needed for laminated structures, see left part of Fig. 9.15. Such approach is equivalent to a h refinement of the discretization with respect to the shell thickness. It is advantageous in case that the solutions – like in elasto-plasticity – do not have a high regularity. The second possibility, to use higher order polynomials for the description of the deformation across the shell thickness, is equivalent to a p refinement. This strategy can be employed for problems in which the solutions can be differentiated sufficiently often.
Integration of the stresses across the shell thickness yields stress resultants which are referred to the shell midsurface. Both, h and p approaches need generally numerical integration over the shell thickness, see e.g. Hughes

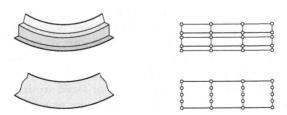

Fig. 9.15 Layer model for shells

and Liu (1981), Parisch (1991) or Büchter et al. (1994). By using the stress directly, an explicit formulation of a flow rule and a yield condition in terms of stress resultants is avoided. Due to its complexity it is often better to use the standard elasto-plastic constitutive equations for the stresses.

When the shell theory is based on stress resultants, constitutive equations for the stress resultants have to be formulated. These constitutive equations represent the stress field which is integrated over the shell thickness and need special care when finite elastic or inelastic deformations have to be considered. In case of elasto-plasticity, there exist formulations which base on the ILYUSHIN yield condition and associated flow rule for the resultants, see e.g. Crisfield (1997). The application of such formulations for shells undergoing finite deformations can be found in Simo and Kennedy (1992). However, while numerically very efficient, these models are questionable since they do not work sufficiently accurate during elastic unloading.

When the stresses from the three-dimensional continuum theory are used within the shell theory, then all constitutive equations, like the ones described in Sect. 3.3, can be employed directly without any change. Application of such formulations are presented in e.g. Büchter et al. (1994), Dvorkin et al. (1995), Seifert (1996) or Miehe (1998). For the special case of thin shells, the assumption of plane stress conditions have to be made. Associated formulations are derived in e.g. Wriggers et al. (1996). Formulations for elasto-plastic material behaviour are documented in e.g. Roehl and Ramm (1996), Wriggers et al. (1996), Soric et al. (1997), Eberlein and Wriggers (1999) and Wagner et al. (2002).

Finite-Element Discretizations. Several new finite element discretization schemes for finite deformation analysis of shells were developed during the last years. The literature regarding this topic is quite extensive, specific approaches can be found in the papers cited so far. Different interpolation schemes were introduced which range from low order interpolations to ansatz spaces with high order.

Low order approximations are selected for problems with low regularity like elasto-plastic deformations since they tend to be more robust when large mesh distortions occur in finite deformation applications. As a further advantage, they lead to a sparse tangent matrix which reduces the solution time for the linear equation system which has to be solved within the incremental solution algorithm of the nonlinear shell problem. One of the main emphases of the finite element method is avoidance of locking effects for low order interpolations.

From linear shell theory, it is well known that the use of C^0 interpolation functions can lead to different stiffening effects which are known as *locking*. In shell problems, different phenomena such as volume-, membrane- and shear-locking can be distinguished.

– **Transverse shear locking** occurs when elements of lower interpolation order cannot reproduce pure bending states without an activation of transverse shear. These transverse shear deformations, which should not be present in bending states, yield contributions to the stiffness matrix and hence lead to additional stiffness. This stiffness can be larger than the stiffness due to bending which results in locking. This is also true for the quasi KIRCHHOFF formulations in (9.199).

− **Membrane locking** occurs due to the same mechanism as transverse shear locking. In that case, bending states cannot be reproduced without activating membrane strains.

− **Volume locking** only occurs when shell theories are used which include thickness changes, for a discussion of this effect see Chap. 10.

An analysis of locking effects from the mathematical standpoint can be found in e.g. Braess (2007) and Brezzi and Fortin (1991). Considerations from the engineering point of view are provided in Andelfinger (1991) and Hauptmann (1997) or in the books of Bathe (1996), Belytschko et al. (2000), Zienkiewicz and Taylor (2000b) and Crisfield (1991).

Locking can be avoided by different measures which are discussed next

− **Use of higher interpolation order:** In general, the order of the ansatz polynomial can be enlarged such that locking disappears totally. The so-called p-version of finite elements with high order interpolation functions was developed in Babuska et al. (1981) but can also be found in Düster et al. (2001). So-called isogeometric ansatz functions which use the same high ansatz order for geometry and interpolation are introduced in Hughes et al. (2005). The polynomial order is very high leading to tangent matrices which are not sparse. Furthermore, this approach assumes a high regularity (smoothness) of the solution of the underlying partial differential equation. Hence application to inelastic problems needs special considerations, see e.g. Düster et al. (2002).

− **Reduced integration:** Within this method, all or selected integrals contributing to the weak form of the shell formulation are integrated not with the necessary integration order but using a reduced order. Such procedure avoids locking but can lead to rank deficiency of the tangent stiffness matrices since zero energy modes occur. This results in so-called *hour-glass* modes appearing within the displacement fields of the finite element solution. Hence stabilization techniques have to be developed to avoid rank deficiency. Reduced integration is applied within shell formulations to the integral containing the transverse shear contributions, for an overview of related methods see e.g. Hughes (1987), Belytschko et al. (2000) and Bischoff et al. (2004).

− **Mixed methods:** A variety of finite element approaches were constructed during the last years based on mixed variational formulations. For continuum problems, these are summarized in Chap. 10. However, mixed variational principles can also be applied to shells. They are usually employed to avoid membrane and volume locking where the last problem only occurs when thickness change is present.[2] One of the widely used methods is the approach based on incompatible modes, see Wilson et al. (1973) and the related variational technique which is known under *enhanced assumed strain* (EAS) method which was developed by Simo and Armero (1992) for nonlinear continuum elements. Application to shell problems can be found in e.g. Andelfinger and Ramm (1993), Büchter et al. (1994), Betsch and Stein (1995), Eberlein and Wriggers (1999) and Gruttmann and Wagner (2005) and will be discussed in the following.

— **Special interpolations:** Special interpolation can be developed for the discretization of shells, e.g. for shear strains, in order to avoid locking. Here the ansatz made by Bathe and Dvorkin (1985) is often used which was also developed in Hughes and Tezduyar (1981) for plates. This ansatz is often employed within nonlinear problems, see e.g. Dvorkin et al. (1995), Miehe (1998) and Eberlein and Wriggers (1999) for finite elasto-plastic deformations of shells.

In the following sections, two shell formulations for finite deformations will be discussed. The first one can be considered as one of the simplest possibilities to model finite deformation problems for elastic and elasto-plastic materials. It is the three-dimensional version of the quasi KIRCHHOFF formulation introduced in Sect. 9.3 where the shear deformations are included but suppressed by a penalty formulation. Such formulation is based on constitutive equations for plane stress states. The associated finite element is very well suited for the analysis of thin shells.

The second shell element is based on a shell theory which also includes thickness changes. Here the rotations are defined by a director vector. Thus the formulation is simpler since it does not need updated formulae for finite rotations. The shell continuum is still a three-dimensional continuum and hance all three-dimensional constitutive equations can be used within this formulation without changes. A comparison of both formulations will be presented at the end of this section.

9.4.2 Kinematics

Shells are three-dimensional curved structural members which have a small extension in thickness direction. Thus a shell is described by introducing a shell midsurface as reference surface. Within such model, it is possible to reduce the description of the shell continuum to two surface parameters ξ^α, which are defined on the shell midsurface $\mathcal{M}$, and to a coordinate ξ in thickness direction. A common choice for the parametrization of the shell midsurface are convective coordinates, which are described in detail in Appendix A.1.2. With the additional assumption that a plane cross section remains

---

[2] For linear problems, the equivalence of mixed methods and reduced integration techniques was shown in Malkus and Hughes (1978) for special problem classes.

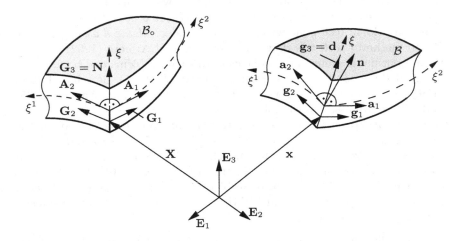

Fig. 9.16 Geometry of the shell continuum

plane during the deformation, the position vector of a point in the current configurations can be formulated by

$$\boldsymbol{\varphi}(\xi^1,\xi^2,\xi,t) = \boldsymbol{\varphi}_M(\xi^\alpha,t) + \xi\,\mathbf{d}(\xi^\alpha,t) \quad \text{with} \quad \xi \in [-\frac{h}{2},+\frac{h}{2}]. \quad (9.182)$$

Here the Greek index α assumes values of 1 and 2. The director $\mathbf{d}$ describes the rotation of the cross section during the deformation and $\boldsymbol{\varphi}_M$ defines the deformation of the shell midsurface $\mathcal{M}$, see Fig. 9.16. Since the deformation of the shell midsurface $\boldsymbol{\varphi}_M$ and as the director $\mathbf{d}$ are represented by three unknown components such formulation is called 6-Parameter theory. A point in the shell continuum is given by the position vector

$$\mathbf{X}(\xi^1,\xi^2,\xi,t) = \mathbf{X}_M(\xi^\alpha,t) + \xi\,\mathbf{N}(\xi^\alpha,t) \quad (9.183)$$

with respect to the initial configuration. Here $\mathbf{N}$ is a vector normal to the shell midsurface $\mathcal{M}$ in the initial configuration.

The co-variant base vectors related to the shell continuum are computed in the initial configuration using (3.36)

$$\begin{aligned} \mathbf{G}_\alpha &= \frac{\partial \mathbf{X}}{\partial \xi^\alpha} = \mathbf{A}_\alpha + \xi\,\mathbf{N}_{,\alpha}\,, \\ \mathbf{G}_3 &= \frac{\partial \mathbf{X}}{\partial \xi} = \mathbf{N}\,. \end{aligned} \quad (9.184)$$

Here the explicit dependencies of the base vectors with respect to the coordinates ξ^α and time t have been neglected to shorten notation. $\mathbf{A}_\alpha$ are tangent vectors at $\mathbf{X}_M(\xi^\alpha, t)$, defined at the shell midsurface $\mathcal{M}$, which follow from the derivative $\mathbf{A}_\alpha = \frac{\partial \mathbf{X}_M}{\partial \xi^\alpha}$. Since the normal vector is perpendicular to $\mathbf{A}_\alpha$, the tangent vector can be used to compute the normal vector

$$\mathbf{N} = \frac{\mathbf{A}_1 \times \mathbf{A}_2}{\| \mathbf{A}_1 \times \mathbf{A}_2 \|}. \tag{9.185}$$

The tangent vectors can be obtained in the same manner with respect to the current configuration. From (9.182), the relations

$$\mathbf{g}_\alpha = \frac{\partial \varphi}{\partial \xi^\alpha} = \mathbf{a}_\alpha + \xi\,\mathbf{d}_{,\alpha},$$

$$\mathbf{g}_3 = \frac{\partial \varphi}{\partial \xi} = \mathbf{d} \tag{9.186}$$

follow and hence the normal vector is presented by the cross product of the tangent vectors $\mathbf{a}_\alpha$

$$\mathbf{n} = \frac{\mathbf{a}_1 \times \mathbf{a}_2}{\| \mathbf{a}_1 \times \mathbf{a}_2 \|}. \tag{9.187}$$

The kinematical relations are now employed to construct the deformation gradient with respect to the shell midsurface $\mathcal{M}$. From the equation (3.39), the deformation gradient

$$\mathbf{F} = \mathbf{F}_{[C]} + \xi\,\mathbf{F}_{[L]} = \mathbf{g}_i \otimes \mathbf{G}^i \tag{9.188}$$

is obtained with

$$\mathbf{F}_{[C]} = \mathbf{a}_\alpha \otimes \mathbf{G}^\alpha + \mathbf{d} \otimes \mathbf{G}^3 \quad \text{and} \quad \mathbf{F}_{[L]} = \mathbf{d}_{,\alpha} \otimes \mathbf{G}^\alpha.$$

Hence the deformation gradient is split with respect to the thickness coordinate ξ into a constant $\mathbf{F}_{[C]}$ and a linear $\mathbf{F}_{[L]}$ part.

Once the deformation gradient is defined, the shell deformations are completely described. All further strain measures follow directly from (9.188) by introduction of the related three-dimensional measures such as, e.g. the right CAUCHY-GREEN tensor from $\mathbf{C} = \mathbf{F}^T\,\mathbf{F}$. Contrary to classical shell theories in which the strains are specified in detail, see also Remark 9.4, the present formulations allow a direct discretization and introduction of the necessary strain measures within the weak form, see Sect. 3.4.1 together with (9.188). Thus all methods used to discretize standard for three-dimensional continuum elements can be employed for shells too.

Remark 9.4: Relation (9.188) can be inserted in (3.15) which yields the GREEN–LAGRANGE strain tensor $\mathbf{E}$ with respect to the shell kinematics (9.182)

$$\mathbf{E} = \frac{1}{2}\,[\,E_{\alpha\beta}\,\mathbf{G}^\alpha \otimes \mathbf{G}^\beta + E_{\alpha 3}\,(\,\mathbf{G}^\alpha \otimes \mathbf{G}^3 + \mathbf{G}^3 \otimes \mathbf{G}^\alpha\,) + E_{33}\,\mathbf{G}^3 \otimes \mathbf{G}^3\,] \tag{9.189}$$

with the components

$$
\begin{aligned}
E_{\alpha\beta} &= (\mathbf{a}_\alpha + \xi\,\mathbf{d}_{,\alpha})\cdot(\mathbf{a}_\beta + \xi\,\mathbf{d}_{,\beta}) - (\mathbf{A}_\alpha + \xi\,\mathbf{N}_{,\alpha})\cdot(\mathbf{A}_\beta + \xi\,\mathbf{N}_{,\beta}), \\
E_{\alpha 3} &= (\mathbf{a}_\alpha + \xi\,\mathbf{d}_{,\alpha})\cdot\mathbf{d}, \\
E_{33} &= \mathbf{d}\cdot\mathbf{d} - 1.
\end{aligned}
\tag{9.190}
$$

Some simplifications are present in these relations since the normal vector $\mathbf{N}$ has length one. This yields $\mathbf{N}\cdot\mathbf{N} = 1$ and $\mathbf{N}\cdot\mathbf{N}_{,\alpha} = 0$. Furthermore, it follows from (9.185) that $\mathbf{A}_\alpha\cdot\mathbf{N} = 0$.

The strain measures are only restricted by the kinematical assumption (9.182) and are valid for arbitrary finite strains. However, relation (9.182) is an approximation for a real three-dimensional continuum. As an example, Eq. $(9.191)_3$ yields a constant strain in thickness direction which is only correct for very special stress states.

An additional constraint in (9.182) leads to a strain in thickness direction which is zero. This constraint is based on the assumption that the director $\mathbf{d}$ follows by a pure rotation $\mathbf{R}$ from the normal vector $\mathbf{N}$

$$
\mathbf{d} = \mathbf{R}\,\mathbf{N}.
\tag{9.191}
$$

In that case, the length of the director will not change during the deformation; hence $\|\mathbf{d}\| = 1$, which leads to E_{33} being zero. This restriction reduces the 6-parameter theory to a 5-parameter model.

9.4.3 Parametrization of the Rotations

As mentioned above, the parametrization of the director $\mathbf{d}$ defines an essential step when the deformation of a shell structure has to be described. Different variants of this parametrization can be found in the literature, see e.g. Argyris (1982) or Betsch et al. (1998). Three different possibilities are selected within the following derivations, leading to distinct shell models.

5-Parameter Model. This parametrization is based of the introduction of an inextensible director field with $\|\mathbf{d}\| = 1$ in (9.182). Such assumption excludes thickness changes of the shell during the deformation. The deformation of the director $\mathbf{d}$ is given by a pure rotation of the normal vector $\mathbf{N}$ in (9.183). The method is very useful for the description of thin shells and produces exact strain measures in the limiting case of vanishing shell thickness.

From many possibilities to define finite rotations of the normal vector, two options are selected:

1. Utilization of two angles as degrees of freedom. This leads to a description of the finite rotation by elementary rotations about fixed axes or by an introduction of spherical coordinates. The formulation has the following properties:
   - additive update of the angles,
   - no storage of variables and
   - not singularity free.
2. Use of linearized rotational degrees of freedom where the finite rotations are rotations about one axis. The properties of this description are:
   - multiplicative update by multiplication of orthogonal matrices,

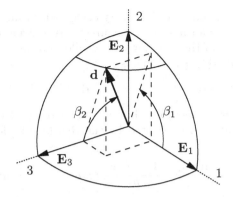

Fig. 9.17 Parametrization of the inextensible director **d**

    – storage of rotations of the last solution step and
    – singularity free for arbitrary rotations.

Both possibilities will be shortly described in the following. A complete overview regarding the parametrization of rotations can be found in Argyris (1982) and in the comparative study in Betsch et al. (1998).

    A simple parametrization, which belongs to 1 and which can also easily be implemented, was firstly applied to finite rotations of shells in Ramm (1976). Within this concept, the director vector **d** is defined as a function of two different independent angles (β_1, β_2) which correspond to spherical coordinates

$$\mathbf{d} = \left\{ \begin{array}{c} \cos \beta_1 \sin \beta_2 \\ \sin \beta_1 \sin \beta_2 \\ \cos \beta_2 \end{array} \right\}. \tag{9.192}$$

The definition of the angles (β_1, β_2) can be found in Fig. 9.17. The angle β_1 is for $\sin \beta_2 = 0 \longrightarrow \beta_2 = (k-1)\pi$, $k \in \mathbb{Z}$ not uniquely defined. In this case, it is necessary to define an angle which is related to β_1 by a constant value to avoid the singularity, for details see Ramm (1976). Otherwise this description of rotations can be applied without further restrictions. The advantage of the description presented in (9.192) lies in the simple additive update of the rotations within incremental solution procedures.

    The shell kinematics of the 5-parameter model are now given in terms of three independent variables related to the deformation of the shell midsurface $\boldsymbol{\varphi}_M$ and the rotation angles (β_1, β_2). Instead of the deformation of the shell midsurface $\boldsymbol{\varphi}_M$, the displacement vector of the shell midsurface $\mathbf{u}_M = \boldsymbol{\varphi}_M - \mathbf{X}_M$ can be introduced. Both formulations are equivalent since the variation or linearization of the coordinates of the midsurface $\mathbf{X}_M$ in its initial position are zero.

    A parametrization of the director field using the approach described in 2 can be obtained for example by application of the RODRIGUES formula. Since the deformation in thickness direction is neglected within the 5-parameter theory ($\|\mathbf{d}\| = 1$), the deformation of the director $\mathbf{d}$ is given by a pure rotation $\mathbf{d} = \mathbf{R}\,\mathbf{N}$. The change of the director vector follows from a time derivative of the scalar product $\mathbf{d} \cdot \mathbf{d} = 1$ and yields

$$\dot{\mathbf{d}} \cdot \mathbf{d} = 0 \quad \Longleftrightarrow \quad \dot{\mathbf{d}} = \boldsymbol{\omega} \times \mathbf{d}. \tag{9.193}$$

Here the axial vector $\boldsymbol{\omega}$ describes the angular velocity of the director. The rotation tensor $\mathbf{R}$ can be represented by the RODRIGUES formula using the axial vector $\boldsymbol{\omega}$

$$\mathbf{R} = \cos\theta\,\mathbf{1} + \frac{\sin\theta}{\theta}\,\hat{\boldsymbol{\omega}} + \frac{1 - \cos\theta}{\theta^2}\,\boldsymbol{\omega} \otimes \boldsymbol{\omega} \quad ; \quad \theta = \|\boldsymbol{\omega}\| \tag{9.194}$$

with the matrices

$$\boldsymbol{\omega} = \left\{ \begin{array}{c} \omega_1 \\ \omega_2 \\ \omega_3 \end{array} \right\} \quad \text{and} \quad \hat{\boldsymbol{\omega}} = \left[\begin{array}{ccc} 0 & -\omega_3 & \omega_2 \\ \omega_3 & 0 & -\omega_1 \\ -\omega_2 & \omega_1 & 0 \end{array} \right]. \tag{9.195}$$

This relation is often applied to describe the dynamics of rigid bodies. A detailed derivation of the RODRIGUES formula can be found in Crisfield (1997). The update formula for the director vector $\mathbf{d}$ is singularity free within this approach. In Eq. (9.194), the three vector components of $\boldsymbol{\omega}$ are used to parameterize $\mathbf{R}$ and thus $\mathbf{d}$. This formulations was applied e.g. in Simo et al. (1989) for the analysis of finite deformations of shells.

6-Parameter Model. Within the shell kinematics of the 6-parameter model, it is assumed that the director vector $\mathbf{d}$ can change its length ($\|\mathbf{d}\| \neq 1$). Hence strains in thickness direction of the shell continuum can be expressed by using this assumption. The main difference to the 5-parameter model lies in the fact that the director is now, like the deformation of the shell midsurface, defined by a vector field. Thus the director can be considered within the 6-parameter model as a vector and no special description of $\mathbf{d}$ by finite rotations is necessary.

As in the 5-parameter model, a displacement vector can be introduced: $\mathbf{u}_M = \boldsymbol{\varphi}_M - \mathbf{X}_M$. Furthermore, instead of the director vector, a difference vector $\mathbf{w}$ is defined leading to

$$\mathbf{d} = \mathbf{N} + \mathbf{w}; \tag{9.196}$$

see Fig. 9.18. Hence within the 6-parameter model, three degrees of freedom are associated with the displacements $\mathbf{u}_M$ of the shell midsurface and three degrees of freedom are associated with the components of the difference vector $\mathbf{w}$. These degrees of freedoms denote the primary variables which have to be discretized within the finite element method.

A further possibility for the description of strains in thickness direction can be found in e.g. Simo et al. (1990B). There a multiplicative decomposition of the director is proposed which yields

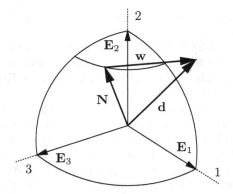

Fig. 9.18 Parametrization of the extensible director **d**

$$\boldsymbol{\varphi} = \boldsymbol{\varphi}_M + \xi\,\lambda\,\mathbf{d}\,. \tag{9.197}$$

Here the director vector is parameterized as in the 5-parameter model with $\|\mathbf{d}\| = 1$. The deformation in thickness direction is then explicitly represented by the additional coordinate λ. Such description can avoid locking which may occur in a finite element discretization based on the 6-parameter model.

9.4.4 Weak Form

The weak form of equilibrium for the 6-parameter model can be formulated in the same way as for the three-dimensional continuum, see (3.289). Using the kinematical relation (9.188) with the variation of the deformation

$$\boldsymbol{\eta} = \boldsymbol{\eta}_M + \xi\,\boldsymbol{\eta}_d,$$

the weak form for the static case is obtained as

$$G\left(\boldsymbol{\varphi}, \boldsymbol{\eta}\right) = \int\limits_{B} \mathbf{P} \cdot \mathrm{Grad}\left(\boldsymbol{\eta}_M + \xi\boldsymbol{\eta}_d\right) dV - \int\limits_{B} \rho_0\,\bar{\mathbf{b}} \cdot \boldsymbol{\eta}\,dV - \int\limits_{\partial B_\sigma} \bar{\mathbf{t}} \cdot \boldsymbol{\eta}\,dA = 0\,.$$

$$\tag{9.198}$$

$\boldsymbol{\eta}_M$ is the variation of the deformation of the midsurface and $\boldsymbol{\eta}_d$ is the variation of the director vector in (9.182). Note that an integration over the complete shell volume has to be performed in (9.198).

This is contrary to classical shell theories where stress resultants are introduced with respect to the shell midsurface. Stress resultants $\mathbf{r}_\sigma$ of the 6-parameter model follow from a direct integration of the stresses over the shell thickness. Due to the nonlinearity of (6.184) with respect to ξ, an explicit integration and hence an explicit representation of the stress resultants is not possible. Thus numerical integration over the thickness has to be performed which is utilized within finite element formulations by GAUSS quadrature, see

Sect. 4.1.1. Strains related to GAUSS points can be associated with different layers, see also Sect. 9.2.3.

Instead of using (3.289), the weak form can also be written in terms of the 2nd PIOLA–KIRCHHOFF stresses (3.292). A detailed description of 6-parameter models including all matrices can be found in Eberlein (1997), see also Bischoff et al. (2004).

When a 5-parameter model of the shell is applied, some peculiarities have to be considered with respect to the three-dimensional formulation. Here –related to the kinematical assumptions – a plain strain condition has to be introduced. However since no stresses are present at the outer surfaces of the shell continuum, a plane stress state can be assumed equally well for the computations. For linear shell theories, it can be shown that this contradiction is consistent and within the kinematical assumptions, see Koiter (1960). For elasto-plastic materials, the plane stress state can be deduced from the three-dimensional form presented in Sect. 6.3.2 where the transverse shear stresses is eliminated. This approach leads to the so-called quasi KIRCHHOFF model which is discussed in Eberlein et al. (1993) for axisymmetric shells. By suppressing the transverse shear strains using a penalty formulation, the internal virtual work of the 5-parameter model is obtained

$$G_i(\mathbf{u}, \boldsymbol{\eta}) = \int_{\mathcal{M}} \int_h S_{\alpha\beta} \, \delta E_{\alpha\beta} \, d\xi d\Omega + c_p \, h \int_{\mathcal{M}} E_{\alpha 3} \, \delta E_{\alpha 3} \, d\Omega \tag{9.199}$$

with $\quad E_{\alpha 3} = E_{3\alpha} = \tfrac{1}{2} \, \mathbf{a}_\alpha \cdot \mathbf{d} = 0 \,.$

The penalty term which suppresses the transverse shear strains $E_{\alpha 3}$, defined in (9.191), can be found in the second term in (9.199) with the penalty parameter c_p. The first integrand contains the 2nd PIOLA-KIRCHHOFF stresses related to the shell surface. Thickness changes do not occur in this formulation. The integration over the shell continuum can be split into an integration over the thickness h and the shell midsurface $\mathcal{M}$, since the stresses cannot be integrated analytically over the thickness in case of nonlinear material response. Due to the constant penalty parameter, the second term in (9.199), however, can be integrated analytically over the thickness.

9.4.5 Constitutive Equations for Shells

Shells can be used in many different applications which depict distinct nonlinear constitutive behaviour, see Sect. 9.3.3. Out of many possibilities, here isotropic elasto-plastic constitutive equations for finite deformations are considered in more detail. Since the main results have already been stated in Sects. 3.3.2 and 6.3, only the additional formulations related to shells will be discussed.

When using the 6-parameter shell, the constitutive equations discussed in Sect. 6.3.2 can be applied directly since general three-dimensional stress states can be modelled within such shell formulation.

In case of the plane stress state which is assumed within the 5-parameter shell model, additional considerations are necessary to derive the relevant constitutive equations and to use the formulations and algorithms developed in Sect. 6.3.2. As long as only small elastic strains are present (see strain energy function (6.158)), the stresses in thickness direction can be eliminated explicitly, which was shown in Wriggers et al. (1995). This elimination leads to efficient algorithms and since small elastic strains are often sufficient to model the elasto-plastic behaviour of shell structures, like in metal forming, a constitutive equation for small elastic strains but finite plastic strains will be formulated for the 5-parameter shell model. However, it should be noted that for special Neo-Hooke type finite strain elasticity it is also possible to analytically eliminate the stresses in thickness direction, related formulations can be found in Campello et al. (2003).

The weak form used for the 5-parameter shell model is related to the initial configuration, see (9.199). Hence Eq. (6.163) has to be pulled back to the initial configuration. This leads to a general eigenvalue problem for the determination of the principal strains, see e.g. Ibrahimbegovic (1994),

$$(\mathbf{C}_n^{p-1} - \lambda_{\alpha e}^{tr\,2}\,\mathbf{C}_{n+1}^{-1})\,\mathbf{N}_\alpha^{tr} = \mathbf{0}\,. \qquad (9.200)$$

Due to the plane stress state, only membrane strains appear in this relation. The shear strains $C_{\alpha 3}$ are assumed to be zero which is in accordance with the kinematical assumption, see also the *penalty* formulation in (9.199). The strain in thickness direction C_{33} is not zero in the plane stress case. This strain, however, can be computed within a postprocessing step, using e.g. the assumption of plastic incompressibility, since it does not occur explicitly in the weak form. A general three-dimensional formulation with respect to the initial configuration can be found in Miehe (1998). This formulation uses the plastic metric analogous to Eq. (6.141).

The algorithm which can be applied to compute the stresses within the plane stress state is mainly equivalent to the algorithms developed in Sect. 6.3.2. It contains the following steps:

1. Assume that the history variables $\mathbf{C}_n^{p-1}$ and α_n are known from the last time step t_n. Furthermore, the right CAUCHY–GREEN tensor $\mathbf{C}_{n+1}$ is known from the solution of the weak form at time t_{n+1}. With these quantities, the eigenvalues or principal strains can be computed from

$$\mathbf{N}^{\alpha\,tr}(\mathbf{C}_{n+1}\,\mathbf{C}_n^{p-1}) = \lambda_{\alpha e}^{tr\,2}\,\mathbf{N}^{\alpha\,tr} \quad \rightsquigarrow \quad \varepsilon_{\alpha e}^{tr} = \ln\lambda_{\alpha e}^{tr}\,. \qquad (9.201)$$

2. The return mapping algorithms, described in Sect. 6.3.2 yields now the strains $\varepsilon_{\alpha e}$, the KIRCHHOFF stresses τ_α and the algorithmic tangent modulus $C_{\alpha\beta}^{ALG}$ with respect to the principal axes.

3. With these quantities, the plastic metric can be determined at time t_{n+1}

$$\mathbf{C}_{n+1}^{p-1} = \sum_{\alpha=1}^{2}\mathbf{N}^\alpha \otimes \mathbf{N}^\alpha \quad \text{with} \quad \mathbf{N}^\alpha = \left(\frac{\lambda_{\alpha e}}{\lambda_{\alpha e}^{tr}}\right)\mathbf{N}^{\alpha\,tr}\,. \qquad (9.202)$$

In the same way, the 2nd PIOLA-KIRCHHOFF stress tensor is obtained as

$$\mathbf{S} = \sum_{\alpha=1}^{2} \frac{\tau_\alpha}{\lambda_{\alpha e}^{tr\,2}} \mathbf{N}^{\alpha\,tr} \otimes \mathbf{N}^{\alpha\,tr}, \tag{9.203}$$

as well as the algorithmic tangent modulus

$$\mathcal{L} = \sum_{\alpha=1}^{2} \sum_{\beta=1}^{2} \frac{C_{\alpha\beta}^{ALG} - \tau_\alpha\,2\,\delta_{\alpha\beta}}{\lambda_{\alpha e}^{tr\,2}\,\lambda_{\beta e}^{tr\,2}} \left(\mathbf{N}^{\alpha\,tr} \otimes \mathbf{N}^{\alpha\,tr} \otimes \mathbf{N}^{\beta\,tr} \otimes \mathbf{N}^{\beta\,tr} \right) +$$

$$\sum_{\alpha\neq\beta} \frac{S_\beta - S_\alpha}{\lambda_{\beta e}^{tr\,2} - \lambda_{\alpha e}^{tr\,2}} \left(\mathbf{N}^{\alpha\,tr} \otimes \mathbf{N}^{\beta\,tr} \otimes \mathbf{N}^{\alpha\,tr} \otimes \mathbf{N}^{\beta\,tr} + \right. \tag{9.204}$$

$$\left. \mathbf{N}^{\alpha\,tr} \otimes \mathbf{N}^{\beta\,tr} \otimes \mathbf{N}^{\beta\,tr} \otimes \mathbf{N}^{\alpha\,tr} \right).$$

The computation of stresses and algorithmic tangent modulus is equivalent to the approach used in Sect. 6.3.2. Note that the associated covariant and contravariant eigenvectors occur in the general eigenvalue problem (9.200). These are related to each other by $\mathbf{N}^{\alpha\,tr} \cdot \mathbf{N}_\beta^{tr} = \delta_\beta^\alpha$. The proposed algorithm does not need the deformation gradient of the defined shell space explicitly since push forward and pull back operations to the related metrics are not necessary when working with the right CAUCHY-GREEN strain tensor. However, the deformation gradient is necessary as kinematical basis for an efficient implementation within the framework of the finite element method.

9.4.6 Finite Element Formulation for the 5-Parameter Model

Shell elements were developed based on different orders of interpolation. The relevant literature review can be found in Sect. 9.4.1. Due to their advantages, only low order finite elements are discussed and a general quadrilateral shell elements with bi-linear shape functions is developed in detail. To avoid the well known locking behaviour of such low order elements, special techniques have to be applied. However, before a detailed discussion, the general isoparametric concept for shells will be considered which is basis for the discretization of general shell geometries.

Isoparametric Concept. In Sect. 9.4.2, curvilinear convective coordinates ξ^α were introduced to parameterize the shell midsurface $\mathcal{M}$. Covariant base vectors follow then from (9.184) or (9.186). Together with (9.189), the components of the strain tensor are obtained for general curvilinear coordinates. At this point, displacement components are introduced in classical shell theories in order to express the base vectors in the current configuration by base vectors related to the initial configuration. In that case, covariant derivatives of the displacements occur in the strain measures, see e.g. Eringen (1962) and Naghdi (1972). The covariant derivatives can be described by using CHRISTOFFEL symbols. Since the CHRISTOFFEL symbols depend on the underlying shell geometry, a general form for a strain–displacement relation for the shell cannot be obtained. However, with finite elements, general geometries are to be described which is not easily possible within the classical approach. Thus the isoparametric concept is directly applied to discretize the shell geometry as well as the displacement and rotation field.

In general, a unique mapping from the reference space of a finite element to the initial and current configuration is defined, see also Sect. 4.1. Within this mapping, the position vector $\mathbf{X}$ to the shell midsurface and the displacement field $\mathbf{u}$ is discretized by the shape function N_I

$$\mathbf{u} = \sum_{I=1}^{n} N_I(\boldsymbol{\zeta}) \, \mathbf{u}_I; \quad \mathbf{X} = \sum_{I=1}^{n} N_I(\boldsymbol{\zeta}) \, \mathbf{X}_I. \qquad (9.205)$$

n denotes the number of nodes of the finite element and $\boldsymbol{\zeta} = \{\zeta^1, \zeta^2\}$ are the convective coordinates. For the interpolation, N_I bilinear shape functions are selected, see (4.28),

$$N_I(\zeta^1, \zeta^2) = \frac{1}{4} \left(1 + \zeta^1 \zeta_I^1\right) \left(1 + \zeta^2 \zeta_I^2\right). \qquad (9.206)$$

The shape functions N_I are referred to the reference configuration $\Omega_\square$, see also (4.28) and Fig. 4.7. Based on (9.205), relation $\mathbf{X} = \mathbf{X}(\boldsymbol{\zeta})$ is obtained. Hence the position vector $\mathbf{X}$, which describes the initial configuration B of the shell continuum, is given by the coordinates $\boldsymbol{\zeta}$ of the reference element $\Omega_\square$. This isoparametric transformation relates initial or current configuration – as in the three-dimensional case – to the reference element $\Omega_\square$, see also (4.54).

The isoparametric concept described in Sect. 4.1 cannot be applied in a direct way to shells since the isoparametric map of a plane reference element into three dimensional space is not singularity free. In order to avoid this singularity, a modified isoparametric description will be used which is related to a local cartesian basis $\{\mathbf{E}_i^{loc}\}$. In general, this basis does not coincide with the global cartesian coordinate system $\{\mathbf{E}_i\}$ of the reference configuration. The local cartesian basis is determined by the basis vectors $\mathbf{G}_\alpha$ of the undeformed initial configuration. For an element, based on isoparametric formulation in (9.184), relation

$$\mathbf{G}_{\zeta^1} = \sum_{I=1}^{4} N_{I,\zeta^1} \, \mathbf{X}_I; \quad \mathbf{G}_{\zeta^2} = \sum_{I=1}^{4} N_{I,\zeta^2} \, \mathbf{X}_I \qquad (9.207)$$

is deduced. The associated normal vector $\mathbf{N}$ is determined by the cross product of $\mathbf{G}_{\zeta^\alpha}$ analogous to (9.185). Now the local cartesian basis $\{\mathbf{E}_i^{loc}\}$ – as depicted in Fig. 9.19 – can be defined by

$$\mathbf{E}_1^{loc} = \frac{\mathbf{G}_{\zeta^1}}{\|\mathbf{G}_{\zeta^1}\|}; \quad \mathbf{E}_3^{loc} = \mathbf{N}; \quad \mathbf{E}_2^{loc} = \mathbf{E}_3^{loc} \times \mathbf{E}_1^{loc}. \qquad (9.208)$$

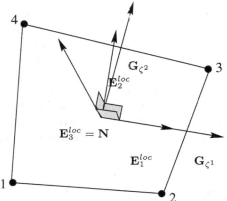

Fig. 9.19 Local cartesian basis system $\{\mathbf{E}_i^{loc}\}$

The gradient $\mathbf{J}$, which belongs to the modified isoparametric map of the shell is given by

$$\mathbf{J} = \text{Grad}_\zeta \mathbf{X} = \frac{\partial \mathbf{X}}{\partial \boldsymbol{\zeta}} = \sum_{I=1}^{4} N_{I,\zeta^\alpha}(\boldsymbol{\zeta}) \, \mathbf{X}_I \otimes \mathbf{E}_{\zeta^\alpha}^{loc} \quad \text{with} \quad \zeta^\alpha = \zeta^1, \zeta^2 \, .$$
(9.209)

With the base vectors defined in equation (9.207)

$$\mathbf{J} = \mathbf{G}_{\zeta^\beta} \otimes \mathbf{E}_{\zeta^\beta}^{loc} = (\mathbf{G}_{\zeta^\beta} \cdot \mathbf{E}_\alpha^{loc}) \, \mathbf{E}_\alpha^{loc} \otimes \mathbf{E}_{\zeta^\beta}^{loc} \quad \text{with} \quad \alpha = 1, 2 \quad (9.210)$$

follows. The coordinates ζ^α can be viewed in the reference configuration as cartesian coordinates with respect to the local basis $\mathbf{E}_\alpha^{loc}$ which was defined in (9.208). By application of the chain rule, the gradient of the interpolation functions follows as

$$\frac{\partial N_I}{\partial \boldsymbol{\zeta}} = \frac{\partial N_I}{\partial \mathbf{X}} \frac{\partial \mathbf{X}}{\partial \boldsymbol{\zeta}} = \frac{\partial N_I}{\partial \mathbf{X}} \mathbf{J}$$
(9.211)

with $\dfrac{\partial N_I}{\partial \boldsymbol{\zeta}} = \nabla_\zeta N_I = N_{I,\zeta^\alpha} \, \mathbf{E}_{\zeta^\alpha}^{loc}$ and $\dfrac{\partial N_I}{\partial \mathbf{X}} = \nabla_{X^{loc}} N_I = N_{I,X_\alpha^{loc}} \, \mathbf{E}_\alpha^{loc}$. Since furthermore the transformation

$$\nabla_{X^{loc}} N_I = \mathbf{J}^{-T} \nabla_\zeta N_I \quad \Longleftrightarrow \quad \left\{ \begin{array}{c} N_{I,1} \\ N_{I,2} \end{array} \right\} = \mathbf{J}^{-T} \left\{ \begin{array}{c} N_{I,\zeta^1} \\ N_{I,\zeta^2} \end{array} \right\}$$
(9.212)

is valid, the derivatives with respect to the local cartesian coordinates X_α^{loc} in the initial configuration can be replaced by the derivatives with respect to ζ^α in the reference configuration.

The idea to use a local cartesian basis is essential for the general formulations of finite shell elements. Due to this choice, all covariant and contravariant derivatives disappear.[3] The transformation to the global cartesian coordinates $\{\mathbf{E}_i\}$ which is necessary for the assembly of residual vectors and tangent matrices will be discussed later.

---

[3] The covariant and contravariant derivatives are related to the change of the base vectors when covarient and contravariant coordinates are selected as basis as done in classical shell theories, see e.g. Simo et al. (1990).

Formulation of the Shell Element. The 5-parameter theory is based on a director field which is inextensible and hence does exclude strains in thickness direction of the shell space. Furthermore, shear deformations are suppressed. With these assumptions, the resulting finite shell element can only be applied for thin shells. The associated parametrization was discussed in Sect. 9.4.3. The constitutive equations can be found in Sect. 9.4.5. Hence the theoretical background is known and this section can concentrate on the development of the finite element discretization of the thin shell.

For the computation of the local deformation gradient $\mathbf{F}^{loc}$, see (9.188), it is advantageous to express $\bar{\mathbf{F}}^{ref}$ completely in terms of the local basis $\{\mathbf{E}_i^{loc}\}$

$$\bar{\mathbf{F}}_{[C]}^{ref} = \mathbf{a}_{\zeta^\alpha} \otimes \mathbf{E}_{\zeta^\alpha}^{loc} + \mathbf{d} \otimes \mathbf{N} \qquad \bar{\mathbf{F}}_{[L]}^{ref} = \mathbf{d}_{,\zeta^\alpha} \otimes \mathbf{E}_{\zeta^\alpha}^{loc} . \tag{9.213}$$

The interpolation of the base vectors $\mathbf{g}_{\zeta^\alpha}$ in (9.213) is provided by

$$\mathbf{g}_{\zeta^\alpha} = \sum_{I=1}^{4} N_{I,\zeta^\alpha} \, \mathbf{x}_I \qquad \text{with} \quad \mathbf{x}_I = x_{I\,i} \, \mathbf{E}_i = \underbrace{x_{I\,i} \, \mathbf{E}_i \cdot \mathbf{E}_{0\,j}^{loc}}_{x_{I\,i}^{loc} \, \delta_{ij}} \, \mathbf{E}_{0\,j}^{loc} . \tag{9.214}$$

Here the local basis $\{\mathbf{E}_{0\,i}^{loc}\}$ is related to the centre of the finite element. From Eq. (9.213) follows a representation of $\bar{\mathbf{F}}^{ref}$ with (9.214) and $x_{I\,i}^{loc}$ which is invariant with respect to rigid body rotations. Hence $\bar{\mathbf{F}}^{ref}$ will reduce to a unit matrix in the initial configuration with respect to $\{\mathbf{E}_{0\,i}^{loc}\}$. The local deformation gradient is computed using the JACOBIAN which stems from the isoparametric map

$$\mathbf{F}^{loc} = \bar{\mathbf{F}}^{ref} \, \boldsymbol{J}^{-1} . \tag{9.215}$$

When using the constitutive equations from Sect. 9.4.5, the stresses have to be transformed to principal directions. Since a plane stress state is considered within the 5-parameter shell model, the transformation between the three remaining stress components $\{\bar{S}_i\}^T = \{S_{11}, S_{22}, S_{12}\}$ and the two principal stresses $\{\bar{S}_j\}^{prin\,T} = \{S_1, S_2, 0\}$ is given by, see also (3.139),

$$\bar{S}_i = T_{ij} \, \bar{S}_j^{prin} \qquad i,j = 1,2,3$$

$$\text{with} \quad T_{ij} = \begin{bmatrix} \cos^2 \varphi & \sin^2 \varphi & -2\sin\varphi\cos\varphi \\ \sin^2 \varphi & \cos^2 \varphi & 2\sin\varphi\cos\varphi \\ \sin\varphi\cos\varphi & -\sin\varphi\cos\varphi & \cos^2 \varphi - \sin^2 \varphi \end{bmatrix} . \tag{9.216}$$

The angle of rotation φ is determined by the components of the right CAUCHY–GREEN tensor since stresses and strains are co-axial for isotropic material behaviour

$$\varphi = \frac{1}{2} \arctan \left(\frac{2\, C_{12}^{loc}}{C_{11}^{loc} - C_{22}^{loc}} \right) . \tag{9.217}$$

The components of the right CAUCHY–GREEN tensor in (9.217) follow from (3.15) by using (9.188), (9.240) and (9.215). Possible singularities in (9.217), which can occur for $C_{11}^{loc} = C_{22}^{loc}$, are avoided by a perturbation of one component by a small number. With these relations, the principal stretches λ_α^{loc} can be stated explicitly

$$(\lambda_\alpha^{loc})^2 = \mathbf{T}^T \, \mathbf{C}^{loc} \, \mathbf{T} \quad \text{with} \quad T_{\alpha\beta} = \begin{bmatrix} \cos\varphi & -\sin\varphi \\ \sin\varphi & \cos\varphi \end{bmatrix}. \tag{9.218}$$

To complete the formulation of the finite shell element, a matrix formulation is introduced for $\bar{\mathbf{F}}^{loc}$ with reference to the local basis $\{\mathbf{E}_i^{loc}\}$ analogous to (9.215)

$$\mathbf{F}^{loc} = \mathbf{F}_{[C]}^{loc} + \xi \, \mathbf{F}_{[L]}^{loc} = \begin{bmatrix} \mathbf{a}_1 & \mathbf{a}_2 & \mathbf{d} \end{bmatrix} + \xi \begin{bmatrix} \mathbf{d}_{,1} & \mathbf{d}_{,2} & \mathbf{0} \end{bmatrix}. \tag{9.219}$$

The approximation of the position vector $\mathbf{X}_M$ and the displacement vector $\mathbf{u}$ with reference to the shell midsurface $\mathcal{M}$ is discussed next. These quantities are described by the local coordinates defined in (9.214)

$$\begin{aligned} \mathbf{X}_M &= \{X_{M1}^{loc}, X_{M2}^{loc}, X_{M3}^{loc}\}^T &= \sum_{I=1}^{4} N_I \, \mathbf{X}_{MI}^{loc} \\ \mathbf{u} &= \{u_1^{loc}, u_2^{loc}, u_3^{loc}\}^T &= \sum_{I=1}^{4} N_I \, \mathbf{u}_I^{loc}. \end{aligned} \tag{9.220}$$

The rotations (β_1, β_2), illustrated in Fig. 9.17, are approximated in the same way within the 5-parameter concept, see e.g. Wagner and Gruttmann (1994). First, the initial angles $(\bar{\beta}_1, \bar{\beta}_2)$ are determined and interpolated in the initial configuration. Then an isoparametric ansatz is chosen to describe the incremental angles (ω_1, ω_2),

$$\bar{\boldsymbol{\beta}} = \{\bar{\beta}_1, \bar{\beta}_2\}^T = \sum_{I=1}^{4} N_I \, \bar{\boldsymbol{\beta}}_I, \qquad \boldsymbol{\omega} = \{\omega_1, \omega_2\}^T = \sum_{I=1}^{4} N_I \, \boldsymbol{\omega}_I. \tag{9.221}$$

The basis vectors $\mathbf{a}_\alpha$ in (9.219) and their variations follow from

$$\mathbf{a}_\alpha = \sum_{I=1}^{4} N_{I,\alpha} \, (\mathbf{X}_{MI}^{loc} + \mathbf{u}_I^{loc}); \qquad \delta\mathbf{a}_\alpha = \sum_{I=1}^{4} N_{I,\alpha} \, \delta\mathbf{u}_I^{loc}. \tag{9.222}$$

Here the derivatives of the interpolation functions $N_{I,\alpha}$ are computed using (9.212). With (9.221), the rotations (β_1, β_2) and their variations are obtained. These are needed to describe the director vector $\mathbf{d}$ in (9.192)

$$\boldsymbol{\beta} = \sum_{I=1}^{4} N_I \, (\bar{\boldsymbol{\beta}}_I + \boldsymbol{\omega}_I); \qquad \delta\boldsymbol{\beta} = \sum_{I=1}^{4} N_I \, \delta\boldsymbol{\omega}_I. \tag{9.223}$$

The weak form leading to the shell element is based on a specialization of (9.199). Thus

$$
G(\boldsymbol{u},\boldsymbol{\alpha},\boldsymbol{\eta}) \;=\; \overset{n_e}{\underset{e=1}{\bigcup}} \sum_{I=1}^{4} \boldsymbol{\eta}_I^T \left\{ \int_{\mathcal{M}_e} \int_h \boldsymbol{B}_I^T \, \bar{\boldsymbol{S}} \, d\xi \, d\Omega \; + \right.
$$

$$
\left. c_p \, h \int_{\mathcal{M}_e} \boldsymbol{B}_I^{pen^T} \left\{ \begin{array}{c} \mathbf{F}_3^{loc^T} \mathbf{F}_1^{loc} \\ \mathbf{F}_3^{loc^T} \mathbf{F}_2^{loc} \end{array} \right\} d\Omega \right\} + '' \textit{Load terms}'' = 0
$$

$$(9.224)$$

is obtained. The vector containing the components of the 2nd PIOLA-KIRCH-HOFF stresses $\bar{\mathbf{S}}$ was already defined in (9.216). The penalty parameter c_p, suppressing shear deformations, has to be selected by the user of the shell element. It depends upon the magnitude of constitutive parameters describing the shell. The vector $\boldsymbol{\eta}_I$ contains the variations $(\delta \boldsymbol{u}_I^{loc}, \delta \boldsymbol{\omega}_I)$. By $\mathbf{q}$, the variation of the *enhanced* parameters $\delta \boldsymbol{\alpha}_\gamma$ is denoted which were introduced in (9.238). The B-matrices $\boldsymbol{B}_I, \boldsymbol{B}_I^{pen}$ have the form

$$
\boldsymbol{B}_I = \begin{bmatrix} \boldsymbol{B}_1^T & B_{11} & B_{12} \\ \boldsymbol{B}_2^T & B_{21} & B_{22} \\ \boldsymbol{B}_3^T & B_{31} & B_{32} \end{bmatrix},
$$

$$(9.225)$$

with $(\alpha, \beta = 1, 2)$

$$
\begin{aligned}
\boldsymbol{B}_\alpha^T &= \mathbf{F}_\alpha^{loc\,T} N_{I,\alpha} \\
\boldsymbol{B}_3^T &= \mathbf{F}_1^{loc\,T} N_{I,2} + \mathbf{F}_2^{loc\,T} N_{I,1} \\
\boldsymbol{B}_{\alpha\beta}^T &= \xi \, \mathbf{F}_\alpha^{loc\,T} (\, \mathbf{d}_\beta^{loc} N_{I,\alpha} + \mathbf{d}_{\beta,\alpha}^{loc} N_I \,) \\
\boldsymbol{B}_{3\alpha}^T &= \xi \, [\, \mathbf{F}_1^{loc\,T} (\, \mathbf{d}_\alpha^{loc} N_{I,2} + \mathbf{d}_{\alpha,2}^{loc} N_I \,) + \mathbf{F}_2^{loc\,T} (\, \mathbf{d}_\alpha^{loc} N_{I,1} + \mathbf{d}_{\alpha,1}^{loc} N_I \,)
\end{aligned}
$$

and

$$
\boldsymbol{B}_I^{pen} = \begin{bmatrix} \mathbf{F}_3^{loc\,T} N_{I,1} & \mathbf{F}_1^{loc\,T} \mathbf{d}_1^{loc} N_I & \mathbf{F}_1^{loc\,T} \mathbf{d}_2^{loc} N_I \\ \mathbf{F}_3^{loc\,T} N_{I,2} & \mathbf{F}_2^{loc\,T} \mathbf{d}_1^{loc} N_I & \mathbf{F}_2^{loc\,T} \mathbf{d}_2^{loc} N_I \end{bmatrix}.
$$

$$(9.226)$$

The vectors $\mathbf{d}_\alpha^{loc}$ stem from the variation of the local director vector which is determined within the 5-parameter theory by (9.192). The variation of these quantities yields

$$
\delta \mathbf{d}^{loc} = \mathbf{d}_\alpha^{loc} \, \delta \beta_\alpha
$$

$$(9.227)$$

with

$$
\mathbf{d}_1^{loc} = \left\{ \begin{array}{c} -\sin\beta_1 \sin\beta_2 \\ \cos\beta_1 \sin\beta_2 \\ 0 \end{array} \right\} \quad \text{and} \quad \mathbf{d}_2^{loc} = \left\{ \begin{array}{c} \cos\beta_1 \cos\beta_2 \\ \sin\beta_1 \cos\beta_2 \\ -\sin\beta_2 \end{array} \right\}.
$$

$$(9.228)$$

The linearization of the discretized weak form (9.224) is needed within the incremental solution procedure using NEWTON'S method. This linearization leads to a system of equations for the incremental displacements $\Delta \mathbf{u}^{loc}$ and the incremental angles of rotation $\Delta \boldsymbol{\omega}$

$$\mathbf{K}_{uu} \left\{ \begin{array}{c} \Delta \mathbf{u} \\ \Delta \boldsymbol{\omega} \end{array} \right\} = -\mathbf{G}_u \qquad (9.229)$$

with $\mathbf{u} = \{u_1^{loc}, u_2^{loc}, u_3^{loc}\}^T$. The tangent matrix in (9.229) is defined by

$$\mathbf{K}_{uu} = \overset{n_e}{\underset{e=1}{\bigcup}} \sum_{I=1}^{4} \sum_{J=1}^{4} \left\{ \int_{\mathcal{M}_e} \int_h \left(\mathbf{B}_I^T \bar{\mathbf{L}} \mathbf{B}_J + \mathbf{G}_{IJ}^1 \right) d\xi \, d\Omega \right. $$
$$\left. + c_p h \int_{\mathcal{M}_e} \left(\mathbf{B}_I^{pen^T} \mathbf{B}_J^{pen} + \mathbf{G}_{IJ}^{pen} \right) d\Omega \right\} . \qquad (9.230)$$

The components of the incremental constitutive tensor $\bar{\mathbf{L}}$ can be computed from (6.185) based on the deformation $\bar{\boldsymbol{\varphi}}$ obtained within the NEWTON scheme. The operator matrices $\mathbf{G}_{IJ}^1$ and $\mathbf{G}_{IJ}^{pen}$ follow by applying the linearization tools described in Sect. 3.5. Explicitly, the matrix $\mathbf{G}_{IJ}^1$ can be written as:

$$\mathbf{G}_{IJ}^1 = \begin{bmatrix} A\,\mathbf{1} & \xi\,\mathbf{b}_1 & \xi\,\mathbf{b}_2 \\ \xi\,\mathbf{c}_1^T & \xi\,(G_{11} + \xi\,H_{11}) & \xi\,(G_{12} + \xi\,H_{12}) \\ \xi\,\mathbf{c}_2^T & \xi\,(G_{21} + \xi\,H_{21}) & \xi\,(G_{22} + \xi\,H_{22}) \end{bmatrix} \qquad (9.231)$$

with

$$\mathbf{b}_\eta = A\,\mathbf{d}_\eta^{lok} + B_\beta\,\mathbf{d}_{\eta,\beta}^{lok} \qquad \mathbf{c}_\eta = A\,\mathbf{d}_\eta^{lok} + C_\alpha\,\mathbf{d}_{\eta,\alpha}^{lok},$$
$$G_{\eta\theta} = \mathbf{F}_\alpha^{lok\,T}\,(\,D_{\alpha\beta}\,\mathbf{d}_{\eta\,6\,\gamma}^{lok}\,\beta_{\gamma,\beta} + (\,B_\alpha + C_\alpha\,)\,\mathbf{d}_{\eta\theta}^{lok}\,),$$
$$H_{\eta\theta} = \mathbf{d}_\eta^{lok\,T}\,(\,A\,\mathbf{d}_\theta^{lok} + B_\beta\,\mathbf{d}_{\theta,\beta}\,) + \mathbf{d}_{\eta,\alpha}^{lok\,T}\,(\,C_\alpha\,\mathbf{d}_\theta^{lok} + D_{\alpha\beta}\,\mathbf{d}_{\theta,\beta}\,)$$

and

$$A = S_{\alpha\beta}\,N_{I,\alpha}\,N_{J,\beta} \qquad B_\beta = S_{\alpha\beta}\,N_{I,\alpha}\,N_J$$
$$C_\alpha = S_{\alpha\beta}\,N_I\,N_{J,\beta} \qquad D_{\alpha\beta} = S_{\alpha\beta}\,N_I\,N_J .$$

Furthermore, $\mathbf{G}_{IJ}^{pen}$

$$\mathbf{G}_{IJ}^{pen} = \begin{bmatrix} \mathbf{0} & A^{pen}\,\mathbf{d}_1^{lok} & A^{pen}\,\mathbf{d}_2^{lok} \\ B^{pen}\,\mathbf{d}_1^{lok\,T} & C_\alpha^{pen}\,\mathbf{F}_\alpha^{lok\,T}\,\mathbf{d}_{11}^{lok} & C_\alpha^{pen}\,\mathbf{F}_\alpha^{lok\,T}\,\mathbf{d}_{21}^{lok} \\ B^{pen}\,\mathbf{d}_2^{lok\,T} & C_\alpha^{pen}\,\mathbf{F}_\alpha^{lok\,T}\,\mathbf{d}_{12}^{lok} & C_\alpha^{pen}\,\mathbf{F}_\alpha^{lok\,T}\,\mathbf{d}_{22}^{lok} \end{bmatrix} \qquad (9.232)$$

follows with

$$A^{pen} = \mathbf{F}_3^{lok\,T}\,\mathbf{F}_\alpha^{lok}\,N_{I,\alpha}\,N_J ,$$
$$B^{pen} = \mathbf{F}_3^{lok\,T}\,\mathbf{F}_\alpha^{lok}\,N_I\,N_{J,\alpha} ,$$
$$C_\alpha^{pen} = \mathbf{F}_3^{lok\,T}\,\mathbf{F}_\alpha^{lok}\,N_I\,N_J .$$

In these relations summation over the indices α, $\beta = 1, 2$ has to be performed. The derivatives and linearizations of the director vector are obtained from (9.222) and (9.227) or from

$$\delta d_{,\alpha}^{lok} = d_\beta^{lok} \, \delta\beta_{\beta,\alpha} + d_{\gamma,\alpha}^{lok} \, \delta\beta_\gamma \quad \text{with} \quad d_{\gamma,\alpha}^{lok} = d_{\gamma\beta}^{lok} \, \beta_{\beta,\alpha}, \qquad (9.233)$$

where the vectors $d_{\alpha\beta}^{lok}$ are given by

$$\begin{aligned} d_{11}^{lok\,T} &= \{ -\cos\beta_1 \sin\beta_2 \quad -\sin\beta_1 \sin\beta_2 \quad 0 \}, \\ d_{12}^{lok\,T} &= \{ -\sin\beta_1 \cos\beta_2 \quad +\cos\beta_1 \cos\beta_2 \quad 0 \}, \qquad (9.234) \\ d_{22}^{lok\,T} &= \{ -\cos\beta_1 \sin\beta_2 \quad -\sin\beta_1 \sin\beta_2 \quad -\cos\beta_2 \}. \end{aligned}$$

The linearization of the variation of the director vector and its derivative yields

$$\Delta\delta d^{lok} = d_{\alpha\beta}^{lok} \, \Delta\beta_\alpha \, \delta\beta_\beta \quad \text{and} \quad \Delta\delta d_{,\alpha}^{lok} = d_{\beta\gamma\delta}^{lok} \beta_{\beta,\alpha} \, \Delta\beta_\gamma \, \delta\beta_\delta, \qquad (9.235)$$

where the newly defined vectors $d_{\beta\gamma\delta}^{lok}$ are computed from

$$\begin{aligned} d_{111}^{lok} &= d_{111}^{lok} = d_{122}^{lok} = d_{212}^{lok} = d_{221}^{lok} = -d_1^{lok} \\ d_{222}^{lok} &= -d_2^{lok} \\ d_{121}^{lok} &= d_{211}^{lok} = -d_{112}^{lok} \quad \text{with} \\ d_{112}^{lok\,T} &= \{ -\cos\beta_1 \cos\beta_2 \quad -\sin\beta_1 \cos\beta_2 \quad 0 \}. \end{aligned}$$

The integration of the shell element with bilinear interpolation is performed with a 2×2 GAUSS quadrature in order to avoid rank deficiency of the tangent matrices. However, this results in shear locking, see e.g. the overview in Andelfinger (1991). Selective reduced integration is a simple method to avoid shear locking, see Sect. 9.4.1; it was developed in Zienkiewicz et al. (1971) for continua and in Hughes et al. (1977a) for plate elements. Using this approach, the penalty term in (9.199) has to be integrated by a one point quadrature. This, however, still results in a rank deficiency of the tangent stiffness matrix which occurs for special boundary conditions. An alternative is provided by the special interpolation for the second term in (9.199) as advocated in Bathe and Dvorkin (1985).

Finally, it has to be realized that the components of the local displacements Δu_{Ii}^{loc}, which are related to one node of the finite element, have to be transformed to the global basis $\{\mathbf{E}_i\}$. This is necessary for the assembly of residual vectors and tangent matrices. Such transformation is given by, compare also (9.214) and Remark 9.1,

$$\Delta u_{Ii}^{loc} \, \mathbf{E}_{0i}^{loc} = \underbrace{\Delta u_{Ii}^{loc} \, \mathbf{E}_{0i}^{loc} \cdot \mathbf{E}_j}_{\Delta u_{Ii} \, \delta_{ij}} \, \mathbf{E}_j. \qquad (9.236)$$

Since all quantities are related to the reference configuration $\Omega_\square$ within the isoparametric concept, it is necessary to transform the area and volume elements in the integrals (9.230) and (9.224) from the initial to the reference configuration. This leads to

$$d\Omega_\circ = \left\| \frac{\partial \mathbf{X}}{\partial \zeta^1} \times \frac{\partial \mathbf{X}}{\partial \zeta^2} \right\| d\zeta^1 \, d\zeta^2 = J \, d\Omega_\square \qquad \text{with} \qquad J = \det \mathbf{J},$$

(9.237)

$$d\Omega = d\xi \, d\mathcal{M}_e = \frac{h}{2} J \, d\hat{\xi} \, d\Omega_\sqsubset.$$

Enhancement of the Membrane Strains. From the methods, mentioned in Sect. 9.4.1, the *enhanced assumed strain* (EAS) method can be selected to avoid membrane locking effects. To suppress shear locking, additionally a selective reduced integration of the penalty term in (9.199) will be applied. Hence the simple element formulation becomes more complex, but based on these additional ingredients bending problems can be described sufficiently accurate.

The EAS method, used for the membrane part of the shell formulation, is described for continua in a detailed way in Sect. 10.5. The basis for the EAS formulation is the variational principle of HU–WASHIZU which was stated for three-dimensional continua in (3.300). Here it will be specified for shells as three-field functional where the displacements $\mathbf{u}$, the displacements gradient $\mathbf{H}$ and the 1st PIOLA-KIRCHHOFF stress tensor $\mathbf{P}$ are selected as independent fields.

The EAS formulation bases on the pioneering work of Simo and Rifai (1990) for linear theory and Simo and Armero (1992) for nonlinear continua. EAS interpolations for the development of finite shell elements were considered in Andelfinger and Ramm (1993) and Betsch (1996). The EAS approach leads to non-physical instabilities at finite deformations in the pressure range as was firstly detected in Wriggers and Reese (1996). This problem is not so serious for shells since the a shell structure will buckle or snap-through before large compressive deformation states are reached. Thus the EAS method can still be applied for shells. Experience with solid elements lead to different EAS formulations. It was shown that the so- called $CG4$ or $Q1/E4T$ interpolations, developed in Korelc and Wriggers (1996a) and Glaser and Armero (1997), are more stable than the original ansatz $Q1/E4$ from Simo and Armero (1992). Thus the $CG4$ interpolation is applied here.

The $CG4$ ansatz leads with (10.173) to the incompatible displacement gradient $\bar{\mathbf{H}}_{CG4}^{ref}$ in the reference configuration $\{\mathbf{E}_{\zeta^\alpha}^{loc}\}$ with the incompatible shape functions M_K

$$\bar{\boldsymbol{H}}_{CG4}^{ref} = \sum_{K=1}^{2} \begin{bmatrix} M_{K,\zeta^1}\,\alpha_1^1 & M_{K,\zeta^1}\,\alpha_2^1 & 0 \\ M_{K,\zeta^2}\,\alpha_1^2 & M_{K,\zeta^2}\,\alpha_2^2 & 0 \\ 0 & 0 & 0 \end{bmatrix} = \begin{bmatrix} \zeta^1\,\alpha_1^1 & \zeta^1\,\alpha_2^1 & 0 \\ \zeta^2\,\alpha_1^2 & \zeta^2\,\alpha_2^2 & 0 \\ 0 & 0 & 0 \end{bmatrix}$$

with $M_K = \dfrac{1}{2}[(\zeta^K)^2 - 1]; \qquad K = 1, 2.$

$$\text{(9.238)}$$

By comparison with the matrix related to the $Q1E4$ interpolation, see (10.79), it can be observed that the ansatz $\bar{\boldsymbol{H}}_{CG4}^{ref}$ is just the transposed of $\bar{\boldsymbol{H}}_{Q1E4}^{ref}$; hence $\bar{\boldsymbol{H}}_{CG4}^{ref} = \bar{\boldsymbol{H}}_{Q1E4}^{ref\,T}$. This means that $\bar{\boldsymbol{H}}_{CG4}^{ref}$ defines no gradient field with respect to ζ^α contrary to $\bar{\boldsymbol{H}}_{Q1E4}^{ref}$. Yet $\bar{\boldsymbol{H}}_{CG4}^{ref}$ is called incompatible displacement gradient.

Since $\bar{\boldsymbol{H}}_{CG4}^{ref}$ in (9.238) is defined with respect to the initial configuration, this gradient has to be transformed to the local cartesian coordinate system $\{\mathbf{E}_i^{loc}\}$. Here a transformation as provided in (9.212) cannot be applied since $\bar{\boldsymbol{H}}_{CG4}^{ref}$ is not a gradient field. Instead, the complete tensor transformation

$$\bar{\boldsymbol{H}}^{loc} = \boldsymbol{J}^{-T}\,\bar{\boldsymbol{H}}_{CG4}^{ref}\,\boldsymbol{J}^{-1} \qquad \text{mit} \quad \boldsymbol{J} = \begin{bmatrix} \mathbf{G}_{\zeta^1}\cdot\mathbf{E}_1^{loc} & \mathbf{G}_{\zeta^2}\cdot\mathbf{E}_1^{loc} & 0 \\ \mathbf{G}_{\zeta^1}\cdot\mathbf{E}_2^{loc} & \mathbf{G}_{\zeta^2}\cdot\mathbf{E}_2^{loc} & 0 \\ 0 & 0 & 0 \end{bmatrix}$$

$$\text{(9.239)}$$

has to be used. This relation is valid for constant JACOBI matrices $\boldsymbol{J}$, see (9.210). To guarantee locking-free behaviour for distorted meshes (9.239) has to be modified

$$\bar{\boldsymbol{H}}^{loc} = \frac{J_0}{J}\,\boldsymbol{J}_0^{-T}\,\bar{\boldsymbol{H}}_{CG4}^{ref}\,\boldsymbol{J}_0^{-1} \qquad \text{with} \quad J_0 = \det \boldsymbol{J}_0; \quad J = \det \boldsymbol{J}. \quad \text{(9.240)}$$

The index 0 denotes an evaluation of the JACOBI matrix at the element centre ($\zeta^1 = \zeta^2 = \xi = 0$). Hence constant stress states can be represented within distorted meshes, see Taylor et al. (1976).

In that case, $\bar{\mathbf{F}}^{ref}$ and $\bar{\boldsymbol{H}}_{CG4}^{ref}$ have the same structure and can be added as described in (10.65). The computation of $\mathbf{F}^{loc}$ follows directly from (9.212) and (9.240)

$$\mathbf{F}^{loc} = \bar{\mathbf{F}}^{ref}\,\boldsymbol{J}^{-1} + \bar{\boldsymbol{H}}^{loc}. \qquad \text{(9.241)}$$

An alternative transformation, which can be used instead of (9.240), is provided in Simo et al. (1993b) and Betsch (1996). It has the form

$$\bar{\boldsymbol{H}} = \frac{J_0}{J}\,\bar{\mathbf{F}}_0\,\boldsymbol{J}_0\,\bar{\boldsymbol{H}}_{Q1E4}^{ref}\,\boldsymbol{J}_0^{-1}. \qquad \text{(9.242)}$$

It is not necessary to evaluate the compatible deformation gradient $\bar{\mathbf{F}}_0$ at the element centre as required in (9.242) since the local form (9.241) of the deformation gradient is used for the description of the shell kinematics. Such notation will now be used to describe $\mathbf{F}^{loc}$ in terms of $\bar{\boldsymbol{H}}^{loc}$ from (9.240) with respect to (9.219)

$$\mathbf{F}^{loc} = \left[\ \bar{\mathbf{F}}^{loc}_{[C]\,1} + \xi\bar{\mathbf{F}}^{loc}_{[L]\,1} + \bar{\mathbf{H}}^{loc}_1 \quad \bar{\mathbf{F}}^{loc}_{[C]\,2} + \xi\bar{\mathbf{F}}^{loc}_{[L]\,2} + \bar{\mathbf{H}}^{loc}_2 \quad \bar{\mathbf{F}}^{loc}_{[C]\,3}\ \right].$$
(9.243)

The weak form of equilibrium follows from the Hu–Washizu principle, see (10.64), by introducing the relations of the 5-parameter theory (9.199). Since the last equation in (10.67) is an orthogonality condition, which is automatically fulfilled by the selected interpolations, only the first two equations have to be discretized

$$G(\boldsymbol{u},\boldsymbol{\alpha},\boldsymbol{\eta}) \;=\; \bigcup_{e=1}^{n_e}\sum_{I=1}^{4}\boldsymbol{\eta}_I^T\left\{\int_{\mathcal{M}_e}\int_h \mathbf{B}_I^T\,\bar{\mathbf{S}}\,d\xi\,d\Omega+\right.$$

$$\left. c_p\,h\int_{\mathcal{M}_e}\mathbf{B}_I^{pen^T}\left\{\begin{array}{c}\mathbf{F}_3^{loc^T}\mathbf{F}_1^{loc}\\\mathbf{F}_3^{loc^T}\mathbf{F}_2^{loc}\end{array}\right\}d\Omega\right\} = 0$$

$$G(\boldsymbol{u},\boldsymbol{\alpha},\mathbf{q}) \;=\; \bigcup_{e=1}^{n_e}\mathbf{q}^T\int_{\mathcal{M}_e}\int_h \mathbf{D}^T\bar{\mathbf{S}}\,d\xi\,d\Omega = 0.$$
(9.244)

Matrix $\mathbf{D}_{4\times3}$ which contains the *enhanced* strains can be stated explicitly as

$$\mathbf{D}=\begin{bmatrix} F_{\alpha1}\,M_{\alpha1} & F_{\alpha1}\,M_{\alpha2} & F_{\alpha1}\,M_{(\alpha+2)1} & F_{\alpha1}\,M_{(\alpha+2)2}\\ F_{\alpha1}\,M_{\alpha3} & F_{\alpha2}\,M_{\alpha4} & F_{\alpha2}\,M_{(\alpha+2)3} & F_{\alpha2}\,M_{(\alpha+2)4}\\ F_{\alpha2}\,M_{\alpha1} & F_{\alpha2}\,M_{\alpha2} & F_{\alpha2}\,M_{(\alpha+2)1} & F_{\alpha2}\,M_{(\alpha+2)2}\\ +F_{\alpha1}\,M_{\alpha3} & +F_{\alpha1}\,M_{\alpha4} & +F_{\alpha1}\,M_{(\alpha+2)3} & +F_{\alpha1}\,M_{(\alpha+2)4}\end{bmatrix},$$
(9.245)

after some algebraic manipulations. Here the components M_{lm} with $l,m = 1,\ldots,4$ stem from the matrix

$$\mathbf{M}=\frac{J_0}{J}\begin{bmatrix} \zeta^1(J_{011}^{-1})^2 & \zeta^1 J_{011}^{-1}J_{021}^{-1} & \zeta^1 J_{011}^{-1}J_{012}^{-1} & \zeta^1 J_{011}^{-1}J_{022}^{-1}\\ \zeta^1 J_{011}^{-1}J_{012}^{-1} & \zeta^1 J_{012}^{-1}J_{021}^{-1} & \zeta^1(J_{012}^{-1})^2 & \zeta^1 J_{012}^{-1}J_{022}^{-1}\\ \zeta^2 J_{011}^{-1}J_{021}^{-1} & \zeta^2(J_{021}^{-1})^2 & \zeta^2 J_{012}^{-1}J_{021}^{-1} & \zeta^2 J_{021}^{-1}J_{022}^{-1}\\ \zeta^2 J_{011}^{-1}J_{022}^{-1} & \zeta^2 J_{021}^{-1}J_{022}^{-1} & \zeta^2 J_{012}^{-1}J_{022}^{-1} & \zeta^2(J_{022}^{-1})^2\end{bmatrix}.$$
(9.246)

$F_{\alpha\beta}$ with $\alpha,\beta = 1,2$ are the components of $\mathbf{F}_\alpha^{loc}$, see (9.219). In (9.245), summation has to be performed over α. The linearization of (9.244) yields

$$\begin{bmatrix}\mathbf{K}_{uu} & \mathbf{K}_{u\alpha}\\ \mathbf{K}_{\alpha u} & \mathbf{K}_{\alpha\alpha}\end{bmatrix}\left\{\begin{array}{c}\Delta\mathbf{u}\\\Delta\boldsymbol{\alpha}\end{array}\right\} = -\left\{\begin{array}{c}\mathbf{G}_u\\\mathbf{G}_\alpha\end{array}\right\}$$
(9.247)

with the unknowns $\Delta\mathbf{u} = \{\Delta u_1^{loc}, \Delta u_2^{loc}, \Delta u_3^{loc}, \Delta\omega_1, \Delta\omega_2\}^T$ at each element node and the incompatible modes $\Delta\boldsymbol{\alpha} = \{\Delta\alpha_1^1, \Delta\alpha_2^1, \Delta\alpha_1^2, \Delta\alpha_2^2\}^T$ per element. The incompatible modes $\Delta\boldsymbol{\alpha}$ can be eliminated at element level. For this a block elimination has to be applied, see equations (10.125) and (10.126) in Sect. 10.5. Hence only the unknowns Δu occur in the global equation system.

The different tangent matrices in (9.247) are provided in (9.230) and

$$
\boldsymbol{K}_{u\alpha} = \bigcup_{e=1}^{n_e} \sum_{I=1}^{4} \int_{\mathcal{M}_e} \int_h \left(\boldsymbol{B}_I^T \, \bar{\boldsymbol{L}} \, \boldsymbol{D} + \boldsymbol{G}_I^2 \right) d\xi \, d\Omega ,
$$

$$
\boldsymbol{K}_{\alpha\alpha} = \bigcup_{e=1}^{n_e} \int_{\mathcal{M}_e} \int_h \left(\boldsymbol{D}^T \, \bar{\boldsymbol{L}} \, \boldsymbol{D} + \boldsymbol{G}^3 \right) d\xi \, d\Omega .
$$

(9.248)

The components of the constitutive tensor $\bar{L}$ can be found in (6.185). The explicit form of the operator matrices $\boldsymbol{G}_I^2$ and $\boldsymbol{G}^3$ is complex. For the 5-parameter theory, they are provided in Gruttmann (1996) and Eberlein (1997).

Enhancement of the Shear Strains. For the improvement of the shear strains, the ansatz developed in Dvorkin and Bathe (1984) and Bathe and Dvorkin (1985) will be discussed. It avoids locking due to the shear term in (9.199) without introducing additional degrees of freedom. This interpolation, known as ANS (*Assumed Natural Strain*) interpolation, allows a 2×2 quadrature of the shear term in (9.199) without leading to rank deficiency as in the selected reduced integration, see (9.229). The ANS ansatz additionally leads to a higher accuracy of solutions for distorted element geometries, see e.g. Betsch (1996).

The BATHE–DVORKIN interpolation approximates the shear strains by a constant in one direction and a linear polynomial in the orthogonal direction. With respect to the reference configuration,

$$
\left\{ \begin{array}{c} \tilde{C}_{13}^{ref} \\ \tilde{C}_{23}^{ref} \end{array} \right\} = \frac{1}{2} \left\{ \begin{array}{c} (1 - \zeta^2) \, \bar{C}_{13\,B}^{ref} + (1 + \zeta^2) \, \bar{C}_{13\,D}^{ref} \\ (1 - \zeta^1) \, \bar{C}_{23\,A}^{ref} + (1 + \zeta^1) \, \bar{C}_{23\,C}^{ref} \end{array} \right\}
$$

(9.249)

follows where the components of the right CAUCHY–GREEN tensor are computed from $\bar{\mathbf{F}}^{ref}$ in (9.213). The collocation points A–D are defined in Fig. 9.20. The components $\bar{C}_{\alpha3}^{ref}$ are given by

$$
\bar{C}_{23\,A,C}^{ref} = \bar{\mathbf{F}}_{2\,A,C}^{ref\,T} \, \bar{\mathbf{F}}_{3\,A,C}^{ref} , \qquad\qquad \bar{C}_{13\,B,D}^{ref} = \bar{\mathbf{F}}_{1\,B,D}^{ref\,T} \, \bar{\mathbf{F}}_{3\,B,D}^{ref}
$$

$$
\text{with} \quad \bar{\mathbf{F}}^{ref} = \left[\bar{\mathbf{F}}_1^{ref} \quad \bar{\mathbf{F}}_2^{ref} \quad \bar{\mathbf{F}}_3^{ref} \right] .
$$

(9.250)

The vectors $\bar{\mathbf{F}}_i^{ref}$, which are evaluated at the collocation points A to D, are defined by standard bilinear shape functions N_I, see (9.206). The transformation of the shear strains $\tilde{C}_{\alpha3}^{ref}$ to the cartesian coordinates X_i^{loc}, defined in (9.208), is computed by the isoparametric map, see (9.212). This yields

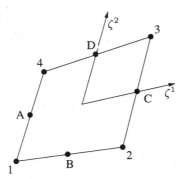

Fig. 9.20 Collocation points for the BATHE–DVORKIN interpolation

$$\left\{ \begin{array}{c} \tilde{C}_{13}^{loc} \\ \tilde{C}_{23}^{loc} \end{array} \right\} = \boldsymbol{J}^{-T} \left\{ \begin{array}{c} \tilde{C}_{13}^{ref} \\ \tilde{C}_{23}^{ref} \end{array} \right\} . \tag{9.251}$$

When the ANS interpolation is used within a finite element model for finite plastic strains, the modified shear strains $\tilde{C}_{\alpha 3}^{ref}$ influence the deformation gradient $\mathbf{F}^{loc}$ in (9.215). This change is only implicitly contained in the formulation. However, for the elasto-plastic formulation, the deformation gradient related to the discretization is needed in order to evaluate relations (9.200) and (9.218). Hence $\mathbf{F}^{loc}$ has to be determined consistent with the ANS interpolation. The associated strategy was pointed out in Dvorkin et al. (1995) and implemented in Eberlein and Wriggers (1999).

Based on the fact that the right CAUCHY–GREEN tensor $\mathbf{C}^{loc}$ is invariant against rigid body rotations, the polar decomposition of the deformation gradient (3.21) can be applied for $\mathbf{F}^{loc}$ and hence this tensor can be split into a rotation part $\mathbf{R}$ and a stretch part $\mathbf{U}$. Since furthermore the right CAUCHY–GREEN tensor is independent on the rotation tensor, see (3.23), the deformation gradient $\tilde{\mathbf{F}}^{loc} = \mathbf{F}_{ANS}^{loc}$ can be introduced which is compatible with the ANS ansatz.

The calculation of $\tilde{\mathbf{F}}^{loc}$ starts from the local CAUCHY–GREEN tensor $\mathbf{C}^{loc} = \mathbf{F}^{loc^T} \mathbf{F}^{loc}$ which is used to obtain the stretch tensor $\mathbf{U}^{loc}$ from the eigenvalue problem

$$(\mathbf{C}^{loc} - \lambda_{(i)}^2 \, \mathbf{1}) \, \mathbf{N}_{(i)} = \mathbf{0} \qquad \Longrightarrow \qquad \mathbf{U}^{loc} . \tag{9.252}$$

Based on this result, the rotation tensor follows from the polar decomposition

$$\mathbf{R} = \mathbf{F}^{loc} \, \mathbf{U}^{loc^{-1}} . \tag{9.253}$$

In the same way, the stretch tensor $\tilde{\mathbf{U}}^{loc}$ is computed via (9.252) from the strain measure $\tilde{\mathbf{C}}^{loc}$ belonging to the ANS interpolation. Now $\tilde{\mathbf{F}}^{loc}$ follows under the assumption that the rotation tensor $\mathbf{R}$ is the same for $\mathbf{C}^{loc}$ and $\tilde{\mathbf{C}}^{loc}$ using (9.253)

$$\tilde{\mathbf{F}}^{loc} = \mathbf{R}\,\tilde{\mathbf{U}}^{loc} = \mathbf{F}^{loc}\,\mathbf{U}^{loc^{-1}}\,\tilde{\mathbf{U}}^{loc}. \tag{9.254}$$

The modified deformation gradient $\tilde{\mathbf{F}}^{loc}$ has then to be applied in (9.200). [4] $\tilde{\mathbf{F}}^{loc}$ can additionally contain the incompatible displacement gradient $\bar{\mathbf{H}}^{loc}$, see (9.239). This is the case when the enhanced strain strategy is applied for the membrane strains.

9.4.7 Shell Intersections

Modelling of real engineering structures needs often the introduction of shell intersections. These arise when shells, consisting of different geometries, meet at a certain zone or when shell parts of the same geometry are differently located in space but act together as one structure. In such cases, the shell midsurface is no longer smooth but has sharp corners. Examples are girders with high webs or shells with a U-shape or L-shape cross section.

The 5-parameter theory only relies on two rotational degrees of freedom which describe the rotation of the shell midsurface $\mathcal{M}$ (the rotation about the shell normal is not relevant). Hence a third component of the rotation tensor is missing when a transformation of the rotational degrees of freedom has to be carried out in order to link the rotations of one midsurface to the other at a shell intersection. This problem can be solved by using a 5/6-parameter concept in which the needed third rotational degree of freedom is introduced at the shell intersection. Within this concept, the shell can still be modelled by the 5-parameter theory.

The 5/6-parameter concept leads to equations in which, instead of the director vectors (9.227), the associated relations of the RODRIGUES formulae (9.193) and (9.194) have to be inserted.

This approach was first presented in Hughes and Liu (1981) and then formulated by Simo (1993) for the 5/6-parameter concept. Generally, a 6-parameter model for the shell intersection is obtained with the three displacement components of $\mathbf{u}$ which can be used to describe the rotations with respect to the axis of the directors.

Rotations around the director axis vanish for smooth shell geometries which can be deducted from the moments of momentum balance when formulated with respect to the normal. Rotations around the normal have to be eliminated for smooth shells to avoid singularity of the resulting equation system. This elimination can be performed by transforming the components of the axial vector $\boldsymbol{\omega}$ to a local cartesian frame $\mathbf{E}_{I\,i}^{loc}$ (defined in (9.208)) at element level. Here index I is related to the nodal number and i denotes the coordinate direction. The coordinate in thickness direction (3-direction) is chosen as fixed coordinate axis within this process. It is now possible to prescribe a boundary condition along this axis in order to eliminate the rotation around the director axis. The following equations summarize this strategy:

---

[4] A linearization of $\tilde{\mathbf{F}}^{loc}$ is not necessary since it does not appear in the weak form, see Eberlein (1997).

– shell intersection:

$$\boldsymbol{\omega} = \omega_{Ii}\,\mathbf{E}_{Ii} = \psi_{Ii}\,,\mathbf{E}_{Ii}^{loc} \tag{9.255}$$

– smooth shell:

$$\boldsymbol{\omega} = \psi_{I\alpha}\,\mathbf{E}_{I\alpha}^{loc} \qquad \text{and} \qquad \psi_{I3} = 0\,. \tag{9.256}$$

This special 5/6-parameter approach for shell intersections reduces to the normal 5-parameter model at all element nodes which belong to smooth shell surfaces.

Note that relation (9.220) does not change within the 5/6-parameter theory. However, the description of shell rotations is not given by (9.223) but through the axial vector $\boldsymbol{\omega}$ which was defined in (9.193). $\boldsymbol{\omega}$ parameterizes the rotation tensor $\mathbf{R}$ based on RODRIGUES formulae (9.194). Hence the rotation angle $\boldsymbol{\beta}$ given in (9.223) approximates the axial vector $\boldsymbol{\omega}$, see (9.256), by

$$\boldsymbol{\omega} \;=\; \sum_{I=1}^{4} N_I\,\boldsymbol{\psi}_I; \qquad \delta\boldsymbol{\omega} \;=\; \sum_{I=1}^{4} N_I\,\delta\boldsymbol{\psi}_I\,. \tag{9.257}$$

within the 5/6-parameter theory. With (9.223) or (9.257), the director field is uniquely defined for the 5- und 5/6-parameter concept.

A special treatment of the shell interactions is not necessary within the 6-parameter model since all three displacements and components of the director vector are used within the formulations. In practical applications however locking is observed. This can be avoided by an introduction of a 6/7-parameter model at a shell intersection. A detailed description of the related concept can be found in e.g. Betsch (1996), where this concept is applied to hyperelastic shells.

9.5 Examples

All numerical simulations in this section are performed by using finite shell elements based on the discussed 5-parameter theory or by using a 6-parameter theory derived in Eberlein (1997), see also Eberlein and Wriggers (1999). Comparison to results gained with different formulations are provided for application in which finite plastic deformations occur. Different shells with smooth surfaces are analyzed when subjected to point or surface loads.

The nonlinear material behaviour necessiates a numerical integration over the shell thickness. It turned out that an integration with five GAUSS-points was sufficient. A Comparison with a larger number of GAUSS-points did not lead to improved results even in the case of finite elasto-plastic deformations. Hence five integration points seem to be adequate for inelastic response of thin shells. The finite element formulation of the 5-parameter quasi-KIRCHHOFF theory is insensitive against the choice of the penalty parameter c_p which occurs in the shear term. In all computations, this parameter is selected such

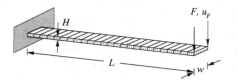

Material data:	
$\mu = 4.6154 \cdot 10^6$	$\Lambda = 6.9231 \cdot 10^6$
$\tau_Y = 2.4 \cdot 10^4$	$K = 1.2 \cdot 10^5$
Geometry:	
$l = 10$	$w = 1$

Fig. 9.21 Bending of a clamped beam

that it has the same magnitude as the average of the maximum values in the tangent matrix. This avoids ill-conditioning of the tangent matrix. The EAS method is employed within all numerical simulations.

9.5.1 Bending of a Clamped Beam

The first example is basically a beam problem. It is selected here to discuss the behaviour of the shell elements in bending situations. The clamped beam is depicted in Fig. 9.21. The numerical simulations are performed for different ratios of length to height of the beam (l/h). Geometry and material data are provided in Fig. 9.21. For $l/h = 100$, simulation results can be found in Dvorkin et al. (1995) which is chosen here for comparison.

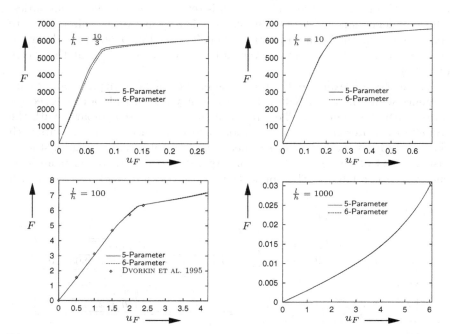

Fig. 9.22 Load–deflection curves

Fig. 9.22 depicts the load-deflection curves for four different ratios of l/h. The solutions (for the case $l/h = 100$) match as well for the 5- as for the 6-parameter theory the results in Dvorkin et al. (1995). The two theories (5- and 6- parameters) yield almost identical results. This is also true for the extremely thin beam with $l/h = 1000$. The element formulation using the 5-parameter theory converges already with 20 elements while formulation based on the 6-parameter theory needs 30 elements. Thus the discussed 5-parameter theory is superior in case of thin shells.

The response of the 5-parameter theory is stiffer for a very thick beam with a length to height ratio of $l/h = 10/3$ since shear deformations are neglected in this model. However, once plastic deformations occur the difference in the solutions between the 5- and 6-parameter model disappears. This is due to the fact that a plastic hinge develops.

9.5.2 Quadratic Plate under Internal Pressure

A quadratic plate is subjected to a constant pressure load of ($p_o = 10^{-2}$) which is increased by a load parameter $\lambda = f$, see Fig. 9.23. The boundary conditions of the plate correspond to a NAVIER plate such that the vertical displacement u_3 is zero at the plate boundary.

Due to symmetry of the problem, only one quarter of the plate is discretized by an unstructured mesh which is refined near the plate boundaries. The load deflection curves of the elasto-plastic response of the thin plate were computed using the 5- and 6-parameter models, see Fig. 9.24. In the diagram, the load parameter λ is plotted versus the displacement $u_{3\,M}$ at the centre of the plate.

As in the first example, the 6-parameter shell element depicts slower convergence while the quasi-KIRCHHOFF element yields already a converged solution with 225 elements. The numerical simulations are compared with the results in Büchter et al. (1994). A good correspondence of these results with the results from the 5- and 6-parameter model is observed in Fig. 9.24 for a displacement of up to $u_{3\,M} \approx 30$. For larger deflections, there exists a difference which is related to the application of the pressure load $p = \lambda\,p_o$ which

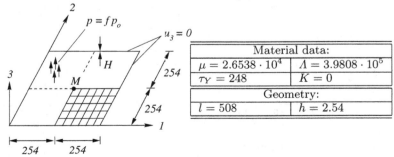

Fig. 9.23 Quadratic plate under internal pressure

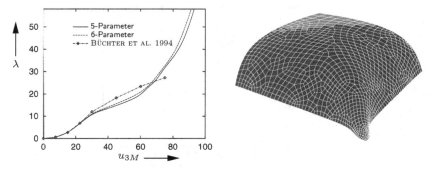

Fig. 9.24 Load-deflection curves and deformed configuration at $\lambda = 70$

is not considered in Büchter et al. (1994) to be deformation dependent. The deformed configuration depicts at $\lambda = 70$ a considerable change of the flat plate surface, see Fig. 9.24.

9.5.3 Pinched cylinder

The last example illustrates the behaviour of a cylindrical shell under point loads. This example has been intensively investigated by many research groups. The first simulation which considered finite deformations can be found in Simo and Kennedy (1992). Simulations with large deformations can also be found e.g. in Wriggers et al. (1996), Miehe (1997) and Soric et al. (1997). Here all numerical simulations are performed with shell elements based on the 5- and 6-parameter theory.

Geometry and initial configuration of the shell can be found in Fig. 9.25. Due to symmetry, only one eighth of the cylinder has to be discretized. The shell is supported at $z = 300$ where all degrees of freedom are fixed besides the displacements in z-direction. The problem is solved by applying

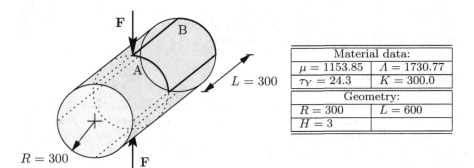

Material data:	
$\mu = 1153.85$	$\Lambda = 1730.77$
$\tau_Y = 24.3$	$K = 300.0$
Geometry:	
$R = 300$	$L = 600$
$H = 3$	

Fig. 9.25 Pinched cylinder

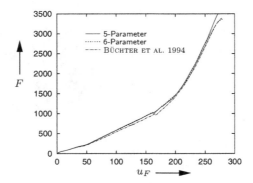

Fig. 9.26 Load-deflection curve

a displacement controlled analysis where a displacement increment of $\Delta u_F \approx$ 2.5 under the load is chosen.[5] This guarantees a stable convergence behaviour within NEWTON'S method. Within this procedure, the total displacement is applied within 100 loading steps.

The load–deflection curves are shown in Fig. 9.26 for a structured mesh with 32×32 elements. They represent the vertical displacement u_F versus the load F. The convergence behaviour of this problem was studied in Wriggers et al. (1996). By comparison with the results in this paper, the solution using the 5-parameter model can be viewed as reference solution. However, the results of the 5- and 6-parameter solution are very close as can be observed from Fig. 9.26. These solutions are furthermore in good agreement with a solution obtained by three-dimensional enhanced strain elements, see Sect. 10.5. Note that the effort to compute the element matrices is considerablly higher when using the 6-parameter model in comparison with the quasi-KIRCHHOFF element (24 against 20 degrees of freedom and five instead of four incompatible modes) and the gain in accuracy is not considerable.

Figure 9.27 depicts the development of the plastic zone for the element with the 5-parameter model. The equivalent plastic strain of the outer layer was plotted. It is interesting that the circular cross section of the cylinder deforms into a rectangular cross section at the loacation of the load which vertices moving outward for increasing load.

9.5.4 Final Remarks

The examples in the previous section lead to the conclusion that the quasi-KIRCHHOFF shell element yields very good results, even somehow unexpected for thick shells. Thus this efficient element can be applied for finite elastic and inelastic deformation problems, like metal sheet forming.

---

[5] Since singularities occur under point loads in shell analysis, these loads have to be distributed on a small fixed area when mesh convergence studies are performed.

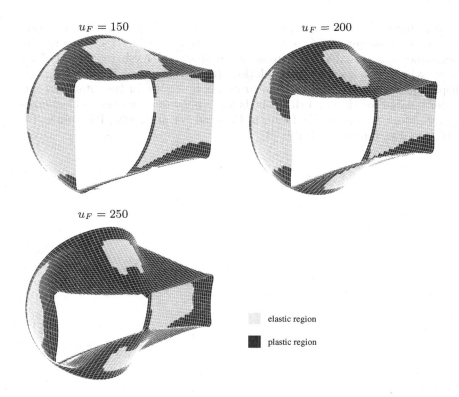

$u_F = 150$ $u_F = 200$

$u_F = 250$

elastic region

plastic region

Fig. 9.27 Development of the plastic zone

However, some problems where elastic material behaviour is considered incorporate boundary layer effects which cannot be modelled correctly by the quasi-KIRCHHOFF theory. This is e.g. the case when resultant shear forces have to be determined for plates with NAVIER boundary conditions. In such cases, the 6-parameter model or higher order interpolations, see Düster et al. (2001) or Hughes et al. (2005) have to be employed. Another possibility would be to use three-dimensional continuum elements. These could also be coupled to finite shell elements since the layer effects are often related only to very small areas. In that case, elements based on the quasi-KIRCHHOFF theory could be used in the undisturbed parts, away from the boundary layer. Such approach would lead to investigations of model adaptivity, see Chap. 8. First investigations for shells can be found in e.g. Han and Wriggers (1998).

The convergence behaviour of the 6-parameter model can be enhanced by additional ANS interpolations in thickness direction, see e.g. Betsch and Stein (1995) and Bischoff and Ramm (1997). This approach is also known as 7-parameter model.

Another discretization technique for shells undergoing finite deformations is based on triangular elements. These have the advantage that more robust techniques are available for automatic meshing and mesh refinement. Elements which have a simple interpolations (quadratic ansatz functions for the displacements and incompatible linear ansatz functions for the rotations) can be formulated for thin and thick shells for hyperelastic and elasto-plastic response. More details can be found in Campello et al. (2003), Pimenta et al. (2004) and Campello et al. (2007).

10. Special Finite Elements for Continua

10.1 Requirements for Continuum Finite Elements

The search for finite elements which can be applied to arbitrary problem classes within solid mechanics has a long history. This can be seen from the numerous scientific papers devoted to this topic. Main target of a development of finite elements is summarized in the following enumeration.

1. *Locking* free behaviour for incompressible materials,
2. good bending performance,
3. no *locking* in thin elements,
4. no sensitivity against mesh distortions,
5. good coarse mesh accuracy,
6. simple implementation of nonlinear constitutive equations and
7. efficiency (e.g. few necessary integration points).

These points result from different demands and can also lead to different element formulations.

The first point is associated with the numerical simulation of a special problem classes which include in solid mechanics rubber like materials and elasto-plastic material equations in the framework of J_2-plasticity. During the last years, different special finite elements were developed for this applications. This results from the fact that classical low order displacement elements, which were described in Chap. 4, are not sufficient. The constraint related to the incompressible behaviour leads even for geometrical linear elements to *locking*, see e.g. Braess (2007), Zienkiewicz and Taylor (1989) and Hughes (1987). Finite elements which are suitable for incompressible materials will be described in detail in Sects. 10.2, 10.4 and 10.5.

The second and third points are of significance when three-dimensional solid elements shall be employed to solve beam- or shell problems since beam and shell structures are often dominated by bending behaviour and are, by construction, thin in one or two spatial coordinated. Using three-dimensional elements allow a simple implementation of three-dimensional constitutive equations which is not so easily possible when classical beam or shell models are used. Furthermore, the treatment of finite rotations are avoided by such formulations, see Sect. 9.4.

The fourth point is essential when modern methods for mesh generation are employed. These methods lead for arbitrary geometries to so-called
unstructured meshes which consist of finite elements shapes with arbitrary
geometry, see e.g. Figs. 8.7, 8.8, 8.9, 8.10, 8.11, 8.12, 8.13, 8.14, 8.15. Another
source for the distortion of finite elements is the change of the nodal coordinates during a nonlinear simulations which can lead to severely deformed
finite elements.

The fifth point is related to the fact that in real engineering applications
often three-dimensional components have to be analysed which size and complexity cannot be modeled using a converged mesh, especially when the simulation is nonlinear. Hence there is still need for elements which depict a good
accuracy, even when used within a coarse mesh. Of course, the importance
of this point will diminish with the increasing computing power, but at the
moment it is still of concern.

The sixth point follows from the fact that more accurate mathematical
and physical models have to be used within the simulation of nonlinear engineering structures. Within this process, new complex nonlinear constitutive
equations have to be implemented. Here a simple interface to the finite element should support the user in order to efficiently change existing finite
elements and to be able to implement new complex constitutive equations.

Finally, it can be mentioned that efficiency is not only related to speed
of the element formulation but also to the memory requirement. The latter
demand is essential when e.g. inelastic problems with several hundred thousand or millions of finite element have to be solved within a given time frame.
This speed of the element formulation is essential when iterative solvers are
applied since in that case the time for the computation of residuals and tangent matrices is of the same order as the time used by the solver within one
iteration.

New developments show that finite elements with a high order of interpolations (so-called p-version of finite elements) can be applied successfully
to finite deformation problems for rubber-like materials, see Heisserer et al.
(2007).

Low order finite elements have been proven to be robust for many nonlinear simulations. This has to do with a low regularity of the analytical
solution which can exclude higher order interpolations, see also Sect. 8.1. A
further fact which supports lower order elements is the sparsity of the global
tangent matrices since low order elements yield a smaller bandwidth. Due
to that the global equation system can be solved more efficiently, which is
crucial for the simulation of large systems. In case of numerical simulations
which include inelastic material behaviour, one or more history variables
have to be stored per integration point. As an example, a problem with J_2
plasticity is considered to obtain the variation of $\boldsymbol{F}_e$ for the ansatz (10.88),
see Sect. 6.2.2, in which six plastic strain components have to be stored
per integration point. This leads to the memory requirement for storing the
history variables when a finite element mesh of a cube with 10^6 finite elements is used which is shown in Table 10.1. The memory requirement is

Table 10.1: Memory requirement for history variables for 10^6 finite elements

Order of interpolation	Number of GAUSS-points	Memory requirement
1	1	56 MByte
1	8	448 MByte
2	27	1512 MByte

larger when iterative solvers or special direct sparse solvers are used, see Sect. 5.2, but is basically of the same order of magnitude. Hence it is advantageous to use elements with a minimum number of integration points in order to optimize memory requirement. Of course, one has to be careful to compare linear and quadratic or other higher order elements since the elements with higher order have a higher order of convergence when the solution has the necessary regularity, see Sect. 8.1. In that case, less finite elements of higher order can be used which yields results with the same accuracy. (Assume that half of the elements per side are sufficient for the discretization of the above cube, then the memory requirement for the history variables reduces for quadratic elements to 189 MByte.) However, in order to compare the finite element discretizazions of different interpolation orders, the total solution time needed to obtain a result with the same accuracy has to be considered.

The memory requirement for history variables play an essential role when explicit integrations schemes are employed to simulate impact or shock problems. Here only the residual has to be stored, see Sect. 6.1.1, which leads to the storage of three values per node. In that case, the storage requirement for the example above is roughly $3 \times 101^3 = 3.091 \times 10^6$ values for the residual vector. The number of history variables for the correct two-point GAUSS-integration in each coordinate direction amounts to $2 \times 2 \times 2 \times 10^3 = 8 \times 10^6$. This is more than double of the storage needed for the residual vector. In order to reduce the overall computing time, all quantities have to be retained in the main memory. In such case, the storage of the history variables is a major concern for explicit computations. Hence most of the explicit finite element codes use specially stabilized finite elements with only one-GAUSS-point. This formulation will be discussed in Sect. 10.4.

It is well known that the pure displacement element with bilinear or trilinear ansatz function has bad convergence behaviour in bending problems, especially if the length in one direction is a lot smaller than in the other ones, e.g. for beam or shell structures. Hence special elements were developed for such problems. With such elements, which is still based on linear ansatz functions, the convergence order cannot be increased with regard to (8.6) or (8.10), but the constant C is reduced considerably. Thus the required accuracy of the finite element solution can be achieved with considerably less elements. In this connection, the ideal element would be an element which is well performing for bending as well as for incompressible problems.

Different formulations have been developed in order to construct finite elements which fulfil all seven requirements stated above. These are:

- techniques which base on a reduced integration of the integrals leading to the element matrices,
- stabilization methods,
- hybrid or mixed variational principles which base on complimentary energy written in terms of the stress field,
- mixed variational principle of HU–WASHIZU type,
- mixed variational principle for rotational fields,
- mixed variational principle for special quantities,
- nodally based elements,
- composite or macro formulations for the element,
- higher order displacement elements and
- formulations based on the COSSERAT point theory.

In the following, different possibilities are summarized and their differences are discussed. After that some of the techniques are presented in detail.

1. **Reduced integration and stabilization.** The most simple method is the "reduced integration" of the integrals leading to the finite element vectors and matrices. It is also very efficient and safes memory for history data storage since less integration points are used. Underintegration or reduced integration means that less GAUSS points are used for the integration of tangent matrices and residual vectors than necessary for the chosen polynomial degree of the shape functions, for first applications see e.g. Zienkiewicz et al. (1971). This reduced integration was developed to avoid *locking* in case of incompressibility. In that case, it is often only applied to the pressure part of the constitutive equation, see e.g. Malkus and Hughes (1978), Hughes (1980) and Sect. 10.2. For reduced integration techniques exist many variants. This stems from the fact that reduced integration is always associated with a rank deficiency of the tangent matrices which is cured by different methods. The related methods are generally known as stabilization techniques. A literature review regarding this topic is presented in Sect. 10.4, in which stabilization techniques from Belytschko et al. (1984) are presented. Using the reduced integration together with stabilization leads to finite elements which fulfil conditions 1, 4, 5, 6 and 7. These elements are *locking* free in case of incompressibility, they have a good coarse mesh accuracy; they are not sensitive against mesh distortions and can be used for arbitrary constitutive equations. The reduced integration provides the most efficient possibility to compute the element residual and the element tangent stiffness (e.g. for an eight-node brick element, only one GAUSS point is needed). However, these elements need the choice of artificial stabilization parameters. In the worst case, e.g. for some bending problems, the finite element solution

can directly depend on the stabilization parameter, see also Sect. 10.4. However, new developments show improvements, see e.g. Reese (2005).

2. **Hybrid or mixed variational principles.** When mixed variational principles are used as basis for finite element discretization, different possibilities exist for the construction of the finite element matrices. This is related to the many different existing mixed forms. Some of them need conforming displacement fields together with non-conforming stress or strain fields, others rely on conforming stress fields but allow non-conforming displacement fields. Theoretical background for linear mixed methods can be found in Washizu (1975) and in various monographs and papers, for the mathematical literature, see e.g. Braess (2007) and Brenner and Scott (2002). For the case of linear elasticity, hybrid elements where first described in Pian (1964) which has lead to many different finite element formulations up to now. Within this approach, Pian and Sumihara (1984) developed a finite element which is efficient and accurate. However due to the need to invert the constitutive equations within the formulation in order to obtain the constitutive equations in terms of the stress field, there are only few elements for ST. VENANT materials known which work for large deformations. A special formulation for Neo–HOOKE materials will be presented in Sect. 10.3.

3. **Enhanced strain elements based on the HU–WASHIZU principle.** Within the enhanced strain formulations, non-conforming strain measures are introduced within the HU–WASHIZU principle. In a first paper, Simo and Rifai (1990) developed *enhanced strain* elements for the geometrical linear theory.

   In follow up work, Simo and Armero (1992) and Simo et al. (1993b) have derived a family of enhanced elements for large deformations and inelastic constitutive equations based in the HU–WASHIZU. This class of elements is related to the incompatible mode elements which were developed by Wilson et al. (1973) and Taylor et al. (1976) for linear problems. The *enhanced strain* elements fulfil point 1 to 6 of the above mentioned requirements. Hence they are well suited for all applications. However, these elements have some disadvantageous. They need a statical condensation on element level. For the two-dimensional case, this leads to the inversion of a 4×4 matrix and, depending on the formulation, in the three-dimensional case a 9×9 or 12×12 matrix has to be inverted. This reduces the efficiency of the enhanced strain elements. Furthermore, storage of the degrees of freedom belonging to the enhanced strains needs additional storage on element level, see also the comments regarding Table 10.1. However, a special efficient formulation has been developed in Puso (2000). A further point which is still under investigation is related to the *hour-glassing* of the enhanced strain elements under pressure. This fact was discovered by Wriggers and Reese (1994), see also Wriggers and Reese (1996). A detailed discussion of this phenomenon can be found in Sect. 10.5. Solutions which partly solve this problem are provided in Korelc and Wriggers (1996a), Glaser and Armero (1997), Reese and Wriggers (2000), Reese (2005) and Mueller-Hoeppe et al. (2008), where different methods have been used to overcome the *hour glassing*, see also Sect. 10.5.4.

It is not possible to enhanced triangular and tetrahedral elements directly. The method is degenerate for triangular and tetrahedral elments, see Reddy and Simo (1995). However, a mixed enhanced approach where ansatz functions for displacements pressures and volume effects are introduced can be employed to generate low order tetrahedral elements which do not lock in incompressibility and perform reasonably well in bending, see e.g. Taylor (1985) and Mahnken et al. (2008).

4. **Mixed variational principles for problems with rotational degrees of freedom.** When not only the momentum is weakly enforced, but also the moment of momentum, which usually leads, see (3.68), to the symmetry of the stress tensor, then rotational degrees of freedom can be introduced as independent field variables. Finite elements which is based on such formulation were constructed in e.g. Hughes and Brezzi (1989). Further applications of such variational formulations can be found for two-dimensional elements which are a basis for shell formulations, see e.g. Ibrahimbegovic et al. (1990), Iura and Atluri (1992) and Gruttmann et al. (1992). A three-dimensional technique using co-rotational formulations for three-dimensional continua was developed in Moita and Crisfield (1996).

5. **Mixed variational principles for special quantities.** Often problems have to be considered which include special constraint conditions. In such cases, it is advantageous to formulate mixed principles which are tailored to fulfil such constraint conditions. Examples are solid elements for plates or shells where, for thin structures, the transverse shear becomes zero in the limit, see also Chap. 9.4. Another example is related to contact problems where the zero gap condition introduces a constraint which has to be considered when deriving associated finite element discretizations, see Chap. 11. The example standing out in solid mechanics is the constraint related to incompressibility. This constraint occurs in rubber elasticity and in case of plastic flow, when the J_2 is applied for mechanical modelling. The related special variational principle relies on a split of the kinematical variables into volumetric and deviatoric parts, details are provided in Sect. 10.2.

The related finite elements fulfil point 1, 4, 5 and 7 of the above mentioned requirements. Due to the kinematical split, the formulation of the constitutive equations is more elaborate than in the standard formulation. This is especially true for large deformations making the linearizations, needed within the NEWTON method, more complex.

6. **Nodally based elements.** Nodally based elements are applied to enhance the bending behaviour of tetrahedral elements and to avoid locking in such cases. Besides a number of other formulations, average nodal pressures or strains can be used to compute average volumetric strains or strains at nodes based on surrounding triangles or tetrahedrals, see Dohrmann et al. (2000) and Bonet and Burton (1998). These types of element have been stabilized by Puso and Solberg (2006) in order to alleviate spurious modes.

7. **Composite or macro elements.** Composite or macro element formulations make use of the possibility to construct finite elements from subelements which use simplified or special shape functions. These type of elements can be developed for triangular and quadrilateral shaped elements. For triangles, this type of formulation is, as well as the nodally based formulation, one of the few possibilities to enhance the element behaviour, see Guo et al. (2000) and Thoutireddy et al. (2002), since triangles cannot be enhanced in the standard way using the HU-WASHIZU principle. This technique is not often employed for quadrilaterals and hexahedral elements since the only gain is a more robust behaviour when the elements are distorted severely at large strain states. Here formulations were developed by Rubin and Jabareen (submitted), based on the COSSERAT point theory, and by Boerner and Wriggers (2008) based on the standard continuum approach.

8. **Higher order displacement elements.** During the last years, finite element discretization schemes were developed which is based on higher order interpolation. These methods depict very good convergence characteristics for finite hyperelastic deformations, see e.g. Düster et al. (2003) but also for elasto-plastic problems undergoing small deformations, see Düster et al. (2002). They can be formulated in an efficient way by hierarchical shape functions using polynomials or NURBS and hence are competitive with respect to low order approximations. However, still special techniques have to be employed for incompressible materials in order to recover optimal convergence rates also for lower order approximations, see e.g. Elguedj et al. (2008) and Heisserer et al. (2007).

9. COSSERAT **point elements.** Lately, elements have been formulated which are based on the COSSERAT point theory. This theory formulates the continuum as a point then director vectors are introduced to account for the deformation modes. For a theoretical background, see Rubin (2000). This formulation transforms directly into a finite element discretization, as was shown in Nadler and Rubin (2003). Furthermore, due to an internal split of the deformation modes it is possible to use linear analytical solution to stabilize the element such that *locking* but also *hour glassing* does not occur. The element has so far superior behaviour

problems with hyperelastic materials for undistorted element geometries, fulfilling points 1-5 and 7, but behaves like pure displacement Q1-element for distorted meshes, see Loehnert et al. (2005). For initially distorted element, geometries approaches to improve the element behaviour are discussed in Boerner et al. (2007) and Rubin and Jabareen (2008).

10.2 Mixed Elements for Incompressibility

Pure is displacement elements are not suitable for problems in which the constitutive behaviour exhibit incompressibility since they tend to *locking*. *Locking* means, in this connection, that the constraint conditions due to incompressibility which are related to the pure volumetric mode (in the elastic case the condition is $J = \det \mathbf{F} = 1$ and for plastic flow the condition $J_p = \det \mathbf{F}_p = 1$ holds) can only be fulfilled with a considerable stiffening of the bending modes, see e.g. Hueck et al. (1994). Thus this behaviour is also called volume locking. Mixed finite element methods can help to avoid locking, see e.g. Zienkiewicz and Taylor (1989) and Brezzi and Fortin (1991).

There exist different possibilities to construct mixed elements. These are shortly discussed in the following.

– **Method of** LAGRANGIAN **multipliers.** Here the constraint condition of incompressibility will be directly introduced via the methods of LAGRANGIAN multipliers. Hence the strain energy

$$W = W_{inkomp} + p\,G(J) \quad \text{with} \quad G(J) = 0 \qquad (10.1)$$

is formulated. The constraint condition is then given for finite deformations as $G(J) = J - 1$ with $J = $ is e.g. given by the MOONEY–RIVLIN material (3.112). Finite elements which are based on this methodology have the disadvantage that contrary to the pure displacement elements additional unknowns occur. These are the LAGRANGIAN multipliers which are equivalent to the pressure p. Furthermore, special techniques are needed to solve the associated incremental equation system for displacements and LAGRANGIAN multipliers

$$\begin{bmatrix} \mathbf{K}_{T\,uu} & \mathbf{B}_{T\,up} \\ \mathbf{B}_{T\,pu}^{T} & \mathbf{0} \end{bmatrix} \left\{ \begin{array}{c} \Delta \mathbf{u} \\ \Delta \mathbf{p} \end{array} \right\} = - \left\{ \begin{array}{c} \mathbf{R}_u \\ \mathbf{R}_p \end{array} \right\} \qquad (10.2)$$

which has zero entries in the diagonal. The sub-matrix $\mathbf{K}_{T\,uu}$ follows from W_{inkomp} while $\mathbf{B}_{T\,up}$ is related to the discretization of the term $p\,G(J)$. Associated finite element formulations can be found in Oden and Key (1970) and Duffet and Reddy (1983).

– **Perturbed** LAGRANGIAN **method.** To have a greater variability for the formulation of ansatz functions, the following strain energy function

$$W = W_{inkomp} + p\,G(J) - \frac{1}{2\,\epsilon}\,p^2 \qquad (10.3)$$

can be introduced. The constraint condition is again given by $G(J) = J - 1$. $\epsilon > 0$ is a perturbation parameter. Choosing now continuous ansatz function for displacements and pressure, the following incremental equation system can be derived

$$\begin{bmatrix} K_{Tuu} & B_{Tup} \\ B^T_{Tpu} & -\frac{1}{\epsilon} K_{pp} \end{bmatrix} \left\{ \begin{array}{c} \Delta u \\ \Delta p \end{array} \right\} = - \left\{ \begin{array}{c} R_u \\ R_{p\epsilon} \end{array} \right\} . \tag{10.4}$$

Here contrary to (10.2), the incremental displacements and pressures can be computed using standard equation solvers. The pressures can be removed from the system by using the SCHUR complement. This leads to

$$\left[K_{Tuu} + \epsilon B_{Tup} K_{pp}^{-1} B^T_{Tpu} \right] \Delta u = -R_u - \epsilon B_{Tup} K_{pp}^{-1} R_{p,\epsilon} . \tag{10.5}$$

When discontinuous ansatz functions are used for the pressure variables then the pressures can be eliminated on element level. This yields an equation in which the inverse of K_{pp} is trivial

$$\left[K_{Tuu} + \epsilon B_{Tup} B^T_{Tpu} \right] \Delta u = -R_u - \epsilon B_{Tup} R_{p,\epsilon} . \tag{10.6}$$

This system of incremental equations is equivalent to a *penalty* formulation for the incompressibility constraint. Note that the solution now depends on the perturbation or penalty parameter. For small values of ϵ, the influence of the constraint condition disappears. For large values of ϵ, the constraint is fulfilled more and more exactly but the condition number of the linear equation system (10.6) will be very large. Then special equation solvers have to be applied. Papers regarding formulation (10.5) have been published for the linear case by Malkus and Hughes (1978) and for the large strain case of rubber elasticity by e.g. Häggblad and Sundberg (1983) and Sussman and Bathe (1987).

– HU–WASHIZU **functional.** In this functional, the incompressibility constraint is introduced as in the penalty method but is formulated via a constitutive equations for the pressure. In that case the functional

$$H(\varphi, p, \theta) = W(\widehat{C}) + K\left[G(\theta) \right]^2 + p\left(J - \theta \right) \tag{10.7}$$

is formulated, see also Sect. 3.4.3. Within the finite element discretization, ansatz functions are selected for the deformation φ, the pressure p and the volumetric strain θ. $G(\theta)$ defines the constitutive equation for the pressure term, here K is the modulus of compression. The formulation of $W(\widehat{C})$ is provided by (3.122). The associated discretization within the finite element method was firstly presented in Simo et al. (1985a).

Finite elements which are derived form mixed methods have to fulfil additional mathematical conditions which guarantee the stability of the element formulation. This condition is known as BB-condition, named after its inventors BABUSKA and BREZZI. Its fulfillment is related to the condition that matrix B_{Tpu} in (10.4) is not rank deficient.

Remark 10.1: With respect to the mathematical formalism, the BB-condition will be stated here for incompressible linear elasticity. With the short hand notation, see also Chap. 8 and Eq. (8.3), the incompressible problem cab be stated for mixed interpolations as

$$a(\mathbf{u}, \boldsymbol{\eta}) + b(p, \boldsymbol{\eta}) = f(\boldsymbol{\eta}) \qquad \forall \boldsymbol{\eta} \in V \qquad (10.8)$$
$$b(q, \mathbf{u}) = 0 \qquad \forall q \in Q$$

where the different terms are given by

$$a(\mathbf{u}, \boldsymbol{\eta}) = 2\mu \int_{\Omega} \mathbf{e}_D(\boldsymbol{\eta}) \cdot \mathbf{e}_d(\mathbf{u}) \, d\Omega ,$$

$$b(p, \boldsymbol{\eta}) = \int_{\Omega} p \operatorname{div} \boldsymbol{\eta} \, d\Omega , \qquad (10.9)$$

$$f(\boldsymbol{\eta}) = \int_{\Omega} \hat{\mathbf{b}} \cdot \boldsymbol{\eta} \, d\Omega + \int_{\Gamma_\sigma} \hat{\mathbf{t}} \cdot \boldsymbol{\eta} \, d\Gamma .$$

The strain deviator $\mathbf{e}_d(\mathbf{u})$, see (3.30), has to be applied in $(10.9)_1$ on order to obtain a clear split between the volumetric strains $\operatorname{div}\mathbf{u}$ and the deviatoric part. The incompressibility condition is described by $\operatorname{div}\mathbf{u} = 0$ in the linear case. It is introduced to the mixed form by the LAGRANGIAN multiplier method.

In the continuous case of solids with sufficiently smooth boundaries, the displacements are in the SOBOLEV space H^1 ($\mathbf{v} \in V = H^1(\Omega)$, for a definition of the spaces see e.g. (8.7)). For the pressure interpolation, the space L^2 ($p \in Q = L^2(\Omega)$) is sufficient since no derivatives of the pressure variable occur in (10.9). With the finite element ansatz functions for the displacements $\mathbf{u}_h \in V_h \subset V$ and for the pressure $p_h \in Q_h \subset Q$, the discretized form of (10.8) follows

$$a(\mathbf{u}_h, \boldsymbol{\eta}_h) + b(p_h, \boldsymbol{\eta}_h) = f(\boldsymbol{\eta}_h) \qquad \forall \boldsymbol{\eta}_h \in V_h \qquad (10.10)$$
$$b(q_h, \mathbf{u}_h) = 0 \qquad \forall q_h \in Q_h.$$

The conditions for existence, uniqueness and stability of the solution are the ellipticity condition and the BB-condition. The first one requires that the ansatz functions $\boldsymbol{\eta}_h$ fulfil for a positive constant $\alpha > 0$, the condition

$$a(\boldsymbol{\eta}_h, \boldsymbol{\eta}_h) \geq \alpha \, \|\boldsymbol{\eta}_h\|_V^2 . \qquad (10.11)$$

The fulfillment of the BB-condition means that a constant $\beta > 0$ exists so that

$$\inf_{q_h \in Q_h} \sup_{\boldsymbol{\eta}_h \in V_h} \frac{b(\boldsymbol{\eta}_h, q_h)}{\|\boldsymbol{\eta}_h\|_{H^1} \|q_h\|_{L^2}} \geq \beta . \qquad (10.12)$$

In case that the ansatz functions fulfil both conditions for incompressible material then the derived finite element method is stable.

For general nonlinear applications, there exists no formulation of the BB-condition. One can apply the condition analogously for the tangent spaces which belong to a given state of deformation and pressure, as e.g. provided in (10.2). The BB-condition has the disadvantage that it cannot be formulated for e single element. One always has to consider a patch of elements, see e.g. B Brezzi and Fortin (1991) or Braess (2007). A numerical method to show fulfillment of the BB-condition was derived in Chapelle and Bathe (1993).

10.2.1 Mixed Q1-P0 Element

In this section, a large deformation finite element is derived which is based on the HU–WASHIZU variational formulation. This element is implemented in many existing finite element codes and uses linear shape functions for the deformation field related to the deviatoric kinematical variables. Additionally, constant ansatz functions are applied to discretize the pressure and volumetric strain.

The continuum mechanical basis for the mixed Q1-P0 element was already discussed in Sect. 3.4.3. Equation (3.308) describes the weak form with respect to the spatial configuration. Inserting the finite element approximation into the weak form yields with (4.94)

$$\nabla^S \boldsymbol{\eta}_e = \sum_{I=1}^{n} \boldsymbol{B}_{0\,I}\, \boldsymbol{\eta}_I \, . \tag{10.13}$$

The virtual strain div $\boldsymbol{\eta}$, related to the change of volume, occurs additionally in (3.308). Discretization of the divergence operator leads to

$$\text{div}\, \boldsymbol{\eta}_e = \sum_{I=1}^{n} \boldsymbol{B}_{V\,I}\, \boldsymbol{\eta}_I \, , \tag{10.14}$$

where the matrix

$$\boldsymbol{B}_{V\,I} = <N_{I,1}, N_{I,2}, N_{I,3}> \tag{10.15}$$

was introduced. The derivatives have to be computed with respect to the current coordinates, as shown in Sect. 4.2.3.

Furthermore, constant ansatz functions are introduced for the pressure $J\,p = \tau_{vol}$, see (3.129), and the volume strain θ_e in Ω_e

$$\tau_{vol\,e} = J\,p_e = J\,\bar{p} \qquad \theta_e = \bar{\theta} \, . \tag{10.16}$$

With these interpolations, the weak form (3.308) can be written as

$$D\Pi(\boldsymbol{\varphi},p,\theta) \cdot \boldsymbol{\eta} \;=\; \bigcup_{e=1}^{n_e} \sum_{I=1}^{n} \boldsymbol{\eta}_I^T \int_{\Omega_e} \{\, (\boldsymbol{B}_{0\,I}^T\, \boldsymbol{\tau}_{iso\,e} + J\,\boldsymbol{B}_{V\,I}^T\, \bar{p} \,\}\, d\Omega - \delta P_{EXT} = 0,$$

$$DII(\boldsymbol{\varphi}, p, \theta)\, \delta p \;\; = \;\; \int_{\Omega_e} \delta\bar{p}\,(\, J_e - \bar{\theta}\,)\, d\Omega = 0, \tag{10.17}$$

$$DII(\boldsymbol{\varphi}, p, \theta)\, \delta\theta \;\; = \;\; \int_{\Omega_e} \delta\bar{\theta}\,\left(\frac{\partial W}{\partial\theta} - \bar{p}\right) d\Omega = 0\,.$$

The integrals are evaluated with respect to the initial configuration. The first equation denotes the weak form of equilibrium where $\boldsymbol{\tau}$ are the KIRCH-HOFF stresses. The second equation is associated with the constraint equation $J_e = \bar{\theta}$ and the third equation yields the constitutive equation for the pressure $\bar{p}$, see also (3.130)$_1$. The last two equations in (10.17) can be fulfilled locally on element level since a discontinuous ansatz was selected for pressure and volume strain. Hence both equations can be solved directly. This leads with (3.12) to

$$\bar{\theta} \;\; = \;\; \frac{1}{\Omega_e} \int_{\Omega_e} J_e\, d\Omega = \frac{\varphi(\Omega_e)}{\Omega_e}$$

$$\bar{p} \;\; = \;\; \frac{1}{\Omega_e} \int_{\Omega_e} \frac{\partial W}{\partial\theta}\, d\Omega = \frac{\partial W}{\partial\theta}(\bar{\theta}) \tag{10.18}$$

The discretization of the weak form (3.308) is now completed and summarized in Eq. (10.17)$_1$ and (10.18). Note that the volumetric variable θ follows simply from the ratio of the element volume in the current configuration $\varphi(\Omega_e)$ to the element volume in the initial configuration Ω_e.

10.2.2 Linearization of the Q1-P0 Element

The linearization of (10.17) yields a matrix form of the Q1-P0 element in which all variables $(\boldsymbol{\varphi}, p, \theta)$ are present. From the first equation of (10.17), the linearization follows with (3.277), (4.112) and (4.113) as

$$D^2 II \cdot \Delta u = \bigcup_{e=1}^{n_e} \sum_{I=1}^{n} \boldsymbol{\eta}_I^T \left[\sum_{K=1}^{n} \bar{\boldsymbol{K}}_{T_{IK}}^{u}\, \Delta\mathbf{u}_K + \bar{\boldsymbol{K}}_{T_I}^{p}\, \Delta\bar{p}\right] \tag{10.19}$$

where the matrices

$$\bar{\boldsymbol{K}}_{T_{IK}}^{u} \;\; = \;\; \int_{\Omega_e} \left[(\nabla_{\bar{x}} N_I)^T\, (\bar{p}\, J\, \mathbf{1} + \bar{\boldsymbol{\tau}}_{iso\,e})\, \nabla_{\bar{x}} N_K \right.$$

$$\left. + \bar{\boldsymbol{B}}_{0\,I}^{T}\, [\,(\mathbf{1} \otimes \mathbf{1} - 2\mathbb{E})\, \bar{p}\, J + \mathbb{C}_{iso}\,]\, \bar{\boldsymbol{B}}_{0\,K} \right] d\Omega, \tag{10.20}$$

$$\bar{\boldsymbol{K}}_{T_I}^{p} \;\; = \;\; \int_{\Omega_e} \boldsymbol{B}_{V\,I}^{T}\, J\, d\Omega\,.$$

occur. The linearization of the second equation (10.17) is derived with the JACOBI determinant, see (3.330), and its associated discretization, see (10.14),

$$\Delta\theta = \frac{1}{\Omega_e} \sum_{K=1}^{n} \int_{\Omega_e} \boldsymbol{B}_{V\,K}\, J\, d\Omega\, \Delta\boldsymbol{u}_K\,. \tag{10.21}$$

The third equation of (10.17) yields the linearization

$$\Delta\bar{p} = \frac{\partial^2 W}{\partial\theta^2}\Delta\theta\,. \tag{10.22}$$

Inserting now (10.21) in (10.22) and using this result in (10.19) and (10.20) leads to the elimination of the variables for pressure $\Delta\bar{p}$ and volumetric strain $\Delta\theta$ on element level. Thus a pure displacement formulation is obtained. Its tangent stiffness matrix has, for the element nodes I and K, the following expression

$$\begin{aligned}
\bar{\boldsymbol{K}}_{T_{I\,K}}^{Q1P0} &= \int_{\Omega_e} \big[\, (\nabla_{\bar{x}}N_I)^T\, (\bar{p}\,J\,\boldsymbol{1} + \bar{\boldsymbol{\tau}}_{iso\,e}\,)\,\nabla_{\bar{x}}N_K \\
&\quad + \bar{\boldsymbol{B}}_{0\,I}^T\,\big[\,(\,\boldsymbol{1}\otimes\boldsymbol{1} - 2\mathbb{E}\,)\,\bar{p}\,J + \mathbb{c}_{iso}\,\big]\,\bar{\boldsymbol{B}}_{0\,K}\,\big]\, d\Omega \\
&\quad + \frac{1}{\Omega_e}\int_{\Omega_e} \boldsymbol{B}_{V\,I}^T\, J\, d\Omega\, \left(\frac{\partial^2 W}{\partial\theta^2}\,\Omega_e\right)\,\frac{1}{\Omega_e}\int_{\Omega_e} \boldsymbol{B}_{V\,K}\, J\, d\Omega\,.
\end{aligned} \tag{10.23}$$

This element does not fulfil the BB-condition in the geometrical linear theory. Thus it can lead to unstable solutions for the pressure when special loading and boundary conditions are given. Often post-processing of the pressures using L^2 smoothing can help. In practical application, it has been observed that this element is quite robust for many problems in solid mechanics which depict quasi-incompressible material behaviour. Hence it is contained in many commercial finite element codes. In case that this element is not sufficient, its high order variant can be used which is the Q2-P1 element with quadratic interpolations for the deformations and linear interpolation for the pressure. It fulfils the BB-condition in case of the linear theory, see e.g. Brezzi and Fortin (1991).

10.3 Mixed Finite Elements for Finite Elasticity

A mixed finite element, based on the Neo–HOOKE material equation (3.119) in Sect. 3.3.1, is developed by using a formulation equivalent to the HELLINGER–REISSNER principle, see e.g. Washizu (1975). The main idea is to use a similar approach as the one advocated by Pian and Sumihara (1984) for the linear case. Within this hybrid approach, the constitutive equation needed to be inverted. Here the Neo–HOOKE material equation is given for the 2nd PIOLA–KIRCHHOFF stress $\boldsymbol{S}$ in terms of the right CAUCHY–GREEN tensor $\boldsymbol{C}$ as

$$\mathbf{S} = \frac{\Lambda}{2} (J^2 - 1) \mathbf{C}^{-1} + \mu (\mathbf{1} - \mathbf{C}^{-1}). \tag{10.24}$$

Under the assumption that it is possible to invert this equation form,

$$\mathbf{C} = \mathbf{f}(\mathbf{S}) \tag{10.25}$$

is obtained.

For the derivation of the mixed hybrid principle in Pian and Sumihara (1984), the classical LEGENDRE transformation $\frac{1}{2}\boldsymbol{\epsilon} \cdot \mathbb{C}[\boldsymbol{\epsilon}] = \boldsymbol{\sigma} \cdot \boldsymbol{\epsilon} - \frac{1}{2}\boldsymbol{\sigma} \cdot \mathbb{C}^{-1}[\boldsymbol{\sigma}]$ was applied. This transformation, a however, a is not valid in the nonlinear case, see e.g. Ogden (1984) and hence cannot be applied the Neo–HOOKE material. Instead, a weak form of the equilibrium G_u and the constitutive relation G_c is formulated which has as primary variables the displacement field and the stress field as follows

$$G_u(\mathbf{u}, \mathbf{S}, \boldsymbol{\eta}) = \int_B \frac{1}{2} \mathbf{S} \cdot \mathbf{C}(\boldsymbol{\eta}) \, dV - \int_B \hat{\mathbf{b}} \cdot \boldsymbol{\eta} \, dV - \int_{\partial B_\sigma} \hat{\mathbf{t}} \cdot \boldsymbol{\eta} \, dA = 0,$$

$$G_c(\mathbf{u}, \mathbf{S}, \mathbf{Q}) = \int_B \frac{1}{2} \mathbf{Q} \cdot [\mathbf{C}(\mathbf{u}) - \mathbf{f}(\mathbf{S})] \, dA = 0. \tag{10.26}$$

Here $\boldsymbol{\eta}$ and $\mathbf{Q}$ are the test functions, $\boldsymbol{\eta}$ is equivalent to the virtual displacement and $\mathbf{Q}$ to the virtual stress. $\mathbf{C}(\boldsymbol{\eta}) = \mathbf{F}^T \operatorname{Grad} \boldsymbol{\eta} + \operatorname{Grad}^T \boldsymbol{\eta} \, \mathbf{F}$ is the virtual strain and $\mathbf{C}(\mathbf{u}) = \mathbf{F}^T \mathbf{F}$ is the right CAUCHY–GREEN tensor; the latter depending only on the displacement field $\mathbf{u}$. This weak form can be viewed as the nonlinear version of the HELLINGER–REISSNER functional.

Note that the inverse (10.25) is not uniquely defined, either locally or globally, see Ogden (1984). However, (10.24) can be inverted by looking at different solution branches. For this (10.24) is rewritten as

$$\hat{\mathbf{A}}(\mathbf{S}) = \hat{\beta}(\mathbf{C}) \, \mathbf{C}^{-1} \tag{10.27}$$

with

$$\hat{\beta} = \alpha - J^2,$$

$$\alpha = 1 + \frac{2\mu}{\Lambda}, \tag{10.28}$$

$$\hat{\mathbf{A}}(\mathbf{S}) = \frac{2}{\Lambda} (\mu\mathbf{1} - \mathbf{S}).$$

Multiplication of (10.27) with $\hat{\mathbf{A}}^{-1}$ from the left side leads to

$$\mathbf{C} = \hat{\beta}(\mathbf{C}) \, \hat{\mathbf{A}}^{-1}(\mathbf{S}). \tag{10.29}$$

Now it remains to compute $\hat{\beta}(\mathbf{C})$ in dependence of $\hat{\mathbf{A}}$. The computation of the determinant of (10.29) yields

$$\det \hat{\mathbf{A}} = \frac{\hat{\beta}^3}{J^2} \tag{10.30}$$

since $J^2 = \det \mathbf{C}$. With the abbreviation $\hat{a} = \det \hat{\mathbf{A}}$ and $(10.28)_1$ this leads to a cubic equation for $\hat{\beta}$:

$$\hat{\beta}^3 + \hat{\alpha}\,\hat{\beta} - \hat{a}\alpha = 0 \,. \tag{10.31}$$

To obtain a simpler solution $\beta = c_1\,\hat{\beta}$, $\mathbf{A} = c_2\,\hat{\mathbf{A}}$ and $a = \det \mathbf{A}$ can be defined. Together with c_1 and c_2

$$c_1 = \frac{2}{3\,\alpha}\,, \qquad c_2 = \frac{1}{3}\left[\frac{4}{\alpha^2}\right]^{\frac{1}{3}}\,, \tag{10.32}$$

which only depend on the LAME constants, this provides the cubic equation for β

$$\beta^3 + 3\,a\,\beta - 2\,a = 0 \,. \tag{10.33}$$

Depending on the discriminant $D = a^3 + a^2$, three different solutions have to be distinguished:

− $D > 0$: Equation (10.33) has only one real solution

$$\beta = r - \frac{a}{r} \quad \text{with} \quad r = \left[a + \sqrt{a^3 + a^2}\right]^{\frac{1}{3}} \,. \tag{10.34}$$

− $D < 0$: This case is equivalent to $a < -1$ and yields three solutions for (10.33)

$$\beta = -2\sqrt{-a}\,\cos\left[\frac{1}{3}\left(\arccos\frac{1}{\sqrt{-a}} + 2\pi k\right)\right]\,, \qquad k = 0,1,2. \tag{10.35}$$

In the physical problem, $J > 0$ has to be fulfilled. From $(10.28)_1$ and $(10.32)_1$ it then follows that $\beta < \frac{2}{3}$. Hence only the solution with $k = 0$ remains under these circumstances

$$\beta = -2\sqrt{-a}\,\cos\left(\frac{1}{3}\arccos\frac{1}{\sqrt{-a}}\right)\,. \tag{10.36}$$

− $D = 0$: Here the determinant a is either $a = 0$ or $a = -1$. For $a = 0$, the only solution is $\beta = 0$ which yields with $(10.28)_1$ and $(10.32)_1$ for the JACOBIAN $J = \sqrt{\alpha}$. For $a = -1$, β can be obtained from (10.36), leading in the limit to $\beta = -2$.

Based on this solution, the expression for the inverse of the Neo–HOOKE material (10.24) can be derived, by employing $(10.28)_1$ and $(10.28)_3$. With

$$\mathbf{A} = \frac{2}{3\,\Lambda}\left[\frac{4}{\alpha^2}\right]^{\frac{1}{3}}(\mu\mathbf{1} - \mathbf{S})\,, \tag{10.37}$$

the final result is obtained

$$\mathbf{C} = \beta \left[\frac{\alpha}{2} \right]^{\frac{1}{3}} \mathbf{A}^{-1} = \frac{3}{4} \beta \alpha \Lambda \left(\mu \mathbf{1} - \mathbf{S} \right)^{-1} . \tag{10.38}$$

Hence it is possible to invert the constitutive equation; the weak formulation (10.26) related to a HELLINGER–REISSNER functional can be used as starting point for the finite element development.

Interpolation has to be selected for the displacement field and the stresses. Here a four-node quadrilateral is derived based on the isoparametric concept. For the displacement field and its variation, the standard shape functions are used

$$\mathbf{u} = \sum_{I=1}^{4} N_I(\xi, \eta) \, \mathbf{u}_I , \qquad \boldsymbol{\eta} = \sum_{I=1}^{4} N_I(\xi, \eta) \, \boldsymbol{\eta}_I . \tag{10.39}$$

As usual, the coordinates are expressed by the same approximation

$$\mathbf{X} = \sum_{I=1}^{4} N_I(\xi, \eta) \, \mathbf{X}_I , \tag{10.40}$$

where the nodal coordinates $\mathbf{X}_I$ are related to the initial configuration and ξ, η are convective coordinates with regard to the reference element. The interpolation function is given by, see Sect. 4.1.2,

$$N_I = \frac{1}{4} \left(1 + \xi \xi_I \right) \left(1 + \eta \eta_I \right) . \tag{10.41}$$

The interpolation introduced by Pian and Sumihara (1984) is chosen for the stress field. It leads to the matrix form

$$\left\{ \begin{array}{c} S^{\xi\xi} \\ S^{\eta\eta} \\ S^{\xi\eta} \end{array} \right\} = \begin{bmatrix} 1 & 0 & 0 & \eta & 0 \\ 0 & 1 & 0 & 0 & \xi \\ 0 & 0 & 1 & 0 & 0 \end{bmatrix} \left\{ \begin{array}{c} \bar{s}_1 \\ \bar{s}_2 \\ \bar{s}_3 \\ \bar{s}_4 \\ \bar{s}_5 \end{array} \right\} \tag{10.42}$$

with respect to the reference element. Note that the stress components are usually contravariant which is in accordance with the stress power $S^{ik} \dot{C}_{ik}$ and the definition of the strain measures via the deformation gradient.

The stresses in the reference element $(\mathbf{S} = S^{\alpha\beta} \, \mathbf{G}_\alpha \otimes \mathbf{G}_\beta)$ have now to be transformed to the global coordinate system which is obtained by

$$\begin{aligned} S^{ik} &= \mathbf{E}^i \cdot \mathbf{S} \, \mathbf{E}^k = \mathbf{E}_i \cdot \left(S^{\alpha\beta} \, \mathbf{G}_\alpha \otimes \mathbf{G}_\beta \right) \mathbf{E}_k \\ &= S^{\alpha\beta} (\mathbf{E}_i \cdot \mathbf{G}_\alpha)(\mathbf{E}_k \cdot \mathbf{G}_\beta) . \end{aligned}$$

For the orthogonal basis, the relation $\mathbf{E}^i = \mathbf{E}_i$ holds. Furthermore, a matrix form of this transformation can be defined which is given by $\mathbf{S}(\mathbf{X}) = \mathbf{T} \, \mathbf{S}(\boldsymbol{\xi}) \, \mathbf{T}^T$. In detail

$$\begin{bmatrix} S^{xx} & S^{xy} \\ S^{yx} & S^{yy} \end{bmatrix} = \begin{bmatrix} T_{11} & T_{12} \\ T_{21} & T_{22} \end{bmatrix} \begin{bmatrix} S^{\xi\xi} & S^{\xi\eta} \\ S^{\eta\xi} & S^{\eta\eta} \end{bmatrix} \begin{bmatrix} T_{11} & T_{21} \\ T_{12} & T_{22} \end{bmatrix} \tag{10.43}$$

is derived, where $T_{i\alpha} = \mathbf{E}_i \cdot \mathbf{G}_\alpha$.

The base vectors can be computed from the isoparametric interpolation since $\mathbf{G}_\alpha = \mathbf{X}_{,\alpha}$

$$\mathbf{G}_\alpha = \sum_{I=1}^{4} N_I(\xi, \eta)_{,\alpha} \, \mathbf{X}_I . \tag{10.44}$$

Hence

$$T_{i\alpha} = \sum_{I=1}^{4} N_I(\xi, \eta)_{,\alpha} \, X_{iI}, \tag{10.45}$$

where $X_{iI} = \mathbf{E}_i \cdot \mathbf{X}_I$. Using the interpolation (10.41), the derivatives are

$$N_{I,\xi} = \frac{\xi_I}{4} (1 + \eta \, \eta_I), \qquad N_{I,\eta} = \frac{\eta_I}{4} (1 + \xi \, \xi_I). \tag{10.46}$$

The transformation matrix will be evaluated at the element centre $\xi = \eta = 0$ leading to

$$N_{I,\xi}^0 = \frac{\xi_I}{4}, \qquad N_{I,\eta}^0 = \frac{\eta_I}{4}. \tag{10.47}$$

and

$$T_{i\xi}^0 = \sum_{I=1}^{4} \frac{\xi_I}{4} X_{iI} \quad \text{and} \quad T_{i\eta}^0 = \sum_{I=1}^{4} \frac{\eta_I}{4} X_{iI}. \tag{10.48}$$

Due to this, the transformation matrix is given by

$$\mathbf{T}^0 = \frac{1}{4} \sum_{I=1}^{4} \begin{bmatrix} \xi_I X_{1I} & \eta_I X_{1I} \\ \xi_I X_{2I} & \eta_I X_{2I} \end{bmatrix}. \tag{10.49}$$

By performing the multiplication in (10.43) and rearranging the components of the stress tensor in VOIGT notation, the stress transformation can be written as

$$\begin{Bmatrix} S^{xx} \\ S^{yy} \\ S^{xy} \end{Bmatrix} = \begin{bmatrix} T_{11}^2 & T_{12}^2 & 2\,T_{11}\,T_{12} \\ T_{21}^2 & T_{22}^2 & 2\,T_{21}\,T_{22} \\ T_{11}\,T_{21} & T_{22}\,T_{12} & T_{12}\,T_{21} + T_{11}\,T_{22} \end{bmatrix} \begin{Bmatrix} S^{\xi\xi} \\ S^{\eta\eta} \\ S^{\xi\eta} \end{Bmatrix}. \tag{10.50}$$

Since the stress interpolation (10.42) is constant for the shear stresses and the transformation matrix is constant element wise, a different representation can be found using the element wise constant matrix $\mathbf{s}^T = \{ s_1, s_2, s_3, s_4, s_5 \}$. This ansatz can be written in global coordinates using the transformation (10.50) at the mid point of the element and leads to the simpler form

$$\begin{Bmatrix} S^{xx} \\ S^{yy} \\ S^{xy} \end{Bmatrix} = \begin{Bmatrix} s_1 \\ s_2 \\ s_3 \end{Bmatrix} + \begin{bmatrix} \eta\,(T_{11}^0)^2 & \xi\,(T_{12}^0)^2 \\ \eta\,(T_{21}^0)^2 & \xi\,(T_{22}^0)^2 \\ \eta\,T_{11}^0\,T_{21}^0 & \xi\,T_{22}^0\,T_{12}^0 \end{bmatrix} \begin{Bmatrix} s_4 \\ s_5 \end{Bmatrix}. \tag{10.51}$$

These interpolations can now be used within the mixed weak form (10.26) and its linearization to derive the matrix form of the associated finite element formulation.

10.4 Stabilized Finite Elements

Stabilized finite elements are formulated in order to obtain efficient elements for which the residual vector and tangent matrix can be computed in a fast way and which need, as few as possible, memory to store history variables related to the chosen constitutive equations. The simplest method to achieve these two goals is to apply reduced integration which is based on a minimum number of GAUSS points and hence has less computational effort and storage requirement for history data. The drawback is that these elements are generally unstable since reduced integration is associated with rank deficiency. Thus underintegrated elements have to be stabilized. Stabilization is performed based on the eigenmodes of the elements. These follow from an eigenvalue analysis of a single finite element matrix. Here zero eigenvalues occur for rigid body modes which naturally do not contribute to the element stiffness. Additional zero eigenvalues have to be stabilized and hence an artificial stiffness has to be introduced to prevent non-physical occurrence of these modes within a finite element analysis.

For the two-dimensional linear elastic case, the eigenvectors computed from the spectral decomposition of the stiffness matrix are depicted in Fig. 10.1, excluding the rigid body modes. The eigenvectors related to the volume change, the elongation and shear can be found in the first row. The second row shows the bending modes of the element. It is well known from the linear theory that the eigenvalues related to the bending modes are zero when reduced integration is applied. In that case, no strain energy is associated with these modes. Hence deformations related to the bending modes can occur in an analysis depending on the loading and boundary conditions. Since two of the bending modes can form an *hour-glass*, these modes are also called *hour-glass* modes, see Fig. 10.2b.

Thus stabilization has to be used to avoid hour-glassing when underintegrated elements are applied within a finite element analysis. In this case,

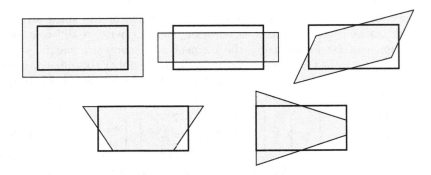

Fig. 10.1 Eigenvectors of the quadrilateral 4-node element

the eigenvectors related to hour-glassing are determined in the initial configuration and then stabilized. This procedure is however not trivial: the mode of a distorted element has to be determined and the magnitude of the stiffness to be added cannot be derived directly from the underlying variational equation.

Two basic approaches are possible.

1. Hour-glass modes can be filtered from the global solution as discussed in Jaquotte and Oden (1986). This however is only possible for elastic problems.
2. The displacement interpolation can be decomposed within a finite element into an linear part and the related orthogonal part. The latter is then used to derive a stabilization matrix. This idea was developed in Kosloff and Frazier (1978) for linear problems. A follow up paper from Belytschko et al. (1984), see also Hughes (1987, p. 251), introduces the so-called γ vectors. Their explicit form can be used to construct the stabilization matrix.

While it is possible to compute the stiffness parameters for the stabilization matrix from the equivalence of mixed methods and stabilized reduced integration procedures, see e.g. the element formulation developed in Pian and Sumihara (1984), this has so far not been achieved for nonlinear problems in a satisfactory way. Approaches can be found in Belytschko and Bindeman (1991), Belytschko and Bindeman (1993), Bonet and Bhargava (1995), Reese et al. (1998) and Reese (2005), see also Sect. 10.5.

The classical stabilization procedure for underintegrated element will be developed in the following for three-dimensional hexahedral elements with linear displacement interpolation. Basically, the tangent stiffness matrix (4.76) which was already derived in Sect. 4.2.2 is evaluated by using a one point GAUSS integration instead of the rank preserving $2 \times 2 \times 2$ integration

$$\bar{K}_{T_{IK}}^{1 \times 1} = \int\limits_{\Omega_e} \left[(\nabla_X N_I)^T \bar{S} \nabla_X N_K + \bar{B}_{LI}^T \bar{D} \bar{B}_{LK} \right] d\Omega \,.$$

The matrix form is provided for the nodal combination I, K of a finite element Ω_e. Within this notation, the sub matrix $\bar{K}_{T_{IK}}$ has the size $n_{dof} \times n_{dof}$ where n_{dof} is the number of degrees of freedom needed to describe the displacement field (for three-dimensional problems $n_{dof} = 3$ holds). The indices I and K are nodes of the element and directly related to the discretization. Summation over all 8 nodes of the hexahedral element yields the tangent matrix for the finite element e: $\bar{K}_{T_e}^{1 \times 1}$. Note that a 1-point-integration requires only one evaluation within the element mid point, see Table 4.1

$$\bar{K}_{T_e}^{1 \times 1} = \bar{K}_{T_e} \big|_{\xi = \eta = \zeta = 0} \,. \tag{10.52}$$

Hence all terms can be neglected which depend on the coordinates ξ, η or ζ. This procedure simplifies the coding of such element and thus leads to a high efficiency.

The stabilization matrix $\bar{K}_{T_e}^{stab}$ is added to (10.52) which leads to

$$\bar{K}_{T_e} = \bar{K}_{T_e}^{1 \times 1} + \bar{K}_{T_e}^{stab} \tag{10.53}$$

with the diagonalized form, see Belytschko et al. (1984),

$$\bar{K}_{T_e}^{stab} = \sum_{k=1}^{12} \alpha_k \, \bar{\gamma}_k \, \bar{\gamma}_k^T \,. \tag{10.54}$$

The scalar parameters $\alpha_k > 0$ can be chosen arbitrarily. However, their magnitude has to be selected such that the parameters avoid hour-glassing on one side and do not influence the solution of the problem on the other side. This however is not always possible, see examples in Reese (1994). Hence the user of stabilized elements has to have sufficient experience when applying this method.

The determination of the $\bar{\gamma}_k$ vectors for stabilization will be presented in the next section. Kosloff and Frazier (1978) have already shown that the diagonal form of $\bar{K}_{T_e}^{stab}$ using 12 scalar parameters, see (10.54), is not sufficient to obtain optimal bending behaviour for generally distorted three-dimensional meshes. Thus the stabilization matrix in (10.54) yields good results for application which do not exhibit bending.

10.4.1 Stabilization Vectors

The isoparametric ansatz functions presented in Sect. 4.1.3 can also be written in an equivalent vector form. This is advantageous when stabilization vectors have to be derived. Instead of (4.40), the interpolation functions are given by

$$N(\xi) = \frac{1}{8} \left[a_1 + \xi \, a_2 + \eta \, a_3 + \zeta \, a_4 + \eta \, \zeta \, a_5 + \xi \, \zeta \, a_6 + \xi \, \eta \, a_7 + \xi \, \eta \, \zeta \, a_8 \right] \tag{10.55}$$

with the constant vectors

$$
\begin{aligned}
a_1^T &= \{\ \ 1,\ \ 1,\ \ 1,\ \ 1,\ \ 1,\ \ 1,\ \ 1,\ \ 1\} \\
a_2^T &= \{-1,\ \ 1,\ \ 1,-1,-1,\ \ 1,\ \ 1,-1\} \\
a_3^T &= \{-1,-1,\ \ 1,\ \ 1,-1,-1,\ \ 1,\ \ 1\} \\
a_4^T &= \{-1,-1,-1,-1,\ \ 1,\ \ 1,\ \ 1,\ \ 1\} \\
a_5^T &= \{\ \ 1,\ \ 1,-1,-1,-1,-1,\ \ 1,\ \ 1\} \\
a_6^T &= \{\ \ 1,-1,-1,\ \ 1,-1,\ \ 1,\ \ 1,-1\} \\
a_7^T &= \{\ \ 1,-1,\ \ 1,-1,\ \ 1,-1,\ \ 1,-1\} \\
a_8^T &= \{-1,\ \ 1,-1,\ \ 1,\ \ 1,-1,\ \ 1,-1\}.
\end{aligned}
$$

By using this notation, the interpolation of the components of the displacement vector $\mathbf{u}_e = u_i \, \mathbf{E}_i$ can be written as

$$u_1 = \mathbf{N}^T \, \mathbf{v}_1 \,, \qquad u_2 = \mathbf{N}^T \, \mathbf{v}_2 \,, \qquad u_3 = \mathbf{N}^T \, \mathbf{v}_3 \,, \qquad (10.56)$$

where the vectors $\mathbf{v}_i$ contain the components of the nodal displacements in coordinate direction i. The stabilization vectors follow from a TAYLOR expansion of the shape functions with respect to the midpoint of the element $\boldsymbol{\xi} = \mathbf{0}$ up to first order terms. This yields for the ansatz functions

$$\mathbf{N} = \mathbf{N}_0 + \left.\frac{\partial \mathbf{N}}{\partial \mathbf{X}}\right|_{\boldsymbol{\xi}=\mathbf{0}} (\mathbf{X} - \mathbf{X}_{|\,0}) + \mathbf{N}_\gamma \qquad (10.57)$$

with a constant term $\mathbf{N}_0$, a linear term and a residual term $\mathbf{N}_\gamma$. $\mathbf{X}_0$ is the position vector of the element midpoint. Since it is not possible to differentiate in (10.57) with respect to $\mathbf{X}$, the chain rule and thus the JACOBI matrix $\mathbf{J}_e$ has to be used, see Sect. 4.1, to obtain

$$
\begin{aligned}
\mathbf{N} &= \mathbf{N}_0 + \left.\left(\frac{\partial \mathbf{N}}{\partial \boldsymbol{\xi}} \, \mathbf{J}_e^{-1}\right)\right|_{\boldsymbol{\xi}=\mathbf{0}} (\mathbf{X} - \mathbf{X}_{|\,0}) + \mathbf{N}_\gamma \\
&= [\,\mathbf{I} - (\mathbf{N}_{,\xi} \, \mathbf{J}_e^{-1})_{|\,0} \, \mathbf{X}_{kn}]\frac{1}{8}\,\mathbf{a}_1 + (\mathbf{N}_{,\xi} \, \mathbf{J}_e^{-1})_{|\,0}\,\mathbf{X} + \mathbf{N}_\gamma \,. \quad (10.58)
\end{aligned}
$$

The matrix $\mathbf{X}_{kn}$ of dimension 3×8 was introduced for a more compact notation. It contains the coordinates $\{\,X_I\,,Y_I\,,Z_I\,\}$ of the position vectors to the element nodes $I = 1, 8$. The index 0 at $\mathbf{J}_{|\,0}$ means that $\mathbf{J}_e$ has to be evaluated at $\boldsymbol{\xi} = \mathbf{0}$. The first two terms in (10.58) represent a vector of the shape functions which is linear in $\mathbf{X}$. Note that this relation is valid for arbitrarily deformed element geometries in the initial configuration. The residual term can now be determined from $\mathbf{N}_\gamma = \mathbf{N} - \mathbf{N}_{lin}$. Within this procedure, the convergence criteria for finite elements have to be fulfilled, see the preliminary remarks in Sect. 8. For this rigid body modes and constant strains have to be recovered for arbitrary element geometries. This requires that vector $\mathbf{N}_\gamma$ has to be orthogonal to the linear part of the shape functions; otherwise it is impossible to obtain constant strain states. This is associated with the classical requirement of the fulfillment of the *patch tests*, see e.g. Bathe (1982) and Hughes (1987). By considering the aforementioned orthogonality, the stabilization vector also called *hour-glass* part

$$
\begin{aligned}
\mathbf{N}_\gamma &= \frac{1}{8}[\,\mathbf{I} - (\mathbf{N}_{,\xi} \, \mathbf{J}_e^{-1})_{|\,0} \, \mathbf{X}_{kn}](\eta\,\zeta\,\mathbf{a}_5 + \xi\,\zeta\,\mathbf{a}_6 + \xi\,\eta\,\mathbf{a}_7 + \xi\,\eta\,\zeta\,\mathbf{a}_8) \\
&= \eta\,\zeta\,\boldsymbol{\gamma}_1 + \xi\,\zeta\,\boldsymbol{\gamma}_2 + \xi\,\eta\,\boldsymbol{\gamma}_3 + \xi\,\eta\,\zeta\,\boldsymbol{\gamma}_4 \qquad (10.59)
\end{aligned}
$$

is derived after some algebraic manipulations, see Belytschko et al. (1984). The 12 stabilization vectors $\bar{\boldsymbol{\gamma}}$ can now be computed from the components of vectors $\boldsymbol{\gamma}_k$ $(k = 1, 4)$ by using four $\boldsymbol{\gamma}$ vectors for each component. Hence the 8 components of the $\boldsymbol{\gamma}$ vectors yield 12 $\bar{\boldsymbol{\gamma}}$ vectors with $3 \times 8 = 24$ components. Explicitly, the vectors are given by

$$
\begin{aligned}
\bar{\boldsymbol{\gamma}}_1 &= \{\gamma_1^1, 0, 0, \gamma_1^2, 0, 0, \ldots, \gamma_1^8, 0, 0\}^T \\
\bar{\boldsymbol{\gamma}}_2 &= \{\gamma_2^1, 0, 0, \gamma_2^2, 0, 0, \ldots, \gamma_2^8, 0, 0\}^T \\
\bar{\boldsymbol{\gamma}}_3 &= \{\gamma_3^1, 0, 0, \gamma_3^2, 0, 0, \ldots, \gamma_3^8, 0, 0\}^T \\
\bar{\boldsymbol{\gamma}}_4 &= \{\gamma_4^1, 0, 0, \gamma_4^2, 0, 0, \ldots, \gamma_4^8, 0, 0\}^T \\
\bar{\boldsymbol{\gamma}}_5 &= \{0, \gamma_1^1, 0, 0, \gamma_1^2, 0, \ldots, 0, \gamma_1^8, 0\}^T \\
\bar{\boldsymbol{\gamma}}_6 &= \{0, \gamma_2^1, 0, 0, \gamma_2^2, 0, \ldots, 0, \gamma_2^8, 0\}^T \\
\bar{\boldsymbol{\gamma}}_7 &= \{0, \gamma_3^1, 0, 0, \gamma_3^2, 0, \ldots, 0, \gamma_3^8, 0\}^T \\
\bar{\boldsymbol{\gamma}}_8 &= \{0, \gamma_4^1, 0, 0, \gamma_4^2, 0, \ldots, 0, \gamma_4^8, 0\}^T \\
\bar{\boldsymbol{\gamma}}_9 &= \{0, 0, \gamma_1^1, 0, 0, \gamma_1^2, \ldots, 0, 0, \gamma_1^8\}^T \\
\bar{\boldsymbol{\gamma}}_{10} &= \{0, 0, \gamma_2^1, 0, 0, \gamma_2^2, \ldots, 0, 0, \gamma_2^8\}^T \\
\bar{\boldsymbol{\gamma}}_{11} &= \{0, 0, \gamma_3^1, 0, 0, \gamma_3^2, \ldots, 0, 0, \gamma_3^8\}^T \\
\bar{\boldsymbol{\gamma}}_{12} &= \{0, 0, \gamma_4^1, 0, 0, \gamma_4^2, \ldots, 0, 0, \gamma_4^8\}^T .
\end{aligned}
$$

Here the terms γ_k^m ($k = 1, 4$ and $m = 1, 8$) are the components of the $\boldsymbol{\gamma}$ vectors defined in (10.59).

10.4.2 Weak Form and Linearization

The weak form for the *hour-glass* stabilized 8-node elements follows from the results derived in Sect. 4.2.1. The matrices and vectors are now evaluated using 1 point GAUSS integration. In detail, the internal virtual work is obtained from (4.54)

$$
\int_B \delta \mathbf{E} \cdot \mathbf{S} \, dV = \bigcup_{e=1}^{n_e} \sum_{I=1}^{8} \boldsymbol{\eta}_I^T \int_{\Omega_\square} (\boldsymbol{B}_{LI}^T \boldsymbol{S}_e)_{|0} \det \boldsymbol{J}_{|0} \, d\square , \tag{10.60}
$$

where index 0 denotes the evaluation of the integrals at the element midpoint $\boldsymbol{\xi} = \mathbf{0}$. The residual term due to the stabilization vectors is given by

$$
G_{stab} = \bigcup_{e=1}^{n_e} \sum_{i=1}^{12} \boldsymbol{\eta}_e^T \alpha_i (\bar{\boldsymbol{\gamma}}_i^T \boldsymbol{u}_e) \bar{\boldsymbol{\gamma}}_i . \tag{10.61}
$$

The vectors $\boldsymbol{\eta}_e$ and $\boldsymbol{u}_e$ contain all 24 components of the test functions and displacements within element $\Omega_{e,}$; hence the sum over all element nodes used in (10.60) disappears. By combining both terms and by evaluating (10.60), using the one-point integration, the residual vector of one finite element is given, see also (4.55),

$$
\begin{aligned}
\boldsymbol{R}_e (\boldsymbol{u}_e) &= \boldsymbol{R}_{e0} (\boldsymbol{u}_e) + \boldsymbol{K}_{stab} \boldsymbol{u}_e \\
&= 8 \sum_{I=1}^{8} \left[\boldsymbol{B}_{LI}^T \boldsymbol{S} \right]_{|0} \det \boldsymbol{J}_{|0} + \sum_{i=1}^{12} \alpha_i (\bar{\boldsymbol{\gamma}}_i^T \boldsymbol{u}_e) \bar{\boldsymbol{\gamma}}_i . \tag{10.62}
\end{aligned}
$$

The linearization of the residual vector yields the tangential stiffness matrix which is needed within the NEWTON method. From (4.76),

$$
\begin{aligned}
\bar{\boldsymbol{K}}_{T_e} &= \bar{\boldsymbol{K}}_{T_{e0}} + \boldsymbol{K}_{stab} \\
&= \sum_{I=1}^{8} \sum_{K=1}^{8} 8 \left[(\nabla_X N_I)^T \, \bar{\boldsymbol{S}} \, (\nabla_X N_K) + \bar{\boldsymbol{B}}_{L\,I}^T \, \bar{\boldsymbol{D}} \, \bar{\boldsymbol{B}}_{L\,K} \right]_{\big|\,0} \det \boldsymbol{J}_{\big|\,0} \\
&\quad + \sum_{i=1}^{12} \alpha_i \, \bar{\boldsymbol{\gamma}}_i \, \bar{\boldsymbol{\gamma}}_i^T
\end{aligned}
\tag{10.63}
$$

is obtained for a one-point integration.

The solution of a problem using the discretized weak form (10.62) depends on the choice of the parameters α_i. The values of α_i do not play a significant role for standard three-dimensional engineering problems in solid mechanics. The parameters can be selected within a certain range and then do not influence the result of the computation. However, when bending dominates the solution behaviour, the solution can depend on the stabilization parameters a_i.

Within the linear theory, it was possible for Kosloff and Frazier (1978) to show that a special choice of the parameter α_i leads to a finite element which is equivalent to the incompatible mode element of Taylor et al. (1976). In that way, a very efficient element with excellent bending behaviour was obtained. For nonlinear problems, there exists no simple way to compute the stabilization parameters. Here the bending solution depends on the parameter α_i, as already shown in Reese (1994) using the example of a simple cantilever under point load. Thus it is desirable to develop a procedure for bending dominated problems in which the parameters α_i can be derived such that a solution dependence disappears. A related method is formulated in Sect. 10.5.3.[1]

10.5 Enhanced Strain Element

It is important to construct finite elements for problems of solid mechanics which can be applied to a wide range of problems. Such elements should be able to model finite strain states for arbitrary elastic and inelastic materials. Furthermore, they should work in the presence of constraints such as incompressibility which lead for standard displacement elements to locking, see Sect. 10.2. Good element performance for bending dominated structural problems is also necessary when arbitrary structural parts have to be discretized and simulated using three-dimensional solids. Last but not least

---

[1] The stabilized finite element formulation derived above for the initial configuration can also be developed with respect to the current configuration. In that case, all quantities have to be mapped to the current configuration using the standard transformations, see Sect. 4.2.3.

elements should be robust when large mesh distortion occur due to large deformations.

In the last twenty years, many different finite elements were developed for finite deformation problems and successfully applied to special problem classes. One example is the Q1-P0 element which is well suited for incompressible materials.

In case of linear elastic applications, there exist many possibilities for the design of finite elements which are locking free, have good bending performance and are robust against mesh distortions, see e.g. the hybrid formulations Pian and Sumihara (1984), or the incompatible mode elements of Taylor et al. (1976). Also the stabilized elements from Kosloff and Frazier (1978) have the same good properties.[2] A variational formulation of the discretization using the incompatible modes was derived in Simo and Rifai (1990). This concept has the advantage that it can also be applied to nonlinear problems like finite elastic or inelastic deformations.

The concept followed in the work by Simo and Armero (1992) and Simo et al. (1993b) is based on the principle of HU–WASHIZU. The finite elements derived by this formulation are called *enhanced strain* or *enhanced assumed strain* (EAS) elements.

While very well suited for linear elastic problem, the enhanced strain elements do not provide a solution for all problem classes mentioned above in nonlinear applications. The elements become instable under compression which was shown for the first time in Wriggers and Reese (1994), see also Wriggers and Reese (1996). Stabilized versions of the enhanced strain elements have been formulated to overcome this disadvantage. However, until lately, these stabilizations could not solve all defects found in Wriggers and Reese (1996) in a satisfactory way. Refined ansatz functions for the enhanced modes solved the instabilities for two-dimensional problems in the compression range, see Korelc and Wriggers (1996a) and Glaser and Armero (1997). But they lead to instabilities in tension states. An in-depth discussion of these phenomena and possible solutions can be found in Sect. 10.5.4.

In the following section, elements based on the enhanced strain concept will be derived, using on one hand the shape functions provided in Simo and Armero (1992) and on the other hand shape functions stemming from a TAYLOR series expansion which was developed in Wriggers and Hueck (1996).

10.5.1 General Concept and Formulation

The development of the nonlinear version of the *enhanced strain* elements is generally based on a mixed variational principle. Following Simo and Armero (1992), HU–WASHIZU'S principle is applied, see Sect. 3.4.3. Here

---

[2] An interesting observation is that the aforementioned formulations can be transferred to each other, see Bischoff et al. (1999a). Thus different mechanical formulations lead to the same element stiffness matrices.

HU–WASHIZU principle is formulated in terms of the deformation φ, the deformation gradient $\mathbf{F}$ and the first PIOLA–KIRCHHOFF stress tensor $\mathbf{P}$ which act as independent variables

$$\Pi(\varphi, \mathbf{F}, \mathbf{P}) = \int_B [\, W(\mathbf{F}) + \mathbf{P} \cdot (\operatorname{Grad}\varphi - \mathbf{F}) \,]\, dV$$

$$- \int_B \varphi \cdot \rho_0\, \hat{\mathbf{b}}\, dV - \int_{\partial B_\sigma} \varphi \cdot \hat{\mathbf{t}}\, dA. \tag{10.64}$$

$W(\mathbf{F})$ denotes the strain energy function of the elastic material under consideration. This formulation is equivalent to the principle provided in (3.300). To simplify notation, the last two terms in (10.64) which describe external forces will be combined and denoted by P_{EXT}.[3]

The variational principle of HU–WASHIZU was formulated in this way in order to be able to additively decompose the deformation gradient, see Simo and Armero (1992). In this decomposition, the local deformation gradient $\operatorname{Grad}\varphi$ is complemented by the independent gradient $\bar{\mathbf{F}}$

$$\mathbf{F} = \operatorname{Grad}\varphi + \bar{\mathbf{F}}. \tag{10.65}$$

Thus the deformation gradient $\mathbf{F}$ is enriched by the *enhanced* gradient $\bar{\mathbf{F}}$ which can be incompatible with the deformation. With Eq. (10.65), relation

$$\Pi(\varphi, \bar{\mathbf{F}}, \mathbf{P}) = \int_B [\, W(\mathbf{F}) - \mathbf{P} \cdot \bar{\mathbf{F}} \,]\, dV - P_{EXT} \tag{10.66}$$

is obtained from (10.64). Its variation yields

$$\int_B \operatorname{Grad}\boldsymbol{\eta} \cdot \frac{\partial W}{\partial \mathbf{F}}\, dV - \delta P_{EXT} = 0,$$

$$\int_B \delta\bar{\mathbf{F}} \cdot \left(-\mathbf{P} + \frac{\partial W}{\partial \mathbf{F}} \right) dV = 0, \tag{10.67}$$

$$\int_B \delta\mathbf{P} \cdot \bar{\mathbf{F}}\, dV = 0.$$

$\operatorname{Grad}\boldsymbol{\eta}$ denotes the variation of the deformation gradient, see (3.289). The first equation denotes the weak form of equilibrium. The second equations

---

[3] It is also possible to formulate the HU–WASHIZU principle in other work conjugate variables. Examples are the 2nd PIOLA–KIRCHHOFF stress tensor and the GREEN–LAGRANGIAN strain tensor $\mathbf{E}$ or the application of the BIOT stress tensor $\mathbf{T}_B$ together with the right stretch tensor $\mathbf{U}$. From the viewpoint of continuum mechanics, these formulations are equivalent. However, due to the fact that the strain measures $\mathbf{F}$, $\mathbf{E}$ and $\mathbf{U}$ are different, their enhancement will lead to different finite element approximations and discretizations.

leads to the constitutive relation. Equation $(10.67)_3$ represents an orthogonality condition between the stress tensor and the variation of the enhanced gradient $\bar{\mathbf{F}}$.

Equations (10.65) and (10.67) provide a variational basis which can be employed to incorporate the incompatible (enhanced) modes in a consistent way to the finite element formulation.

An efficient implementation of the enhanced element can be obtained by transforming all quantities in (10.67) to the current configuration, see also Sect. 4.2.3. By computing the deformation gradient from (10.65), relation

$$\int_B \nabla^S \boldsymbol{\eta} \cdot \left(2\,\mathbf{F}\,\frac{\partial W}{\partial \mathbf{C}}\,\mathbf{F}^T \right) dV - \delta P_{EXT} = 0,$$

$$\int_B \delta\,\bar{\mathbf{h}}^S \cdot \left(-\boldsymbol{\tau} + 2\,\mathbf{F}\,\frac{\partial W}{\partial \mathbf{C}}\,\mathbf{F}^T \right) dV = 0, \qquad (10.68)$$

$$\int_B \delta\,\boldsymbol{\tau} \cdot \bar{\mathbf{h}}\,dV = 0$$

is deduced based on (10.67). Here the gradient $\nabla^S \boldsymbol{\eta} = \mathrm{sym}\,[\mathrm{Grad}\,\boldsymbol{\eta}\,\mathbf{F}^{-1}]$ is the symmetric part of the variation of the deformation gradient with respect to the current configuration. The tensor $\mathbf{C} = \mathbf{F}^T \mathbf{F}$ is the right CAUCHY–GREEN strain tensor, see (3.15). The enhanced gradient in the current configuration is computed from $\bar{\mathbf{h}} = \bar{\mathbf{F}}\mathbf{F}^{-1}$. The symmetric KIRCHHOFF stress tensor follows with the 1st PIOLA–KIRCHHOFF stresses from $\boldsymbol{\tau} = \mathbf{P}\,\mathbf{F}^T$ or with (3.84) from the 2nd PIOLA–KIRCHHOFF stress tensor: $\boldsymbol{\tau} = \mathbf{F}\,\mathbf{S}\,\mathbf{F}^T$. Since $\boldsymbol{\tau}$ is a symmetric tensor, only the symmetric parts of the deformation gradient $\nabla^S \boldsymbol{\eta}$ and the enhanced gradient $\bar{\mathbf{h}}^S$ contribute to the scalar product with the KIRCHHOFF stresses in Eq. (10.68).

To complete the model, a strain energy function W is needed. Different variants can be found for hyperelastic materials in Sect. 3.3.1. The KIRCHHOFF stresses follow then from the 2nd PIOLA–KIRCHHOFF stresses via (3.104).

10.5.2 Discretization of the Enhanced Strain Element

An isoparametric ansatz, see (4.4), is introduced to discretize the displacement field and the geometry of the current configuration in (10.65)

$$\mathbf{x}_e = \mathbf{X}_e + \mathbf{u}_e = \sum_{I=1}^{n} N_I(\boldsymbol{\xi})\,\mathbf{x}_I \qquad \text{with} \quad x_I = X_I + u_I. \qquad (10.69)$$

In the two-dimensional case, the bilinear shape functions (4.28) are applied. In case of three-dimensional discretizations, the shape functions (4.40) are used.

The conforming part of the deformation gradient can now be determined from (10.69). With (4.8) and (4.11), the deformation gradient follows

$$\text{Grad}\,\boldsymbol{\varphi}_e = \sum_{I=1}^{n} \mathbf{x}_I \otimes \nabla_X N_I(\boldsymbol{\xi}) = \sum_{I=1}^{n} \mathbf{x}_I \otimes \mathbf{J}_e^{-T} \nabla_\xi N_I(\boldsymbol{\xi})\,. \tag{10.70}$$

For the enhanced part of the deformation gradient, an interpolation has to be selected which even can be incompatible. Following Glaser and Armero (1997), a product form is defined for the enriched part $\bar{\mathbf{F}}$

$$\bar{\mathbf{F}} = \mathbf{F}_0\,\bar{\mathbf{M}}\,\boldsymbol{\alpha}\,. \tag{10.71}$$

$\boldsymbol{\alpha}$ denote the *enhanced* parameters, $\bar{\mathbf{M}}$ contains the interpolation functions. $\mathbf{F}_0$ is the constant part of the conform deformation gradient (10.70), which is evaluated at the element midpoint

$$\mathbf{F}_0 = \sum_{I=1}^{n} \mathbf{x}_I \otimes \nabla_X N_I(\mathbf{0})\,. \tag{10.72}$$

The ansatz (10.71) fulfils the requirements for objectivity of the enhanced element formulation for arbitrary interpolations $\bar{\mathbf{M}}$, see Glaser and Armero (1997).[4]

The interpolations of the enriched part $\bar{\mathbf{M}}$ are related to the initial configuration of a finite element Ω_e. Since the incompatible interpolations have to be formulated with respect to the reference configuration $\Omega_\square$, like the isoparametric interpolations, $\bar{\mathbf{M}}$ has to be transformed to $\Omega_\square$ (for the relevant notation, see Fig. 4.3). This is performed by using the tensor transformation

$$\bar{\mathbf{M}} = \frac{j_0}{j}\,\mathbf{J}_0\,\mathbf{M}(\boldsymbol{\xi})\,\mathbf{J}_0^{-1}\,. \tag{10.73}$$

Here $\mathbf{J}_0$ defines the mapping between Ω_e and $\Omega_\square$, see (4.7), which is evaluated at the element midpoint ($\boldsymbol{\xi} = \mathbf{0}$). The determinant of the transformation is denoted by $j = \det \mathbf{J}_e$. Its evaluation at the element midpoint is denoted by $j_0 = \det \mathbf{J}_0$.

Now the interpolation for the enhanced modes have to be selected. These can be incompatible since no derivatives of the enriched deformation gradient appear in (10.67). In general, the ansatz

$$\mathbf{M}(\boldsymbol{\xi})\boldsymbol{\alpha} = \sum_{L=1}^{n_{enh}} \mathbf{M}_L(\boldsymbol{\xi})\,\alpha_L \tag{10.74}$$

---

[4] This representation deviates from the form advocated in Simo and Armero (1992) in such a way that $\bar{\mathbf{M}}$ was introduced as a gradient and hence could be interpolated without using $\mathbf{F}_0$, see also Exercise 10.1.

can be introduced with n_{enh} interpolations for the additional incompatible modes. The ansatz can be written for two-dimensional elements in the compact form

$$\mathbf{M}(\boldsymbol{\xi})\,\boldsymbol{\alpha} = \begin{bmatrix} M_1\,(\xi,\eta)\,\alpha_1 & M_2\,(\xi,\eta)\,\alpha_2 \\ M_3\,(\xi,\eta)\,\alpha_3 & M_4\,(\xi,\eta)\,\alpha_4 \,. \end{bmatrix}. \tag{10.75}$$

The interpolations M_L have to obey the orthogonality condition $(10.67)_3$ within the element

$$\int_{\Omega_e} \delta\mathbf{P}_e \cdot \bar{\mathbf{F}}_e\, d\Omega = 0\,. \tag{10.76}$$

By assuming constant stresses in Ω_e, condition

$$\int_{\Omega_e} \bar{\mathbf{M}}\, d\Omega = 0 \tag{10.77}$$

is obtained based on (10.71). It yields with (10.73)

$$\int_{\Omega_\square} \mathbf{M}(\boldsymbol{\xi})\, d\square = 0\,. \tag{10.78}$$

The interpolations M_L in (10.75) have to fulfil this condition which is the case for polynomials with uneven exponents. Hence the simplest interpolation with four enhanced or incompatible modes is given by

$$\mathbf{M}(\boldsymbol{\xi})^{2D}\,\boldsymbol{\alpha} = \sum_{L=1}^{4} \mathbf{M}(\boldsymbol{\xi})_L^{2D}\,\alpha_L = \begin{bmatrix} \xi\,\alpha_1 & \eta\,\alpha_2 \\ \xi\,\alpha_3 & \eta\,\alpha_4 \end{bmatrix}. \tag{10.79}$$

The finite element based on this ansatz is called Q1/E4 element, see Simo and Armero (1992). This element is equivalent in the linear case with the incompatible mode element by Taylor et al. (1976).

The corresponding interpolation for the three-dimensional case leads to an ansatz for the enhanced deformation gradient with nine modes

$$\mathbf{M}^{3D}\,\boldsymbol{\alpha} = \sum_{L=1}^{9} \mathbf{M}(\boldsymbol{\xi})_L^{3D}\,\alpha_L = \begin{bmatrix} \xi\,\alpha_1 & \eta\,\alpha_2 & \zeta\,\alpha_3 \\ \xi\,\alpha_4 & \eta\,\alpha_5 & \zeta\,\alpha_6 \\ \xi\,\alpha_7 & \eta\,\alpha_8 & \zeta\,\alpha_9 \end{bmatrix}. \tag{10.80}$$

It is simply the extension of the two-dimensional interpolation and yields the so-called Q1/E9 element. As already shown in Simo et al. (1993b), this ansatz is not sufficient to prevent locking. Thus additional enhanced modes have to be introduced in order to prevent volume locking. The related element has 12 incompatible modes and hence is called Q1/E12 element.

The matrix formulation of the enhanced finite element is based on a description of the deformation gradient in vector form. It will be developed here for the two-dimensional case. A different formulation is provided in Exercise 10.1 in detail.

Basis of the implementation is the mixed form (10.67) with respect to the initial configuration, but form (10.68) could also be employed which is referred to the current configuration. Essential for an efficient implementation is the use of formulations which lead to sparse matrices. As already discussed in the standard formulation of isoparametric elements, see Sects. 4.2.2 and 4.2.4, the formulation (10.68) with respect to the current configurations provides the most efficient variant.

The quantities $\nabla^S \boldsymbol{\eta}$ and $\delta \mathbf{h}^S$ have to be discretized in (10.68). Furthermore, the KIRCHHOFF stresses are computed from $\boldsymbol{\tau} = 2 \mathbf{F} \frac{\partial W}{\partial C} \mathbf{F}^T$ using (10.65). The variation of the deformation dependent part of the deformation gradient follows with (3.32) within the element Ω_e in the current configuration as

$$\nabla \boldsymbol{\eta}_e = \mathrm{Grad}\, \boldsymbol{\eta}_e \, \mathbf{F}_e^{-1} = \left[\sum_{I=1}^{n} \boldsymbol{\eta}_I \otimes \nabla_X N_I(\boldsymbol{\xi}) \right] \mathbf{F}_e^{-1}. \qquad (10.81)$$

Here the enriched deformation gradient has to be introduced for $\mathbf{F}$, see (10.65). The symmetrical part follows as in (4.94). Its matrix form is given by

$$\nabla^S \boldsymbol{\eta}_e = \sum_{I=1}^{n} \begin{bmatrix} N_{I,1} & 0 \\ 0 & N_{I,2} \\ N_{I,2} & N_{I,1} \end{bmatrix} \left\{ \begin{matrix} \eta_1 \\ \eta_2 \end{matrix} \right\}_I = \sum_{I=1}^{n} \mathbf{B}_I \, \boldsymbol{\eta}_I, \qquad (10.82)$$

where the derivatives have to be determined using (10.81). For the enhanced modes, the vector form is given by

$$\delta \bar{\mathbf{h}}_e^S = \sum_{L=1}^{n_{enh}} \begin{bmatrix} M_{11}^L \\ M_{22}^L \\ M_{12}^L + M_{21}^L \end{bmatrix} \delta \alpha_L = \sum_{I=L}^{n_{enh}} \mathbf{G}_L \, \delta \alpha_L. \qquad (10.83)$$

In this relation, the components $M_{11}^L, M_{12}^L, M_{21}^L$ and M_{22}^L have to be computed based on (10.73) and (10.74) from

$$\bar{\mathbf{M}}_L = \begin{bmatrix} M_{11}^L & M_{12}^L \\ M_{21}^L & M_{22}^L \end{bmatrix} = \mathbf{F}_0 \, \frac{j_0}{j} \, J_0 \, \mathbf{M}(\boldsymbol{\xi})_L \, \mathbf{J}_0^{-1} \, \mathbf{F}_e^{-1}. \qquad (10.84)$$

$\mathbf{M}(\boldsymbol{\xi})_L$ denotes the Lth mode, see (10.79). The weak form (10.68) can now be rewritten as

$$\bigcup_{e=1}^{n_e} \left[\sum_{I} \delta \boldsymbol{\eta}_I{}^T \int_{\Omega_e} \mathbf{B}_I{}^T \boldsymbol{\tau}_e \, d\Omega \right] - \delta P_{EXT} = 0$$

$$\sum_{L} \delta \alpha_L \int_{\Omega_e} \mathbf{G}_L{}^T \boldsymbol{\tau}_e \, d\Omega = 0. \qquad (10.85)$$

Note that $(10.68)_3$ is directly fulfilled by construction of the enhanced interpolations, see (10.73).

The solution of this nonlinear algebraic equation system will be obtained by NEWTON'S method. Hence the linearization of (10.85) has to be derived. Analogous to the procedure given in Sect. 4.2.4, the incremental equation system

$$\begin{bmatrix} \boldsymbol{K}_{uu} & \boldsymbol{K}_{u\alpha} \\ \boldsymbol{K}_{\alpha u} & \boldsymbol{K}_{\alpha\alpha} \end{bmatrix} \begin{Bmatrix} \Delta u \\ \Delta \alpha \end{Bmatrix} = - \begin{Bmatrix} \boldsymbol{G}_u \\ \boldsymbol{G}_\alpha \end{Bmatrix} \tag{10.86}$$

can be deduced. In this form, the sub matrices are given by

$$\boldsymbol{K}_{uu} = \bigcup_{e=1}^{n_e} \sum_{I=1}^{n} \sum_{K=1}^{n} \int_{\Omega_e} [\boldsymbol{B}_I^T \, \boldsymbol{D}^{MR} \, \boldsymbol{B}_K + (\overline{\nabla}_x N_I)^T \, \boldsymbol{\tau}_e \, \overline{\nabla}_x N_K] \, d\Omega,$$

$$\boldsymbol{K}_{u\alpha} = \bigcup_{e=1}^{n_e} \sum_{I=1}^{n} \sum_{M=1}^{n_{enh}} \int_{\Omega_e} [\, \boldsymbol{B}_I^T \, \boldsymbol{D}^{MR} \, \boldsymbol{G}_M \tag{10.87}$$

$$+ (\overline{\nabla}_x N_I)^T \, \boldsymbol{\tau}_e \, \boldsymbol{G}_M + (\overline{\nabla}_x N_I|_0)^T \, \boldsymbol{\tau}_e \, \boldsymbol{G}_M] \, d\Omega,$$

$$\boldsymbol{K}_{\alpha\alpha} = \bigcup_{e=1}^{n_e} \sum_{L=1}^{n_{enh}} \sum_{M=1}^{n_{enh}} \int_{\Omega_e} [\, \boldsymbol{G}_L^T \, \boldsymbol{D}^{MR} \, \boldsymbol{G}_M + \bar{\boldsymbol{M}}_L \boldsymbol{\tau} \cdot \bar{\boldsymbol{M}}_M] \, d\Omega.$$

The residuals $\boldsymbol{G}_u$ and $\boldsymbol{G}_\alpha$ follow directly from (10.85). As in the previous equations, the derivatives have to be determined with respect to x via (10.81). The definition of $\boldsymbol{D}^{MR}$ can be found in (4.113). $(\overline{\nabla}_x N_I|_0)$ denotes the evaluation of the gradient at the element midpoint, see also (10.72). For the solution of equation system (10.86), block elimination can be employed. It provides an efficient implementation since $\boldsymbol{K}_{\alpha\alpha}$ can be inverted directly on element level due to the incompatible interpolation functions. This procedure is explicitly shown in Exercise 10.1.

Exercise 10.1: Derive the discretization and resulting matrix formulation for a two-dimensional 4-node element based on the HU–WASHIZU principle. Use for the derivatives of the shape functions and for the interpolation of the enhanced modes a TAYLOR series expansion up to order 2 with respect to the element mid point. The element has to be constructed for finite elastic deformations.

Solution: Within the element Ω_e, the displacements will be approximated by isoparametric shape functions. The use of a TAYLOR series expansion of order 2 for the standard shape functions and enhanced mode interpolations leads to explicit expressions for the gradients. Within the range of small strains, it was shown in Hueck and Wriggers (1995) that this method can be applied to all terms which are associated with the enhanced element.

In case of finite deformations explicit expressions are developed for the standard and the enhanced displacement gradients in (10.65). These gradients are related to the initial configuration $\mathbf{X}$. After that the equations will be transformed to the current configuration $\mathbf{x}$.

The bilinear isoparametric form functions (4.28) are used for the interpolation

$$N_I(\xi, \eta) = \frac{1}{4}(1 + \xi\,\xi_I)(1 + \eta\,\eta_I) = \frac{1}{4}(1 + \xi_I\,\xi + \eta_I\,\eta + \xi_I\eta_I\,\xi\eta). \quad (10.88)$$

ξ_I and η_I are the coordinates of node I in the ξ–η reference configuration of the element. The coordinates within the element are given by

$$
\begin{aligned}
X &= a_0 + a_1\,\xi + a_2\,\xi\eta + a_3\,\eta \\
Y &= b_0 + b_1\,\xi + b_2\,\xi\eta + b_3\,\eta,
\end{aligned} \quad (10.89)
$$

where the constants a_i are defined as follows

$$
\begin{aligned}
a_0 &= \frac{1}{4}\sum_{I=1}^{4} X_I, & a_1 &= \frac{1}{4}\sum_{I=1}^{4} \xi_I\,X_I, \\
a_2 &= \frac{1}{4}\sum_{I=1}^{4} \xi_I\,\eta_I\,X_I, & a_3 &= \frac{1}{4}\sum_{I=1}^{4} \eta_I\,X_I.
\end{aligned}
$$

The constants b_i are computed in an analogous way where X_I is exchanged by Y_I. The deformation gradient within the element follows with (10.88)

$$\operatorname{Grad}\varphi_e = \sum_{I=1}^{4}\begin{bmatrix} N_{I,X}\,x_I & N_{I,Y}\,x_I \\ N_{I,X}\,y_I & N_{I,Y}\,y_I \end{bmatrix}. \quad (10.90)$$

In this relation, x_I and y_I are the coordinates of node I in the current configuration. In Eq. (10.90), the derivatives of the form functions have to be computed with respect to X and Y. A TAYLOR series expansion up to order 1 yields with respect to the element midpoint $\xi = \eta = 0$

$$N_I = N_I|_0 + \left.\frac{\partial N_I}{\partial X}\right|_0 (X - X_0) + \left.\frac{\partial N_I}{\partial Y}\right|_0 (Y - Y_0) + N_{\gamma I}. \quad (10.91)$$

The remaining higher order terms are denoted by $N_{\gamma I}$. From (10.88) $N_I|_0 = 1/4$ is obtained. The evaluation of the chain rule at the element midpoint leads to

$$\left\{ \begin{array}{c} \left.\dfrac{\partial N_I}{\partial X}\right|_0 \\[2mm] \left.\dfrac{\partial N_I}{\partial Y}\right|_0 \end{array} \right\} = \boldsymbol{J}_0^{-1} \left\{ \begin{array}{c} \left.\dfrac{\partial N_I}{\partial \xi}\right|_0 \\[2mm] \left.\dfrac{\partial N_I}{\partial \eta}\right|_0 \end{array} \right\}. \quad (10.92)$$

$\boldsymbol{J}_0$ is the JACOBI matrix $\boldsymbol{J}$ evaluated at the element midpoint. The derivatives of the shape functions can be computed at the element mid point by, as shown in Hueck and Wriggers (1995),

$$\left. \frac{\partial N_I}{\partial X} \right|_0 = \frac{1}{4\,j_0} \left(b_3\, \xi_I - b_1\, \eta_I \right), \tag{10.93}$$

$$\left. \frac{\partial N_I}{\partial Y} \right|_0 = \frac{1}{4\,j_0} \left(-a_3\, \xi_I + a_1\, \eta_I \right). \tag{10.94}$$

The determinant $\mathbf{J}$ is given by $j = j_0 + j_1\,\xi + j_2\,\eta$ with

$$j_0 = a_1\, b_3 - a_3\, b_1, \qquad j_1 = a_1\, b_2 - a_2\, b_1 \qquad \text{and} \qquad j_2 = a_2\, b_3 - a_3\, b_2.$$

This leads to $\det \mathbf{J}_0 = j_0$. The solution of (10.91) yields with (10.93) and (10.94) after some algebraic manipulations the higher order term

$$N_{\gamma I} = \gamma_I\, \xi\eta \qquad \text{with} \quad \gamma_I = \frac{1}{4} \left(\xi_I\eta_I - \frac{j_2}{j_0}\, \xi_I - \frac{j_1}{j_0}\, \eta_I \right), \tag{10.95}$$

where the so-called stabilization- or γ-vector has been introduced, see also Sect. 10.4 and Belytschko et al. (1984).

The interpolation functions for the enhanced gradient $\bar{\mathbf{F}}$ in (10.65) are determined analogous to (10.90). Wilson et al. (1973) have introduced the classical incompatible modes by

$$M_1 = (1 - \xi^2), \qquad M_2 = (1 - \eta^2). \tag{10.96}$$

These represent a discontinuous interpolation between different elements Ω_e. An expansion using the TAYLOR series around the element midpoint yields, for the incompatible modes,

$$M_L = M_L|_0 + \left. \frac{\partial M_L}{\partial X} \right|_0 (X - X_0) + \left. \frac{\partial M_L}{\partial Y} \right|_0 (Y - Y_0) + M_{\gamma L}. \tag{10.97}$$

The constant term is $M_I|_0 = 1$. By the chain rule, it can be shown that all terms of first order are zero in (10.97). The remaining terms of higher order in (10.97) are

$$M_{\gamma 1} = -\xi^2 \qquad \text{and} \quad M_{\gamma 2} = -\eta^2. \tag{10.98}$$

Now the higher order terms in Eqs. (10.95) and (10.96) will be expanded in X and Y by a TAYLOR series of second order with respect to the element midpoint. To simplify notation, the terms are combined in $\mathbf{q}^T = \{ q_1, q_2, q_3 \} = \{ \xi^2, \xi\eta, \eta^2 \}$. TAYLOR series expansion yields

$$\mathbf{q} = \frac{1}{2} \left(\left. \frac{\partial^2 \mathbf{q}}{\partial X^2} \right|_0 \Delta X^2 + 2 \left. \frac{\partial^2 \mathbf{q}}{\partial X \partial Y} \right|_0 \Delta X\, \Delta Y + \left. \frac{\partial^2 \mathbf{q}}{\partial Y^2} \right|_0 \Delta Y^2 \right) + \mathbf{r}_3 \tag{10.99}$$

with $\Delta X = X - X_0$ and $\Delta Y = Y - Y_0$. Constant terms and terms of first order do not appear in this equation since $\mathbf{q}$ only consists of terms of higher order, which appear as remainders in the expansion of N_I (10.91) and M_L (10.97). The term $\mathbf{r}_3$ contains terms of third order and will be neglected in the

following derivations. The computation of the second derivatives in (10.99)
are described in detail in Hueck and Wriggers (1995). They lead to the form

$$N_{\gamma I} = -\frac{1}{\bar{j}_0^2} [b_1 b_3 \Delta X^2 - (a_1 b_3 + a_3 b_1) \Delta X \Delta Y + a_1 a_3 \Delta Y^2] \gamma_I ,$$

$$M_{\gamma 1} = -\frac{1}{\bar{j}_0^2} [b_3^2 \Delta X^2 - 2 a_3 b_3 \Delta X \Delta Y + a_3^2 \Delta Y^2] , \qquad (10.100)$$

$$M_{\gamma 2} = -\frac{1}{\bar{j}_0^2} [b_1^2 \Delta X^2 - 2 a_1 b_1 \Delta X \Delta Y + a_1^2 \Delta Y^2] .$$

The shape functions and the incompatible interpolations can be approximated by these equations and by (10.93) and (10.94). Finally, with Eqs. (10.91) and (10.100), the derivatives of the shape functions with respect to X and Y yield

$$\begin{aligned} N_{I,X} &= N_{I,X}|_0 + N_{I\gamma,X} \\ &= \frac{1}{4\,j_0} (b_3 \xi_I - b_1 \eta_I) - \frac{1}{\bar{j}_0^2} [2 b_1 b_3 \Delta X - (a_1 b_3 + a_3 b_1) \Delta Y] \gamma_I . \end{aligned}$$
$$(10.101)$$

Since Eq. (10.89) leads to $\Delta X = a_1 \xi + a_2 \xi \eta + a_3 \eta$ and $\Delta Y = b_1 \xi + b_2 \xi \eta + b_3 \eta$, explicit expressions can be derived for the derivatives of N_I with respect to X

$$N_{I,X} = \frac{1}{4\,j_0} (b_3 \xi_I - b_1 \eta_I) + \frac{1}{j_0} \left[-b_1 \xi + \frac{1}{j_0} (j_1 b_3 - j_2 b_1) \xi \eta + b_3 \eta \right] \gamma_I .$$
$$(10.102)$$

Analogously the derivatives of N_I with respect to Y follow as

$$N_{I,Y} = \frac{1}{4\,j_0} (a_1 \eta_I - a_3 \xi_I) + \frac{1}{j_0} \left[a_1 \xi + \frac{1}{j_0} (j_2 a_1 - j_1 a_3) \xi \eta - a_3 \eta \right] \gamma_I .$$
$$(10.103)$$

The derivatives of the incompatible modes are obtained using (10.97) and (10.100)

$$M_{1,X} = -\frac{2}{j_0} b_3 \left(\xi + \frac{j_2}{j_0} \xi \eta \right) , \qquad M_{1,Y} = \frac{2}{j_0} a_3 \left(\xi + \frac{j_2}{j_0} \xi \eta \right) ,$$

$$M_{2,X} = \frac{2}{j_0} b_1 \left(\eta + \frac{j_1}{j_0} \xi \eta \right) , \qquad M_{2,Y} = -\frac{2}{j_0} a_1 \left(\eta + \frac{j_1}{j_0} \xi \eta \right) .$$
$$(10.104)$$

It is possible to compute the gradients (10.90) and (10.105) with respect to the reference configuration $\mathbf{X}$ by using expressions (10.102) to (10.104). For the enhanced deformation, gradient $\bar{\mathbf{F}}$ follows

$$\bar{\mathbf{F}}_e = \sum_{L=1}^{2} \boldsymbol{\alpha}_L \bar{\mathbf{G}}_L^T \quad \text{with} \quad \boldsymbol{\alpha}_L = \left\{ \begin{array}{c} \alpha_L \\ \phi_L \end{array} \right\} \quad \text{and} \quad \bar{\mathbf{G}}_L = \left\{ \begin{array}{c} M_{L,X} \\ M_{L,Y} \end{array} \right\} ,$$
$$(10.105)$$

where α_L and ϕ_L are the variables with respect to the coordinate directions related to the enhanced modes.

Remark 10.2: In Eqs. (10.102), (10.103) and (10.104) only the constant term J_0 of the JACOBI determinant appears in the denominator. This expression is proportional to the element area and cannot become zero or negative, even when an element is highly distorted. Hence this formulation is more robust against geometric mesh distortion.

By using (10.68) as a basis for the nonlinear finite element formulation, the gradients have to be transformed to the current configuration. The standard displacement gradient is transformed by $\nabla \mathbf{u} = (\mathrm{Grad}\,\mathbf{u})\,\mathbf{F}^{-1}$ to the spatial displacement gradient

$$\nabla \mathbf{u}_e = \sum_{I=1}^{4} \begin{bmatrix} N_{I,x}\,u_I & N_{I,y}\,u_I \\ N_{I,x}\,v_I & N_{I,y}\,v_I \end{bmatrix} \tag{10.106}$$

with the nodal displacements u_I and v_I. The derivatives of the shape functions with respect to $\mathbf{x}$ follow for the two-dimensional case in explicit form

$$\begin{Bmatrix} N_{I,x} \\ N_{I,y} \end{Bmatrix} = \frac{1}{\det \mathbf{F}_e} \begin{Bmatrix} F_{22}\,N_{I,X} - F_{21}\,N_{I,Y} \\ -F_{12}\,N_{I,X} + F_{11}\,N_{I,Y} \end{Bmatrix}. \tag{10.107}$$

Here F_{ik} are the components of the deformation gradient $\mathbf{F}$, see (10.65).

At the same time the enhanced gradient is transformed to the current configuration. This yields – as for the displacement gradient – $\bar{\mathbf{h}} = \bar{\mathbf{F}}\,\mathbf{F}^{-1}$. Together with (10.105), it follows

$$\bar{\mathbf{h}}_e = \sum_{L=1}^{2} \boldsymbol{\alpha}_L\,\bar{\mathbf{g}}_L^T \quad \text{with} \quad \bar{\mathbf{g}}_L = \mathbf{F}_e^{-T}\,\bar{\mathbf{G}}_L, \tag{10.108}$$

where

$$\bar{\mathbf{g}}_L = \begin{Bmatrix} M_{L,x} \\ M_{L,y} \end{Bmatrix} = \frac{1}{\det \mathbf{F}} \begin{Bmatrix} F_{22}\,M_{L,X} - F_{21}\,M_{L,Y} \\ -F_{12}\,M_{L,X} + F_{11}\,M_{L,Y} \end{Bmatrix} \tag{10.109}$$

is valid. Thus the enhanced gradient is transformed to the current configuration in a similar way as the displacement gradient.

The discretization of the weak form (10.68) requires, for plane strain, the matrices

$$\boldsymbol{\tau} = \begin{Bmatrix} \tau_{11} \\ \tau_{22} \\ \tau_{12} \end{Bmatrix}, \mathbf{b} = \begin{Bmatrix} b_{11} \\ b_{22} \\ b_{12} \end{Bmatrix}, \nabla^S \boldsymbol{\eta} = \begin{Bmatrix} \eta_{,x} \\ \eta_{,y} \\ \eta_{,y} + \eta_{,x} \end{Bmatrix}, \delta\bar{\mathbf{h}}^S = \begin{Bmatrix} \delta h_{11} \\ \delta h_{22} \\ \delta h_{12} + \delta h_{21} \end{Bmatrix}. \tag{10.110}$$

From the constitutive relation (3.120), the KIRCHHOFF stresses

$$\boldsymbol{\tau}_e = \begin{Bmatrix} \tau_{11} \\ \tau_{22} \\ \tau_{12} \end{Bmatrix}_e = \frac{\Lambda}{2}\,[\,J^2 - 1\,] \begin{Bmatrix} 1 \\ 1 \\ 0 \end{Bmatrix} + \mu \left[\begin{Bmatrix} b_{11} \\ b_{22} \\ b_{12} \end{Bmatrix} - \begin{Bmatrix} 1 \\ 1 \\ 0 \end{Bmatrix} \right] \tag{10.111}$$

can be deduced. In this expression, the discrete approximation for the left CAUCHY–GREEN tensor $\mathbf{b}$ is given by

$$\mathbf{b}_e = \left\{ \begin{array}{c} (F_{11})^2 + (F_{12})^2 \\ (F_{22})^2 + (F_{21})^2 \\ F_{11}\,F_{21} + F_{12}\,F_{22} \end{array} \right\}. \tag{10.112}$$

The components of the deformation gradient $\mathbf{F}$ are computed from (10.65) together with (10.90) and (10.105)

$$\left[\begin{array}{cc} F_{11} & F_{12} \\ F_{21} & F_{22} \end{array} \right]_e = \sum_{I=1}^{4} \left[\begin{array}{cc} N_{I,x}\,x_I & N_{I,y}\,x_I \\ N_{I,x}\,y_I & N_{I,y}\,y_I \end{array} \right] + \sum_{L=1}^{2} \left[\begin{array}{cc} M_{L,x}\,\alpha_L & M_{L,y}\,\alpha_L \\ M_{L,x}\,\phi_L & M_{L,y}\,\phi_L \end{array} \right] \tag{10.113}$$

within an element Ω_e. The variation of the symmetric displacement gradient is provided in Ω_e by

$$\nabla^S \boldsymbol{\eta}_e = \sum_{I=1}^{4} \mathbf{B}_I\,\boldsymbol{\eta}_I = \sum_{I=1}^{4} \left[\begin{array}{cc} N_{I,x} & 0 \\ 0 & N_{I,y} \\ N_{I,y} & N_{I,x} \end{array} \right] \left\{ \begin{array}{c} \eta_{x\,I} \\ \eta_{y\,I} \end{array} \right\}. \tag{10.114}$$

This defines the $\mathbf{B}$-matrix, see also (4.94). The derivatives of the shape functions are computed from (10.107) with (10.102) and (10.103) with respect to the current configuration. Analogously, the variation of the enhanced displacement gradient $\bar{\mathbf{h}}$ is given with (10.108) and (10.105) by

$$\delta\bar{\mathbf{h}}_e^S = \sum_{L=1}^{2} \mathbf{G}_L\,\delta\boldsymbol{\alpha}_L = \sum_{L=1}^{2} \left[\begin{array}{cc} M_{L,x} & 0 \\ 0 & M_{L,y} \\ M_{L,y} & M_{L,x} \end{array} \right] \left\{ \begin{array}{c} \delta\alpha_L \\ \delta\phi_L \end{array} \right\}. \tag{10.115}$$

The relations (10.104) and (10.109) have to be applied in (10.115) to compute $M_{L,x}$ and $M_{L,y}$.

The discretization of (10.68) yields with (10.110) to (10.115) the residuals of the enhanced element

$$\left. \begin{array}{l} \bigcup\limits_{e=1}^{n_e} \left\{ \sum\limits_{I=1}^{4} \boldsymbol{\eta}_I^T \displaystyle\int_{\Omega_e} \mathbf{B}_I^T\,\boldsymbol{\tau}_e\,d\Omega \right\} - \delta P_{EXT} = 0 \\[18pt] \sum\limits_{L=1}^{2} \delta\boldsymbol{\alpha}_L^T \displaystyle\int_{\Omega_e} \mathbf{G}_L^T\,\boldsymbol{\tau}_e\,d\Omega = 0 \end{array} \right\} \Rightarrow \begin{array}{l} \mathbf{g}_u(\,\mathbf{u},\,\boldsymbol{\alpha}\,) = \mathbf{0} \\[6pt] \mathbf{g}_\alpha^e(\,\mathbf{u},\,\boldsymbol{\alpha}\,) = \mathbf{0}, \end{array}$$

$$\tag{10.116}$$

where the abbreviations $\mathbf{g}_u = \mathbf{0}$ and $\mathbf{g}_\alpha^e = \mathbf{0}$ were introduced for the first and second equation. The last equation in (10.68) has only to be fulfilled on element level. This follows from the fact that the interpolation functions for the enhanced modes are discontinuous over the element domains. Furthermore, the interpolation of the stress field can be selected in (10.67) and (10.68) such that (10.68)$_3$ is automatically fulfilled, see Simo and Armero (1992).

Remark 10.3: Equation $(10.116)_2$ leads for constant stresses to the condition $\int_{\Omega_e} \bar{\mathbf{G}}_L \, dV = 0$. This has to be considered in order to fulfil the patch test for piecewise constant stress fields. With the enhanced functions in (10.104), this condition is fulfilled exactly when the following approximation is used for the integration

$$\int_B f(x,y)\, dV = \int_{-1}^{1} \int_{-1}^{1} f(\xi,\eta)\, j \, d\xi \, d\eta \approx \int_{-1}^{1} \int_{-1}^{1} f(\xi,\eta)\, j_0 \, d\xi \, d\eta.$$

Due to this the use of j_0 instead of j is important for the mapping onto the reference configuration within this element formulation, see also (10.78).

NEWTON's method is usually applied to solve the nonlinear algebraic equation system (10.116) for the unknown displacements $\boldsymbol{u}$ and the enhanced variables $\boldsymbol{\alpha}$, see Sect. 5.1.1. This iterative scheme requires the linearization of (10.116). As was shown in Sect. 3.5.3, Eq. (10.68) is transformed for this operation to the initial configuration

$$
\begin{aligned}
G_u &= \int_B \operatorname{Grad}\boldsymbol{\eta} \cdot \left(2\,\mathbf{F}\,\frac{\partial W}{\partial \mathbf{C}} \right) dV - \delta P_{EXT} = 0\,, \\
G_\alpha &= \int_B \delta\bar{\mathbf{F}} \cdot \left(2\,\mathbf{F}\,\frac{\partial W}{\partial \mathbf{C}} \right) dV = 0\,.
\end{aligned}
\tag{10.117}
$$

The linearization will be denoted by $\Delta(\bullet)$, as introduced in Sect. 3.5.3.

The linearization of $\mathbf{F}$ yields with (10.65) $\Delta\mathbf{F} = \operatorname{Grad}\Delta\mathbf{u} + \Delta\bar{\mathbf{F}}$. This relation is used to linearize $\mathbf{C} = \mathbf{F}^T\mathbf{F}$

$$\Delta\mathbf{C} = \Delta(\mathbf{F}^T\mathbf{F}) = [\,(\operatorname{Grad}\Delta\mathbf{u})^T + \Delta\bar{\mathbf{F}}^T\,]\,\mathbf{F} + \mathbf{F}^T\,[\operatorname{Grad}\Delta\mathbf{u} + \Delta\bar{\mathbf{F}}\,]\,. \tag{10.118}$$

Use of the linearized kinematical quantities in (10.117) leads to

$$
\begin{aligned}
\Delta G_u &= \int_B \operatorname{Grad}\boldsymbol{\eta} \cdot 2\left[(\operatorname{Grad}\Delta\mathbf{u} + \Delta\bar{\mathbf{F}})\,\frac{\partial W}{\partial \mathbf{C}} + \mathbf{F}\,\frac{\partial^2 W}{\partial \mathbf{C}\,\partial \mathbf{C}}\Delta\mathbf{C} \right] dV = 0, \\
\Delta G_\alpha &= \int_B \delta\bar{\mathbf{F}} \cdot 2\left[(\operatorname{Grad}\Delta\mathbf{u} + \Delta\bar{\mathbf{F}})\,\frac{\partial W}{\partial \mathbf{C}} + \mathbf{F}\,\frac{\partial^2 W}{\partial \mathbf{C}\,\partial \mathbf{C}}\Delta\mathbf{C} \right] dV = 0\,.
\end{aligned}
\tag{10.119}
$$

This result is pushed forward to the current configuration. Employing the relation between the 2nd PIOLA–KIRCHHOFF stress tensor $\mathbf{S}$ and the KIRCHHOFF stress tensor $\boldsymbol{\tau} = \mathbf{F}\,\mathbf{S}\,\mathbf{F}^T$, see (3.84), and using the incremental material tensor in the current configuration $\mathbf{c}$, see (3.245), relations

$$
\begin{aligned}
\Delta g_u &= \int_B \left\{ \nabla^S \boldsymbol{\eta} \cdot \mathbf{c} \left[\nabla^S (\Delta \mathbf{u}) \right] + \nabla^S \boldsymbol{\eta} \, \nabla^S (\Delta \mathbf{u}) \cdot \boldsymbol{\tau} \right\} dV \\
&\quad + \int_B \left\{ \nabla^S \boldsymbol{\eta} \cdot \mathbf{c} \left[\Delta \bar{\mathbf{h}} \right] + \nabla^S \boldsymbol{\eta} \, \Delta \bar{\mathbf{h}} \cdot \boldsymbol{\tau} \right\} dV = 0, \\
\Delta g_\alpha &= \int_B \left\{ \delta \bar{\mathbf{h}}^S \cdot \mathbf{c} \left[\nabla^S (\Delta \mathbf{u}) \right] + \bar{\mathbf{h}}^S \, \nabla^S (\Delta \mathbf{u}) \cdot \boldsymbol{\tau} \right\} dV \\
&\quad + \int_B \left\{ \delta \bar{\mathbf{h}}^S \cdot \mathbf{c} \left[\Delta \bar{\mathbf{h}} \right] + \delta \bar{\mathbf{h}} \, \Delta \bar{\mathbf{h}} \cdot \boldsymbol{\tau} \right\} dV = 0
\end{aligned}
\tag{10.120}
$$

are deduced after some algebraic manipulations.

In case of a plane strain state, the explicit expression for the constitutive tensor (3.120), see also (3.271), is given by

$$
\boldsymbol{D} = \begin{bmatrix} e_1 & e_2 & 0 \\ e_2 & e_1 & 0 \\ 0 & 0 & g \end{bmatrix} \quad \text{with} \quad
\begin{aligned}
e_1 &= \mu + \Lambda \\
e_2 &= \Lambda J^2 \\
g &= \mu - \frac{\Lambda}{2} \left[J^2 - 1 \right].
\end{aligned}
\tag{10.121}
$$

The operators for determining $\nabla^S \boldsymbol{\eta}$ and $\delta \bar{\mathbf{h}}^S$ are stated in discrete form in (10.114) and (10.115). The same operators can also be applied for the determination of $\nabla^S (\Delta \mathbf{u})$ and $\Delta \bar{\mathbf{h}}^S$. With this notation, the following tangent matrices are defined as

$$
\boldsymbol{K}_{uu} = \bigcup_{e=1}^{n_e} \sum_{I=1}^{4} \sum_{J=1}^{4} \int_{\Omega_e} \left[\boldsymbol{B}_I^T \boldsymbol{D} \boldsymbol{B}_J + G_{IJ}^1 \, \mathbf{I}_{2 \times 2} \right] d\Omega
$$

$$
\boldsymbol{K}_{u\alpha} = \bigcup_{e=1}^{n_e} \sum_{I=1}^{4} \sum_{L=1}^{2} \int_{\Omega_e} \left[\boldsymbol{B}_I^T \boldsymbol{D} \boldsymbol{G}_L + G_{IL}^2 \, \mathbf{I}_{2 \times 2} \right] d\Omega
\tag{10.122}
$$

$$
\boldsymbol{K}_{\alpha\alpha} = \bigcup_{e=1}^{n_e} \sum_{L=1}^{2} \sum_{M=1}^{2} \int_{\Omega_e} \left[\boldsymbol{G}_L^T \boldsymbol{D} \boldsymbol{G}_M + G_{LM}^3 \, \mathbf{I}_{2 \times 2} \right] d\Omega
\tag{10.123}
$$

with

$$
\begin{aligned}
G_{IJ}^1 &= <N_{I,x} , N_{I,y}> \begin{bmatrix} \tau_{11} & \tau_{12} \\ \tau_{21} & \tau_{22} \end{bmatrix} \begin{Bmatrix} N_{J,x} \\ N_{J,y} \end{Bmatrix} \\
G_{IL}^2 &= <N_{I,x} , N_{I,y}> \begin{bmatrix} \tau_{11} & \tau_{12} \\ \tau_{21} & \tau_{22} \end{bmatrix} \begin{Bmatrix} M_{L,x} \\ M_{L,y} \end{Bmatrix} \\
G_{LM}^3 &= <M_{L,x} , M_{L,y}> \begin{bmatrix} \tau_{11} & \tau_{12} \\ \tau_{21} & \tau_{22} \end{bmatrix} \begin{Bmatrix} M_{M,x} \\ M_{M,y} \end{Bmatrix}.
\end{aligned}
$$

Since the interpolation functions for the enhanced strains are discontinuous, it is possible to invert the matrix $\boldsymbol{K}_{\alpha\alpha}$ on element level. By writing the equation system for one element as

$$\boldsymbol{K}^e_{uu}\,\Delta\boldsymbol{u}^e + \boldsymbol{K}^e_{u\alpha}\,\Delta\boldsymbol{\alpha}^e = -\boldsymbol{g}^e_u$$
$$\boldsymbol{K}^e_{\alpha u}\,\Delta\boldsymbol{u}^e + \boldsymbol{K}^e_{\alpha\alpha}\,\Delta\boldsymbol{\alpha}^e = -\boldsymbol{g}^e_\alpha \tag{10.124}$$

a block elimination technique, as employed in Simo and Rifai (1990), is efficient in combination with NEWTON'S method to solve (10.124). Within this procedure, the variables $\boldsymbol{\alpha}^e$ are eliminated on element level

$$\Delta\boldsymbol{\alpha}^e = -\boldsymbol{K}^{e\,-1}_{\alpha\alpha}\,(\,\boldsymbol{K}^e_{\alpha u}\Delta\boldsymbol{u}^e + \boldsymbol{g}^e_\alpha\,)\,. \tag{10.125}$$

This leads to the displacement formulation

$$(\,\boldsymbol{K}^e_{uu} - \boldsymbol{K}^e_{u\alpha}\,\boldsymbol{K}^{e\,-1}_{\alpha\alpha}\,\boldsymbol{K}^e_{\alpha u}\,)\Delta\boldsymbol{u}^e = -\boldsymbol{g}^e_u + \boldsymbol{K}^e_{u\alpha}\,\boldsymbol{K}^{e\,-1}_{\alpha\alpha}\,\boldsymbol{g}^e_\alpha \tag{10.126}$$

and hence to the definition of the element residual and tangent matrix for the enhanced element

$$\hat{\boldsymbol{g}}_u = \boldsymbol{g}^e_u - \boldsymbol{K}^e_{u\alpha}\,\boldsymbol{K}^{e\,-1}_{\alpha\alpha}\,\boldsymbol{g}^e_\alpha \quad \text{and} \quad \hat{\boldsymbol{K}}_{uu} = \boldsymbol{K}^e_{uu} - \boldsymbol{K}^e_{u\alpha}\,\boldsymbol{K}^{e\,-1}_{\alpha\alpha}\,\boldsymbol{K}^e_{\alpha u}\,. \tag{10.127}$$

An efficient implementation which avoids the storage of $\boldsymbol{K}^{e\,-1}_{\alpha\alpha}\,\boldsymbol{K}^e_{\alpha u}$ and $\boldsymbol{K}^{e\,-1}_{\alpha\alpha}\,\boldsymbol{g}_\alpha$ on element level can be found in Simo et al. (1993b).

10.5.3 Combination of Enhanced Formulation and Hour-Glass Stabilization

A possibility in which the advantage of the stabilized *hour-glass* elements of Belytschko et al. (1984) (high efficiency) is combined with the advantage of the *enhanced strain* elements (*locking* free behaviour) was developed in Reese et al. (1998) and has been refined since then in Reese (2003) and Reese (2005). Staring point of this development are the relations (10.62) and (10.63). These lead after assembly to the nonlinear equation

$$\boldsymbol{R}_0 + \boldsymbol{K}_{stab}\,\boldsymbol{v} = \boldsymbol{P} \tag{10.128}$$

and its linearization

$$(\,\boldsymbol{K}_{T0} + \boldsymbol{K}_{stab}\,)\,\Delta\boldsymbol{v} = \boldsymbol{P} - \boldsymbol{R}_0 - \boldsymbol{K}_{stab}\,\boldsymbol{v}\,. \tag{10.129}$$

In order to derive the explicit form of $\boldsymbol{K}_{stab}$ for this formulation, the deformation gradient $\boldsymbol{F}$ and its enhanced part $\bar{\boldsymbol{F}}$ in (10.65) is written as

$$
\begin{aligned}
\boldsymbol{F}_e &= \boldsymbol{B}\,\boldsymbol{x}_e & \text{Grad}\,\boldsymbol{\eta}_e &= \boldsymbol{B}\,\boldsymbol{\eta}_e \quad \text{and} \\
\bar{\boldsymbol{F}}_e &= \boldsymbol{G}\,\boldsymbol{\alpha}_e & \delta\bar{\boldsymbol{F}}_e &= \boldsymbol{G}\,\delta\boldsymbol{\alpha}_e\,,
\end{aligned} \tag{10.130}
$$

where vector notation is introduced.

In the two-dimensional case, the explicit form

$$\text{Grad}\,\boldsymbol{\eta}_e = \left\{\begin{array}{c} \eta_{1,1} \\ \eta_{1,2} \\ \eta_{2,1} \\ \eta_{2,2} \end{array}\right\} = \sum_{I=1}^{4} \boldsymbol{B}_I\,\boldsymbol{\eta}_I = \sum_{I=1}^{4} \left[\begin{array}{cc} N_{I,X} & 0 \\ N_{I,Y} & 0 \\ 0 & N_{I,X} \\ 0 & N_{I,Y} \end{array}\right] \left\{\begin{array}{c} \eta_{X\,I} \\ \eta_{Y\,I} \end{array}\right\} \quad (10.131)$$

is obtained for the variation of $\boldsymbol{F}_e$ using the ansatz (10.88). This relation can be written in a compact way as

$$\text{Grad}\,\boldsymbol{\eta}_e = [\,\boldsymbol{B}_1\,,\boldsymbol{B}_2\,,\boldsymbol{B}_3\,,\boldsymbol{B}_4\,] \left\{\begin{array}{c} \eta_{X\,1} \\ \eta_{Y\,1} \\ \ldots \\ \eta_{X\,4} \\ \eta_{Y\,4} \end{array}\right\} = \boldsymbol{B}\,\boldsymbol{\eta}_e\,. \quad (10.132)$$

Starting from the TAYLOR series expansion of the shape functions, see (10.91), the $\boldsymbol{B}$-matrix can be split into linear and *hour-glass* parts. This leads after Reese and Wriggers (2000) to

$$\boldsymbol{B} = \boldsymbol{j}\,(\,\boldsymbol{B}_{lin}\,\boldsymbol{M}_{lin} + \boldsymbol{B}_{hg}\,\boldsymbol{M}_{hg}\,)\,. \quad (10.133)$$

For two-dimensions, the matrices in (10.133) have the form

$$\boldsymbol{j} = \left[\begin{array}{cccc} \frac{\partial\xi}{\partial X} & \frac{\partial\eta}{\partial X} & 0 & 0 \\ 0 & 0 & \frac{\partial\xi}{\partial Y} & \frac{\partial\eta}{\partial Y} \\ \frac{\partial\xi}{\partial Y} & \frac{\partial\eta}{\partial Y} & 0 & 0 \\ 0 & 0 & \frac{\partial\xi}{\partial X} & \frac{\partial\eta}{\partial X} \end{array}\right], \quad (10.134)$$

$$\boldsymbol{B}_{lin} = \left[\begin{array}{cccccc} 0 & 1 & 0 & 0 & 0 & 0 \\ 0 & 0 & 1 & 0 & 0 & 0 \\ 0 & 0 & 0 & 0 & 1 & 0 \\ 0 & 0 & 0 & 0 & 0 & 1 \end{array}\right], \quad \boldsymbol{B}_{hg} = \left[\begin{array}{cc} \eta & 0 \\ \xi & 0 \\ 0 & \eta \\ 0 & \xi \end{array}\right], \quad (10.135)$$

and

$$\boldsymbol{M}_{lin}^T = \left[\begin{array}{cccccc} \boldsymbol{N}_0 & \boldsymbol{N}_{,X\,0} & \boldsymbol{N}_{,Y\,0} & \boldsymbol{O} & \boldsymbol{O} & \boldsymbol{O} \\ \boldsymbol{O} & \boldsymbol{O} & \boldsymbol{O} & \boldsymbol{N}_0 & \boldsymbol{N}_{,X\,0} & \boldsymbol{N}_{,Y\,0} \end{array}\right],$$

$$\boldsymbol{M}_{hg}^T = \left[\begin{array}{cc} \boldsymbol{\gamma} & \boldsymbol{O} \\ \boldsymbol{O} & \boldsymbol{\gamma} \end{array}\right]. \quad (10.136)$$

The components $N_I|_0$, $\frac{\partial N_I}{\partial X}\big|_0$ and $\frac{\partial N_I}{\partial Y}\big|_0$, computed in (10.91), (10.93) and (10.94), are contained in vectors $\boldsymbol{N}_0$, $\boldsymbol{N}_{,X\,0}$ and $\boldsymbol{N}_{,Y\,0}$. In vector $\boldsymbol{\gamma}$, the components of the $\boldsymbol{\gamma}$ vector, see (10.95), are assembled.

The enhanced strain parts will now be specified based on the application of the ansatz (10.96). The variation of the enhanced strain gradient in (10.65) follows analogously to (10.131)

$$\delta \bar{\mathbf{F}}_e = \left\{ \begin{array}{c} \delta \bar{F}_{11} \\ \delta \bar{F}_{12} \\ \delta \bar{F}_{21} \\ \delta \bar{F}_{22} \end{array} \right\} = \sum_{L=1}^{2} \boldsymbol{G}_L \, \delta \boldsymbol{\varphi}_L = \sum_{I=1}^{4} \begin{bmatrix} M_{L,X} & 0 \\ M_{L,Y} & 0 \\ 0 & M_{L,X} \\ 0 & M_{L,Y} \end{bmatrix} \left\{ \begin{array}{c} \delta \varphi_L \\ \delta \phi_L \end{array} \right\} .$$

(10.137)

This can be written in compact form

$$\delta \bar{\mathbf{F}}_e = [\, \boldsymbol{G}_1 \,,\, \boldsymbol{G}_2 \,] \left\{ \begin{array}{c} \delta \varphi_1 \\ \delta \phi_1 \\ \delta \varphi_2 \\ d \phi_2 \end{array} \right\} = \boldsymbol{G} \, \delta \boldsymbol{\alpha}_e \,.$$

(10.138)

Since the approximation for the enhanced strain term does not contain constant and linear parts, the $\boldsymbol{G}$ matrix can be expressed by a TAYLOR series expansion

$$\boldsymbol{G} = j \, \hat{\boldsymbol{G}} \quad \text{with} \quad \hat{\boldsymbol{G}} = \begin{bmatrix} \xi & 0 & 0 & 0 \\ 0 & \eta & 0 & 0 \\ 0 & 0 & \xi & 0 \\ 0 & 0 & 0 & \eta \end{bmatrix} .$$

(10.139)

The variational equation (10.67) follows from the HU–WASHIZU principle. Its discretization uses the above defined matrices

$$\bigcup_{e=1}^{n_e} \boldsymbol{\eta}_e^T \int_{\Omega_e} [j \, (\boldsymbol{B}_{lin} \, \boldsymbol{M}_{lin} + \boldsymbol{B}_{hg} \, \boldsymbol{M}_{hg})]^T \, \boldsymbol{P}_e \, d\Omega - \delta P_{EXT} = 0$$

$$\delta \boldsymbol{\alpha}_e^T \int_{\Omega_e} (j \, \hat{\boldsymbol{G}})^T \, \boldsymbol{P}_e \, d\Omega = 0. \quad (10.140)$$

For the fulfillment of the last equation in (10.67), the ansatz $\boldsymbol{G} = \frac{j_0}{j} \, j_0 \, \hat{\boldsymbol{G}}$ has to be selected for distorted element geometry. This leads with the incremental constitutive matrix $\boldsymbol{A} = \frac{\partial^2 W}{\partial F \, \partial F}$, for the linearization of the second equation of (10.140), to

$$\delta \boldsymbol{\alpha}_e^T \left[\int_{\Omega_e} \hat{\boldsymbol{G}}^T \hat{\boldsymbol{A}} \boldsymbol{B}_{lin} \, d\Omega \, \boldsymbol{M}_{lin} \, \varDelta \, \mathbf{u}_e + \int_{\Omega_e} \hat{\boldsymbol{G}}^T \hat{\boldsymbol{A}} \boldsymbol{B}_{hg} \, d\Omega \, \boldsymbol{M}_{hg} \, \varDelta \, \mathbf{u}_e \right.$$

$$\left. + \int_{\Omega_e} \hat{\boldsymbol{G}}^T \hat{\boldsymbol{A}} \hat{\boldsymbol{G}} \, d\Omega \, \varDelta \, \boldsymbol{\alpha}_e \right] = -\delta \boldsymbol{\alpha}_e^T \int_{\Omega_e} \hat{\boldsymbol{G}}^T \, \hat{\boldsymbol{P}}_e \, d\Omega \,.$$

(10.141)

Here the abbreviation $\hat{\boldsymbol{A}} = \boldsymbol{j}^T \boldsymbol{A} \boldsymbol{j}$ was introduced together with the abbreviation for the 1st PIOLA–KIRCHHOFF stress tensor $\hat{\boldsymbol{P}}_e = \boldsymbol{j}^T \boldsymbol{P}_e$. The form (10.141) can now be simplified by assuming that $\hat{\boldsymbol{A}}$, $\hat{\boldsymbol{P}}$ and $j \, dV$ are constant within an element Ω_e. These assumptions are approximations for arbitrary element geometries. The assumption of constant stress states and rhomboidal element forms enables an exact evaluation of (10.141). Hence the solution converges for arbitrary meshes when a sufficient number of finite elements is used. In that case, the stress fields in the elements are nearly constant. Due to this simplification, the first integral in (10.141) disappears, since $\boldsymbol{B}_{lin}$ is constant and $\hat{\boldsymbol{G}}$ is linear in ξ and η. With the definitions

$$\boldsymbol{K}_{\alpha u} = \int_{\Omega_e} \hat{\boldsymbol{G}}^T \hat{\boldsymbol{A}}_0 \boldsymbol{B}_{hg} \, d\Omega_0 \quad \text{and} \quad \boldsymbol{K}_{\alpha \alpha} = \int_{\Omega_e} \hat{\boldsymbol{G}}^T \hat{\boldsymbol{A}} \hat{\boldsymbol{G}} \, d\Omega_0, \qquad (10.142)$$

the matrix relation

$$\Delta \boldsymbol{\alpha} = -\boldsymbol{K}_{\alpha \alpha}^{-1} \boldsymbol{K}_{\alpha u} \, \boldsymbol{M}_{hg} \, \Delta \boldsymbol{v} \qquad (10.143)$$

follows for the incremental enhanced variables $\Delta \boldsymbol{\alpha}$ on element level. In (10.142), the index $()_0$ denotes evaluation of a quantity at element midpoint (this is equivalent to a 1-point-integration). The increments of the gradients follow from (10.133) and (10.139) with (10.143)

$$\Delta \boldsymbol{F} + \Delta \bar{\boldsymbol{F}} = j \left(\boldsymbol{B}_{lin} \, \boldsymbol{M}_{lin} + \boldsymbol{B}_{stab} \, \boldsymbol{M}_{hg} \right) \Delta \boldsymbol{v}, \qquad (10.144)$$

where the new $\boldsymbol{B}$-matrix, $\boldsymbol{B}_{stab}$, is defined by

$$\boldsymbol{B}_{stab} = \boldsymbol{B}_{hg} - \hat{\boldsymbol{G}} \, \boldsymbol{K}_{\alpha \alpha}^{-1} \boldsymbol{K}_{\alpha u}. \qquad (10.145)$$

This relation can be inserted in the linearized form of $(10.140)_1$. It leads to the tangent matrix, by noting that $\int_{\Omega_e} \boldsymbol{B}_{lin}^T \hat{\boldsymbol{A}}_0 \boldsymbol{B}_{stab} \, d\Omega_0$ and $\int_{\Omega_e} \hat{\boldsymbol{G}}^T \hat{\boldsymbol{A}}_0 \boldsymbol{B}_{stab} \, d\Omega_0$ are zero,

$$\boldsymbol{K}_T = \boldsymbol{M}_{lin}^T \, \boldsymbol{K}_0 \, \boldsymbol{M}_{lin} + \boldsymbol{M}_{hg}^T \, \boldsymbol{K}_{stab} \, \boldsymbol{M}_{hg} \qquad (10.146)$$

with

$$\boldsymbol{K}_0 = \int_{\Omega_e} \boldsymbol{B}_{lin}^T \hat{\boldsymbol{A}}_0 \boldsymbol{B}_{lin} \, d\Omega_0 \quad \text{and} \quad \boldsymbol{K}_{stab} = \int_{\Omega_e} \boldsymbol{B}_{stab}^T \hat{\boldsymbol{A}}_0 \boldsymbol{B}_{stab} \, d\Omega_0. \qquad (10.147)$$

Since $\boldsymbol{B}_{lin}$ is constant, $\boldsymbol{K}_0$ is integrated exactly by a 1-point-GAUSS integration. $\boldsymbol{K}_{stab}$ can be integrated analytically. Thus an efficient computation of the tangent matrix $\boldsymbol{K}_T$ is possible. A further advantage is that the constitutive tensor has to be evaluated only at the element midpoint. Since the element volume is also computed using the element midpoint, the element is insensitive against mesh distortions.

Since the first matrix in (10.146) is equivalent to matrix $\boldsymbol{K}_{T0}$ (10.129), the second matrix in (10.146) can be interpreted as stabilization matrix which is here computed by using the enhanced formulation. With this all matrices in (10.129) are known.

Since a constant stabilization matrix is used within the concept of stabilization, see Sect. 10.4, the stabilization matrix in (10.146) also has to be kept constant during the NEWTON iterations within a load step. For large

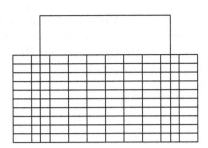

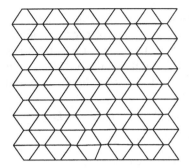

Fig. 10.2a Homogeneous deformation   **Fig. 10.2b** *Hourglass* eigenvector

load steps, however, the stabilization matrix in (10.146) is not optimal in the sense of the enhanced strain method, since e.g. the incremental constitutive tensor may change. Then a post-iteration is required to update the matrix according to the computed deformation and stress state. Such procedure can be viewed as an UZAWA algorithm known from optimization, see e.g. Luenberger (1984). In a recent paper, Reese (2005) presented a new formulation which basically overcomes this problem.

10.5.4 Instabilities Related to Enhanced Elements

Enhanced strain elements were developed over the last 15 years for finite strain problems which include bending dominated response or incompressible behaviour. The advantage of this element formulation is its relatively simple implementation in which complex constitutive equations for finite elastic and inelastic strains can be included. An additional advantage is a good coarse mesh accuracy for different applications. However, there is one disadvantage which is, in the classical formulation of enhanced strain elements, related to instability, see Wriggers and Reese (1994) and Wriggers and Reese (1996). In these papers, it was shown that a block under homogeneous pressure state will lead to a non-physical instability which is related to the enhanced element formulation, see Fig. 10.2a for the problem definition. This instability occurs at a finite deformation state, independently on the constitutive equation.[5]

By applying loading and boundary conditions as depicted in Fig. 10.2a, the loss of uniqueness of the solution occurs and the tangent matrix becomes singular. The eigenmode related to the zero eigenvalue of the tangent stiffness has the form shown in Fig. 10.2b which is well known as hour-glass mode.

---

[5] Enhanced elements derived for geometrical linear elastic problems are known to be stable for all strain and stress states. However, a even a geometrically linear formulation of the enhanced strain element with an inelastic constitutive equation can depict such instable behaviour.

Thus the instability has nothing to do with the stability problems as discussed in Chap. 7. It can be shown, see below, that the enhanced strain element is rank deficient for this deformation state. This loss of rank can of course also be observed under more complex loading states where pressure occurs locally or for different types of material behaviour.

Interesting enough this phenomena will even occur for a single element. Hence an analytical investigation of this element behaviour is feasable in which all matrices can be presented in closed form.

Here one element will be investigated, assuming hyperelastic constitutive behaviour. A compressible Neo–HOOKE material is selected where the strain energy, after (3.116) and (3.118), is given in terms of the principal strains λ_i^2 of the right CAUCHY–GREEN tensor, see (3.15), as

$$W = \frac{1}{2}\mu\,[\,(\lambda_1^2 + \lambda_2^2 + \lambda_3^2) - 3\,] - \mu \ln J + \frac{\Lambda}{4}\,(\,J^2 - 1 - 2\ln J\,). \quad (10.148)$$

$J = \lambda_1\,\lambda_2\,\lambda_3$ denotes the JACOBI determinant of the deformation gradient.

A homogeneous plain strain state is considered in a rectangular plate, see Fig. 10.2a. Thus it is possible to perform the analysis with respect to the principal strains since the principal directions coincide in this case with the cartesian coordinates. From the strain energy, the 1st PIOLA–KIRCHHOFF stresses $\mathbf{P} = \sum_{i=1}^{3} P_i\,\mathbf{n}_i \otimes \mathbf{N}_i$, can be computed, see e.g. Ogden (1984), as

$$P_i = \frac{\partial W}{\partial \lambda_i} = \frac{1}{\lambda_i}\left[\mu\,(\lambda_i^2 - 1) + \frac{\Lambda}{2}\,(J^2 - 1)\right]. \quad (10.149)$$

Furthermore, the coefficients of the incremental constitutive tensor related to a formulation using $\mathbf{P}$ are needed. After some algebra and analogous to the derivation in (3.265), the incremental constitutive tensor follows with (10.149) from $\mathbb{A}_{iJkL} = \partial P_{iJ}\,/\,\partial F_{kL}$. Hence the non-zero elements of this tensor are given with respect to the principal strains with $(i,j = 1,2$ and $i \neq j)$ as

$$\mathbb{A}_{iiii} = \mu\left(1 + \frac{1}{\lambda_i^2}\right) + \frac{\Lambda}{2\lambda_i^2}\,(J^2 + 1)$$

$$\mathbb{A}_{iijj} = \Lambda\,J \qquad\qquad (10.150)$$

$$\mathbb{A}_{ijij} = \mu$$

$$\mathbb{A}_{ijji} = \frac{1}{\lambda_i\,\lambda_j}\left[\mu + \frac{\Lambda}{2}\,(1 - \lambda_i^2\,\lambda_j^2)\right],$$

where it is not necessary to distinguish between derivations with respect to the initial- and current configuration.

As already mentioned, it is sufficient to show the rank deficiency for a single finite element. Here an isoparametric bilinear element will be considered which is chosen such that the local ξ,η-axis coincide with the global X,Y-axis in Fig. 4.2. Hence initial- and reference configuration are the same, see

Fig. 10.3a. For such a discretization all vectors and matrices can be presented explicitly in closed form. The bilinear shape functions are given in this special case in terms of the cartesian coordinates

$$N_I(X,Y) = \frac{1}{4}\,(1 + X\,X_I)(1 + Y\,Y_I)\,. \tag{10.151}$$

This leads directly to the derivatives needed for the computation of the deformation gradient after (10.131)

$$N_{I,X} = \frac{X_I}{4}\,(1 + Y_I\,Y)\quad\text{and}\quad N_{I,Y} = \frac{Y_I}{4}\,(1 + X_I\,X)\,. \tag{10.152}$$

Hence the $\boldsymbol{B}_I$ matrix in (10.131) is linear in X and Y.

The deformation gradient is enhanced in Eq. (10.65) by $\bar{\mathbf{F}}$. Using the interpolation of the incompatible modes $M_L(X,Y)$, see Taylor et al. (1976) and (10.96), it follows for the derivatives in the enhanced gradient (10.137)

$$M_{1,X} = -X\,,\quad M_{1,Y} = 0\,,\quad M_{2,X} = 0\quad\text{and}\quad M_{2,Y} = -Y\,. \tag{10.153}$$

Now the first two equations of the mixed formulation (10.67) can be formulated with this interpolation. Since a plain stress state is assumed ($P_{33} = 0$), the four stress components

$$\boldsymbol{P}^T = \{\,P_{11}\,,P_{12}\,,P_{21}\,,P_{22}\,\} \tag{10.154}$$

have to be determined. These components of $\boldsymbol{P}$ can be obtained from (10.149) using the deformation gradient from (10.65). This leads to the weak form of the single element Ω_e in Fig. 10.3a

$$\sum_{I=1}^{4} \boldsymbol{\eta}_I^T \int_{\Omega_e} \boldsymbol{B}_I^T\,\boldsymbol{P}\,d\Omega - \delta P_{EXT} = 0$$

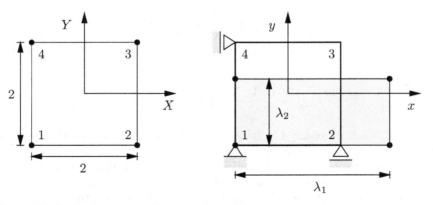

Fig. 10.3 Finite element and homogeneous deformation

$$\sum_{L=1}^{2} \delta\varphi_L^T \int_{\Omega_e} G_L^T \, P \, d\Omega = 0 \qquad (10.155)$$

which is referred to the initial configuration. In case of a homogeneous stress field, the stress P_{22} is constant and $P_{11} = P_{12} = P_{21} = 0$. This yields

$$P_{22} \int_{-1}^{1} \int_{-1}^{1} \sum_{I=1}^{4} N_{I,Y} \, \eta_{YI} \, t \, dX \, dY - \delta P_{EXT} = P_{22} \sum_{I=1}^{4} (-Y_I) \, \eta_{YI} \, t - \delta P_{EXT}$$

$$= \boldsymbol{\eta}^T \, \boldsymbol{G}_u = \boldsymbol{\eta}^T (\, \boldsymbol{R}_u - \boldsymbol{P}_{EXT} \,) = 0 \qquad (10.156)$$

with $\boldsymbol{R}_u^T = \{ \, 0, -P_{22}, 0, -P_{22}, 0, P_{22}, 0, P_{22} \, \} t$.

In the same way, the explicit form of Eq. $(10.155)_2$ is given for $P_{22} = $ const.

$$P_{22} \int_{-1}^{1} \int_{-1}^{1} \sum_{L=1}^{2} M_{L,Y} \, \delta\phi_L \, t \, dX \, dY = \delta\boldsymbol{\alpha}^T \, \boldsymbol{G}_\alpha = 0 \,. \qquad (10.157)$$

In this special situation $\boldsymbol{G}_\alpha^T = \{ 0,0,0,0 \}$ follows from (10.153).

The solution of the nonlinear equations (10.156) and (10.157) follows usually by employing NEWTON's method which needs the tangent matrix of the weak form. It is obtained from the general form, see e.g. (10.124), and can be stated explicitly for the square element Ω_e

$$\boldsymbol{K}_{uu} = \begin{bmatrix} \boldsymbol{K}_{uu}^1 & \boldsymbol{K}_{uu}^2 \\ \boldsymbol{K}_{uu}^{2\,T} & \boldsymbol{K}_{uu}^1 \end{bmatrix} \quad \text{with}$$

$$\boldsymbol{K}_{uu}^1 = \begin{bmatrix} 2a+2e & c+d & -2a+e & c-d \\ c+d & 2b+2e & -c+d & b-2e \\ -2a+e & -c+d & 2a+2e & -c-d \\ c-d & b-2e & -c-d & 2b+2e \end{bmatrix} t$$

$$\boldsymbol{K}_{uu}^2 = \begin{bmatrix} -a-e & -c-d & a-2e & -c+d \\ -c-d & -b-e & c-d & -2b+e \\ a-2e & c-d & -a-e & c+d \\ -c+d & -2b+e & c+d & -b-e \end{bmatrix} t$$

$$\boldsymbol{K}_{\alpha u} = \begin{bmatrix} 0 & \frac{4}{3}c & 0 & -\frac{4}{3}c & 0 & \frac{4}{3}c & 0 & -\frac{4}{3}c \\ \frac{4}{3}d & 0 & -\frac{4}{3}d & 0 & \frac{4}{3}d & 0 & -\frac{4}{3}d & 0 \\ 0 & \frac{4}{3}d & 0 & -\frac{4}{3}d & 0 & \frac{4}{3}d & 0 & -\frac{4}{3}d \\ \frac{4}{3}c & 0 & -\frac{4}{3}c & 0 & \frac{4}{3}c & 0 & -\frac{4}{3}c & 0 \end{bmatrix} t = \boldsymbol{K}_{u\alpha}^T$$

$$\boldsymbol{K}_{\alpha\alpha} = \begin{bmatrix} 8a & 0 & 0 & 0 \\ 0 & 8e & 0 & 0 \\ 0 & 0 & 8e & 0 \\ 0 & 0 & 0 & 8b \end{bmatrix} t \,. \qquad (10.158)$$

The coefficients in these matrices are given by

$$a = \frac{\mathbb{A}_{1111}}{6}; \qquad b = \frac{\mathbb{A}_{2222}}{6}; \qquad c = \frac{\mathbb{A}_{1122}}{4} = \frac{\mathbb{A}_{2211}}{4}$$

$$d = \frac{\mathbb{A}_{1221}}{4} = \frac{\mathbb{A}_{2112}}{4}; \qquad\qquad e = \frac{\mathbb{A}_{1212}}{6} = \frac{\mathbb{A}_{2121}}{6}.$$

Using block elimination within the solution of the linear equation system (10.124), the enhanced variables $\boldsymbol{\alpha}$ can be eliminated. With $\boldsymbol{K} = \boldsymbol{K}_{uu} - \boldsymbol{K}_{u\alpha}\,\boldsymbol{K}_{\alpha\alpha}^{-1}\,\boldsymbol{K}_{u\alpha}^T$, an equation system for the unknown displacements can be written as

$$\boldsymbol{K} = \boldsymbol{K}_{uu} - \begin{bmatrix} f & 0 & -f & 0 & f & 0 & -f & 0 \\ 0 & g & 0 & -g & 0 & g & 0 & -g \\ -f & 0 & f & 0 & -f & 0 & f & 0 \\ 0 & -g & 0 & g & 0 & -g & 0 & g \\ f & 0 & -f & 0 & f & 0 & -f & 0 \\ 0 & g & 0 & -g & 0 & g & 0 & -g \\ -f & 0 & f & 0 & -f & 0 & f & 0 \\ 0 & -g & 0 & g & 0 & -g & 0 & g \end{bmatrix} t \qquad (10.159)$$

where

$$f = \frac{2}{9}\left(\frac{d^2}{e} + \frac{c^2}{b}\right) \qquad g = \frac{2}{9}\left(\frac{d^2}{e} + \frac{c^2}{a}\right).$$

This relation constitutes the explicit structure of the element matrix for the homogeneous stress field with $P_{22} = \text{const.}$ at finite deformations. Specification of boundary conditions which are related to the homogeneous deformation, see Fig. 10. 3b, yields a further reduced matrix system.

For the computation of the eigenvector which is associated with the rank deficiency of the enhanced strain element, it is sufficient to consider only the nodal displacements (u_2, u_3). The vertical displacements $(v_3 = v_4)$ follow from condition $P_{11} = 0$. Hence they are known values within the analysis. These considerations yield the nodal displacement vector for the element depicted in Fig. 10.3b: $\mathbf{v} = \{0, 0, u_2, 0, u_3, v_3, 0, v_4\}$. Additionally, in case of a homogeneous stress state, it can be concluded: $u_2 = u_3$. However, both unknowns u_2 and u_3 have to be kept within the analysis; otherwise the *hour glass* form of the eigenvector cannot be detected.

The unknown increments of the enhanced variables $\Delta\boldsymbol{\alpha}$ follow from (10.125). Since $\boldsymbol{G}_{\alpha} = \boldsymbol{0}$ can be deduced from (10.157), it follows from (10.125) and the special structure of $\boldsymbol{K}_{u\,\alpha}$, see (10.158), that $\boldsymbol{\alpha}$ is generally zero for a homogeneous stress state.

The reduced form of (10.159) results from the specification of the boundary conditions

$$\boldsymbol{K} = \begin{bmatrix} 2a + 2e - f & a - 2e + f \\ a - 2e + f & 2a + 2e - f \end{bmatrix} t = \begin{bmatrix} A - f & B + f \\ B + f & A - f \end{bmatrix} t. \qquad (10.160)$$

A rank deficiency of K is present once the eigenvalue of the matrix are less or equal zero. The eigenvalues can be computed from $K - \omega I$ and are determined from

$$\frac{1}{t^2} \det (K - \omega I) = \omega^2 + 2\omega (f - A) + K, \quad \text{with } K = A^2 - B^2 - 2f(A + B).$$
(10.161)

This yields

$$\omega_{1,2} = A - f \pm \sqrt{(A - f)^2 - K} \quad \Longrightarrow \quad \left\{ \begin{array}{ll} \omega_1 = & A + B \\ \omega_2 = & A - B - 2f. \end{array} \right.$$
(10.162)

The coefficients A, B and f depend upon the coefficients of the constitutive tensor (10.150) and also on the principal stretches λ_1 and λ_2. Since the normal stress P_{11} is equal to zero, which is also true for the principal stress P_1, it is possible to determine the stretch λ_2 as a function of λ_1

$$P_1 = 0 = \frac{1}{\lambda_1} \left[\mu (\lambda_1^2 - 1) + \frac{\Lambda}{2} (J^2 - 1) \right] \longrightarrow \lambda_2 = \frac{1}{\lambda_1} \sqrt{1 - \frac{2\mu}{\Lambda} (\lambda_1^2 - 1)}.$$
(10.163)

Now the eigenvalues ω_1 and ω_2 of the tangent matrix K_T can be written as a function depending on λ_1. Since

$$A + B = \frac{\mathbb{A}_{1111}}{2} = \mu + \frac{\Lambda}{2\lambda_1^2},$$
(10.164)

the eigenvalue ω_1 is for $\mu > 0$ and $\Lambda \geq 0$ always positive. Thus the hour glass instability can only be observed by looking at the second eigenvalue

$$\hat{\omega}_2(\lambda_1) = A - B - 2f = \frac{1}{6}\mathbb{A}_{1111} + \frac{2}{3}\mathbb{A}_{1212} - \frac{1}{6}\left(\frac{\mathbb{A}_{1221}^2}{\mathbb{A}_{1212}} + \frac{\mathbb{A}_{1122}^2}{\mathbb{A}_{2222}}\right) < 0.$$
(10.165)

The function $\omega_2 = \hat{\omega}_2(\lambda_1)$ is shown in Fig. 10.4 for a value of the LÁME constant $\Lambda = 100.000$ and the shear modulus $\mu = 20$. As can be seen in Fig. 10.4, a negative eigenvalue ω_2 occurs for a stretch $\lambda_1 > 1.6344$. From (10.163), it follows that $\lambda_2 < 0.6116$.

The eigenvector associated with $\omega_2 = 0$ can be computed from $(K_T - \omega_2 1) \phi_2 = 0$. With (10.158), (10.161) and (10.162), the eigenvector

$$\phi_2^T = \{\phi^{uT}, \phi^{\alpha T}\} = \{1, -1, 0, \alpha_\alpha, 0, \beta_\alpha\}$$
(10.166)

is obtained. In this result, the first two components are the displacements ϕ^u in X-direction. The last four components belong to the enhanced modes ϕ^α with

$$\alpha_\alpha = \frac{d}{3e} = \frac{1}{2}\frac{\mathbb{A}_{1221}}{\mathbb{A}_{1212}} = \frac{1}{2J}\left[1 + \frac{\Lambda}{2\mu}(1 - J^2)\right]$$

$$\beta_\alpha = \frac{c}{3b} = \frac{1}{2}\frac{\mathbb{A}_{1122}}{\mathbb{A}_{2222}} = \frac{1}{2}\frac{\lambda_2^2 J}{\frac{\mu}{\Lambda}(\lambda_2^2 + 1) + \frac{1}{2}(J^2 + 1)}.$$
(10.167)

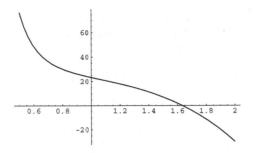

Fig. 10.4 Eigenvalue as function of the stretch λ_1

For the above selected values of Λ and μ, the eigenvectors follow which are, depicted in Fig. 10.5.

Remark 10.4:

1. In the linear elastic case, the stretches are $\lambda_1 \approx 1$, $\lambda_2 \approx 1$. Then it follows from (10.165)

$$\hat{\omega}_2(1) = \frac{1}{6}\left(\Lambda - \frac{\Lambda^2}{\Lambda + 2\mu}\right) + \frac{5}{6}\mu\,.$$

   The eigenvalue is for $\mu > 0$ always positive. Thus the hour-glassing described above does not occur.

2. The eigenvectors of the pure Q1-displacement element can be determined in the same way. In that case, f in (10.165) is equal to zero which yields

$$\omega_{1,2} = A \pm B \longrightarrow \begin{cases} \omega_1 = \frac{\mu}{2}(1 + \frac{1}{\lambda_1^2}) + \frac{\Lambda}{4\lambda_1^2}(J^2 + 1) > 0 \\ \omega_2 = \frac{\mu}{6}(5 + \frac{1}{\lambda_1^2}) + \frac{\Lambda}{12\lambda_1^2}(J^2 + 1) > 0\,. \end{cases}$$

   Also in this case the hour-glass instability does not occur for parameters of the LÁME constants ($\mu > 0$, $\Lambda \geq 0$) which make physically sense.

3. It can be shown that the hour-glass instability does not depend on the material model. In Reese (1994) and Glaser and Armero (1997), the same effects were

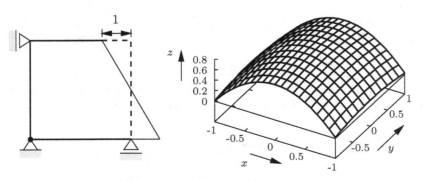

Fig. 10.5a X-component $\boldsymbol{\phi}^u$ **Fig. 10.5b** Y-component $\boldsymbol{\phi}^\alpha$

observed for OGDEN materials. Furthermore, rank deficiency of the enhanced strain element was observed in de Souza Neto et al. (1995) for elasto-plastic deformations.

4. For the class of enhanced strain interpolations discussed in this section, hourglass instabilities are only observed for pressure states since for $0 < \lambda_1 \leq 1$ no zero eigenvalue occurs, see Fig. 10.4.

The hour-glass modes discussed above can also be found when using standard enhanced strain elements for inhomogeneous stress states. The rank deficiency occurs only for elements which are situated in areas where compressive stresses occur.

10.5.5 Stabilization of the Enhanced Strain Formulation

Once the phenomenon was detected, different research groups started to work on methods to overcome instable behaviour of the enhanced strain elements. Within this research work different methods were developed. One method is related to classical hour-glass stabilization, as discussed in Sect. 10.4. Another technique is related to the choice of a different interpolation of the enhanced strains. A third method uses different strain energies within the enhanced formulation. These methods are discussed below.

Hour-Glass Stabilization. The *hour-glass* stabilization is performed as well for the displacements as for the enhanced modes. For the displacements, the stabilization can be obtained using γ-vectors as defined in (10.59). The two-dimensional form is presented in (10.95) explicitly. When additionally the eigenvectors, related to the enhanced modes, are stabilized then in the two-dimensional case the stabilization vectors

$$
\begin{aligned}
\bar{\gamma}_1{}^T &= \{\gamma_1, 0, \dots, \gamma_4, 0, 0, \alpha_\alpha, 0, \beta_\alpha\}, \\
\bar{\gamma}_2{}^T &= \{0, \gamma_1, \dots, 0, \gamma_4, \alpha_\alpha, 0, \beta_\alpha, 0\}
\end{aligned}
\tag{10.168}
$$

are obtained. Here α_α and β_α are defined by (10.167). The last four terms define the stabilization of the enhanced modes.

With these stabilization vectors, the incremental equation system for the unknowns $\boldsymbol{v}^T = \{\boldsymbol{u}^T, \boldsymbol{\alpha}^T\}$ can be written as

$$
\left(\boldsymbol{K}_T + \sum_{s=1}^{2} c_s \, \bar{\boldsymbol{\gamma}}_s \, \bar{\boldsymbol{\gamma}}_s^T \right) \Delta \boldsymbol{v} = -\boldsymbol{G} - \sum_{s=1}^{2} c_s \, \bar{\boldsymbol{\gamma}}_s (\bar{\boldsymbol{\gamma}}_s^T \, \boldsymbol{v}),
\tag{10.169}
$$

see also (10.124) and (10.63). This equation system is solved as (10.126) by block-elimination.

This stabilization is only used when negative eigenvalues are found within an element. This requires for general quadrilaterals a generalized computation of the eigenvalues, see e.g. Glaser and Armero (1997). Here the problem is that the components of the eigenvector belonging to the enhanced modes depend upon the deformations, see (10.167). A simplified version for the determination of the constants α_α and β_α follows from the computation of the constants for $\lambda_i \longrightarrow 1$ and $J \longrightarrow 1$

$$\alpha_\alpha = \frac{1}{2} \qquad \beta_\alpha = \frac{1}{2}\frac{\Lambda}{2\mu + \Lambda}. \qquad (10.170)$$

This procedure was implemented in a two-dimensional Q1E4 element. For the basic formulation, see Simo and Armero (1992) or Wriggers and Hueck (1996) and Exercise 10.1. The resulting enhanced elements are rank deficient in compression states.

In order to investigate the influence of the described stabilization, a block under compression is considered under plane strain conditions. Its initial configuration is depicted in Fig. 10.6a for a finite element mesh with 16 × 16 elements. At the upper side, a constant vertical displacement is applied such that a constant stress state occurs. The constitutive parameters were selected as $\Lambda = 100.000$ and $\mu = 20$. The first physical eigenvector is shown in Fig. 10.6b. Convergence of the solution is obtained for a discretization with 64 × 64 elements. For this mesh the critical stretch, belonging to the physical eigenvector, is $\lambda_2 = 0.575$. Several computations were performed in order to investigate the dependency of the stabilization parameter on the solution. This also included a convergence study regarding the necessary mesh refinement. The computation of the system depicted in Fig. 10.6a yields a stretch λ_2 which belongs to the first physical eigenvector. This value is provided for the parameter c_s depending on the mesh refinement in Table 10.2. For comparison, the stretch which belongs to the first singularity of the non-stabilized enhanced element is documented in the first row of Table 10.2. The stretch belonging to the nonphysical hour-glass mode, see also Fig. 10.2b, is $\lambda_2 = 0.695$. This result is independent on mesh refinement, since the rank deficiency is a local phenomenon, see above. For the discretization with one element, the solution from Fig. 10.5a was used. It is different since in this special case different boundary conditions were employed.

It is clear from the values reported in Table 10.2 that the stabilization procedure avoids the hour-glass instability of the enhanced element. Furthermore, the solution only depends slightly upon the stabilization parameter c_s. The formulation converges to the stretch $\lambda_2 = 0.575$. For comparison reasons, the solution of the Q1-displacement element is reported too.

Table 10.2 Stretch λ_2 belonging to singularity

FEM	1x1	8x8	16x16	32x32	64x64
$c_s = 0$	0.612	0.695	0.695	0.695	0.695
$c_s = 10\,\mu$	0.260	0.475	0.555	0.575	0.580
$c_s = 100\,\mu$	0.245	0.470	0.550	0.570	0.575
$c_s = 1000\,\mu$	0.245	0.470	0.550	0.570	0.575
$c_s = 10000\,\mu$	0.245	0.470	0.550	0.570	0.575
Q1-Element	—	0.040	0.085	0.165	0.370

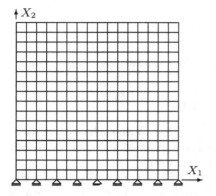

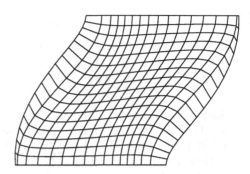

Fig. 10.6a FEM discretization   **Fig. 10.6b** 1st physical eigenvector

However it shows the locking of the displacement element which results from the quasi-incompressible material behaviour.

10.5.6 Special Interpolation of the Enhanced Modes

The rank deficiency occurs as well for distorted as for undistorted element geometries within the enhanced strain formulation. Hence a finite element formulation has to be developed which does not degenerate to a Q1E4 element.[6] In some cases, an interpolation of the enhanced modes can be constructed such that negative eigenvalues are avoided in (10.165). The associated formulation was presented in Korelc and Wriggers (1996b) and Glaser and Armero (1997). Starting from the two-dimensional formulation (10.79), the interpolation of the incompatible modes can be written in more general form as

$$\widehat{\mathbf{M}}(\boldsymbol{\xi})^{2D}\,\boldsymbol{\alpha} = \sum_{L=1}^{4} \mathbf{M}(\boldsymbol{\xi})^{2D}_{L}\,\alpha_L = \begin{bmatrix} \xi\,\alpha_1 & M_2(\xi,\eta)\,\alpha_2 \\ M_3(\xi,\eta)\,\alpha_3 & \eta\,\alpha_4 \end{bmatrix}. \qquad (10.171)$$

The interpolation on the main diagonal of $\boldsymbol{M}^{2D}$ cannot be changed in order to avoid volume locking. Korelc and Wriggers (1996a) developed orthogonality conditions for the ansatz polynomials M_{12} and M_{21} which resulted from the eigenvalue analysis (10.161). These were designed to avoid negative eigenvalues and hence rank deficiency. The conditions are

$$\int_{\Omega_e} M_2(\xi,\eta)\,M_3(\xi,\eta)\,d\Omega \;=\; 0\,,$$

$$\int_{\Omega_e} M_2(\xi,\eta)\,d\Omega \;=\; 0\,, \qquad (10.172)$$

---

[6] Note that higher order integration does not solve the problem.

$$\int_{\Omega_e} M_3(\xi, \eta)\, d\Omega \;=\; 0\,.$$

The application of these conditions yield in the simplest case

$$\widehat{M}(\xi)^{2D}\, \boldsymbol{\alpha} = \sum_{L=1}^{4} M(\xi)_L^{2D}\, \alpha_L = \begin{bmatrix} \xi\,\alpha_1 & \xi\,\alpha_2 \\ \eta\,\alpha_3 & \eta\,\alpha_4 \end{bmatrix} = [M(\xi)^{2D}]^T\,, \qquad (10.173)$$

which is the transpose of the interpolation in (10.79). With such interpolation, called CG4 or Q1/E4T, no instabilities occur in compression states, see Korelc and Wriggers (1996a) and Glaser and Armero (1997).[7]

However, this formulation (CG4 or Q1/E4T) is not totally free of singularities which can occur in the case of large elasto-plastic deformations in tension states and hence lead to rank deficiency of these enhanced strain formulations. This is also true for the three-dimensional formulation (CG9), which is based on the transposed of the interpolation matrix (10.80), see Korelc and Wriggers (1996b).

Additionally, several other approaches were proposed to stabilize the enhanced element formulation when applied to the numerical simulation of the finite deformation problems. Some of these methods are discussed below.

1. A possibility to avoid rank deficiency of the Q1/E4 element is provided by a change in the continuum formulation. Crisfield et al. (1995) have used the right stretch tensor $\mathbf{U}$ in the HU–WASHIZU functional instead of the deformation gradient $\mathbf{F} = \mathbf{R}\,\mathbf{U}$. This however is not sufficient to prevent instabilities. Hence the authors have additionally evaluated the rotation tensor $\mathbf{R}$ only at the element mid point as in a co-rotational formulation. This element does not hour-glass in compression states. However, the formulation is quite complex since the rotation and the stretch tensor have to be determined and all constitutive equations must be provided for BIOT stresses, see also Exercise 3.10. Furthermore, it seems that this formulation tends to lock in some applications.

2. de Souza Neto et al. (1996) developed an element which is based on an interpolation of the strains using a constant deformation gradient. This element depicts no rank deficiency but has several drawbacks. First, the formulation results in a non-symmetric tangent matrix - even for elastic materials - and second it locks in bending situations.

3. Another stabilization technique was developed by Glaser and Armero (1997) based on the Q1/E4T element. In this formulation, the authors add to the functional (10.64), after elimination of the stresses a stabilization term which acts on the volumetric part of the deformation

$$\Pi_\alpha(\boldsymbol{\varphi}, \mathbf{F}) = \Pi(\boldsymbol{\varphi}, \mathbf{F}) + \int_B \frac{\alpha}{2}\,[\det \mathbf{F} - 1\,]^2\, dV\,. \qquad (10.174)$$

---

[7] The CG4 and Q1/E4 interpolations cannot be distinguished in the linear theory, see Korelc and Wriggers (1996b).

This approach avoids hour glassing of the Q1/E4T or CG4 formulation under tension states if a small value for α / μ is selected, see Glaser and Armero (1997). A scheme, how to determine α depending on a problem at hand, is however not provided by the authors.

4. Bischoff et al. (1999b) employ least square methods using stabilization concepts known from work in the area of numerical flow simulations. With such techniques, the authors circumvent hour - glassing of the enhanced elements at finite deformation states. The stabilization is obtained via a deformation dependent function; however in the cited paper all results are valid only for rectangular elements.

5. Reese and Wriggers (2000) use the stability analysis discussed in the last section to develop a technique which automatically changes the element formulation such that hour glassing does not occur. In this approach, it is necessary to do the stability analysis for elements with arbitrary distorted geometries. Once an eigenvalue in (10.165) is equal zero or negative, the element formulation is changed such that instability is circumvented. Different cases have to be distinguished, for details see Reese and Wriggers (2000). These techniques have been successfully employed for three-dimensional simulations of finite elasto-plastic problems, see Reese (2003) and Reese (2005).

10.5.7 Special One Point Integration and Enhanced Stabilization

The enhanced variational methods provide a high flexibility for the generation of different finite elements. This will be shown by the following formulation in which an element will be derived which can be applied successfully to solid problems of finite elasticity and is based on a split of the element deformation into a homogeneous and inhomogeneous part, as introduced in Nadler and Rubin (2003) for the COSSERAT point element. It does not depict, as well as the COSSERAT point element, any nonphysical instabilities and does not rely on any analytical solutions. The difference is that simply the inhomogeneous part of the deformation is enhanced, as it is responsible for the locking behavior. Within this formulation, the deformation gradient $\mathbf{F}$, see (3.14), is additively split into its homogeneous and inhomogeneous part

$$\mathbf{F} = \bar{\mathbf{F}} + \widehat{\mathbf{F}}, \tag{10.175}$$

with

$$\bar{\mathbf{F}} = \frac{1}{V} \int_\Omega \mathbf{F} \, dV \tag{10.176}$$

being the volume average of the displacement gradient and V the element volume in the initial configuration. The strain energy density function is split accordingly, leading to

$$W(\mathbf{F}) = W_H(\bar{\mathbf{F}}) + W_I(\widehat{\mathbf{F}}). \tag{10.177}$$

For the homogeneous part of the deformation, a compressible Neo–Hooke material introduced for the strain energy function W_H, see (3.116) with $g(J) = \frac{A}{2}\left(\bar{J} - 1\right)^2$. Substituting the overall deformation measures by their homogeneous parts yields for the 1st PIOLA–KIRCHHOFF stresses

$$\bar{\mathbf{P}} = \frac{\partial W_H}{\partial \bar{\mathbf{F}}} = \Lambda \bar{J}\left(\bar{J} - 1\right)\bar{\mathbf{F}}^{-T} + \mu\left(\bar{\mathbf{F}} - \bar{\mathbf{F}}^{-T}\right), \tag{10.178}$$

where

$$\bar{J} = \det\left(\bar{\mathbf{F}}\right) \text{ and } \bar{\mathbf{C}} = \bar{\mathbf{F}}^T \bar{\mathbf{F}}$$

are the volume averaged values of the deformation gradient, the JACOBIAN and the right CAUCHY–GREEN tensor, respectively.

For the inhomogeneous part of the element deformation, the strain energy density function is defined by a linear elastic model model, see (3.121), since the inhomogeneous deformation part consists mainly of bending and torsion deformations, see Nadler and Rubin (2003), which can be described well by this model.

$$W_I(\widehat{\mathbf{H}}) = \frac{1}{2}\widehat{\mathbf{H}} \cdot \mathbf{C}_0\left[\widehat{\mathbf{H}}\right] \tag{10.179}$$

with a constant elasticity tensor $\mathbf{C}_0$, see (3.272). The 1st PIOLA–KIRCHHOFF stress tensor is then given by

$$\widehat{\mathbf{P}} = \Lambda\mathrm{tr}(\widehat{\mathbf{H}})\mathbf{1} + \mu\left(\widehat{\mathbf{H}} + \widehat{\mathbf{H}}^T\right). \tag{10.180}$$

Now the inhomogeneous part of the displacement gradient is enhanced such that

$$\widehat{\mathbf{H}} = \tilde{\mathbf{H}} + \hat{\mathbf{H}} \tag{10.181}$$

where

$$\tilde{\mathbf{H}} = \mathbf{H}\left(\varphi\right) - \bar{\mathbf{H}} \tag{10.182}$$

is the displacement gradient following from the deformation and $\hat{\mathbf{H}}$ is the enhanced displacement gradient.

By using the above definitions, the HU–WASHIZU functional (10.64) can be rewritten as

$$\Pi\left(\varphi, \widehat{\mathbf{H}}, \mathbf{P}\right) = \int_\Omega \left[W_H(\widehat{\mathbf{H}}) + W_S(\widehat{\mathbf{H}}) - \widehat{\mathbf{P}} \cdot \hat{\mathbf{H}}\right] dV - P_{ext} = 0. \tag{10.183}$$

The variation of Eq. (10.183) w.r.t. its independent variables φ, $\widehat{\mathbf{H}}$ and $\mathbf{P}$ leads to

$$\int_\Omega \delta\bar{\mathbf{H}} \cdot \frac{\partial W_H}{\partial \bar{\mathbf{F}}}\, dV + \int_\Omega \delta\tilde{\mathbf{H}} \cdot \frac{\partial W_S}{\partial \widehat{\mathbf{H}}}\, dV - \delta P_{ext} = 0,$$

$$\int_\Omega \delta\hat{\mathbf{H}} \cdot \left(\frac{\partial W_S}{\partial \widehat{\mathbf{H}}} - \widehat{\mathbf{P}}\right) dV = 0,$$

$$\int_\Omega \delta\widehat{\mathbf{P}} \cdot \hat{\mathbf{H}}\, dV = 0. \tag{10.184}$$

Equation (10.184)$_1$ yields the standard weak form of the equilibrium and Eq. (10.184)$_2$ the constitutive equation. In a weak sense, Eq. (10.184)$_3$ provides an orthogonality condition between the stress tensor and the displacement gradient. Hence the finite element interpolations for the enhanced displacement gradient have to be chosen such that the orthogonality conditions

$$\int_\Omega \delta \hat{\mathbf{H}} \cdot \hat{\mathbf{P}} \, dV = 0, \qquad \int_\Omega \delta \hat{\mathbf{P}} \cdot \hat{\mathbf{H}} \, dV = 0 \qquad (10.185)$$

are fulfilled. Then, Eqs. (10.184) become

$$\int_\Omega \delta \bar{\mathbf{H}} \cdot \frac{\partial W}{\partial \bar{\mathbf{F}}} \, dV + \int_\Omega \delta \tilde{\mathbf{H}} \cdot \frac{\partial W}{\partial \widehat{\mathbf{H}}} \, dV - \delta P_{ext} = 0,$$

$$\int_\Omega \delta \hat{\mathbf{H}} \cdot \frac{\partial W}{\partial \widehat{\mathbf{H}}} \, dV = 0. \qquad (10.186)$$

These equations are the basis for the subsequent development of the enhanced finite element formulation.

Finite Element Discretization. A standard finite element discretization is employed within a single element Ω_e where the position of a material point in the current configuration φ is approximated by trilinear isoparametric shape functions

$$\varphi^h = \sum_{I=1}^n N_I \varphi_I = \sum_{I=1}^n N_I \left(\mathbf{X}_I + \mathbf{u}_I \right), \qquad (10.187)$$

where $\mathbf{X}_I$ are the positions of the nodes of Ω_e with respect to the initial configuration and $\mathbf{u}_I$ are the nodal displacements. The formulation is presented here for an eight-node brick element as shown in Fig. 4.8. For the interpolation functions N_I standard trilinear shape functions, defined in (4.40), are used.

With the transformation, see also (4.44),

$$\frac{\partial N_I}{\partial \mathbf{X}} = \mathbf{J}^{-T} \frac{\partial N_I}{\partial \xi}, \qquad (10.188)$$

where $\mathbf{J} = \frac{\partial \mathbf{X}}{\partial \xi}$ is the standard JACOBIAN of the isoparametric map and (ξ, η, ζ) are the coordinates of the point ξ in the reference configuration, the displacement gradient can be written as

$$
\mathbf{H}^h =
\begin{bmatrix}
H_{11}^h \\
H_{22}^h \\
H_{33}^h \\
H_{12}^h \\
H_{21}^h \\
H_{23}^h \\
H_{32}^h \\
H_{13}^h \\
H_{31}^h
\end{bmatrix}
= \sum_{I=1}^{8} \boldsymbol{B}_I \, \mathbf{u}_I \quad \text{with} \quad
\boldsymbol{B}_I =
\begin{bmatrix}
N_{I,X} & 0 & 0 \\
0 & N_{I,Y} & 0 \\
0 & 0 & N_{I,Z} \\
N_{I,Y} & 0 & 0 \\
0 & N_{I,X} & 0 \\
0 & N_{I,Z} & 0 \\
0 & 0 & N_{I,Y} \\
N_{I,Z} & 0 & 0 \\
0 & 0 & N_{I,X}
\end{bmatrix} .
$$

$$(10.189)$$

The discrete form of the homogeneous part of the displacement gradient is obtained by inserting Eq. (10.189) into Eq. (10.176), leading to

$$
\bar{\mathbf{H}}^h = \frac{1}{\Omega_e} \int_{\Omega_e} \mathbf{H}^h \, d\Omega = \sum_{I=1}^{8} \frac{1}{\Omega_e} \int_{\Omega_e} \boldsymbol{B}_I \, d\Omega \, \mathbf{u}_I = \sum_{I=1}^{8} \bar{\boldsymbol{B}}_I \, \mathbf{u}_I . \tag{10.190}
$$

Note that a numerical integration over the element volume can be avoided in this equation by using the ansatz functions introduced by Belytschko et al. (1984) which allow an analytical integration.

With Eqs. (10.189) and (10.190), the discrete form of the inhomogeneous part of the displacement gradient $\tilde{\mathbf{H}}$ is written as

$$
\tilde{\mathbf{H}}^h = \sum_{I=1}^{8} \left(\boldsymbol{B}_I - \bar{\boldsymbol{B}}_I \right) \mathbf{u}_I = \sum_{I=1}^{8} \tilde{\boldsymbol{B}}_I \, \mathbf{u}_I . \tag{10.191}
$$

For the enhanced displacement gradient $\tilde{\mathbf{H}}$, the ansatz functions have to be chosen such that they fulfil the orthogonality condition given in Equations (10.185). Here, three quadratic functions are used to interpolate the enhanced modes, as introduced in Wilson et al. (1973)

$$
M_1 = \left(1 - \xi^2 \right) \quad M_2 = \left(1 - \eta^2 \right) \quad M_3 = \left(1 - \zeta^2 \right) . \tag{10.192}
$$

Then, the enhanced displacement gradient can be discretized on the element level as

$$
\hat{\mathbf{H}}^h =
\begin{bmatrix}
\hat{H}_{11}^h \\
\hat{H}_{22}^h \\
\hat{H}_{33}^h \\
\hat{H}_{12}^h \\
\hat{H}_{21}^h \\
\hat{H}_{23}^h \\
\hat{H}_{32}^h \\
\hat{H}_{13}^h \\
\hat{H}_{31}^h
\end{bmatrix}
= \sum_{L=1}^{3} \mathbf{G}_L \boldsymbol{\alpha}_L \quad \text{with} \; \mathbf{G}_L =
\begin{bmatrix}
M_{L,X} & 0 & 0 \\
0 & M_{L,Y} & 0 \\
0 & 0 & M_{L,Z} \\
M_{L,Y} & 0 & 0 \\
0 & M_{L,X} & 0 \\
0 & M_{L,Z} & 0 \\
0 & 0 & M_{L,Y} \\
M_{L,Z} & 0 & 0 \\
0 & 0 & M_{L,X}
\end{bmatrix}
$$

$$(10.193)$$

and $\boldsymbol{\alpha}_L$ are the enhanced variables,

$$\boldsymbol{\alpha}_L^T = [\alpha_{1L}, \alpha_{2L}, \alpha_{3L}]. \tag{10.194}$$

Linearization and Solution Procedure. With the help of Eqs. (10.190), (10.191) and (10.193), the discrete form of the variational equations (10.186) is given by

$$\bigcup_{e=1}^{n_e} \left(\sum_{I=1}^{8} \delta \boldsymbol{u}_I^T \bar{\boldsymbol{B}}_I^T \bar{\boldsymbol{P}} \, \Omega_e + \sum_{I=1}^{8} \delta \boldsymbol{u}_I^T \int_{\Omega_e} \tilde{\boldsymbol{B}}_I^T \widehat{\boldsymbol{P}} \, d\Omega \right) - \delta P_{\text{EXT}} = 0$$

$$\sum_{K=1}^{3} \delta \boldsymbol{\alpha}_K^T \int_{\Omega_e} \boldsymbol{G}_K^T \, \mathbb{C}_0 \left[\tilde{\boldsymbol{H}}^h \right] d\Omega = 0 \tag{10.195}$$

where Eq. $(10.195)_2$ is defined on the element level. This leads to the nodal residual vectors within an element Ω_e

$$\boldsymbol{R}_I^u = \bar{\boldsymbol{B}}_I^T \bar{\boldsymbol{P}}^h \, \Omega_e + \int_{\Omega_e} \tilde{\boldsymbol{B}}_I^T \widehat{\boldsymbol{P}}^h \, d\Omega - \boldsymbol{P}_I^{EXT},$$

$$\boldsymbol{R}_L^\alpha = \int_{\Omega_e} \tilde{\boldsymbol{B}}_I^T \mathbb{C}_0 \tilde{\boldsymbol{H}}^h \, d\Omega, \tag{10.196}$$

where $\boldsymbol{P}_I^{EXT}$ is the nodal vector related to the external loads. The linearization of equations (10.195) yields on element level

$$\boldsymbol{K}_{IJ}^{uu} = \bar{\boldsymbol{B}}_I^T \mathbb{D} \bar{\boldsymbol{B}}_J \, \Omega_e + \int_{\Omega_e} \tilde{\boldsymbol{B}}_I^T \mathbb{C}_0 \tilde{\boldsymbol{B}}_J \, d\Omega$$

$$\boldsymbol{K}_{IL}^{u\alpha} = \int_{\Omega_e} \tilde{\boldsymbol{B}}_I^T \mathbb{C}_0 \boldsymbol{G}_L \, d\Omega$$

$$\boldsymbol{K}_{KJ}^{\alpha u} = \int_{\Omega_e} \boldsymbol{G}_K^T \mathbb{C}_0 \tilde{\boldsymbol{B}}_J \, d\Omega$$

$$\boldsymbol{K}_{KL}^{\alpha\alpha} = \int_{\Omega_e} \boldsymbol{G}_K \mathbb{C}_0 \boldsymbol{G}_L \, d\Omega \ . \tag{10.197}$$

With Eqs. (10.195) and (10.197), the system of linear equations which has to be solved in every NEWTON iteration can be constructed by standard assembly, see Sect. 4.2. Here, as discussed already in Sect. 10.5.2, a block elimination of the variables $\boldsymbol{\alpha}$ can be obtained based on the linear system on element level, for details see Exercise 10.1.

This element is called Q1/EI9 due to the fact that standard tri-linear ansatz functions are used to interpolate the volume averaged and the enhanced part and that additionally nine enhanced modes are applied to describe the small strain elastic enhanced stabilization part.

10.6 Examples

The performance of the different elements is shown by means of examples suitable to point out important properties of the different elements such as high coarse mesh accuracy, low mesh distortion sensitivity and locking free response for bending and incompressibility dominated problems. Furthermore, it can be shown that some of the elements do not hour-glass for arbitrary problem classes and loading.

The element formulation which are compared in this section are standard isoparametric as well as special elements for good bending performance and for incompressible problems. The following elements were selected:

− two standard elements Q1 and Q2 which use tri-linear and tri-quadratic interpolations, respectively, see Sect. 4.2,
− the mixed Q1/P0 element as proposed by Simo et al. (1985a) for finite deformations, see also Sect. 10.2.1,
− the classical enhanced element QM1/E12, developed in Simo et al. (1993b), and
− the Q1/EI9 element described in the last section.

All elements use a hyperelastic material model, see (3.116) with $g(J) = \frac{A}{2}\left(\bar{J} - 1\right)^2$.

10.6.1 Patch Test

The patch test proposed by MacNeal and Harder (1985) is used for a displacement patch test. The the finite element mesh is shown in Fig. 10.7(a). Boundary conditions are set such that a rotation around the x_3-axis is possible, but no other rigid body motion. A displacement in the x_2-direction is applied at point P, resulting in a rotation around the x_3-axis. All elements fulfil this patch test; hence the computed stresses are zero. For the traction controlled patch test, the same mesh is used. The nodes at $x_1 = 0$, $x_2 = 0$ and $x_3 = 0$ are fixed in the x_1-direction, x_2-direction and x_3-direction, respectively. A surface load is applied in the x_1-direction, see Fig. 10.7(b). This configuration should lead to a uniform stress σ_{11}, while all other stresses should be zero. The force patch test is fulfilled by all elements except for the QM1/E12.

10.6.2 Beam with Distorted Mesh

A cantilever beam of length l, width $2w$ and height h is loaded with an equally distributed shear force $F = 12\,\mathrm{N}$ at its free end, as shown in Fig. 10.8. The boundary conditions are such that the clamped end is fixed in the

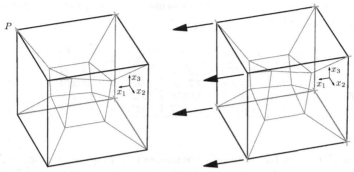

Fig. 10.7 (a) displacement and (b) force patch test

x_1-direction. Additionally, the node at $x_1 = x_2 = x_3 = 0$ is fixed in the x_2-direction to avoid rigid body motion. All nodes at $x_3 = 0$ are fixed in the x_3-direction. Due to the loading, the rectangular cross section and the boundary conditions, symmetry conditions can be enforced. In order to circumvent a stress singularity at the clamped end, the shear load is applied there in the opposite direction instead of fixing these points in the x_2-direction. The geometry and the material data of the beam as well as the load applied and the boundary conditions are provided in Fig. 10.8.

The convergence of the deflection v_P in x_2-direction of point P depicted in Fig. 10.8 is investigated for the Q1/EI9 element as well as the Q1, Q2 and QM1/E12 element. Four different meshes are used, with $16 \times 4 \times 2$, $32 \times 8 \times 4$, $64 \times 16 \times 8$ and $128 \times 32 \times 16$ elements, respectively.

In Table 10.3, the displacement v_P in x_2-direction of point P is depicted for all elements, as a function of the number of degrees of freedom. As expected, the Q1 element locks. Both enhanced strain elements are softer than the Q2 element, where the QM1/E12 is closer to the Q2 element than the Q1/EI9 element. This shows the good coarse mesh accuracy of the enhanced elements. As expected, all elements converge to the same solution. For better visualization, only the results for the elements which do not lock are shown in Fig. 10.9 where the displacement v_P is plotted for the Q1/EI9, the Q2 and

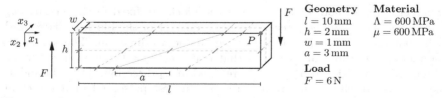

Fig. 10.8 Beam: system, load and material data

Geometry $l = 10\,\text{mm}$, $h = 2\,\text{mm}$, $w = 1\,\text{mm}$, $a = 3\,\text{mm}$ **Material** $\Lambda = 600\,\text{MPa}$, $\mu = 600\,\text{MPa}$ **Load** $F = 6\,\text{N}$

Table 10.3 Beam with distorted mesh: Displacement v_P [mm] for the Q1/EI9, Q1, Q2 and QM1/E12 element

Degrees of freedom	Q1/EI9	Q1	Q2	QM1/E12
664	1.0379	0.5778	1.0128	1.0299
4112	1.0314	0.7840	1.0257	1.0279
28576	1.0283	0.9358	1.0270	1.0273
28576	1.0275	1.0007	1.0271	1.0272

the QM1/E12 with respect to the number of elements. The pure displacement element Q2 converges from below, as the mathematical theory predicts. Both mixed elements, QM1/E12 and Q1/EI9, converge from above.

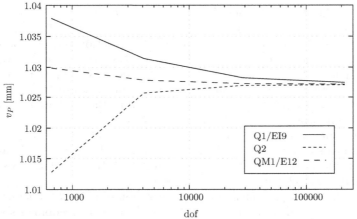

Fig. 10.9 Beam with distorted mesh: Displacement v_P for the Q1/EI9, Q2 and QM1/E12 element

10.6.3 Nearly Incompressible Block

A nearly incompressible block of length l, width w and height h is loaded by an equally distributed surface load q at its top centre, as shown in Fig. 10.10. Furthermore, all nodes on the top of the block are fixed in the x_1- and x_2-directions. For symmetry reasons, only a quarter of the block is discretized. The bottom face of the block is fixed in the x_3-direction. The symmetry boundary conditions are set such that nodes at $x_1 = 0.5\,w$ are fixed in x_1-direction and nodes at $x_2 = 0.5\,l$ are fixed in x_2-direction. These boundary conditions are chosen according to a similar test presented in Reese et al. (2000).

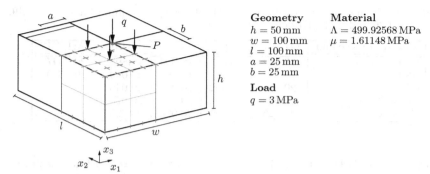

Geometry	Material
$h = 50\,\mathrm{mm}$	$\Lambda = 499.92568\,\mathrm{MPa}$
$w = 100\,\mathrm{mm}$	$\mu = 1.61148\,\mathrm{MPa}$
$l = 100\,\mathrm{mm}$	
$a = 25\,\mathrm{mm}$	
$b = 25\,\mathrm{mm}$	

Load

$q = 3\,\mathrm{MPa}$

Fig. 10.10 Nearly incompressible block: system, load and material data

The geometry and the material as well as the applied load and the boundary conditions are provided in Fig. 10.10. The convergence of the vertical displacement w_P in x_3-direction at the point P, in Fig. 10.10, is investigated for the Q1/EI9 and the Q1, Q2, Q1P0 and the QM1/E12 element for regular meshes with $4 \times 4 \times 4$, $8 \times 8 \times 8$, $16 \times 16 \times 16$, $32 \times 32 \times 32$ and $64 \times 64 \times 64$ elements.

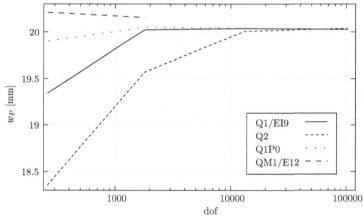

Fig. 10.11 Nearly incompressible block: Displacement w_P for the Q1/EI9, Q2, Q1P0 and QM1/E12 element

In Table 10.4, the vertical displacement w_P in x_3-direction of point P is shown as a function of the number of degrees of freedom for all elements. It can be observed that the Q1 element locks, as can be expected for this nearly incompressible problem. Both enhanced strain elements and the Q1P0 element are softer than the Q2 element. Thus still mild locking occurs for the higher order quadratic displacement element. Again, all elements except the QM1/E12 element converge to the same solution. For the QM1/E12 element,

Table 10.4 Nearly incompressible block: Displacement w_P [mm] for the Q1/EI9, Q1, Q2, Q1P0 and QM1/E12 element

Degrees of freedom	Q1/EI9	Q1	Q2	Q1P0	QM1/E12
260	19.342	7.656	18.354	19.898	20.2097
1800	20.023	13.083	19.569	20.049	20.1549
13328	20.038	17.492	20.008	20.040	
102432	20.028	19.493	20.040	20.028	
802880	20.025	19.951	20.026	20.025	

solutions can only be obtained for the two coarsest meshes. For finer mesh resolutions, the QM1/E12 element depicts nonphysical hour-glass instabilities. The displacement w_P are plotted for the Q1/EI9, Q2, Q1P0 and QM1/E12 element in Fig. 10.11 to visualize the results for the elements that are known to perform well for this test. It can be seen that the Q1P0 and the Q1/EI9 element perform extremely well, even for very coarse meshes. The Q2 element converges slower which is related to a mild locking of this element.

11. Contact Problems

The numerical treatment of contact problems requires the formulation of kinematical relations and constraints, constitutive equations at the contact interface, variational equations and its discretization using special finite elements within the contact area. The mathematical formulation of contact problems leads to variational inequalities. Hence special algorithms have to be constructed for the solution of such problems.

Based on today's possibilities to model engineering applications using very refined discretizations, it is often necessary to resolve contact boundary conditions also. Hence the application of contact within industrial and engineering problems ranges from forming processes, via tyre computations and car-crash simulations to general bearing problems, gears or bio-mechanical problems like teeth implants.

This chapter will address all aspects summarized above. However, not all topics can be discussed in detail. The reader can find an in-depth treatment in the monographs by Laursen (2002) and Wriggers (2006).

11.1 Contact Kinematics

This section summarizes the relations needed to formulate the geometrical contact conditions. In more detail, the distance or penetration function is needed for the formulation of normal contact. Furthermore relative tangential velocities and the associated displacements have to be derived for frictional contact. More detailed derivations of these kinematical relations can be found in Wriggers (2006).

The formulation of the contact constraints is based on the assumption that the contact solids undergo finite deformations. Both solids are described in the initial configuration by B^γ, where $\gamma = 1, 2$ denotes one of the bodies. The deformation $\boldsymbol{\varphi}^\gamma$ maps point in the initial configuration $\mathbf{X}^\gamma \in B^\gamma$ onto points of the deformed configuration $\mathbf{x}^\gamma = \boldsymbol{\varphi}^\gamma(\mathbf{X}^\gamma)$.

In order to define the distance function between two contacting solids, the approach of two surfaces Γ_c^γ is described in the deformed configuration $\boldsymbol{\varphi}^\gamma(\Gamma_c^\gamma)$. In this formulation, the possible contact surface is given by $\Gamma_c^\gamma \subset \partial B^\gamma$, see Fig. 11.1. It is assumed that these surfaces are convex, for the non-convex case, see e.g. Wriggers (2006).

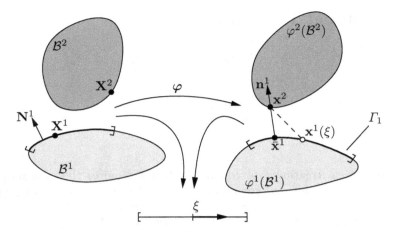

Fig. 11.1 Contact geometry and geometric approach

For the mathematical formulation of the distance or penetration function, the surfaces which come into contact have to be distinguished. In this context, one surface is denoted as *slave* and the other as *master* surface. For every point on the slave surface, the distance to the master surface will be computed. The slave surface will now be associated with body B^1 which, in a sense, is arbitrary. One can, without problems, exchange the rolls of master and slave without changing the final result, as will be discussed later. With these preliminary results, the deformed slave surface is denoted by $\varphi^1(\Gamma_c^1)$ and the master surface by $\varphi^2(\Gamma_c^2)$, the latter is reference surface, see Fig. 11.1, which however can deform itself.

We parameterize now the master surface Γ_c^2 in its initial and current configuration by convective coordinates ξ^1, ξ^2. Hence the material surfaces are described in the initial configuration by $\mathbf{X}^2 = \hat{\mathbf{X}}^2(\xi^1, \xi^2)$ and in the current configuration by $\mathbf{x}^2 = \hat{\mathbf{x}}^2(\xi^1, \xi^2)$. The associated tangent vectors are given by $\mathbf{A}_\alpha^2 = \hat{\mathbf{X}}_{,\alpha}^2(\xi^1, \xi^2)$ and $\mathbf{a}_\alpha^2 = \hat{\mathbf{x}}_{,\alpha}^2(\xi^1, \xi^2)$ where $(\ )_{,\alpha}$ is the derivative with respect to the convective coordinates ξ^α.

Point $\bar{\mathbf{x}}^2 = \mathbf{x}^2(\bar{\boldsymbol{\xi}})$ on the master surface which has minimal distance to a fixed point $\mathbf{x}^1$ on the slave surface is determined from

$$\| \mathbf{x}^1 - \bar{\mathbf{x}}^2 \| = \min_{\mathbf{x}^2 \subseteq \Gamma^2} \| \mathbf{x}^1 - \mathbf{x}^2(\boldsymbol{\xi}) \| , \qquad (11.1)$$

see Fig. 11.1 which illustrated the two-dimensional case. Relation (11.1) leads to the condition

$$\frac{d}{d\xi^\alpha} \| \mathbf{x}^1 - \mathbf{x}^2(\xi^1, \xi^2) \| = \frac{\mathbf{x}^1 - \mathbf{x}^2(\xi^1, \xi^2)}{\| \mathbf{x}^1 - \mathbf{x}^1(\xi^1, \xi^2) \|} \cdot \mathbf{x}_{,\alpha}^2(\xi^1, \xi^2) = 0 . \qquad (11.2)$$

In this relation, the tangent vector $\mathbf{x}_{,\alpha}^2 = \mathbf{a}_\alpha^2$ occurs which is related to the master surface. The tangent vector is in the solution point $\bar{\mathbf{x}}^2$ of (11.1) perpendicular to vector $\mathbf{x}^1 - \mathbf{x}^2(\xi^1, \xi^2)$. Hence the latter vector is normal to the master surface, see Fig. 11.1.

Once point $\bar{\mathbf{x}}^2$ is known an inequality condition can be written which describes the non-penetration of the solids. For this the distance, function $g_N = [\mathbf{x}^1 - \mathbf{x}^2(\bar{\xi})] \cdot \mathbf{n}^2(\bar{\xi})$ will be defined. It defines the following states in the contact area. Hence the constraint condition to exclude penetration is given as

$\begin{aligned} g_N &> 0 & &\text{no contact,} \\ g_N &= 0 & &\text{perfect contact,} \\ g_N &< 0 & &\text{penetration} \end{aligned}$

by

$$g_N = (\mathbf{x}^1 - \bar{\mathbf{x}}^2) \cdot \bar{\mathbf{n}}^2 \geq 0. \tag{11.3}$$

For some numerical algorithms, the definition of a penetration function is needed

$$g_N^- = \begin{cases} (\mathbf{x}^1 - \bar{\mathbf{x}}^2) \cdot \bar{\mathbf{n}}^2 & \text{falls } (\mathbf{x}^1 - \bar{\mathbf{x}}^2) \cdot \bar{\mathbf{n}}^2 < 0 \\ 0 & \text{otherwise} \end{cases} \tag{11.4}$$

on the slave surface $\varphi^1(\Gamma_c^1)$. The penetration function (11.4) provides the following information:

1. g_N^- can be used to check the contact state. This yields condition:

$$\text{contact} \iff g_N < 0. \tag{11.5}$$

2. g_N^- is used for $g_N^- < 0$ as local kinematical variable in a constitutive relation for the contact pressure.

In case of contact, the variation δg_N of the penetration function can be computed. With (11.4),

$$\delta g_N = [\boldsymbol{\eta}^1 - \hat{\boldsymbol{\eta}}^2(\bar{\xi})] \cdot \bar{\mathbf{n}}^2 \tag{11.6}$$

is obtained where $\boldsymbol{\eta}$ denotes the test function or virtual displacement.

In case that one body slides on the surface of another body then this relative tangential movement can be expressed by the change of the minimal distance at the solution point $(\bar{\xi}^1, \bar{\xi}^2)$. This yields the sliding distance of a point $\mathbf{x}^1$ on the deformed master Fl"ache

$$g_T = \int_{t_0}^{t} \|\dot{\bar{\xi}}^\alpha \, \bar{\mathbf{a}}_\alpha^2\| \, dt. \tag{11.7}$$

t is the time used to parameterize the path of point $\mathbf{x}^1$. t_0 is the time of the first contact of point $\mathbf{x}^1$ with the master surface. In order to evaluate the integral in (11.7), the time derivative of ξ^α at the projection point $\bar{\mathbf{x}}^2$ have to be computed. It can be determined from (11.2)

$$\frac{d}{dt}\left[\mathbf{x}^1 - \bar{\mathbf{x}}^2(\bar{\xi}^1, \bar{\xi}^2)\right] \cdot \bar{\mathbf{a}}_\alpha^2 = \left[\mathbf{v}^1 - \bar{\mathbf{v}}^2 - \bar{\mathbf{a}}_\beta \dot{\bar{\xi}}^\beta\right] \cdot \bar{\mathbf{a}}_\alpha^2 + \left[\mathbf{x}^1 - \bar{\mathbf{x}}^2\right] \cdot \dot{\bar{\mathbf{a}}}_\alpha^2 = 0. \quad (11.8)$$

With $\dot{\bar{\mathbf{a}}}_\alpha^2 = \bar{\mathbf{v}}_{,\alpha}^2 + \hat{\mathbf{x}}_{,\alpha\beta}^2 \dot{\bar{\xi}}^\beta$, the relation

$$\bar{H}_{\alpha\beta}\,\dot{\bar{\xi}}^\beta = \bar{R}_\alpha \quad (11.9)$$

can be derived, where

$$\bar{H}_{\alpha\beta} = \left[\bar{a}_{\alpha\beta} + g_N\,\bar{b}_{\alpha\beta}\right], \qquad \bar{R}_\alpha = \left[\mathbf{v}^1 - \hat{\mathbf{v}}^2(\bar{\boldsymbol{\xi}})\right] \cdot \bar{\mathbf{a}}_\alpha^2 + g_N\,\bar{\mathbf{n}}^2 \cdot \hat{\mathbf{v}}_{,\alpha}^2(\bar{\boldsymbol{\xi}}). \quad (11.10)$$

The tensors $\bar{a}_{\alpha\beta}$ and $\bar{b}_{\alpha\beta}$ are well known from differential geometry as metric- and curvature tensors of the deformed surface, respectively. Another formulation using a complete convective description can be found in Konyukhov and Schweizerhof (2006).

With these preliminary results, the relative tangential velocity is defined on the current slave surface $\varphi^1(\Gamma_c^1)$

$$\dot{\mathbf{g}}_T = \dot{\bar{\xi}}^\alpha\,\bar{\mathbf{a}}_\alpha^2. \quad (11.11)$$

From this equation, the slip in tangential direction is determined by integration. Hence (11.11) is an evolution equation for the slip $\mathbf{g}_T$. As can be seen in the next section, the tangential velocity $\dot{\mathbf{g}}_T$ enters as kinematical variable in the constitutive relations for the tangential contact stress.

Remark 11.1:

1. The second term on the right hand side of (11.9) depends on the gap function g_N. In case that the non-penetration condition ($g_N = 0$) is strongly enforced, this term disappears and (11.9) reduces to

$$\dot{\bar{\xi}}^\beta = \left[\mathbf{v}^1 - \bar{\mathbf{v}}^2\right] \cdot \bar{\mathbf{a}}^{2\,\beta}. \quad (11.12)$$

Additionally $\mathcal{L}_v\,\mathbf{g}_T$ in (11.11) is determined by the projection of the spatial velocities onto the tangent plane at the current contact point

$$\dot{\mathbf{g}}_T = \left(\bar{\mathbf{a}}_\alpha^2 \otimes \bar{\mathbf{a}}^{2\,\alpha}\right)\left[\mathbf{v}^1 - \bar{\mathbf{v}}^2\right]. \quad (11.13)$$

2. For a plane contact surface, the curvature tensor $\bar{b}_{\alpha\beta}$ is zero. Thus relation (11.9) simplifies.
3. For two-dimensional contact, equation (11.9) can be specified as

$$\dot{\bar{\xi}} = \frac{1}{\bar{a}_{11} + g_N\,\bar{b}_{11}} \left\{\left[\mathbf{v}^1 - \mathbf{v}^2(\bar{\xi})\right] \cdot \mathbf{x}^2,_\xi(\bar{\xi}) + g_N\,\bar{\mathbf{n}}^2 \cdot \mathbf{v}^2,_\xi(\bar{\xi})\right\} \quad (11.14)$$

with the metric $\bar{a}_{11} = \mathbf{x}^2,_\xi(\bar{\xi}) \cdot \mathbf{x}^2,_\xi(\bar{\xi})$ and the curvature $\bar{b}_{11} = \mathbf{x}^2,_{\xi\xi}(\bar{\xi}) \cdot \bar{\mathbf{n}}^2$. Now the relation (11.7) for the computation of the total slip simplifies to

$$g_T = \int\limits_{t_0}^{t} \|\dot{\bar{\xi}}\, \mathbf{\bar{x}}_{,\xi}^2\|\, dt = \int\limits_{\xi_0}^{\bar{\xi}} \sqrt{\bar{a}_{11}}\, d\xi \,. \tag{11.15}$$

Note that the integral is re-parameterized by the surface coordinate such that there is no time dependency.

By exchanging the velocity field by the variation or test function in (11.11), the virtual change of the relative tangential slip is given by

$$\delta \mathbf{g}_T = \delta \bar{\xi}^\alpha\, \bar{\mathbf{a}}_\alpha^2 \,. \tag{11.16}$$

11.2 Constitutive Equations at the Contact Interface

Depending on the accuracy which is needed to describe the mechanical behaviour at the contact interface there exist, in the literature, different approaches to model constitutive equations for the contact zone. While the tangential relative velocity enters the frictional constitutive equations, the behaviour in normal direction is either described by micro-mechanical relations or by the kinematical constraint of non-penetration.

11.2.1 Normal Contact

The first, classical mathematical modeling relates to the constraint equation (11.3) for the normal components of the displacements which prevents the penetration of one solid into the other. Within this formulation, the normal contact force follows is a reaction force. The mathematical formulation leads to the so-called Kuhn-Tucker-Karush condition

$$g_N \geq 0\,, \quad p_N \leq 0\,, \quad p_N\, g_N = 0\,, \tag{11.17}$$

where p_N is the contact pressure which follows as a reaction due to the constraint $g = 0$ in case of contact.

A second approach introduces constitutive relations within the contact interface which describe the contact pressure in terms of the approach of both surfaces in normal direction. This modeling is based on micro-mechanical considerations in which the surface roughness plays a significant role. Based on statistical considerations, the constitutive equations can be derived, see e.g. Kragelsky et al. (1982), which include the micro-mechanical behaviour in the contact area. This micro-mechanical behaviour depends upon physical parameters like hardness but also on geometrical parameters like the surface roughness. However, in reality, these interactions are far more complex, see e.g. Kragelsky and Alisin (2001), and only constitutive equations can be derived which model the essential phenomena. In general, such constitutive relation has the form

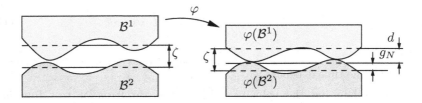

Fig. 11.2 Physical approach in the contact area Γ_c: initial and deformed configuration

$$p_N = f(d) \quad \text{or} \quad d = h(p_N) . \tag{11.18}$$

f and h are nonlinear functions of the approach of the deformed middle planes of the rough contact surfaces d or of the contact pressure p_N.

The approach of the deformed middle planes of the rough contact surfaces can be expressed by the geometrical approach or distance function g_N (11.4)

$$g_N = \zeta - d , \tag{11.19}$$

where ζ is the distance between the middle planes of the rough contact surface in the initial configuration. This is defined as the distance of the middle planes which occurs when both rough surface touch each other in Γ_c, see Fig. 11.2.1.

A possible constitutive equation to describe the approach of the solids in the contact area has, after Kragelsky et al. (1982), the form

$$p_N = c_N (\zeta - g_N)^{\alpha_N} . \tag{11.20}$$

The constitutive parameters c_N and α_N have to be determined from experiments.

Note that relation (11.20) can be interpreted as a law for a nonlinear elastic spring. Due to the fact that the approach within the contact area is extremely small when compared to the other deformations of the solid, it means that the spring stiffness is very large. This is, from the numerical point of view, a big disadvantage since it will lead to badly conditioned equation systems. Here special solution techniques are needed to overcome this problem, see e.g. Wriggers and Zavarise (1993).

11.2.2 Tangential Contact

The constitutive behaviour in tangential direction is very complex. It depends on many factors such as surface roughness, magnitude of the normal pressure, tangential relative velocities, contaminants or humidity, etc. This complexity can be reflected by the constitutive equations which then depend on many material parameters. Since it is not easy to determine such parameters, the most simplest constitutive relation, the so-called COULOMB law is employed in many engineering applications which only depends upon one material parameter, the coefficient of friction. However, depending on the material pairing, there exist many variants of this constitutive relation for the tangential stresses. The general expression is provided by

$$\mathbf{t}_T = -\hat{f}(p_N, \dot{g}_T, g_T, \theta, \ldots) \frac{\dot{\mathbf{g}}_T}{\|\dot{\mathbf{g}}_T\|}, \tag{11.21}$$

where the function $\hat{f}$ depends upon the contact pressure p_N, the total relative tangential velocity $\dot{g}_T = \|\dot{\mathbf{g}}_T\|$, the total slip g_T, see (11.7) and other parameters such as temperature θ. The stress due to friction acts always opposite to the relative tangential velocity $\dot{\mathbf{g}}_T$. The following table summarizes some constitutive equations, for details see Wriggers (2006), for frictional response. There exist several other constitutive relations which consider micromechanical aspects, see e.g. Woo and Thomas (1980) or the overview in Oden and Martins (1986). The physical background for the fictional behaviour can be found in e.g. Tabor (1981) and Kragelsky and Alisin (2001).

Since, in tangential contact, besides sliding also stick phenomena are observed, the stick phase has to be defined. It can be described using (11.11) as a constraint condition for the relative tangential movement

$$\dot{\mathbf{g}}_T = \mathbf{0} \Leftrightarrow \dot{\xi}^\alpha = 0. \tag{11.22}$$

The stick/slip behaviour in the contact interface yields, for the simplest constitutive equation in Table 11.1, a response which is non-smooth, as can be seen in Fig. 11.2.2. This leads to mathematical difficulties which also effect the algorithms used to solve frictional contact using the finite element method. Due to that there exist mainly two possibilities to overcome the algorithmic problems due to non-smoothness. One is to regularize the constitutive equations for friction in (11.21) and the other is to formulate the friction problem like an elasto-plastic problem, see Sect. 6.2.

Table 11.1 Frictional constitutive equations

Material pairing	Constitutive equation	Material parameters
General	$\hat{f} = \mu\, p_N$	μ
Rubber	$\hat{f} = \mu_0 + c_1\,\theta\,[\ln \dot{g}_T - \ln(c_2\,\theta)]$	μ_0, c_1, c_2
Rubber	$\hat{f} = \mu_0(p_N) + c_1 \ln \dfrac{\dot{g}_T}{v_1} - c_2 \ln \dfrac{\dot{g}_T}{v_2}$	$\mu_0, c_1, c_2, v_1, v_2$
Metal	$\hat{f} = \mu_D + (\mu_S - \mu_D)\,e^{-c\,\dot{g}_T}$	μ_D, μ_S, c

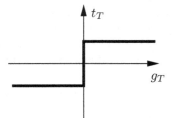

Fig. 11.3 COULOMB'S friction law

The regularisation can be performed by introducing a smooth function such that f in (11.21) has the form

$$\hat{f} = \mu \, \tanh\left(\frac{\dot{g}_T}{\varepsilon}\right) \mid p_N \mid, \qquad (11.23)$$

in which ε is the regularization parameter. For $\varepsilon \to 0$, the constitutive relation of the first line in Table 11.1 is recovered. Note that such constitutive equation describes the stick–slip motion only approximately and might reproduce, for large values of ε, non-physical result.

The other method which has been applied during the last years extensively to solve frictional contact problems is based on a split of the tangential motion into an elastic (stick) and plastic (slip) part.

$$\mathbf{g}_T^e = \mathbf{g}_T - \mathbf{g}_T^s . \qquad (11.24)$$

Here the elastic part is given by $\mathbf{g}_T^e$, which approximates or regularizes the stick part, and the plastic part $\mathbf{g}_T^s$ denotes the irreversible part of the tangential relative motion, see Fig. 11.2.2. When using this split and interpretation of the frictional behaviour at the contact interface, all known algorithms from plasticity can be applied to describe the stick–slip motion.

In this context, many different constitutive relations can be formulated to characterize the frictional contact behaviour, see e.g. Michalowski and Mroz (1978), Curnier (1984) and Wriggers (2006). The elasto-plastic analogy was first used in Fredriksson (1976) in the context of finite element analysis for contact in order to describe softening frictional behaviour. The biggest advantage of this analogy lies, however, in the possibility to apply the projection methods developed in Simo and Taylor (1985) for plasticity which was firstly used in the context of frictional problems in Wriggers (1987) for the COULOMB law. Further formulations can be found for small deformations in Giannokopoulos (1989) and for large deformations in Wriggers et al. (1990).

Note that with the split in (11.24) the stick part can now be computed using a constitutive equation which can be interpreted in a way that elastic micro-displacements occur in the relative tangential motion. The physical interpretation would be that the asperities of the surface roughness behave

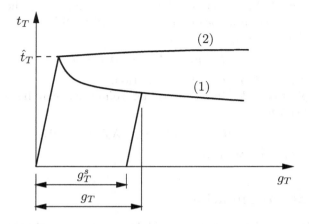

Fig. 11.4 Constitutive equation for stick and slip

elastically. Hence the simplest constitutive model is based on the assumption of isotropic linear elastic behaviour which yields

$$\mathbf{t}_T = c_T\,\mathbf{g}_T^e \tag{11.25}$$

with the elastic constant c_T. Since the relative elastic displacement $\mathbf{g}_T^e$ is usually very small (it has to approximate the stick behaviour), quite large values for c_T have to be selected which can lead to bad conditioning of the tangent matrix.

The "plastic" slip g_T^s is described by a set of constitutive evolution equations

$$
\begin{aligned}
\dot{\mathbf{g}}_T^s &= \lambda\,\frac{\partial \hat{f}_s(\mathbf{t}_T)}{\partial \mathbf{t}_T} = \lambda\,\mathbf{n}_T \qquad \text{with} \quad \mathbf{n}_T = \frac{\mathbf{t}_T}{\|\mathbf{t}_T\|}, \\
\dot{g}_T &= \lambda,
\end{aligned}
\tag{11.26}
$$

in which $\hat{f}_s$ is the so-called slip function which characterizes the "elastic" stick region which depends, in general, on the total slip g_T. A general form of the slip function is

$$\hat{f}_s(\mathbf{t}_T) = \|\,\mathbf{t}_T\,\| - h(\,p_N\,,g_T\,,\ldots) \le 0\,. \tag{11.27}$$

It depends on the contact pressure in the simplest case, but can also depend on the total slip g_T, the relative tangential velocity $\dot{g}_T$ or the temperature θ, see also the definition of the direction constitutive equation for the tangential stresses in (11.21). which depends. As depicted in Fig. 11.2.2, hardening or softening behaviour can be modelled within this approach. The internal variable is defined in (11.15). It accumulates the sliding path which is seen by the material point $\mathbf{x}^1$. Note that this definition is equivalent to the introduction of an equivalent strain in elasto-plasticity, see Sect. 6.2.

The special choice

$$\hat{f}_s(\mathbf{t}_T) = \| \, \mathbf{t}_T \, \| - \mu \, p_N \leq 0 \tag{11.28}$$

yields the classical COULOMB law of friction.

As in plasticity, loading– unloading conditions can be formulated in KUHN-TUCKER form

$$\lambda \geq 0 \, , \quad \hat{f}_s(\mathbf{t}_T) \leq 0 \, , \quad \lambda \, \hat{f}_s(\mathbf{t}_T) = 0 \, , \tag{11.29}$$

which yield the slip parameter λ.

11.3 Weak Formulation

The introduction of the inequality (11.1) which represents the contact condition yields with (3.296) to a variational inequality of the form

$$\sum_{\gamma=1}^{2} \int_{\Omega^\gamma} \boldsymbol{\tau}^\gamma \cdot \nabla^S \left(\boldsymbol{\eta}^\gamma - \boldsymbol{\varphi}^\gamma \right) dV \geq \sum_{\gamma=1}^{2} \int_{\Omega^\gamma} \bar{\mathbf{f}}^\gamma \cdot (\boldsymbol{\eta}^\gamma - \boldsymbol{\varphi}^\gamma) \, dV - \int_{\Gamma_\sigma{}^\gamma} \bar{\mathbf{t}}^\gamma \cdot (\boldsymbol{\eta}^\gamma - \boldsymbol{\varphi}^\gamma) \, dA \, .$$

$$\tag{11.30}$$

In this variational inequality, the integration has, contrary to (3.296), to be performed with respect to the region Ω^γ which is assumed by $\mathcal{B}^\gamma$ in the initial configuration. Due to that the KIRCHHOFF stress $\boldsymbol{\tau}$ appears in (11.30) instead of the CAUCHY stress. However $\boldsymbol{\tau}$ is, as also the gradient operator "$\nabla^S()$", related to the current configuration.

Based on (11.30), the deformation $(\boldsymbol{\varphi}^1 , \boldsymbol{\varphi}^2) \in \mathbf{K}$ of both in bodies being in frictionless contact has to be determined for both bodies being in frictionless contact. The set $\mathbf{K}$ is then defined by

$$\mathbf{K} = \{ \, (\boldsymbol{\eta}^1 , \boldsymbol{\eta}^2) \in \mathbf{V} \, | \, [\, \boldsymbol{\eta}^1 - \hat{\boldsymbol{\eta}}^2(\bar{\xi}^1, \, \bar{\xi}^2)] \cdot \bar{\mathbf{n}}^2 \geq 0 \, \} \, . \tag{11.31}$$

In case of finite elastic deformations, the existence of solutions of (11.30) can be shown, see e.g. Ciarlet (1988) and Curnier et al. (1992). For that the strain energy function which describes the constitutive behaviour has to be polyconvex, see also the remarks in Sect. 3.3.1.

Remark 11.2: In the geometrically linear theory, (11.30) can be written as variational inequality of the form

$$a(\mathbf{u}, \mathbf{v} - \mathbf{u}) \geq f(\mathbf{v} - \mathbf{u}) \, . \tag{11.32}$$

The operators $a(\mathbf{u} , \mathbf{v})$ and $(\mathbf{u})$ are defined by

$$a(\mathbf{u}, \mathbf{w}) = \int_\Omega \boldsymbol{\varepsilon}(\mathbf{u}) \cdot \mathbb{C}_0[\boldsymbol{\varepsilon}(\mathbf{w})] \, d\Omega \, ,$$

$$f(\mathbf{w}) = \int_\Omega \hat{\mathbf{b}} \cdot \mathbf{w} \, d\Omega + \int_{\Gamma_\sigma} \hat{\mathbf{t}} \cdot \mathbf{w} \, . \, d\Gamma.$$

Furthermore, the total region occupied by both bodies is $\Omega = \cup_\gamma \mathcal{B}^\gamma$. $\mathbb{C}_0$ is the elasticity tenor of the linear theory, see (3.273). The linear strain tensor is given by $\boldsymbol{\varepsilon}(\mathbf{u}) = \frac{1}{2}(\nabla\mathbf{u} + \nabla^T\mathbf{u})$. Due to the contact constraints, the variational equation is nonlinear. One has to determine the displacement $\mathbf{u} \in \mathbf{K}$ such that (11.32) is fulfilled for all test functions $\mathbf{v} \in \mathbf{K}$

$$\mathbf{K} = \{\mathbf{v} \in \mathbf{V} \,|\, (\mathbf{v}^1 - \bar{\mathbf{v}}^2) \cdot \bar{\mathbf{n}}^2 + g_0 \geq 0 \text{ on } \Gamma_c\}. \tag{11.33}$$

The mathematical structure of the variational equation (11.32) is discussed in depth in Duvaut and Lions (1976) and Kikuchi and Oden (1988).

Algorithms for the solution of variational inequalities exist in a large variety. Most of them stems form optimisation theory, see e.g. the overview in Luenberger (1984), Bertsekas (1984) and Bazaraa et al. (1993). For the simulation of contact problems using the finite element method, algorithms have to be selected which can handle a large number of inequality constraints in an efficient way. The most popular algorithms are not only the penalty and augmented Lagrangian algorithms, but also newer techniques like projected gradient algorithms, see e.g. Dostal (2003), and interior point methods, see e.g. Wright (1997), are available.

When using the finite element method together with the penalty or Lagrange multiplier method, the inequality constraints are formulated are split into active and inactive constraints which change their number within the solution. Hence, instead of the variational inequality (11.30), a variational equation can be written including the active constraints in the contact area γ_c^{akt}

$$\sum_{\gamma=1}^{2}\left\{ \int_{\Omega^\gamma} \boldsymbol{\tau}^\gamma \cdot \operatorname{grad}\boldsymbol{\eta}^\gamma\, dV - \int_{\Omega^\gamma} \bar{\mathbf{f}}^\gamma \cdot \boldsymbol{\eta}^\gamma\, dV - \int_{\Gamma_\sigma{}^\gamma} \bar{\mathbf{t}}^\gamma \cdot \boldsymbol{\eta}^\gamma\, dA \right\}$$

$$+ \quad ''contact\ contributions'' = 0. \tag{11.34}$$

For the contact between two bodies, the *contact contributions* can now be formulated in the framework of the LAGRANGE multiplier or the penalty method. This leads for the active part of the contact area Γ_c^{akt} to:

1. Method of LAGRANGE multipliers:

$$\int_{\Gamma_c^{akt}} (\lambda_N\, \delta g_N + \boldsymbol{\lambda}_T \cdot \delta\mathbf{g}_T)\, dA \tag{11.35}$$

λ_N is the LAGRANGE multiplier related to the constraint $g_N = 0$ which can be identified as contact pressure p_N. δg_N is the variation of the distance function in normal direction (11.6). The term $\boldsymbol{\lambda}_T \cdot \delta\mathbf{g}_T$ describes the constraint in tangential direction. If this is given by (11.22), then $\boldsymbol{\lambda}_T$ is the reaction due to stick. If sliding occurs, it is not possible to interpret $\boldsymbol{\lambda}_T$ as reaction. In that case, it is the stress vector $\mathbf{t}_T$ in tangential direction which follows from the constitutive equations (11.25), (11.26), (11.27) and (11.29).

2. **Penalty method:** The contact constrain $g_N = 0$ is introduced in this case via a penalty term in the weak form. This yields

$$\int_{\Gamma_c} \epsilon_N \, g_N \, \delta g_N \, dA \,, \quad \epsilon_N > 0 \,. \tag{11.36}$$

One can show, see Luenberger (1984), that for $\epsilon_N \rightarrow \infty$ the result of the LAGRANGE multiplier method is recovered. However, the choice of a large penalty parameter leads to a badly conditioned tangential stiffness matrix. As in the LAGRANGE multiplier approach, stick and slip in tangential direction has to be distinguished. For stick, the penalty constraint is introduced in tangential direction also

$$\int_{\Gamma_c} \left(\epsilon_N \, g_N \, \delta g_N + \epsilon_T \, \mathbf{g}_T \cdot \delta \mathbf{g}_T \right) dA \,, \quad \epsilon_N > 0 \,, \epsilon_T > 0 \,. \tag{11.37}$$

In case of sliding, the variational form change to

$$\int_{\Gamma_c} \left(\epsilon_N \, g_N \, \delta g_N + \mathbf{t}_T \cdot \delta \mathbf{g}_T \right) dA \,, \quad \epsilon > 0, \tag{11.38}$$

where again the constitutive relations (11.25), (11.26), (11.27) and (11.29) have to be applied to compute $\mathbf{t}_T$.

In equations (11.35), (11.36), (11.37) and (11.38), the variation of the distance function g_N occurs which follows from (11.6). The variation of the relative tangential displacement is provided in (11.16).

Remark 11.3:
1. When the constitutive parameter c_T in (11.25) is exchanged by the penalty parameter ϵ_T, then a regularisation of the frictional constitutive equation for stick is obtained, see e.g. Ju and Taylor (1988) and Curnier and Alart (1988).
2. A further method to include contact constraints into (11.34) is provided by direct elimination. Then the constraint condition on Γ_c^{akt} leads to $g_N = 0 \longrightarrow$ $\mathbf{x}^1 \cdot \bar{\mathbf{n}}^2 = \hat{\mathbf{x}}_t^2 \cdot \bar{\mathbf{n}}^2$ and can be used to eliminate the associate displacements of either body $\mathcal{B}^1$ or $\mathcal{B}^2$. This method is implemented in some commercial codes; a more detailed description can be found in Wohlmuth (2000) and Wriggers (2006).
3. A further technique to include contact constraints is provided by the barrier method. This adds the term

$$\int_{\Gamma_c} \epsilon_N \frac{1}{g_N^2} \delta g_N \, d\Gamma$$

to the weak form instead of (11.36). This function introduces a repellent effect between the bodies which fades with distance quadratically. Due to that always all constrains are active, however the solution has always to stay in the feasible region. For this safeguard algorithms are needed, see e.g. Bazaraa et al. (1993). More general schemes based on barrier functions are provided by interior point methods, see e.g. Wright (1997).

4. A technique which combines penalty and barrier method can be found in Zavarise et al. (1998). By a special formulation, all distance functions are active – as in the pure barrier method – and the penalty methods acts as safeguard function which prevents penetration.

5. A so-called (*perturbed*) LAGRANGE formulation can be applied to combine penalty and LAGRANGE multiplier method within a mixed method, see e.g. Oden (1981) and Simo et al. (1985b). Using this formulation, special mixed finite element discretization for contact problems can be developed.

6. The main problem when using the penalty method is the bad condition of the tangent stiffness matrix. A method which can be applied to circumvent this problem is the *augmented* LAGRANGE formulation, see e.g. Glowinski and Le Tallec (1984). This method was also applied to contact problem to see e.g. Wriggers et al. (1985), Kikuchi and Oden (1988) and Laursen and Simo (1993a). The UZAWA-algorithm, which is related to the *augmented* LAGRANGE formulation, is based on the idea to fix the LAGRANGE multiplier within an iterative solution step and then compute the next value of the LAGRANGE multiplier by an update formula. This leads to the following weak form

$$\sum_{\gamma=1}^{2} \left\{ \int_{\mathcal{B}^\gamma} \boldsymbol{\tau}^\gamma \cdot \operatorname{grad} \boldsymbol{\eta}^\gamma \, dV - \int_{\mathcal{B}^\gamma} \bar{\mathbf{f}}^\gamma \cdot \boldsymbol{\eta}^\gamma \, dV - \int_{\Gamma_\sigma{}^\gamma} \bar{\mathbf{t}}^\gamma \cdot \boldsymbol{\eta}^\gamma \, dA \right\}$$

$$+ \int_{\Gamma_c} [\, \bar{\lambda}_N + \epsilon_N \, g_N^L \,) \, \delta g_N + \mathbf{t}_T \cdot \delta \mathbf{g}_T \,] \, dA = 0 \qquad (11.39)$$

with the update $\bar{\lambda}_{N_{new}} = \bar{\lambda}_{N_{old}} + \epsilon_N \, g_{N_{new}}$. Note that this update is only first order, which leads to more iterations within the overall algorithm. Methods to improve the order of this update can be found, in general, in Bertsekas (1984) or in the context of finite element methods for contact problems in Alart and Curnier (1991).

11.4 Discretization

A general formulation of a contact undergoing finite deformations using the finite element method has to allow finite sliding of a contact (slave) point of the entire (master) surface of the other body. This possibility is represented within the discretization by a so-called *node-to-segment* contact element, see Fig. 11.4.1 for a more detailed description. Die resulting matrix formulation including the tangential stiffness matrices was developed for the two-dimensional firstly in Wriggers and Simo (1985) for frictionless contact. The extension for frictional contact can be found in Wriggers et al. (1990). Three-dimensional discretization is provided in Parisch (1989) for the frictionless and in Peric and Owen (1992) and Laursen and Simo (1993b) for the frictional case.

In the most general case of finite deformations, the element nodes do not match at the contact interface as in classical finite element discretizations, see Fig. 11.4. First implementations for such case were provided by Hallquist (1979) and Hughes et al. (1977b). Other formulations can be found

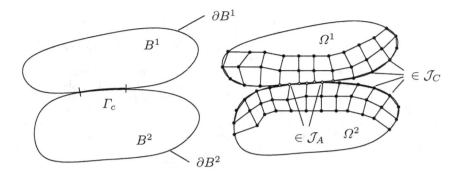

Fig. 11.5 Discretization of contact for finite deformations

in Bathe and Chaudhary (1985) or Hallquist et al. (1985). Today many commercial codes can handle such problems, even with self-contact, see e.g. Hallquist et al. (1992). However, there is still a vivid development of contact discretization for finite deformation contact such as smooth, see e.g. Pietrzak and Curnier (1999) and Wriggers et al. (2001) and Krstulovic-Opara et al. (2002), and mortar discretization schemes, see e.g. Puso (2004), Puso and Laursen (2004) and Fischer and Wriggers (2006).

Within the actual implementation of contact schemes, one has to distinguish between points on the contact surface which are in contact and others which are not in contact. For this purpose, the set of all possible nodes $\mathcal{J}_C \in \Gamma$ is defined. The active nodes are given by $\mathcal{J}_A$. In the following matrix formulations, only the active nodes are considered in the weak form (11.34). The main difference between the LAGRANGE multiplier (11.35) and the penalty method (11.36) are that different interpolations can be used for the LAGRANGE multipliers λ_N and the displacement field which appears in g_N

$$\int_{\Gamma_c} \lambda_N \, \delta g_N \, d\Gamma \longrightarrow \int_{\Gamma_c^h} \lambda_{N\,c} \, \delta g_{N\,c} \, d\Gamma \,. \tag{11.40}$$

The ansatz function for $\lambda_{N\,c}$ and $\delta g_{N\,c}$ are given by

$$\lambda_{N\,c} = \sum_K M_K(\xi)\,\lambda_{NK} \quad \text{and} \quad \delta g_{N\,c} = \sum_I N_I(\xi)\,\delta g_{NI}\,. \tag{11.41}$$

Note that the interpolations in (11.41) have to be chosen such that the BABUSKA–BREZZI condition is fulfilled which guarantees the stability of the mixed method, see Remark 10.1 and also Kikuchi and Oden (1988).

When using the penalty method, only the displacement field has to be discretized

$$\int_{\Gamma_c} \epsilon_N \, g_N \, \delta g_N \, d\Gamma \longrightarrow \int_{\Gamma_c^h} \epsilon_N \, g_{N\,c} \, \delta g_{N\,c} \, d\Gamma\,. \tag{11.42}$$

Here the same ansatz is chosen for the displacements entering the distance function and the test functions entering the variation of the distance function

$$g_{Nc} = \sum_I N_I(\xi)\, g_{NI} \quad \text{and} \quad \delta g_{Nc} = \sum_I N_I(\xi)\, \delta g_{NI}. \qquad (11.43)$$

Also the ansatz function for the penalty method have implicitly fulfil the BABUSKA–BREZZI condition. This is related to the equivalence of both formulations which can be established by the *perturbed* LAGRANGE formulation, see Remark 11.3 *Nr. 5*. It is equivalent to the incompressibility constraint in solid mechanics which was discussed in Malkus and Hughes (1978). Due to this reason, a reduced integration for the integrals in (11.42) has to be used eventually. Proper choices for the numerical integration schemes can be found in e.g. Oden (1981).

11.4.1 NTS-Discretization

Since this book is aimed at finite deformations, only discretization schemes are considered which can handle large sliding in the contact interface. These exist for two- and three-dimensional applications. Here a two-dimensional discretization will be considered using the node-to-segment (NTS) formulation. Related three-dimensional formulation can be found in Laursen (2002) and Wriggers (2006). The NTS-discretization is the most simple possibility to treat large sliding at contact interfaces. Hence it can be found in many commercial finite element codes. For contact discretization using the NTS element, it is assumed that a *slave* node (s), given by the position vector $\mathbf{x}_s^1$ in $\varphi(B^1)$, comes into contact with the *master* segment (1)–(2), which is described by the position vectors $\mathbf{x}_1^2$ and $\mathbf{x}_2^2$ with respect to $\varphi(B^2)$, see Fig. 11.4.1. The kinematical relations for this discretization follow directly from the continuum formulation, see Sect. 11.1.

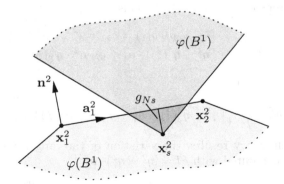

Fig. 11.6 Node-to-Segment contact element

With a linear interpolation of the master surface described by the surface coordinate ξ,

$$\hat{\mathbf{x}}^2(\xi) = \mathbf{x}_1^2 + (\mathbf{x}_2^2 - \mathbf{x}_1^2)\,\xi, \qquad 0 \le \xi \le 1, \tag{11.44}$$

the tangent vectors

$$\bar{\mathbf{a}}_1^2 = \hat{\mathbf{x}}^2(\xi),_1 = (\mathbf{x}_2^2 - \mathbf{x}_1^2) \tag{11.45}$$

can be computed. The tangent vector can be normalized which yields $\mathbf{a}_1^2 = \bar{\mathbf{a}}_1^2 / l$ with the length of the master element $l = \| \mathbf{x}_2^2 - \mathbf{x}_1^2 \|$. With the unit tangent vector $\mathbf{a}_1^2$, the unit normal vector of the segment (1)–(2) is obtained by $\mathbf{n}^2 = \mathbf{e}_3 \times \mathbf{a}_1^2$.

$\bar{\xi}$ and g_{Ns} follow from solutions of (11.1) and (11.10) which denotes a projection of the slave node $\mathbf{x}_s$ in (s) onto the master segment (1)–(2)

$$\bar{\xi} = \frac{1}{l}(\mathbf{x}_s^1 - \mathbf{x}_1^2) \cdot \mathbf{a}_1^2 \quad \text{and} \quad g_{Ns} = \| \mathbf{x}_s^1 - (1 - \bar{\xi})\mathbf{x}_1^2 - \bar{\xi}\mathbf{x}_2^2 \|. \tag{11.46}$$

From these equations and the continuous formulation (11.6), the variation of the distance function g_{Ns} can be derived

$$\delta g_{Ns} = [\,\boldsymbol{\eta}_s - (1 - \bar{\xi})\,\boldsymbol{\eta}_1^2 - \bar{\xi}\,\boldsymbol{\eta}_2^2\,] \cdot \mathbf{n}^2. \tag{11.47}$$

With the interpolation for the coordinate $\hat{\mathbf{x}}^2(\xi) = \mathbf{x}_1^2 + \xi\,(\mathbf{x}_2^2 - \mathbf{x}_1^2)$ on the master segment (1)–(2), relation

$$g_{Ts} = \int_{\xi_0}^{\bar{\xi}} l\,d\xi = (\bar{\xi} - \xi_0) \tag{11.48}$$

is obtained from (11.15) which leads to the discrete form of the variation of the tangential slip distance

$$\delta g_{Ts} = l\,\delta\bar{\xi} + (\bar{\xi} - \xi_0)\,\delta l. \tag{11.49}$$

Inserting the interpolation into (11.9) yields

$$\begin{aligned} \bar{H}_{11} &= (\bar{a}_{11} + g_N\,\bar{b}_{11}) = a_{11} = l^2 \\ \bar{R}_1 &= [\,\boldsymbol{\eta}^1 - \hat{\boldsymbol{\eta}}^2(\bar{\xi})\,] \cdot \bar{\mathbf{a}}_1^2 + g_N\,\mathbf{n}^2 \cdot \hat{\boldsymbol{\eta}}_{,\xi}^2(\bar{\xi}) \end{aligned}$$

and hence the variation of $\bar{\xi}$

$$\delta\bar{\xi} = \frac{1}{l^2}\left\{ [\,\boldsymbol{\eta}^1 - \hat{\boldsymbol{\eta}}^2(\bar{\xi})\,] \cdot \bar{\mathbf{a}}_1^2 + g_N\,\mathbf{n}^2 \cdot \hat{\boldsymbol{\eta}}_{,\xi}^2(\bar{\xi}) \right\}. \tag{11.50}$$

Using these preliminary results, the variation of the tangential slip for the NTS element is computed with $\delta l = [\,\boldsymbol{\eta}_2^2 - \boldsymbol{\eta}_1^2\,] \cdot \mathbf{a}_1^2$

$$\delta g_{Ts} = [\,\boldsymbol{\eta}_s^1 - (1 - \bar{\xi})\,\boldsymbol{\eta}_1^2 - \bar{\xi}\,\boldsymbol{\eta}_2^2\,] \cdot \mathbf{a}_1^2 + \frac{g_{Ns}}{l}\,[\,\boldsymbol{\eta}_2^2 - \boldsymbol{\eta}_1^2\,] \cdot \mathbf{n}^2 + \frac{g_{Ts}}{l}\,[\,\boldsymbol{\eta}_2^2 - \boldsymbol{\eta}_1^2\,] \cdot \mathbf{a}_1^2. \tag{11.51}$$

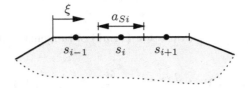

Fig. 11.7 Multiple slave nodes on one segment

Equations (11.47) and (11.51) represent the essential kinematical relations for the contact element depicted in Fig. 11.4.1. These can now be used in the weak form (11.37) where the integral has to be evaluated with respect to the deformed contact surface. This yields, with the assumption of a constant contact pressure,

$$\int_{\varphi(\Gamma_c)} (p_N \, \delta g_N + t_T \, \delta g_T) \, d\gamma \longrightarrow \sum_{s=1}^{n_c} (P_{N\,s} \, \delta g_{Ns} + T_{T\,s} \, \delta g_{Ts}), \qquad (11.52)$$

where all contributions of the active contact segments have to be summed up. The nodal normal force P_{Ns} is given by $P_{Ns} = p_{Ns} \, a_s$ at the slave node (s). For the penalty method, $P_{Ns} = \epsilon_N \, g_{Ns} \, a_s$ is obtained. The tangential force is given by $T_{Ts} = t_{Ts} \, a_s$. In case of stick, it follows from the penalty method $T_{Ts} = \epsilon_T \, g_{Ts} \, a_s$. In case of sliding, it will be determined from the integration of the friction law, see next section. a_s is the area of the contact element which is equal to l in the two-dimensional case.

Remark 11.4: If more than one slave node is in contact with the same segment, see Fig. 11.4.1, then the area a_s related to one slave node is no longer given by total surface of the master segment. In such case, a_s can be computed from the midpoints between the projections $\bar{\xi}$ of the neighbouring slave nodes. The counter i describes the slave node s_i and the neighbouring nodes s_{i-1} and s_{i+1}. With these definitions, the area belonging to s_i is given by

$$a_{s_i} = \frac{l}{2} \, (\bar{\xi}_{i+1} - \bar{\xi}_{i-1}). \qquad (11.53)$$

There are special cases which also include the boundary of the segment; they will not be discussed here in detail.

11.4.2 Matrix Form of Contact Residual

With equations (11.37) and (11.52), the discrete weak form which is associated with a node (s) is given by

$$\delta g_{Ns} \, P_{N\,s} + \delta g_{Ts} \, T_{T\,s}. \qquad (11.54)$$

This equation will now be presented in matrix form. For the first part in (11.54), the variation (11.47) of the distance function is obtained as

$$\delta g_{Ns} = \boldsymbol{\eta}_s^T \mathbf{N}_s . \tag{11.55}$$

Using similar notation, the variation (11.51) of tangential sliding by matrices can be described by

$$\delta g_{Ts} = \boldsymbol{\eta}_s^T \left(\mathbf{T}_s + \frac{g_{Ns}}{l} \mathbf{N}_{0\,s} + \frac{g_{T\,s}}{l} \mathbf{T}_{0\,s} \right) = \boldsymbol{\eta}_s^T \, \hat{\mathbf{T}}_s . \tag{11.56}$$

The following matrices were used in these equation

$$\boldsymbol{\eta}_s = (\, \eta_s^1 \quad \eta_1^2 \quad \eta_2^2 \,)^T , \tag{11.57}$$

$$\mathbf{N}_s = \left\{ \begin{array}{c} \mathbf{n}^2 \\ -(1 - \bar{\xi})\,\mathbf{n}^2 \\ -\bar{\xi}\,\mathbf{n}^2 \end{array} \right\}_s , \qquad \mathbf{N}_{0\,s} = \left\{ \begin{array}{c} \mathbf{0} \\ -\mathbf{n}^2 \\ \mathbf{n}^2 \end{array} \right\}_s , \tag{11.58}$$

and

$$\mathbf{T}_s = \left\{ \begin{array}{c} \mathbf{a}_1^2 \\ -(1 - \bar{\xi})\,\mathbf{a}_1^2 \\ -\bar{\xi}\,\mathbf{a}_1^2 \end{array} \right\}_s , \qquad \mathbf{T}_{0\,s} = \left\{ \begin{array}{c} \mathbf{0} \\ -\mathbf{a}_1^2 \\ \mathbf{a}_1^2 \end{array} \right\}_s . \tag{11.59}$$

With this the weak form of the contact contribution of one slave node yields $\boldsymbol{\eta}^T \mathbf{G}_s$ with the element residual

$$\mathbf{G}_s = P_{N\,s}\,\mathbf{N}_s + T_{T\,s}\,\hat{\mathbf{T}}_s . \tag{11.60}$$

11.4.3 Integration of the Friction Law

The integration of the constitutive equation for friction is based on (11.24), (11.25), (11.26), (11.27), (11.28) and (11.29). It yields an algorithmic update of the tangential stresses $\mathbf{t}_{T\,n+1}$. Since the differential equations governing the evolution of the sliding are stiff, an implicit EULER method will be selected, see also Sect. 6.2. Such procedure was first suggested in Wriggers (1987) and Giannokopoulos (1989) for frictional contact. The results due to the integration procedure are summarized for a time increment $\Delta t_{n+1} = t_{n+1} - t_n$.

The increment of the sliding within a time step Δt_{n+1} is given by

$$\Delta \mathbf{g}_{T\,n+1} = (\, \bar{\xi}_{n+1}^\alpha - \bar{\xi}_n^\alpha \,)\bar{\mathbf{a}}_{\alpha\,n+1} . \tag{11.61}$$

This total increment has to be subdivided into an elastic (stick) and plastic (slip) part, see (11.24). By

$$\mathbf{t}_{t\,n+1}^{tr} = c_T \left(\mathbf{g}_{T\,n+1} - \mathbf{g}_{T\,n}^s \right) = \mathbf{t}_{T\,n} + c_T \, \Delta \mathbf{g}_{T\,n+1}, \tag{11.62}$$

a *trial* stress is defined which is the stress computed by assuming only stick. This stress is now inserted in the slip condition

$$f_{s\,n+1}^{tr} = \| \mathbf{t}_{T\,n+1}^{tr} \| - \mu\,p_{N\,n+1} . \tag{11.63}$$

In case that the state computed with (11.62) is elastic ($f_{s\,n+1}^{tr} \leq 0$), no sliding takes place and the tangential stress at t_{n+1} is given by $\mathbf{t}_{t\,n+1} = \mathbf{t}_{t\,n+1}^{tr}$. In case that the slip condition is not fulfilled in the time increment Δt_{n+1}, $f_{s\,n+1}^{tr} > 0$, then the tangential stresses have to be projected onto the admissible region. By using the implicit EULER scheme, it follows

$$
\begin{aligned}
\mathbf{g}_{T\,n+1}^{s} &= \mathbf{g}_{T\,n}^{s} + \lambda\,\mathbf{n}_{T\,n+1}\,, \\
g_{v\,n+1} &= g_{v\,n} + \lambda\,.
\end{aligned}
\tag{11.64}
$$

On basis of the formulations and algorithms described in Chap. 6.2, the projected stresses

$$
\begin{aligned}
\mathbf{t}_{T\,n+1} &= \mathbf{t}_{t\,n+1}^{tr} - \lambda\,c_T\,\mathbf{n}_{T\,n+1}\ \text{with} \\
\mathbf{n}_{T\,n+1} &= \mathbf{n}_{T\,n+1}^{tr}
\end{aligned}
\tag{11.65}
$$

are obtained. Multiplication of (11.65) by $\mathbf{n}_{T\,n+1}$ leads to an equation for the still unknown parameter λ

$$
\kappa(\lambda) = \|\,\mathbf{t}_{T\,n+1}^{tr}\| - \hat{g}_s(p_{N\,n+1}\,,\theta\,,g_{v\,n+1}) - c_T\,\lambda = 0,
\tag{11.66}
$$

where $\hat{g}_s$ is a nonlinear function of λ. This means that, in general, an iterative method like NEWTON's method has to be applied to solve $\kappa(\lambda) = 0$. For the special case of COULOMB's models, (11.66) can be solved explicitly for λ

$$
\lambda = \frac{1}{c_T}\,(\,\|\,\mathbf{t}_{t\,n+1}^{tr}\| - \mu\,p_{N\,n+1}\,)\,.
\tag{11.67}
$$

This result can now be inserted in (11.65). With this the tangential stresses are known. The slip within one increment is given by Eq. (11.64). For COULOMB's law this yields the model

$$
\begin{aligned}
\mathbf{t}_{T\,n+1} &= \mu\,p_{N\,n+1}\,\mathbf{n}_{T\,n+1}^{tr}\,, \\
\mathbf{g}_{T\,n+1}^{s} &= \mathbf{g}_{T\,n}^{s} + \frac{1}{c_T}\,(\,\|\,\mathbf{t}_{t\,n+1}^{tr}\| - \mu\,p_{N\,n+1}\,)\,\mathbf{n}_{T\,n+1}^{tr}\,.
\end{aligned}
\tag{11.68}
$$

11.4.4 Algorithms

General algorithms for contact have to include search procedures which determine bodies which possibly can come into contact. Once possible contact is detected, the local contact conditions based on the penetration function (11.4) have to be established. The total number of nodes which are in contact state is denoted by n_c. For this number of nodes, the problem will be solved within one increment.

Before the related algorithm is stated, the matrix formulation of the global problem is provided

$$\mathbf{G}_c^p(\mathbf{v}) = \mathbf{G}(\mathbf{v}) + \sum_{s=1}^{n_c} \mathbf{G}_s^c(\mathbf{v}) = \mathbf{0}, \qquad (11.69)$$

where $\mathbf{G}(\mathbf{v})$ describes the contribution of the body due the weak form (11.34). n_c are the active contact segments, $\mathbf{G}_s^c$ was defined in (11.60).

Out of many possible algorithms for the solution of contact problems, a penalty scheme is discussed which is used in most finite element codes. It is based on the definition of active sets

$-$ Initialize algorithms
$-$ Set: $\mathbf{v}_1 = \mathbf{0}$
  $-$ LOOP over iterations: $i = 1, ..,$ until convergence
    $\bullet$ Test for contact: $g_{N\,s\,i} \leq 0 \rightarrow$ active node
    $\bullet$ Solve: $\mathbf{G}_c(\mathbf{v}_i) = \mathbf{G}(\mathbf{v}_i) + \cup_{s=1}^{n_c} \mathbf{G}_s^c(\mathbf{v}_i) = \mathbf{0}$
    $\bullet$ Convergence test: $\|\mathbf{G}_c(\mathbf{v}_i)\| \leq TOL \Rightarrow$ END LOOP
  $-$ END LOOP

This algorithm can be shortened by evaluating the distance function directly within the iterative solution of the nonlinear equations (11.69). This leads to:

$-$ Initialize algorithms
$-$ Set: $\mathbf{v}_1 = \mathbf{0}$
  $-$ LOOP over iterations: $i = 1, ..,$ until convergence
    $\bullet$ Test for contact: $g_{N\,s\,i} \leq 0 \rightarrow$ active node
    $\bullet$ Compute new displacement increment:
      $[\, D\mathbf{G}(\mathbf{v}_i) + \cup_{s=1}^{n_c} D\mathbf{G}_s^c(\mathbf{v}_i)]\Delta\mathbf{v}_i = -\mathbf{G}_c(\mathbf{v}_{i-1})$
    $\bullet$ Convergence test: $\|\mathbf{G}_c(\mathbf{v}_i)\| \leq TOL \Rightarrow$ END LOOP
  $-$ END LOOP

Here $[\, D\mathbf{G}(\mathbf{v}_i) + \cup_{s=1}^{n_c} D\mathbf{G}_s^c(\mathbf{v}_i)]$ defines the tangent matrix including the contact contributions needed in the NEWTON method. In case that an ill-conditioned system occurs due to the chosen penalty parameter, the UZAWA algorithm can be applied, see Remark 11.3 *Nr. 6.*

11.4.5 Linearization of the Contact Residual

NEWTON'S method is applied within the above described algorithm for the solution of contact problems. This requires the computation of tangent matrices. For the discretization derived in Sect. 11.4.1, these matrices can be computed analytically. They are summarized below, for a more detailed derivation, see e.g. Wriggers and Simo (1985) for the frictionless contact and Wriggers (1995) for frictional contact problems. Further details which include three-dimensional discretizations can also be found in the textbooks of Laursen (2002) and Wriggers (2006).

The tangent matrix for the normal component of contact follows from the term $\delta g_{Ns}\, P_{N\,s}$ in (11.54). In the linearization of the first term (11.47), the dependency of $\bar{\xi}$ from the current displacements has to be considered as well as the change of the normal vector $\mathbf{n}^1$. For the penalty method, the tangent matrix

$$\mathbf{K}_{Ns}^c = \epsilon_N \left[\mathbf{N}_s \mathbf{N}_s^T - \frac{g_{Ns}}{l} \left(\mathbf{N}_{0s} \mathbf{T}_s^T + \mathbf{T}_s \mathbf{N}_{0s}^T + \frac{g_{Ns}}{l} \mathbf{N}_{0s} \mathbf{N}_{0s}^T \right) \right] \quad (11.70)$$

is derived with $P_{Ns} = \epsilon_N g_{Ns}$. All matrices which enter (11.70) are already defined in (11.58) and (11.59). Note that all terms disappear in (11.70) which are multiplied by g_{Ns}. This yields the simple matrix structure $\mathbf{K}_{Ns}^{Lc} = \epsilon_N \mathbf{N}_s \mathbf{N}_s^T$.

In order to determine the tangential part of the tangent matrix for one contact segment, the term $\delta g_{Ts} T_{Ts}$ has to be linearized. For stick, this leads to

$$\mathbf{K}_{Ts}^c = c_T \left\{ [\hat{\mathbf{T}}_s \hat{\mathbf{T}}_s^T + \frac{g_{Ts}}{l} \left[\mathbf{N}_s \mathbf{N}_{0s}^T + \mathbf{N}_{0s} \mathbf{N}_s^T \right. \right.$$
$$\left. \left. - \frac{g_{Ns}}{l} (\mathbf{T}_{0s} \mathbf{N}_{0s}^T + \mathbf{N}_{0s} \mathbf{T}_{0s}^T) + \frac{g_{Ts}}{l} \mathbf{N}_{0s} \mathbf{N}_{0s}^T \right] \right\} . \quad (11.71)$$

In case of a geometrically linear theory, all terms multiplied by g_{Ns} and g_{TS} disappear. Hence (11.71) reduces to $\mathbf{K}_{Ts}^{Lc} = c_T \mathbf{T}_s \mathbf{T}_s^T$.

The case of sliding yields an additional part for (11.71). It follows from the linearization of the algorithmic update formula (11.68) for COULOMB'S law. This leads, for a node-to-segment element with (11.68), to the matrix form

$$\mathbf{K}_{Ts}^{Sc} = \mathbf{K}_{Ts}^c + \mu \epsilon_N \left(\mathbf{T}_s + \frac{g_{Ns}}{l} \mathbf{N}_{0s} \right) \mathbf{N}_s^T. \quad (11.72)$$

Note that matrix $\mathbf{K}_{Ts}^{Sc}$ is non-symmetric. This is a result of the non-associative character of the frictional constitutive equations.

12. Automation of the Finite Element Method by J. Korelc

Nowadays, the use of advanced software technologies – especially symbolic and algebraic systems – problem solving environments and automatic differentiation tools influence directly how the mechanical problem and corresponding numerical model are formulated mathematically and solved, leading to the automation of the finite element method. Automation of the finite element method has attracted attention of researches from the field of mathematics, computer science and computational mechanics, resulting in a variety of approaches and available software tools. Alternative approaches are discussed in the first section of this chapter, while an emphasis is given to the automatic generation of the finite element codes using the computer algebra systems. In order to formulate nonlinear finite elements symbolically in a general but simple way, a clear mathematical formulation is needed at the highest abstract level possible. Appropriate problem descriptions for the fully implicit analysis of non-linear, path-dependent problems and a symbolic input for the generation of a finite strain elasto-plastic element are presented at the end of this chapter.

12.1 Advanced Software Tools and Techniques

Most of the existing numerical methods for solving partial differential equations can be subdivided into two classes: finite difference (FD) and related methods and finite element (FE) and related methods. In the last years, various approaches to the automation of the two methods were studied extensively. In many ways, the present stage of the automation of the finite difference method is more elaborated and more general than the automation of the FE method. Various transformations, differentiation, matrix operations, and a large number of degrees of freedom involved in the derivation of characteristic FE quantities often lead to exponential growth of the expressions in space and time, see e.g. Fritzson and Fritzson (1984). This makes automation of the FE method more complex than automation of the finite difference method.

A complete finite element simulation can be, from the aspect of the level of automation, decomposed into the following steps:

1. formulation of the strong form of an initial boundary-value problem;
2. transformation of the strong form into a weak form or variational functional;
3. definition of the discretization of the domain and approximation of the field variables and their virtual counterparts (test functions);
4. derivation and solution of additional algebraic equations or differential equations defined at the element level (e.g. plastic evolution equations);
5. derivation of algebraic equations that describe the contribution of one element to the global internal force vector and to the global tangential stiffness matrix;
6. coding of the derived equations in the required computer language;
7. generation of a finite element mesh and its boundary conditions;
8. solution of the global problem;
9. presentation and analysis of results.

Alternatively, one can also start from the free HELMHOLTZ energy of a problem, see Sect. 3.2.3, and derive element equations directly as a gradient of the free energy. This approach is especially appealing for the automation due to the numerical efficiency of the solution when the gradient is obtained by the reverse mode of automatic differentiation.

As demonstrated throughout this book, there are almost countless ways of how a particular problem can be solved by the FE method. If the automation of all nine steps is chosen, then only very specific subsets of possible formulations can be covered. Usually, only the standard spatial discretization (see also Chap. 4) is considered as presented in Logg (2007). On the other hand, the standard discretization is of little use for problems involving coarse meshes, locking phenomena and distorted element shapes where highly problem-specific formulations described in Chaps. 10 and 11 have to be used. As usual in science, the high uniqueness of a specific formulation renders the whole concept of automation questionable. Making templates or deriving objects for something that is used only once simply does not pay off. This may be the main reason why the complete automation of the FE method is still not used within the commercial FE environments. More often, the level of automation used involves only steps that are from the numerical aspect deterministic (e.g. various correctness preserving symbolic manipulations, differentiation and automatic code generation) while the true decisions are left to the researcher.

The following techniques, which are result of the rapid development in computer science in the last decades, are particularly relevant for the description of nonlinear finite element models on a high abstract level, while preserving the numerical efficiency.

12.1.1 Symbolic and Algebraic Computational Systems

Computer algebra (CA) systems are tools for the manipulation of mathematical expressions in symbolic form. Widely used CA systems such as

Mathematica (www.wolfram.com) or *Maple* (www.maplesoft.com) have
become an integrated computing environment that covers all aspects of computational processes, including numerical analysis and graphical presentation
of the results. The general CA systems are also one of the most complex
software systems ever developed and the CA system *Mathematica* is often
described as the "world's single largest consumer of algorithms". In case of
complex mechanical models, the direct use of CA systems is not possible
due to several reasons. For the numerical implementation, CA systems cannot keep up with the run-time efficiency of programming languages such as
FORTRAN and C and by no means with highly problem-oriented and efficient numerical environments used for finite element analysis. However, CA
systems can be used for the automatic derivation of appropriate formulas
and generation of numerical codes. The FE method is usually implemented
as an additional package or toolbox within the general CA systems such as
AceFEM (www.fgg.uni-lj.si/symech/) for *Mathematica*.

The major limitation of the symbolic systems, when applied to complex
engineering problems, as pointed out before by many authors (see e.g. Wang
(1986), Fritzson and Fritzson (1984), Korelc (1997) and Korelc (2002)) is an
uncontrollable growth of expressions and consequently redundant operations
and inefficient codes. This is especially problematic when a CA system is
used to derive formulas needed in numerical procedures such as the finite
element method where the numerical efficiency of the derived formulas and
the generated code are of utmost priority. The problem of expression growth
is discussed in more detail in Sect. 12.3.

12.1.2 Automatic Differentiation Tools

Differentiation is an arithmetic operation that plays a crucial role in the development of new numerical procedures. Often it is difficult to obtain the
exact analytical derivatives, which is the reason for using instead numerical differentiation. Automatic differentiation (AD) represents an alternative
solution to the numerical differentiation as well as to the symbolic differentiation performed either manually or by a computer algebra system. With
the AD technique, one can avoid the problem of expression growth that is
associated with the symbolic differentiation performed by the CA system.
The AD technique is explained in more detail in Sect. 12.2 due to the central
role of AD in the automation of the finite element method.

12.1.3 Problem Solving Environments

Problem solving environments (PSE) are automatic code generators with libraries containing routines for various numerical solution methods. These
routines form templates for the generated program codes. The system libraries include a variety of numerical solution methods available in such

systems. They are meant to solve problems, in particular ordinary differential equations or partial differential equations, in an already established way. Several problem-solving environments for a high level abstract description of partial differential equations have been derived based on finite difference method, such as *SciNapse* (Akers et al. 1998) and *Ctadel* (van Engelen et al. 1995). A comprehensive overview can be found in Gallopoulos et al. (1994). Additionally, to the general problem solving environment, there are also tools that support only numerical operations such as compiled numerical libraries (e.g. *NAG*, www.nag.co.uk), numerical matrix languages (e.g. *MATLAB*, www.mathworks.com) and high-level object oriented languages with object libraries.

General finite element environments, such as commercial codes like *ABA-QUS* (www.hks.com) and *ANSYS* (www.ansys.com) or research codes like *FEAP* (www.ce.berkeley.edu/rlt/feap/), can also be viewed as a specialized PSE. The general finite element environments can handle, regardless of the type of finite elements, all phases of a typical finite element simulation: preprocessing of the input data, manipulation and organization of the data related to nodes and elements, material characteristics, displacements and stresses, construction of the global matrices by invoking different elements subroutines, solution of the system of equations, post-processing and analysis of the results.

12.1.4 Hybrid Approaches

The level of automation of finite element method can be greatly increased by combining several approaches and tools. Some possible combinations are discussed below.

Hybrid Object-Oriented Approach. The object-oriented approach has brought a new perspective for the development of complex software; hence in the past decade, numerous object-oriented FE environments were developed. While the object-oriented approach deals primarily with the high level of data abstraction and organization, its principles can be extended also to the complete automation of the finite element method. An overview of object-oriented hybrid symbolic-numerical approach can be found in Eyheramendy and Zimmermann (2000) and in Beall and Shephard (1999). Modern hybrid object-oriented (HOO) systems, such as FEniCS, see Logg (2007), provide tools for automation of all FE simulation steps, spanning the arc from the strong form of a given PDE to the solution and the presentation of the results. A typical HOO system introduces its own domain-specific languages and uses built-in C++ libraries for symbolic manipulation. The HOO systems are, in general, restricted to a particular type of formulations where the general knowledge of the appropriate procedure that leads from a strong form to the element equations has already been established. This also reduces the expression growth problem since the symbolic code derivation is used only for sub-problems.

Hybrid Symbolic-Numerical Approach HSN. The disadvantage of the hybrid object oriented approach is the loss of generality and flexibility compared to a general computed algebra systems. Only a small fraction of symbolic manipulation capabilities of a general CA systems is presented in specialized finite element C++ libraries for symbolic manipulation. While the hybrid object-oriented systems tends to offer complete FE solution, the idea behind the hybrid symbolic-numeric (HSN) approach is to use a general CA system for the derivation of the characteristic element quantities and the automatic code generation of user subroutines at the level of one finite element. The automatically generated code is then incorporated into an existing finite element environment (one or possibly more) and used within the global numerical solution procedures. The hybrid symbolic-numerical approach is explained in more detail in Sect. 12.3.

12.2 Automatic Differentiation

Automatic differentiation techniques are based on the fact that every computer program executes a sequence of elementary operations with known derivatives, thus allowing evaluation of exact derivatives via the chain rule for an arbitrary complex formulation.

12.2.1 Principles of Automatic Differentiation

If a computer code is given which allows to evaluate a function f and needs to compute the gradient ∇f of f with respect to arbitrary variables, then the automatic differentiation tools, see e.g. Griewank (2000), Griewank and Walther (2008), Bartholomew-Biggs et al. (2000) and Bischof et al. (2002), can be applied to generate the appropriate program code. There are two approaches for the automatic differentiation of a computer program, often recalled as the forward and the reverse mode of automatic differentiation. The procedures are illustrated by means of a simple example of the function f defined by

$$f = b\,c \quad \text{with} \quad b = \sum_{l=1}^{n} a_l^2 \quad \text{and} \quad c = \sin(b), \tag{12.1}$$

where $a_1, a_2, ..., a_n$ are n independent variables. The forward mode accumulates the derivatives of intermediate variables with respect to the independent variables as follows

$$
\begin{aligned}
\nabla b &= \left\{ \tfrac{db}{da_l} \right\} = \left\{ 2\,a_l \right\} & l &= 1, 2, ..., n \\
\nabla c &= \left\{ \tfrac{dc}{da_l} \right\} = \left\{ \cos(b)\,\nabla b_l \right\} & l &= 1, 2, ..., n \\
\nabla f &= \left\{ \tfrac{df}{da_l} \right\} = \left\{ \nabla b_l\, c + b\,\nabla c_l \right\} & l &= 1, 2, ..., n
\end{aligned}
\tag{12.2}
$$

In contrast to the forward mode, the reverse mode propagates adjoints $\bar{x} = \frac{\partial f}{\partial x}$, which are the derivatives of the final values, with respect to intermediate variables:

$$
\begin{aligned}
\bar{f} &= \frac{df}{df} = 1 && 1 \\
\bar{c} &= \frac{df}{dc} = \frac{\partial f}{\partial c}\bar{f} = b\,\bar{f} && 1 \\
\bar{b} &= \frac{df}{db} = \frac{\partial f}{\partial b}\bar{f} + \frac{\partial c}{\partial b}\bar{c} = c\bar{f} + \cos(b)\,\bar{c} && 1 \\
\nabla f &= \{\bar{a}_l\} = \left\{\frac{\partial b}{\partial a_l}\,\bar{b}\right\} = \{2\,a_l\,\bar{b}\} && l = 1, 2, ..., n.
\end{aligned}
\tag{12.3}
$$

The numerical efficieny of the differentiation can be measured by a numerical work ratio

$$
wratio(f) = \frac{numerical\_cost(f(a_1, a_2, a_3, ..., a_n), \nabla f = \frac{\partial f}{\partial a_i})}{numerical\_cost(f(a_1, a_2, a_3, ..., a_n))} .
\tag{12.4}
$$

The numerical work ratio is defined as the ratio between the numerical cost of the evaluation of function f together with its gradient ∇f and the numerical cost of evaluation of function f alone. The ratio is proportional to the number of independent variables $O(n)$ in the case of forward mode and constant in the case of reverse mode. The upper bound for the ratio in the case of reverse mode is $wratio(f) \leq 5$ and is usually around 1.5 if care is taken in handling quantities that are common to the function and gradient, see e.g. Griewank (2000) and Griewank and Walther (2008). Although numerically superior, the reverse mode requires potential storage of a large amount of intermediate data during evaluation of the function f that can be as high as the number of numerical operations performed. Additionally, a complete reversal of the program flow is required. This is because the intermediate variables are used in reverse order when related to their computation.

There exist many strategies how the automatic differentiation procedure can be implemented, see e.g. Bischof et al. (2002). The simplest approach is to use operator overloading and, during the evaluation of function f, create a trace of all numerical operations and their arguments, later used to evaluate the gradient in forward or reverse mode. The operator overloading strategy is computationally too inefficient to be used within finite element procedures. More efficient is a source-to-source transformation strategy that transforms the source code for computing a function into the source code for computing the derivatives of the function. The AD tools based on source-to-source transformation have been developed for most of the programming languages, e.g. *ADIFOR* (www-unix.mcs.anl.gov/autodiff/ADIFOR/) for Fortran, *ADOL-C* (www.math.tu-dresden.de/adol-c/)[1] for C, MAD for

---

[1] ADOL-C includes operator overloding also which can be used efficiently for higher order derivatives, see Griewank and Walther (2008).

Matlab (www.amorg.co.uk/AD/MAD/) and *AceGen* (www.fgg.uni-lj.si/ symech) for *Mathematica.*

12.2.2 Automatic Differentiation and FEM

The tools for automatic differentiation (AD) were primarily developed for the evaluation of the gradient of an objective function used within the Newton-type optimization procedures where the Hessian of objective function is needed. The objective function are often defined by a large, complex program composed of many subroutines. Thus AD tools can be applied directly within the complete FE environment, including all subroutines, to obtain the required derivatives when the evaluation of the objective function involves finite element simulations. The AD tools have been successfully applied to get gradients of residuals defined by FE environments with several hundred thousend lines of code, see e.g. Bischof et al. (2003).

The AD technology can also be used for the evaluation of specific quantities that appear as part of a finite element simulation. It would be difficult and computationally inefficient to apply the AD tools within large FE systems to get e.g. the global stiffness matrix of large-scale problem directly. This is especially problematic when a fully implicit Newton type procedure is used to solve nonlinear, transient and coupled problems involving various types of elements, complicated continuation or arc-length methods and adaptive procedures.

However, one can still use automatic differentiation at the single element level to evaluate element specific quantities in an efficient way such as:

– strain and stress tensors,
– nonlinear coordinate transformations,
– consistent tangent stiffness matrix,
– residual vector and
– sensitivity pseudo-load vector.

A direct use of automatic differentiation tools for the development of nonlinear finite elements turns out to be complex and not straightforward; furthermore the numerical efficiency of the resulting codes is poor. One solution, followed mostly in hybrid object-oriented systems, is to apply problem specific solutions to evaluate the local tangent matrix in an optimal way, see. e.g. Kirby et al. (2005). Another solution, followed in hybrid symbolic-numeric systems, see e.g. Korelc (2002), is to combine a general computer algebra system and the AD technology.

The implementation of the automatic differentiation procedure has to fulfil specific requirements in order to develop element source codes automatically that are as efficient as manually written codes. Some basic requirements are:

- The AD procedure can be initiated at any time and at any point of the derivation of the formulae and as many times as required (e.g. in the example at the end the AD is used 13 times during the generation of an element subroutine). The recursive use of standard AD tools on the same code, if allowed at all, leads to numerically inefficient source code. This requirement limits the use of standard AD tools. An alternative approach is implemented in Korelc (2002) where the source-to-source transformation strategy is replaced by a method that consistently enhances the existing code rather than producing a new one.
- The storage of the intermediate variables is not a limitation when the differentiation in reverse mode is used at the single element level. Finite element formulations at the single element level involve a relatively small set of independent and intermediate variables.
- For the reasons of efficiency, the results of all previous applications of automatic differentiation have to be accounted for, when automatic differentiation is used several times inside the same subroutine.
- The user has to be able to employ all the capabilities of the symbolic system within the final and the intermediate results of the AD procedure.
- The AD procedure must offer a mechanism for the descriptions of various mathematical formalisms applied within a finite element formulation.

The mathematical formalisms that are part of the traditional FE formulation are e.g. partial derivatives $\frac{\partial(\bullet)}{\partial(\bullet)}$, total derivatives $\frac{D(\bullet)}{D(\bullet)}$ or directional derivatives. They can all be represented by an AD procedure if possible exceptions are treated in a proper way. However, the result of AD procedure may not automatically correspond to any of the above mathematical formalisms. Hence let us define a "conditional derivative" by the following formalism

$$\nabla f = \left. \frac{\partial f(\mathbf{a}, \mathbf{b}(\mathbf{a}))}{\partial(\mathbf{a})} \right|_{\frac{\partial(\mathbf{b})}{\partial(\mathbf{a})} = \mathbf{M}}, \tag{12.5}$$

where function f depends upon a set of mutually independent variables $\mathbf{a}$ and a set of mutually independent intermediate variables $\mathbf{b}$. The above formalism has to be viewed in an algorithmic way. It represents the automatic differentiation of function f with respect to variables $\mathbf{a}$. During the AD procedure, the total derivatives of intermediate variables $\mathbf{b}$ with respect to independent variables are set to be equal to matrix $\mathbf{M}$. Some situations that typically appear in the formulation of finite elements are presented in Table 12.1.

In case A, there exists an explicit algorithmic dependency on $\mathbf{b}$ with respect to $\mathbf{a}$, hence the derivatives can be obtained in principle automatically, without intervention by the user, simply by the chain rule. However, there also exists a profound mathematical relationship that enables evaluation of derivatives in a more efficient way. This is often the case when the evaluation of $\mathbf{b}$ involves iterative loops, inverse matrices, etc.

Case B represents the situation when variables $\mathbf{b}$ are independent variables and variables $\mathbf{a}$ implicitly depend on $\mathbf{b}$. This implicit dependency has

Table 12.1 Automatic differentiation exceptions

Type	Formalism	Schematic *AceGen* input
A	$\Delta f = \dfrac{\partial f(a,b(a))}{\partial(a)}\bigg\|_{\frac{\partial(b)}{\partial(a)}=M}$	a ⊢ SMSReal[a$$] b ⊢ SMSFreeze[f_b[a]] δf ⊢ SMSD[f[a, b], a, "Implicit" → {b, a, M}]
B	$\Delta f = \dfrac{\partial f(b)}{\partial(a(b))}\bigg\|_{\frac{\partial(b)}{\partial(a)}=M}$	b ⊢ SMSReal[b$$] a ⊢ SMSFreeze[f_a[b]] δf ⊢ SMSD[f[b], a, "Implicit" → {b, a, M}]
C	$\Delta f = \dfrac{\partial f(a,b(a))}{\partial(a)}\bigg\|_{\frac{\partial(b)}{\partial(\bullet)}=0}$	a ⊢ SMSReal[a$$] b ⊢ f_b[a] δf ⊢ SMSD[f[a, b], a, "Constant" → b]
D	$\dfrac{\partial(\bullet)}{\partial(\bullet)}\bigg\|_{\frac{\partial(b)}{\partial(a)}=M}$	a ⊢ SMSReal[a$$] b ⊢ SMSFreeze[f_b[a], "Dependency" → {a, M}] ... δf_i ⊢ SMSD[f_i[a, b], a]

to be considered for the differentiation. In this case, automatic differentiation would not provide the correct result without the user intervention. A typical example for this situation is a differentiation that involves a transformation of coordinates. Usually the numerical integration procedures as well as interpolation functions require additional reference coordinate system (for details, see Chap. 4). An exception for automatic differentiation of type B is then introduced to properly handle differentiation involving coordinate transformations from initial $\mathbf{X}$ to reference coordinates ξ as follows:

$$\frac{\partial(\bullet)}{\partial\mathbf{X}} \Rightarrow \frac{\partial(\bullet)}{\partial\mathbf{X}}\bigg\|_{\frac{\partial\xi}{\partial\mathbf{X}}=\left[\frac{\partial\mathbf{X}}{\partial\xi}\right]^{-1}} \tag{12.6}$$

In case C, there exists an explicit dependency between variables $\mathbf{b}$ and $\mathbf{a}$ that has to be neglected for differentiation. The status of the dependent variable $\mathbf{b}$ is thus temporary. For the duration of the AD procedure, it is changed into an independent variable. The situation frequently appears in the formulation of mechanical problems where instead of the total variation some arbitrary variation of a given quantity has to be evaluated.

The exceptions of cases A, B and C are imposed within automatic differentiation only during the execution of the particular call of the AD procedure. Case D is equal to case A with an AD exception defined globally; thus valid for every call of the AD procedure during the derivation of the problem. When in collision, then exeptions of type A, B and C overrule the D type exception.

12.3 Hybrid Symbolic-Numerical Approach

The real power of the symbolic approach for the development, testing and application of new, unconventional ideas is provided by general purpose CA systems. However, there use is limited for problems which lead to large systems like finite element simulations. Furthermore, the use of largescale commercial finite element environments for analyzing a variety of problems is an everyday practice of engineers. The hybrid symbolic-numerical (HSN) approach is a way to combine both.

Although large FE environments often offer a possibility to incorporate user defined elements and material modes, it is time consuming to develop and test these user defined new pieces of software. Practice shows that, at the research stage of the derivation of a new numerical model, different languages and different platforms are the best means for the assessment of specific performances and, of course, failures of the numerical model. The basic tests, which are performed on a single finite element or on a small patch of elements, can be done most efficiently by using general CA systems.

Many design flaws of nonlinear finite elements, such as element instabilities or poor convergence properties, can be easily identified, if the element quantities are investigated on a symbolic level. Unfortunately, a standalone CA system becomes very inefficient once there is a larger number of nonlinear finite elements to process or if iterative numerical procedures have to be executed. In order to assess element performances under real conditions, the easiest way is to run the necessary test simulations on sequential machines with good debugging capabilities and with an open source FE environment designed for research purposes, e.g. FEAP (www.ce.berkeley.edu/ rlt/feap/), *AceFEM* (www.fgg.uni-lj.si/symech/) or Diffpack (www.diffpack.com). At the end, for real industrial simulations involving complex geometries, a large commercial FE environment has to be used.

In order to meet all these demands in an optimal way, an approach is needed that would offer multi-language and multi-environment generation of numerical codes. The automatically generated code is then incorporated into the FE environment that is most suitable for the specific step of the research process. The structure of the hybrid symbolic-numerical system *AceGen* for multi-language and multi-environment code generation introduced by Korelc (2002) is presented in Fig. 12.1. Using the classical approach, re-coding of the element in different languages would be time consuming and is rarely done. With the general computer algebra systems, re-coding comes practically for free, since the code can be automatically generated for several languages and for several platforms from the same basic symbolic description. An advantage of using a general computer algebra system is also that it provides well known and defined description language for the derivation of FE equations, generation of FE code and the possibly also for a complete FE analysis, as opposed to the hybrid object oriented systems which introduce their own domain-specific language.

When the symbolic approach is used in a standard way to describe complex-engineering problems, the common experience of computer algebra users is an uncontrollable swell of expression, as pointed out before,

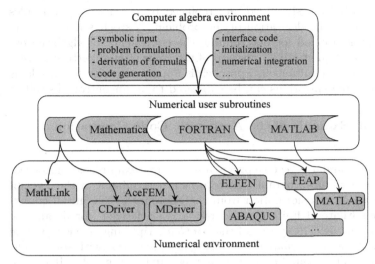

Fig. 12.1 Multi-language and multi-environement FE code generation

leading to inefficient or even unusable codes. Not many attempts have been undertaken to design a general FE code generator where this key issue, controlling the expression growth within the FE code generation, is treated within the automatic procedure. Techniques such as the use of the symmetric properties of the formulae, the automatic introduction of intermediate variables and pattern search were only used within specialized systems, see Wang (1986).

The general computer algebra systems come with the built in code optimization capabilities, see e.g. *Maple*, or additional packages for code optimization, such as *AceGen* (www.fgg.uni-lj.si/symech/) for *Mathematica*. The classical way of optimizing expressions in a computer algebra system is searching for common sub-expressions after all formulae have been derived and before the numerical code is generated. This seems to be insufficient for the general nonlinear mechanical problems and only relatively simple finite elements can be derived within this approach.

An alternative approach for automatic code generation is employed in *AceGen* and called Simultaneous Stochastic Simplification of numerical code, see Korelc (1997). This approach avoids the problem of expression swell by combining the following techniques:

– symbolic and algebraic capabilities of the general computer algebra system *Mathematica*,
– automatic differentiation techniques and
– simultaneous optimization of expressions with automatic selection and introduction of appropriate intermediate variables.

Formulae are optimized, simplified and replaced by the auxiliary variables simultaneously with the derivation of the problem. A stochastic evaluation of the formulae is applied for determining the equivalence of algebraic expressions, see e.g. Gonnet (1986), instead of the conventional pattern matching techniques. The simultaneous approach is appropriate also for problems where intermediate expressions can be subjected to an uncontrolled swell.

12.3.1 Typical Example of the Automatic Code Generation Procedure

To illustrate the standard *AceGen* procedure, a simple example is considered. A typical numerical sub-program that returns a determinant of the JACOBI matrix of nonlinear transformation from the reference to the initial configuration for quadrilateral finite element topology is derived (for details see Eq. (4.35) in Chap. 4). The syntax of the *AceGen* script language is the same as the syntax of the *Mathematica* script language, however, with some additional functions. The input for *AceGen* is presented in Fig. 12.2. It can be divided into six characteristic steps:

- At the beginning of the session the *SMSInitialize* function initializes the system.
- The *SMSModule* function defines the input and output parameters of the subroutine "DetJ".
- The *SMSReal* function assigns the input parameters $X\$\$$ and $k\$\$$ and $e\$\$$ of the subroutine to the standard *Mathematica* symbols. Double $\$$ characters indicate that the symbol is an input or output parameter of the generated subroutine.
- During the description of the problem, the special operators ($\vdash, \dashv, \models$) are used to perform the simultaneous optimization of expressions and the creation of new intermediate variables. The *SMSD* function performs an automatic differentiation of one or several expressions with respect to the arbitrary variable or the vector of variables by simultaneously enhancing the already derived code.

```
<< AceGen`;
SMSInitialize["DetJ", "Language" -> "C"];
SMSModule["DetJ",
   Real[X$$[2, 4], k$$, e$$, J$$]];
{ξ, η} ⊢ SMSReal[{k$$, e$$}];
{Xi, Yi} ⊢ SMSReal[Array[X$$, {2, 4}]]];
Ni ⊢ {(1 - ξ) (1 - η), (1 + ξ) (1 - η),
     (1 + ξ) (1 + η), (1 - ξ) (1 + η)} / 4;
J ⊢ SMSD[{Ni.Xi, Ni.Yi}, {ξ, η}];
SMSExport[Det[J], J$$];
SMSWrite[];
```

Fig. 12.2 Typical *AceGen* input

```
SUBROUTINE DetJ(v,X,k,e,J)
IMPLICIT NONE
include 'sms.h'
DOUBLE PRECISION v(5001),X(2,4),k,e,J
v(20)=((-1d0)+k)/4d0
v(21)=((-1d0)-k)/4d0
v(22)=(1d0+e)/4d0
v(19)=((-1d0)+e)/4d0
  J=(v(19)*(X(1,1)-X(1,2))+v(22)*(X(1,3)-X(1,4)))*(v(21)*(X(2,2)
&-X(2,3))+v(20)*(X(2,4)))-(v(21)*(X(1,2)-X(1,3))+v(20)*(X
&(1,1)-X(1,4)))*(v(19)*(X(2,1)-X(2,2))+v(22)*(X(2,3)-X(2,4)))
END
```

```
#include "sms.h"
void DetJ(double v[5001],double X[2][4],double
(*k),double (*e),double (*J)){
v[20]=(-1e0+(*k))/4e0;
v[21]=(-1e0-(*k))/4e0;
v[22]=(1e0-(*e))/4e0;
v[19]=(-1e0+(*e))/4e0;
(*J)=(v[19]*(X[0][0]-X[0][1])+v[22]*(X[0][2]-X[0][3]))*
(v[21]*(X[1][1]-X[1][2])+v[20]*(X[1][0]
 -X[1][3]))-(v[21]*(X[0][1]-X[0][2])+v[20]*(X[1][0]
 (X[0][0]-X[0][3]))*(v[19]*(X[1][0]-X[1][1])+v[22]*
 (X[1][2]-X[1][3]));
};
```

Fig. 12.3 Typical automatically generated subroutine in FORTRAN and C language.

- The results of the derivation are assigned to the output parameter *J$$* of the subroutine by the *SMSExport* function.
- At the end of the session, the *SMSWrite* function writes the contents of the vector of the generated formulae to the file in a prescribed language format. The generated subroutines, in C and FORTRAN language, are presented in Fig. 12.3.

12.4 Abstract Symbolic Formulations in Computational Mechanics

The true benefit using symbolic tools is not about the development of a theory what is normally done manually on a sheet of paper using a pencil, or if a computer shall be used a simple word processor is adequate for such task. The advantage of the symbolic approach in computational mechanics becomes apparent only when the description of the problem, which means that the basic equations are written down, is appropriate for the symbolic description. Unfortunately, some of the traditional descriptions, used in computational mechanics, are not appropriate for the symbolic description. The symbolic formulation of the computational mechanics problems differs often from the classical formulations described in detail in other chapters of this book, and thus brings up the need for rethinking and reformulating of known and traditional ways. Despite that, there exist strong arguments why, at the end, symbolic formulations are indeed beneficial, i.e.:

- A symbolic formulation is more compressed and thus provides fewer possibilities for an error.
- Algebraic operations, such as differentiation, are done automatically.
- Automatically generated codes are highly efficient and portable.
- The multi-language and multi-environment capabilities of symbolic systems enable generation of numerical codes for various numerical environments from the same symbolic description.
- An available collection of prepared symbolic inputs for a broad range of finite elements can be easily adjusted for the user specific problem leading to the on-demand numerical code generation.
- The multi-field and multi-physic problems can be easily implemented. For example, the symbolic inputs for mechanical and thermal analysis can be combined into a new symbolic input that would create a finite element for fully coupled and quadratically convergent thermo-mechanical analysis.

For example, the standard formulation of the first term of the tangential stiffness matrix $\boldsymbol{B}^T \boldsymbol{D} \boldsymbol{B}$ (for details see Chap. 4, e.g. Eq. (4.83)) can be easily repeated using the symbolic tools. Having in mind that the tangential stiffness matrix of a finite element is either the JACOBIAN of the resulting system of discrete algebraic equations or the HESSIAN of the variational functional, then automatic differentiation should be sufficient tool for obtaining the tangent matrix. The work of implementing the $\boldsymbol{B}^T \boldsymbol{D} \boldsymbol{B}$ formulation and the efficiency of the resulting code is inferior to the approach when the tangent matrix is derived by the reverse automatic differentiation. The latter approach requires, regardless of the complexity of the topology and the material model, a single line of symbolic input. The standard $\boldsymbol{B}^T \boldsymbol{D} \boldsymbol{B}$ formulation would require much more input for the same result.

It should be pointed out that the symbolic differentiation is one of the algebraic operations prone to severe expression growth and it can result even for relatively simple nonlinear elements in hundreds of pages of code. Thus, the use of a hybrid system that combines the symbolic tool with the automatic differentiation technique is essential for the high abstract symbolic formulation of nonlinear finite element models. To increase the numerical efficiency of the generated code and to limit the physical size of the generated code, it is essential to minimize the number of calls to the automatic differentiation procedure. In the reverse mode of automatic differentiation, the expression $SMSD\,[a, \mathbf{c}] + SMSD\,[b, \mathbf{c}]$ can result in a code that is twice as large and twice slower than the code produced by the equivalent expression $SMSD\,[a + b, \mathbf{c}]$.

In this section, an abstract symbolic formulation is described which is needed to obatain the contribution of a single element Ω_e to the internal force vector $\boldsymbol{R}$ and to the tangential stiffness matrix $\boldsymbol{K}_T$. The formulation follows the basic equations of continuum mechanics provided in Chap. 3 and spatial discretization techniques given in Chap. 4. As pointed out in Sect. 3.4, the variational functional approach and the weak form approach are the two basic possibilities open for the derivation of variational formulation of equilibrium equations and their linearizations.

12.4.1 Variational Principle

In case of hyperelastic material responses, a principle of stationary elastic potential can be formulated, see Sect. 3.4.3. Within this formulation, the functional of the strain energy density function W can be formulated, by using the discretization techniques in Chap. 4, as a function of N generalized displacement parameters $\boldsymbol{u}_e$ of the element. The contribution of one element to the residual $\boldsymbol{R}$ and the tangent matrix $\boldsymbol{K}_T$ is then obtained by automatic differentiation as first and second derivative of W integrated over the domain Ω of the element

$$\boldsymbol{R}_e = \int_{\Omega_e} \frac{\partial W}{\partial \boldsymbol{u}_e} d\Omega\,,$$

$$\boldsymbol{K}_{Te} = \frac{\partial \boldsymbol{R}_e}{\partial \boldsymbol{u}_e}\,. \tag{12.7}$$

The use of automatic differentiation is straightforward in this case and there is no need to derive any additional intermediate quantities such as the traditional $\boldsymbol{B}$-matrices and the incremental constitutive matrix $\boldsymbol{D}$. When the reverse mode of automatic differentiation is applied, then the evaluation also becomes optimal from the point of numerical efficiency of the evaluation of $\boldsymbol{R}_e$ and $\boldsymbol{K}_{Te}$. The evaluation of the residual $\boldsymbol{R}_e$ starts from the scalar strain energy function W; thus only one reversal of the program flow and only one construction of the adjoining variables is required. The numerical cost ratio for the evaluation of $\boldsymbol{R}_e$ is then $wratio, (\boldsymbol{R}_e) < 5$. The evaluation of $\boldsymbol{K}_{Te}$ starts from the residual, a vector of N functions; thus one reversal of the program flow and a construction of N adjoining variables is required. The numerical cost ratio for the evaluation of $\boldsymbol{K}_{Te}$ is then proportional to N, making the total cost of the evaluation of the element contribution proportional to the number of degrees of freedom of the element $(O(N))$.

12.4.2 Weak Form

Another approach which can be persued to derive the finite element discretizations starts from the weak form of equilibrium, as introduced in Sect. 3.4.1. The part of the weak form of equilibrium describing the stress divergent term (without the load contribution) is provided in Eq. (3.292)

$$\int_{\Omega_e} \mathbf{S} \cdot \delta \mathbf{E}\, d\Omega \tag{12.8}$$

where $\mathbf{S}$ is the 2nd PIOLA–KIRCHHOFF stress tensor and $\mathbf{E}$ the GREEN–LAGRANGIAN strain tensor. Other work conjugated stress–strain pairs can be considered as well as described in Sects. 3.4.1 and 3.4.2. The symbolic formulation of the weak form is not straightforward, since the variation $\delta \mathbf{E}$ is not a real but rather factitious quantity and automatic differentiation cannot be applied directly. Of course, a general computer algebra system can be used for building the necessary apparatus to deal with the variations in a traditional way and then automatic code generation can be applied on the results. But then the elegance of using automatic differentiation would then be lost. However, the automatic differentiation can be applied directly after the variation is discretized.

The discretization of the variation $\delta \mathbf{E}$ leads to

$$\delta \mathbf{E} = \sum_{i=1}^{N} \frac{\partial \mathbf{E}}{\partial u_{ei}} \delta u_{ei}, \tag{12.9}$$

where the u_{ei} are the N generalized displacements parameters of the element. The finite element approximation of the weak form, appropriate for symbolic description, and its linearization can then easily be obtained. This leads to a form for the element contribution $\mathbf{R}_e$ and $\mathbf{K}_{Te}$ to the residual and tangent matrix which is relevant for the symbolic formulation

$$\mathbf{R}_e = \int_{\Omega_e} \mathbf{S} \cdot \frac{\partial \mathbf{E}}{\partial \mathbf{u}_e} \, d\Omega \, ,$$
$$\mathbf{K}_{Te} = \frac{\partial \mathbf{R}_e}{\partial \mathbf{u}_e} \, . \tag{12.10}$$

While the variational functional formulation starts from one scalar quantity, the weak form formulation starts from the six or nine scalar quantities (the six components of the symmetric strain tensor $\mathbf{E}$ or nine components, if the chosen strain measure is nonsymmetric). The evaluation of the weak form $\mathbf{R}_e$ by reverse automatic differentiation is thus theoretically six times more expensive when $(12.10)_1$ is used. Additionally, the stress tensor $\mathbf{S}$ has to be evaluated. The actual difference is usually much smaller due to the code optimization procedures. The evaluation of $\mathbf{K}_{Te}$ starts again from a vector containing N functions which leads to a numerical cost ratio for the evaluation of $\mathbf{K}_{Te}$ proportional to N. The total cost of the evaluation of the element contribution is proportional to N. This is roughly the same as for the formulation based on the variational functional, see (12.7).

12.4.3 Symbolic Formulation of Elasto-Plastic Problems

In this section, a general method is presented for the automatic derivation of $\mathbf{R}_e$ and $\mathbf{K}_{Te}$ in the case of arbitrary elasto-plastic problems as introduced in Sects. 3.3.2, 6.2 and 6.3. Let $\mathbf{u}_{e\,n+1}$ be a vector of generalized displacement parameters of the element, $\mathbf{p}_{n+1}$ a vector of unknowns at GAUSS point level and $\boldsymbol{p}_n$ is a vector of history values at GAUSS point level from the previous time step.

The elasto-plastic problem is defined by a hyperelastic strain energy density function W, a yield condition f and a set of algebraic constraints to be fulfilled at GAUSS point level $\boldsymbol{Q}_{n+1}(\boldsymbol{u}_{n+1}, \boldsymbol{p}_{n+1}, \boldsymbol{p}_n)$ that have to be solved for unknowns $\boldsymbol{p}_{n+1}$ when the material point is in a plastic state. In general, the vector $\boldsymbol{p}_{n+1}$ is composed of an appropriate measure of plastic strains (or stresses in the small deformation case), the hardening variables and the consistency parameter λ where the $\boldsymbol{Q}_{n+1}$ are composed of the corresponding set of discretized evolution equations that describe the evolution of plastic strains and hardening variables and the consistency condition $f = 0$. No restriction is imposed on the form of the algebraic equations at GAUSS point level at this point. For more details about possible formulations at GAUSS point level, see Sects. 6.2 and 6.3.

Due to the fact that the evolution equations are stiff, an implicit EULER integration is chosen. As discussed in Sects. 6.2 and 6.3, the yield condition is evaluated for a the trial state by freezing the state variables as follows

$$f^{tr} = f(\boldsymbol{u}_{n+1}, \boldsymbol{p}_n). \tag{12.11}$$

Due to the dissipative nature of elasto-plastic problems, the variational functional does not exist and the weak form (12.10) has to be used. A material point is in the elastic domain for $f^{tr} \leq 0$. The stress tensor $\boldsymbol{S}$ and the unknowns at the GAUSS point $\boldsymbol{p}_{n+1}$ are given for an elastic state by

$$\boldsymbol{S} = 2\frac{\partial W(\boldsymbol{u}_{n+1}, \boldsymbol{p}_n)}{\partial \mathbf{C}}, \tag{12.12}$$

$$\boldsymbol{p}_{n+1} = \boldsymbol{p}_n. \tag{12.13}$$

The set of algebraic equations $\boldsymbol{Q}_{n+1}$ has to be fulfilled at each GAUSS point of the finite element discretization for which the yield condition is violated: $f^{tr} > 0$. The associated nonlinear equations are solved by the iterative NEWTON method using an additional iterative loop at each GAUSS point, as already discussed in Sects. 6.2 and 6.3. The related algorithm is presented in Box 12.1, where $\tilde{\boldsymbol{p}}_{n+1}$ denotes a vector of the local unknowns at GAUSS point level within the iterative loop. $\boldsymbol{A}_{n+1}$ is a matrix that follows from the linearization of the nonlinear equation set $\boldsymbol{Q}_{n+1}$. Due to the iterative loop needed to solve $\boldsymbol{Q}_{n+1}$, the variables $\boldsymbol{p}_{n+1}$ depend now implicitly upon the generalized displacement parameters $\boldsymbol{u}_{e\,n+1}$. The direct application of the automatic differentiation procedure to obtain $\boldsymbol{S}$ would consider this algorithmic dependency and the evaluated stress tensor would not be correct. With the use of the type C exception of the auomatic differentiation procedure (see Table 12.1), the correct stress tensor $\boldsymbol{S}$ can be expressed as follows

$$\boldsymbol{S} = 2\frac{\partial W}{\partial \mathbf{C}}\bigg|_{\frac{\partial \boldsymbol{p}}{\partial(\bullet)} = 0} \tag{12.14}$$

summation over integration points

$\mathbf{x} := \mathbf{X}(\xi) + \mathbf{u}(\mathbf{u}_{e\,n+1}, \xi)$

use AD exception of type B for coordinate transformation

$\mathbf{F} := \frac{\partial \mathbf{x}}{\partial \mathbf{X}}\big|_{\frac{\partial \xi}{\partial \mathbf{X}} = [\frac{\partial \mathbf{X}}{\partial \xi}]^{-1}}$

$\tau^{trial} := \tau(\mathbf{u}_{e\,n+1}, \mathbf{p}_n)$

$f(\tau^{trial}) \leq 0 \;\{\mathbf{p}_{n+1} := \mathbf{p}_n$

$f(\tau^{trial}) > 0 \;\begin{cases} \quad \text{local Newton loop} \\ \quad \tilde{\mathbf{p}}_{n+1} := \mathbf{p}_n \\ \quad \text{repeat} \\ \qquad \mathbf{A}_{n+1} := \frac{\partial \mathbf{Q}_{n+1}(\mathbf{u}_{n+1}, \tilde{\mathbf{p}}_{n+1}, \mathbf{p}_n)}{\partial \tilde{\mathbf{p}}_{n+1}} \\ \qquad \Delta\tilde{\mathbf{p}}_{n+1} := -\mathbf{A}_{n+1}^{-1}\mathbf{Q}_{n+1}(\mathbf{u}_{n+1}, \tilde{\mathbf{p}}_{n+1}, \mathbf{p}_n) \\ \qquad \tilde{\mathbf{p}}_{n+1} := \tilde{\mathbf{p}}_{n+1} + \Delta\tilde{\mathbf{p}}_{n+1} \\ \quad \text{until } \|\Delta\tilde{\mathbf{p}}_{n+1}\| < TOL \\ \quad \mathbf{p}_{n+1} := \tilde{\mathbf{p}}_{n+1} \\ \quad \text{define AD exception of type D for } \mathbf{p}_{n+1} \\ \qquad \frac{\partial(\cdot)}{\partial(\cdot)}\big|\frac{\partial \mathbf{p}_{n+1}}{\partial \mathbf{u}_{e\,n+1}} = -\mathbf{A}_{n+1}^{-1}\frac{\partial \mathbf{Q}_{n+1}(\mathbf{u}_{n+1}, \mathbf{p}_{n+1}, \mathbf{p}_n)}{\partial \mathbf{u}_{e\,n+1}} \end{cases}$

use AD exception of type C

$\mathbf{R}_e := \frac{\partial W(\mathbf{u}_{e\,n+1}, \mathbf{P}_{n+1})}{\partial \mathbf{u}_{e\,n+1}}\big|_{\frac{\partial \mathbf{P}_{n+1}}{\partial(\bullet)} = 0}$

$\mathbf{K}_{Te} := \frac{\partial \mathbf{R}_e}{\partial \mathbf{u}_{e\,n+1}}$

end loop

Box 12.1 Algorithm for the abstract symbolic description of elasto-plastic problems

From the definition of the GREEN–LAGRANGIAN strain tensor $\mathbf{E}$, noting that Eq. (12.14) implies Eq. (12.12) and by assuming the same discretization for displacements and the variation of the displacements (test functions) a final "basic equation of the symbolic plasticity", is derived

$$\mathbf{R}_e = \int_{\Omega_e} 2\frac{\partial W}{\partial \mathbf{C}}\bigg|_{\frac{\partial \mathbf{p}}{\partial(\bullet)} = 0} \cdot \frac{1}{2}\frac{\partial \mathbf{C}}{\partial \mathbf{u}_{e\,n+1}}\, d\Omega$$
$$= \int_{\Omega_e} \frac{\partial W}{\partial \mathbf{u}_{\epsilon\,n+1}}\bigg|_{\frac{\partial \mathbf{p}}{\partial(\bullet)} = 0}\, d\Omega\,. \tag{12.15}$$

An efficient and accurate numerical solution of the corresponding coupled nonlinear system of algebraic equations requires quadratically convergent numerical procedure. For this the linearization of (12.15) is needed which leads to the tangent stiffness matrix. This matrix can be derived for a finite element by directly applying the automatic differentiation procedure leading to

$$\mathbf{K}_{Te} = \frac{\partial \mathbf{R}_e}{\partial \mathbf{u}_{e\,n+1}}\,. \tag{12.16}$$

Tangent stiffness matrix derived in this way is already "consistent" with the algorithm used for plasticity. Hence no additional procedures to derive a consistent tangent modulus are required. However, this involves the differentiation of the complete iterative NEWTON procedure at GAUSS point level. This can be avoided if additionally the type D exception of automatic differentiation is used. The derivatives of the variables at GAUSS point level with respect to the generalized displacements can be obtained in a more efficient way by solving the following sensitivity problem

$$
\boldsymbol{A}_{n+1}\frac{\partial \boldsymbol{p}_{n+1}}{\partial \boldsymbol{u}_{e\,n+1}} = -\frac{\partial \boldsymbol{Q}_{n+1}(\boldsymbol{u}_{n+1}, \boldsymbol{p}_{n+1}, \boldsymbol{p}_n)}{\partial \boldsymbol{u}_{e\,n+1}}. \tag{12.17}
$$

The derivatives are then defined by the following type D exception of automatic differentiation

$$
\frac{\partial(\cdot)}{\partial(\cdot)}\bigg|\frac{\partial \boldsymbol{p}_{n+1}}{\partial \boldsymbol{u}_{e\,n+1}} = -\boldsymbol{A}_{n+1}^{-1}\frac{\partial \boldsymbol{Q}_{n+1}(\boldsymbol{u}_{n+1}, \boldsymbol{p}_{n+1}, \boldsymbol{p}_n)}{\partial \boldsymbol{u}_{e\,n+1}}. \tag{12.18}
$$

The exception (12.18) of automatic differentiation effectively bypasses the true algorithm used to calculate $\boldsymbol{p}_{n+1}$. In some cases, a closed form solution for all or a part of the variables at GAUSS point level can be derived, improving the overall numerical efficiency of the procedure. Several advantages of the formulation (12.15) can be observed with respect to the standard formulation of elasto-plastic problems, see Sects. 6.2 and 6.3:

- Equations (12.15) and (12.16) unify the elastic and plastic state; thus only two calls to automatic differentiation procedure are needed (one to evaluate $\boldsymbol{R}_e$ and one to evaluate $\boldsymbol{K}_{Te}$).
- The formulation starts from the scalar quantity W which is optimal for the automatic differentiation and automatic code generation.
- The stress and the strain tensors do not appear explicitly in (12.15), thus the question of choosing the optimal stress–strain pair does not arise at all. The only free parameters of the formulation are the strain energy function, the yield condition, the evolution equations and the discretization of the domain and displacements.
- The presented formulation is expressed with respect ot the initial configuration. The spatial formulation can also be derived. Due to the use of automatic differentiation, the $\boldsymbol{B}$ matrix does not appear explicitly as a part of the formulation. Consequently, the advantage of the spatial formulation (sparse $\boldsymbol{B}$ matrix, see Remark 4.5) does not materialize.
- Equation (12.15) can be obtained also directly as a gradient of the free energy function.
- The described formulation can be employed to derive small strain as well as finite strain plasticity models, like multi-surface plasticity, non-associate plasticity models, compressible plasticity models, etc. Also various finite

element discretization techniques (standard displacement elements, enhanced strain elements, underintegrated formulations, etc. can be incorporated).

12.5 Finite Strain Plasticity Example

In this section, an abstract symbolic formulation for the contribution of one finite element Ω_e to the internal force vector $\boldsymbol{R}_e$ and to the tangential stiffness matrix $\boldsymbol{K}_{Te}$ is presented for a problem undergoing finite plastic strains.

12.5.1 Formulation

The formulation employs the general implicit EULER scheme for the integration of the finite strain inelastic constitutive equations as stated in Sect. 6.3. The used NEO–HOOKE strain energy W is defined in Sect. 3.3.1, see e.g. Eq. (3.116). The parts of the finite strain plasticity model necessary for the abstract symbolic description are briefly summarized in Box 12.2. The derivation is here described for a two-dimensional quadrilateral finite element with four nodes. The spatial discretization of domain and displacement is based on the standard isoparametric concept as presented in Chap. 4. The plane strain condition is enforced. The vector of the variables at GAUSS point level contains the components of the plastic strains $\mathbf{F}_{n+1}^{p-1} - \mathbf{1}$ at time t_{n+1} and the consistency parameter λ_{n+1} which is related to the same time

$$p_{n+1}^T = \left\{ (F_{n+1}^{p-1})_{11} - 1, (F_{n+1}^{p-1})_{12}, (F_{n+1}^{p-1})_{21}, (F_{n+1}^{p-1})_{22} - 1, (F_{n+1}^{p-1})_{33} - 1, \lambda_{n+1} \right\}. \tag{12.19}$$

It is worth noticing that inaccurate integration of the plastic evolution equations leads to a loss of volume in case of incompressible plasticity to

$$
\begin{aligned}
&\mathbf{F}_{n+1}^e = \mathbf{F}_{n+1} \mathbf{F}_{n+1}^{p-1} \\
&\mathbf{C}_{n+1}^e = \mathbf{F}_{n+1}^{eT} \mathbf{F}_{n+1}^e \\
&J^2 = \det(\mathbf{C}_{n+1}^e) \\
&W = \frac{\mu}{2}(tr(\mathbf{C}_{n+1}^e) - 3 - \ln(J^2)) + \frac{\lambda}{4}(J^2 - 1 - \ln(J^2)) \\
&\boldsymbol{\tau} = 2\mathbf{F}_{n+1}^e \frac{\partial W}{\partial \mathbf{C}_{n+1}^e} \mathbf{F}_{n+1}^{eT} \\
&\mathbf{s} = \boldsymbol{\tau} - \frac{tr(\boldsymbol{\tau})}{3}\mathbf{1} \\
&\alpha = \sqrt{\frac{2}{3}}\,\lambda \\
&f = \sqrt{\mathbf{s}\cdot\mathbf{s}} - \sqrt{\frac{2}{3}}\,(Y_0 + H\alpha) \\
&\mathbf{Q}_{n+1} = \left\{ \begin{array}{l} \mathbf{F}_{n+1}^e - exp(-(\lambda_{n+1} - \lambda_n)\frac{\partial f}{\partial \boldsymbol{\tau}})\mathbf{F}_{n+1}\mathbf{F}_n^{p-1} = 0 \\ f = 0 \end{array} \right\}
\end{aligned}
$$

Box 12.2 Summary of the finite strain plasticity model

a non-symmetric global tangent matrix in case of isotropic plasticity, and hence to a loss of objectivity of the resulting finite element. The problems can be avoided by an exact exponential approximation of the evolution of the plastic deformation gradient, see e.g. Simo (1998). An exponential approximation requires a reliable evaluation of the matrix exponential and its derivatives which has proved to be a difficult task. Thus a numerical approximation is used instead, see e.g. Itskov (2003) and Lu (2004). With the *AceGen* function *SMSMatrixExp* the exact, closed form solution of a matrix exponent is obtained by automatic differentiation of an appropriate scalar function presented in Lu (2004). The automatically generated closed form solution of a matrix exponential yields accurate results up to machine precision. This is also the case for multiple eigenvalues, and hence significantly improves the reliability of the finite strain plasticity formulation.

12.5.2 *AceGen* Input

Now the structure of the input for *AceGen* is presented using, as an example, the two-dimensional version of the finite strain plasticity element as described above.

– Step 1: **Initialization**
   Here the *AceGen* is initialized and the element characteristics necessary for the automatic creation of the interface between the automatically generated code and the chosen finite element environment are defined. The *SMSStandardModule* command starts the definition of the user subroutine for the calculation of the tangent matrix and the residual vector. After that the loop over the GAUSS points is initiated.

```
<< AceGen`;
SMSInitialize["FpW", "Environment" -> "AceFEM"];
nhistory = 7; nstate = 6;
SMSTemplate["SMSTopology" -> "Q1",
 "SMSNoTimeStorage" -> nhistory es$$["id", "NoIntPoints"],
 "SMSGroupDataNames" -> {
   "E -elastic modulus", "v -poisson ratio",
   "Y0 -initial yield stress", "H -hardening coefficient"},
 "SMSSymmetricTangent" -> True]

SMSStandardModule["Tangent and residual"]

SMSDo[IpIndex, 1, SMSInteger[es$$["id", "NoIntPoints"]]];
```

– Step 2: **Interface to the input data of the user element subroutine**
   Here the coordinates of the current integration points ξ, η, ζ, the integration point weights w_g, the coordinates of the element nodes Xi, Yi, the current values of the displacements ui, vi and the material properties of the element are taken from the supplied arguments of the subroutine. All global degrees of freedom are then collected in one vector $un1 \equiv \boldsymbol{u}_{n+1}$, such that the proper degree of freedom ordering is established. The variable hi

defines the location of the variables $pn \equiv \boldsymbol{p}_n$ at GAUSS point level within the field of the element history variables ($ed\$\$[hp, ...]$) for the $IpIndex$ -th integration point.

```
{ξ, η, ζ, wg} ⊦ Array[SMSReal[es$$["IntPoints", #1, IpIndex]] &, 4];
Xi ⊧ Array[SMSReal[nd$$[#1, "X", 1]] &, 4];
Yi ⊧ Array[SMSReal[nd$$[#1, "X", 2]] &, 4];
ui ⊧ Array[SMSReal[nd$$[#1, "at", 1]] &, 4];
vi ⊧ Array[SMSReal[nd$$[#1, "at", 2]] &, 4];
un1 ⊦ Flatten[Transpose[{ui, vi}]];
{Em, ν, Y0, H } ⊧ Array[SMSReal[es$$["Data", #1]] &, 4];
hi ⊧ SMSInteger[(IpIndex - 1) * nhistory];
pn ⊧ SMSReal[Array[ed$$["hp", hi + #1] &, nstate]];
```

- Step 3: **Definition of the trial functions and kinematic equations**
  This defines the shape functions N_i, the interpolation of the physical coordinates X, Y and the displacements u, v within the element, the JACOBI matrix of the isoparametric mapping, the displacement gradient $Dn1 \equiv \boldsymbol{H}_{n+1}$, the deformation gradient $Fn1 \equiv \boldsymbol{F}_{n+1}$ and the inverse plastic deformation gradient $Fpin \equiv \boldsymbol{F}_n^{p-1}$.

```
     1
Ni ⊧ ─ { (1 - ξ) (1 - η), (1 + ξ) (1 - η), (1 + ξ) (1 + η), (1 - ξ) (1 + η) };
     4
X ⊧ SMSFreeze[Ni.Xi]; Y ⊧ SMSFreeze[Ni.Yi];
Z ⊧ SMSFreeze[ζ];
Jm ⊧ SMSD[{X, Y, Z}, {ξ, η, ζ}];
u ⊧ Ni.ui; v ⊧ Ni.vi;
Dn1 ⊧ SMSD[{u, v, 0}, {X, Y, Z},
    "Implicit" → {{{ξ, η, ζ}, {X, Y, Z}, Inverse[Jm]}}];
Fn1 ⊧ IdentityMatrix[3] + Dn1;
```

	pn[[1]]	pn[[2]]	0
Fpin = IdentityMatrix[3] +	pn[[3]]	pn[[4]]	0
	0	0	pn[[5]]

;

- Step 4: **Definition of the constitutive model dependent quantities**
  Here the fWA function is defined. It returns, the value of the input parameter $task$, the yield condition f, the strain energy function W or the evolution equations Q at a GAUSS point. They are evaluated for the given values of the variables pt, at GAUSS point level. Thus, the function returns, with respect to the supplied parameters, either trial values or the iterative values. Note that the function $SMSMatrixExp$ returns an exact, closed form solution of the matrix exponent.

```
fWA[pt_, task_] := Block[{},
```

$$
\text{Fpin1 = IdentityMatrix[3] +}
\begin{array}{|c|c|c|}
\hline
\text{pt[[1]]} & \text{pt[[2]]} & 0 \\
\hline
\text{pt[[3]]} & \text{pt[[4]]} & 0 \\
\hline
0 & 0 & \text{pt[[5]]} \\
\hline
\end{array}
\; ;
$$

```
Fe ⊧ Fn1.Fpin1;
Ce ⊧ SMSFreeze[Transpose[Fe].Fe, "KeepStructure" → True];
{λ, μ} ⊧ SMSHookeToLame[Em, ν];
Je2 ⊧ Det[Ce];
W = Simplify[ μ / 2 (Tr[Ce] - 3 - Log[Je2]) + λ / 4 ( Je2 - 1 - Log[Je2])];
If[task == "W", Return[W]];
τ ⊧ Simplify[2 Fe.SMSD[W, Ce, "IgnoreNumbers" → True,
      "Symmetric" → True].Transpose[Fe]];
τF ⊧ SMSFreeze[τ, "KeepStructure" → True];
{λ, λn} = {pt[[6]], pn[[6]]};
```

$$
\text{s = τF} - \frac{1}{3}\ \text{IdentityMatrix[3] Tr[τF];}
$$

$$
\alpha = \sqrt{2/3}\ \lambda;
$$

```
f = SMSSqrt[Total[s s, 2]] - √(2 / 3)  (Y0 + H α);
If[task == "f", Return[f]];
𝒜 = Simplify[SMSD[f, τF, "IgnoreNumbers" → True,
      "Symmetric" → True]];
𝒵 ⊧ Simplify[Fe - SMSMatrixExp[- (λ - λn) 𝒜].Fn1.Fpin];
Q ⊧ {𝒵[[1, 1]], 𝒵[[1, 2]], 𝒵[[2, 1]], 𝒵[[2, 2]], 𝒵[[3, 3]], f};
Return[Q];
]
```

− Step 5: **Elastic part**

The state at a GAUSS point is stored as an additional history variable. The stored information is used within the first global NEWTON iteration in order to improve the convergence radius of the global NEWTON iteration used to solve the nonlinear weak form of the problem at hand. For $pt = pn$, the fWA function returns the trial yield condition $ftr \equiv f^{tr}$. The history variables $pn1 \equiv \boldsymbol{p}_{n+1}$ at the GAUSS point are not changed with respect to the values at the previous time step t_n in the case of elastic state $ftr \leq 0$:
$$\boldsymbol{P}_{n+1} = \boldsymbol{P}_n.$$

```
ftr ⊧ fWA[pn, "f"];
SMSIf[ (iter == 1 && SMSInteger[ed$$["hp", hi + nstate + 1]] == 0)  ||
```

$$
\left(\text{iter > 1 && ftr} < \frac{1}{10^8} \right)];
$$

```
pn1 ⊣ pn;
SMSExport[0, ed$$["ht", hi + nstate + 1]];

SMSElse[];
```

- Step 6: **Plastic part**

  For plastic deformations ($ftr > 0$), the local NEWTON iterative loop is implemented, according to Box 12.1, for the vector of variables $pln1 \equiv \tilde{\mathbf{p}}_{n+1}$ at the GAUSS point. The derivatives of the variables at GAUSS point level with respect to the generalized displacements $\delta p\delta u \equiv \frac{\partial \mathbf{p}_{n+1}}{\partial \mathbf{u}_{e\,n+1}}$ are evaluated and the automatic differentiation exception of type D is defined for variable $pn1$. The error flag is set, if the NEWTON iterative loop could not converge within 30 iterations. Note that the function *SMSLUFactor* performs full symbolic factorization of the system of linear equations and the function *SMSLUSolve* full symbolic back substitution.

```
pt ⊣ pn;
SMSDo[i, 1, 30, 1, pt];
 Q ⊨ fWA[pt, "Q"];
 A ⊨ SMSD[Q, pt];
 LU ⊣ SMSLUFactor[A];
 Δp ⊨ SMSLUSolve[LU, -Q];
 pt ⊣ pt + Δp;
 SMSIf[Sqrt[Δp.Δp] < 1 / 10^9];
  δpδu ⊣ SMSLUSolve[LU, -SMSD[Q, un1, "Constant" → pt]];
  SMSBreak[];
 SMSEndIf[];
 SMSIf[i == 29];
  SMSExport[2, idata$$["ErrorStatus"]];
  SMSBreak[];
 SMSEndIf[];
SMSEndDo[pt, δpδu];
pn1 ⊣ SMSFreeze[pt, "Dependency" → {un1, δpδu}];
SMSExport[1, ed$$["ht", hi + nstate + 1]];
SMSEndIf[pn1];
```

- Step 7: **Element tangent stiffness matrix and internal force vector**

  Here the strain energy W, the part of the tangent stiffness matrix $KeTij$ associated with a nodal value ij and the internal force vector Rei associated with a nodal value i, are evaluated for the final values of the variables $pn1$ at a GAUSS point. The vectors, containing the quantities $KeTij$, the Rei and the $pn1$, are exported by *SMSExport* to the output parameters of the user element subroutine.

```
W ⊨ fWA[pn1, "W"];
SMSDo[i, 1, 8];
 Rei ⊨ Det[Jm] wg SMSD[W, un1, i, "Constant" → pn1];
 SMSExport[Rei, p$$[i], "AddIn" → True];
 SMSDo[j, 1, 8];
  KeTij = SMSD[Ri, un1, j];
  SMSExport[KeTij, s$$[i, j], "AddIn" → True];
 SMSEndDo[];
SMSEndDo[];
SMSExport[pn1, ed$$["ht", hi + #] &];
```

– Step 8: **Code generation**
This is the end of the integration loop. The element source code is generated and written in "J2C.c" file in C language.

```
SMSEndDo[];
SMSWrite[];
```

12.5.3 Efficiency of Automatically Generated Codes

Following the procedures described in previous sections, various elements can be derived. It is essential for the use of the symbolic approach that the automatically generated elements are efficient with respect to evaluation time when compared to finite elements which were coded manually. The following different finite elements are investigated:

– Q1: the standard two-dimensional displacement, quadrilateral, isoparametric element for plane strain problems, see e.g. Exercise 4.3 in Sect. 4.2.1,
– Q1E4: the two-dimensional, quadrilateral, enhanced assumed strain element (EAS) with four enhanced modes, introduced by Simo and Rifai (1990), for plane strain problems, see Sect. 10.5,
– H1: the standard three-dimensional displacement, hexahedral, isoparametric element, see Sect. 4.2.1,
– H1E9: the three-dimensional hexahedral enhanced assumed strain element (EAS) with nine enhanced modes, introduced by Simo and Armero (1992), for plane strain problems, see Sect. 10.5.

Each element is derived and analysed for four different cases: linear elasticity, hyperelasticity, small strain elasto-plasticity and finite strain elasto-plasticity. A 2×2 GAUSS integration is used for all two-dimensional and a $3 \times 3 \times 3$ GAUSS integration is employed for all three-dimensional elements.

In Table 12.2, characteristic data related to the generated output of *Ace-Gen* are compared for the different element summarized above:

– The size of the code of the automatically generated user-subroutine that evaluates $\boldsymbol{R}_e$ and $\boldsymbol{K}_{Te}$.
– The time needed for the numerical evaluation of the $\boldsymbol{R}_e$ and $\boldsymbol{K}_{Te}$. The time is normalized for the two-dimensional case with respect to linear elastic Q1 element since it varies for different hardware platforms. In the same way, the evaluation time is scaled with respect to the linear elastic H1 element in the three-dimensional case. It is interesting that the evaluation time for the H1 element is 6.1 times larger than for the Q1 element while a simple comparison related to the number of GAUSS points would give with $27 / 4 = 6.75$ a larger value, not to mention the handling of larger $\boldsymbol{B}$-matrix in manually coded elements. This underlines the advantage of

Table 12.2 Comparison of the code size and the numerical efficiency

Element	Constitutive model	Code size (Kbytes)	Evaluation time (normalized)	*AceGen* time (normalized)
Q1	linear elastic	9	1	1
Q1	hyperelastic	9	1.6	1.3
Q1	small strain elasto-plastic	24	3.0	7.4
Q1	finite strain elasto-plastic	48	9.5	25
Q1E4	linear elastic	10	1.6	2.11
Q1E4	hyperelastic	15	3.4	3.5
Q1E4	small strain elasto-plastic	27	3.7	12
Q1E4	finite strain elasto-plastic	66	11.8	49
H1	linear elastic	18	1	4.2
H1	hyperelastic	21	1.5	4.5
H1	small strain elasto-plastic	46	2.2	23.2
H1	finite strain elasto-plastic	105	6.9	69.0
H1E9	linear elastic	25	1.9	10.6
H1E9	hyperelastic	46	4.3	16.5
H1E9	small strain elasto-plastic	53	3.4	40.5
H1E9	finite strain elasto-plastic	134	10.0	117.8

using *AceGen*. For elasto-plastic models, the comparison is controlled by the extent of plastification and the number of GAUSS point iterations; thus it depends upon the actual example solved. The presented comparison is based on an example where a rectangular bar is stretched, thus all the GAUSS points are either in elastic or in plastic state.
– The time needed for the generation of the element code. The time is normalized with respect to the linear elastic Q1 element.

The code size is in the range from 9 to 66 Kbytes. The normalized evaluation times scale basically for two- and three-dimensional elements in the same way. This is remarkable since matrix operations, especially when using the EAS elements, have to be performed which result in the 2D case in the inversion of a 4 × 4 matrix while the three-dimensional element needs the inversion of a 9 × 9 matrix at element level when manually coded.

The generation of the finite strain elasto-plastic element on a 2GHz PC takes with *AceGen* code generator approximately 160 seconds. Thus, all other elements in Table 12.2 can be generated faster, e.g. the hyperelastic Q1 element is obtained within only one second.

Both, the code size and the derivation time are small enough to allow "real time" automatic derivation of complex nonlinear finite elements.

A. Vectors and Tensors

This section summarizes sum rules of tensor algebra and tensor analysis which can be found in e.g. Eringen (1967) and Marsden and Hughes (1983). This summary is not meant to be complete, but it should help in understanding some mathematical derivations and results provided in the previous chapters. A more complete treatment can be found in textbooks on tensor algebra and analysis.

A.1 Tensor Algebra

A.1.1 Definition of a Tensor

A tensor is defined as linear map between two vector spaces $\mathcal{V}$ and $\mathcal{W}$. This yields

$$
\begin{aligned}
\mathbf{T} : \mathcal{V} &\mapsto \mathcal{W} \\
\mathbf{v} &\mapsto \mathbf{w} = \mathbf{T}\,\mathbf{v} \qquad \mathbf{v} \in \mathcal{V}, \mathbf{w} \in \mathcal{W}, \\
\mathbf{T}(\mathbf{u} + \mathbf{v}) &= \mathbf{T}\,\mathbf{u} + \mathbf{T}\,\mathbf{v}, \\
\mathbf{T}(\alpha\,\mathbf{u}) &= \alpha\,\mathbf{T}\,\mathbf{u}.
\end{aligned}
$$

A special tensor is the dyad which consists of vectors defined in the spaces $\mathcal{V}$ and $\mathcal{W}$

$$
\mathbf{T} = \mathbf{a} \otimes \mathbf{b}, \qquad \mathbf{a} \in \mathcal{W}, \mathbf{b} \in \mathcal{V}.
$$

A linear map of a vector $\mathbf{c} \in \mathcal{V}$ to the space $\mathcal{W}$ is obtained with the dyad following the rule

$$
(\mathbf{a} \otimes \mathbf{b})\,\mathbf{c} = (\mathbf{b} \cdot \mathbf{c})\,\mathbf{a} \qquad \mathbf{c}, \mathbf{b} \in \mathcal{V} \quad \mathbf{a} \in \mathcal{W}.
$$

Note that the vector spaces have to be chosen such that the scalar product $\mathbf{b} \cdot \mathbf{c}$ is defined which means that $\mathbf{b}$ and $\mathbf{c}$ are in the same vector space, here $\mathcal{V}$.

A.1.2 Vectors and Tensors in a Base System

The vectors and tensors defined in Appendix A.1.1 have to be written with respect to a basis. This can be a cartesian coordinate system, defined by orthogonal base vectors $\mathbf{E}_1, \mathbf{E}_2, \mathbf{E}_3$ or $\mathbf{e}_1, \mathbf{e}_2, \mathbf{e}_3$. A more general basis is provided by a convective coordinate systems which yields co-variant, $\mathbf{g}_i$, and contra-variant, $\mathbf{g}^i$, base vectors. The associated convective coordinates are denoted by $\{\Theta^j\}$. One can assume that these coordinates are inscribed in the bodies, see Fig. A.1.2. Hence the convective coordinates are deformed as well when the body is deformed under the action of loads. Let us assume that the cartesian coordinates of the initial and current configuration $\{X_A\}$ and $\{x_i\}$, see Chap. 3, can be written as a function of the convective coordinates $\{\Theta^j\}$ as

$$X_A = \hat{X}_A\left(\Theta^1, \Theta^2, \Theta^3\right), \qquad x_i = \hat{x}_i\left(\Theta^1, \Theta^2, \Theta^3\right).$$

In short, the transformation reads: $\mathbf{X} = \hat{\mathbf{X}}\left(\Theta^j\right)$ and $\mathbf{x} = \hat{\mathbf{x}}\left(\Theta^j\right)$. When convective coordinates are used, the base vectors are different at each point of a body in the initial and current configuration. The so-called covariant base vectors are given for a point $\mathbf{X}$ in the initial configuration B of a body by

$$\mathbf{G}_j = \frac{\partial \mathbf{X}}{\partial \Theta^j} = \mathbf{X}_{,j}.$$

In an analogous way, the co-variant base vector for a point $\boldsymbol{\varphi}\left(\mathbf{X}, t\right)$ in the current configuration $\varphi(B)$ is obtained

$$\mathbf{g}_j = \frac{\partial \boldsymbol{\varphi}\left(\mathbf{X}, t\right)}{\partial \Theta^j} = \boldsymbol{\varphi}_{,j}.$$

The co-variant base vectors are tangents to the convective coordinates, see Fig. A.1.2.

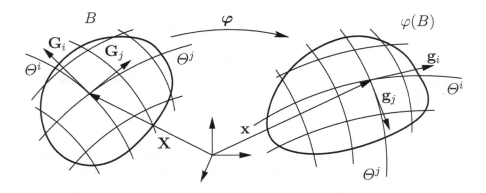

Fig. A.1 Convective coordinates of configurations B and $\varphi(B)$

Now vectors and tensors can be described with respect to the convective coordinate system. For the following details, the basis $\{\mathbf{g}_i\}$ is selected. When using convective coordinates additional base vectors, so-called contra-variant base vectors, have to be defined. These are given by

$$\mathbf{g}_i \cdot \mathbf{g}^k = \delta_i^k$$

with the KRONECKER symbol

$$\delta_k^i = \begin{cases} 1 & \text{for } i = k \\ 0 & \text{for } i \neq k \end{cases}.$$

From this definition, it is observed that the contra-variant base vectors are orthogonal to the co-variant base vectors (e.g. $\mathbf{g}^1 \perp \mathbf{g}_2$, $\mathbf{g}_3$). Now vectors and tensors can be defined with respect to this base system:

- The form of a vector $\mathbf{u}$ with respect to co-variant and contra-variant basis is given by

$$\mathbf{u} = u^i\,\mathbf{g}_i, \qquad \mathbf{v} = v_i\,\mathbf{g}^i,$$

where contra-variant u^i and the co-variant v_i components are defined by

$$u^i = \mathbf{u} \cdot \mathbf{g}^i, \qquad v_i = \mathbf{v} \cdot \mathbf{g}_i.$$

Note further that the co-variant and contra-variant basis vectors can be transformed by the **metric tensor**

$$\begin{aligned} \mathbf{g} &= g_{ik}\,\mathbf{g}^i \otimes \mathbf{g}^k, \\ \mathbf{g}^{-1} &= g^{ik}\,\mathbf{g}_i \otimes \mathbf{g}_k, \\ \mathbf{g}^{-1}\mathbf{g} &= \mathbf{1}. \end{aligned}$$

This yields

$$\begin{aligned} \mathbf{g}^i &= g^{ik}\,\mathbf{g}_k, \\ \mathbf{g}_i &= g_{ik}\,\mathbf{g}^k. \end{aligned}$$

Note that in the special case of orthogonal cartesian coordinates X_1, X_2, X_3, condition $\mathbf{E}_i \cdot \mathbf{E}_k = \delta_{ik}$ holds and hence $g_{ik} = \delta_{ik}$ with the KRONECKER symbol of the cartesian basis

$$\delta_{ik} = \begin{cases} 1 & \text{for } i = k \\ 0 & \text{for } i \neq k \end{cases}.$$

Thus the metric tensor is equal to the unit tensor.
- Different forms can be found for the representation of a second order tensor with respect to a co-variant and contra-variant basis

$$\mathbf{S} = S^{ik}\,\mathbf{g}_i \otimes \mathbf{g}_k\,,$$
$$\mathbf{T} = T^i_{.k}\,\mathbf{g}_i \otimes \mathbf{g}^k\,,$$
$$\mathbf{U} = U_{ik}\,\mathbf{g}^i \otimes \mathbf{g}^k\,.$$

The related components follow from

$$S^{ik} = \mathbf{g}^i \cdot \mathbf{S}\,\mathbf{g}^k\,,$$
$$T^i_{.k} = \mathbf{g}^i \cdot \mathbf{T}\,\mathbf{g}_k\,,$$
$$U_{ik} = \mathbf{g}_i \cdot \mathbf{U}\,\mathbf{g}_k$$

with the co-variant $\mathbf{g}_i$ and the contra-variant $\mathbf{g}^i$ base vectors.

In the special case of a cartesian coordinate system co-variant and contra-variant bases are identical. Hence the component form of a vector is given by

$$\mathbf{u} = u_i\,\mathbf{E}_i,$$

while a tensor has the form

$$\mathbf{T} = T_{ik}\,\mathbf{E}_i \otimes \mathbf{E}_k\,,$$

where $\mathbf{E}_i$ denotes the cartesian basis. Basically all formulae presented in the following, which are written in either co-variant or contra-variant form, can be reduced to cartesian basis by using indices as subscripts.

The linear map of a vector by a tensor has for arbitrary tensors of second order and for a dyadic the following representation in components

$$\mathbf{T}\,\mathbf{u} = (T^{ik}\,\mathbf{g}_i \otimes \mathbf{g}_k)\,u_l\,\mathbf{g}^l = T^{ik}\,u_l(\mathbf{g}_k \cdot \mathbf{g}^l)\,\mathbf{g}_i$$
$$= T^{ik}\,u_l\,\delta^l_k\,\mathbf{g}_i = T^{ik}\,u_k\,\mathbf{g}_i\,,$$
$$(\mathbf{a} \otimes \mathbf{b})\,\mathbf{c} = (\mathbf{b} \cdot \mathbf{c})\,\mathbf{a} = (b^i\,c_i)\,a^m\,\mathbf{g}_m.$$

A.1.3 Operations with Vectors and Tensors

Vectors and tensors can be combined using different operations. Some important possibilities are summarized below:

1. Scalar product of vectors and tensors:

$$\mathbf{a} \cdot \mathbf{b} = a^i\,b_i\,,$$
$$(\mathbf{a} \otimes \mathbf{b}) \cdot (\mathbf{c} \otimes \mathbf{d}) = (\mathbf{a} \cdot \mathbf{c})\,(\mathbf{b} \cdot \mathbf{d})\,,$$
$$\mathbf{S} \cdot \mathbf{T} = (S^{ik}\mathbf{g}_i \otimes \mathbf{g}_k) \cdot (T^{lm}\mathbf{g}_l \otimes \mathbf{g}_m)$$
$$= S^{ik}\,T^{lm}\,(\mathbf{g}_i \cdot \mathbf{g}_l)\,(\mathbf{g}_k \cdot \mathbf{g}_m) = S^{ik}\,T^{lm}\,g_{il}\,g_{km}$$
$$= S^{ik}\,T_{ik}\,.$$

The scalar product can be defined between two vectors or tensors of same order. It always yields a single number, the scalar.

2. Cross product of vectors in a cartesian basis:

$$\mathbf{a} \times \mathbf{b} = e_{ikl}\, a_i\, b_k \mathbf{E}_l \,,$$

with the permutation symbol

$$e_{ikl} = \begin{cases} 0 & \text{for } i = k\,, i = l\,, k = l\,, i = k = l \\ +1 & \text{for } ikl = 123\,, = 312\,, = 231 \\ -1 & \text{for } ikl = 321\,, = 213\,, = 132 \end{cases}.$$

The cross product of two vectors yields a vector perpendicular to both vectors.

3. Product of two second order tensors:

$$\begin{aligned} (\mathbf{a} \otimes \mathbf{b})(\mathbf{c} \otimes \mathbf{d}) &= (\mathbf{b} \cdot \mathbf{c})\,\mathbf{a} \otimes \mathbf{d}\,, \\ \mathbf{T}\,\mathbf{S} &= (T^{ik}\,\mathbf{g}_i \otimes \mathbf{g}_k)(S^{lm}\,\mathbf{g}_l \otimes \mathbf{g}_m) \\ &= T^{ik}\,S_{lm}\,(\mathbf{g}_k \cdot \mathbf{g}_l)\,\mathbf{g}_i \otimes \mathbf{g}_m \\ &= T^{ik}\,S_{lm}\,g_{kl}\,\mathbf{g}_i \otimes \mathbf{g}_m = T^{ik}\,S_k^l\,\mathbf{g}_i \otimes \mathbf{g}_l \,. \end{aligned}$$

The result of this product is again a second order tensor.

A.1.4 Special Forms of Tensors

Transposed Tensor.

$$\begin{aligned} (\mathbf{a} \otimes \mathbf{b})^T &= \mathbf{b} \otimes \mathbf{a}\,, \\ \mathbf{T}^T &= T^{ik}\,\mathbf{g}_k \otimes \mathbf{g}_i\,, \\ \mathbf{u} \cdot \mathbf{T}\mathbf{v} &= \mathbf{v} \cdot \mathbf{T}^T \mathbf{u} \\ \mathbf{a} \cdot (\mathbf{b} \otimes \mathbf{c})\,\mathbf{d} &= \mathbf{d} \cdot (\mathbf{c} \otimes \mathbf{b})\,\mathbf{a}\,. \end{aligned}$$

Inverse Tensor.

$$\mathbf{T}\,\mathbf{T}^{-1} = \mathbf{1}\,.$$

A tensor multiplied by its inverse tensor yields the unit tensor.

SHERMAN–MORRISON formula for the inverse of the sum of an arbitrary second order tensor and a dyadic product

$$[\mathbf{T} + \mathbf{a} \otimes \mathbf{b}]^{-1} = \mathbf{T}^{-1} - \frac{\mathbf{T}^{-1}\mathbf{a} \otimes \mathbf{b}\,\mathbf{T}^{-1}}{1 + \mathbf{b} \cdot \mathbf{T}^{-1}\mathbf{a}}\,.$$

Unit Tensor.

$$\mathbf{1} = \delta_k^i\,\mathbf{g}_i \otimes \mathbf{g}^k = \mathbf{g}_i \otimes \mathbf{g}^i\,.$$

The unit tensor related to a cartesian basis is given by

$$\mathbf{1} = \delta_{ik}\,\mathbf{E}_i \otimes \mathbf{E}_k = \mathbf{E}_i \otimes \mathbf{E}_i\,.$$

Axial Vector t_A.
$$\mathbf{T}_A \, \mathbf{v} = \mathbf{t}_A \times \mathbf{v}.$$
The axial vector represents a vector obtained by the linear map of an arbitrary skew symmetric tensor $\mathbf{T}_A$ and an arbitrary vector $\mathbf{v}$.

special case: $\mathbf{T}_A = \frac{1}{2} \left(\mathbf{a} \otimes \mathbf{b} - \mathbf{b} \otimes \mathbf{a} \right) \Rightarrow \mathbf{t}_A = \frac{1}{2} \left(\mathbf{b} \times \mathbf{a} \right).$

Orthogonal tensor preserves the scalar product
$$(\mathbf{Q}\,\mathbf{a}) \cdot (\mathbf{Q}\,\mathbf{b}) = \mathbf{a} \cdot \mathbf{b} = \mathbf{b} \cdot (\mathbf{Q}^T \, \mathbf{Q}\,\mathbf{a}) \Longrightarrow \mathbf{Q}^T \, \mathbf{Q} = \mathbf{1}.$$
Thus an orthogonal tensor has the properties
$$\begin{aligned}
\mathbf{Q}^{-1} &= \mathbf{Q}^T, \\
\mathbf{Q}\,\mathbf{Q}^T &= \mathbf{1}.
\end{aligned}$$
The orthogonal tensor $\mathbf{Q}$ has in an orthogonal basis $\{\mathbf{r}, \mathbf{s}, \mathbf{t}\}$, the representation
$$\mathbf{Q} = \mathbf{r} \otimes \mathbf{r} + (\,\mathbf{s} \otimes \mathbf{s} + \mathbf{t} \otimes \mathbf{t}\,) \cos\theta - (\,\mathbf{s} \otimes \mathbf{t} - \mathbf{t} \otimes \mathbf{s}\,) \sin\theta.$$
Hence the orthogonal tensor represents a rotation about axis $\mathbf{r}$.

A.1.5 Eigenvalues and Invariants of Tensors

Before the invariants and eigenvalues of second order tensors are discussed, it is useful to define the trace and determinant of a second order tensor.

Trace of a Tensor.
$$\begin{aligned}
\operatorname{tr}\mathbf{T} &= \mathbf{1} \cdot \mathbf{T} = (\mathbf{g}_i \otimes \mathbf{g}^i) \cdot (\,T^{lm}\,\mathbf{g}_l \otimes \mathbf{g}_m) \\
&= (\mathbf{g}_i \cdot \mathbf{g}_l)(\mathbf{g}^i \cdot \mathbf{g}_m)\,T^{lm} = g_{il}\,\delta^i_m\,T^{lm} = T^l_l, \\
\operatorname{tr}(\mathbf{a} \otimes \mathbf{b}) &= \mathbf{a} \cdot \mathbf{b}
\end{aligned}$$
with the properties
$$\begin{aligned}
\operatorname{tr}\mathbf{T}^T &= \operatorname{tr}\mathbf{T} \\
\operatorname{tr}(\mathbf{S} + \mathbf{T}) &= \operatorname{tr}\mathbf{S} + \operatorname{tr}\mathbf{T} \\
\operatorname{tr}(\mathbf{S}\,\mathbf{T}) &= \operatorname{tr}(\mathbf{T}\,\mathbf{S}) \\
\operatorname{tr}(\mathbf{S}\,\mathbf{T}) &= \mathbf{S} \cdot \mathbf{T} = S_{ik}\,T^{ik} = \operatorname{tr}(\mathbf{S}\,\mathbf{T}^T) = \mathbf{T}^T \cdot \mathbf{S}^T.
\end{aligned}$$

Determinant.
$$\begin{aligned}
\det(\alpha\,\mathbf{T}) &= \alpha^3 \det\mathbf{T}, \\
\det(\mathbf{S}\,\mathbf{T}) &= \det\mathbf{S} \det\mathbf{T}, \\
\det(\mathbf{T}^{-1}) &= \frac{1}{\det\mathbf{T}}.
\end{aligned}$$

With respect to an orthogonal cartesian coordinate system, relation

$$\det \mathbf{T} = e_{ikl}\, T_{i1}\, T_{k2}\, T_{l3}$$

is obtained with the permutation symbol e_{ikl}.

Eigenvalues of a Tensor.

From the special eigenvalue problem,

$$(\mathbf{T} - \lambda\,\mathbf{1})\,\boldsymbol{\varphi} = \mathbf{0}$$

follows a cubic equation for the eigenvalues, known as characteristic polynomial,

$$\det(\mathbf{T} - \lambda\,\mathbf{1}) = \lambda^3 - I_T\,\lambda^2 + II_T\,\lambda - III_T = 0$$

with the three invariants of the tensor $\mathbf{T}$:

$$
\begin{aligned}
I_T &= \operatorname{tr}\mathbf{T} = T_i^i\,, \\
II_T &= \frac{1}{2}\left[(\operatorname{tr}\mathbf{T})^2 - \operatorname{tr}(\mathbf{T}^2)\right] = \frac{1}{2}\left[(T_i^i)^2 - T_m^i T_i^m\right]\,, \\
III_T &= \det\mathbf{T} = \frac{1}{6}\left[(\operatorname{tr}\mathbf{T})^3 - 3\operatorname{tr}\mathbf{T}\operatorname{tr}(\mathbf{T}^2) + 2\operatorname{tr}(\mathbf{T}^3)\right]\,.
\end{aligned}
$$

It is possible to compute the eigenvalues λ_i and eigenvectors $\boldsymbol{\varphi}_i$ of a symmetrical second order tensor from the invariants. One scheme, which can also be found in Simo and Hughes (1998), is given by

$$
\begin{aligned}
r &= \frac{1}{54}\left(-2\,I_T + 9\,I_T\,II_T - 27\,III_T\right) \\
q &= \frac{1}{9}\left(I_T^2 - 3\,II_T\right) \\
\theta &= \arccos\left(r/\sqrt{q^3}\right) \\
\lambda_1 &= -2\sqrt{q}\,\cos[\theta/3] + \frac{1}{3}\,I_T \\
\lambda_2 &= -2\sqrt{q}\,\cos[(\theta + 2\pi)/3] + \frac{1}{3}\,I_T \\
\lambda_3 &= -2\sqrt{q}\,\cos[(\theta - 2\pi)/3] + \frac{1}{3}\,I_T.
\end{aligned}
$$

From these eigenvalues, the eigenvectors follow. Here three cases have to be distinguished.

1. All eigenvalues are different $\lambda_1 \neq \lambda_2 \neq \lambda_3$:

$$\boldsymbol{\varphi}_i \otimes \boldsymbol{\varphi}_i = \frac{\lambda_i}{2\,\lambda_i^3 - I_T\,\lambda_i^2 + III_T}\left(\mathbf{T}^2 - (I_T - \lambda_i)\,\mathbf{T} + \frac{III_T}{\lambda_i}\,\mathbf{1}\right).$$

2. Two eigenvalues are equal $\lambda_i \neq \lambda_j = \lambda_k$:

$$\boldsymbol{\varphi}_j \otimes \boldsymbol{\varphi}_j = \mathbf{1} - \boldsymbol{\varphi}_i \otimes \boldsymbol{\varphi}_i.$$

3. All eigenvalues are equal $\lambda_i = \lambda_j = \lambda_k$:

$$\boldsymbol{\varphi}_i \otimes \boldsymbol{\varphi}_i = \mathbf{1}.$$

When knowing the eigenvalues, the invariants

$$
\begin{aligned}
I_T &= \lambda_1 + \lambda_2 + \lambda_3, \\
II_T &= \lambda_1 \lambda_2 + \lambda_2 \lambda_3 + \lambda_3 \lambda_1, \\
III_T &= \lambda_1 \lambda_2 \lambda_3
\end{aligned}
$$

can be computed in a simpler way. A generalization yields instead of the characteristic polynomial a characteristic equation for the tensor itself. This is known as CAYLEY–HAMILTON theorem.

$$\mathbf{T}^3 - I_T \mathbf{T}^2 + II_T \mathbf{T} - III_T \mathbf{1} = \mathbf{0}.$$

This equation can be used to e.g. compute the inverse of a tensor by multiplying with the inverse

$$\mathbf{T}^{-1} = \frac{1}{III_T} \left[\mathbf{T}^2 - I_T \mathbf{T} + II_T \mathbf{1} \right].$$

For skew symmetric tensors $\mathbf{T}_A$ with their axial vector $\mathbf{t}_a$ the invariants are

$$
\begin{aligned}
I_{T_A} &= \operatorname{tr} \mathbf{T}_A = 0, \\
II_{T_A} &= \| \mathbf{t}_A \|^2, \\
III_{T_A} &= \det \mathbf{T}_A = 0.
\end{aligned}
$$

Thus the eigenvalues can be computed for $\mathbf{T}_A$ from $\lambda^2 + \| \mathbf{t}_A \|^2 = 0$.

Spectral Decomposition of a Tensor.

Once the eigenvalues and eigenvectors of a symmetric tensor $\mathbf{S}$ are known, it can be represented via a spectral decomposition

$$\mathbf{S} = \sum_{\alpha=1}^{3} \lambda_\alpha \boldsymbol{\varphi}_\alpha \otimes \boldsymbol{\varphi}_\alpha.$$

Based on this result, powers and logarithms of tensors can be defined as

$$
\begin{aligned}
\ln \mathbf{S} &= \sum_{\alpha=1}^{3} \ln \lambda_\alpha \, \boldsymbol{\varphi}_\alpha \otimes \boldsymbol{\varphi}_\alpha \\
\mathbf{S}^{\frac{1}{2}} &= \sum_{\alpha=1}^{3} \sqrt{\lambda_\alpha} \, \boldsymbol{\varphi}_\alpha \otimes \boldsymbol{\varphi}_\alpha.
\end{aligned}
$$

A.1.6 Tensors of Higher Order

Tensors of higher order will be defined in terms of dyadic products. The derived rules can then be applied to the bases of arbitrary tensors. Here only tensors of third and fourth order are considered.

1. Third order dyadic product:

$$(\mathbf{a} \otimes \mathbf{b}) \otimes \mathbf{c} = \mathbf{a} \otimes \mathbf{b} \otimes \mathbf{c}$$
$$(\mathbf{a} \otimes \mathbf{b} \otimes \mathbf{c})(\mathbf{d} \otimes \mathbf{e}) = (\mathbf{b} \cdot \mathbf{d})(\mathbf{c} \cdot \mathbf{e})\,\mathbf{a}.$$

2. Fourth order dyadic product:

$$(\mathbf{a} \otimes \mathbf{b}) \otimes (\mathbf{c} \otimes \mathbf{d}) = \mathbf{a} \otimes \mathbf{b} \otimes \mathbf{c} \otimes \mathbf{d}$$

$$(\mathbf{a} \otimes \mathbf{b} \otimes \mathbf{c} \otimes \mathbf{d})(\mathbf{f} \otimes \mathbf{g}) = (\mathbf{c} \cdot \mathbf{f})(\mathbf{d} \cdot \mathbf{g})(\mathbf{a} \otimes \mathbf{b}).$$

Rules:

$$\begin{aligned}
(\mathbf{T} \otimes \mathbf{c})\,\mathbf{v} &= (\mathbf{c} \cdot \mathbf{v})\,\mathbf{T}, \\
(\mathbf{a} \otimes \mathbf{T})\,\mathbf{R} &= (\mathbf{T} \cdot \mathbf{R})\,\mathbf{a}, \\
(\mathbf{a} \otimes \mathbf{b} \otimes \mathbf{c})\,\mathbf{1} &= (\mathbf{b} \cdot \mathbf{c})\,\mathbf{a}, \\
(\mathbf{T} \otimes \mathbf{v})(\mathbf{a} \otimes \mathbf{b}) &= (\mathbf{b} \cdot \mathbf{v})\,\mathbf{T}\,\mathbf{a}, \\
(\mathbf{T} \otimes \mathbf{v})\,\mathbf{R} &= (\mathbf{T}\,\mathbf{R})\,\mathbf{v}, \\
(\mathbf{T} \otimes \mathbf{v})\,\mathbf{1} &= \mathbf{T}\,\mathbf{v}, \\
(\mathbf{T} \otimes \mathbf{R})\,\mathbf{S} &= (\mathbf{R} \cdot \mathbf{S})\,\mathbf{T}, \\
(\mathbf{T} \otimes \mathbf{R})\,\mathbf{v} &= \mathbf{T} \otimes \mathbf{R}\,\mathbf{v}, \\
(\mathbf{T} \otimes \mathbf{R})\,\mathbf{1} &= (\mathrm{tr}\mathbf{R})\,\mathbf{T}.
\end{aligned}$$

In general, the representation of a fourth order tensor is given by

$$\mathbb{C} = C^{ijkl}\,\mathbf{g}_i \otimes \mathbf{g}_j \otimes \mathbf{g}_k \otimes \mathbf{g}_l.$$

A tensor of fourth order can be applied to define a linear mapping between two tensors of second order,

$$\begin{aligned}
\mathbf{U} &= \mathbb{C}[\mathbf{V}] \\
U^{ij}\,\mathbf{g}_i \otimes \mathbf{g}_j &= (C^{ijkl}\,\mathbf{g}_i \otimes \mathbf{g}_j \otimes \mathbf{g}_k \otimes \mathbf{g}_l)(V_{mn}\,\mathbf{g}^m \otimes \mathbf{g}^n) \\
&= C^{ijkl} V_{mn} \delta_k^m \delta_l^n\,\mathbf{g}_i \otimes \mathbf{g}_j = C^{ijkl} V_{kl}\,\mathbf{g}_i \otimes \mathbf{g}_j.
\end{aligned}$$

A.2 Tensor Analysis

In this section, scalars, vectors and tensors are discussed which are functions of a position vector $\mathbf{X}$ and time t. For that the following fields are defined:

- Scalar field: $\alpha(\mathbf{X}, t)$
- Vector field: $\mathbf{v}(\mathbf{X}, t)$
- Tensor field: $\mathbf{T}(\mathbf{X}, t)$

Examples for scalar fields are density, pressure or temperature. Displacements, velocities or momentum can be described by vector fields. Stresses or strains are represented by tensor fields.

A.2.1 Differentiation with Respect to a Real Variable

The differentiation with respect to a real variable, e.g. the time, is based on the following definitions

$$\text{Definition:} \quad \dot{\mathbf{v}}(\mathbf{X}, t) = \frac{\partial \mathbf{v}(\mathbf{X}, t)}{\partial t}$$

For scalar-, vector- and tensor valued fields the following rules apply.
Rules:

$$
\begin{aligned}
(\lambda \mathbf{v})^{\cdot} &= \dot{\lambda}\mathbf{v} + \lambda\dot{\mathbf{v}}, \\
(\mathbf{u} \otimes \mathbf{v})^{\cdot} &= \dot{\mathbf{u}} \otimes \mathbf{v} + \mathbf{u} \otimes \dot{\mathbf{v}}, \\
(\mathbf{u} \cdot \mathbf{v})^{\cdot} &= \dot{\mathbf{u}} \cdot \mathbf{v} + \mathbf{u} \cdot \dot{\mathbf{v}}, \\
(\mathbf{u} \times \mathbf{v})^{\cdot} &= \dot{\mathbf{u}} \times \mathbf{v} + \mathbf{u} \times \dot{\mathbf{v}}, \\
(\mathbf{T}\mathbf{v})^{\cdot} &= \dot{\mathbf{T}}\mathbf{v} + \mathbf{T}\dot{\mathbf{v}}, \\
(\mathbf{T}\mathbf{S})^{\cdot} &= \dot{\mathbf{T}}\mathbf{S} + \mathbf{T}\dot{\mathbf{S}}, \\
(\mathbf{T} \cdot \mathbf{S})^{\cdot} &= \dot{\mathbf{T}} \cdot \mathbf{S} + \mathbf{T} \cdot \dot{\mathbf{S}}, \\
(\mathbf{T}^{T})^{\cdot} &= (\dot{\mathbf{T}})^{T}, \\
(\mathbf{T}^{-1})^{\cdot} &= -\mathbf{T}^{-1}(\dot{\mathbf{T}})\mathbf{T}^{-1}.
\end{aligned}
$$

A.2.2 Gradient of a Field

The gradient of a field yields always a field which is one order higher. Hence a gradient of a scalar field yields a vector field and so forth.

$$
\mathbf{v} = \operatorname{Grad}\alpha(\mathbf{X}, t) = \frac{\partial \alpha}{\partial \mathbf{X}} = \frac{\partial \alpha}{\partial X_i}\mathbf{G}^i,
$$

$$
\mathbf{T} = \operatorname{Grad}\mathbf{v}(\mathbf{X}, t) = \frac{\partial \mathbf{v}}{\partial \mathbf{X}} = \frac{\partial \mathbf{v}}{\partial X_i} \otimes \mathbf{G}^i.
$$

Often the NABLA operator ∇ is used instead of the gradient operator Grad. Then

$$
\begin{aligned}
\operatorname{grad}\alpha &= \nabla\alpha, \\
\operatorname{grad}\mathbf{u} &= \nabla\mathbf{u}
\end{aligned}
$$

is written.

Rules:

$$
\begin{aligned}
\mathrm{Grad}\,(\alpha\,\beta) &= (\mathrm{Grad}\,\alpha)\,\beta + \alpha\,\mathrm{Grad}\,\beta\,, \\
\mathrm{Grad}\,(\alpha\,\mathbf{v}) &= \mathbf{v}\otimes\mathrm{Grad}\,\alpha + \alpha\,\mathrm{Grad}\,\mathbf{v}\,, \\
\mathrm{Grad}\,(\alpha\,\mathbf{T}) &= \mathbf{T}\otimes\mathrm{Grad}\,\alpha + \alpha\,\mathrm{Grad}\,\mathbf{T}\,, \\
\mathrm{Grad}\,(\mathbf{u}\cdot\mathbf{v}) &= (\mathrm{Grad}\,\mathbf{u})^{T}\mathbf{v} + (\mathrm{Grad}\,\mathbf{v})^{T}\mathbf{u}
\end{aligned}
$$

or in index notation

$$
\begin{aligned}
(\alpha\,\beta)_{,i} &= \alpha_{,i}\,\beta + \alpha\,\beta_{,i}\,, \\
(\alpha\,v_i)_{,k} &= v_i\,\alpha_{,k} + \alpha\,v_{i,k}\,, \\
(\alpha\,T_{ik})_{,m} &= \alpha_{,m}\,T_{ik} + \alpha\,T_{ik,m}\,, \\
(u_i\,v_i)_{,k} &= u_{i,k}\,v_i + u_i\,v_{i,k}\,.
\end{aligned}
$$

In case that the scalar variable α is given as function of a vector valued field $\mathbf{u}(\mathbf{X}\,t)$ or the vector valued variable $\mathbf{u}$ given as function of a scalar field $\alpha(\mathbf{X},t)$, the gradient

$$
\begin{aligned}
\mathrm{Grad}\,\alpha\{\mathbf{u}(\mathbf{X},t)\} &= (\mathrm{Grad}\,\mathbf{u})^{T}\frac{\partial\alpha}{\partial\mathbf{u}}\,, \\
\mathrm{Grad}\,\mathbf{u}\{\alpha(\mathbf{X},t)\} &= \frac{\partial\mathbf{u}}{\partial\alpha}\otimes\mathrm{Grad}\,\alpha
\end{aligned}
$$

is obtained.

The differentiation of a symmetric tensor $\mathbf{T}$ with respect to itself yields a fourth order tensor. In index notation, this reads

$$
\left(\frac{\partial\mathbf{T}}{\partial\mathbf{T}}\right)_{iklm} = \frac{1}{2}\left(\delta_{il}\,\delta_{km} + \delta_{im}\,\delta_{kl}\right).
$$

In an analogous way, the differentiation of the inverse is given by

$$
\left(\frac{\partial\mathbf{T}^{-1}}{\partial\mathbf{T}}\right)_{iklm} = \frac{1}{2}\left(T_{il}^{-1}\,T_{mk}^{-1} + T_{im}^{-1}\,T_{lk}^{-1}\right).
$$

Special cases are

$$
\begin{aligned}
\frac{\partial\mathbf{T}^{-1}}{\partial\mathbf{T}}\,[\mathbf{V}] &= -\mathbf{T}^{-1}\,\mathbf{V}\,\mathbf{T}^{-1}\,, \\
\frac{\partial\mathbf{T}^{-1}}{\partial\mathbf{T}}\,[\mathbf{T}]\otimes\mathbf{T}^{-1} &= -\mathbf{T}^{-1}\otimes\mathbf{T}^{-1}
\end{aligned}
$$

or in index notation

$$
\begin{aligned}
\left(\frac{\partial\mathbf{T}^{-1}}{\partial\mathbf{T}}\right)_{iklm} V_{lm} &= -T_{ij}^{-1}\,V_{jn}\,T_{nk}^{-1}\,, \\
\left(\frac{\partial\mathbf{T}^{-1}}{\partial\mathbf{T}}\right)_{iklm} T_{lm}\,T_{no}^{-1} &= -T_{ik}^{-1}\,T_{no}^{-1}\,.
\end{aligned}
$$

With the product rule, it further follows

$$\frac{\partial(\alpha \mathbf{S})}{\partial \mathbf{T}} = \mathbf{T} \otimes \frac{\partial \alpha}{\partial \mathbf{T}} + \alpha \frac{\partial \mathbf{S}}{\partial \mathbf{T}} \,.$$

A.2.3 Divergence of a Field

The computation of the divergence of a field reduces the order of the field by one. Thus the divergence of e.g. a tensor field yields a vector field.

$$\begin{aligned}
\operatorname{Div} \mathbf{v}(\mathbf{X}, t) &= \operatorname{Grad} \mathbf{v}(\mathbf{X}, t) \cdot \mathbf{1}\,, \\
\operatorname{Div} \mathbf{T}(\mathbf{X}, t) &= \operatorname{Grad} \mathbf{T}(\mathbf{X}, t)\, \mathbf{1}\,.
\end{aligned}$$

Rules:

$$\begin{aligned}
\operatorname{Div}(\alpha\, \mathbf{v}) &= \mathbf{v} \cdot \operatorname{Grad} \alpha + \alpha \operatorname{Div} \mathbf{v}\,, \\
\operatorname{Div}(\alpha\, \mathbf{T}) &= \mathbf{T} \operatorname{Grad} \alpha + \alpha \operatorname{Div} \mathbf{T}\,, \\
\operatorname{Div}(\mathbf{T}\, \mathbf{v}) &= \mathbf{T}^T \cdot \operatorname{Grad} \mathbf{v} + \operatorname{Div} \mathbf{T}^T \cdot \mathbf{v}\,, \\
\operatorname{Div}(\mathbf{u} \otimes \mathbf{v}) &= (\operatorname{Grad} \mathbf{u})\, \mathbf{v} + (\operatorname{Div} \mathbf{v})\, \mathbf{u}\,, \\
\operatorname{Div}(\mathbf{u} \times \mathbf{v}) &= (\operatorname{Grad} \mathbf{u} \times \mathbf{v}) \cdot \mathbf{1} - (\operatorname{Grad} \mathbf{v} \times \mathbf{u})\,.
\end{aligned}$$

A.2.4 Rotation of a Vector Field

The rotation of a vector field is defined as follows:

$$\operatorname{Rot} \mathbf{v}(\mathbf{X}, t) = e_{ijk} \frac{\partial \mathbf{v}}{\partial X_k} \mathbf{G}^j$$

with the permutation symbol e_{ijk}.

Rules:

$$\begin{aligned}
\operatorname{Rot}(\mathbf{u} \times \mathbf{v}) &= \operatorname{Div}(\mathbf{u} \otimes \mathbf{v} - \mathbf{v} \otimes \mathbf{u})\,, \\
\operatorname{Rot}(\mathbf{T}_A\, \mathbf{v}) &= [(\operatorname{Div} \mathbf{t}_A)\, \mathbf{1} - \operatorname{Grad} \mathbf{t}_A]\, \mathbf{v}\,, \\
\operatorname{Rot} \mathbf{T} \cdot \mathbf{1} &= 0 \quad \text{for} \quad \mathbf{T} = \mathbf{T}^T\,.
\end{aligned}$$

A.2.5 Derivation of an Invariant with Respect to a Tensor

The invariants of a symmetric tensor $\mathbf{T}$: I_T, II_T, III_T have been defined above. The derivation of these quantities with respect to the tensor yield

$$\begin{aligned}
\frac{\partial I_T}{\partial \mathbf{T}} &= \mathbf{1}\,, \\
\frac{\partial II_T}{\partial \mathbf{T}} &= I_T\, \mathbf{1} - \mathbf{T}\,, \\
\frac{\partial III_T}{\partial \mathbf{T}} &= III_T\, \mathbf{T}^{-1}\,.
\end{aligned}$$

By application of the CAYLEY–HAMILTON theorem $\mathbf{T}^3 - I_T\,\mathbf{T}^2 + II_T\,\mathbf{T} - III_T\,\mathbf{1} = \mathbf{0}$, relation

$$\frac{\partial III_T}{\partial \mathbf{T}} = \mathbf{T}^2 - I_T\,\mathbf{T} + II_T\,\mathbf{1}$$

is deduced. For an invertible tensor $\mathbf{A}$ of second order, it yields

$$\frac{\partial \mathrm{tr}\,\mathbf{A}}{\partial \mathbf{A}} = \mathbf{1}\,,$$

$$\frac{\partial \mathrm{tr}\,(\mathbf{A}^2)}{\partial \mathbf{A}} = 2\,\mathbf{A}^T\,,$$

$$\frac{\partial \det \mathbf{A}}{\partial \mathbf{A}} = \det \mathbf{A}\,\mathbf{A}^{-T}\,.$$

A.2.6 Pull Back and Push Forward Operations

In this section, the *pull back* (φ^*) and *push forward* (φ_*) operations will be discussed. The co-variant basis vectors $\{\mathbf{g}_i\}$, see Fig. A.1, are related to the tangent space while the contr-variant basis vectors $\{\mathbf{g}^i\}$ can be denoted as one forms. These bases have different behaviour when pulled back from the current to the initial configuration and when pushed forward from the initial configuration to the current one. The following table depicts the different behaviour during transformation

$$\mathbf{g}_i = \mathbf{F}\,\mathbf{G}_i\,, \qquad \mathbf{g}^i = \mathbf{F}^{-T}\,\mathbf{G}^i\,,$$

$$\mathbf{G}_i = \mathbf{F}^{-1}\,\mathbf{g}_i\,, \qquad \mathbf{G}^i = \mathbf{F}^T\,\mathbf{g}^i\,.$$

As in Chap. 3, small letters are associated with quantities measured in the current configuration and capital letters are related to quantities in the initial configuration. Thus basis vectors $\mathbf{G}_i$ refer to the initial configuration B and basis vectors $\mathbf{g}_i$ refer to the current configuration $\varphi(B)$.

The transformation behaviour of the divergence operator is given by

$$\mathrm{div}\,\mathbf{v} = \tfrac{1}{J}\,\mathrm{Div}\,\mathbf{v}\,, \quad \mathrm{Div}\,\mathbf{v} = J\,\mathrm{div}\,\mathbf{v}\,.$$

In analogous way, the gradients can be transferred:

$$\mathrm{grad}\,\alpha = \mathbf{F}^{-T}\,\mathrm{Grad}\,\alpha\,, \quad \mathrm{Grad}\,\alpha = \mathbf{F}^T\,\mathrm{grad}\,\alpha\,,$$

$$\mathrm{grad}\,\mathbf{v} = \mathrm{Grad}\,\mathbf{v}\,\mathbf{F}\,, \quad \mathrm{Grad}\,\mathbf{v} = \mathrm{grad}\,\mathbf{v}\,\mathbf{F}^{-1}\,.$$

Tensors can, as well as gradients, be referred to base systems of the initial or current configuration. Here this will be exemplarily performed for the CAUCHY stress tensor $\boldsymbol{\sigma}$, which is defined with respect to the current configuration. With the representation of $\boldsymbol{\sigma}$ with respect to co-variant and contra variant bases

$$\boldsymbol{\sigma}^\flat = \sigma_{ik}\,\mathbf{g}^i \otimes \mathbf{g}^k\,,$$

$$\boldsymbol{\sigma}^{\sharp} = \sigma^{ik}\, \mathbf{g}_i \otimes \mathbf{g}_k\,,$$

$$\boldsymbol{\sigma}_1 = \sigma^i_{\cdot k}\, \mathbf{g}_i \otimes \mathbf{g}^k\,,$$

the pull back and push forward operations are

$$\text{push forward} \qquad\qquad \text{pull back}$$

$$\boldsymbol{\sigma}^{\flat} = \mathbf{F}^{-T}\, \boldsymbol{\Sigma}^{\flat}\, \mathbf{F}^{-1} \qquad \boldsymbol{\Sigma}^{\flat} = \mathbf{F}^T\, \boldsymbol{\sigma}^{\flat}\, \mathbf{F}$$

$$\boldsymbol{\sigma}^{\sharp} = \mathbf{F}\, \boldsymbol{\Sigma}^{\sharp}\, \mathbf{F}^T \qquad \boldsymbol{\Sigma}^{\sharp} = \mathbf{F}^{-1}\, \boldsymbol{\sigma}^{\sharp}\, \mathbf{F}^{-T}$$

$$\boldsymbol{\sigma}_1 = \mathbf{F}\, \boldsymbol{\Sigma}_1\, \mathbf{F}^{-1} \qquad \boldsymbol{\Sigma}_1 = \mathbf{F}^{-1}\, \boldsymbol{\sigma}_1\, \mathbf{F}.$$

Here $\boldsymbol{\Sigma}$ denotes the stress tensor referred to the initial configuration B.

A.2.7 Lie-Derivative of Stress Tensors

The LIE derivative is a derivative of a spatial tensor with respect to time. It is defined by

$$L_{\mathbf{v}}(t) = \Phi_{t*}\left[\frac{d}{dt}\Phi_t^*(t)\right].$$

This means that one has first to pull the tensor back to initial configuration; then perform the time derivative and after that push the result forward to the spatial or current configuration.

It can be applied to tensors which are defined with respect to the current configuration, see e.g. Sect. 3.1.4. The LIE derivative yields the flux related to the used stress tensor. From the definition of the pull back and push forward operations, it is clear that the selected tensor basis (co-variant or contra-variant) has influence on the result of the LIE derivative. Since stress tensors are mostly written with respect to a co-variant basis, this basis is used for the following considerations.

Application of the LIE derivative to the KIRCHHOFF stress tensor $\boldsymbol{\tau}$ yields the so-called OLDROYD stress flux or stress rate

$$L_{\mathbf{v}}(\boldsymbol{\tau}^{\sharp}) = \mathbf{F}\frac{d}{dt}\left[\mathbf{F}^{-1}\boldsymbol{\tau}\mathbf{F}^{-t}\right]\mathbf{F}^T.$$

With (3.47), the result $\mathbf{F}\,\dot{\mathbf{F}}^{-1} = -\mathbf{l}$ is obtained from $\frac{d}{dt}(\mathbf{F}\,\mathbf{F}^{-1}) = \mathbf{0}$ and hence the final expression for the OLDROYD stress rate is

$$L_{\mathbf{v}}(\boldsymbol{\tau}^{\sharp}) = \dot{\boldsymbol{\tau}}^{\sharp} - \mathbf{l}\boldsymbol{\tau}^{\sharp} - \boldsymbol{\tau}^{\sharp}\mathbf{l}^T.$$

The TRUESDELL stress rate $L_{\mathbf{v}}^J(\boldsymbol{\sigma}^{\sharp})$ follows from the relation between the 2nd PIOLA–KIRCHHOFF stress tensor and the CAUCHY stress tensor

$$\begin{aligned} L_{\mathbf{v}}^J(\boldsymbol{\sigma}^{\sharp}) &= J^{-1}\mathbf{F}\frac{d}{dt}[J\mathbf{F}^{-1}\boldsymbol{\sigma}^{\sharp}\mathbf{F}^{-T}]\mathbf{F}^T \\ &= \dot{\boldsymbol{\sigma}}^{\sharp} - \mathbf{l}\boldsymbol{\sigma}^{\sharp} - \boldsymbol{\sigma}^{\sharp}\mathbf{l}^T + \boldsymbol{\sigma}^{\sharp}\,tr(\mathbf{d}). \end{aligned}$$

These stress rates are objective, see Sect. 3.2.6.

Further known stress rates can be derived in an analogous way. Let us remark that the addition of two objective stress rates yield again an objective stress rate.

A.2.8 Integral Theorems

The integral theorems are subdivided into two categories. The first one is related to the transformation of area into volume integrals and the second one is associated with the transformation of line or curve integrals into area integrals.

Transformation of area integrals to volume integrals:

$$\int_{\partial B} \mathbf{u} \cdot \mathbf{n}\, da = \int_B \operatorname{Div} \mathbf{u}\, dv\,,$$

$$\int_{\partial B} \mathbf{T}\mathbf{n}\, da = \int_B \operatorname{Div} \mathbf{T}\, dv\,,$$

$$\int_{\partial B} (\mathbf{u} \times \mathbf{T}\mathbf{n})\, da = \int_B (\mathbf{u} \times \operatorname{Div} \mathbf{T} + \operatorname{Grad} \mathbf{u} \times \mathbf{T})\, dv\,,$$

$$\int_{\partial B} \mathbf{n} \times \mathbf{u}\, da = \int_B \operatorname{Rot} \mathbf{u}\, dv\,.$$

Since $\operatorname{Div} \mathbf{x} = 3$,

$$V = \int_B dv = \frac{1}{3} \int_{\partial B} \mathbf{x} \cdot \mathbf{n}\, da$$

can be deducted from the first expression.

Transformation of line or curve integrals into area integrals:

$$\oint_C \Phi\, d\mathbf{x} = \int_{\partial B} \mathbf{n} \times \operatorname{Grad} \Phi\, da\,,$$

$$\oint_C \mathbf{u} \times d\mathbf{x} = \int_{\partial B} (\operatorname{Div} \mathbf{u}\, \mathbf{1} - \operatorname{Grad}^T \mathbf{u})\, \mathbf{n}\, da\,,$$

$$\oint_C \mathbf{u} \cdot d\mathbf{x} = \int_{\partial B} \operatorname{Rot} \mathbf{u} \cdot \mathbf{n}\, da\,.$$

Bibliography

Ainsworth M. and Oden J.T. A procedure for a posteriori error estimation for h-p finite element methods. *Computer Methods in Applied Mechanics and Engineering*, 101:73–96 (1992).

Akers R., Baffes P., Kant E., Randall C. and Steinberg R.Y. Automatic synthesis of numerical codes for solving partial differential equations. *Mathematics and Computers in Simulation*, 45:3–22 (1998).

Alart P. and Curnier A. A mixed formulation for frictional contact problems prone to Newton like solution methods. *Computer Methods in Applied Mechanics and Engineering*, 92:353–375 (1991).

Altenbach J. and Altenbach H. *Einführung in die Kontinuumsmechanik*. Teubner-Verlag, Stuttgart (1994).

Andelfinger U. *Untersuchungen zur Zuverlässigkeit Hybrid-Gemischter Finiter Elemente für Flächentragwerke*. Dissertation, Institut für Baustatik der Universität Stuttgart (1991). Bericht Nr. 13.

Andelfinger U. and Ramm E. EAS-elements for two-dimensional, three-dimensional plate and shell structures and their equivalence to HR-elements. *International Journal for Numerical Methods in Engineering*, 36:1311–1337 (1993).

Archer G.C. A technique for the reduction of dynamic degrees of freedom. *Earthquake Engineering and Structural Dynamics*, 30:127–145 (2001).

Argyris J.H. An excursion into large rotations. *Computer Methods in Applied Mechanics and Engineering*, 32:85–155 (1982).

Argyris J.H., Doltsinis J.S., Pimenta P.M. and Wuestenberg H. Thermomechanical response of solids at high strains - natural approach. *Computer Methods in Applied Mechanics and Engineering*, 32:3–57 (1982).

Argyris J.H. and Kleiber M. Incremental formulation in nonlinear mechanics and large strain elasto-plasticity - Natural approach. I. *Computer Methods in Applied Mechanics and Engineering*, 11:215–247 (1977).

Argyris J.H., Pister K.S., Szimmat J. and Willam K.J. Unified concepts of constitutive modelling and numerical solution methods for concrete creep problems. *Computer Methods in Applied Mechanics and Engineering*, 10:199–246 (1976).

Arnold D.N., Falk R.S. and Winther R. Preconditioning discret approximations of the reissner–mindlin plate model. *Mathematical Modelling and Numerical Analysis*, 31:517–557 (1997).

Atluri S.N. On constitutive relations at finite strain: hypo-elasticity and elasto-plasticity with isotropic and kinematic hardening. *Computer Methods in Applied Mechanics and Engineering*, 43:137–171 (1984).

Axelsson O. *Iterative Solution Methods*. Cambridge University Press, Cambridge (1994).

Axelsson O. and Barker V.A. *Finite Element Solution of Boundary Value Problems: Theory and Computation*. Society for Industrial and Applied Mathematics, Philadelphia, PA, USA (2001).

Babuska I. and Rheinboldt W. Error estimates for adaptive finite element computations. *SIAM Journal on Numerical Analysis*, 15:736–754 (1978).

Babuska I., Strouboulis T., Upadhyay C.S., Gangaraj S.K. and Copps K. Validation of a posteriori error estimators by numerical approach. *International Journal for Numerical Methods in Engineering*, 37:1073–1123 (1994).

Babuska I., Szabo B.A. and Katz I.N. The p-Version of the Finite Element Method. *SIAM Journal on Numerical Analysis*, 18(3):515–545 (1981).

Balay S., Buschelman K., Eijkhout V., Gropp W.D., Kaushik D., Knepley M.G., McInnes L.C., Smith B.F. and Zhang H. PETSc users manual. *Technical Report ANL-95/11 – Revision 2.1.5*, Argonne National Laboratory, Argonne (2004).

Balay S., Buschelman K., Gropp W.D., Kaushik D., Knepley M.G., McInnes L.C., Smith B.F. and Zhang H. PETSc Web page (2001). Http://www.mcs.anl.gov/petsc.

Bank R.E. PLTMG: A software package for solving elliptic partial differential equations. *Technical Report Vol. 7*, Society for Industrial and Applied Methematics, Philadelphia (1990).

Bartels S. and Carstensen C. Each averaging technique yields reliable a posteriori error control in fem on unstructured grids. Part ii: High order fem. *Mathematics of Computation*, 71:971–994 (2002).

Bartholomew-Biggs M., Brown S., Christianson B. and Dixon L. Automatic differentiation of algorithms. *Journal of Computational and Applied Mathematics*, 124:171–190 (2000).

Basar Y. and Ding Y. Finite-rotation elements for nonlinear analysis of thin shell structures. *International Journal of Solids & Structures*, 26:83–97 (1990).

Basar Y. and Ding Y. Shear deformation models for large strain shell analysis. *International Journal of Solids & Structures*, 34:1687–1708 (1996).

Bass J.M. and Oden J.T. Adaptive finite element methods for a class of evolution problems in viscoplasticity. *International Journal of Engineering Science*, 25:623–653 (1987).

Bastian P. and Wittum G. On robust and adaptive multi-grid methods. In P.W. Hemker et al., editors, *Multigrid methods IV. Proceedings of the fourth European multigrid conference*, pages 1–17. Birkhaeuser, ISNM, Int. Ser. Numer. Math. Volume 116, Amsterdam (1994).

Bathe K.J. *Finite Element Procedures in Engineering Analysis*. Prentice-Hall, Englewood Cliffs, New Jersey (1982).

Bathe K.J. *Finite-Elemente-Methoden, Matrizen und lineare Algebra. Die Methode der finiten Elemente. L"osung von Gleichgewichtsbedingungen und Bewegungsgleichungen; Deutsche "Ubersetzung von P. Zimmermann*. Springer-Verlag, Berlin-Heidelberg-New York (1986).

Bathe K.J. *Finite Element Procedures*. Prentice-Hall, Englewood Cliffs, New Jersey (1996).

Bathe K.J. and Bolourchi S. Large displacement analysis of three-dimensional beam structures. *International Journal for Numerical Methods in Engineering*, 14:961–986 (1979).

Bathe K.J. and Chaudhary A.B. A solution method for planar and axisymmetric contact problems. *International Journal for Numerical Methods in Engineering*, 21:65–88 (1985).

Bathe K.J. and Dvorkin E.N. A four-node plate bending element based on Mindlin/Reissner plate theory and a mixed interpolation. *International Journal for Numerical Methods in Engineering*, 21:367–383 (1985).

Bathe K.J. and Gracewski S. On nonlinear dynamic analysis using substructuring and mode superposition. *Computers and Structures*, 13:699–707 (1981).

Bathe K.J., Ramm E. and Wilson E.L. Finite element formulation for large deformation analysis. *International Journal for Numerical Methods in Engineering*, 9:353–386 (1975).

Baumann M., Klarmann R. and Schweizerhof K. Algorithmen zur Optimierung von Gleichungssystemen bei Finite-Element-Berechnungen. *Technical Report 2/1990*, Institut für Baustatik, Karlsruhe (1990).

Bazant Z.P. and Cedolin L. *Stability of Structures*. Oxford University Press, New York (1991).

Bazant Z.P. and Cedolin L. *Stability of Structures: Elastic, Inelastic, Fracture, and Damage Theories*. Dover Publications, New York (2003).

Bazaraa M.S., Sherali H.D. and Shetty C.M. *Nonlinear Programming, Theory and Algorithms*. Wiley, Chichester, second edition (1993).

Beall M. and Shephard M. Object-oriented framework for reliable numerical simulations. *Engineering with Computers*, 15:61–72 (1999).

Becker A. Berechnung ebener Stabtragwerke nach der Fließgelenktheorie II. Ordnung unter Ber"ucksichtigung der Normal- und Querkraftinteraktion mit Hilfe der Methode der Finiten Elemente. *Technical report*, Diplomarbeit am Institut für Baumechanik und Numerische Mechanik der Universität Hannover (1985).

Becker E. and Bürger W. *Kontinuumsmechanik*. B.G. Teubner, Stuttgart (1975).

Becker R. and Rannacher R. A feed-back approach to error control in finite element methods: Basic analysis and examples. *EAST-WEST Journal of Numerical Mathematics*, 4:237–264 (1996).

Belytschko T. and Bindeman L.P. Assumed strain stabilization of the 4-node quadrilateral with 1-point quadrature for nonlinear problems. *Computer Methods in Applied Mechanics and Engineering*, 88(3):311–340 (1991).

Belytschko T. and Bindeman L.P. Assumed strain stabilization of the eight node hexahedral element. *Computer Methods in Applied Mechanics and Engineering*, 105(2):225–260 (1993).

Belytschko T., Chiapetta T. and Bartel R.L. Efficient large-scale non-linear transient analysis by finite elements. *International Journal for Numerical Methods in Engineering*, 10:579–596 (1976).

Belytschko T., Liu W.K. and Moran B. *Nonlinear finite elements for continua and structures*. Wiley, Chichester (2000).

Belytschko T., Ong J.S.J., Liu W.K. and Kennedy J.M. Hourglass control in linear and nonlinear problems. *Computer Methods in Applied Mechanics and Engineering*, 43:251–276 (1984).

Bergan P.G., Horrigmoe G., Krakeland B. and Soreide T.H. Solution techniques for non-linear finite element problems. *International Journal for Numerical Methods in Engineering*, 12:1677–1696 (1978).

Bertsekas D.P. *Constrained Optimization and Lagrange Multiplier Methods*. Academic Press, New York (1984).

Besseling J.F. and van der Giessen E. *Mathematical Modelling of Inelastic Deformations*. Chapman & Hall, London (1994).

Betsch P. *Statische und dynamische Berechnungen von Schalen endlicher elastischer Deformationen mit gemischten Finiten Elementen*. Dissertation, Institut für Baumechanik und Numerische Mechanik der Universität Hannover (1996). Bericht Nr. F 96/4.

Betsch P., Gruttmann F. and Stein E. A 4-node finite shell element for the implementation of an assumed general hyperelastic 3d-elasticity at finite strains. *Computer Methods in Applied Mechanics and Engineering*, 130:57–79 (1996).

Betsch P., Meyer L. and ·Stein E. On the parametrization of finite rotations in computational mechanics: A classification of concepts with application to smooth shells. *Computer Methods in Applied Mechanics and Engineering*, 155:273–305 (1998).

Betsch P. and Stein E. An assumed strain approach avoiding artificial thickness straining for a nonlinear 4-node shell element. *Communications in Applied Numerical Methods*, 11:899–909 (1995).

Betsch P. and Steinmann P. Conservation properties of a time finite element method. Part i: Time-stepping schemes for n-body problems. *International Journal for Numerical Methods in Engineering*, 49:599–638 (2000).

Bidmon W. *Zum Weiterreißverhalten von beschichteten Geweben*. Dissertation, Institut für Werkstoffe im Bauwesen der Universität Stuttgart (1989). Mitteilung 1989/2.

Bischof C., Buecker H.M., Lang B., Rasch A. and Risch J.W. Extending the functionality of the general-purpose finite element package sepran by automatic differentiation. *International Journal for Numerical Methods in Engineering*, 58:2225–2238 (2003).

Bischof C., Hovland P. and Norris B. Implementation of automatic differentiation tools. In: C. Norris and J.J.B. Fenwick, editors, *Proceedings of the ACM SIGPLAN Workshop on Partial Evaluation and Semantics-Based Program Manipulation*. ACM Press, New York (2002).

Bischoff M. and Ramm E. Shear deformable shell elements for large strains and rotations. *International Journal for Numerical Methods in Engineering*, 40: 4427–4449 (1997).

Bischoff M., Ramm E. and Braess D. A class of equivalent enhanced assumed strain and hybrid stress finite elements. *Computational Mechanics*, 22:443–449 (1999a).

Bischoff M., Wall W.A., Bletzinger K.U. and Ramm E. Models and finite elements for thin-walled structures. In E. Stein, R. de Borst and T.J.R. Hughes, editors, *Encyclopedia of Computational Mechanics*, pages 59–137. Wiley, Chichester (2004).

Bischoff M., Wall W.A. and Ramm E. Stabilized enhanced assumed strain elements for large strain analysis without artificial kinematic modes. In W. Wunderlich, editor, *ECCM 99*, pages 1–19. Lehrstuhl für Statik, München (1999b).

Bodner S.R. and Partom Y. Constitutive equations for elastic viscoplastic strain hardening materials. *Journal of Applied Mechanics*, 42:385–389 (1975).

Boerner E., Loehnert S. and Wriggers P. A new finite element based on the theory of a cosserat point – extension to initially distorted elements for 2d plane strain. *International Journal for Numerical Methods in Engineering*, 71:454–472 (2007).

Boerner E. and Wriggers P. A macro-element for incompressible finite deformations based on a volume averaged deformation gradient. *Computational Mechanics* 42:407–416 (2008).

Boersma A. and Wriggers P. An algebraic multigrid solver for finite element computations in solid mechanics. *Engineering Computations*, 14:202–215 (1997).

Bolotin V. and Armstrong H. The dynamic stability of elastic systems. *American Journal of Physics*, 33:752 (1965).

Bonet J. and Bhargava P. A uniform deformation gradient hexahedron element with artificial hourglass control. *International Journal for Numerical Methods in Engineering*, 38:2809–2828 (1995).

Bonet J. and Burton A.J. A simple average nodal pressure tetrahedral element for incompressible and nearly incompressible dynamic explicit application. *Communications in Numerical Methods in Engineering*, 14:437–449 (1998).

Braess D. *Finite Elements: Theory, Fast Solvers, and Applications in Solid Mechanics*. Cambridge University Press, Cambridge (2007).

Brandt A. Algebraic multigrid theory: The symmetric case. *Applied Mathematics and Computation*, 19:23–56 (1986).

Brandt A., McCormick S.F. and Ruge J.W. Algebraic multigrid (amg) for sparse matrix equations. In E.D. J., editor, *Sparsity and Its Applications*. Cambridge University Press, Cambridge (1985).

Brank B., Briseghella L., Tonello N. and Damjanic F.B. On non-linear dynamics of shells: implementation of energy-momentum conserving algorithm for a finite rotation shell model. *International Journal for Numerical Methods in Engineering*, 42:409–42 (1998).

Bremer C. Algorithmen zum effizienten Einsatz der Finite-Element-Methode. *Technical Report 86–48*, Institut für Statik, Braunschweig (1986).

Brendel B. and Ramm E. Nichtlineare stabilit"atsuntersuchungen mit der methode der finiten elemente. *Ingenieur-Archiv*, 51:337–362 (1982).

Brenner S.C. and Scott L.R. *The Mathematical Theory of Finite Element Methods*. Springer, New York (2002).

Brezzi F. and Fortin M. *Mixed and Hybrid Finite Element Methods*. Springer, Berlin, Heidelberg, New York (1991).

Briseghella L., Majorana C. and Pellegrino C. Dynamic stability of elastic structures: A finite element approach. *Computers and Structures*, 69:11–25 (1998).

Bucher C. Stabilization of explicit time integration by modal reduction. In W.A. Wall, K.U. Bletzinger and K. Schweizerhof, editors, *Trends in Computational Structural Mechanics,*, pages 429–437. Lehrstuhl für Statik, Barcelona, Spain (2001).

Büchter N. and Ramm E. Shell theory versus degeneration – A comparison in large rotation finite element analysis. *International Journal for Numerical Methods in Engineering*, 34:39–59 (1992).

Büchter N., Ramm E. and Roehl D. Three-dimensional extension of non-linear shell formulation based on the enhanced assumed strain concept. *International Journal for Numerical Methods in Engineering*, 37:2551–2568 (1994).

Bufler H. Pressure loaded structures under large deformations. *Zeitschrift für angewandte Mathematik und Mechanik*, 64:287–295 (1984).

Campello E.M.B., Pimenta P.M. and Wriggers P. A triangular finte shell element based on a fully nonlinear shell formulation. *Computational Mechanics*, 31: 505–518 (2003).

Campello E.M.B., Pimenta P.M. and Wriggers P. Elastic-plastic analysis of metallic shells at finite strains. *to appear: Computational Mechanics* (2008).

Carstensen C. and Bartels S. Each averaging technique yields reliable a posteriori error control in fem on unstructured grids. Part i: Low order conforming, nonconforming and mixed fem. *Mathematics of Computation*, 71:945–969 (2002).

Carstensen C. and Funken S. Averaging technique for fe – a posteriori error control in elasticity. Part i: Conforming fem. *Computer Methods in Applied Mechanics and Engineering*, 190:2483–2498 (2001).

Carstensen C., Scherf O. and Wriggers P. Adaptive finite elements for elastic bodies in contact. *SIAM Journal of Scientific Computing*, 20:1605–1626 (1999).

Castanier M.P., Tan Y.C. and Pierre C. Characteristic constraint modes for component mode synthesis. *AIAA Journal*, 39:1182–1187 (2001).

Chadwick P. *Continuum Mechanics, Concise Theory and Problems*. Dover Publications, Mineola (1999).

Chadwick P. and Ogden R.W. A theorem of tensor calculus and its application to isotropic elasticity. *Archives of Rational Mechanics*, 44:54–68 (1971).

Chapelle D. and Bathe K.J. The inf-sup test. *Computers and Structures*, 47: 537–545 (1993).

Chuong C.J. and Fung Y. Three-dimensional stress distribution in arteries. *Journal of Biomechanical Engineering*, 105:268–274 (1983).

Ciarlet P.G. *Mathematical Elasticity I: Three-dimensional Elasticity.* North-Holland, Amsterdam (1988).

Ciarlet P.G. *Introduction to Numerical Linear Algebra and Optimization, English Translation of 1982 Edition in Cambridge Texts in Applied Mathematics.* Cambridge University Press, Cambridge (1989).

Cirak F., Ortiz M. and Schröder P. Subdivision surfaces: A new paradigm for thin shell finite-element analysis. *International Journal for Numerical Methods in Engineering*, 47:2039–2072 (2000).

Crisfield M.A. A fast incremental/iterative solution prodedure that handles snap through. *Computers and Structures*, 13:55–62 (1981).

Crisfield M.A. *Non-linear Finite Element Analysis of Solids and Structures*, volume 1. Wiley, Chichester (1991).

Crisfield M.A. *Non-linear Finite Element Analysis of Solids and Structures*, volume 2. Wiley, Chichester (1997).

Crisfield M.A., Moita G.F., Jelenic G. and Lyons L.P.R. Enhanced lower-order element formulations for large strains. *Computational Mechanics*, 17:62–73 (1995).

Crisfield M.A. and Shi J. A review of solution procedures and path-following techniques in relation to the non-linear finite element analysis of structures. In P. Wriggers and W. Wagner, editors, *Computational Methods in Nonlinear Mechanics*. Springer, Berlin (1991).

Crisfield M.A. and Shi J. A co-rotational element/time-integration strategy for non-linear dynamics. *International Journal for Numerical Methods in Engineering*, 37:1897–1913 (1994).

Crisfield M.A. and Shi J. An energy conserving co-rotational procedure for non-linear dynamics with finite elements. *Nonlinear Dynamics*, 9(1):37–52 (1996).

Cuitino A. and Ortiz M. A material-independent method for extending stress update algorithms from small-strain plasticity to finite plasticity with multiplicative kinematics. *Engineering Computations*, 9:437–451 (1992).

Curnier A. A theory of friction. *International Journal of Solids & Structures*, 20:637–647 (1984).

Curnier A. and Alart P. A generalized Newton method for contact problems with friction. Journal of Theoretical and Applied Mechanics, 7:67–82 (1988).

Curnier A., He Q.C. and Telega J.J. Formulation of unilateral contact between two elastic bodies undergoing finite deformation. *C. R. Academy Science Paris*, 314:1–6 (1992).

Cuthill E. and McKee J. Reducing the bandwith of sparse symmetric matrices. *ACM Publications P-69*, pages 157–172 (1969).

Dafalias Y.F. The plastic spin. *Journal of Applied Mechanics*, 52:865–871 (1985).

Davis T.A. and Duff I.S. A combined unifrontal/multifrontal method for unsymmetric sparse matrices. *ACM Transactions on Mathematical Software*, 25:1–19 (1999).

Dennis J.E. and Schnabel R.B. *Numerical Methods for Unconstrained Optimization and Nonlinear Equations.* Prentice-Hall, Englewood Cliffs, New Jersey (1983).

Desai C.S. and Siriwardane H.J. *Constitutive Laws for Engineering Materials.* Prentice-Hall, Englewood Cliffs, New Jersey (1984).

Dhatt G. and Touzot G. *The Finite Element Method Displayed.* Wiley, Chichester (1985).

Doblare M. Non-linear dynamics of three-dimensional rods: exact energy and momentum conserving algorithms. *International Journal for Numerical Methods in Engineering*, 38:1431–1473 (1995).

Dohrmann C., Heinstein M., Jung J., Key S.W. and Witkowski W. Node-based uniform strain elements for three-node triangular and four-node tetrahedral mesh. *International Journal for Numerical Methods in Engineering*, 47: 1549–1568 (2000).

Doll S. and Schweizerhof K. On the development of volumetric strain energy functions. *Transactions of the ASME - E - Journal of Applied Mechanics*, 67:17–21 (2000).

Dostal Z. A proportioning based algorithm for bound constraint quadratic programming with the rate of convergence. *Numerical Algorithms*, 34:293–302 (2003).

Douglas C., Haase G. and Langer U. *A Tutorial on Elliptic PDE Solvers and their Parallelization*. SIAM, Philadelphia (2003).

Drucker D.C. and Prager W. Soil mechanics and plastic analysis or limit design. *Quarterly Appl. Math.*, 10:157–165 (1952).

Duff I.S. Ma57 – a new code for the solution of sparse symmetric definite and indefinite systems. *ACM Transactions on Mathematical Software*, 30:118–154 (2004).

Duff I.S., Erisman A.M. and Reid J.K. *Direct Methods for Sparse Matrices*. Clarendon Press, Oxford (1989).

Duffet G. and Reddy B.D. The analysis of incompressible hyperelastic bodies by the finite element method. *Computer Methods in Applied Mechanics and Engineering*, 41:105–120 (1983).

Düster A., Bröker H. and Rank E. The p-version of the finite element method for three-dimensional curved thin walled structures. *International Journal for Numerical Methods in Engineering*, 52:673–703 (2001).

Düster A., Hartmann S. and Rank E. p-FEM applied to finite isotropic hyperelastic bodies. *Computer Methods in Applied Mechanics and Engineering*, 192: 5147–5166 (2003).

Düster A., Niggl A., Nübel V. and Rank E. A numerical investigation of high-order finite elements for problems of elastoplasticity. *Journal of Scientific Computing*, 17(1):397–404 (2002).

Duvaut G. and Lions J.L. *Inequalities in Mechanics and Physics*. Springer Verlag, Berlin (1976).

Dvorkin E.N. and Bathe K.J. A continuum mechanics based four-node shell element for general nonlinear analysis. *Engineering Computations*, 1:77–88 (1984).

Dvorkin E.N., Onate E. and Oliver J. On a nonlinear formulation for curved Timoshenko beam elements considering large displacement/rotation increments. *International Journal for Numerical Methods in Engineering*, 26:1597–1613 (1988).

Dvorkin E.N., Pantuso D. and Repetto A. A formulation of the MITC4 shell element for finite strain elasto-plastic analysis. *Computer Methods in Applied Mechanics and Engineering*, 125:17–40 (1995).

Eberlein R. *Finite-Elemente-Konzepte für Schalen mit großen elastischen und plastischen Verzerrungen*. Dissertation, Institut für Mechanik IV der Technischen Hochschule Darmstadt (1997).

Eberlein R. and Wriggers P. Finite element concepts for finite elastoplastic strains and isotropic stress response in shells: Theoretical and computational analysis. *Computer Methods in Applied Mechanics and Engineering*, 171:243–279 (1999).

Eberlein R., Wriggers P. and Taylor R. A fully non-linear axisymmetrical quasi-kirchhoff-type shell element for rubberlike materials. *International Journal for Numerical Methods in Engineering*, 36:4027–4043 (1993).

Ehrlich D. and Armero F. Finite element methods for the analysis of softening plastic hinges in beams and frames. *Computational Mechanics*, 35(4):237–264 (2005).

Ekh M., Johansson A., Thorberntsson H. and Josefson B. Models for Cyclic Ratchetting Plasticity–Integration and Calibration. *Journal of Engineering Materials and Technology*, 122:49 (2000).

Elguedj T., Bazilevs Y., Calo V.M. and Hughes T.J.R. $\bar{B}$ and $\bar{F}$ projection methods for nearly incompressible linear and nonlinear elasticity and plasticity using higher-order NURBS elements. *Computer Methods in Applied Mechanics and Engineering* (2008).

Elman H.C., Silvester D.J. and Wathen A.J. *Finite Elements and Fast Iterative Solvers with Applications in Incompressible Fluid Dynamics*. Oxford University Press, Oxford (2005).

van Engelen R.A., Wolters L. and Cats G. Ctadel: A generator of efficient code for PDE-based scientific applications. *Technical report*, Leiden Institute of Advanced Computer Science, http://www.liacs.nl/TechRep/1995/tr95-26.ps.gz, (1995).

Eriksson E. On some path-related measures for non-linear structural f. e. problems. *International Journal for Numerical Methods in Engineering*, 26:1791–1803 (1988).

Eringen A. *Nonlinear Theory of Continuous Media*. McGraw-Hill, New York, London (1962).

Eringen A. *Mechanics of Continua*. Wiley, New York, London, Sidney (1967).

Eschenauer H. and Schnell W. *Elastizitätstheorie*. BI Wissenschaftsverlag, Mannheim, dritte edition (1993).

Eyheramendy D. and Zimmermann T. Object-oriented symbolic derivation and automatic programming of finite elements in mechanics. *Engineering with Computers*, 15(1):12–36 (2000).

Farhat C. and Roux F. A method of finite element tearing and interconnecting and its parallel solution algorithm. *International Journal for Numerical Methods in Engineering*, 32(6):1205–1227 (1991).

Felippa C.A. A historical outline of matrix structural analysis: A play in three acts. *Technical Report CU-CAS-00-13*, Center for Aerospace Structures, University of Colorado (2000).

Feucht M. *Ein gradientenabhängiges Gursonmodell zur Beschreibung duktiler Schädigung mit Entfestigung*. Dissertation, Institut für Mechanik der TU Darmstadt (1999). Bericht Nr. D 17.

Findley W.N., Lai J.S. and Onaran K. *Creep and Relaxation of Nonlinear Viscoelastic Materials*. Dover Publications, New York (1989).

Fischer K.A. and Wriggers P. Mortar based frictional contact formulation for higher order interpolations using the moving friction cone. *Computer Methods in Applied Mechanics and Engineering*, 195:5020–5036 (2006).

Fletcher R. Conjugated gradient methods for indefinite systems. *Lecture Notes in Mathematics*, 506:773–789 (1976).

Flory P. Thermodynamic relations for high elastic materials. *Transactions of the Faraday Society*, 57:829–838 (1961).

Fourment L. and Chenot J.L. Error estimators for viscoplastic materials: Applications to forming processes. *International Journal for Numerical Methods in Engineering*, 38:469–490 (1995).

Fredriksson B. Finite element solution of surface nonlinearities in structural mechanics with special emphasis to contact and fracture mechanics problems. *Computers and Structures*, 6:281–290 (1976).

Fried I. Orthogonal trajectory accession to the nonlinear equilibrium curve. *Computer Methods in Applied Mechanics and Engineering*, 15:283–297 (1984).

Fritzson P. and Fritzson D. The need for high-level programming support in scientific computing applied to mechanical analysis. *Computers and Structures*, 45:387–395 (1984).

Fujii F. and Ramm E. Computational bifurcation theory: path-tracing, pinpointing and path-switching. *Engineering Structures*, 19:385–392 (1997).

Gear C.W. *Numerical Initial Value Problems in Ordinary Differential Equations.* Prentice-Hall, Englewood Cliffs (1971).

Geradin M. and Rixen D. *Mechanical Vibrations.* Wiley, Chichester, second edition (1997).

Giannokopoulos A.E. The return mapping method for the integration of friction constitutive relations. *Computers and Structures*, 32:157–168 (1989).

Glaser S. and Armero F. On the formulation of enhanced strain finite elements in finite deformations. *Engineering Computations*, 14:759–791 (1997).

Glowinski R. and Le Tallec P. Finite element analysis in nonlinear incompressible elasticity. In *Finite Element, Vol. V: Special Problems in Solid Mechanics.* Prentice-Hall, Englewood Cliffs, New Jersey (1984).

Golub G. and Ortega J.M. *Scientific Computing, Eine Einführung in das wissenschaftliche Rechnen und Parallele Numerik.* Teubner-Verlag, Stuttgart (1996).

Golub G.H. and van Loan C.F. *Matrix Computations.* John Hopkins University Press, Baltimore (1989).

Gonnet G. New results for random determination of equivalence of expression. In B.W. Char, editor, *Proceedings of 1986 ACM Symposium on Symbolic and Algebraic Computation*, pages 127–131. ACM, Waterloo (1986).

Gonzalez O. Exact energy and momentum conserving algorithms for general models in nonlinear elasticity. *Computer Methods in Applied Mechanics and Engineering*, 190(13–14):1763–1783 (2000).

Gould N.I.M., Hu Y. and Scott J.A. A numerical evaluation of sparse direct solvers for the solution of large sparse, symmetric linear systems of equations. *Technical Report RAL-TR-2005-00*, Computational Science and Engineering Department, Atlas Centre, Rutherford Appleton Laboratory, Oxfordshire (2005).

Govindjee S. and Simo J.C. Mullin's effect and the strain amplitiude dependence of the storage modulus. *International Journal of Solids & Structures*, 29:1737–1751 (1992).

Griewank A. *Evaluating Derivatives: Principles and Techniques of Algorithmic Differentiation.* SIAM, Philadelphia (2000).

Griewank A. and Walther A. *Evaluating Derivatives: Principles and Techniques of Algorithmic Differentiation.* SIAM, Philadelphia, second edition (2008).

Gross D., Hauger W., Schnell W. and Wriggers P. *Technische Mechanik 4.* Springer, Berlin, third edition (1999).

Gruttmann F. *Theorie und Numerik dünnwandiger Faserverbundstrukturen.* Habilitation, Institut für Baumechanik und Numerische Mechanik der Universität Hannover (1996). Bericht Nr. F 96/1.

Gruttmann F., Sauer R. and Wagner W. A geometrical non-linear eccentric 3D-beam element with arbitrary cross-sections. *Computer Methods in Applied Mechanics and Engineering*, 160:383–400 (1998).

Gruttmann F., Sauer R. and Wagner W. Theory and numerics of three-dimensional beams with elastoplastic material behaviour. *International Journal for Numerical Methods in Engineering*, 48:1675–1702 (2000).

Gruttmann F. and Stein E. Tangentiale Steifigkeitsmatrizen bei Anwendung von Projektionsverfahren in der Elastoplastizitätstheorie. *Ingenieur-Archiv*, 58:15–24 (1988).

Gruttmann F. and Wagner W. A linear quadrilateral shell element with fast stiffness computation. *Computer Methods in Applied Mechanics and Engineering*, 194 (39-41):4279–4300 (2005).

Gruttmann F., Wagner W. and Wriggers P. A nonlinear quadrilateral shell element with drilling degrees of freedom. *Ingenieur Archiv*, 62:474–486 (1992).

Guo Y., Ortiz M., Belytschko T. and Repetto E.A. Triangular composite finite elements. *International Journal for Numerical Methods in Engineering*, 47: 287–316 (2000).

Gurson A.L. Continuum theory of ductile rupture by void nucleation and growth, Part i. *Journal Engineering Material Technology*, 99:2–15 (1977).

Haase G., Langer U., Reitzinger S. and Schöberl J. Algebraic multigrid methods based on element preconditioning. *International Journal of Computer Mathematics*, 78(4):575–598 (2001).

Habraken A. and Cescotto S. An automatic remeshing technique for finite element simulation of forming processes. *International Journal for Numerical Methods in Engineering*, 30:1503–1525 (1990).

Hackbusch W. *Iterative Solution of Large Sparse Systems*. Springer, New York (1994).

Hackbusch W. *Multi-Grid Methods and Applications*. Springer, New York (2003).

Häggblad B. and Sundberg J.A. Large strain solutions of rubber components. *Computers and Structures*, 17:835–843 (1983).

Hallquist J.O. Nike2d: An implicit, finite-deformation, finite element code for analysing the static and dynamic response of two-dimensional solids. *Technical Report UCRL-52678*, University of California, Lawrence Livermore National Laboratory (1979).

Hallquist J.O., Goudreau G.L. and Benson D.J. Sliding interfaces with contact-impact in large-scale lagrange computations. *Computer Methods in Applied Mechanics and Engineering*, 51:107–137 (1985).

Hallquist J.O., Schweizerhof K. and Stillman D. Efficiency refinements of contact strategies and algorithms in explicit fe programming. In D.R.J. Owen, E. Hinton and E. Onate, editors, *Proceedings of COMPLAS III*, pages 359–384. Pineridge Press (1992).

Han C.S. *Eine h-adaptive Finite-Element-Methode für elasto-plastische Schalenprobleme in unilateralem Kontakt*. Dissertation, Institut für Baumechanik und Numerische Mechanik der Universität Hannover (1999). Bericht Nr. F 99/2.

Han C.S. and Wriggers P. A simple local a posteriori bending indicator for axisymmetrical membrane and bending shell elements. *Engineering Computations*, 15:977–988 (1998).

Hart E.W. Constitutive relations for the nonelastic deformation of metals. *Trans. ASME, Journal of Engineering Materials and Technology*, 98:193–201 (1976).

Hassler M. and Schweizerhof K. On the static interaction of fluid and gas loaded multi-chamber systems in large deformation finite element analysis. *Computer Methods in Applied Mechanics and Engineering*, 197:1725–1749 (2008).

Hauptmann R. *Strukturangepasste geometrisch nichtlineare finite Elemente für Flächentragwerke*. Dissertation, Institut für M der Universität Fredericiana Karlsruhe (1997). Bericht Nr. M 97/3.

Hauptmann R. and Schweizerhof K. A systematic development of 'solid-shell' element formulation for linear and nonlinear analyses employing only displacement degree of freedom. *International Journal for Numerical Methods in Engineering*, 42:49–69 (1998).

Heisserer U., Hartmann S., Yosibash Z. and Düster A. On volumetric locking-free behavior of p-version finite elements under finite deformations. *Communications in Numerical Methods in Engineering*. DOI: 10.1002/cnm.1008 (in press) (2007).

Henning A. Traglastberechnung ebener Rahmen – Theorie II. Ordnungschweizerhof und Interaktion. *Technical Report 75-12*, Institut für Statik, Braunschweig (1975).

Hilber H., Hughes T.R.J. and Taylor R.L. Improved numerical dissipation for time integration algorithms in structural dynamics. *Earthquake Engineering and Structural Dynamics*, 5:283–292 (1977).

Hill R. *The Mathematical Theory of Plasticity*. Clarendon Press, Oxford (1950).

Hill R. A general theory of uniqueness and stability in elasto-plastic solids. *Journal of Mechanics and Physics of Solids*, 6:236–249 (1958).

Hill R. Constitutive inequalities for isotropic elastic solids under finite strain. *Proceedings of the Royal Society, London*, A314:457–472 (1970).

Hinton E. and Owen D.R.J. *An Introduction to Finite Element Computations*. Pineridge Press, Swansea (1979).

Hinton E., Rock T. and Zienkiewicz O.C. A note on mass lumping and related processes in the finite element method. *Earthquake Engineering and Structural Dynamics*, 4:245–249 (1976).

Hofstetter G. and Mang H.A. *Computational Mechanics of Reinforced Concrete Structures*. Vieweg, Berlin (1995).

Hoger A. The stress conjugate to logarithmic strain. *International Journal of Solids & Structures*, 23:1645–1656 (1987).

Hohenemser K. and Prager W. Über die ansätze der mechanik isotroper kontinua. *Zeitschrift für angewandte Mathematik und Mechanik*, 12:216–226 (1932).

Hoit M. and Wilson E.L. An equation numbering algorithm based on a minimum front criteria. *Computers and Structures*, 16:225–239 (1983).

Holmes P., Lumley J.L. and Berkooz G. *Turbulence, Coherent Structures, Dynamical Systems and Symmetry*. Cambridge University Press, Cambridge (1996).

Holzapfel G., Eberlein R., Wriggers P. and Weizsäcker H. Large strain analysis of soft biological and rubber-like membranes: Formulation and finite element analysis. *Computer Methods in Applied Mechanics and Engineering*, 132:45–61 (1996a).

Holzapfel G., Eberlein R., Wriggers P. and Weizsäcker H. A new axisymmetrical membrane element for anisotropic, finite strain analysis of arteries. *Communications in Applied Numerical Methods*, 12:507–517 (1996b).

Holzapfel G.A. *Nonlinear Solid Mechanics*. Wiley, Chichester (2000).

Hueck U., Reddy B. and Wriggers P. On the stabilization of the rectangular four-node quadrilateral element. *Communications in Applied Numerical Methods*, 10:555–563 (1994).

Hueck U. and Wriggers P. A formulation for the four-node quadrilateral element, Part i: Plane element. *International Journal for Numerical Methods in Engineering*, 38:3007–3037 (1995).

Hughes T.J.R. Generalization of selective integration procedures to anisotropic and nonlinear media. *International Journal for Numerical Methods in Engineering*, 15:1413–1418 (1980).

Hughes T.J.R. and Brezzi F. On drilling degrees of freedom. *Computer Methods in Applied Mechanics and Engineering*, 72:105–121 (1989).

Hughes T.J.R., Cottrell J.A. and Bazilevs Y. Isogeometric analysis: CAD, finite elements, NURBS, exact geometry, and mesh refinement. *Computer Methods in Applied Mechanics and Engineering*, 194(39-41):4135–4195 (2005).

Hughes T.J.R. and Liu W.K. Nonlinear Finite Element Analysis of Shells: Part I. Threedimensional Shells. *Computer Methods in Applied Mechanics and Engineering*, 26:331–362 (1981).

Hughes T.J.R., Taylor R.L. and Kanoknukulchai W. A Simple and Efficient Finite Element for Plate Bending. *International Journal for Numerical Methods in Engineering*, 11:1529–1547 (1977a).

Hughes T.J.R. and Tezduyar T.E. Finite elements based upon mindlin plate theory with particular reference to the four-node bilinear isoparametric element. *Journal of Applied Mechanics*, 48:587–596 (1981).

Hughes T.R.J. *The Finite Element Method*. Prentice Hall, Englewood Cliffs, New Jersey (1987).

Hughes T.R.J., Taylor R.L. and Kanoknukulchai W. A finite element method for large displacement contact and impact problems. In K.J. Bathe, editor, *Formulations and Computational Algorithms in FE Analysis*, pages 468–495. MIT Press, Boston (1977b).

Ibrahimbegovic A. Finite Elastoplastic Deformations of Space-Curved Membranes. *Computer Methods in Applied Mechanics and Engineering*, 119:371–394 (1994).

Ibrahimbegovic A. A finite element implementaion of geometrically nonlinear Reissner's beam theory: Three-dimensional curved beam elements. *Computer Methods in Applied Mechanics and Engineering*, 122:11–26 (1995).

Ibrahimbegovic A., Taylor R.L. and Wilson E.L. A robust quadrilateral membrane element with drilling degrees of freedom. *International Journal for Numerical Methods in Engineering*, 30:445–457 (1990).

Idelsohn S.R. and Cardona A. Reduction methods and explicit time integration. *Advances in Engineering Software*, 6:36–44 (1984).

Idelsohn S.R. and Cardona A. A reduction method for nonlinear structural dynamic analysis. *Computer Methods in Applied Mechanics and Engineering*, 49:253–279 (1985).

Irons B. Quadrature rules for brick based finite elements. *International Journal for Numerical Methods in Engineering*, 3:293–294 (1971).

Irons B. and Ahmad S. *Techniques of Finite Elements*. Ellis Horwood, Chichester, U.K. (1986).

Isaacson E. and Keller H.B. *Analysis of Numerical Methods*. Wiley, London (1966).

Itskov M. Computation of the exponential and other isotropic tensor functions and their derivatives. *Computer Methods in Applied Mechanics and Engineering*, 192:3985–3999 (2003).

Iura M. and Atluri S.N. Formulation of a membrane finite element with drilling degrees of freedom. *Computational Mechanics*, 39:417–428 (1992).

Jaquotte O.P. and Oden J.T. An accurate and efficient a posteriori control of hourglass instabilities in underintegrated linear and nonlinear elasticity. *Computer Methods in Applied Mechanics and Engineering*, 55:105–128 (1986).

Jelenic G. and Saje M. A kinematically exact space finite strain beam model - finite element formulations by generalised virtual work principle. *Computer Methods in Applied Mechanics and Engineering*, 120:131–161 (1995).

Jepson A. and Spence A. Folds in solutions of two parameter systems amd theri calculation, Part i. *SIAM Journal on Numerical Analysis*, 22:347–368 (1985).

Joe B. Quadrilateral mesh generation in polygonal regions. *Computer Aided Design*, 27:209–222 (1995).

Johansson G., Ekh M. and Runesson K. Computational modeling of inelastic large ratcheting strains. *International Journal of Plasticity*, 21(5):955–980 (2005).

Johnson C. *Numerical Solution of Partial Differential Equations by the Finite Element Method*. Cambridge University Press, Cambridge (1987).

Johnson C. and Hansbo P. Adaptive finite element methods in computational mechanics. *Computer Methods in Applied Mechanics and Engineering*, 101:143–181 (1992).

Ju W. and Taylor R.L. A perturbed lagrange formulation for the finite element solution of nonlinear frictional contact problems. *Journal of Theoretical and Applied Mechanics*, 7:1–14 (1988).

Jung M. and Langer U. *Methode der finiten Elemente für Ingenieure*. Teubner (2001).

Kahn R. Finite-element-berechnungen ebener stabwerke mit flissgelenken und grossen verschiebungen. *Technical Report F 87/1*, Forschungs- und Seminarberichte aus dem Bereich der Mechanik der Universität Hannover (1987).

Kane C., Marsden J.E. and Ortiz M. Symplectic-energy-momentum preserving variational integrators. *Journal of Mathematical Physics*, 40(7):3353 (1999).

Kappus R. Zur Elastizitätstheorie endlicher Verschiebungen. *Zeitschrift für angewandte Mathematik und Mechanik*, 19:271–361 (1939).

Keller H.B. Numerical solution of bifurcation and nonlinear eigenvalue problems. In P. Rabinowitz, editor, *Application of Bifurcation Theory*, pages 359–384. Academic Press, New York (1977).

Khan A.S. and Huang S. *Continuum Theory of Plasticity*. Wiley, Chichester, New York (1995).

Kickinger F. Algebraic multigrid solver for discrete elliptic second order problems. *Technical Report 96-5*, Department of Mathematics, Johannes Kepler University, Linz (1996).

Kikuchi N. and Oden J.T. *Contact Problems in Elasticity: A Study of Variational Inequalities and Finite Element Methods*. SIAM, Philadelphia (1988).

Kirby R.C., Knepley M., Logg A. and Scott L.R. Optimizing the evaluation of finite element matrices. *SIAM Journal on Scientific Computing*, 27:741–758 (2005).

Kirsch U., Bogomolni M. and Sheinman I. Nonlinear dynamic reanalysis for structural optimization. In van CampenD. H., L.M. D. and van den Oever W. P. J . M, editors, *Proceedings of the Fifth EUROMECH Nonlinear Dynamics Conference (ENOC)*. Eindhoven (2005).

Koiter W.T. A consistent first approximation in the general theory of thin elastic shells. In W.T. Koiter, editor, *The Theory of Thin Elastic Shells*, pages 12–33. North-Holland, Amsterdam (1960).

Konyukhov A. and Schweizerhof K. A special focus on 2d formulations for contact problems using a convective description. *International Journal for Numerical Methods in Engineering*, 66:1432–1465 (2006).

Korelc J. Automatic generation of finite-element code by simultaneous optimization of expressions. *Theoretical Computer Science*, 187:231–248 (1997).

Korelc J. Multi-language and multi-environment generation of nonlinear finite element codes. *Engineering with Computers*, 18:312–327 (2002).

Korelc J. and Wriggers P. Consistent gradient formulation for a stable enhanced strain method for large deformations. *Engineering Computations*, 13:103–123 (1996a).

Korelc J. and Wriggers P. Improved enhanced strain 3-d element with Taylor expansion of shape functions. *Computational Mechanics*, 19:30–40 (1996b).

Korneev V.G., Langer U. and Xanthis L. On fast domain decomposition solving procedures for hp-discretizations of 3d elliptic problems. *Computational Methods in Applied Mathematics*, 3:536–559 (2003).

Kosloff D. and Frazier G.A. Treatment of hourglass pattern in low order finite element codes. *International Journal for Numerical and Analytical Methods in Geomechanics*, 2:57–72 (1978).

Kočvara M. and Mande l.J. A multigrid method for three dimensional elasticity and algebraic convergence estimates. *Applied Mathematics and Computation*, 23:121–135 (1987).

Kragelsky I.V. and Alisin V.V. *Tribology - Lubrication , Friction, and Wear*. Professional Engineering Publishing (2001).

Kragelsky I.V., Dobychin M.N. and Kombalov V.S. *Friction and Wear - Calculation Methods (Translated from The Russian by N. Standen)*. Pergamon Press, Oxford (1982).

Krätzig W.B. 'Best' transverse shearing and stretching shell theory for non-linear finite element simulations. *Computer Methods in Applied Mechanics and Engineering*, 103:135–160 (1993).

Krempl E., McMahon J. and Yao D. Viscoplasticity based on overstress with a differential growth law for the equilibrium stress. *Mechanics of Materials*, 5: 35–48 (1986).

Kreuzer E. and Kust O. Analysis of long torsional strings by proper orthogonal decomposition. *Archive of Applied Mechanics*, 184:68–80 (1995).

Krstulovic-Opara L., Wriggers P. and Korelc J. A c1-continuous formulation for 3d finite deformation frictional contact. *Computational Mechanics*, 29:27–42 (2002).

Krysl P., Lall S. and Marsden J.E. Dimensional model reduction in non-linear finite element dynamics of solids and structures. *International Journal for Numerical Methods in Engineering*, 51:479–504 (2001).

Kühborn A. and Schoop H. A Nonlinear Theory for Sandwich Shells Including the Wrinkling Phenomenon. *Ingenieur-Archiv*, 62:413–427 (1992).

Kuhl D. and Crisfield M.A. Energy-conserving and decaying algorithms in nonlinear structural dynamics. *International Journal for Numerical Methods in Engineering*, 45(5):569–599 (1999).

Kuhl D. and Ramm E. Constraint energy momentum algorithm and its application to nonlinear dynamics of shells. *Computer Methods in Applied Mechanics and Engineering*, 136:293–315 (1996).

Ladeveze P. Constitutive relation error estimators for time-dependent nonlinear f.e. analysis. In S. Idelsohn, E. Onate and E. Dvorkin, editors, *Computational Mechanics*. CIMNE, Barcelona (1998).

Ladeveze P. and Leguillon D. Error estimate procedure in the finite element method and applications. *SIAM Journal on Numerical Analysis*, 20:485–509 (1983).

Ladeveze P. and Pelle J.P. *Mastering Calculations in Linear and Nonlinear Mechanics*. Springer, Mechanical Engineering Series, New York (2005).

Langer U. Multigrid – methoden. *Technical report*, Institut für Mathematik, Johannes Kepler Universität Linz (1996).

Larsson R., Runesson K. and Ottosen N. Discontinuous displacement approximation for capturing plastic localization. *International Journal for Numerical Methods in Engineering*, 36:2087–2105 (1993).

Lasry D. and Belytschko T. Localization limiters in transient problems. *International Journal of Solids & Structures*, 24:581–597 (1988).

Laursen T.A. *Computational Contact and Impact Mechanics*. Springer, Berlin, New York, Heidelberg (2002).

Laursen T.A. and Meng X.N. A new solution procedure for application of energy-conserving algorithms to general constitutive models in nonlinear elastodynamics. *Computer Methods in Applied Mechanics and Engineering*, 190(46-47): 6309–6322 (2001).

Laursen T.A. and Simo J.C. Algorithmic symmetrization of Coulomb frictional problems using augmented Lagrangians. *Computer Methods in Applied Mechanics and Engineering*, 108:133–146 (1993a).

Laursen T.A. and Simo J.C. A continuum-based finite element formulation for the implicit solution of multibody, large deformation frictional contact problems. *International Journal for Numerical Methods in Engineering*, 36:3451–3485 (1993b).

Leblond J.B., Perrin G. and Devaux J. Bifurcation effects in ductile metals with nonlocal damage. *Journal of Applied Mechanics*, 61:236–242 (1994).

Lee E.H. and Liu D.T.   Finite strain elasto-plastic theory with applications to plane-wave analysis. *Journal of Applied Mechanics*, 38:19–27 (1967).

Leger P. Mode superposition methods. In M. Papadrakakis, editor, *Solving Large-scale Problems in Mechanics*, pages 225–257. Wiley, New York (1993).

Leppin C. and Wriggers P.   Numerical simulations of the behaviour of cohesionless soil. In D.R.J. Owen, E. Hinton and E. Onate, editors, *Proceedings of COMPLAS 5*. CIMNE, Barcelona (1997).

Libai A. and Simmonds J.G.   Large-strain constitutive laes for the cylindrical deformation of shells. *International Journal of Nonlinear Mechanics*, 16:91–103 (1992).

Loehnert S., Boerner E., Rubin M. and Wriggers P.   Response of a nonlinear elastic general cosserat brick element in simulations typically exhibiting locking and hourglassing. *Computational Mechanics*, 36:255–265 (2005).

Logg A. Automating the finite element method. *Archives of Computational Methods in Engineering*, 14:93–138 (2007).

Löhner R. Progress in grid generation via the advancing front technique. *Engineering Computations*, 12:186–210 (1996).

Lu J. Exact expansions of arbitrary tensor functions and their derivatives. *International Journal of Solids & Structures*, 41:337–349 (2004).

Lubliner J. A model of rubber viscoelasticity. *Mechancis Research Communications*, 12:93–99 (1985).

Lubliner J. *Plasticity Theory*. MacMillan, London (1990).

Luenberger D.G. *Linear and Nonlinear Programming*. Addison-Wesley, Reading, MA, second edition (1984).

Lumpe G. Geometrisch nichtlineare berechnung von räumlichen stabwerken. *Technical Report 28*, Institut für Statik, Universität Hannover (1982).

MacNeal R.H. and Harder R.L. A Proposed Standard Set of Problems to Test Finite Element Accuracy. *Finite Elements in Analysis and Design*, 1:3–20 (1985).

Mahnken R., Caylaka I. and Laschet G. Two mixed finite element formulations with area bubble functions for tetrahedral elements. *Computer Methods in Applied Mechanics and Engineering*, 197:1147–1165 (2008).

Mäkinen J. Total lagrangian Reissner's geometrically exact beam element without singularities. *International Journal for Numerical Methods in Engineering*, 70:1009–1048 (2007).

Malkus D.S. and Hughes T.J.R.   Mixed finite element methods – Reduced and selective integration techniques: A unification of concepts. *Computer Methods in Applied Mechanics and Engineering*, 15:63–81 (1978).

Malvern L.E. *Introduction to the Mechanics of a Continuous Medium*. Prentice-Hall, Englewood Cliffs, New Jersey (1969).

Mandel J. Thermodynamics and Plasticity. In *Foundations of Continuum Thermodynamics*, Delgado Domingers, J. J. and Nina, N. R. and Whitelaw, J. H. (Eds.). MacMillan, London (1974). 283–304.

Marsden J.E. and Hughes T.J.R. *Mathematical Foundations of Elasticity*. Prentice-Hall, Englewood Cliffs, NJ (1983).

Matthies H. and Strang G. The solution of nonlinear finite element equations. *International Journal for Numerical Methods in Engineering*, 14:1613–1626 (1979).

Maugin G.A. *The Thermomechanics of Plasticity and Fracture*. Cambridge University Press, Cambridge, New York (1992).

Meisel M. and Meyer A. Implementierung eines parallelen vorkonditionierten Schur-Komplement CG-Verfahrens in das Programmpaket FEAP. *Technical Report SPC 95-2*, Institut für Mathematik, TU Chemnitz-Zwickau (1995).

Meyer A. A parallel preconditioned conjugate gradient method using domain decomposition and inexact solvers on each subdomain. *Computing*, 45 (1990).

Meyer M. and Matthies H. Efficient model reduction in non-linear dynamics using the Karhunen-Loeve expansion and dual-weighted-residual methods. *Computational Mechanics*, 31:179–191 (2003).

Meynen S., Boersma A. and Wriggers P. Application of a parallel algebraic multigrid method for the solution of elasto-plastic shell problems. *Numerical Linear Algebra with Application*, 4:223–238 (1997).

Michalowski R. and Mroz Z. Associated and non-associated sliding rules in contact friction problems. *Archives of Mechanics*, 30:259–276 (1978).

Miehe C. Kanonische modelle multiplikativer elasto-plastizit"at. thermodynamische formulierung und numerische implementation. *Technical Report F 93/1*, Forschungs- und Seminarberichte aus dem Bereich der Mechanik der Universität Hannover (1993).

Miehe C. Aspects of the Formulation and Finite Element Implementation of Large Strain Isotropic Elasticity. *International Journal for Numerical Methods in Engineering*, 37:1981–2004 (1994).

Miehe C. A formulation of finite elastoplasticity in shells based on dual co- and contra-variant eigenvectors normalized with respect to a plastic metric. In D.R.J. Owen, E. Onate and E. Hinton, editors, *Computational Plasticity, Fundamentals and Applications*. CIMNE, Barcelona (1997). 1922–1929.

Miehe C. A theoretical and computational model for isotropic elastoplastic stress analysis in shells at large strains. *Computer Methods in Applied Mechanics and Engineering*, 155:193–233 (1998).

Miehe C. and Schröder J. Post-critical discontinous localization analysis of small-strain softening elastoplastic solids. *Archive of Applied Mechanics*, 64:267–285 (1994).

Mittelmann H.D. and Weber H. Numerical methods for bifurcation problems – a survey and classification. In H.D. Mittelmann and H. Weber, editors, *Bifurcation Problems and their Numerical Solution, ISNM 54*, pages 1–45. Birkhäuser, Basel, Boston, Stuttgart (1980).

Moita G.F. and Crisfield M.A. A finte element formulation for 3-d continua using the co-rotational technique. *International Journal for Numerical Methods in Engineering*, 39(22):3775–3792 (1996).

Mooney M. A theory of large elastic deformations. *Journal for Applied Physics*, 11:582–592 (1940).

Moreau J.J. Application of convex analysis to the treatment of elastoplastic strucures. In P. Germain and B. Nayroles, editors, *Applications of Methods of Functional Analysis to Problems in Mechanics*. Springer-Verlag, Berlin (1976).

Morman K.N. The generalized strain measure with applications to non-homogeneous deformations in rubber-like solids. *Journal of Applied Mechanics*, 53:726–728 (1987).

Mueller-Hoeppe D.S., Loehnert S. and Wriggers P. A finite deformation brick element with inhomogeneous mode enhancement, on-line first. *International Journal for Numerical Methods in Engineering* (2008).

Nadler B. and Rubin M. A new 3-d finite element for nonlinear elasticity using the theory of a cosserat point. *International Journal of Solids and Structures*, 40:4585–4614 (2003).

Naghdi P.M. *The Theory of Shells*, volume VIa/2 of *Handbuch der Physik, Mechanics of Solids II*. Springer, Berlin (1972).

Nagtegaal J.C. On the implementation of inelastic constitutive equations with special reference to large deformation problems. *Computer Methods in Applied Mechanics and Engineering*, 33(1-3):469–484 (1982).

Nagtegaal J.C., Parks D.M. and Rice J.C. On numerically accurate finite element solutions in the fully plastic range. *Computer Methods in Applied Mechanics and Engineering*, 4: 153–177 (1990).

Needleman A. A numerical study of necking in circular cylindrical bars. *Journal of Mechanics and Physics of Solids*, 20:111–127 (1972).

Needleman A. and Rice J.R. Limits to ductility set by plastic flow localization. In D.P. Koistinen and N.-W. Wang, editor, *Mechanics of Sheet Metal Forming*, pages 237–267. Plenum Press, New York (1978).

Newmark N.M. A method of computation for structural dynamics. *Proceedings of ASCE, Journal of Engineering Mechanics*, 85:67–94 (1959).

Nguyen Q. *Stability and Nonlinear Solid Mechanics*. Wiley, New York (2000).

Nickell R.E. Nonlinear dynamics by mode superposition. *Computer Methods in Applied Mechanics and Engineering*, 7:107–129 (1976).

Noor A.K. Recent advances and applications of reduction methods. *Applied Mechanics Review*, 47:125–146–94 (1994).

Oden J.T. Exterior penalty methods for contact problems in elasticity. In W. Wunderlich, E. Stein and K.J. Bathe, editors, *Nonlinear Finite Element Analysis in Structural Mechanics*. Springer, Berlin (1981).

Oden J.T. and Key J.E. Numerical analysis of finite axisymmetrical deformations of incompressible elastic solids of revolution. *International Journal of Solids & Structures*, 6:497–518 (1970).

Oden J.T. and Martins J.A.C. Models and computational methods for dynamic friction phenomena. *Computer Methods in Applied Mechanics and Engineering*, 52:527–634 (1986).

Oden J.T., Zohdi T. and Rodin G.J. Hierarchical modelling of heterogeneous bodies. *Computer Methods in Applied Mechanics and Engineering*, 138:273–298 (1996).

Ogden R.W. Large deformation isotropic elasticity: On the correlation of theory and experiment for incompressible rubberlike solids. *Proceedings of the Royal Society of London*, 326:565–584 (1972).

Ogden R.W. Elastic deformations of rubberlike solids. In H.G. Hopkins and M.J. Sewell, editors, *Mechanics of Solids, The Rodney Hill 60th Anniversary Volume*, pages 499–537. Pergamon Press, Oxford (1982).

Ogden R.W. *Non-Linear Elastic Deformations*. Ellis Horwood and John Wiley, Chichester (1984).

Oliver J. Continuum modeling of strong discontinuities in solid mechanics using damage models. *Computational Mechanics*, 17:49–61 (1995).

Oliver J., Huespe A.E., Blanco S. and Linero D.L. Stability and robustness issues in numerical modeling of material failure with the strong discontinuity approach. *Computer Methods in Applied Mechanics and Engineering*, 195: 7093–7114 (2006).

Onate E. and Cervera M. Derivation of thin plate bending elements with one degree of freedom per node: A simple three node triangle. *Engineering computations*, 10(6):543–561 (1993).

Oran C. and Kassimali A. Large deformations of framed structures under static and dynamic loads. *Computers and Structures*, 6:539–547 (1976).

Ortega J. and Rheinboldt W. *Iterative Solution of Nonlinear Equations in Several Variables*. Academic Press, New York (1970).

Ortiz M. and Quigley J.J. Adaptive mesh refinement in strain localization problems. *Computer Methods in Applied Mechanics and Engineering*, 90:781–804 (1991).

Owen D.R.J. and Hinton E. *Finite Elements in Plasticity: Theory and Practice*. Pineridge Press, Swansea, U.K. (1980).

Owen S.J. *A Survey of Unstructured Mesh Generation Technology*. http://www.andrew.cmu.edu/user/sowen/survey/index.html (1999).

Pamin J. *Gradient-Dependent Plasticity in Numerical Simulation of Localization Phenomena.* Ph.D. thesis, Delft University of Technology, Delft (1994).

Papadrakakis M. *Solving Large-Scale Linear Problems in Solid and Structural Mechanics, in Solving Large-Scale Problems in Mechanics.* Wiley & Sons, Chichester (1993).

Parisch H. A consistent tangent stiffness matrix for three-dimensional non-linear contact analysis. *International Journal for Numerical Methods in Engineering*, 28:1803–1812 (1989).

Parisch H. An investigation of a finite rotation four node assumed strain element. *International Journal for Numerical Methods in Engineering*, 31:127–150 (1991).

Peric D., Hochard C., Dutko M. and Owen D.R.J. Transfer operators for evolving meshes in small strain elasto-plasticity. *Computer Methods in Applied Mechanics and Engineering*, 137:331–344 (1996).

Peric D. and Owen D.R.J. Computational model for contact problems with friction based on the penalty method. *International Journal for Numerical Methods in Engineering*, 36:1289–1309 (1992).

Peric D. and Owen D.R.J. On error estimates and adaptivity in elastoplastic solids: Applications to the numerical simulation of strain localization in classical and cosserat continua. *International Journal for Numerical Methods in Engineering*, 37:1351–1379 (1994).

Peric D. and Owen D.R.J. Finite-element applications to the nonlinear mechanics of solids. *Reports on Progress in Physics*, 61:1495–1574 (1997).

Peric D., Owen D.R.J. and Honnor M.E. A model for finite strain elasto-plasticity based on logarithmic strains: Computational issues. *Computer Methods in Applied Mechanics and Engineering*, 94(1):35–61 (1992).

Perić D., Vaz M. and Owen D.R.J. On adaptive strategies for large deformations of elasto-plastic solids at finite strains: Computational issues and industrial applications. *Computer Methods in Applied Mechanics and Engineering*, 176 (1-4):279–312 (1999).

Perzyna P. The constitutive equations for rate sensitive plastic materials. *Quarterly Applied Mathematics*, 20:321–332 (1963).

Perzyna P. Fundamental problems in viscoplasticity. *Advances in Applied Mechanics*, 9:243–377 (1966).

Petersen C. *Statik und Stabilität der Baukonstruktionen.* Vieweg & Sohn, Berlin (1980).

Petryk H. and Thermann K. On discretized plasticity problems with bifurcations. *International Journal of Solids & Structures*, 29:745–765 (1992).

Pflüger A. *Stabilitätsprobleme in der Elastostatik.* Springer-Verlag, Berlin, Heidelberg, New York, dritte edition (1975).

Pian T.H.H. Derivation of element stiffness matrices by assumed stress distributions. *AIAA-J. 2*, 7:1333–1336 (1964).

Pian T.H.H. and Sumihara K. Rational approach for assumed stress finite elements. *International Journal for Numerical Methods in Engineering*, 20: 1685–1695 (1984).

Pietraszkiewicz W. Geometrically nonlinear theories of thin elastic shells. *Technical Report 55*, Mitteilungen des Instituts für Mechanik der Ruhr-Universität Bochum (1978).

Pietrzak G. and Curnier A Large deformation frictional contact mechanics: continuum formulation and augmented lagrangean treatment. *Computer Methods in Applied Mechanics and Engineering*, 177:351–381 (1999).

Pimenta P. and Yojo T. Geometrically exact analysis of spatial frames. *Applied Mechanics Review*, 46:113–128 (1993).

Pimenta P.M., Campello E.M.B. and Wriggers P. A fully nonlinear multi-parameter shell model with thickness variation and a triangular shell finite element formulation. *Computational Mechanics*, 34:181–193 (2004).

Planinc I. and Saje M. A quadratically convergent algorithm for the computation of stability points: The application of the determinant of the tangent stiffness matrix. *Computer Methods in Applied Mechanics and Engineering*, 169:89–105 (1999).

Prager W. *Probleme der Plastizitätstheorie*. Birkäuser, Basel, Stuttgart (1955).

Prager W. *Einführung in die Kontinuumsmechanik*. Birkäuser, Basel, Stuttgart (1961).

Puso M.A. A highly efficient enhanced assumed strain physically stabilized hexahedral element. *International Journal for Numerical Methods in Engineering*, 49:1029–1064 (2000).

Puso M.A. A 3D mortar method for solid mechanics. *International Journal for Numerical Methods in Engineering*, 59(3):315–336 (2004).

Puso M.A. and Laursen T.A. A mortar segment-to-segment contact method for large deformation solid mechanics. *Computer Methods in Applied Mechanics and Engineering*, 193:601–629 (2004).

Puso M.A. and Solberg J. A stabilized nodally integrated tetrahedral. *International Journal for Numerical Methods in Engineering*, 67:841–867 (2006).

Ramm E. Geometrisch nichtlineare Elastostatik und Finite Elemente. *Technical Report Nr. 76-2*, Institut für Baustatik der Universität Stuttgart (1976).

Ramm E. Strategies for tracing the nonlinear response near limit points. In W. Wunderlich, E. Stein and K.J. Bathe, editors, *Nonlinear Finite Element Analysis in Structural Mechanics*. Springer, Berlin, Heidelberg, New York (1981).

Ramm E. and Cirak F. Adaptivity for nonlinear thin-walled structures. In D.R.J. Owen, E. Hinton and E. Onate, editors, *Proceedings of COMPLAS 5*, pages 145–163. CIMNE, Barcelona (1997).

Ramm E., Rank E., Rannacher R., Schweizerhof K., Stein E., Wendland W., Wittum G., Wriggers P. and Wunderlich W. *Error-Controlled Adaptive Finite Elements in Solid Mechanics*. Wiley, Chichester (2003).

Raniecki B. and Bruhns O. Bounds to bifurcation stresses in solids with non-associated plastic flow law at finite strain. *Journal of the Mechanics and Physics of Solids*, 29:153–172 (1981).

Rank E., Schweingruber M. and Sommer M. Adaptive mesh generation and transformation of triangular to quadrilateral meshes. *Communications in Numerical Methods in Engineering*, 9:121–129 (1993).

Rankin C.C. and Brogan F.A. An element independent corotational procedure for the treatment of large roations. In L.H. Sobel and K. Thomas, editors, *Collapse Analysis of Structures*, pages 85–100. ASME, New York (1984).

Rannacher R. and Suttmeier F.T. A feed back approach to error control in finite element methods: Application to linear elasticity. *Computational Mechanics*, 19(5):434–446 (1997a).

Rannacher R. and Suttmeier F.T. A posteriori error control in finite element methods via duality techniques: Application to perfect plasticity. *Technical Report 97-16*, Institut for Applied Mathematics, SFB 359, University of Heidelberg (1997b).

Rannacher R. and Suttmeier F.T. A posteriori error control in finite element methods via duality techniques: Application to perfect plasticity. *Computational Mechanics*, 21(2):123–133 (1998).

Rannacher R. and Suttmeier F.T. A posteriori error estimation and mesh adaptation for finite element models in elasto-plasticity. *Computer Methods in Applied Mechanics and Engineering*, 176(1-4):333–361 (1999).

Reddy B.D. and Simo J.C. Stability and convergence of a class of enhanced strain methods. *SIAM Journal of Numerical Analysis*, 32:1705–1728 (1995).

Reese S. Theorie und Numerik des Stabilitätsverhalten hyperelastischer Festkörper. *Technical Report D 17*, Institut für Mechanik der TH Darmstadt (1994).

Reese S. On a consistent hourglass stabilization technique to treat large inelastic deformations and thermo-mechanical coupling in plane strain problems. *International Journal for Numerical Methods in Engineering*, 57:1095–1127 (2003).

Reese S. On a physically stabilized one point finite element formulation for three-dimensional finite elasto-plasticity. *Computer Methods in Applied Mechanics and Engineering*, 194:4685–4715 (2005).

Reese S. and Govindjee S. A theory of finite viscoelasticity and numerical aspects. *International Journal of Solids & Structures*, 35:3455–3482 (1998).

Reese S. and Wriggers P. A finite element method for stability problems in finite elasticity. *International Journal for Numerical Methods in Engineering*, 38: 1171–1200 (1995).

Reese S. and Wriggers P. A new stabilization concept for finite elements in large deformation problems. *International Journal for Numerical Methods in Engineering*, 48:79–110 (2000).

Reese S., Wriggers P. and Reddy B.D. A new locking-free brick element formulation for continuous large deformation problems. In *Proceedings of WCCM IV in Buenos Aires* (1998).

Reese S., Wriggers P. and Reddy B.D. A new locking-free brick element technique for large deformation problems in elasticity. *Computers and Structures*, 75: 291–304 (2000).

Rehle N. Adaptive Finite Element Verfahren bei der Analyse von Flächentragwerken. *Technical Report Nr. 20*, Institut für Baustatik der Universität Stuttgart (1996).

Reissner E. On one-dimensional finite strain beam theory, the plane problem. *Journal of Applied Mathematics and Physics*, 23:795–804 (1972).

Reitinger R. and Ramm E. Buckling and imperfection sensitivity in the optimization of shell structures. *Thin-Walled Structures*, 23:159–177 (1995).

Rheinboldt W. Numerical analysis of continuation methods for nonlinear structural problems. *Computers and Structures*, 13:103–113 (1981).

Rheinboldt W. *Methods for Solving Systems of Nonlinear Equations*. Society for Industrial and Applied Mathematics, Philadelphia (1984).

Rheinboldt W. Error estimates for nonlinear finite element computations. *Computers and Structures*, 20:91–98 (1985).

Ribó R., Bugeda G. and Oñate E. Some algorithms to correct a geometry in order to create a finite element mesh. *Computers and Structures*, 80(16–17):1399–1408 (2002).

Riccius J., Schweizerhof K. and Baumann M. Combination of adaptivity and mesh smoothing for the finite element analysis of shells with intersections. *International Journal for Numerical Methods in Engineering*, 40:2459–2474 (1997).

Riks E. The application of Newtons method to the problem of elastic stability. *Journal of Applied Mechanics*, 39:1060–1066 (1972).

Riks E. Some computational aspects of stability analysis of nonlinear structures. *Computer Methods in Applied Mechanics and Engineering*, 47:219–260 (1984).

Riks E., Rankin C. and Brogan F. On the solution mode of jumping phenomena in thin-walled shell structures. *Computer Methods in Applied Mechanics and Engineering*, 136:59–92 (1996).

Rivlin R.S. Large elastic deformations of isotropic materials. *Proc. of the Royal Society of London*, 241:379–397 (1948).

Roehl D. and Ramm E. Large elasto-plastic finite element analysis of solids and shells with the enhanced assumed strain concept. *International Journal of Solids & Structures*, 33:3215–3237 (1996).

Romero I. and Armero F. An objective finite element approximation of the kinematics of geometrically exact rods and its use in the formulation of an energy–momentum conserving scheme in dynamics. *International Journal for Numerical Methods in Engineering*, 54:1683–1716 (2002).

Rubin M. *Cosserat Theories: Shells, Rods and Points*, volume 79. Kluwer Academic Publishers, Dordrecht (2000).

Rubin M. and Jabareen M. An improved 3-d brick cosserat point element for irregular shaped element. *Computational Mechanics* 40:979–1004 (2007).

Rubin M. and Jabareen M. A 3d brick macro-element for non-linear elasticity. *Private communication* (2007).

Ruge J.W. Amg for problems of elasticity. *Applied Mathematics and Computation*, 19:293–309 (1986).

Saad Y. Practical use of polynomial preconditionings for the conjugate gradient method. *SIAM Journal on Scientific and Statistical Computing*, 4 (1985).

Saad Y. *Iterative Methods for Sparse Linear Systems*. SIAM, Philadelphia (2003).

Saad Y. and Schultz M.H. Gmres: A generalized residual algorithm for solving non-symmetric linear systems. *SIAM Journal on Scientific and Statistical Computing*, 7:856–869 (1986).

Sansour C. A theory and finite element formulation of shells at finite deformations involving thickness change: Circumventing the use of a rotation tensor. *Ingenieur-Archiv*, 65:194–216 (1995).

Sansour C., Sansour J. and Wriggers P. A finite element approach to the chaotic motion of geometrically exact rods undergoing plane deformations. *Nonlinear Dynamics*, 11:189–212 (1996).

Sansour C., Wriggers P. and Sansour J. Nonlinear dynamics of shells: Theory, finite element formulation and integration schemes. *Nonlinear Dynamics*, 13:279–305 (1997).

Schenk O. and Gärtner K. Solving unsymmetric sparse systems of linear equations with pardiso. *Journal of Future Generation Computer Systems*, 20:475–487 (2004).

Scherf O. Kontinuumsmechanische modellierung nichtlinearer kontaktprobleme und ihre numerische analyse mit adaptiven finite-element-methoden. *Technical Report D 17*, Institut für Mechanik der TH Darmstadt (1997).

Schöberl J. *Robust Multigrid Methods for Parameter Deoendent Problems*. Dissertation, Institut für Analysis und Numerik, Johannes Kepler Universität Linz (1999).

Schoop H. Oberflächenorientierte Schalentheorien endlicher Verschiebungen. *Ingenieur-Archiv*, 56:427–437 (1986).

Schreyer H.L. and Neilsen M.K. Analytical and numerical tests for loss of material stability. *International Journal for Numerical Methods in Engineering*, 39:1721–1736 (1996).

Schwarz H.R. *FORTRAN-Programme zur Methode der finiten Elemente*. Teubner, Stuttgart (1981).

Schweizerhof K. Nichtlineare Berechnung von Tragwerken unter verformungsabhängiger Belastung mit finiten Elementen. *Technical Report 82-2*, Institut für Baustatik, Stuttgart (1982).

Schweizerhof K. and Ramm E. Displacement dependent pressure loads in nonlinear finite element analysis. *Computers and Structures*, 18 (1984).

Schweizerhof K., Vielsack P., Rottner T. and Ewert E. Stability and sensitivity investigations of thin-walled shell structures using transient finite element analysis.

In H. Mang, editor, *5 thWorld Congress on Computational Mechanics-WCCM V*. Vienna (2002).

Schweizerhof K. and Wriggers P. Consistent linearization for path following methods in nonlinear fe-analysis. *Computer Methods in Applied Mechanics and Engineering*, 59:261–279 (1986).

Schwetlick H. and Kretschmar H. *Numerische Verfahren für Naturwissenschaftler und Ingenieure*. Fachbuchverlag, Leipzig (1991).

Seifert B. *Zur Theorie und Numerik Finiter Elastoplastischer Deformationen von Schalenstrukturen*. Dissertation, Institut für Baumechanik und Numerische Mechanik der Universität Hannover (1996). Bericht Nr. F 96/2.

Sewell M.J. On configuration-dependent loading. *Archives of Rational Mechanics*, 23:321–351 (1967).

Sheng D. and Sloan S.W. Load stepping schemes for critical state models. *International Journal for Numerical Methods in Engineering*, 50:67–93 (2001).

Shepard M.S. and Georges M.K. Three-dimensional mesh generation by finite octree technique. *International Journal for Numerical Methods in Engineering*, 32: 709–749 (1991).

Simitses G. *Dynamic Stability of Suddenly Loaded Structures*. Springer-Verlag, Berlin (1990).

Simo J., Oliver J. and Armero F. Analysis of strong dicontinuities in rate independent softening materials. *Computational Mechanics*, 12:277–296 (1993a).

Simo J., Taylor R. and Wriggers P. A note on finite element implementation of pressure boundary loading. *Communications in Applied Numerical Methods*, 7:513–525 (1991).

Simo J.C. A finite strain beam formulation. The three-dimensional dynamic problem. Part I. *Computer Methods in Applied Mechanics and Engineering*, 49:55–70 (1985).

Simo J.C. On a fully three-dimensional finite-strain viscoelastic damage model: Formulation and computational aspects. *Computer Methods in Applied Mechanics and Engineering*, 60:153–173 (1987).

Simo J.C. A framework for finite strain elastoplasticity based on the multiplicative decomposition and hyperelastic relations. Part ii: Computational aspects. *Computer Methods in Applied Mechanics and Engineering*, 67:1–31 (1988).

Simo J.C. Algorithms for static and dynamic multiplicative plasticity that preserve the classical return mapping schemes of the infinitesimal theory. *Computer Methods in Applied Mechanics and Engineering*, 99:61–112 (1992).

Simo J.C. On a stress resultant geometrically exact shell model. Part VII. Shell intersections with 5/6-DOF finite element formulations. *Computer Methods in Applied Mechanics and Engineering*, 108:319–339 (1993).

Simo J.C. Numerical analysis and simulation of plasticity. In P.G. Ciarlet and J.L. Lions, editors, *Handbook of Numerical Analysis*, volume 6, pages 179–499. North-Holland, Amsterdam (1998).

Simo J.C. and Armero F. Geometrically non-linear enhanced strain mixed methods and the method of incompatible modes. *International Journal for Numerical Methods in Engineering*, 33:1413–1449 (1992).

Simo J.C., Armero F. and Taylor R.L. Improved versions of assumed enhanced strain tri-linear elements for 3D finite deformation problems. *Computer Methods in Applied Mechanics and Engineering*, 110:359–386 (1993b).

Simo J.C., Fox D.D. and Rifai M.S. On a stress resultant geometrical exact shell model. Part I: Formulation and optimal parametrization. *Computer Methods in Applied Mechanics and Engineering*, 72:267–304 (1989).

Simo J.C., Fox D.D. and Rifai M.S. On a stress resultant geometrical exact shell model. Part III: Computational aspects of the nonlinear theory. *Computer Methods in Applied Mechanics and Engineering*, 79:21–70 (1990).

Simo J.C., Hjelmstad K.D. and Taylor R.L. Numerical formulations for finite deformation problems of beams accounting for the effect of transverse shear. *Computer Methods in Applied Mechanics and Engineering*, 42:301–330 (1984).

Simo J.C. and Hughes T.J.R. *Computational Inelasticity*. Springer, New York, Berlin (1998).

Simo J.C. and Kennedy J.G. On a stress resultant geometrically exact shell model. Part V. Nonlinear plasticity: Formulation and integration algorithms. *Computer Methods in Applied Mechanics and Engineering*, 96:133–171 (1992).

Simo J.C. and Miehe C. Associative coupled thermoplasticity at finite strains: formulation, numerical analysis and implementation. *Computer Methods in Applied Mechanics and Engineering*, 98:41–104 (1992).

Simo J.C. and Ortiz M. A unified approach to finite deformation elastoplasticity based on the use of hyperelastic constitutive equations. *Computer Methods in Applied Mechanics and Engineering*, 49:201–215 (1985).

Simo J.C. and Pister K.S. Remarks on rate constitutive equations for finite deformation problems. *Computer Methods in Applied Mechanics and Engineering*, 46:201–215 (1984).

Simo J.C. and Rifai M.S. A class of assumed strain methods and the method of incompatible modes. *International Journal for Numerical Methods in Engineering*, 29:1595–1638 (1990).

Simo J.C., Rifai M.S. and Fox D.D. On a stress resultant geometrically exact shell model. Part IV. Variable thickness shells with through-the-tickness stretching. *Computer Methods in Applied Mechanics and Engineering*, 81:91–126 (1990B).

Simo J.C. and Tarnow N. The discrete energy–momentum method. Conserving algorithms for nonlinear elastodynamics. *Zeitschrift für angewandte Mathematik und Physik*, 43:757–792 (1992).

Simo J.C. and Taylor R.L. Consistent tangent operators for rate-independent elastoplasticity. *Computer Methods in Applied Mechanics and Engineering*, 48:101–118 (1985).

Simo J.C. and Taylor R.L. A return mapping algorithm for plane stress elastoplasticity. *International Journal for Numerical Methods in Engineering*, 22:649–670 (1986).

Simo J.C. and Taylor R.L. Quasi-incompressible finite elasticity in principal stretches continuum basis and numerical algorithms. *Computer Methods in Applied Mechanics and Engineering*, 85:273–310 (1991).

Simo J.C., Taylor R.L. and Pister K.S. Variational and projection methods for the volume constraint in finite deformation elasto-plasticity. *Computer Methods in Applied Mechanics and Engineering*, 51:177–208 (1985a).

Simo J.C. and Vu-Quoc L. Three dimensional finite strain rod model. Part II: computational aspects. *Computer Methods in Applied Mechanics and Engineering*, 58:79–116 (1986).

Simo J.C., Wriggers P. and Taylor R.L. A perturbed Lagrangian formulation for the finite element solution of contact problems. *Computer Methods in Applied Mechanics and Engineering*, 50:163–180 (1985b).

Skeie G., Astrup O.C. and Bergan P. Application of adapted nonlinear solution strategies. In N.E. Wiberg, editor, *Advances in Finite Element Technology*, pages 212–236. CIMNE, Barcelona (1995).

Sloan S.W. A fast algorithm for constructing delaunay triangularization in the plane. *Advances in Engineering Software*, 9:34–55 (1987a).

Sloan S.W. Substepping schemes for the numerical integration of elastoplastic stress-strain relations. *International Journal for Numerical Methods in Engineering*, 24:893–911 (1987b).

Sloan S.W. A fast algorithm for generating constrained delaunay triangulations. *Computers and Structures*, 47:441–450 (1993).

Sloan S.W., Abbo A.J. and Sheng D. Refined explicit integration of elastoplastic models with automatic error control. *Engineering Computations*, 18:121–154 (2001).

Sluys L.J. *Wave propagation, localization and dispersion in softening solids*. Ph.D. thesis, Delft University of Technology, Delft (1992).

Soric J., Montag U. and Krätzig W.B. An efficient formulation of integration algorithms for elastoplastic shell analysis based on Layered finite element approach. *Computer Methods in Applied Mechanics and Engineering*, 148:315–328 (1997).

de Souza Neto E.A., Peric D., Dutko M. and Owen D.R.J. Design of simple lower-order finite elements for large-deformation analysis of nearly incompressible solids. *International Journal of Solids & Structures*, 33:3277–3296 (1996).

de Souza Neto E.A., Peric D., Huang G.C. and Owen D.R.J. Remarks on stability of enhanced strain elements in finite elasticity and elastoplasticity. In D.R.J. Owen, E. Hinton and E. Onate, editors, *Proceedings of COMPLAS 4*, volume 1, pages 361–372. Pineridge Press, Swansea (1995).

Spence A. and Jepson A.D. The numerical calculation of cusps, bifurcation points and isola formation points in two parameter problems. In T. Küpper, H.D. Mittelmann and H. Weber, editors, *Numerical Methods for Bifurcation Problems, ISNM 70*, pages 502–514. Birkhäuser, Basel, Boston, Stuttgart (1984).

Spiess H. Reduction methods in finite element analysis of nonlinear structural dynamics. *Technical Report F06/2*, Forschungs- und Seminarberichte aus dem Bereich der Mechanik der Universität Hannover (2006).

Stein E. and Ohnimus S. Dimensional adaptivity in linear elasticity with hierarchical test-spaces for h- and p-refinement processes. *Engineering Computations*, 12:107–119 (1996).

Stein E., Steinmann P. and Miehe C. Instability phenomena in plasticity: Modelling and computation. *Computational Mechanics*, 17:74–87 (1995).

Steinmann P., Larsson R. and Runesson K. On the localization properties of multiplicative hyperelasto-plastic continua with strong discontinuities. *International Journal of Solids & Structures*, 34:969–990 (1997).

Stoer J. and Bulirsch R. *Numerische Mathematik 2*. Springer Verlag, Berlin, Heidelberg, Wien, dritte edition (1990).

Strang G. and Fix G.J. *An Analysis of the Finite Element Method*. Prentice-Hall, Inc., Englewood Cliffs (1973).

Stueben K. Algebraic multigrid (amg) experiences and comparisons. *Applied Mathematics and Computation*, 13:419–451 (1983).

Sussman T. and Bathe K.J. A finite element formulation for nonlinear incompressible elastic and inelastic analysis. *Computers and Structures*, 26:357–409 (1987).

Szabó I. *Höhere Technische Mechanik*. Springer, Berlin, Heidelberg, Wien, fünfte edition (1977).

Tabor D. Friction – The present state of our understanding. *Journal Lubrication Technology*, 103:169–179 (1981).

Taylor R.L. Solution of linear equations by a profile solver. *Engineering Computations*, 2:334–350 (1985).

Taylor R.L. A mixed-enhanced formulation for tetrahedral finite elements. *International Journal for Numerical Methods in Engineering*, 47:205–227 (2000).

Taylor R.L., Beresford P.J. and Wilson E.L. A non-conforming element for stress analysis. *International Journal for Numerical Methods in Engineering*, 10: 1211–1219 (1976).

Taylor R.L., Pister K.S. and Goudreau G.L. Thermomechanical analysis of viscoelastic solids. *International Journal for Numerical Methods in Engineering*, 2:45–79 (1970).

Taylor R.L., Wilson E.L. and Sackett S.J. Direct solution of equations by frontal and varaible band, active column methods. In E.S. W. Wunderlich and K.J. Bathe, editors, *Nonlinear Finite Element Analysis in Structural Mechanics*, pages 33–107. Springer-Verlag, Berlin, Heidelberg (1981).

Thoutireddy P., Molinari J.F., Repetto E.A. and Ortiz M. Tetrahedral composite finite elements. *International Journal for Numerical Methods in Engineering*, 53:1337–1351 (2002).

Treloar L.R.G. Stress–strain data for vulcanized rubber under various types of deformation. *Transactions of the Faraday Society*, 40:59–70 (1944).

Truesdell C. and Noll W. The nonlinear field theories of mechanics. In S. Flügge, editor, *Handbuch der Physik III/3*. Springer, Berlin, Heidelberg, Wien (1965).

Truesdell C. and Toupin R. The classical field theories. In *Handbuch der Physik III/1*. Springer, Berlin, Heidelberg, Wien (1960).

Tvergaard V. Material failure by void growth to coalescence. *Advances in Applied Mechanics*, 27:83–151 (1989).

Tvergaard V. and Needleman A. Analysis of the cup-cone fracture in a round tensile bar. *Archives of Mechanics*, 32:157–169 (1984).

Vainberg M.M. *Variational Methods for the Study of Nonlinear Operators*. Holden Day, San Francisco (1964).

Verfürth R. *A Review of A Posteriori Error Estimation and Adaptive Mesh-Refinement Techniques*. Wiley, Teubner, Chichester, New York, Stuttgart, Leipzig (1996).

Vogel U. *Die Traglastberechnung st"ahlerner Rahmentragwerke nach der Plastizit"atstheorie II. Ordnung*. Stahlbau Verlag, K"oln (1965).

Vogel U. Calibrating frames, vergleichsrechnungen an verschiedenen rahmen. *Der Stahlbau*, 10:295–301 (1985).

den Vorst H.A.V. Bi-cgstab: A fast and smoothly converging variant of bi-cg for the solution of non-symmetric linear systems. *SIAM Journal on Scientific and Statistical Computing*, 13:631–644 (1992).

Vukazich M., Mish K. and Romstad K. Nonlinear dynamic response of frames using Lanczos modal analysis. *Journal of Structural Engineering*, 122:1418–1426 (1996).

Wagner W. A finite element model for non-linear shells of revolution with finite rotations. *International Journal for Numerical Methods in Engineering*, 29: 1455–1471 (1990).

Wagner W. Zur Behandlung von Stabilitätsproblemen mit der Methode der Finiten Elemente. *Technical Report F91/1*, Forschungs- und Seminarberichte aus dem Bereich der Mechanik der Universität Hannover (1991).

Wagner W. and Gruttmann F. A simple finite rotation formulation for composite shell elements. *Engineering Computations*, 11:145–176 (1994).

Wagner W., Klinkel S. and Gruttmann F. Elastic and plastic analysis of thin-walled structures using improved hexahedral elements. *Computers and Structures*, 80 (9–10):857–869 (2002).

Wagner W. and Wriggers P. A simple method for the calculation of secondary branches. *Engineering Computations*, 5:103–109 (1988).

Wang P.S. Finger: A symbolic system for automatic generation of numerical programs in finite element analysis. *Journal of Symbolic Computation*, 2:305–316
(1986).

Washizu K. *Variational Methods in Elasticity and Plasticity*. Pergamon Press,
Oxford, second edition (1975).

Weber G. and Anand L. Finite defformation constitutive equations and a time integration procedure for isotropic hyperelastic-viscoelastic solids. *Computer Methods in Applied Mechanics and Engineering*, 79:173–202 (1990).

Wempner G. Finite elements, finite roataions and small strains of flexible shells.
International Journal of Solids & Structures, 5:117–153 (1969).

Werner B. and Spence A. The computation of symmetry-breaking bifurcation
points. *SIAM Journal on Numerical Analysis*, 21:388–399 (1984).

Wiberg N.E., Abdulwahab F. and Ziukas S. Enhanced superconvergent patch recovery incorporating equilibrium and boundary conditions. *International Journal
for Numerical Methods in Engineering*, 36:3417–3440 (1994).

Wilson E.L. and Dovey H.H. Solution or reduction of equilibrium equations for
large complex structural systems. *Advances in Engineering Software*, 1:19–25
(1978).

Wilson E.L., Taylor R.L., Doherty W.P. and Ghaboussi J. Incompatible displacements models. In *Numerical and Computer Models in Structural Mechanics*,
Fenves S. J., Perrone N., Robinson A. R. and Schnobrich W. C. (Eds.). Academic Press, New York (1973). 43–57.

Wilson E.L., Yuan M. and Dickens J.M. Dynamic analysis by direct superposition
of ritz vectors. *Earthquake Engineering and Structural Dynamics*, 10:813–821
(1982).

Windels R. Traglasten von Balkenquerschnitten beim Angriff von Biegemoment,
Längs- und Querkraft. *Der Stahlbau*, 39:10–16 (1970).

Wohlmuth B.I. *Discretization Methods and Iterative Solvers based on Domain Decomposition*. Springer Verlag, Berlin, Heidelberg, New York (2000).

Woo K.L. and Thomas T.R. Contact of rough surfaces: A review of experimental
works. *Wear*, 58:331–340 (1980).

Wood W.L. *Practical Time-stepping Schemes*. Clarendon Press, Oxford (1990).

Wood W.L., Bossak M. and Zienkiewicz O.C. An alpha modification of Newmark's method. *International Journal for Numerical Methods in Engineering*, 15:
1562–1566 (1981).

Wriggers P. On consistent tangent matrices for frictional contact problems. In
G. Pande and J. Middleton, editors, *Proceedings of NUMETA 87*. M. Nijhoff
Publishers, Dordrecht (1987).

Wriggers P. Finite element algorithms for contact problems. *Archive of Computational Methods in Engineering*, 2:1–49 (1995).

Wriggers P. *Computational Contact Mechanics*. Springer, Berlin, Heidelberg,
New York (2006).

Wriggers P. and Boersma A. A parallel algebraic multigrid solver for problems
in solid mechanics discretized by finite elements. *Computers and Structures*,
69:129–137 (1998).

Wriggers P. and Carstensen C. An efficient algorithm for the computation of stability points of dynamical systems under step load. *Engineering Computations*,
9:669–679 (1992).

Wriggers P., Eberlein R. and Gruttmann F. An axisymmetrical quasi-Kirchhoff-
type shell element for large plastic deformations. *Archiv of Applied Mechanics*,
65:465–477 (1995).

Wriggers P., Eberlein R. and Reese S. Comparison between shell and 3d-elements in finite plasticity. *International Journal of Solids & Structures*, 33:3309–3326 (1996).

Wriggers P. and Gruttmann F. Large deformations of thin shells: Theory and finite-element discretization, analytical and computational models of shells. In A.K. Noor, T. Belytschko and J.C. Simo, editors, *ASME, CED-Vol.3*, volume 135-159 (1989).

Wriggers P. and Gruttmann F. Thin shells with finite rotations formulated in biot stresses theory and finite-element-formulation. *International Journal for Numerical Methods in Engineering*, 36:2049–2071 (1993).

Wriggers P. and Hueck U. A formulation of the enhanced qs6-element for large elastic deformations. *International Journal for Numerical Methods in Engineering*, 39:3039–3053 (1996).

Wriggers P., Krstulovic-Opara L. and Korelc J. Smooth c1-interpolations for two-dimensional frictional contact problems. *International Journal for Numerical Methods in Engineering*, 51:1469–1495 (2001).

Wriggers P. and Meynen S. Parallele algorithmen und hardware für berechnungsverfahren des ingenieurbaus. In E. Ramm, E. Stein and W. Wunderlich, editors, *Finite Elemente in der Baupraxis*. Ernst & Sohn, München (1995).

Wriggers P., Miehe C., Kleiber M. and Simo J. A thermomechanical approach to the necking problem. *International Journal for Numerical Methods in Engineering*, 33:869–883 (1992).

Wriggers P. and Reese S. A note on enhanced strain methods for large deformations. *Technical Report 3/94*, Bericht des Instituts für Mechanik (1994).

Wriggers P. and Reese S. A note on enhanced strain methods for large deformations. *Computer Methods in Applied Mechanics and Engineering*, 135:201–209 (1996).

Wriggers P., Rieger A. and Scherf O. Comparison of different error measures for adaptive finite element techniques applied to contact problems involving large elastic strains. *Computer Methods in Applied Mechanics and Engineering* (2000).

Wriggers P. and Scherf O. An adaptive finite element method for elastoplastic contact problems. In D.R.J. Owen, E. Hinton and E. Onate, editors, *Proceedings of COMPLAS 4*. Pineridge Press, Swansea (1995).

Wriggers P. and Simo J. A note on tangent stiffnesses for fully nonlinear contact problems. *Communications in Applied Numerical Methods*, 1:199–203 (1985).

Wriggers P. and Simo J. A general procedure for the direct computation of turning and bifurcation points. *International Journal for Numerical Methods in Engineering*, 30:155–176 (1990).

Wriggers P., Simo J. and Taylor R. Penalty and augmented lagrangian formulations for contact problems. In J. Middleton and G. Pande, editors, *Proceedings of NUMETA Conference*. Balkema, Rotterdam (1985).

Wriggers P. and Taylor R. A fully nonlinear axisymmetrical membrane element for rubberlike materials. *Engineering Computations*, 7:303–310 (1990).

Wriggers P., Van T.V. and Stein E. Finite-element-formulation of large deformation impact-contact-problems with friction. *Computers and Structures*, 37:319–333 (1990).

Wriggers P., Wagner W. and Miehe C. A quadratically convergent procedure for the calculation of stability points in finite element analysis. *Computer Methods in Applied Mechanics and Engineering*, 70:329–347 (1988).

Wriggers P. and Zavarise G. On the application of augmented lagrangian techniques for nonlinear constitutive laws in contact interfaces. *Communications in Applied Numerical Methods*, 9:815–824 (1993).

Wright S.J. *Primal-Dual Interior-Point Methods*. SIAM, Philadelphia (1997).

Yagawa G., Yoshimura S. and Nakao K. Automatic mesh generation of complex geometries based on fuzzy knowledge processing and computational geometry. *Integrated Computer-Aided Engineering*, 2(4):265–280 (1995).

Yosibash Z., Hartmann S., Heisserer U., Düster A., Rank E. and Szanto M. Axisymmetric pressure boundary loading for finite deformation analysis using p-FEM. *Computer Methods in Applied Mechanics and Engineering*, 196: 1261–1277 (2007).

Zavarise G., Wriggers P. and Schrefler B.A. A method for solving contact problems. *International Journal for Numerical Methods in Engineering*, 42:473–498 (1998).

Zhu J.Z., Zienkiewicz O.C., Hinton E. and Wu J. A new approach to the development of automatic quadrilateral mesh generation. *International Journal for Numerical Methods in Engineering*, 32:849–866 (1991).

Zienkiewicz O.C., Bauer J., Morgan K. and Onate E. A simple and efficient element for axisymmetrical shells. *International Journal for Numerical Methods in Engineering*, 11:1545–1558 (1977).

Zienkiewicz O.C. and Cheung Y.K. *The Finite Element Method in Structural and Soild Mechanics*. McGraw Hill, London (1967).

Zienkiewicz O.C. and Taylor R.L. *The Finite Element Method, 4th Ed.*, volume 1. McGraw Hill, London (1989).

Zienkiewicz O.C. and Taylor R.L. *The Finite Element Method, 4th Ed.*, volume 2. McGraw Hill, London (1991).

Zienkiewicz O.C. and Taylor R.L. *The Finite Element Method*, volume 1. Butterworth-Heinemann, Oxford, UK, 5th edition (2000a).

Zienkiewicz O.C. and Taylor R.L. *The Finite Element Method*, volume 2. Butterworth-Heinemann, Oxford, UK, 5th edition (2000b).

Zienkiewicz O.C., Taylor R.L. and Too J.M. Reduced integration technique in general analysis of plates and shells. *International Journal for Numerical Methods in Engineering*, 3:275–290 (1971).

Zienkiewicz O.C. and Zhu J.Z. A simple error estimator and adaptive procedure for practical engineering analysis. *International Journal for Numerical Methods in Engineering*, 24:337–357 (1987).

Zienkiewicz O.C. and Zhu J.Z. The superconvergent patch recovery (SPR) and adaptive finite element refinement. *Computer Methods in Applied Mechanics and Engineering*, 101:207–224 (1992).

Index